DIFFERENTIAL CALCULUS

For B.A. & B.Sc. Classes as per UGC Model Syllabus

SHANTI NARAYAN

*Formerly, Dean of Colleges, Delhi University, Delhi
and Principal, Hans Raj College, Delhi*

Revised by

Dr. P.K. MITTAL

M.Sc., Ph.D.
*Head of Mathematics Department
Govt. Post Graduate College
Rishikesh (Uttaranchal)*

SHYAMLAL CHARITABLE TRUST

RAM NAGAR, NEW DELHI-110 055

First Edition 1942
Subsequent Editions and Reprints 1974, 75, 77, 78, 79, 80, 81, 82, 83, 84, 85 (Twice), 86, 87, 88, 90, 91, 92, 93, 94, 96, 97, 98, 99, 2000, 2001, 2002, 2003, 2004, 2005 (Twice), 2007 (Twice), 2008 (Thrice), 2009 (Thrice), 2010, 2011; Revised and Enlarged Edition 2012; Reprint 2013
Reprint 2014 (Twice)

ISBN : 81-219-0471-4   Code : 14A 049

PRINTED IN INDIA
By Goyal Offset Printers, A-60/1, G.T. Karnal Road, Indil. Area, Delhi-110 033
and published by Shyam Lal Charitable Thrust, 7361, Ram Nagar, New Delhi -110 055.

PREFACE TO THE FIFTEENTH EDITION

This book originally written about 62 year ago, has during the intervening period, been revised and reprinted several times. Since its last revision, tremendous changes in the trend of studies have taken place which necessiated thorough revision of the book. But, it was really a very difficult task to revise this perfect well written book of late Dr. Shanti Narayan. Since a long time undoubtedly, it was a book loved by teachers and students alike.

I took up the task to revise the book with great devotion to meet the rapidly changing demands of students interested in self-study and appearing in different competitive examination. Keeping this in view, a large variety of illustrative solved examples have been included in every chapter. Keeping in mind the syllabi of different Universities, specially that introduced by U.G.C. certain topics have been included which were not present in the original text.

This book, in the present form, is a humble effort to make it more useful to students and teachers.

I owe my special gratitude to Shri Ravindra Kumar Gupta, Managing Director and Sri Himanshu Gupta, Director of S. Chand & Co. Ltd., for giving me opportunity to revise the books of late Dr. Shanti Narayan, an eminent Indian Mathematician.

I would like to express, heart felt gratitude to my friend Mr. Naveen Joshi, General Manager (Sales & Marketing) of S. Chand & Co. Ltd. without whose continuous encouragement and persuation the present book may have not reached to your hands. My sincere thanks are also due to Mr. Shishir Bhatnagar of S. Chand & Co. Ltd.

The final tribute and greatest appreciation, however, is reserved for my wife Prabha, without whose persuation, cooperation and timely help, this book would not have seen the light of the day.

As the need for improvement is never ending, I will look forward to receive valuable and useful suggestions from our users for further improvement in the book.

— Dr. P.K. MITTAL

PREFACE TO THE FIRST EDITION

This book is meant for students preparing for the B.A. and B.Sc. examinations of our universities. Some topics of the Honours standard have also been included. They are given in the form of appendices to the relevant chapters. The treatment of the subject is rigorous but no attempt has been made to state and prove the theorems in generalised forms and under less respective conditions as in the case with the modern Theory of Functions. It has also been a constant endeavour of the author to see that the subject is not presented just as a body of formulae. This is to see that the student does not form an unfortunate impression that the study of Calculus consists only in acquiring a skill to manipulate some formulae through constant drilling.

The book opens with a brief outline of the development of Real Numbers, their expression as infinite decimals and their representation by points along a line. This is followed by a discussion of the graphs of the elementary functions x^n, log x, e^x, sin x, $\sin^{-1} x$ etc. Some of the difficulties of inverse functions have also been illustrated by precise formulation of Inverse trigonometrical functions. It is suggested that the teacher in the class needs to refer only a few salient points of this part of the book. The student should, on his part, go through the same in complete details to acquire a sound grasp of the basics of the subject. This part is so present that a student would have no difficulty in an independent study of the same.

While the first part of the book is analytical in character, the latter part deals with the geometrical applications of the subject. But this order of the subject is by no means suggested to be rigidly followed in the class. A different order may usefully be adopted at the direction of the teacher.

An analysis of the 'Layman's' concept has frequently been made to serve basis for the precise formulation of the corresponding 'Scientist's' concepts. This specially relates to the two concepts of *Continuity* and *Curvature*.

Geometrical interpretation of results analytically obtained have been given for the benefit of the students. A chapter on '*Some Important Curves',* has been given before dealing with geometrical applications. This will enable the student to get familiar with the names and shapes of some of the important curves. It is felt that a student will have better understanding of the properties of a curve if he knows how the curve looks like. This chapter will also serve as a useful introduction to the subject of *Double points* of a curve.

Asymptote of a curve has been defined as a line so that the distance of any point on the curve from this line tends to zero, as the point tends to infinity along the curve. It is believed that, of all the definitions of an asymptote, this is the one which is most natural. It embodies the idea to which the concept of asymptotes owes its importance. Moreover, the definition gives rise to a simple method for determining the asymptotes.

The various principles and methods have been profusely illustrated by means of a large number of solved examples.

I am indebted to Prof. Sita Ram Gupta, M.A., P.E.S., formerly of the Government College, Lahore who very kindly went through the manuscript and made a number of suggestions. My thanks are also due to my old pupils and friends, Professors Jagan Nath M.A., Vidya Sagar M.A., and Om Parkash M.A., for the help which they rendered in preparing this book.

Suggestions for improvement will be thankfully acknowledged.

—SHANTI NARAYAN

CONTENTS

CHAPTER 1
REAL NUMBERS

1.1. Introduction

The subject of Differential Calculus takes its stand upon the set of real numbers and concerns itself with the various properties of the same. It specially introduces and deals with what is called the *Limit operation*, in addition to being concerned with the Algebraic operations of Addition and Multiplication and their Inverses, Subtraction and Division.

The subject is a development of the important notion of *Instantaneous rate of change*, which is itself a limit idea, as distinguished from the average rate of change and as such, it finds applications to all those branches of knowledge which deal with the same. Thus, it is applied to Geometry, Mechanics and other branches of Theoretical Physics and also to Social Sciences, such as Economics and Psychology.

It may be noted here that this application is essentially based on the notion of measurement, whereby we employ real numbers to measure the particular entity which is the object of investigation in any particular department of knowledge. In Mechanics, for instance, we are concerned with the notion of time and, therefore, in the application of Calculus to Mechanics the first step is to correlate the two notions of Time and Real Numbers, *i.e.* to measure time in terms of real numbers. Similar is the case with other notions such as amount of Heat, Intensity of Light, Force, Demand, Intelligence, etc. The formulation of an entity in terms of real numbers, *i.e.*, measurement must, of course take note of the properties which we intuitively associate with the same. This remark will later on be illustrated with reference to the concepts of Velocity, Acceleration, Curvature, etc.

The importance of real numbers for the study of the subject in hand being thus clear, we will, in some of the following articles, see how starting from the set of natural numbers, we arrive at the set of *real numbers*.

It is, however, not intended to give here any logically connected account of the development of the set of real numbers and only a very brief reference to some well-known salient facts will suffice for our purpose. An account of the logical development of number systems is given in the book *"Number Systems"* by the author.

It may also be mentioned here that even though it satisfies a deep philosophical need to base the theory part of Calculus on the notion of number alone to the entire exclusion of every physical basis, but a rigid insistence on the same is not within the scope of the present book and intuitive geometrical notions of Point, Distance etc. will often be applied to for securing simplicity in the context of aims of this book.

1

1.2. Rational Numbers and their Representation by Points along a Straight Line

1.2.1. Natural Numbers. It was the numbers, 1, 2, 3, 4 etc. that we were first introduced through the process of counting certain sets of objects. The totality of these numbers is known as the set of Natural numbers.

While the operations of addition and multiplication are unrestrictedly possible in relation to the set of natural numbers, this is not the case in respect of the inverse operations of subtraction and division. Thus, for example, the symbols 2–3, 2 ÷ 3 are meaningless insofar as the set of natural numbers is concerned. Natural numbers are also referred to as positive integers.

The set {1, 2, 3, 4, ...} of natural numbers will be denoted as **N**.

1.2.2. Fractional Numbers. At a later stage, another set of numbers like $p/q, \left(e.g., \dfrac{1}{2}, \dfrac{3}{4} \right)$ where p and q are numbers, was adjoined to the set N of natural numbers. This new set is known as the set of Fractions and it obviously included natural numbers as a sub-set ; q being equal to 1 in this case.

The introduction of Fractional numbers is motivated from an abstract point of view, to render Division unrestrictedly possible and from the concrete point of view, to render numbers serviceable for practical needs of measurement also in addition to counting.

1.2.3. Rational Numbers. Still later the set of fractions was enlarged by incorporation in it the set of negative fractions and zero. The entire set of these numbers is known as the set of Rational numbers and is denoted as **Q**.

The introduction of negative numbers is motivated from an abstract point of view, to render subtraction always possible and, from a concrete point of view, to facilitate a unified treatment of oppositely directed pairs of entities, such as gain and loss ; rise and fall, etc.

The natural numbers 1, 2, 3, 4, ... are called positive integers and the numbers –1, –2, ... are called negative integers.

The set {0, –1, 1, –2, 2, –3, 3 ....}, called the set of Integers, is denoted as **I**.

Every non-zero rational number is expressible in the form p/q where p and q are any two integers and q is not zero.

1.2.4. Fundamental operations on rational numbers. An important property of the set of rational numbers is that the operations of addition, multiplication, subtraction and division can be performed upon any two such numbers (with one *exception viz.,* Division by zero which is considered below in (§. 1.2.5) and the number obtained as the result of these operations is again a rational number.

This property is expressed by saying that *the set of rational numbers is* **closed** *with respect to the four fundamental operations (division by zero being excluded).*

1.2.5. Meaningless operation of division by zero. It is important to note that the only exception to the above property is 'Division by zero' which is a meaningless operation. This may be seen as follows :-

To divide a by b amounts to determining a number c such that $bc = a$, and the division will be intelligible, if and only if, the determination of c is *uniquely possible*.

Now, there is no number which when multiplied by zero produces a number other than zero and as such $a/0$ is *no* number when $a \neq 0$. Also *any* number when multiplied by zero produces zero so that $0/0$ may be any number.

On account of this *Impossibility* in one case and *Indefiniteness* in the other, the operation of division by zero *must be always avoided.*

A disregard of this exception often leads to absurd results as is illustrated below in (*i*).

(*i*)   Let $x = 6$, we have
$$x = 6 \Rightarrow x^2 - 36 = x - 6 \Rightarrow (x - 6)(x + 6) = x - 6.$$

Dividing both sides by $(x - 6)$, we get

$$x + 6 = 1 \Rightarrow 6 + 6 = 1 \Rightarrow 12 = 1$$

which is clearly absurd.

Division by $x - 6$, which is zero, *here*, is responsible for this absurd conclusion.

(*ii*) We may also remark in this connection that

$$\frac{x^2 - 36}{x - 6} = \frac{(x - 6)(x + 6)}{(x - 6)} = x + 6, \text{ only when } x \neq 6 \qquad \qquad ..(i)$$

For $x = 6$, the left hand expression $(x^2 - 36)/(x - 6)$ is meaningless whereas the right hand expression $x + 6$, is equal to 12 so that equality ceases to hold for $x = 6$.

The equality (*i*) above is obtained on dividing the numerator and denominator of the expression $(x^2 - 36)/(x - 6)$ by $(x - 6)$ and this operation of division is possible only when the divisor $(x - 6) \neq 0$, *i.e.*, when $x \neq 6$. This explains the restricted character of the equality (*i*).

Ex. 1. Show by examples that the set of natural numbers is not closed with respect to the operations of subtraction and division. Also show that the set of positive fractions is not closed with respect to the operation of subtraction.

Ex. 2. Show that every rational number is expressible as a terminating or a recurring decimal.

To decimalise p/q, we have first to divide p by q and then each remainder, after multiplicaton with 10, is to be divided by q to obtain the successive figures in the decimal expression of p/q. The decimal expression will be terminating if at some stage, the remainder vanishes. Otherwise, the process will be unending. In the latter case, the remainder will always be one of the finite set of numbers, 1, 2, ..., $q - 1$ and so must repeat itself at some stage. From this stage onward, the quotients will also repeat themselves and the decimal expression will, therefore, be recurring.

For a proper understanding of the argument, the student may actually express some fractional numbers, say $\dfrac{3}{7}, \dfrac{3}{13}, \dfrac{31}{123}$, in decimal notation.

Ex. 3. For what values of x are the following equalities *not* valid :

(*i*) $\dfrac{x}{x} = 1$,

(*ii*) $\dfrac{x^2 - a^2}{x - a} = x + a$,

(*iii*) $\dfrac{1 - x}{1 - \sqrt{x}} = 1 + \sqrt{x}$,

(*iv*) $\dfrac{1 - \cos x}{\sin x} = \tan \dfrac{x}{2}$.

ANSWERS

3.   (*i*) 0,   (*ii*) *a*,   (*iii*) 1,   (*iv*) $n\pi : \in I$

1.2.6. Representation of rational numbers by points along line or by segments of a line. The mode of representing rational numbers by points along a line or by segments of a line, which may be known as *number line* will now be explained.

We start with marking an arbitrary point O on the line and calling it the *Origin* or zero point. The number zero will be represented by the point O.

The point O divides the line into two parts or sides. Any one of these may be called positive and the other, then called negative. Usually, the line is drawn

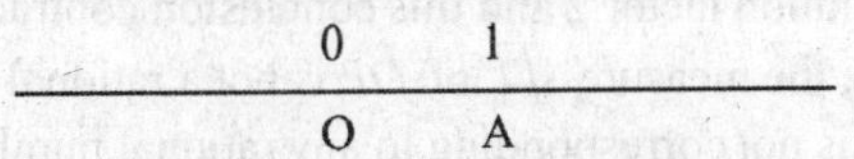

Fig. 1.1

parallel to the printed lines of the page and the right hand side of O is termed positive and the left hand side of O is negative.

On the positive side, we take an arbitrary length OA, and call it the *unit* length.

We say that the number 1 is represented by the point A. After having fixed an *origin, Positive sense* and a *unit length* on the number line in the manner indicated here, we are in a position to determine a point representing any given rational number as explained below :

Positive Integers. Firstly we consider any positive integer m. We take a point on the positive side of the line such that its distance from O is m times the unit length OA. This point will be reached by measuring successively m steps each equal to OA starting from O. This point, then, is said to represent the positive integer m.

Negative Integers. To represent a negative integer, $- m$, we take point on the negative side of O such that its distance from O is m times the unit length OA. This point represents the negative integer, $- m$.

Rational Numbers. Finally, let p/q be any rational number; q being a positive integer. Let OA be divided into q equal parts; OB being one of them. We take a point on the positive or negative side of O according as p is positive or negative such that its distance from O is p times (or, $- p$ times if p is negative) the distance OB. The point so obtained represents the rational number p/q.

1.3. Irrational Numbers

We have seen in the last article that every rational number can be represented by a point of a line. Also, it is easy, to see that we can cover the line with such points as closely as we like. The natural question now arises, "Is the converse true ?" "Is it possible to assign a rational number to every point on the number line ?" A simple consideration, as detailed below, will show that it is not so.

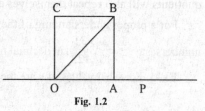

Fig. 1.2

Construct a square with one of its sides as OA of unit length and take a point P on the line such that OP is equal in length to the diagonal of this square. It will now be shown that the point P *cannot* correspond to a rational number, *i.e.,* the length of OP cannot have a rational number as its measure.

If possible, let its measure be a rational number p/q so that by the Pythagoras' theorem, we have

$$(p/q)^2 = 1^2 + 1^2 = 2 \Rightarrow p^2 = 2q^2.$$

We suppose that the natural numbers p and q have no common factor, for such factors, if any, can be cancelled to begin with.

Firstly we notice that

$$(2n)^2 = 4n^2, \qquad (2n + 1)^2 = (4n^2 + 4n) + 1$$

so that the square of an even number is even and that of an odd number is odd.

From the equation (i), we see that p^2 is an even number. Therefore, p itself must be even.

Let then, p be equal to $2n$ where n is an integer. We have $4n^2 = 2q^2 \Rightarrow q^2 = 2n^2 \Rightarrow q^2$ is even $\Rightarrow q$ is even.

Hence, p and q have a common factor 2 and this conclusion contradicts the hypothesis that they have no common factor. Thus, the measure $\sqrt{2}$ of *OP* is not a rational number. There exists, therefore, a point on the number axis not corresponding to any rational number.

Again, we take a point L on the line such that the length OL is any rational multiple say, p/q, of OP.

The length OL cannot have rational measure. For if possible, let m/n be the measure of OL so that,

$$\frac{p}{q}\sqrt{2} = \frac{m}{n} \Rightarrow \sqrt{2} = \frac{mq}{np}$$

which states that $\sqrt{2}$ is a rational number, being equal to mq/np. Thus we arrive at contradiction.

Hence L cannot correspond to a rational number.

Thus we see that there exist an unlimited number of points on the number line which do not correspond to any rational number.

If we now require that our set of numbers should be such that after the choice of unit length on the line, every point of the line should have a number corresponding to it (or that *every length should be capable of measurement*), we are obliged to extend our system of numbers further by the introduction of *Irrational numbers*.

We will thus associate an irrational number to every point of the line which does not correspond to a rational number.

Real number. *A number, rational or irrational, is called a real number.*

The set of rational and irrational numbers is thus the set of real numbers.

This set is denoted as **R.**

Each real number is represented by some point of the number axis and each point of the number axis has some real number, rational or irrational, corresponding to it.

The positive real numbers correspond to points on the right of O and the negative real numbers to points on the left of O.

We might say that each positive real number is the measure of some length OP and that the set of positive real numbers is adequate to measure every length.

A method of representing irrational numbers in the decimal notation is given in Article 1.4.

1.3.1. Number and point. If any real number, say x, is represented by a point P, then we usually say that *the point* P is x.

Thus, the terms, number and point, are generally used in an indistinguishable manner.

In the following, unless otherwise stated by a number will be meant a real number.

1.3.2. Closed and open intervals. Let a, b be two given numbers such that $a < b$. Then the set of numbers x such that

$$a \leq x \leq b$$

is called a **closed** interval denoted by the symbols $[a, b]$.

We generally describe this situation as follows :

$$[a, b] = \{x : a \leq x \leq b\},$$

so that the closed interval denoted by $[a, b]$ is the set of numbers x satisfying the condition $a \leq x \leq b$. Thus

$$x \in [a, b] \Leftrightarrow a \leq x \leq b.$$

The set of numbers x such that

$$a < x < b$$

is called an **open** interval denoted by the symbol $] a, b [$.

Thus $] a, b [= \{x : a < x < b\}$

We may also have **semi-closed** or **semi-open** intervals $] a, b]$, $[a, b[$ such that,

$$] a, b] = \{x : a < x \leq b\} \text{ and } [a, b [= \{x : a \leq x < b\}.$$

The number $b - a$ is referred to as the length of each of intervals

$$[a, b],] a, b [,] a, b], [a, b [.$$

1.4. Decimal Representation of Real Numbers.

Let P be any given point of the number line. We now seek to obtain the decimal representation of the number associated with the point P.

To start with, we suppose that the point P lies on the positive side of O.

Let the points corresponding to integers be marked on the number line so that the whole line is divided into intervals of length one each.

Now if P coincides with someone of these points of division, it corresponds to an integer and we need proceed no further. In case P falls *between* two points of division, say $a, a + 1$, we sub-divide the interval $[a, a + 1]$ into 10 equal parts so that the length of each part is 1/10. The points of division, now, are

$$a, a + \frac{1}{10}, a + \frac{2}{10}, \ldots\ldots\ldots, a + \frac{9}{10}, a + 1.$$

If *P coincides with anyone of these points of division, then it corresponds to a rational number.* In the alternative case, it falls between two points of division, say

$$a + \frac{a_1}{10}, a + \frac{a_1 + 1}{10}, i.e., a . a_1, a . (a_1 + 1),$$

where a_1, is anyone of the integers 0, 1, 2,3,............9.

We again sub-divide the interval

$$\left[a + \frac{a_1}{10}, a + \frac{a_1 + 1}{10} \right]$$

into 10 equal parts so that the length of each part is $1/10^2$.

These points of division, now, are

$$a + \frac{a_1}{10}, a + \frac{a_1}{10} + \frac{1}{10^2}, a + \frac{a_1}{10} + \frac{2}{10^2}, \ldots\ldots, a + \frac{a_1}{10} + \frac{9}{10^2}, a + \frac{a_1 + 1}{10}.$$

The point *P* will either coincide with one of the above points of division (in which case it corresponds to a rational number) or will lie between two points of division, say

$$a + \frac{a_1}{10} + \frac{a_2}{10^2}, a + \frac{a_1}{10} + \frac{a_2 + 1}{10^2}, i.e., a . a_1 a_2, a . a_1 (a_2 + 1),$$

where a_2 is one of the integers 0, 1, 2, ..., 9

We again sub-divide this last interval and continue to repeat the process. After a number of steps, say n, the point P will either be found to coincide with some point of division (in this case it corresponds to a rational number) or lie between two points of the form

$$a + \frac{a_1}{10} + \frac{a_2}{10^2} + \ldots + \frac{a_n}{10^n}, a + \frac{a_1}{10} + \frac{a_2}{10^2} + \ldots + \frac{a_n + 1}{10^n},$$

i.e., $\qquad a . a_1 a_2 \ldots\ldots\ldots a_n, a . a_1, a_2 a_3 \ldots\ldots\ldots (a_n + 1),$

the distance between which is $1/10^n$ and which gets smaller and smaller as n increases.

The process can clearly be continued indefinitely.

The successive intervals in which P lies go on shrinking in length and will clearly close up to the point P. This point P is then represented by the infinite decimal

$$a . a_1, a_2, a_3 \ldots\ldots\ldots$$

Conversely, consider any infinite decimal

$$a . a_1, a_2, a_3 \ldots\ldots\ldots a_n \ldots\ldots\ldots$$

and construct the series of intervals.

$$[a, a + 1], [a . a_1, a. (a_1 + 1)], [a . a_1 a_2, a . a_1 (a_2 + 1)], \ldots\ldots\ldots$$

Each of these intervals lies within the preceding one. Their lengths go on diminishing and by taking n sufficiently large, we can make the length as near to zero as we like. We thus see that these intervals close up to a point which belongs to each of these intervals. This fact is related to the intuitively perceived aspect of the continuity of a straight line.

Thus, there is one and only one point common to this series of intervals and this is the point represented by the decimal.

$$a . a_1 a_2 a_3 \ldots\ldots$$

Combining the results of this article with that of Ex. 2, § 1.2.5, p. 3, we see that *every decimal, finite or infinite, denotes a number which is rational if the decimal is terminating or recurring and irrational in the contrary case.*

Let now, P lie on the negative side of O. Then the number representing it is

$$-a . a_1 a_2 \ldots\ldots\ldots a_n \ldots\ldots$$

where

$$a . a_1 a_2 \ldots\ldots\ldots a_n \ldots\ldots\ldots$$

is the number representing the point P′ on the positive side of O such that PP′ is bisected at O.

Illustrations. The numbers

$$2. 12122122212222\ldots\ldots$$

$$0. 01234001234000123400001234\ldots\ldots$$

are irrational.

Ex. 1. Calculate the cube root of 2 to three decimal places. We have,

$$1^3 = 1 < 2 \text{ and } 2^3 = 8 > 2$$

$\Rightarrow \qquad\qquad 1 < \sqrt[3]{2} < 2.$

We consider the numbers

$$1, 1\cdot1, \cdot1\cdot2, \ldots\ldots\ldots1\cdot9, 2$$

which divide the interval [1 , 2] into 10 equal parts and find two successive numbers such that the cube of the first is < 2 and that of second is > 2. We find that

$$(1.2)^3 = 1.728 < 2$$

and $\qquad\qquad (1.3)^3 = 2.197 > 2$

$\Rightarrow \qquad\qquad 1.2 < \sqrt[3]{2} < 1.3.$

Again consider the numbers

$$1\cdot2, 1\cdot21, 1\cdot22, \ldots\ldots\ldots, 1\cdot29, 1\cdot3$$

which divide the interval [1.2. 1.3] into 10 equal parts. We find that

$$(1.25)^3 = 1.953125 < 2$$

and $\qquad\qquad (1.26)^3 = 2.000376 > 2$

$\Rightarrow \qquad\qquad 1.25 < \sqrt[3]{2} < 1.26$

Again, consider the numbers

$$1.25, 1.251, 1.252, \ldots\ldots\ldots, 1.259, 1.26$$

which divide the interval, [1.25, 1.26] into 10 equal parts. We find that

$$(1.259)^3 = 1.995616979 < 2$$

and $\qquad\qquad (1.26)^3 = 2.000376 > 2$

$\Rightarrow \qquad\qquad 1.259 < \sqrt[3]{2} < 1.26.$

Hence $\qquad\qquad \sqrt[3]{2} = 1.259\ldots\ldots$

Thus, to three decimal places, we have

$$\sqrt[3]{2} = 1.259.$$

Ex. 2. Calculate the cube root of 5 to 2 decimal places.

Ex. 3. Give any four different irrational numbers in decimal notation and arrange them in order of magnitude.

Ex. 4. Consider the following two irrational numbers

$$0 \cdot 10201022010222010222201022222201 \ldots\ldots,$$

$$0 \cdot 120121201212120\ldots\ldots$$

which of these two numbers is greater than the other. Insert any three irrational numbers between them.

Ex. 5. Consider the following numbers :

$$x = \cdot 430340430340430$$

$$y = \cdot 234223422234222234.$$

Show that $.6 < x + y < .8,$ $.66 < x + y < .68,$

$$.08 < xy < .15,   .09 < xy < .1.$$

Note. The method described above in Ex. 1, which is indeed very cumbersome has only been given to illustrate the basic and elemenrary nature of the problem. In actual practice, however, other methods involving infinite series or other limiting processes are employed.

ANSWERS

2.   1.7104,   4. Second > first.

1.5. The Modulus of a Real Number

Let x be any given real number. We have the following possibilities :

$$x > 0,   x < 0,   x = 0.$$

We write $|x| = \begin{cases} x \text{ if } \geq 0, \\ -x \text{ if } x < 0, \end{cases}$

and call $|x|$ the *Modulus* or the *Absolute value* of the real number x.

Thus $|3| = 3, |-3| = -(-3) = 3, |0| = 0,$

• $|5 - 7| = |7 - 5| = 2.$

The modulus of the difference between two numbers is the measure of the distance between the corresponding points on the number line.

It would be seen that $\forall\ x \in \mathbf{R},$ we have

$$|x| \geq 0,$$

$$|x| \geq x, |x| \geq -x.$$

In fact $|x| = \max \{x, -x\},$

where $\max \{x, -x\}$ denotes the greater of the two numbers x and $-x$.

We may also see that $|x|$ denotes the positive square root of x^2 so that we have

$$|x| = \sqrt{(x^2)}.$$

Some results involving moduli. We now state some simple and useful results involving the moduli of numbers.

1.5.1. $|a+b| \le |a|+|b|$

i.e., the modulus of the sum of two numbers is less than or equal to the sum of their moduli.

We have $\qquad a \le |a|, \quad b \le |b|$

$\Rightarrow \qquad a+b \le |a|+|b|.$

Also $\qquad -a \le |a|, -b \le |b|$

$\Rightarrow \qquad -(a+b) \le |a|+|b|$

Now $\qquad |a+b| = \max \{(a+b), -(a+b)\}.$

Thus $\qquad |a+b| \le |a|+|b|.$

In fact, we have

$\qquad |a+b| = |a|+|b|,$ if a, b have the same sign,

and $\qquad |a+b| < |a|+|b|,$ if a, b have opposite signs.

For example.

$$|7+3| = |7|+|3|$$
$$|-7-3| = |-7|+|-3|$$
$$4 = |7-3|$$
$$= |7+(-3)|$$
$$< |7|+|-3|$$
$$= 10$$

1.5.2. $|ab| = |a||b|$

i.e. the modules of the product of two numbers is equal to the product of their moduli.

We have $\qquad |a| = \sqrt{(a^2)}, \qquad |b| = \sqrt{(b^2)}$

$\qquad |a||b| = \sqrt{(a^2)}\sqrt{(b^2)} = \sqrt{a^2 b^2} = \sqrt{(ab)^2} = |ab|.$

For example,

$$|4.3| = 12 = |4|.|3|,$$
$$|(-4)(-3)| = 12 = |-4|.|-3|,$$
$$|(-4).(3)| = 12 = |-4|.|3|.$$

1.5.3. $|a-b| \ge ||a|-|b||$

we have

$$|a| = |(a-b)+b| \le |a-b|+|b|$$

$\Rightarrow \qquad |a|-|b| \le |a|-|b| \qquad\qquad\qquad\qquad\qquad \dots(i)$

Again

$$|b| = |(b-a)+a| \le |b-a|+a|$$

$\Rightarrow \qquad |b|-|a| \le |b-a| = |a-b| \qquad\qquad\qquad\qquad \dots(ii)$

Since

$$||a|-|b|| = \max \{|a|-|b|, -(|a|-|b|)\}$$

We have from (i) and (ii)

$$||a|-|b|| \le |a-b|$$

1.5.4. $|x-a| < 1 \Leftrightarrow a-1 < x < a+1$ $\qquad\qquad\qquad\qquad\qquad \dots(A)$

The result can be stated in another form as follows :

$$|x-a| < l \Leftrightarrow x \in]a-l, a+l[$$

We have

$$|x-a| = \max\{(x-a), (a-x)\} < l$$
$$\Leftrightarrow x-a<l \wedge a-x<l$$
$$\Leftrightarrow x<a+l \wedge x>a-l$$
$$\Leftrightarrow a-l<x<a+l.$$

Note. It is easy to see the truth of the result (A) by means of the Number line. The inequality $|x-a| < l$ implies that the numerical difference between

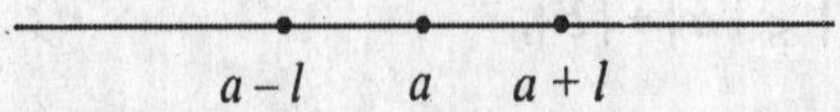

$$\begin{array}{ccc} \bullet & \bullet & \bullet \\ a-l & a & a+l \end{array}$$

Fig. 1.3

a and x must be less than l, so that the point x (which may lie to the right or to the left of a) can be only at a distance less than l from the point a.

Now, from the figure, we see that this is possible, if and only if, x lies between $a-l$ and $a+l$. It follows that

$$|x-a| \le l \Leftrightarrow a-l \le x \le a+l.$$

EXERCISES

1. Give the equivalents of the following inequalities in terms of the modulus notation :

    (*i*)　$-1 \le x \le 3$,　　　(*ii*)　$2<x<5$,　　　(*iii*)　$-3 \le x \le 7$,　　　(*iv*)　$l-\varepsilon<x<l+\varepsilon$.

2. Give the equivalents of the following by doing away with the modulus notation :

    (*i*)　$|x-2|<3$,　　　(*ii*)　$|x+1| \le 2$,　　　(*iii*)　$0<|x-1|<2$.

3. Which of the following statements are true :

    (*i*)　$\{x: |x-3|<4\} = \{x: -1<x<7\}$

    (*ii*)　$\{x: |4-x|<1\} = \{x: 3<x<5\}$

    (*iii*)　$\{x: |1-x|<2\} = \{x: 1<x<3\}$

4. Show that :

    (*i*)　$\left|\dfrac{1}{x}\right| = \dfrac{1}{|x|}$ if $x \ne 0$　　(*ii*)　$\left|\dfrac{x}{y}\right| = \dfrac{|x|}{|y|}$ if $y \ne 0$　(*iii*)　$|x|<\varepsilon \Leftrightarrow -\varepsilon<x<\varepsilon$

5. If　$y=|x|+|x-1|$, show that :

    $$y = \begin{cases} 1-2x, & \text{for } x \le 0, \\ 1, & \text{for } 0<x<1. \\ 2x-1, & \text{for } x \ge 1. \end{cases}$$

6. Show that :

    $$\left.\begin{array}{l} |a-b| <l \\ |b-c| <m \end{array}\right\} \Rightarrow |a-c|<l+m.$$

    [We have $|a-c| = |a-b+b-c| \le |a-b|+|b-c|<l+m.$]

ANSWERS

1. (*i*)　$|x-1| \le 2$.　　(*ii*)　$\left|x-\dfrac{7}{2}\right| < \dfrac{3}{2}$.　　(*iii*)　$|x-2| \le 5$.　　　　(*iv*)　$|x-l|<e$.

2. (*i*)　$-1<x<5$.　　(*ii*)　$-3 \le x \le 1$.　　(*iii*)　$-1<x<3, x \ne 1$

3. (*i*),　(*ii*)

CHAPTER
2
FUNCTIONS AND GRAPHS– ELEMENTARY FUNCTIONS

2.1 Functions : Domain and Range of a function

In this chapter, we shall introduce the notion of '*Function*' which is the cornerstone of the study of all mathematics including Calculus.

Let A and B be any two sets and let f denotes a rule which associates to each member of A a member of B. We say that f is *a function from A into B*. Also A is said to be **Domain** of this function.

If x denotes a member of the set A, then the member of the set B, which the function f associates to $x \in A$, is denoted by $f(x)$ called the value of the function f for x or at x. The function may be described as $x \to f(x)$, or $y = f(x)$ where $x \in A$ and $y \in B$.

Range of a function : Let f be a function from a set A into a set B so that the domain of f is the set A. Then the set of all the *function values* is called the **Range** of f. Thus,

$$\text{Range of } f = \{f(x) : x \in A\}.$$

While we shall be exclusively concerned with functions for which the sets A and B are some subsets of the set **R** of real numbers, the notion of functions may also be seen to be present in non-number situations as illustrated below.

1. Let A denote the set of students in some class of a school. Suppose that the class is given a test in some subject, say, Mathematics.

Here we have a situation in which we associate to each student member of the class the marks obtained by him in the test. Let us denote this association by f. If x denotes a student member of the class, then $f(x)$ denotes the marks obtained by the student x in the test.

2. Consider the set of financial years beginning from say, 1948- 49 to 1992- 93. During each of these financial years, India exported goods of certain value. Thus here we have a situation in which we associate to each of these financial years the value of the exports during that year. If f denotes the association refered to here and x denotes anyone of the financial years, in question, then $f(x)$ denotes the value of exports during the financial year x.

We shall now consider some examples of functions from the set of real numbers into the set of real numbers. Such functions are called Real valued functions of a real variable. Before giving these examples, however, we shall indicate the usual types of domains of functions.

Note. We generally refer $f(x)$ as a function rather than f. This terminology is rather erroneous in as much as $f(x)$ denotes the number which the function f associates to the number x and not the function f.

11

2.1.1 Some Types of Domains of Functions : Usually the domain of a function is an interval, open, closed, semi-closed or semi-open.

Domains may also extend without bound in one or the other directions *i.e.* the domain may be an interval of any of the following types :

$$]-\infty, b\,] = \{x : x \le b\}, \qquad [a, \infty\,[= \{x : x \ge \infty\},$$
$$]-\infty, b\,[= \{x : x < b\}, \qquad]a, \infty\,[= \{x : x > a\}.$$

The domain of a function may be **R** itself.

The set **R** is sometimes written as $]-\infty, \infty\,[$.

Sometimes when a function is defined by means of some expression, it may not be necessary to indicate the domain of the function. In that case, the domain is supposed to consist of all those numbers for which the expression defining the function has a meaning, unless otherwise stated.

2.1.2 Independent and Dependent Variables : The symbol which denotes a member of a domain of a function is called an *Independent Variable*, and the symbol denoting a member of the range of a function is called a *Dependent Variable*. Also if x denotes the independent variable, we may refer to the function as a function of x.

EXAMPLES

1. Consider numbers x and y connected by the relation $y = \sqrt{(1 - x^2)}$, where we take only the positive value of the square root.

We observe that there is no real number whose square is negative and hence, so far as real numbers are concerned, the square root of a negative number does *not* exist.

Now $1 - x^2$, is non - negative, if and only if, x^2 is less than or equal to 1 *i.e.* if x satisfies the relation

$$-1 \le x \le 1 \Leftrightarrow x \in [-1, 1].$$

Thus the given equation associates a real number y only to such of the real numbers x which belong to $[-1, 1]$. We say that the given equation defines a real valued function of a real variable or simply a function, with domain $[-1, 1]$. This function may be described as $y = \sqrt{(1 - x^2)}$, $x \in [-1, 1]$.

2. Consider the numbers x and y connected by the relation

$$y = (x - 1)/(x - 2).$$

The determination of y for $x = 2$ involves the meaningless operation of division by zero and, therefore, the relation does not assign any value to y corresponding to $x = 2$.

Thus the relation $y = (x - 1)/(x - 2)$ defines a function whose domain is the set of real numbers excluding 2 *i.e.*, the set **R** ~ {2}.

The function defined by the relation $y = (x - 1)/(x - 2)$ may be described as $y = (x - 1)/(x - 2)$, $x \in \mathbf{R} \sim \{2\}$.

3. Consider the two numbers x and y with their relationships defined by the equations

$y = x^2$	when	$x < 0$	...(*i*)
$y = x$	when	$0 \le x \le 1,$	...(*ii*)
$y = 1/x$	when	$x > 1$	...(*iii*)

These relations assign a definite value to y corresponding to every value of x, although the value is not determined by a single expression, as was the case in Ex. 1 and Ex. 2. In order to determine a value of y corresponding to a given value of x, we have to select one of the three equations depending upon the value of x in question.

For instance

$$x = -2 \qquad \Rightarrow \qquad y = (-2)^2 = 4 \qquad\qquad \text{[Equation } (i) \because -2 < 0$$

$$x = 1/2 \qquad \Rightarrow \qquad y = 1/2 \qquad\qquad \text{[Equation } (ii) \because 0 < \tfrac{1}{2} < 1$$

$$x = 3 \qquad \Rightarrow \qquad y = 1/3 \qquad\qquad \text{[Equation } (iii) \because 3 > 1$$

Here we have a function whose domain is **R**.

This example illustrates an important point that it is not necessary that only one expression should be used to determine a function. What is required is simply the existence of a law or laws which associates a number to each member of the domain of the function.

4. The domain of the function

$$y = 1/\sqrt{(1-x)(x-2)}$$

is the open interval] 1, 2 [.

For $x = 1$ and 2 the denominator $(1-x)(x-2)$ becomes zero. Also for $x < 1$ and $x > 2$ the expression $(1-x)(x-2)$ under the radical sign becomes negative.

5. The domain of the function $y = \sqrt{(1-x)(x-2)}$ is the closed interval [1, 2].

6. The domains of the functions

$$y = 1/\sqrt{x}, \qquad y = 1/\sqrt{(1-x)}$$

are] 0, ∞ [and] – ∞, 1 [respectively.

7. The domain of the function $y = x$! is the set of Natural Numbers.

8. The domain of the function $y = |x|$ is **R**.

9. The domain and range of $y = x^2$ is **R** and [0, ∞ [respectively.

10. The domain and range of $y = \sin x$ is **R** and [–1, 1] respectively.

11. Let $y = \dfrac{1}{q}$ when x is a rational number $\dfrac{p}{q}$ in its lowest terms

$= 0$, when x is irrational.

The domain of this function is **R** – {0}. What is the range ?

EXERCISES

1. Give the domains of the following functions :

(i) $y = \dfrac{2x}{2x+7}$,

(ii) $y = \dfrac{1}{x^3 - x}$,

(iii) $y = \sqrt{(x^3 - x)}$,

(iv) $y = \sqrt{\left(\dfrac{1}{x} - 2\right)}$.

(v) $y = \sqrt{\left(x - \dfrac{x}{1-x}\right)}$,

(vi) $y = \sqrt{\left(\dfrac{1}{\sqrt{x}} - \sqrt{(x+1)}\right)}$.

(vii) $y = 1/(1 + \cos x)$,

(viii) $y = \sqrt{(1 + 2\sin x)}$.

2. Give the domains of the following functions :

(i) $f(x) = \dfrac{x^2 - 3x - 1}{x - 2}$

(ii) $f(x) = \sqrt{(x^2 - 1)}$

(iii) $f(x) = \sqrt{\dfrac{1 - |x|}{2 - |x|}}$

(iv) $f(x) = \sqrt{(-x^2 + 6x - 9)}$ (v) $f(x) = \dfrac{x + 3}{\sqrt{(x^2 - 5x + 4)}}$ (vi) $f(x) = \dfrac{x^2 + 3}{x^2 + |x|}$

(vii) $f(x) = \sqrt{(1 - |x|)}$,   (viii) $f(x) = \dfrac{1}{\sqrt{(-x^2 + 6x - 9)}}$

3. Give the differences in the domains of the following pairs of functions :

(i) $y = \sqrt{(1 - x)}$ and $y = \sqrt{(1 - x^2)}$,

(iii) $y = \dfrac{1}{1 - x}$ and $y = \dfrac{1}{1 - x^2}$.

4. What is the difference between the domains of the following pairs of functions :

(i) $y = \dfrac{x^2}{x}$ and $y = x$.

(ii) $y = \sqrt{[x^2 (x + 1)]}$ and $y = x \sqrt{(x + 1)}$.

(iii) $y = \sqrt{(1 + x^2)}$ and $y = \sqrt{(x^2 - 1)}$.

(iv) $y = x - 5$ and $y = \sqrt{(x - 5)^2}$.

ANSWERS

1. (i) $\mathbf{R} \sim \left\{ -\frac{7}{2} \right\}$ (ii) $\mathbf{R} \sim \{-1, 0, 1\}$ (iii) $[-1, 0] \cup [1, +\infty[$

(iv) $]0, \frac{1}{2}]$ (v) $]1, +\infty[$ (vi) $[0, \frac{1}{2}(-1 + \sqrt{5})]$

(vii) $\mathbf{R} \sim \{(2n+1)\pi : n \in I\}$ (vii) $\left\{ \left[\left(2n - \frac{1}{6}\right) \pi, \left(2n + \frac{7}{6}\right) \pi \right] : n \in \mathbf{I} \right\}$

2. (i) $\mathbf{R} \sim \{2\}$,   (ii) $]-\infty, -1] \cup [1, +\infty[$

(iii) $[-1, 1] \cup]2, +\infty[\cup]-\infty, -2[$ (iv) $\{3\}$

(v) $]-\infty, 1[\cup]4, +\infty[$ (vi) $\mathbf{R}$ (vii) $[-1, 1]$

(viii) For no value of x.

3. (i) $]-\infty, 1], [-1, 1]$.

The domain of $y = \sqrt{(1 - x^2)}$ is a proper sub-set of that of $y = \sqrt{(1 - x)}$.

(ii) $\mathbf{R} \sim \{1\}$, $\mathbf{R} \sim \{-1, 1\}$

While -1 is a member of the domain of $y = 1/(1 - x)$, -1 is not a member of that of $y = 1/(1 - x^2)$.

4. (i) $\mathbf{R} \sim \{0\}$, $\mathbf{R}$.   (ii) Same domain.

(iii) $[-1, 1]$, $\mathbf{R} \sim]-1[$.   (iv) $\mathbf{R}, \mathbf{R}$.

2.2 Graphs of functions

Let f be a function with domain $[a, b]$ We have

$$y = f(x), \qquad x \in [a, b].$$

The function f associates to each $x \in [a, b]$ a number denoted by $f(x)$.

We shall here introduce cartesian graphs of functions.

To represent the function graphically, we take two straight lines X'OX and Y'OY, called co-ordinate axes, at right angles to each other, as in Plane Analytical Geometry. We take O as origin for both the axes and select unit intervals on OX, OY (usually of the same lengths).

To a number $x \in [a, b]$ corresponds a point M on X-axis such that OM = x. Again the function f associates to $x \in [a, b]$ a number $f(x)$. We write $f(x) = y$. We now have a point N on Y-axis such that ON = $f(x) = y$.

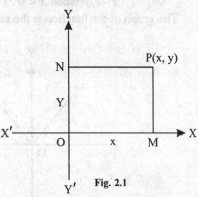

Completing the rectangle OMPN, we obtain a point P which is said to correspond to the pair of numbers $[x, f(x)]$ i.e., (x, y).

The set of points (x, y) obtained by giving different values to x, is said to be the **graph** of the function f. Also $y = f(x)$ is said to be the equation of the graph.

Fig. 2.1

While we are considering here **Cartesian graphs,** we shall later on introduce the notion of **Polar Graphs.** The graph of a function f is also said to be the graph of the equation $y = f(x)$.

EXAMPLES

1. The graph of the function

$y = x^2, x < 0; y = x, x \in [0, 1]; y = 1/x, x \in [1, \propto [\text{ is Fig. 2.2.}$

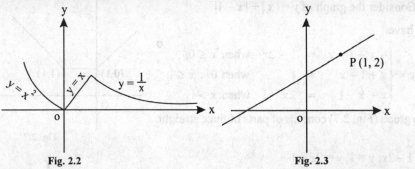

Fig. 2.2   Fig. 2.3

2. The graph (Fig. 2.3) of the function $y = (x^2 - 1)/(x - 1)$ is the straight line $y = x + 1$, excluding the point P (1, 2).

3. The graph of $y = x$ consists of the set of points

(1, 1), (2, 2) (3, 6), (4, 24), etc.

4. The graph of the function f defined as follows :

$$f(x) = \begin{cases} x, & \text{when } 0 \le x < \frac{1}{2} \\ 1, & \text{when } x = \frac{1}{2} \\ 1 - x & \text{when } \frac{1}{2} < x \le 1, \end{cases}$$

is as given in Fig. 2.4.

5. The graph of the function which associates to each real number x the positive square root of x^2 is given in Fig. 2.5

As $\sqrt{x^2} = x$ or $-x$ according as x is positive or negative,

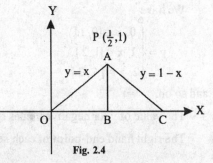

Fig. 2.4

we have

$$f(x) = \begin{cases} x, & \text{when } x \ge 0 ; \\ -x, & \text{when } x < 0 . \end{cases}$$

The graph of the function is the same as that of $y = |x|$.   *(Calicut 2004)*

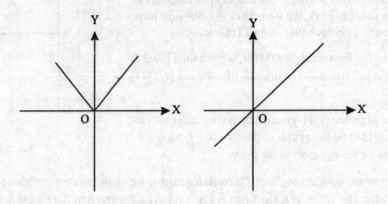

Fig. 2.5   Fig. 2.6

The student should compare the graph (Fig. 2.5) of
$y = \sqrt{x^2}$ with the graph (Fig. 2.6) of $y = x$.

6. Consider the graph of $y = |x| + |x - 1|$

We have

$$y = \begin{cases} -x + 1 - x & = 1 - 2x & \text{when } x \le 0, \\ x + 1 - x & = 1 & \text{when } 0 < x \le 1, \\ x + x - 1 & = 2x - 1 & \text{when } x > 1 \end{cases}$$

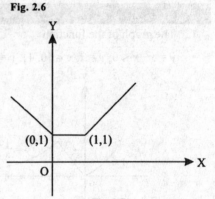

(0,1)   (1,1)

The graph (Fig. 2.7) consists of parts of three straight
lines,

Fig. 2.7

$y = 1 - 2x; y = 1, y = 2x - 1$

corresponding to the intervals

$] - \infty , 0],] 0, 1],] 1, \infty [.$

7. The graph of $y = [x]$ where $[x]$ denotes the greatest
integer not greater than x is given in Fig. 2.8.

We have

$$y = \begin{cases} 0, x \in [0, 1 [\\ 1, x \in [1, 2 [\\ 2, x \in [2, 3 [\end{cases}$$

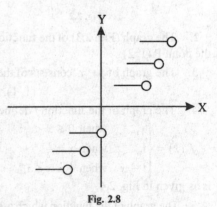

and so on.

Fig. 2.8

The value of y for negative values of x can also be similarly given.

The right hand end-**point** of each segment is *not* a **point** of the graph.

1. Draw the graphs of the functions with the function values as given :

(i) $f(x) = \begin{cases} 1, & \text{when } x \leq 0 \\ -1, & \text{when } x > 0 \end{cases}$

(ii) $f(x) = \begin{cases} x, & \text{when } 0 \leq x < \frac{1}{2} \\ 1 - x, & \text{when } \frac{1}{2} \leq x \leq 1 \end{cases}$

(iii) $f(x) = \begin{cases} x, & \text{when } 0 \leq x \leq \frac{1}{2} \\ 2 - x, & \text{when } \frac{1}{2} < x < 1 \end{cases}$

(iv) $f(x) = \begin{cases} x, & \text{when } 0 \leq x < \frac{1}{2} \\ 1, & \text{when } \quad x = \frac{1}{2} \\ 1 - x, & \text{when } \frac{1}{2} < x < 1 \end{cases}$

(v) $f(x) = \begin{cases} x^2, & \text{when } x \leq 0 \\ \sqrt{x}, & \text{when } x > 0 \end{cases}$

(vi) $f(x) = \begin{cases} 1/x, & \text{when } x < 0 \\ 0, & \text{when } x = 0 \\ -1/x, & \text{when } x > 0 \end{cases}$

2. Give the graph of the following functions :

(i) $y = \dfrac{\sqrt{x^2}}{x}$,

(ii) $y = x + \dfrac{\sqrt{(x-1)^2}}{x-1}$,

(iii) $y = x + \dfrac{\sqrt{(x-1)^2}}{x-1} + \dfrac{\sqrt{(x-2)^2}}{x-2}$,

(iv) $y = x + \dfrac{\sqrt{(x-1)^2}}{x-1} + \dfrac{\sqrt{(x-2)^2}}{x-2} + \dfrac{\sqrt{(x-3)^2}}{x-3}$.

The positive value of the square root is to be taken in each case.

Give also the domain of the function in each case.

3. Give the graphs of the following functions :

(i) $y = |x|$,

(ii) $y = |x| + |x + 1|$,

(iii) $y = 2|x - 1| + 3|x + 2|$,

(iv) $y = [x]^2$,

(v) $y = [x] + [x + 1]$.

4. Draw the graphs of the following functions :

(i) $y = [2x]$,

(ii) $y = 2[x]$,

(iii) $y = 2x + [x - 1]^2$,

(iv) $x = x^2 + [x]^2$.

5. Give the graphs of the following functions for the domain [0, 8]. Give also the range in each case :

(i) $y = \left[x + \dfrac{1}{2}\right]$,

(ii) $y = [x] + \frac{1}{2}$,

(iii) $y = x - [x]$, *(Calicut 2004)*

(iv) $y = x + \lfloor x \rfloor$,

(v) $y = [x]^2$,

(vi) $y = [x^2]$,

(vii) $y = x \cdot [x]$,

(viii) $y = \left[\frac{1}{2} x\right] + 1$,

(ix) $y = x^2 + [x]$,

(x) $y = x^2 - [x]$,

(xi) $y = 1/3^{[x]}$,

(xii) $y = 3^{[3x]}$,

(xiii) $y = [|x|]$,

(xiv) $y = |[x]|$.

6. Sketch the graph of the following function $y = \text{sig}(x)$ where sig (x) is defined as follows :

$$\text{sig}(x) = \begin{cases} 1 & \text{for } x > 0, \\ 0 & \text{for } x = 0, \\ -1 & \text{for } x < 0. \end{cases}$$

7. Give the graphs of the following equations :

     (*i*) $y = \text{sig}|x|$,   (*ii*) $y = \text{sig}[x]$,

     (*iii*) $y = \text{sig}(x - |x|)$.

ANSWERS

2.   (*i*) $R \sim \{0\}$ (*ii*) $R \sim \{1\}$

    (*iii*) $R \sim \{1, 2\}$ (*iv*) $R \sim \{1, 2, 3\}$

2.3 Operations on Functions

Let f, g be two functions with domains denoted by D_f, D_g respectively. We now proceed to give meanings to the symbols.

$$f + g, f - g, fg, f \div g.$$

To define $f + g$, we have to give a meaning to $(f + g)(x)$ where x belongs to the domain of $f + g$. A natural way of defining $(f + g)(x)$ is as follows :

$$(f + g)(x) = f(x) + g(x).$$

Since we talk of $f(x) + g(x)$, the number x must necessarily belong to the domain of f, as also to that of g and as such x must be a member of the intersection $D_f \cap D_g$ of the domains of the functions f and g.

Thus we have

$$(f + g)(x) = f(x) + g(x) \;\; \forall x \in D_f \cap D_g$$

Similarly, we have the following definitions :

$$(f - g)(x) = f(x) - g(x) \;\; \forall \; x \in D_f \cap D_g.$$

$$(fg)(x) = f(x) g(x) \;\; \forall \; x \in D_f \cap D_g.$$

While we define $f \div g$ as follows :

$$(f \div g)(x) = f(x) \div g(x),$$

it is necessary to exclude those of the members of the domain D_g of the function g for which $g(x)$ is 0. Thus the domain of the function $f \div g$ is the set

$$D_f \cap [D_g \sim \{x : g(x) = 0\}].$$

Ex. Given that the functions f and g are defined as follows :

     (*i*) $f(x) = \sqrt{x},$ $g(x) = 1/x^2,$

     (*ii*) $f(x) = x^2 - 3x + 1,$ $g(x) = 1/(x - 2),$

     (*iii*) $f(x) = \sqrt{(x^2 - 5x + 4)},$ $g(x) = x + 3,$

     (*iv*) $f(x) = \sqrt{(\sin x)},$ $g(x) = 1/(1 - \sin x),$

what are the domains of the functions fg and $f \div g$ in each case.

    (*i*) $]0, +\infty[,]0, +\infty[$

    (*ii*) $\mathbf{R} \sim \{2\}, \mathbf{R} \sim \{2\}$.

    (*iii*) $\mathbf{R}, \mathbf{R} \sim \{3\}$

    (*vi*) $\{2n\pi + (2n+1)\pi\} : n \in \mathbf{I}\} \sim \{2n\pi + \pi/2\} : n \in \mathbf{I}\}$

          $\{[2n\pi + (2n+1)\pi] : n \in \mathbf{I}\} \sim \{(2n\pi + \pi/2) : n \in \mathbf{I}\}$.

2.3.1. Polynomial functions and Rational functions.

A function of the form

$$f(x) = a_0 x^n + a_1 x^{n-1} + \ldots\ldots\ldots + a_{n-1} x + a_n$$

where $a_0, a_1, a_2, \ldots\ldots\ldots, a_n$ are given real numbers and $a_0 \neq 0$ is called a *Polynomial function* of degree n. The domain of every polynomial function is $\mathbf{R}$.

A function of the from

$$g(x) = \frac{a_0 x^n + a_1 x^{n-1} + \ldots + a_{n-1} x + a_n}{b_0 x^m + b_1 x^{m-1} + \ldots + b_{m-1} x + b_m}$$

where $a_0 x^n + a_1 x^{n-1} + \ldots + a_{n-1} x + a_n, b_0 x^m + b_1 x^{m-1} + \ldots + b_{m-1} x + b_m$, are polynomials is called a *Rational function*.

The domain of a rational function is the set of all those real numbers for which the value of the polynomial in the denominator is *not zero*.

2.3.2. Constant functions and the Identity function.

It is interesting to see that every polynomial function and every rational function can be obtained as a result of the operations of addition, subtraction, multiplication and division performed on the *Constant functions* and *Identity function* which we now define.

Let a be any given number. Then the function $f(x) = a \; \forall \; x \in \mathbf{R}$ is called a *Constant function*, also denoted by a permitting an abuse of language.

The function $f(x) = x, \; \forall \; x \in \mathbf{R}$ is called the *Identity function* which may be denoted by I.

Illustration. The polynomial function $f(x) = x^2 + 2x + 3$ can be described as $I^2 + 2 \, I + 3$ where I denotes the Identity function $I^2 = I$. I and 2, 3 stand for the Constant functions $\forall \, x \in \mathbf{R}$ respectively.

Ex. Consider any four rational functions and describe them as obtained through the operations of addition, subtraction, multiplication and division performed on the Identity function and the Constant functions.

2.4 One-one Functions.

Consider a function

$$y = f(x)$$

with domain A and range B respectively.

To each $x \in A$, there corresponds a $y \in B$. Again if y is a given member of the range B of the function, there will exist *at least* one member x of the domain A such that $f(x) = y$. It is possible, however, that in any given case, there may exist more than one member of A, say x_1, x_2 such that while $x_1 \neq x_2$ we have

$$f(x_1) = f(\mathrm{x}_2).$$

Def. *A function is said to be one-one if each member of the range of the function arises for one and only one member of the domain of the function.*

Illustrations.

1. The function $y = x^3$ with domain $\mathbf{R}$ is one-one (Fig. 2.9).
2. The function $y = x^2$ with domain $\mathbf{R}$ is not one-one. (Fig. 2.10).

The range of this function is the set of all non-negative numbers and each non-zero member of the range arises from two different members of the domain. As for example, the number 9 of the range arises from two different members – 3 and 3 of the domain (Fig. 2.10).

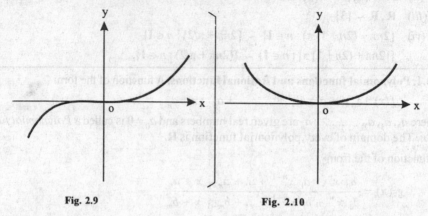

Fig. 2.9   Fig. 2.10

3. The function $y = x^2$, the domain the set of all non-negative real numbers *only,* is one-one (Fig. 2.11).

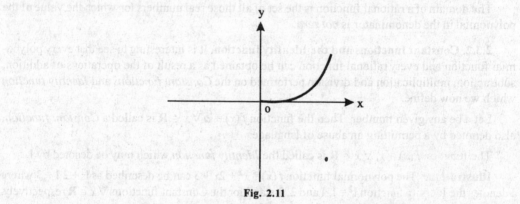

Fig. 2.11

4. The function $y = \sin x$ with domain **R** is *not* one-one. The range of the function is [–1, 1] and each member of the range arises from an infinite number of members of the domain. For example, the member, 0, of the range arises from each member of the infinite set $\{n\pi \,; n \in \mathbf{I}\}$ (Fig. 2.12).

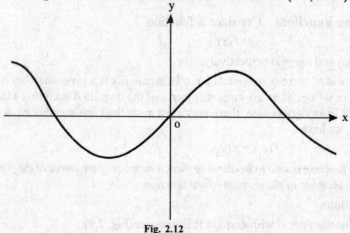

Fig. 2.12

5. The function $y = \sin x$ with domain $[-\pi/2, \pi/2]$ is one-one (Fig. 2.13).

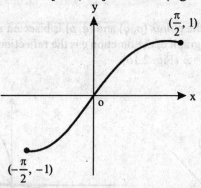

Fig. 2.13

2.4.1. Graph of one-one functions. It may be seen that a function is one-one, if and only if, no line parallel to the x-axis meets the graph of the function in more than one point.

For example, the function represented by the graph on the left (Fig. 2.14) is one-one and that represented by the graph on the right (Fig. 2.15) is *not* one-one.

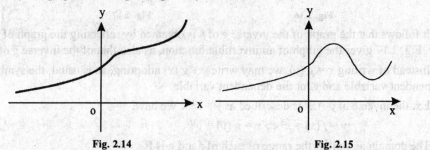

Fig. 2.14   **Fig. 2.15**

2.4.2. Invertible functions. Consider a one-one function with domain A and range B. Let $y \in B$. The function f being one-one, the member $y \in B$ arises from one and only one member $x \in A$ such that $f(x) = y$.

Thus we define a new function, say g, such that

$$g(y) = x \Leftrightarrow f(x) = y.$$

Also the domain of the function g is the range of the given function f and *vice-versa*.

The function g is said to be an **Inverse** of the function f. We also say that f is an **Invertible** function.

We thus see that *one-one functions are Invertible.*

Clearly if g is the inverse of f, then f is also the inverse of g.

Illustrations. The function :

(*i*) $y = x^2$ with domain **R** is invertible.

(*ii*) $y = x^2$ with domain **R** is not invertible.

(*iii*) $y = x^2$ with domain $[0, \infty[$ is invertible.

(*iv*) $y = \sin x$ with domain **R** is not invertible.

(*v*) $y = \sin x$ with domain $[-\pi/2, \pi/2]$ is invertible.

2.4.3. Graph of the inverse of an invertible function. Let $y = f(x)$ be an invertible function and let g denote the inverse of f, we have

$$y = f(x) \Leftrightarrow x = g(y).$$

Let (p, q) be a point on the graph of the function f. Then (q, p) is the corresponding point on the graph of the function g.

The segment joining the points (p, q) and (q, p) is bisected at right angles by the line $y = x$ so that the point (q, p) on the graph of the function g is the reflection of the point (p, q) on the graph of the function f in the line $y = x$ (Fig. 2.16).

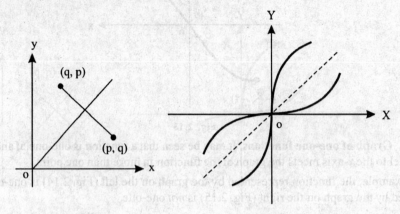

Fig. 2.16   Fig. 2.17

It follows that the graph of the inverse g of f is obtained by reflecting the graph of f in the line $y = x$. Fig. 2.17 gives the graph of an invertible function, as also that of the inverse g of f.

Instead of writing $x = g(y)$, we may write $y = g(x)$ adopting, as is usual, the symbol x for the independent variable and y for the dependent variable.

Ex. the inverse of $y = x^3$ is described as $y = x^{1/3}$. We have

$$y = f(x) = x^3 \Leftrightarrow x = g(y) = y^{1/3}.$$

The domain as well as the range of each of f and g is **R**.

2.4.4. Criteria for invertibility. Increasing and decreasing functions.

We now introduce the notion of increasing and decreasing functions on the basis of which it is possible to give a simple criterion for the invertibility of a function.

Def. *A function f is called an increasing function in its domain D if*

$$x_2 > x_1 \Rightarrow f(x_2) \geq f(x_1), x_1 \in D, x_2 \in D.$$

*A function f is called **strictly** increasing if we do not have the equality so that*

$$x_2 > x_1 \Rightarrow f(x_2) > f(x_1).$$

Def. *A function is called a decreasing functon in D if*

$$x_2 > x_1 \Rightarrow f(x_2) \leq f(x_1) ; x_1 \in D, x_2 \in D.$$

*A function f is called **strictly** decreasing if*

$$x_2 > x_1 \Rightarrow f(x_2) < f(x_1).$$

Increasing and Decreasing functions are also called **Monotonic functions**.

Illustrations : The function

1. $y = x^2$ is strictly decreasing in $] - \infty, 0]$ and strictly increasing in $[0, \infty]$ (Fig. 2.18).

2. $y = x^3$ is strictly increasing in $] -, \infty [$ (Fig. 2.19).

3. $y = 1/x^2$ is strictly increasing in $] - \infty, 0[$ and strictly decreasing in $[0, \infty]$ (Fig. 2.20).

4. $y = 1/x$ is strictly decreasing in $] - \infty, 0[$ and $] 0, \infty [$ (Fig. 2.21).

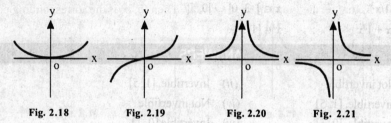

Fig. 2.18   Fig. 2.19   Fig. 2.20   Fig. 2.21

5. $y = [x]$ is increasing but *not* strictly increasing in **R**.

6. $y = \tan x$ is strictly increasing in each of $]-\pi/2, \pi/2[,]\pi/2, 3\pi/2[,]3\pi/2, 5\pi/2[$ etc. (Fig. 2.22).

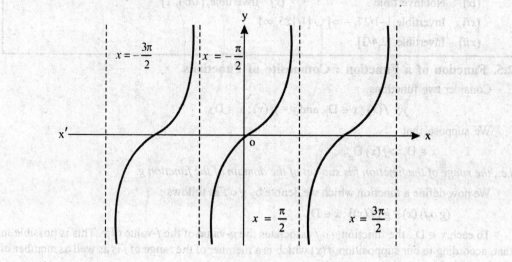

Fig. 2.22

2.4.5. Invertibility of strictly monotonic functions. Let a function f be strictly monotonic in its domain A and B be its range.

Because of the strictly monotonic character of the function f each member of its range B arises from *one and only* one member of its domain A. Thus the function f is one-one and therefore invertible.

It follows that *every strictly monotonic function is invertible.*

Ex. Which of the following functions are invertible? Give also the domains of the inverse of the invertible functions :

(i) $y = 1 + x^2$; $x \in [-2, 2]$,

(ii) • $y = 1 + x^2$; $x \in [-2, 0]$,

(iii) $y = 1 + x^2$; $x \in [0, 2]$,

(iv) $y = |x|$; $x \in [-1, 3]$,

(v) $y = |x|$; $x \in [0, 3]$,

(vi) $y = |x|$; $x \in [-1, 0]$,

(vii) $y = x^4$; $x \in]-\infty, \infty[$,

(viii) $y = x^4$; $x \in [0, \infty]$,

(ix) $y = 1/x^4$; $x \in]-\infty, 0[\cup]0, \infty[$,

(x) $y = 1/x^4$; $x \in [1, 8]$,

(xi) $y = 1/x^3$; $x \in \,]\text{--}3, 0[\,\cup\,]0, 5[,$

(xii) $y = x + 1/x$; $x \in [1, 3],$

ANSWERS

Ex.   (i)   Not invertible   (ii)   Invertible, $[1, 5]$

(iii)   Invertible, $[1, 5]$ (iv)   Not invertible

(v)   Invertible, $[0, 3]$ (vi)   Invertible $[0, 1]$

(vii) Not invertible   (viii) Invertible, $[0, \infty[$

(ix)   Not invertible   (x)   Invertible, $[1/64, 1]$

(xi)   Invertible, $[-1/27, -\infty[\,\cup\, [1/125, \infty[$

(xii)   Invertible, $[2, 4/3]$.

2.5. Function of a Function : Composite of Functions.

Consider two functions

$$y = f(x) ; x \in D_f, \text{ and } y = g(x) ; x \in D_g.$$

We suppose that

$$x \in D_f \Rightarrow f(x)\, D_g,$$

i.e., the range of the function f is sub-set of the domain of the function g.

We now define a function which we denote by $g \circ f$ as follows :

$$(g \circ f)(x) = g[f(x)] ; x \in D_f.$$

To each $x \in D_f$, the function $g \circ f$ associates the g-value of the f-value of x. This is possible in that, according to our supposition, $f(x)$ which is a member of the range of f is as well as member of the domain of g.

The function $g \circ f$ is called the *Composite* of the functions g and f taken in this order. It is also called the *function of a function.*

Illustrations :

1. Consider the two functions

$$f(x) = x^2, g(x) = \sin x.$$

The domain of the function f is **R**. Also the range $[0, \infty[$ of f is a sub-set of the domain **R** of g. Thus we have

$$(g \circ f)(x) = g[f(x)] = g(x^2) = \sin x^2, x \in \mathbf{R}$$

Let us also consider $f \circ g$ We have

$$(f \circ g)(x) = f[g(x)] = (\sin x^2), x \in \mathbf{R}.$$

The range of the function g being $[-1, 1]$ we have to restrict the domain of f to $[-1, 1]$ only.

It should be noticed that the functions $g \circ f$ and $f \circ g$ may not be same.

2. Consider the function

$$y = \sqrt{(x^2 - 3x + 2)}.$$

We shall now express this function as the composite of two functions.

Consider

$$f(x) = x^2 - 3x + 2. \qquad g(x) = \sqrt{x}.$$

Surely we have

$$(g \circ f)(x) = g[f(x)] = \sqrt{(x^2 - 3x + 2)}.$$

We now consider the domain of the composite $g \circ f$ of g and f. The domain of f is **R**. Now only that part of the domain **R** of f can be considered such that the corresponding range is a sub-set of the domain $[0, \infty[$ of g. We have

$$x^2 - 3x + 2 = (x - 1)(x - 2).$$

Now $(x - 1)(x - 2) \geq 0 \Leftrightarrow x \in]\infty -, 1] \cup [2, \infty[= D$, say.

Thus we see that the domain of the composite $g \circ f$ is the set

$$]-\infty, 1] \cup [2, \infty[= \mathbf{R} \sim]1, 2[$$

We may also see that $(f \circ g)(x) = (x - 3)\sqrt{x} + 2$ with domain $[0, \infty[$.

EXERCISES

1. Express the following functions as the composites of appropriate functions :

   (i) $y = \sqrt{(\sin x)},$ (ii) $y = \sin\sqrt{(x^2 + 1)}.$

   (iii) $y = \tan(\tan x).$

2. Show that the function $f(x) = (2x + 3)/(x - 3)$ in invertible. Find the inverse g of f and verify that $g \circ f$ is the identity function.

3. Repeat the above question with the function $f(x) = (4x + 5)/(5x + 4)$.

4. Given that f is an invertible function with its inverse g, show that gof and fog are identity functions.

5. Consider the three functions :

$$f(x) = x^2, g(x) = \sin x, \ h(x) = \sqrt{x}.$$

Compute

$[(g \circ f) \circ h](x), [g \circ (fo\ h)](x), [(fo\ g)\ oh](x), [fo\ (g\ o\ h)](x)$. What do you notice?

ANSWERS

1. (i) $f(x) = \sin x, g(u) = \sqrt{u}, (gof)(x) = \sqrt{\sin x}$

   (ii) $f(x) = x^2 + 1, g(u) = \sqrt{u}$

            $h(v) = \sin v, (hog)\ of = \sin\sqrt{(x^2 + 1)}.$

   (iii) $f(x) = \tan x, (fof)(x) = \tan(\tan x).$

2. (i) $g(x) = \dfrac{3 + 3x}{x - 2}, x \neq 2$

3. (ii) $g(x) = \dfrac{5 - 4x}{5x - 4}, x \neq \dfrac{4}{5}$

4. $((gof)\ oh)(x) = (go\ (foh))(x) = \sin x$

   $((fog)\ oh)(x) = (fo\ (goh))(x) = \sin^2\sqrt{x}$

In the following, we shall study the graphs of the exponential function $y = a^x$ and its inverse $y = \log_a x$ as also the graphs of the trigonometrical functions $y = \sin x, y = \cos x,$ $y = \tan x, y = \cot x, y = \sec x, y = \csc x$ and their inverses in appropriately chosen domains.

The exponential function will be introduced in stages.

2.6. The function $y = x^{1/n}$; n being any integer.

The consideration will be given to the four cases which arise when n is a

    (i) *positive even integer,*   *(ii)* *positive odd integer,*

    (iii) *negative even integer,*   *(iv)* *negative odd integer.*

In *(i)* the function $y = x^n$ is strictly decreasing in $]-\infty, 0]$ and striclty increasing in $[0, \infty]$ (Fig. 2.23).

In *(ii)* the function $y = x^n$ is strictly increasing in $]\infty - 0[$ (Fig. 2.24).

If follows that the function $y = x^n$ is strictly increasing in $[0, \infty[$ when n is a positive integer, even or odd and as such the function is invertible. We write

$$y = x^n \Leftrightarrow x = y^{1/n}, x \, [0, \infty[.$$

Thus the inverse of the function $y = x^n$ is the function $y = x^{1/n}$, the domain and range of either function being $[0, \infty[.$

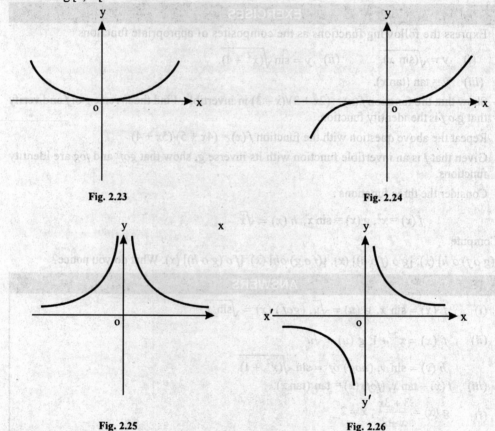

    Fig. 2.23   **Fig. 2.24**

    Fig. 2.25   **Fig. 2.26**

In *(iii)* the function $y = x^n$ is strictly increasing in $]-\infty, 0[$ and strictly decreasing in $]0, \infty[$ (Fig. 2.25).

In *(iv)*, the function $y = x^n$ is strictly decreasing in $]-\infty, 0[$ and also strictly decreasing in $]0, \infty[$ (Fig. 2.26).

If follows that the function $y = x^n$ is strictly decreasing in $]0, \infty[$ when n is a negative integer, even or odd and as such the function is invertible.

We write $y = x^n \Leftrightarrow x = y^{1/n}, x \in]0, \infty[$

so that the inverse of the function $y = x^n$ is the function $y = x^{1/n}$, the domain and range of either function being $]\,0, \infty\,[$.

It is important to note that in each case $x^{1/n}$, *denotes the positive nth root of the positive number x.*

2.6.1. Meaning of x^a when a is any real number.

Case I. Let a be a rational number m/n where m and n are integers and $n \neq 0$. Then we have

$$x^{m/n} = (x^m)^{1/n}.$$

We assume that the reader is already familar with the various properties of the power function $y = x^a$ such $x^a . x^b = x^{a+b}$ where a, b are arbitrary rational numbers.

Case II. Let a be an irrational number, say $b.a_1 a_2 a_3 a_4 \ldots$

The number a can be approximated by the sequence of rational numbers

$$b, b.a_1, b.a_2, b.a_1 a_2 a_3 \ldots\ldots$$

This suggests that x^a can be thought of as approximated by the sequence of numbers

$$x^b, x^{b.a_1}, x^{b.a_1 a_2}, x^{b.a_1 a_2 a_3},$$

the indices being rational numbers.

What we have stated here is only an indication of what appears to be natural. The actual treatment requires very sophisticated considerations which is beyond the scope of this book.

Power Function. The function $y = x^a, x \in \,]\,0, \infty\,[$ is called the General power function, a being any given real number.

2.6.2. Exponential Functions. The function $y = a^x$ where the base a is a constant and the index, x a variable is called an *Exponential function.* In the context of what is stated in the preceding, we are not, in this book, concerned with the rigorous justification for the statements which we shall be making. The only basis for these statements will be our natural expectations.

To consider the graph of $y = a^x$, we have to consider two cases (*i*) $a > 1$, (*ii*) $0 < a < 1$.

Let us take $a = 2$.

To have an indication of the nature of variation of the function $y = a^x$, we consider the case when x varies over the sequence

$$\ldots\ldots\ldots -3, -2, -1, 0, 1, 2, 3, \ldots\ldots\ldots$$

of integers and the corresponding sequence of the values of a^x is

$$\ldots\ldots\ldots 1/8, 1/4, 1/2, 1, 2, 4, 8, 16, \ldots\ldots\ldots$$

Every member of this sequence is positive. Also the numbers on the left can be made as near 0 as we like and those on the right as large as we like.

Now take $a = 1/2 < 1$. The sequence of values of a^x now obtained is

$$\ldots\ldots\ldots, 16, 8, 4, 2, 1, 1/2, 1/4, 1/8, \ldots\ldots\ldots$$

In this case the number on the left can be made as large as we like and those on the right as near 0 as we like.

The above statements *suggest* conclusions which we shall now formulate for drawing the graph of the function $y = a^x$.

I. $y = a^x; a > 1$.

    (*i*)   The domain of the function is $]-\infty, \infty\,[$ and the range is $]\,0, \infty\,[$.

    (*ii*)   The domain is strictly increasing.

We have the graph as given in Fig. 2.27.

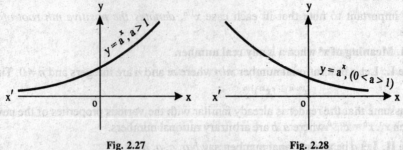

| Fig. 2.27 | Fig. 2.28 |

It will be seen that if $a > 1$, the curve $y = a^x$ approaches the x-axis *viz.* the line $y = 0$ nearer and nearer if x, while taking negative values becomes greater and greater in absolute value. Also the perpendicular distance of the points of the curve $y = a^x$ from the line $y = 0$ can be made as small as we like if only $|x|$ is sufficiently large so that the curve $y = a^x$ approaches the lines $y = 0$ asymptotically for large negative values of x. We also say that $y = 0$ is the **asymptote** of $y = a^x$.

    II. $y = a^x; 0 < a < 1$.

        (*i*)   The domain of the function is $] -\infty, \infty [$ and the range is $] 0, \infty [$.

        (*ii*)   The function is strictly decreasing.

We have the graph as given in Fig. 2.28.

Ex. Describe the relationship of the curve $y = a^x$ and the line $y = 0$ if $0 < a < 1$.

The logarithmic function y = log $_a$x.

We have seen that $y = a^x$ is strictly increasing when $a > 1$ and strictly decreasing when $0 < a < 1$. Thus this function is invertible. The inverse of this function is denoted by log a^x we write.

$$y = a^x \Leftrightarrow x = \log_a y,$$

where $x \in] -\infty, \infty [$ and $y \in] 0, \infty [$.

Writing $y = \log_a x$ in place of $x = \log_a y$, we have the graphs of $y = \log_a x$ as shown in Figs. 2.29, 2.30.

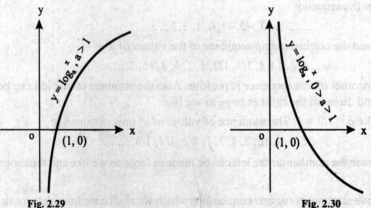

| Fig. 2.29 | Fig. 2.30 |

It will be seen that the domain of $y = \log_a x$ is $]0, \infty [$ and range $] -\infty, \infty[$. Also $a^o = 1 \Leftrightarrow \log_a 1 = 0$.

Comparing the relationship of the curve $y = a^x$ with the line $y = 0$ with that of the curve $y = \log_a x$ with the line $x = 0$, we say that the line $x = 0$ is the asymptote of $y = \log_a x$.

2.7. Trigonometric and Inverse trigonometric functions.

1. The functions $y = \sin x$, $y = \sin^{-1} x$.

The domain of $y = \sin x$ is **R** and range $[-1, 1]$. It increases strictly from -1 to 1 as x increases from $-\pi/2$ to $\pi/2$, decreases strictly from 1 to -1 as x increases from $\pi/2$ to $3\pi/2$ and so on.

We have the graph of $y = \sin x$ as given in Fig. 2.31 :

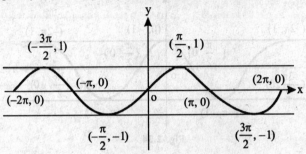

Fig. 2.31

$\sin x = 0$ when $x \in \{0, \pm\pi, \pm 2\pi, \pm 3\pi, \ldots\ldots\}$.

It follows that $y = \sin x$ is invertible in the domain $\left[-\dfrac{\pi}{2}, \dfrac{\pi}{2} \right]$ We write

$$y = \sin x \Leftrightarrow x = \sin^{-1} y \; ; x \in \left[-\dfrac{\pi}{2}, \dfrac{\pi}{2} \right], \, y \in [-1, 1].$$

We give below the graphs of the function $y = \sin x$ and its inverse $y = \sin^{-1} x$. (Fig. 2.32 and 2.33 respectively)

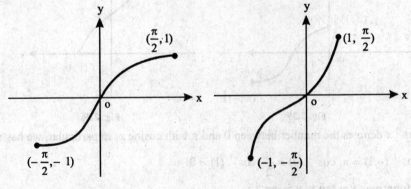

Fig. 2.32   **Fig. 2.33**

We notice that $\sin^{-1} x$ denotes the number lying between $\left[-\dfrac{\pi}{2}, \dfrac{\pi}{2} \right]$ whose sin is x.

Also $\sin^{-1} x$ strictly increases from $-\pi/2$ to $+\pi/2$ as x increases from -1 to 1. In particular, we have

$$\sin^{-1} 0 = 0, \; \sin^{-1} (-1) = -\dfrac{\pi}{2}, \; \sin^{-1} (1) = \dfrac{\pi}{2}.$$

Ex. Give other possible ways of defining $y = \sin^{-1} x$.

2. The functions y = cos x, y = cos⁻¹ x.

The domain of $y = \cos x$ is **R** and range $[-1, 1]$. Also the function decreasing strictly from 1 to –1 as x increases from 0 to π, increases strictly from – 1 to 1 as x increases from π to 2π and so on.

We have the graph of $y = \cos x$ as given in Fig. 2.34.

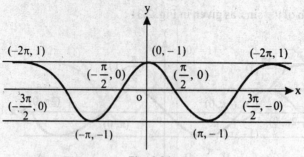

Fig. 2.34

$$\cos x = 0 \text{ when } x \in \left\{\pm \tfrac{\pi}{2}, \pm \tfrac{3\pi}{2}, \pm \tfrac{5\pi}{2}, ...\right\}$$

It will thus be seen that in the domain $[0, \pi]$ the function $y = \cos x$ is invertible. We write

$$y = \cos x \Leftrightarrow x = \cos^{-1} y; x \in [0, \pi], y \in [-1, 1].$$

We give below the graphs of the functions $y = \cos x$ and its inverse $y = \cos^{-1} x$. (Fig. 2.35 and 2.36 respectively)

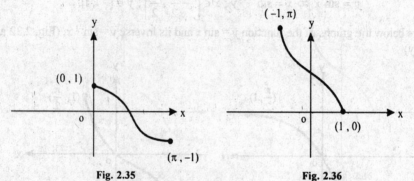

Fig. 2.35   **Fig. 2.36**

Thus $\cos^{-1} x$ denotes the number between 0 and π with cosine x. In particular, we have

$$\cos^{-1} (-1) = \pi, \cos^{-1} (0) = \frac{\pi}{2}, \cos^{-1} (1) = 0.$$

3. The functions y = tan x, y = tan⁻¹ x.

The domain of the function $y = \tan x$ is

$$\mathbf{R} \sim \{\pm \pi/2, \pm 3\pi/2, \pm 5\pi/2, ...\}.$$

So that we exclude from **R** the numbers for which $\cos x = 0$.

The range is $]-\infty, \infty[$.

The graph of $y = \tan x$ is given in Fig. 2.37

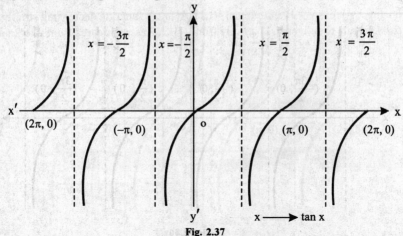

Fig. 2.37

The function $y = \tan x$ increases strictly from $-\infty$ to $+\infty$ as x increases in $\left]-\dfrac{\pi}{2}, \dfrac{\pi}{2}\right[$ and is as such invertible.

We write

$$y = \tan x \Leftrightarrow x = \tan^{-1} y; x \in \left]-\frac{\pi}{2}, \frac{\pi}{2}\right[\ y \in \ ]-\infty, \infty[.$$

We give below the graphs of $y = \tan x, y = \tan^{-1} x$.

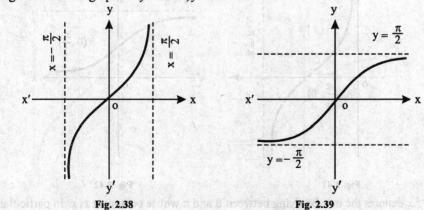

Fig. 2.38   **Fig. 2.39**

as shown in Fig. 2.38 and 2.39 respectively, $\tan^{-1} x$ denotes the number between $-\pi/2$ and $\pi/2$ whose tangent is x. In particular, we have

$$\tan^{-1}(0) = 0, \tan^{-1}(1) = \pi/4, \tan^{-1}(-1) = -\pi/4.$$

Note. $x = -\pi/2, x = \pi/2, x = -3\pi/2, x = 3\pi/2$ etc. are the asymptotes of $y = \tan x$. Also $y = \pi/2$ and $y = -\pi/2$ are the asymptotes of $y = \tan^{-1} x$.

4. The functions $y = \cot x$, $y = \cot^{-1} x$.

The domain of $y = \cot x$ is

$$\mathbf{R} \sim [0, \pm \pi, \pm 2\pi, \pm 3\pi, \ldots \ldots].$$

so that we exclude from $\mathbf{R}$ the set of numbers for which $\sin x = 0$. The range is $]-\infty, \infty[$.

We give below the graph of $y = \cot x$.

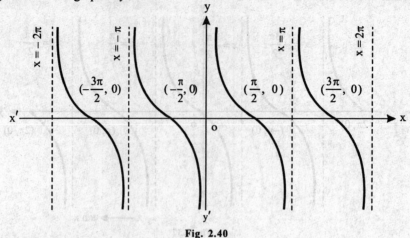

Fig. 2.40

The function $y = \cot x$ decreases strictly from $+ \infty$ to $- \infty$ as x increases in $] \, 0, \pi \, [$ so that the function is invertible in $] \, 0, \pi \, [$.

$$y = \cot x \Leftrightarrow x = \cot^{-1} y; x \in \,] \, 0, \pi \, [, y \in \,] -\infty, \infty \, [.$$

We give below the graphs of

$y = \cot x$ and $y = \cot^{-1} x$. (Fig. 2.41 and 2.42 respectively)

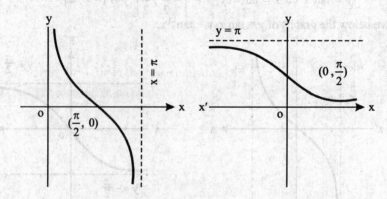

Fig. 2.41   **Fig. 2.42**

$\cot^{-1} x$ denotes the number lying between 0 and π whose cotangent is x. In particular,

$\cot^{-1} (0) = \pi/2$, $\cot^{-1} (1) = \pi/4$, $\cot^{-1} (-1) = 3\pi/4$.

Ex. Give the asymptotes of

$\quad y = \cot x$ and $y = \cot^{-1} x$.   [**Ans.** $\{x = n\pi$ such that $n \in I\}, y = \pi, y = 0.]$

5. The functions y = sec x, y = sec⁻¹ x.

The domain of $y = \sec x$ is

$$\mathbf{R} \sim \{\pm \pi/2, \pm 3\pi/2, \pm 5\pi/2, ...\}$$

so that we have excluded those of the numbers for which $\cos x = 0$.

The range is $] -\infty, -1] \cup [1, \infty \, [\, [= \mathbf{R} \sim \,] -1, 1 \, [.$

Given below is the graph of $y = \sec x$.

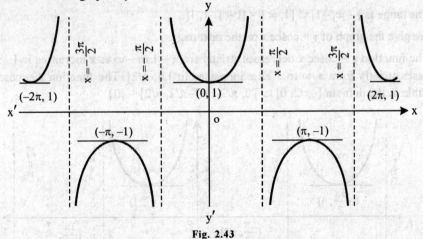

Fig. 2.43

The function $y = \sec x$ is strictly increasing from 1 to ∞ as x increases in $[0, \pi/2[$ and strictly increasing form $-\infty$ to -1 as x increases in $]\pi/2, \pi]$.

Thus $y = \sec x$ is invertible in the domain

$$[0, \pi/2 [\cup] \pi/2, \pi] = [0, \pi] \sim \{\pi/2\}.$$

We write

$$y = \sec x \Leftrightarrow x = \sec^{-1} y; x \in [0, \pi] \sim \{\pi/2\}, y \, \mathbf{R} \,]{-1}, 1[.$$

We give below the graphs of the functions

$$y = \sec x \text{ and } y = \sec^{-1} x. \text{ (Figs. 2.44 and 2.45 respectively)}$$

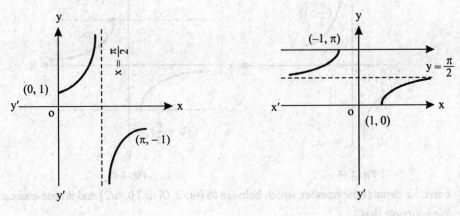

Fig. 2.44   **Fig. 2.45**

$\sec^{-1} x$ denotes the number which $\in [0, \pi/2 [\cup] \pi/2, \pi]$ whose secant is x. In particular, we have

$$\sec^{-1} (1) = 0, \sec^{-1} (-1) = \pi.$$

6. The functions y = cosec x, y = cosec⁻¹ x.

The domain of $y = \text{cosec } x$ is

$$\mathbf{R} \sim \{0, \pm \pi, \pm 2\pi, \pm 3\pi, \dots\dots\dots\}$$

so that we have excluded the set of numbers for which $\sin x = 0$.

The range is $]-\infty, -1] \cup [1, \infty [= \mathbf{R} \sim]-1, 1[$.

We give the graph of $y = \operatorname{cosec} x$ on the next page.

The function $y = \operatorname{cosec} x$ decreases strictly from -1 to $-\infty$ as x increases in $[-\pi/2, 0[$ and decreases strictly from $+\infty$ to 1 as x increases in $]\,0, \pi/2]$. The function $y = \operatorname{cosec} x$ is thus invertible in the domain $[-\pi/2, 0[\cup]\,0, \pi/2,] = [-\pi/2, \pi/2] \sim \{0\}$.

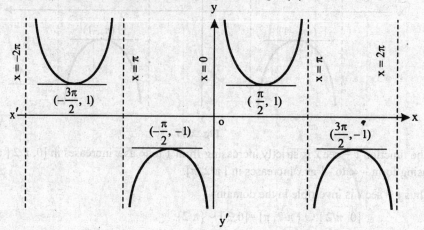

Fig. 2.46

We give below the graphs of

$$y = \operatorname{cosec} x, \ y = \operatorname{cosec}^{-1} x.$$

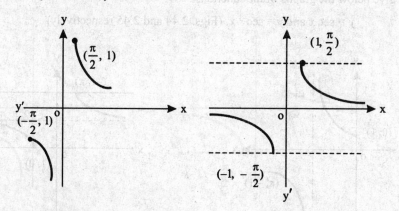

Fig. 2.47   Fig. 2.48

$\operatorname{cosec}^{-1} x$ denotes the number which belongs to $[-\pi/2, 0[\cup]\,0, \pi/2]$ and whose cosecant is x. We may note that

$$\operatorname{cosec}^{-1}(-1) = -\pi/2, \ \operatorname{cosec}^{-1} 1 = \pi/2.$$

Ex. Give the asymptotes of

$$y = \sec x, \ y = \operatorname{cosec} x, \ y = \sec^{-1} x, \ y = \operatorname{cosec}^{-1} x.$$

[**Ans.** $\{x = (2n + 1) \pi/2 : x \in I\} \{x = n\pi : n \in I\}, y = \pi/2, y = 0\}]$

Asymptotes of a curve. While we have introduced in this chapter the notion of asymptotes of a curve in the context of special curves, we shall take us the general discussion of asymptotes in a later chapter.

Note : The following table gives the domains and ranges of the inverse trigonometric functions:

Functions	Domains	Ranges
$y = \sin^{-1} x$	$[-1, 1]$	$[-\pi/2, \pi/2]$
$y = \cos^{-1} x$	$[-1, 1]$	$[0, \pi]$
$y = \tan^{-1} x$	$\mathbf{R}$	$]-\pi/2, \pi/2[$
$y = \cot^{-1} x$	$\mathbf{R}$	$]\,0, \pi[$
$y = \sec^{-1} x$	$\mathbf{R} -]-1, 1[$	$[0, \pi] \sim \{\pi/2\}$
$y = \operatorname{cosec}^{-1} x$	$\mathbf{R} -]-1, 1[$	$[\,\pi/2, \pi/2] - \{0\}$

A useful inequality

In dealing with Trigonometric functions, we shall find it useful to know that

$$|\sin x| < |x| \text{ when } x \in \,]-\pi/2, \pi/2\,[$$

except for $x = 0$ when $\sin x = x$; both being zero.

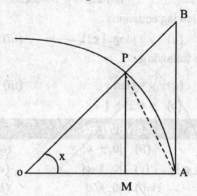

Fig. 2.49

Consider a circle with centre O and radius 1.

Let $\angle AOP = x$ where $0 < x < \pi/2$. (Fig. 2.49) We have

Area if $\triangle OAP <$ Area of sector $OAP <$ Area of $\triangle OAB$.

Area of $\triangle OAP = \frac{1}{2} OA. OP \sin x = \frac{1}{2} \sin x$.

Area of sector $OAP = \frac{1}{2} .1 . x = \frac{1}{2} x$.

Area of $OAB = \frac{1}{2} .OA . AB = \frac{1}{2} \tan x$,

Thus $\sin x < x < \tan x$.

We have thus seen that $\sin x < x$ when $x \in \,]0, \pi/2[$ or equivalently $|\sin x| < |x|$ in that $\sin x$ and x are both positive when $x \in \,] 0, \pi/2[$.

Let $x \in \,]-\pi/2, 0[$ so that $\sin x$ and x are both negative. We have

$$\sin (-x) < -x \text{ for } -x \in \,] 0, \pi/2 \,[$$

$\Rightarrow \qquad |\sin (-x)| < |-x| \, |\sin x| < |x|.$

Thus we have seen that $|\sin x| < |x|$ when

$$x \in \,]-\pi/2, 0] \cup [0, \pi/2[$$

Note. The inequality is also generally given in books on Elementary Trigonometry.

1. Give the graphs of the following functions :

(i) $y = 2^{|x|}$,   (ii) $y = \left(\frac{1}{2}\right)^{|x|}$,   (iii) $y = \log_2 |x|$,

(iv) $y = \log_{1/2} |x|$,   (v) $y = |\log_2 x|$,   (vi) $y = |\cos x|$,

(vii) $y = |\tan x|$,   (viii) $y = |\operatorname{cosec} x|$,   (ix) $y = |\sin^{-1} x|$,

(x) $y = |\tan^{-1} x|$,   (xi) $y = |\sec^{-1} x|$

2. Give the domains of the following functions :

(i) $y = \sin(1/x)$,   (ii) $y = \log_2 (1/x)$,   (iii) $y = \log_3 (1/\sqrt{(x-1)})$,

(iv) $y = \log_{1/2} \sin x$,   (v) $y = \sin \log_2 x$,   (vi) $y = \log_2 \sin^{-1} x$,

(vii) $y = \log_3 \tan^{-1} x$,   (viii) $y = \log_4 (1/\tan^{-1} x)$,   (ix) $y = \log_{1/3} (1 + x)$,

3. Give the graphs of the functions with the following function values :

$\tan^2 x$, 2^{x^2}, $2^{1/x^2}$, $2^{-1/x^2}$

$\sin 2x$, $\cos 3x$, $\tan 2x$, $\cot 3x$.

4. Draw the graphs of the following equations :

(i) $y = x \sin x$,   (ii) $y = |\log_2 |x||$,   (iii) $y = \tan(\tan^{-1} x)$.

5. Give the asymptotes of the following :

(i) $y = \tan 2x$,   (ii) $y = |\tan x|$,   (iii) $y = \cot \frac{1}{2} x$,

(iv) $y = 3^{x/2}$,   (v) $y = 2^x + 1$.

ANSWERS

2. (i) $R \sim \{0\}$ (ii) $]0, +\infty[$ (iii) $]1, +\infty[$

(iv) $\{]2n\pi, (2n+1)\pi[: n \in I\}$ (v) $]0, +\infty[$ (vi) $]0, \pi/2[$

(vii) $]0, \pi/2[$ (viii) $]0, \pi/2[$ (ix) $]-1, +\infty[$

5. (i) $\{x = \left(n + \frac{1}{2}\right)\pi : n \in I\}$ (ii) $\{x = (2n+1)\pi/2 : n \in I\}$

(iii) $\{x = 2n\pi : n \in I\}$ (iv) $y = 0$ (v) $y = 1$.

2.8 Implicitly Defined Functions

The relations $y = f(x)$ is said to be the equation of the graph of the function f. Sometimes, however, we come across equations in x and y which are not of the form $y = f(x)$ so that to some values of x correspond more than one value of y. For example, for the equation

$$x^3 + y^3 + xy + 2y^2 + x + y = 0$$

$x = 0$ corresponds to $y = 0$ and $y = -1$.

Consider now the equations

$$y^2 = 4ax, \quad \frac{x^2}{a^2} + \frac{y^2}{b^2} = 1, \quad \frac{x^2}{a^2} - \frac{y^2}{b^2} = 1 \qquad \ldots(i)$$

which represent a parabola, an ellipse and a hyperbola respectively. These equations are solvable for y and when, so solved, we obtain

$$y = \pm 2\sqrt{(ax)}, \quad y = \pm \frac{b}{a}\sqrt{(a^2 - x^2)}, \quad y = \pm \frac{b}{a}\sqrt{(x^2 - a^2)}$$

so that, in general, to each admissible value of x correspond two values of y. Thus in each case, the given equation defines two different functions.

Each point on the graph of the two functions.

$$y = 2 \sqrt{(ax)} \qquad\qquad y = -2 \sqrt{(ax)}$$

is a point on the graph of the equation

$$y^2 = 4ax.$$

We may similarly describe the situation in the case of the other two equations also.

The graphs of the given equations (i) are given in figures 2.50, 2.51, 2.52 respectively.

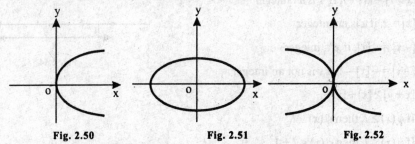

Fig. 2.50   Fig. 2.51   Fig. 2.52

The thick parts of these graphs lying above x-axis are those of the functions

$$y = 2 \sqrt{(ax)}, \qquad y = \frac{b}{a} \sqrt{(a^2 - x^2)}, \qquad y = \frac{b}{a} \sqrt{(x^2 - a^2)}$$

and those lying below x-axis are the graphs of the functions

$$y = -2 \sqrt{(ax)}, \qquad y = -\frac{b}{a} \sqrt{(a^2 - x^2)}, \qquad y = -\frac{b}{a} \sqrt{(x^2 - a^2)}$$

We say that the two functions $y = 2 \sqrt{(ax)}$, $y = -2 \sqrt{(ax)}$ are implicitly defined by the equation $y^2 = 4ax$. Similar statements may be made about the other equations also.

The above illustraions do not exhibit the general situation in that while the given equations (i) were explicitly solvable for y in terms of x, it may not be so in every given case. The discussion of the general question as to whether or not to a given equation in x and y correspond one or more functions such that every point on their graphs is a point on the graph of the given equation and *vice-versa* is beyond the scope of this book. In any given case of an equation in x and y, we shall only *assume* that the equation determines one or more functions. Each of these functions will be said to be implicitly defined by the given equation.

2.9 Some Special Functions

(a) **Absolute value function (or modulus function) :**

$$y = |x| = \begin{cases} x, & x \geq 0 \\ -x, & x < 0 \end{cases}$$

It is the numerical value of x.

Properties

(i) $|x| \leq a \Rightarrow -a \leq x \leq a; (a \geq 0)$

(ii) $|x| \geq a \Rightarrow x \leq -a$ or $x \geq a; (a \geq 0)$

(iii) $|x \pm y| \geq ||x| - |y||$

(iv) $|x \pm y| \leq |x| + |y|$

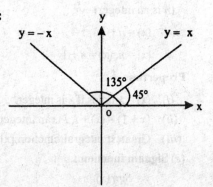

Fig. 2.53

(*b*) Greatest Integer Function :

[*x*] indicates the integral part of *x* which is nearest and smaller integer for *x*. It is also known as *floor of x*.

Thus, $[3.43201] = 3, [0.325] = 0,$

$$[7] = 7, [-7.392] = -8, [-0.8] = -1.$$

In general $n \le x < n + 1$ (*n* is an integer)

i.e. $[x] = n.$

Properties

(*i*) $[x + I] = [x] + I$, if *I* is an integer

(*ii*) $[x] = x$, if *x* is an integer

(*iii*) $[-x] = -[x]$, if $x \in$ integer

(*iv*) $[-x] = -[x] - 1$ if *x* is not an integer

(*v*) $[x + y] \ge [x] + [y]$

(*vi*) If $\phi(x) \ge I$, then $\phi(x) \ge I$

$\quad\quad$ If $\phi(x) \le I$, then $\phi(x) < I + 1.$

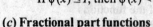

Fig. 2.54

(*c*) Fractional part functions

$\quad$ $y = \{x\}$. It indicates the fractional part of *x*. if

$\quad$ $x = I + f, I = [x]$ and $f = \{x\}$

Then $y = \{x\} = x - [x].$

Properties

(*i*) $\{x\} = 0$ if *x* is an integer

(*ii*) $\{x\} = x$ if $0 \le x < 1$

(*iii*) $\{-1\} = 1 - \{x\}$ if *x* is not an integer

(*d*) Least integer functions

$\quad$ $y = (x)$ indicates the integral part of *x* which is nearest and greater integer of *x*.

$\quad$ It is also known as *ceiling of x*.

$\quad$ If $n < x \le n + 1$

$\quad$ (*n* is an integer)

i.e. $(x) = n + 1$

$\quad$ $[x] = n, (x) = n + 1$

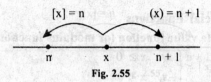

$[x] = n \qquad (x) = n + 1$

Fig. 2.55

Properties

(*i*) $(x) = x$ holds if *x* is integer.

(*ii*) $(x + 1) = (x) + I$, *I* is an integer

(*iii*) Greatest integral function [*x*] converts $x = I + f$ into (*I*) while [*x*] converts to $I + 1.$

(*e*) Signum function :

$\quad$ $y = Sgn(x).$

It is defined by

$$y = Sgn\,(x) = \begin{cases} \dfrac{|x|}{x} & \text{or} \quad \dfrac{x}{|x|},\ x \neq 0 \\ 0 & \text{if} \quad x = 0 \end{cases}$$

$$= \begin{cases} 1 & \text{if} \quad x > 0 \\ -1 & \text{if} \quad x < 0 \\ 0 & \text{if} \quad x = 0 \end{cases}$$

MISCELLANCEOUS EXAMPLES

1. *Find domain of* $f(x) = \log_{10} \log_{10}(1 + x^3)$.

$f(x) = \log_{10} \log_{10}(1 + x^3)$ exists

If $\qquad \log_{10}(1 + x^3) > 0$

or $\qquad 1 + x^3 > 1$

or $\qquad x^3 > 0$

or $\qquad x \in (0, \infty)$. Thus, domain of given function exists if $x > 0$.

2. *Find the domain for* $y = \sin^{-1}\left(\log_2\left(\dfrac{x^2}{2}\right)\right)$

For $\qquad y$ to be defined, $\dfrac{x^2}{2} > 0$ ...(*i*)

and $\qquad -1 \le \log_2\left(\dfrac{x^2}{2}\right) \le 1$ ...(*ii*)

Form (*i*) $\qquad x \in R - \{0\}$

From (*ii*) $\qquad 2^{-1} \le \dfrac{x^2}{2} \le 2^1 \Rightarrow 1 \le x^2 \le 4$

$\Rightarrow \qquad -2 \le x \le -1$ or $1 \le x \le 2$

Thus we have

$x \in [-2, -1] \cup [1, 2]$.

3. *Given* : $y = 2\,[x] + 3$ *and* $y = 3\,[x - 2] + 5$, *than find value of* $[x + y]$.

we have $\qquad 2\,[x] + 3 = 3\,[x - 2] + 5$

$\Rightarrow \qquad 2\,[x] + 3 = 3\,[x] - 6 + 5$

$\Rightarrow \qquad [x] = 4 \Rightarrow 4 \le x < 5$

or $\qquad x = 4 + f,\ f$ is the fractional part

$\therefore \qquad y = 2\,[x] + 3 = 11$.

Hence $\qquad [x + y] = [4 + f + 11] = [15 + f] = 15$.

4. *Solve* : $\qquad 4\,\{x\} = x + [x]$

we know that $\qquad x = [x] + \{x\}$

$\therefore \qquad 4\,[x] = [x] + \{x\} + [x]$

$\Rightarrow$ $$\{x\} = \frac{2\,[x]}{3}$$ ...(i)

But $0 \le \{x\} \le 1$

So, $0 \le \dfrac{2\,[x]}{3} \le 1$

$\Rightarrow$ $0 \le [x] \le \frac{3}{2} \Rightarrow [x] = 0$ or 1.

If $[x] = 1$, then $\{x\} = \dfrac{2}{3}$ from (i)

Thus $x = \dfrac{5}{3}$

If $[x] = 0, \{x\} = 0$

So, $x = 0$

Thus, solutions of the given equation are

$$x \in \left\{0, \frac{5}{3}\right\}$$

5. *If* $[x]$ = *the greatest integer less than or equal to* x, *and* (x) = *the least integer greater than or equal to* x, *and* $[x]^2 + (x)^2 > 25$, *then find* x.

Let $x = I + f$ where I is an integer and f is fractional part.

$\therefore$ $[x]^2 + (x)^2 > 25$

$\Rightarrow$ $[I + f]^2 + (I + f)^2 > 25$

$\Rightarrow$ $I^2 + \{I + 1\}^2 > 25$

$\Rightarrow$ $2\,I^2 + 2\,I - 24 > 0$

$\Rightarrow$ $(I + 4)\,(I - 3) > 0$

$\therefore$ $I < -4$ or $I > 3$

Here, $x = I + f$

So, $x < -4 + f$ or $x > 3 + f$.

Since, $0 \le f < 1$

$\therefore$ $x \le -4$ and $x \ge 4$

$\therefore$ $x \in \,]-\infty, -4] \cup [4, \infty[$.

6. *Find domain and range of the function*

$$y = \log_e (3x^2 - 4x + 5).$$

y is defined if $3x^2 - 4x + 5 > 0$

Here discriminant is $16 - 4\,(3)\,(5) < 0$

and coefficient of $x^2 = 3 > 0$

Hence, $(3x^2 - 4x + 5) > 0 \;\forall \in x \in R$

$\therefore$ Domain $\in R$.

Now, $\qquad y = \log_e (3x^2 - 4x + 5)$

$\Rightarrow \qquad 3x^2 - 4x + 5 = e^y$

$\Rightarrow \qquad 3x^2 - 4x + (5 - e^y) = 0$

Since x is real thus,

$$(-4)^2 - 4(3)(5 - e^y) \geq 0$$

$\Rightarrow \qquad e^y \geq 11/3$

$\Rightarrow \qquad y \geq \log_e (11/3)$

Hence, range is $\left[\log \left(\dfrac{11}{3} \right), \infty \right[$.

7. *Find out whether the following function is even, odd or neither even nor odd,*

$$f(x) = \begin{cases} x|x| & x \leq -1 \\ [1+x] + [1-x], & -1 < x < 1 \\ -x|x|, & x \geq 1 \end{cases}$$

where $||$ and $[\ ]$ represent modulus and greatest integer function.

The given function can be written as (using definitions of modulus and greatest integer function).

$$f(x) = \begin{cases} -x^2, & x \leq -1 \\ 2 + [x] + [-x], & -1 < x < 1 \\ -x^2, & x \geq 1 \end{cases}$$

$$\Rightarrow \quad f(x) = \begin{cases} -x^2, & x \leq -1 \\ 2 - 1 + 0, & -1 < x < 0 \\ 2, & x = 0 \\ 2 + 0 - 1, & 0 < x < 1 \\ -x^2, & x \geq 1 \end{cases}$$

$$\Rightarrow \quad f(x) = \begin{cases} -x^2, & x \leq -1 \\ 1, & -1 < x < 0 \\ 2, & x = 0 \\ 1 & 0 < x < 1 \\ -x^2, & x \geq 1 \end{cases}$$

which is clearly even as if $f(-x) = f(x)$

Hence, the given function is even.

8. *If k be a positive real number such that $f(x + k) + f(x) = 0$ for all $x \in R$. Prove that $f(x)$ is a periodic function with period $2k$.*

We have $\qquad f(x + k) + f(x) = 0, \forall x \in R$

$\Rightarrow \qquad f(x + k) = -f(x), \forall x \in R$

$\Rightarrow \qquad f(x + 2k) = -f(x + k) = f(x), \forall x \in R$

Hence $f(x)$ is periodic with period $2k$.

9. *Find the period of* $f(x) = |\sin x| + |\cos x|$

    $|\sin x|$ has period π, $|\cos x|$ has period π,

    Here, $f(x)$ is even, thus period of

    $$f(x) = \frac{1}{2} | \text{LCM of } \pi \text{ and } \pi]$$

    $$= \pi/2$$

    $\therefore$ Period of $f(x) = \pi/2$.

10. *Let* $f(x) = \log_{x^2} 25$ *and* $g(x) = \log_x 5$ *then* $f(x) = g(x)$ *holds, then find the interval for x.*

    Domain of $f \in R - \{1, 0\}$

    Domain of $g \in (0, \infty) - \{1\}$

    for $f(x) = g(x)$,

    domain of f = domain of g

    i.e. $f(x) = g(x)$, if $x \in (0, \infty) - \{1\}$

11. *Let* $f : R \to R$ *be defined by* $f(x) = \dfrac{e^x - e^{-x}}{2}$, *Is* $f(x)$ *invertible? If so, find its inverse.*

    To check the invertibility, we has

    (*i*)   one-one : $f'(x) = \dfrac{e^x + e^{-x}}{2}$

    $$= \frac{e^{2x} + 1}{2 e^x} \text{ which is strictly increasing as } e^{2x} > 0 \text{ for all } x, \text{ Thus one-one}$$

    (*ii*) on to ; Let $y = f(x)$

    $$y = \frac{e^x + e^{-x}}{2} \text{ where } y \text{ is strictly monotonic.}$$

    Hence range of $f(x) =]f(-\infty), f(\infty)[$

    $\Rightarrow$ range of $f(x) =]-\infty, \infty[$

    $\Rightarrow$ range of $f(x)$ = co-domain

    Hence $f(x)$ is one-one and onto, so invertible.

    To find the inverse, we have

    $$y = \frac{e^{2x} - 1}{2 e^x}$$

    $\Rightarrow$ $e^{2x} - 2e^x y - 1 = 0$

    $\Rightarrow$ $e^x = \dfrac{2 y \pm \sqrt{4y^2 + 4}}{2}$

    $\Rightarrow$ $x = f^{-1}(y) = \log(y \pm \sqrt{(y^2 + 1)})$

    Neglecting negative sign,

    $$f^{-1}(x) = \log(x + \sqrt{x^2 + 1})$$

12. *Find the domain of single valued function* $y = f(x)$ *given by the equation* $10^x + 10^y = 10$.

we have

$$10^y = 10 - 10^x$$

$\Rightarrow \qquad\qquad y = \log_{10}(10 - 10^x)$

Now. v is defined if,

$$10 - 10^x > 0$$

$\Rightarrow \qquad\qquad 1 > x.$

Hence domain of $y = f(x)$ is $x < 1$

$\therefore \qquad\qquad x \in \ ]-\infty, 1[$

13. *If* $x \in (0, \pi/2)$, *find the solution of the function*

$$f(x) = \frac{1}{\sqrt{-\log_{\sin x} \tan x}}$$

Here $x \in (0, \pi/2)$

$\Rightarrow \qquad\qquad 0 < \sin x < 1 \qquad\qquad\qquad\qquad ...(i)$

we know that

$$\log_a x < b \Rightarrow x > a^b, \text{ if } 0 < a < 1$$

$$x < a^b, \text{ if } a > 1 \qquad\qquad ...(ii)$$

Thus, $\qquad f(x) = \dfrac{1}{\sqrt{-\log_{\sin x} \tan x}}$ exists if, $-\log_{\sin x}(\tan x) > 0$

$\Rightarrow \qquad\qquad \tan x > (\sin x)^0$

$\Rightarrow \qquad\qquad \tan x > 1$

$\Rightarrow \qquad\qquad x \in \left(\dfrac{\pi}{4}, \dfrac{\pi}{2} \right)$

$\therefore \qquad\qquad x \in \left] \dfrac{\pi}{4}, \dfrac{\pi}{2} \right[$

$\therefore \left(\dfrac{\pi}{4}, \dfrac{\pi}{2} \right)$ is the required solution .

14. *Find all possible values of* x *satisfying,*

$$\frac{[x]}{[x-2]} - \frac{[x-2]}{[x]} = \frac{8\{x\}+12}{[x-2][x]}$$

where [ ] *donotes the greatest integer function and* { } *is fractional part.*

We have $\qquad \dfrac{[x]}{[x-2]} - \dfrac{[x-2]}{[x]} = \dfrac{8\{x\}+12}{[x-2][x]}$

$\Rightarrow \qquad \dfrac{[x]^2 - [x-2]^2}{[x-2][x]} = \dfrac{8\{x\}+12}{[x][x-2]}$

$\Rightarrow \qquad ([x]-[x-2])([x]+[x-2]) = 8\{x\}+12$

$\Rightarrow \qquad ([x]-[x]+2)([x]+[x]-2) = 8\{x\}+12 \qquad\qquad [\because \{x+1\} = \{x\}+1\}$

$\Rightarrow \qquad 4([x]-1) = 8\{x\}+12$

$\Rightarrow \qquad [x]-4 = 2\{x\} \qquad\qquad\qquad\qquad\qquad\qquad ...(i)$

Now, we know that

$$0 \le \{x\} < 1 \Rightarrow 0 \le 2\{x\} < 2$$

$$\Rightarrow \quad 0 \le [x] - 4 < 2 \Rightarrow 4 \le [x] < 6$$

$$\Rightarrow \quad [x] = 4, 5$$

If, $[x] = 4 \Rightarrow 2\{x\} = [x] - 4$

become $\{x\} = 0$ ...(ii)

and if, $[x] = 5 \Rightarrow 2\{x\} = [x] - 4$

become $\{x\} = \frac{1}{2}$ ...(iii)

From (ii) and (iii) we have

$$x = [x] + \{x\}$$

i.e., $x = 4 + 0 = 4$

and $x = 5 + \frac{1}{2} = \frac{11}{2}$

$$\Rightarrow \qquad x \in \left\{4, \frac{11}{2}\right\}$$

15. *Find the domain and range for*

$$f(x) = \left[\log\left(\sin^{-1}\sqrt{(x^2 + 3x + 2)}\right)\right] \text{ where [ ] denotes the greatest integer function.}$$

For domain,

$$0 < x^2 + 3x + 2 \le 1$$

Now, $x^2 + 3x + 2 > 0$

$$\Rightarrow \qquad (x+1)(x+2) > 0$$

$$\Rightarrow \qquad x \in (-\infty, -2) \cup (-1, \infty) \qquad \qquad ...(i)$$

and $x^2 + 3x + 2 \le 1$

$$\Rightarrow \qquad x^2 + 3x + 1 \le 0$$

$$\Rightarrow \qquad x \in \left[\frac{-3 - \sqrt{5}}{2}, \frac{-3 + \sqrt{5}}{2}\right] \qquad \qquad ...(ii)$$

∴ From (i) and (ii)

$$x \in \left[\frac{-3 - \sqrt{5}}{2}, -2\right[\cup \left]-2, \frac{-3 + \sqrt{5}}{2}\right]$$

This is the domain for $f(x)$.

For range,

$$0 < \sin^{-1}\sqrt{(x^2 + 3x + 2)} \le \frac{\pi}{2}$$

$$\Rightarrow \qquad -\infty < \log\sin^{-1}\sqrt{(x^2 + 3x + 2)} \le \log\left(\frac{\pi}{2}\right)$$

$$\Rightarrow \qquad -\infty < \log\sin^{-1}\sqrt{(x^2 + 3x + 2)} \le \text{(a value less than 1)}$$

$$\Rightarrow \qquad \log\sin^{-1}\sqrt{(x^2 + 3x + 2)} \text{ can take all non-positive integers.}$$

∴   Range of $f(x)$ = set of all non-positive intergers

16. *The function $f(x)$ is defined in $[0, 1]$. Find the domain of $f(\tan x)$.*

$f(x)$ is defined in $[0, 1]$

$\Rightarrow \qquad\qquad x \in [0, 1]$

For $f(\tan x)$ to be defined, we must have

$$0 \le \tan x \le 1$$

i.e., $\qquad\qquad n\pi \le x \le n\pi + \pi/4$

or $\qquad\qquad 0 \le x \le \pi/4$

$\therefore \qquad$ domain for $f(\tan x) \in \left[n\pi,\; n\pi + \dfrac{\pi}{4} \right]$.

17. *If $f(x)$ satisfies the relation, $f(x + y) = f(x) + f(y)$ for all $x, y \in R$ and $f(1) = 5$, find $\displaystyle\sum_{n=1}^{p} f(n)$. Also prove that $f(x)$ is odd.*

we have $\qquad f(r) = f(r-1) + f(1)$

$\Rightarrow \qquad\qquad f(r) = f(r-1) + 5$

$\qquad\qquad f(r) = [f(r-2) + 5] + 5$

$\qquad\qquad\qquad = f(r-2) + 2.5$

$\qquad\qquad\qquad = f(r-3) + 3.5$

$\qquad\qquad\qquad = f(1) + (r-1).5$

$\qquad\qquad\qquad = 5 + (r-1).5$

$\qquad\qquad\qquad = 5r$

$\therefore \displaystyle\sum_{n=1}^{p} f(n) = \sum_{n=1}^{p} (5n) = \dfrac{5.p\,(p+1)}{2}$

Putting $x = 0, y = 0$ in the given function, we have

$$f(0 + 0) = f(0) + f(0)$$

$\Rightarrow \qquad\qquad f(0) = 0.$

Putting $(-x)$ for y, we have

$$f(x - x) = f(x) + f(-x)$$

$\Rightarrow \qquad f(x) + f(-x) = 0$

$\Rightarrow \qquad\qquad f(-x) = -f(x),$

Hence $f(x)$ is odd.

18. *If the functions f anu g be defined as $f(x) = e^x$ and $g(x) = 3x - 2$, where $f : R \to R$ and $g : R \to R$. Then find the functions fog and gof. Also find the domain of $(fog)^{-1}$ and $(gof)^{-1}$.*

we have $\qquad (fog)(x) = f\{ g(x) \}$

$\qquad\qquad\qquad = f(3x - 2) = e^{3x-2}$

and $\qquad (gof)(x) = g\{f(x)\} = g(e^x)$

$\qquad\qquad\qquad = 3\,e^x - 2$

Now, let $\qquad (fog)(x) = y = e^{3x-2}$

$\Rightarrow \qquad\qquad\qquad x = \dfrac{\log y + 2}{3}$

$$\Rightarrow \qquad (fog)^{-1}(x) = \frac{\log x + 2}{3}$$

and domain of $(fog)^{-1}(x)$ is $x > 0$,

i.e., $\qquad\qquad x \in]0, \infty[.$

Again let $\qquad (gof)(x) = z = 3e^x - 2$

$$\Rightarrow \qquad x = \log \frac{z+2}{3}$$

$$\Rightarrow \qquad (gof)^{-1}(x) = \log\left(\frac{x+2}{3}\right)$$

and domain of $(gof)^{-1}$ is $\frac{x+2}{3} > 0$

i.e., $\qquad x > -2, i.e., x \in]-2, \infty[,$

19. Let $f(x+p) = 1 + [2 - 3f(x) + 3(f(x))^2 - (f(x))^3]^{1/3} \; \forall \, x \in R$, where $p > 0$, prove that $f(x)$ is periodic and find its period.

We have $\qquad f(x+p) = 1[1 + \{1 - f(x)^3\}^{1/3}]$

$\Rightarrow \qquad f(x+p) - 1 = [1 - \{f(x) - 1\}^3]^{1/3}$

$\Rightarrow \qquad g(x+p) = [1 - \{g(x)\}^3]^{1/3}$

where $\qquad g(x) = f(x) - 1.$

Then $\qquad g(x+2p) = [1 - \{g(x+p)\}^3]^{1/3}$

$\Rightarrow \qquad g(x+2p) = [1 - \{\{1 - \{g(x)\}^3\}^{1/3}\}^3]^{1/3}$

$\Rightarrow \qquad g(x+2p) = [1 - \{1 - \{g(x)\}^3\}]^{1/3}$

$\Rightarrow \qquad g(x+2p) = [\{g(x)\}^3]^{1/3} = g(x)$

$\Rightarrow \qquad f(x+2p) = f(x).$

Hence, $f(x)$ is periodic with period $2p$.

20. If the terms of the A.P. $\sqrt{a-x}, \sqrt{x}, \sqrt{a+x}, \; ..... $ are all integers where $a, x > 0$, then find the least composite odd integral value of a.

Since $\sqrt{a-x}, \sqrt{x}, \sqrt{a+x},....$ are in A.P.,

hence $\qquad 2\sqrt{x} = \sqrt{a-x} + \sqrt{a+x}$

$\Rightarrow \qquad 4x = 2a + 2\sqrt{a^2 - x^2}$

or $\qquad \sqrt{a^2 - x^2} = 2x - a$

$\Rightarrow \qquad a^2 - x^2 = (2x - a)^2$

$\Rightarrow \qquad x = 0 \text{ or } \frac{4a}{5}.$

∴ The terms of the arithmetic progression are :

$$\sqrt{\frac{a}{5}}, 2\sqrt{\frac{a}{5}}, 3\sqrt{\frac{a}{5}} \; ......$$

which are integers.

∴ $\qquad a = 5n^2 ; n \in N.$

For $\qquad n = 1, a = 5$ which is not composite

for $\qquad n = 2, a = 20$, which is composite but not odd.

for $\qquad n = 3, a = 45$ which is the least composite odd.

EXERCISES

1. Find the domain of the function

$$f(x) = \frac{1}{\log_{10}(1-x)} + \sqrt{x+2}$$

2. Find the domain of the function

$$f(x) = \log\left\{\log_{|\sin x|}(x^2 - 8x + 23) - \frac{3}{\log_2|\sin x|}\right\}$$

3. Find the domain of defintion of $f(x) = \dfrac{\log_2(x+3)}{x^2 + 3x + 2}$.

4. Find domain for $f(x) = \sin^{-1}\left(\dfrac{1+x^2}{2x}\right)$.

5. Find the domain of following :

   (i) $^{16-x}C_{2x-1} + {}^{20-3x}P_{4x-5}$

   (ii) $f(x) = \sqrt{e^{\cos^{-1}(\log_4 x^2)}}$

   (iii) $f(x) = \log_e |\log_e x|$

   (iv) $f(x) = \sqrt{\dfrac{\log_{0.3}(x-1)}{x^2 - 2x - 8}}$

   (v) $f(x) = \cos^{-1}\sqrt{\log_{[x]}\dfrac{|x|}{x}}$, where [ ] denotes the greatest integer.

   (vi) $f(x) = \sqrt{x^2 + 4x}C_{2x^2 + 3}$

   (vii) $f(x) = \sin^{-1}\left[\log_2\left(\dfrac{x^2}{2}\right)\right]$ where [.] denotes the greatest integer.

6. Find domain for $f(x) = [\sin x]\cos\left(\dfrac{\pi}{[x-1]}\right)$

7. Find the solution set of $(x)^2 + (x+1)^2 = 25$, where (x) is the least integer greater than equal to x.

8. Find range of $y = \sqrt{x-1} + \sqrt{5-x}$.

9. Find the range of following :

   (i) $f(x) = \log_3(5 + 4x - x^2)$

   (ii) $f(x) = \log_{[x-1]}\sin x$, where [.] denotes greatest integer.

   (iii) $f(x) = \sqrt{[\sin 2x] - [\cos 2x]}$, where [.] denotes greatest integer.

   (iv) $f(x) = \sqrt{\log(\cos(\sin x))}$

   (v) $f(x) = \dfrac{\sin x}{\sqrt{1 + \tan^2 x}} - \dfrac{\cos x}{\sqrt{1 + \cot^2 x}}$

   (vi) $f(x) = \dfrac{\tan\left(\pi\,[x^2 - x]\right)}{1 + \sin\,(\cos x)}$

   (vii) $f(x) = [|\sin x| + |\cos x|]$, where [.] denotes the greatest integer function.

10.   If f is an even function; find the real values of x satisfying the equation $f(x) = \left(\dfrac{x+1}{x+2}\right)$.

11.   Prove that $f(x) = x - [x]$ is periodic function. Also, find its period.

12.   Find the period if $f(x) = \sin x + \{x\}$, where $\{x\}$ is fractional part of x.

13.   Find period of $f(x) = \cos\,(\cos x) + \cos\,(\sin x)$.

14.   Find the period of following functions :

   (i) $f(x) = \dfrac{1}{2}\left\{\dfrac{|\sin x|}{\cos x} + \dfrac{|\cos x|}{\sin x}\right\}$

   (ii) $f(x) = e^{\cos^4 \pi x - [x] + \cos^2 \pi x}$

   (iii) $f(x) = \sin\dfrac{\pi x}{n!} - \cos\dfrac{\pi x}{(n+1)!}$

   (iv) $f(x) = e^{\ln(\sin x)} + \tan^3 x - \csc\,(3x - 5)$.

15.   Consider the real valued function satisfying $2f(\sin x) + f(\cos x) = x$. Find the domain and range of $f(x)$.

16.   Let $f(x) = \begin{cases} 1 + x & 0 \le x \le 2 \\ 3 - x, & 2 < x \le 3 \end{cases}$, find $(fof)\,(x)$.

17.   Let $f : [\frac{1}{2}, \infty) \to [\frac{3}{4}, \infty)$, where $f(x) = x^2 - x + 1$. Find the inverse of $f(x)$.

18.   Find the set of all solutions of the equation
$2^{|y|} - |2^{y-1} - 1| = 2^{y-1} + 1$.

19.   Find the domain and range of, $f(x) = \log\left[\cos|x| + \dfrac{1}{2}\right]$; where [.] denotes the greatest integer function.

20.   Find the domain and range of $f(x) = \sin^{-1}(\log\,[x]) = \log\,(\sin^{-1}\,[x]$; where [.] denotes the greatest integer function.

21.   If $f : R \to R$ is defined by $f(x) = x^2 + 1$, then find value of $f^{-1}(17)$ and $f^{-1}(-3)$.

22.   If a function is defined as $g(x) = |\sin x| + \sin x$, $\phi(x) = \sin x + \cos x$, $0 \le x \le \pi$, then find domain for

$$f(x) = \sqrt{\log_{\phi(x)}\,g(x)}\,.$$

23. $f(x) = \begin{cases} x - 1, & -1 \le x \le 0 \\ x^2, & 0 \le x \le 1 \end{cases}$ and $g(x) = \sin x$.

   Find $h(x) = f(|g(x)|) + |f(g(x))|$.

24.   Solve $\dfrac{1}{[x]} + \dfrac{1}{[2x]} = \{x\} + \dfrac{1}{3}$, where [.] denotes greatest integral function and {.} denotes fractional part of x.

25.   Find the solution of $|x^2 - 1 + \sin x| = |x^2 - 1| + |\sin x|$ belonging to the interval $[-2\pi, 2\pi]$.

26. Find the range of : $f(x) = 3 \sin \sqrt{\dfrac{\pi^2}{9} - x^2}$.

27. If the function $f : [1, \infty[\to [1, \infty[$ is defined by $f(x) = 2^{x(x-1)}$ then find $f^{-1}(x)$.

28. If $f(x)$ be a polynomiol function satisfying

$$f(x) . f\left(\frac{1}{x}\right) = f(x) + f\left(\frac{1}{x}\right) \text{ and } f(4) = 65. \text{ Then find } f(6).$$

29. Let $f(x)$ be defined on $[-2, 2]$ and is given by

$$f(x) - \begin{cases} -1, & -2 \le x \le 0 \\ x - 1, & 0 \le x \le 2 \end{cases} \text{ and } g(x) = f(|x|) + |f(x)|. \text{ Then find } g(x).$$

30. Let two functions are defined as :

$$g(x) = \begin{cases} x^2, & -1 \le x < 2 \\ x + 2, & 2 \le x \le 3 \end{cases} \text{ and } f(x) = \begin{cases} x + 1, & x \le 1 \\ 2x + 1, & 1 < x \le 2. \end{cases}$$

    find gof.

31. Let $f(x) = x^2 - 2x, x \in R$ and $g(x) = f(f(x) - 1) + f(5 - f(x))$. Show that $g(x) \ge 0 \; \forall \; x \in R$.

32. If f is a polynomial function satisfying $2 + f(x) . f(y) = f(x) + f(y) + f(xy) \; \forall \; x, y \in R$ and if $f(2) = 5$, then find $f(f(2))$.

33. Check whether the function defined by

$$f(x + \lambda) = 1 + \sqrt{2 f(x) - f^2(x)} \; \forall \; x \in R \text{ is periodic or not. If yes, then find its period.}$$

34. Let $f(x, y)$ be a periodic function satisfying the condition $f(x, y) = f((2x + 2y), (2y - 2x)) \; \forall \; x, y \in R$. Define a function g by $g(x) = f(2^x, 0)$. Then prove that, $g(x)$ is periodic function, find its period.

35. Find all the solutions of the equation

$$\sin\left(x - \frac{\pi}{4}\right) - \cos\left(x + \frac{3\pi}{4}\right) = 1$$

    which satisfy the inequality $\dfrac{2 \cos 7x}{\cos 3 + \sin 3} > 2^{\cos 2x}$

36. Solve the equation :

$$10^{(x+1)(3x+4)} - 2 . 10^{(x+1)(x+2)} = 10^{1-x-x^2}$$

37. Prove that $a^2 + b^2 + c^2 + 2abc < 2$, where a, b, c are the sides of triangle ABC such that $a + b + c = 2$.

38. Show that there exists no polynomial $f(x)$ with integral coefficients which satisfy $f(a) = b$, $f(b) = c, f(c) = a$, where a, b, c are distinct integers.

39. Consider a real valued function $f(x)$ satisfying $2 f(xy) = \{f(x)\}^y + \{f(y)\}^x$ for all $x, y \in R$ and $f(1) = a$, where $a \ne 1$.

    Prove that $(a - 1) \displaystyle\sum_{i=1}^{n} f(x) = a^{n+1} - a$.

40. Let $f : R - \{2\} \to R$ function satisfies the following functional equation :

$$2 f(x) + 3 f\left(\frac{2x + 29}{x - 2}\right) = 100x + 80, \forall \; x \in R - \{2\}. \text{ Determine } f(x).$$

ANSWERS

1. $[-2, 0[\cup]0, 1[$;

2. $]3, \pi[\cup \left] \pi, \dfrac{3\pi}{2} \right[\cup \left[\dfrac{3\pi}{2}, 5 \right[$;

3. $]-3, \infty[- \{-1, -2\}$;

4. $\{-1, 1\}$;

5.   (i) $x = \{2, 3\}$ (ii) $x \in [-2, -\tfrac{1}{2}] \cup [\tfrac{1}{2}, 2]$

     (iii) $x \in]0, 1[\cup]1, \infty[$;   (iv) $x \in]2, 4[$

     (v) $x \in]2, \infty[$;   (iv) $]-\sqrt{8} - 1[\cup [1, \sqrt{8}]$;

6. $R - [1, 2]$;   7. $x \in]-5, -4] \cup [2, 3]$;   8. $[2, \sqrt[2]{2}]$;

9.   (i)   Range $\in]-\infty, \log_3 9[$ (ii) $[-\infty, 0]$;

     (iii) $\{0, 1\}$;   (iv) $\{0\}$;

     (v) $[-1, 1]$;   (vi) $\{0\}$;

     (vii) $\{1\}$;

10. $x = \left\{ \dfrac{-1 \pm \sqrt{5}}{2}, \dfrac{-3 \pm \sqrt{5}}{2} \right\}$;   11. period $= 1$;

12. not periodic ;   13. $\pi/2$;

14.   (i) 2π ;   (ii) 1 ;

     (iii) $2(n+1)|$;   (iv) 2π ;

15. Domain $\in [-1, 1]$, Range $\in \left[-\dfrac{2\pi}{3}, \dfrac{\pi}{3} \right]$;   16. $\begin{cases} 2 + x, & 0 < x \le 1 \\ 2 - x & 1 < x \le 2 \\ 4 - x, & 2 < x \le 3, \end{cases}$

17. $x = 1$;   18. $\{y : y \ge 1 \cup y = -1\}$;

19. Domain : $x \in \bigcup\limits_{n=1} \left[(2n\pi - \dfrac{\pi}{3}), (2n\pi + \dfrac{\pi}{3}) \right]$

     Range : $\{0\}$

20. Domain : $[1, 2[$; Range of $f(x) \in \left\{ \log \dfrac{\pi}{2} \right\}$ 21. $x = \{+4, -4\}$;

22. Domain of $f(x)$ is ; $\left[\dfrac{\pi}{6}, \dfrac{\pi}{2} \right[$;

23. $h(x) = \begin{bmatrix} \sin^2 x - \sin x + 1, & -1 < x < 0 \\ 2\sin^2 x, & 0 \le x \le 1 \end{bmatrix}$ 24. $\dfrac{29}{12}, \dfrac{19}{6}, \dfrac{97}{24}$;

25. $x \in [-2\pi, -\pi] \cup [-1, 0] \cup [1, \pi] \cup \{2\pi\}$;

26. range of $f(x) \in \left[0, \dfrac{3\sqrt{3}}{2} \right]$ 27. $\dfrac{1 \pm \sqrt{1 + 4\log_2 x}}{2}$;

28. 217 ;   29. $g(x) = \begin{cases} -x, & -2 \le x \le 0 \\ 0, & 0 \le x \le 1 \\ 2(x-1) & 1 \le x \le 2 \end{cases}$

30. $(g \circ f)(x) = \{(x+1)^2, -2 \le x \le 1\}$ 32. $f(f(2)) = 26$

33. 2λ ;   34. period $= 12$;

35. $x = 2n\pi + \dfrac{3\pi}{4}$; **36.** $x = -1 \pm \sqrt{\dfrac{1}{2}\log_{10}(1 + \sqrt{11})}$

40. $f(x) = 16 - 40x - 60\,\dfrac{(2x + 29)}{(x - 2)}$.

2.10 Polar Co-ordinates and Graphs

While we have so far considered the cartesian system of co-ordinates, we now intoroduce to *Polar System*. In the polar system of co-ordinates, we start with a fixed line called the *Initial line* and a fixed point on it called the *Pole*.

Let OX be the initial line and O the pole. Let P be a given point. Then the distance OP, called the radius vector, and the $\angle XOP$, called the vectorial angle, are called the polar co-ordinates of the point P. Writing $OP = r$, $\angle XOP, = \theta$, we may refer to the point P as (r, θ). (Fig. 2.56)

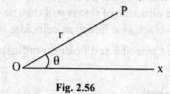

Fig. 2.56

Graph of a function $r = f(\theta)$ is the set of points $[\theta, f(\theta)]$ where θ belongs to the domain of the given function f. Also we say that $r = f(\theta)$ is the polar equation of the graph.

Unrestricted variation of polar co-ordinates. If we were concerned with assigning polar co-ordinates to only *individual* points in the plane, then it would be enough to consider the radius vector to have positive values and the vectorial angle to lie between 0 and 2π only. While considering graphs of given functions or equivalents it becomes necessary, however, to remove this restriction and to consider both r and θ as varying in the interval $[-\infty, \infty]$. The necessary conventions for this will be introduced now.

The angle, θ, will be regarded as the measure of rotation of a line which starting from OX revolves round it, the measure being positive or negative according as the rotation is counter-clockwise or clockwise.

To find the point (r, θ) where r, is negative and, θ has any value, we proceed as follows :

Let the revolving line starting from OX revolve through, θ, Then produce this final position of the revolving line backwards through O

The point P on this produced line such that $OP = |r|$ is the required point (r, θ).

Thus if r is negative, the point (r, θ) is the same as the point $(|r|, \theta + \pi)$.

The positions of the points $(1, 9\pi/4)$, $(1, -9\pi/4)$, $(-1, 9\pi/4)$, $(-1, -9\pi/4)$ have been marked in the figures 2.57 to 2.60 given below :

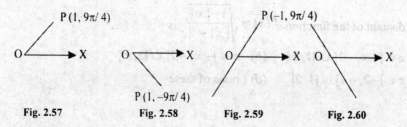

| Fig. 2.57 | Fig. 2.58 | Fig. 2.59 | Fig. 2.60 |

We may notice that of the two points (r, θ), $(r, -\theta)$ each is the reflection of the other in the initial line. Thus the graph of the function f or equivalently that of the equation $r = f(\theta)$ will be symmetrical about the initial line, if we have

$$f(-\theta) = f(\theta) \,\forall\, \theta.$$

In case, therefore, the graph of the function is known to be symmetrical about the initial line, we may, in the first instance, draw only that part of it which corresponds to $\theta \in [0, \pi]$. The complete graph will then consist of the union of the part already drawn and its reflection in the initial line.

Transformation of co-ordinates. It is often found necessary and useful to express a given cartesian equation as a polar equation and *vice-versa*. For this purpose, we need relations between the cartesian co-ordinates (x, y) and the polar co-ordinates (r, θ) of the same point. These relations will naturally assume a simpler from if the reference systems for the two systems of co-ordinates are chosen so as to be related in some suitable manner.

We suppose that the initial line of the polar system is the positive direction of x-axis and the pole is the origin. The positive direction of y-axis will then be such that the initial line after revolving through $\pi/2$ in the counter-clockwise direction coincides with it.

Let (x, y) and (r, θ) be the Cartesian and Polar co-ordinates respectively of a point P.

From the $\triangle OMP$, we get

$OM/OP = \cos\theta \Rightarrow \mathbf{x = r\cos\theta}$ ...(i)

$MP/OP = \sin\theta \Rightarrow \mathbf{y = r\sin\theta}$ ...(ii)

The equations (i) and (ii) determine the cartesian co-ordinates (x, y) of the point P in terms of its polar-co-ordinates (r, θ) and *vice-versa*.

We give below the cartesian forms of a few equations and the corresponding forms obtained by means of the ralations.

$$x = r\cos\theta, \ y = r\sin\theta$$

Cartesian form	Polar form
$x^3 + y^3 = 3axy$	$r = \dfrac{3a\sin\theta\cos\theta}{\cos^3\theta + \sin^3\theta}$
$x^4 + y^4 = 4a^2xy$	$r^2 = \dfrac{4a\sin\theta\cos\theta}{\sin^4\theta + \cos^4\theta}$
$y^4 - x^4 + xy = 0$	$r^2 = \dfrac{1}{2}\tan 2\theta$
$x^4 + y^4 = a^2(x^2 - y^2)$	$r^2 = \dfrac{a^2\cos 2\theta}{\cos^4\theta + \sin^4\theta}$

OBJECTIVE QUESTIONS

Note : *For each of the following questions four alternatives are given for the answer. Only one of them is correct alternative. Choose the correct alternative.*

1. The domain of the function $f(x) = \sqrt{\dfrac{1 - |x|}{|x| - 2}}$ is

   (a) $x \in \,]-\infty, -1[\, \cup\,]1, \infty[$ (b) $x \in \,]-\infty, -2[\, \cup\,]2, \infty[$

   (c) $x \in \,]-2, -1]\, \cup\, [1, 2[$ (d) none of these

2. The number of roots of the equation

$$\cot x = \frac{\pi}{2} + x \text{ in } \left[-\pi, \frac{3\pi}{2}\right] \text{ is}$$

(a) 3   (b) 2   (c) 1   (d) infinite

3. The solution of the equation

$|2^x - 1| + |4 - 2^x| \le 3$ is,

(a) $(-\infty, 0)$ (b) $(0, 2)$ (c) $(\infty -, \infty)$ (d) none of these

4. The domain of $f(x) = \sqrt{1 - \left|\dfrac{x^2}{x-1}\right|}$ is

(a) $]-\infty, \infty[- \{1\}$

(b) $\left[\dfrac{-1-\sqrt{5}}{2}, \dfrac{-1+\sqrt{5}}{2}\right]$

(c) $\left[\dfrac{-1-\sqrt{5}}{2}, \dfrac{-1+\sqrt{5}}{2}\right] \cup]1, \infty[$

(d) $\left[\dfrac{-1+\sqrt{5}}{2}, 1\right[$

5. Period of the function $\left|\sin^3 \dfrac{x}{2}\right| + \left|\cos^5 \dfrac{x}{5}\right|$ is

(a) 2π (b) 10π (c) 8π (d) 5π

6. The domain of $f(x) = \sqrt{\left(\log_2 \dfrac{x}{[x]}\right)}$; where [.] denotes the greatest integer function is

(a) $]-\infty, \infty[- [0,1[$ (b) $]-\infty, 0[$ (c) $[1, \infty[$ (d) none of these

7. The domain of $f(x) = \sqrt{(\tan x - 1)(\tan x - 2)}$ is

(a) $\left]n\pi + \dfrac{\pi}{4}, n\pi + \tan^{-1} 2\right[$

(b) $\left[\dfrac{\pi}{4}, \tan^{-1} 2\right[$

(c) $R - \left]n\pi + \dfrac{\pi}{4}, n\pi + \tan^{-1} 2\right[$

(d) none of these

8. The solution set of the equation

$$\tan^{-1} \sqrt{x(x+1)} + \sin^{-1} \sqrt{x^2 + x + 1} = \frac{\pi}{2} \text{ is}$$

(a) $]-1, 0[$ (b) $[-1, 0]$ (c) $\{-1, 0\}$ (d) none of these

9. If $f(x) = \dfrac{1}{\sqrt{1 + x^2}}$, then $(f \circ f \circ f)(x)$ is

(a) $\dfrac{3x}{\sqrt{1 + x^2}}$ (b) $\dfrac{x}{\sqrt{1 + 3x^2}}$ (c) $\dfrac{3x}{\sqrt{1 - x^2}}$ (d) none of these

10. Let $f\left(x + \dfrac{1}{x}\right) = x^2 + \dfrac{1}{x^2}$, where $x \ne 0$; then $f(x)$ is

(a) $x^2 - 2, x \ne 0$ (b) $x^2 - 2, |x| < 2$ (c) $x^2 - 2, |x| \ge 2$ (d) none of these

11. The period of $e^{\cos^4 \pi x + x - [x]} + \cos^2 \pi x$ is

(a) 0   (b) 1   (c) $\dfrac{\pi}{2}$ (d) 2π

12. The range of the function $f(x) = \sin(\sin^{-1}\{x\})$ where $\{\,.\,\}$ is fractional part of x, is

(a) $[0, 1[$ (b) $[0, 1]$ (c) $]-1, 1[$ (d) none of these

13. If $f(x) = \left[\sin \dfrac{x}{r}\right], x, r \in\,]0, \pi[$, then range of $f(x)$ is

(a) $]0, 1]$ (b) $[0, 1]$ (c) $\{0, 1\}$ (d) $\{0\}$

14. Let $f : \left[-\dfrac{\pi}{3}, \dfrac{2\pi}{3}\right] \to [0, 4]$ be a function defined as $f(x) = \sqrt{3}\, \sin x - \cos x + 2$. Then $f^{-1}(x)$ is given by

(a) $\sin^{-1}\left(\dfrac{x-2}{2}\right) - \dfrac{\pi}{6}$ (b) $\sin^{-1}\left(\dfrac{x-2}{2}\right) + \dfrac{\pi}{6}$

(c) $\dfrac{2\pi}{3} + \cos^{-1}\left(\dfrac{x-2}{2}\right)$ (d) none of these

15. If f is a function such that $f(0) = 2, f(1) = 3$ and $f(x+2) = 2f(x) - f(x+1)$ for every real x then $f(5)$ is

(a) 7   (b) 13   (c) 1   (d) 3

16. If $f(x) = \begin{cases} x^2 & \text{for } x \geq 0 \\ x, & \text{for } x < 0 \end{cases}$ then $f o f(x)$ is given by

(a) x^2 for $x \geq 0$, x for $x < 0$ (b) x^4 for $x \geq 0$, x^2 for $x < 0$

(c) x^4 for $x \geq 0$, $-x^2$ for $x < 0$ (d) x^4 for $x \geq 0$, x for $x < 0$

17. If $f(x) + 2f(1 - x) = x^2 + 2, \ \forall\, x \in R$, then $f(x)$ is given as

(a) $\dfrac{(x-2)^2}{3}$ (b) $x^2 - 2$ (c) 1   (d) none of these

18. If $f : I \to I$ be defined by $f(x) = [x + 1]$, where $[\,.\,]$ denotes the greatest integer function, then $f^{-1}(x)$ is equal to

(a) $x - 1$ (b) $[x + 1]$ (c) $\dfrac{1}{[x-1]}$ (d) $\dfrac{1}{x+1}$

19. Domain of $f(x)$ satisfying $2^x + 2^{f(x)} = 2$ is

(a) $]-1, 1[$ (b) $]-\infty, -1[$ (c) $]-\infty, 1[$ (d) none of these

20. Let $f(x) \begin{cases} x^3 - 1, & x < 2 \\ x^2 + 3, & x \geq 2 \end{cases}$, then $f^{-1}(x)$ is

(a) $\begin{cases} (x+1)^{1/3}, & x < 2 \\ (x-3)^{1/2}, & x \geq 2 \end{cases}$ (b) $\begin{cases} (x+1)^{1/3}, & x < 7 \\ (x-3)^{1/2}, & x \geq 7 \end{cases}$ (c) $\begin{cases} (x+1)^{1/3}, & x < 1 \\ (x-3)^{1/2}, & x \geq 1 \end{cases}$ (d) does not exist.

ANSWERS

1.	(c)	2.	(a)	3.	(b)	4.	(b)	5.	(b)	6.	(c)	7.	(c)
8.	(c)	9.	(b)	10.	(c)	11.	(b)	12.	(a)	13.	(c)	14.	(b)
15.	(a)	16.	(d)	17.	(a)	18.	(a)	19.	(c)	20.	(b)		

CHAPTER 3

CONTINUITY AND LIMIT

3.1. Introduction

The statement that a function is defined in a certain interval means that to each number belonging to the interval, there corresponds a value of the function. *The value of the function for any particular number may be quite independent of the value of the function for another number* and no relationship in the various values of a function corresponding to the members of the domain is implied in the definition of a function as such.

Thus, if x_1, x_2 are any two members of the domain of a function f so that $f(x_1)$, $f(x_2)$ are the corresponding values of the function, then $|f(x_2) - f(x_1)|$ may be large even though $|x_2 - x_1|$ is small. It is because the value $f(x_2)$ assigned to the function for $x = x_2$ is quite independent of the value $f(x_1)$ of the function for $x = x_1$.

We now propose to study the change $|f(x_2) - f(x_1)|$ relative to the change $|x_2 - x_2|$ by introducing the notion of *Continuity* and *Discontinuity* of a function.

In the next article, we first analyse the intuitive notion of *Continuity* and then state its precise meaning in the form of a definition.

3.1.1. Continuity of a function at an interior point of an interval.

Intuitively, *continuous* variation implies absence of *sudden* changes so that in order to arrive at a suitable definition of continuity, we have to examine the precise meanings of its implication in relation to a function which is defined in an interval $[a, b]$. Let c, be a point of $]a, b[$, *i.e.*, an interior point of $[a, b]$.

The function f will vary continuously at c if, as x changes from c to either side of it, the change in the value of the function is not sudden, *i.e.*, the change in the value $f(x)$ of the function is *small* if only the change in the value of x is *small*.

We consider a value $c + h$ of x belonging to the interval $[a, b]$. Here, h, which is the changes in x, may be positive or negative. Then $f(c + h) - f(c)$, is the corresponding change in the value of the function f, which, again, may be positive or negative.

For continuity, we require that $f(c + h) - f(c)$ should be numerically small, if h is numerically small. This means that

$$|f(c + h) - f(c)|$$

can be made as small as we like by taking $|h|$ sufficiently small.

The precise definition of continuity should not involve the use of the word *small*, whose meaning is indefinite, as there exists no absolute standard of smallness. Such a definition would now be given.

A function f is continuous at c if, to any positive number, ε, arbitrarily assigned, there corresponds a positive number δ such that

$$|f(c + h) - f(c)| < ε$$

for all values of, h, such that

$$|h| < δ \qquad\qquad\qquad\qquad (Avadh\ 2000)$$

This means that $f(c + h)$ lies between $f(c) - ε$ and $f(c) + ε$ for all those values of h, which lie between $-δ$ and $δ$.

Alternatively, replacing $c + h$ by x, we can say that *a function f is continuous at c, if there exists an interval $]c - δ, c + δ[$ around c such that, for all values of x which belong to this interval, we have*

$$f(c) - ε < f(x) < f(c) + ε,$$

ε being any positive number arbitrarily assigned.

A function which is continuous at c is also said to be continuous for c.

Meaning of continuity explained graphically. Consider the point $P[c, f(c)]$ on the graph of the function f. draw the lines

$$y = f(c) - ε, \qquad\qquad y = f(c) + ε \qquad\qquad\qquad ....(i)$$

which lie on different sides of the point P and are parallel to x-axis.

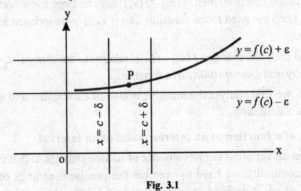

Fig. 3.1

Here, ε which is an arbitrarily assigned positive number, measures the degree of closeness of the lines (*i*) from each other.

The continuity of the function f at c, then, requires that we should be able to draw two lines

$$x = c - δ, \qquad\qquad x = c + δ \qquad\qquad\qquad ....(ii)$$

which are parallel to y-axis and which lie around the line $x = c$, such that every point of the graph between the two lines (*ii*) lies also between the two lines (*i*).

3.1.2 Continuity at end points. We say that a function f is continuous at the left end-point a of its domain $[a, b]$, if, given any positive number ε, there exists a positive number δ such that

$$|f(a + h) - f(a)| < ε \text{ when } 0 ≤ h < δ.$$

Similarly f is continuous at the right end-point b of its domain if given $ε > 0$ there exists $δ > 0$ such that

$$|f(b - h) - f(b)| < ε \text{ when } 0 ≤ h < δ.$$

3.1.3 Continuity of a function in an interval. In the preceding discussion, we have dealt with the definition of continuity of a function at a point of its domain. We now extend this definition to *Continuity in an interval* and say that :

A functions is continuous in an interval if it is continuous at every point thereof.

3.1.4 Discontinuity. *A function which is* **not** *continuous at c is said to be discontinuous at c.* This concept will be discussed in detail later on.

The notion of continuity and discontinuity will now be illustrated by means of some simple examples.

EXAMPLES

1. *Show that* $f(x) = 3x + 1$ *is continuous at* $x = 1$.

We have $f(1) = 4$

and $f(x) - f(1) = 3x + 1 - 4 = 3(x - 1)$.

We will now attempt to make the numerical value of this difference smaller than any pre-assigned positive number, *say* .001.

(*i*) Let $x > 1$ so that $3(x - 1)$ is positive and is, therefore, the numerical value of $f(x) - f(1)$. Now

$$|f(x) - f(1)| = 3(x - 1) < .001$$

if $\qquad x - 1 < .001/3$, *i.e.*, if $x < 1 + .001/3$ ...(*i*)

(*ii*) Let $x < 1$, so that $3(x - 1)$ is negative and the numerical value of $f(x) - f(1)$ is $3(1 - x)$. Now

$$|f(x) - f(1)| = 3(1 - x) < .001$$

if $\qquad 1 - x < .001/3$, *i.e.*, if $1 - .001/3 < x$ ...(*ii*)

Combining (*i*) and (*ii*), we see that

$$|f(x) - f(1)| < .001$$

for all those values of x for which

$$1 - .001/3 < x < 1 + .001/3.$$

The test of continuity for 1 is thus satisfied for the particular value .001 of ε. We may similarly show that the test is true for other particular values of ε also.

The *complete* argument is, however, as follows :

Let ε be a given positive number. We have

$$|f(x) - f(1)| = 3|x - 1|.$$

Now

$$3|x - 1| < \varepsilon \Leftrightarrow |x - 1| < \varepsilon/3 \Leftrightarrow 1 - \varepsilon/3 < x < 1 + \varepsilon/3.$$

Thus, there exists an interval $]1 - \varepsilon/3, 1 + \varepsilon/3[$ around 1 such that for every x in this interval, the numerical value of the difference between $f(x)$ and $f(1)$ is less than the pre-assigned positive number ε. Here $\delta = \varepsilon/3$.

Hence f is continuous at 1.

Note. It may be shown that the function $f(x) = 3x + 1$, is continuous in **R** which is the domain of the function.

2. *Show that* $f(x) = 3x^2 + 2x - 1$ *is continuous at* 2.

Let ε be a given positive number. We have $f(2) = 15$. Also

$$|f(x) - f(2)| = |3x^2 + 2x - 1 - 15| = |x - 2||3x + 18|$$

We suppose that x lies between 1 and 3. For values of x between 1 and 3, $3x + 8$ is positive and less than $(3.3 + 8) = 17$.

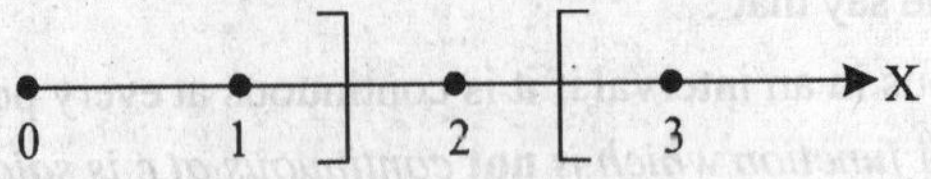

Fig. 3.2

Thus for $\qquad 1 < x < 3$
we have $|f(x) - f(2)| < 17|x - 2|$

Now $\qquad 17|x - 2| < \varepsilon \qquad$ if $\qquad |x - 2| < \varepsilon/17$

Thus, we see that there exists a positive number $\varepsilon/17$ such that

$$|f(x) - f(2)| < \varepsilon \text{ when } |x - 2| < \varepsilon/17.$$

Therefore, f is continuous for $x = 2$.

Note. It may be shown that $f(x) = 3x^2 + 2x - 1$, is continuous for every value of x.

3. *Prove that $f(x) = \sin x$ is continuous for every value of x.*

It will be shown that $f(x) = \sin x$ is continuous for any given value of x ; say c.

Let ε be an *arbitrarily assigned positive number.* We have

$$|f(x) - f(c)| = |\sin x - \sin c|$$
$$= \left| 2 \cos \frac{x + c}{2} \sin \frac{x - c}{2} \right|$$
$$= \left| 2 \cos \frac{x + c}{2} \right| \left| \sin \frac{x - c}{2} \right|.$$

Now $\left| \cos \dfrac{x + c}{2} \right| \le 1$ for every value of x and c.

Also $\qquad \left| \sin \dfrac{x - c}{2} \right| \le \left| \dfrac{x - c}{2} \right| \qquad\qquad$ (From *Elementary Trigonometry*)

Thus, we have

$$|\sin x - \sin c| = 2 \left| \cos \frac{x + c}{2} \right| \left| \sin \frac{x - c}{2} \right|$$
$$\le 2.1. \left| \frac{x - c}{2} \right| = |x - c|.$$

Thus, $\qquad |\sin x - \sin c| < \varepsilon$ when $|x - c| < \varepsilon$.

There exists, therefore, an interval $]\,c - \varepsilon, c + \varepsilon\,[$ around c such that for every value of x in this interval

$$|\sin x - \sin c| < \varepsilon.$$

It follows that $f(x) = \sin x$ is continuous for $x = c$ and, therefore, also for every value of x ; c being any number.

4. *Show that $f(x) = \sin^2 x$ is continuous for every value of x.* $\qquad$ (*Manipur 2002*)

Let ε be any given positive number. We have

$$|f(x) - f(c)| = |\sin^2 x - \sin^2 c|$$
$$= |\sin(x + c)||\sin(x - c)|$$
$$\le |\sin(x - c)| \le |x - c|.$$

Thus, $\qquad |f(x) - f(c)| < \varepsilon \qquad$ if $\qquad |x - c| < \varepsilon$.

Hence, $f(x) = \sin^2 x$ is continuous for $x = c$, and, therefore, also for every value of x ; c being any number.

5. *Show that the function f, as defined below, is discontinuous at* $x = \dfrac{1}{2}$.

$$f(x) = \begin{cases} x, & \text{when } 0 \le x < \dfrac{1}{2} \\[2mm] 1, & \text{when } = \dfrac{1}{2} \\[2mm] 1 - x, & \text{when } \dfrac{1}{2} < x < 1 \end{cases}$$

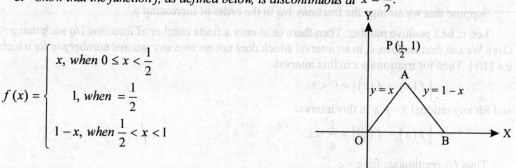

Fig. 3.3

The argument will be better grasped by considering the graph of the function.

The graph consists of the point $P\,(1/2, 1)$ and the line segments OA, AB excluding the point A. Our intuition immediately suggests that there is a discontinuity at A, there being a gap in the graph at A.

Also, analytically, we can see that for any value of x, about $x = \dfrac{1}{2}$, $f(x)$ differs from $f\left(\dfrac{1}{2}\right)$

which is equal to 1, by a number greater than $\dfrac{1}{2}$ so that there is no question of making the difference

between $f(x)$ and $f\left(\dfrac{1}{2}\right)$ less than *any* positive number arbitrarily assigned.

Therefore, f is discontinuous at $x = \left(\dfrac{1}{2}\right)$.

6. *Discuss the function f defined by*

$$f(x) = \begin{cases} 1/q, & \text{when } x \text{ is a rational } p/q \ne 0 \text{ in lowest terms,} \\ 0, & \text{when } x \text{ is irrational and } x = 0 \end{cases}$$

as regards its continuity.

It will be shown that f is continuous for irrational values of x and for $x = 0$ and discontinuous for *non-zero* rational values of x.

Let $p/q \ne 0$ be a rational number so that

$$f\left(\frac{p}{q}\right) = \frac{1}{q}.$$

We know that in every interval there lie an infinite number of irrational numbers. Accordingly in every neighbourhood of p/q, there exist irrational numbers, such that for these numbers, x, we have

$$\left| f(x) - f\left(\frac{p}{q}\right) \right| = \left| 0 - \frac{1}{q} \right| = \frac{1}{q}.$$

Thus, if, ε, be a positive number $< 1/q$, there cannot exist a neighbourhood of p/q for every point x of which

$$\left| f(x) - f\left(\frac{p}{q}\right) \right| < \varepsilon$$

Hence, f is discontinuous for $x = p/q$.

Let, now, c, be an irrational number, so that $f(c) = 0$.

Suppose that we arrange the fractions $1/q$ in the order of increasing q.

Let, ε, be a positive number. Then there exist only a finite number of fractions $1/q$ such that $q < (1/\varepsilon)$. We can thus enclose, c, in an interval which does not enclose any rational number p/q for which $q < (1/\varepsilon)$. Then for irrationals x in this interval

$$\left| f(x) - f(c) \right| = 0 < \varepsilon.$$

and for any rational $x = p/q$ in this interval

$$\left| f(x) - f(c) \right| = \frac{1}{q} < \varepsilon.$$

Thus f is continuous for $x = c$.

EXERCISES

1. Show that every constant function and the Identity function are continuous in **R**.

2. Show that the following functions are constinuous at the given values:

   (i) $f(x) = 1/x$ at 3   (ii) $f(x) = (x + 2)(x - 3)$ at 5

   (iii) $f(x) = 2x^2 + 1$ at 0   (iv) $f(x) = 1/(x^2 + 1)$ at 0

   (v) $f(x) = x + \sin x$ at 0   (vi) $f(x) = \sqrt{x}$ at 0 and 1.

3. Give the set of points of discontinuity of $f(x) = [x]$.

4. Show that f defined by $f(x) = \sqrt{(1 - x)} \; \forall \; x \leq 1$ is continuous in $[0, 1]$.

5. Given that f is defined by

   $$f(x) = 2 \text{ if } x \in \,]\,1, 2[, \; f(x) = -1 \text{ if } x \in \,]\,2, 3[, \; f(x) = 1/2 \text{ if } x \in \,]\,3, 4[$$

   and $f(1) = 5, f(2) = f(3) = 0, f(4) = 1/2,$

   Investigate if f is continuous or discontinuous for 2, 3.

6. Consider the function which associates to the number x the following values, whenever they exist, and establish the corresponding statements.

   (i) $f(x) = 1 + x + |x|,$ f is continuous at $-1, 0, 1$

   (ii) $f(x) = x^2 + x + 1,$ f is continuous at 0 and 1

   (iii) $f(x) = x^2 - 5x + 4,$ f is continuous at 1 and 4

   (iv) $f(x) = x^3 - 3x,$ f is continuous at 0

   (v) $f(x) = (x + 3)/(x - 1),$ f is continuous at 2

7. Show that

   (i) $f(x) = (ax + b)/(cx + d) \,; c \neq 0$ is continuous $\forall \; x \in R \sim \{-d/c\}$.

   (ii) $f(x) = (x^2 + 4)/(x^2 - x - 2)$ is continous $\forall \; x \in R \sim \{-1/2\}$.

8. Examine the continuity at 0 of f such that

   (i) $f(x) = (x^2 - 9)/(x - 3)$ when $x \neq 3$ and $f(3) = 6$.

   (ii) $f(x) = \sin(1/x)$ when $x \neq 0$ and $f(0) = 0$.

9. Prove that $f(x) = \cos x$ is continuous for every value of x.

10. Show that $f(x) = \cos^2 x$ and $f(x) = 2 + x + x^2$ are continuous for every value of x.

11. Draw the graph of the function g defined as follows:

$$g(x) = 0, x \in \,] -\infty, 0\,[, g(x) = 1, x \in [\,0, 1\,[,$$
$$g(x) = 2, x \in [\,1, \infty\,[,$$

and show that it has two points of discontinuity.

ANSWERS		
3. 1;	**5.**	Discontinuous at 2 and 3;
8. (*i*) Continuous	(*ii*)	Discontinuous.

3.2. Limit.

The important concept of the limit of a function will now be introduced. For the continuity of a function f at c, the values of the function for values of x near c lie near $f(c)$. In general.

(*i*) the value $f(c)$ of the function for c and

(*ii*) the values $f(x)$ for values of x near c

are independent. It may, in fact, sometimes happen that the values of the function for values of x and c lie near a number l which is not equal to $f(c)$ or that the values do not lie near any number at all.

Thus, for example, we have seen in Ex. 5, page 74 that the values of the function for values of x near 1/2 lie near 1/2 which is different from the value, 1, of the function for $x = 1/2$. In fact this inequality itself was the cause of discontinuity of this function for $x = 1/2$.

The above remarks lead us to introduce the notion of limit as follows:

Def. *A function f is said to tend to a limit, l, as x tends to c, if, to any positive ε, arbitary assigned, there corresponds a positive number δ, such that for every $x \in \,]\,c - \delta, c + \delta\,[\, \sim \{c\}, f(x)$ differs from l numerically by a number which is less than ε, i.e.*

$$|f(x) - l| < \varepsilon,$$

for evry value of x, such that

$$0 < |x - c| < \delta,$$

or equivalently

$$l - \varepsilon < f(x) < l + \varepsilon \; \forall \, x \in \,]\,c - \delta, c\,[\,\cup\,]\,c, c + \delta\,[. \qquad \textbf{\textit{(Kanpur 2001)}}$$

In this case, we write $\lim\limits_{x \to c} f(x) = l.$

3.2.1 Right handed and left handed limits.

$$\lim\limits_{x \to (c+0)} f(x) = l \qquad\qquad \lim\limits_{x \to (c-0)} f(x) = l$$

A function f is said to tend to a limit, l, as x tends to c, **from the right,** *if to any positive number, ε, arbitrarily assigned there corresponds a positive number, δ, such that*

$$|f(x) - l| < \varepsilon$$

for every value of $x \in \,]\,c, c + \delta\,[.$

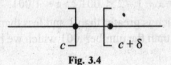

Fig. 3.4

In this case, we write

$$\lim_{x \to (c-0)} f(x) = l$$

and say that, l is the *right handed limit* of $f(x)$.

A function f is said to tend to a limit, l, as x tends to c **from the left,** *if to any positive number, ε, arbitrarily assigned there corresponds a positive number, δ, such that*

$$|f(x) - l| < \varepsilon \text{ for every value of } x \in \,] \, c - \delta, c \, [$$

In this case, we wirte

$$\lim_{x \to (c-0)} f(x) = l$$

and say that, l, is the *left handed limit* of $f(x)$.

From above it at once follows that

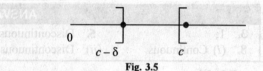

Fig. 3.5

$$\lim_{x \to c} f(x) = l \Leftrightarrow \lim_{x \to (c+0)} f(x) = l = \lim_{x \to (c-0)} f(x).$$

3.2.2. Non-existence of limit. It is important to remember that a limit may not exist.

An example of this possibility is given in Ex. 3 page 64.

Remarks: 1. In order that a function may tend to a limit, it is necessary and sufficient that corresponding to every positive number ε, a choice of δ is possible. The function will not either approach a limit or at any rate will not have the limit l, if for *some* ε, a corresponding δ does not exist.

2. The question of the limit of $f(x)$ as x approaches 'c' does not take any note of the value of the function for c. The function may not even be defined for $x = c$.

From the above it appears that in ordre that the statement

$$\lim_{x \to c} f(x) = l$$

has a meaning, it is necessary, that the function f is defined in a neighbourhood of c except possibly at c.

EXAMPLES

1. *Examine the limit of the function $y = (x^2 - 1)/(x - 1)$ as x tends to 1*

The function is defined for every value of x other than 1 and

$$y = \frac{x^2 - 1}{x - 1} = x + 1, \text{ when } x \neq 1.$$

Case I. Firstly consider the behaviour of the values of y for values of x greater than 1. Clearly, y is greater than 2 when x is greater than 1.

If, x, while remaining greater than 1, takes up values whose difference from 1 constantly diminishes, then y, while remaining greater than 2, takes up values whose difference from 2 constantly diminishes also.

In fact, difference between y and 2 can be made as small as we like by taking x sufficiently near 1.

For instance, consider the number .001.

Then $\qquad |y - 2| = y - 2 = x + 1 - 2 < .001 \Leftrightarrow x < 1.001.$

Thus, for every value of x which is greater than 1 and less than 1.001, the absolute value of the difference between y and 2 is less than the number .001 which we had arbitrarily selected.

Instead of the particular number .001, we now consider any positive number ε. Then

$$y - 2 = x - 1 < \varepsilon \Leftrightarrow x < 1 + \varepsilon.$$

Thus, there exists an interval]1, 1 + ε[, such that the value of y, for any value of x in this interval, differs from 2 numerically, by a number which is smaller than the positive number ε, selected arbitrarily.

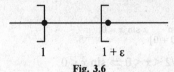

Fig. 3.6

Thus the limit of y as x approaches 1 from the right is 2 and we have

$$\lim_{x \to (1 + 0)} y = 2.$$

Case II. We now consider the behaviour of the values of y for values of x less than 1.

When x is less than 1, y is less than 2.

If, x, while remaining less than 1, takes up values whose difference from 1 constantly diminishes, then y, while remaining less than 2, takes up values whose difference from 2 constantly diminishes also.

Let, now, ε be any arbitrarily assigned positive number, however small.

We then have

$$|y - 2| = 2 - y = 2 - (x + 1)$$
$$= 1 - x < \varepsilon$$

Fig. 3.7

so that for every value of x less than 1 but > 1 − ε, the absolute value of the difference between y and 2 is less than the number ε.

Thus there exists *an interval*]1 − ε, 1[*such that the value of y, for any x in the interval, differs from 2 numerically by a number which is smaller than the arbitrarily selected positive number* ε.

Thus the limit of y, as x approaches 1, from the left is 2 and we write

$$\lim_{x \to (1 - 0)} y = 2.$$

Case III. Combining the conclusions arrived at in the last two cases, we see that corresponding to any arbitrarily assigned positive number ε, there exists an interval] 1 − ε, 1 + ε[around 1, such that for every value of x in this interval, other than 1 where the function is not defined, y differs from 2 numerically by a number which is less than ε, *i.e.*, we have

$$|y - 2| < \varepsilon$$

for any x, other than 1, such that $|x - 1| < \varepsilon$

Thus $\lim_{x \to 1} y = 2 \Leftrightarrow y \to 2$ as $x \to 1$.

2. *Examine* $\lim_{x \to 0} (x \sin x)$

Case 1. Let x > 0. Now

$$0 < x < \pi/2 \Rightarrow \sin x > 0.$$

For 0 < x < π/2, we have

$$\sin x < x \Rightarrow x \sin x < x^2.$$

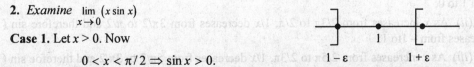

Fig. 3.8

Let ε be an arbitrarily assigned positive number.

For values of x which are positive and less than $\sqrt{\varepsilon}$, we have $x^2 < \varepsilon$. Thus

$$0 < x \sin x < \varepsilon \text{ when } 0 < x \sqrt{\varepsilon}.$$

It follows that

$$\lim_{x \to (0 + 0)} x \sin x = 0.$$

Case II. Let $x < 0$. Now $-\pi/2 < x < 0 \Rightarrow \sin x < 0$.

The values of the function for two values of x which are equal in magnitude but opposite in signs are equal. Hence, as in case I, we see that for any value of x in the interval the numerical value of the difference between $x \sin x$ and 0 is less than ε. Thus

$$\lim_{x \to (0 - 0)} x \sin x = 0$$

Case III. Combining the conclusions arrived at in the last two cases, we see that corresponding to any positive number ε arbitrarily assigned, there exists an interval $]-\sqrt{\varepsilon}, \sqrt{\varepsilon}[$ around 0, such that for any x belonging to this interval, the numerical value of of the difference between $x \sin x$ and 0 is $< \varepsilon$, i.e.,

$$|x \sin x - 0| < \varepsilon.$$

Thus
$$\lim_{x \to 0} x \sin x = 0.$$

Remarks. It will be seen that the inequality $|x \sin x - 0| < \varepsilon$ is satisfied even for $x = 0$. But it should be carefully noted that no difference would arise as to the conclusion

$$\lim_{x \to 0} x \sin x = 0,$$

even if $x = 0$ were an exception.

Again, we see that for $x = 0$, the value of the function is $0 \sin 0 = 0$ which is also its limit as x approaches 0. Thus in this case, the limit of the function is the same as its value so that function is continuous at 1 is irrelevant.

3. *Examine the limit of sin $(1/x)$ as x approaches* 0.

Let $\qquad\qquad y = \sin(1/x)$. $\qquad\qquad\qquad\qquad\qquad\qquad$ *(Kanpur 2001)*

The function is defined for every value of x, other than 0; *i.e.*, the domain of the function is $\mathbf{R} \sim \{0\}$.

The graph of this function will enable the student to understand the argument better.

To draw the graph, we note the following points :

(*i*) As x increases from $2/\pi$ to ∞, $1/x$ decreases from $\pi/2$ to 0 and therefore sin $(1/x)$ decreases from 1 to 0;

(*ii*) As x increases from $2/3\pi$ to $2/\pi$, $1/x$ decreases from $3\pi/2$ to $\pi/2$ and therefore sin $(1/x)$ increases from -1 to 1;

(*iii*) As x increases from $2/5\pi$ to $2/3\pi$, $1/x$ decreases from $5\pi/2$ to $3\pi/2$ and therefoe sin $(1/x)$ decreases from $+1$ to -1; and so on.

Thus the positive values of x can be divided in an infinite number of intervals

$$\cdots\cdots\cdots\left[\frac{2}{7\pi},\frac{2}{5\pi}\right],\left[\frac{2}{5\pi},\frac{2}{3\pi}\right],\left[\frac{2}{3\pi},\frac{2}{\pi}\right],\left[\frac{2}{\pi},\infty\right[$$

such that the function decreases from 1 to 0 in the first interval on the left and oscillates from -1 to 1 from 1 to -1 alternately in the others beginning from the second interval on the right.

It may be similarly seen that the negative values of x also divide themselves in an infinite set of intervals

$$\left]-\infty,\frac{-2}{\pi}\right],\left[\frac{-2}{\pi},\frac{-2}{3\pi}\right],\left[\frac{-2}{3\pi},\frac{-2}{5\pi}\right],\cdots\cdots\cdots$$

such that the function decreases from 0 to -1 in the first interval on the left and oscillates from -1 to 1 and from 1 to -1 alternately in the others beginning from the second interval on the left.

Hence, we have the graph (Fig. 3.9) as drawn

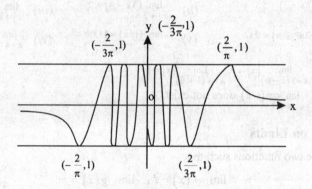

Fig. 3.9

The function oscillates between -1 and 1 more and more rapidly as x approaches nearer and nearer zero from either side. If we take any interval enclosing 0, however small it may be, then for an infinite number of points of this interval the function assumes the values 1 and -1.

There can, therfore, exist no number which differs from sin $1/x$ by a number less than an arbitrarily assigned positive number for values of x near 0.

Hence $\lim\limits_{x\to 0}(\sin 1/x)$ does *not* exist.

This example illustrates an important fact that a **limit may not always exist.**

4. *Examine* $\lim\limits_{x\to 0}[x\sin 1/x].$

Obviously the function is not defined for $x = 0$, so that 0 does not belong to the domain of the function.

Now, for non-zero values of x, we have

$$\left|x\sin\frac{1}{x}-0\right|=|x|\left|\sin\frac{1}{x}\right|\le|x|<\varepsilon,$$

when $|x-0|=|x|<\varepsilon.$

Thus, we see that if, ε, be any positive number, there exists an interval $]-\varepsilon,\varepsilon[$ around 0, such that for every value of x in this interval, with the sole exception of 0, $x\sin(1/x)$ differs from 0 by a number less than ε. Hence

$$\lim\limits_{x\to 0}\left[x\sin(1/x)\right]=0.$$

In fact, as may be easily seen, the graph of $y = x \sin(1/x)$ oscillates between $y = x$ and $y = -x$ as x tends to zero.

EXERCISES

1. Given that
$$x < 3 \Rightarrow f(x) = x + 1 \; ; x > 3 \Rightarrow f(x) = 2(5-x), f(3) = 5, \text{ show that } \lim f(x) = 4 \text{ when } x \to 3$$

2. Consider $f(x) = \left(2x^2 + |x|\right) \big/ x, \quad x \in R \sim \{0\}$ and show that
$$\lim_{x \to (0-0)} f(x) = -1, \quad \lim_{x \to (0+0)} f(x) = 1.$$

3. Show that $\displaystyle \lim_{x \to (1+0)} \frac{\sqrt{x}-1}{x-1} = \frac{1}{2}.$

4. Show that the following limits exist and have the values as given:

   (i) $\displaystyle \lim_{x \to 1} (2x+3) = 5.$ (ii) $\displaystyle \lim_{x \to 2} (3x-4) = 2.$ (iii) $\displaystyle \lim_{x \to 1} \left(x^2+3\right) = 4.$

   (iv) $\displaystyle \lim_{x \to 3} \left(2x^2+x\right) = 21.$ (v) $\displaystyle \lim_{x \to 3} (1/x) = 1/3.$ (vi) $\displaystyle \lim_{x \to 0} 1/(2x/3) = 1/3.$

5. Show that $\displaystyle \lim_{x \to (1-0)} [x] = 0, \quad \lim_{x \to (1+0)} [x] = 1.$

6. Show that $\displaystyle \lim_{x \to 0} \cos(1/x)$ does not exist.

3.3. Theorems on Limits

Let f and g be two functions such that
$$\lim_{x \to a} f(x) = l, \quad \lim_{x \to a} g(x) = m.$$

Then

(i) $\displaystyle \lim_{x \to a} (f+g)(x) = \lim_{x \to a} [f(x) + g(x)]$
$$= \lim_{x \to a} f(x) + \lim_{x \to a} g(x) = l + m$$

i.e., the limit of the sum of two functions is equal to the sum of their limits ;

(ii) $\displaystyle \lim_{x \to a} (f-g)(x) = \lim_{x \to a} [f(x) - g(x)]$
$$= \lim_{x \to a} f(x) - \lim_{x \to a} g(x) = l - m$$

i.e., the limit of the difference of two functions is equal to the difference of their limits :

(iii) $\displaystyle \lim_{x \to a} (f \cdot g)(x) = \lim_{x \to a} [f(x) \cdot g(x)]$
$$= \lim_{x \to a} f(x) \cdot \lim_{x \to a} g(x) = l \cdot m$$

i.e., the limit of the product of two functions is equal to the product of their limits ;

(iv) $\displaystyle \lim_{x \to a} (f \div g)(x) = \lim_{x \to a} [f(x) \div g(x)]$
$$= \lim_{x \to a} f(x) \div \lim_{x \to a} g(x) = l \div m, \, (m \neq 0)$$

*i.e., the limit of the quotient of two functions is equal to the quotient of their limits provided the limit of the divisor is **not** zero.*

These results are of fundamental importance but their formal proofs will *not* be given in this book.

These results can also be extended to the case of any finite number of functions.

3.3.1. Infinite limits and variables tending to infinity. Meanings of

(*i*) $\lim\limits_{x \to c} f(x) = \infty.$ (*ii*) $\lim\limits_{x \to c} f(x) = -\infty.$ *(Kanpur 2001)*

A function f is said to tend ∞ as x tends to c if to any positive number G, however large, there corresponds a positive number δ such that for all values of $x \in \left]c - \delta, c\right[\cup\left]c, c + \delta\right[$,

$$f(x) > G.$$

A function f is said to tend to $-\infty$ as x tends to c if to any positive number G, however large, there corresponds a positive number δ such that for all values of $x \in \left]c - \delta, c\right[\cup\left]c, c + \delta\right[$,

$$f(x) < -G.$$

Right handed and left handed limits can be defined as in § 3.2.1.

(*iii*) $\lim\limits_{x \to \infty} f(x) = l.$ (*iv*) $\lim\limits_{x \to -\infty} f(x) = l.$

A function f is said to tend to l as x tends to ∞, if to any given $\varepsilon > 0$ there corresponds $G > 0$ such that

$$\left| f(x) - l \right| < \varepsilon \; \forall \; x > G.$$

A function f is said to tend to l as x tends to $-\infty$, if to any given $\varepsilon > 0$, there corresponds $G > 0$ such that

$$\left| f(x) - l \right| < \varepsilon \; \forall \; x > -G.$$

(*v*) $\lim\limits_{x \to \infty} f(x) = \infty,$ (*vi*) $\lim\limits_{x \to -\infty} f(x) = \infty,$

A function f is said to tend to ∞, as x tends to if to given $\Delta > 0$, there corresponds $G > 0$ such that

$$f(x) > \Delta \; \forall \; x > G.$$

A function f is said to tend to $+\infty$ as x tends to $-\infty$ if to given $\Delta > 0$, there corresponds $G > 0$ such that

$$f(x) > \Delta \; \forall \; x < -G.$$

We may similarly give meanings to

$$\lim\limits_{x \to \infty} f(x) = -\infty, \quad \lim\limits_{x \to -\infty} f(x) = -\infty,$$

Instead of ∞, we may find it more useful to write $+\infty$.

EXAMPLES

1. *Show that*

$$\lim\limits_{x \to 0} \left(1/x^2\right) = \infty.$$

The function is not defined for 0, *i.e.*, 0 does not belong to the domain of the function $f(x) = 1/x^2$.

We write $y = 1/x^2$.

Case I. Let $x > 0$. Consider any positive number, say 10^6.

Now $y = 1/x^2 > 10^6$,   if $0 < x < 1/10^3$.

Instead of the particular number 10^6, we may consider any positive number G. Then

$$1/x^2 > G \text{ if } 0 < x < 1/\sqrt{G}.$$

Thus, there exists an interval $]0, 1/\sqrt{G}[$, such that for every value of x belonging to it, y is greater than the arbitrarily assigned positive number G, Hence

$$\lim_{x \to (0+0)} (1/x^2) = \infty.$$

Case II. Let $x < 0$ be any arbitrarily assigned positive number. Then,

$$\forall x \in]-1/\sqrt{G}, 0[\text{ we have } 1/x^2 > G$$

and as such $\lim_{x \to (0-0)} (1/x^2) = \infty.$

Case III. Combining the conclusions arrived at in the last two cases, we see that corresponding to *any* arbitrarily assigned positive number G, there exists an interval $]-1/\sqrt{G}, 1/\sqrt{G}[$, around 0, such that for every value of x (other than 0) belonging to this interval, we have $1/x^2 > G$. Thus

$$\lim_{x \to 0} (1/x^2) = \infty \text{ or } (1/x^2) \to \infty \text{ as } x \to 0.$$

2. *Show that*

$$\lim_{x \to (0+0)} \frac{1}{x} = \infty, \quad \lim_{x \to (0-0)} \frac{1}{x} = -\infty$$

and $\lim_{x \to 0} \dfrac{1}{x}$ *does not exist.*

The function $f(x) = 1/x$ is not defined $x = 0$. We write $y = 1/x$.

Case I. Let $x > 0$ so that y is positive. If G be any positive number taken arbitrarily, then

$$1/x > G, \text{ if } 0 < x < 1/G$$

$\Rightarrow$ $\lim (1/x) = \infty$ when $x \to (0+0)$.

Case II. Let $x < 0$ so that $y = 1/x$ is negative.

If G be any positive number taken arbitrarily, then

$1/x < -G$ if $-1/G < x < 0$

$\Rightarrow$ $\lim (1/x) = -\infty$ when $x \to (0-0)$.

Case III. Clearly when $x \to 0$, $\lim (1/x)$ does not exist.

3. *Consider* $f(x) = (2x + 1)/(x - 3)$ *and show that*

$$\lim_{x \to \infty} f(x) = 2, \quad \lim_{x \to -\infty} f(x) = 2.$$

Let $\varepsilon > 0$ be given. We have

$$\left| \frac{2x+1}{x-3} - 2 \right| = \frac{7}{|x-3|} < \varepsilon \text{ for } x > \frac{7}{\varepsilon} + 3$$

so that $\displaystyle \lim_{x \to \infty} f(x) = 2$.

Again, we have

$$\left| \frac{2x+1}{x-3} - 2 \right| = \frac{7}{|x-3|} < \varepsilon \text{ for } x < \frac{-7}{\varepsilon} + 3$$

so that

$$\lim_{x \to -\infty} f(x) = 2.$$

Combining the two results, we see that

$$\lim_{|x| \to \infty} f(x) = 2.$$

EXERCISES

Show that :

(i) $\displaystyle \lim_{x \to (2-0)} 1/(x-2) = -\infty,$ $\qquad \displaystyle \lim_{x \to (2+0)} 1/(x-2) = +\infty.$

(ii) $\displaystyle \lim_{x \to (3-0)} (5x+2)/(x-3) = -\infty,$ $\qquad \displaystyle \lim_{x \to (3+0)} (5x+2)/(x-3) = +\infty.$

(iii) $\displaystyle \lim_{x \to (0-0)} (5x^2+3)/x^3 = -\infty,$ $\qquad \displaystyle \lim_{x \to (0+0)} (5x^2+3)/x^3 = +\infty.$

(iv) $\displaystyle \lim_{x \to +\infty} x^2 = +\infty,$ $\qquad \displaystyle \lim_{x \to -\infty} x^2 = +\infty.$

(v) $\displaystyle \lim_{x \to c+\infty} x^3 = +\infty,$ $\qquad \displaystyle \lim_{x \to \infty} (x^3 - x^2) = +\infty.$

3.4 Extension of Operations on Limits

Corresponding to the theorems on limits as given in § 3.3. page 66, we also have results pertaining to infinite limits and variables tending to infinity. We give these results without proof in the following tables. These results are valid when x tends to

$$c, \qquad c-0. \quad c+0, \quad +\infty \quad \text{or} \quad -\infty$$

	If $\lim f$ is	If $\lim g$ is	Then $\lim f + g$ is
1	l	$+\infty$	$+\infty$
2	l	$-\infty$	$-\infty$
3	$+\infty$	$+\infty$	$+\infty$
4	$-\infty$	$-\infty$	$-\infty$
5	$+\infty$	$-\infty$	No general conclusion

	If $\lim |f|$ is	If $\lim |g|$ is	Then $\lim |fg|$ is
1	$l \neq 0$	$+\infty$	$+\infty$
2	0	$+\infty$	No general conclusion
3	$+\infty$	$+\infty$	$+\infty$

	If $\lim |g|$ is	Then $\lim 1/|g|$ is
1	0	$+\infty$
2	$+\infty$	0

	If $\lim |f|$ is	If $\lim |g|$ is	Then $\lim |f| 1 |g|$ is
1	$l \neq 0$	$+\infty$	$+\infty$
2	0	0	No general conclusion
3	l	$+\infty$	0
4	$+\infty$	l	$+\infty$
5	$+\infty$	$+\infty$	No general conclusion

Indeterminate Forms. From the first table, it is seen that if $\lim f(x) = +\infty$ and $\lim g(x) = -\infty$, we cannot deduce any *specific* conclusion about $\lim [f(x) + g(x)]$. The fact is that in such a case the behaviour of $f(x) + g(x)$ just does not depend upon the individual behaviours of $f(x)$ and $g(x)$ but on their behaviours relatively to each other.

We know consider a few examples to illustrate the different possibilities.

EXAMPLES

1. $\lim\limits_{x \to \infty} \sqrt{(x+1)} = \infty, \quad \lim\limits_{x \to \infty} \sqrt{(x-1)} = \infty.$

We examine $\qquad \lim\limits_{x \to \infty} \left[\sqrt{(x+1)} - \sqrt{(x-1)} \right].$

We have

$$\sqrt{(x+1)} - \sqrt{(x-1)} = \frac{(x+1) - (x-1)}{\sqrt{(x+1)} + \sqrt{(x-1)}}$$

$$= \frac{2}{\sqrt{(x+1)} + \sqrt{(x-1)}} \to 0 \text{ as } x \to \infty.$$

Thus while

$$\lim\limits_{x \to \infty} \sqrt{(x+1)} = \infty, \quad \lim\limits_{x \to \infty} \sqrt{(x-1)} = \infty.$$

we have $\lim\limits_{x \to \infty} \left[\sqrt{(x+1)} - \sqrt{(x-1)} \right] = 0.$

2. $\lim\limits_{x \to \infty} \sqrt{(x^2+1)} = \infty, \quad \lim\limits_{x \to \infty} \sqrt{(x+1)} = \infty.$

Now $\sqrt{(x^2+1)} - \sqrt{(x+1)} = \dfrac{(x^2+1)-(x+1)}{\sqrt{(x^2+1)} + \sqrt{(x+1)}}$

$$= \dfrac{x(x-1)}{\sqrt{(x+1)} + \sqrt{(x+1)}}$$

$$= (x-1)\dfrac{1}{\sqrt{\left(1+\dfrac{1}{x^2}\right)} + \sqrt{\left(\dfrac{1}{x}+\dfrac{1}{x^2}\right)}}$$

so that $\displaystyle \lim_{x \to \infty} \left[\sqrt{(x^2+1)} - \sqrt{(x+1)} \right] = \infty.$

Thus while

$$\lim_{x \to \infty} \sqrt{(x^2+1)} = \infty, \ \lim_{x \to \infty} \sqrt{(x+1)} = \infty,$$

we have $\displaystyle \lim_{x \to \infty} \left[\sqrt{(x^2+1)} - \sqrt{(x+1)} \right] = \infty.$

3. $\displaystyle \lim_{x \to 0} \dfrac{1}{x^2} = \infty, \ \lim_{x \to 0} \dfrac{1}{x^4} = \infty,$

$$\lim_{x \to 0} \left(\dfrac{1}{x^2} - \dfrac{1}{x^4} \right) = \lim_{x \to 0} \dfrac{x^2-1}{x^4} = -\infty.$$

4. $\displaystyle \lim_{x \to 0} \left(2+\dfrac{1}{x^2} \right) = \infty, \ \lim_{x \to 0} \left(1+x+\dfrac{1}{x^2} \right) = \infty,$

$$\lim_{x \to 0} \left[\left(2+\dfrac{1}{x^2} \right) - \left(1+x+\dfrac{1}{x^2} \right) \right] = \lim_{x \to 0} (1-x) = 1.$$

In case, we have

$$\lim_{x \to a} f(x) = \infty, \ \lim_{x \to a} g(x) = \infty,$$

we say that $f(x) - g(x)$ assumes the Indeterminate form $\infty - \infty$ when $x \to a$.

5. $\displaystyle \lim_{x \to 0} \left[\cot x \right], \ \lim_{x \to \infty} \left[\cot^{-1} x \right]$

To check $\displaystyle \lim_{x \to 0} \left[\cot x \right],$ put $\cot x = t$, now as $x \to 0$; $\cot x$ exhibits two values for $x \to (0+0)$ and

$x \to (0-0)$, *i.e.*, $\cot x \to +\infty$ and $\cot x \to -\infty$ respectively.

$\therefore$ we should apply right hand and left hand limit;

i.e., $\displaystyle \lim_{x \to (0+0)} \left[\cot x \right], = \lim_{x \to +\infty} \left[t \right] = \infty$

$$\left[\because \cot x = t \Rightarrow t \to +\infty \text{ as } x \to (0+0) \right]$$

and $\displaystyle \lim_{x \to (0-0)} \left[\cot x \right] = \lim_{x \to \infty} \left[t \right] = -\infty$

$$\left[\because \cot x = t \Rightarrow t \to -\infty \text{ as } x \to (0-0) \right]$$

$\therefore$ limit does not exist.

To find $\qquad \lim_{x \to \infty} \left[\cot^{-1} x \right] \Rightarrow \lim_{t \to [0+0]} \left[t \right]$

$$\left[\because \cot^{-1} x = t \Rightarrow t \to (0+0) \text{ as } x \to +\infty \right]$$

$$\Rightarrow \left[0 + h \right] = 0.$$

Other Indeterminate Forms $0 \cdot \infty, \dfrac{0}{0}, \dfrac{\infty}{\infty}.$

In the context of the tables 2 and 4, we make the following statements:

1. If $f(x) = 0$ and $\lim g(x) = \infty$ then $f(x) g(x)$ is said to assume the Indeterminate form $0.\infty$.

2. If $\lim f(x) = 0$ and $\lim g(x) = 0$, then $f(x) \div g(x)$ is said to assume the indeterminate form $\dfrac{0}{0}$.

3. If $\lim f(x) = \infty$, $\lim g(x) = \infty$, then $f(x) \div g(x)$ is said to assume the indeterminate form $\dfrac{\infty}{\infty}$.

In the following exercises, we shall give several cases of the determination of the limits of expressions which assume one or the other types of the Indeterminate forms.

The general procedure for determining the limits of expressions assuming indeterminate forms will be considered in Chapter IX.

EXAMPLES

1. *Find* $\lim_{x \to +\infty} \left(7x^3 + 8x^2 + 5x - 7 \right).$

We have $\qquad 7x^3 + 8x^2 + 5x - 7 = x^3 \left(7 + \dfrac{8}{x} + \dfrac{5}{x^2} - \dfrac{7}{x^3} \right).$

Now $\qquad \lim_{x \to +\infty} \dfrac{8}{x} = \lim_{x \to +\infty} \dfrac{5}{x^2} = \lim_{x \to +\infty} \dfrac{-7}{x^3} = 0$

so that $\qquad \lim_{x \to \infty} \left(7 + \dfrac{8}{x} + \dfrac{5}{x^2} - \dfrac{7}{x^3} \right) = 7.$

Also $\qquad \lim_{x \to \infty} x^3 = \infty.$

Thus $\qquad \lim_{x \to \infty} \left(7x^3 + 8x^2 + 5x - 7 \right) = \infty.$

2. *Obtain* $\lim_{x \to -\infty} \left(7x^3 + 8x^2 + 5x - 7 \right).$

We have $7x^3 + 8x^2 + 5x - 7 = x^3 \left(7 + \dfrac{8}{x} + \dfrac{5}{x^2} - \dfrac{7}{x^3} \right).$

Now $\qquad \lim_{x \to -\infty} \dfrac{8}{x} = \lim_{x \to -\infty} \dfrac{5}{x^2} = \lim_{x \to -\infty} \dfrac{7}{x^3} = 0.$

Also $\qquad \lim_{x \to -\infty} x^3 = -\infty.$

It follows that $\lim_{x \to -\infty} \left(7x^3 + 8x^2 + 5x - 7 \right) = -\infty,$

3. *Evaluate* : $\lim\limits_{x \to \infty} \left(\sqrt{x^2 + x + 1} - \sqrt{x^2 + 1} \right)$

$$\lim_{x \to \infty} \left(\sqrt{x^2 + x + 1} - \sqrt{x^2 + 1} \right) = \lim_{x \to \infty} \frac{\left(\sqrt{x^2 + x + 1} - \sqrt{x^2 + 1} \right)}{1} \times \frac{\left(\sqrt{x^2 + x + 1} + \sqrt{x^2 + 1} \right)}{\left(\sqrt{x^2 + x + 1} + \sqrt{x^2 + 1} \right)}$$

$$= \lim_{x \to \infty} \frac{x}{\sqrt{x^2 + x + 1} + \sqrt{x^2 + 1}}$$

$$= \lim_{x \to \infty} \frac{1}{\sqrt{1 + \dfrac{1}{x} + \dfrac{1}{x^2}} + \sqrt{1 + \dfrac{1}{x^2}}}$$

$$= \frac{1}{1 + 1} \qquad \left[\text{as } \frac{1}{x} \to 0 \text{ as } x \to \infty \right]$$

$$= \frac{1}{2}.$$

4. *Evaluate* $= \lim\limits_{n \to \infty} \dfrac{(n + 2)! + (n + 1)!}{(n + 2)! - (n + 1)!}$

$$G iven\ limit = \lim_{x \to \infty} \frac{(n + 1)\,![n + 2 + 1]}{(n + 1)\,![n + 2 - 1]} = \lim_{x \to 0} \frac{n + 3}{n + 1}$$

$$\lim_{n \to \infty} \frac{1 + 3/n}{1 + 1/n} = \frac{1 + 0}{1 + 0} = 1 \qquad \left[\text{as } \frac{1}{n} \to 0 \text{ as } n \to \infty \right]$$

5. *Solve* : $\lim\limits_{x \to \infty} \dfrac{- 3n + (- 1)^n}{4n - (- 1)^n}$

while evaluating the given limit, we come across two cases : when n is even or n is odd.

Case 1. *n* is even, say $n = 2k$

$$\Rightarrow \qquad \lim_{k \to \infty} \frac{- 6k + (- 1)^{2k}}{8k - (- 1)^{2k}}$$

$$\Rightarrow \qquad \lim_{k \to \infty} \frac{- 6k + 1}{8k - 1} = \lim_{k \to \infty} \frac{- 6 + \dfrac{1}{k}}{8 - \dfrac{1}{k}}$$

$$= \frac{- 6}{8} = \frac{- 3}{4}.$$

Case 11 : *n* is odd, say $n = 2k + 1$

$$\Rightarrow \qquad \lim_{k \to \infty} \frac{- 3\,(2k + 1) - (- 1)^{2k + 1}}{4\,(2k + 1) - (- 1)^{2k + 1}} = \lim_{k \to \infty} \frac{- 6k - 2}{8k + 5}$$

$$= \lim_{k \to \infty} \frac{-6 - \dfrac{2}{k}}{8 + \dfrac{5}{k}}$$

$$= \frac{-3}{4}$$

$$\therefore \quad \lim_{n \to \infty} \frac{-3n + (-1)^n}{4n - (-1)^n} = \frac{-3}{4}.$$

6. *Find* $\displaystyle \lim_{n \to \infty} \left[\sum_{r=1}^{n} \frac{1}{2^r} \right]$, *where* $[.]$ *denotes the greatest integer.*

$$\sum_{r=1}^{n} \frac{1}{2^r} = \frac{\dfrac{1}{2}(1 - \dfrac{1}{2^n})}{1 - 1/2} = 1 - \left(\frac{1}{2}\right)^n$$

This tends to one as $n \to \infty$ but always remains less than 1.

Thus

$$\left[\sum_{2=1}^{n} \frac{1}{2^r} \right] \to 0 \text{ as } n \to \infty$$

$$\therefore \qquad \lim_{n \to \infty} \left[\sum_{r=1}^{n} \frac{1}{2^r} \right] = 0.$$

7. *Find* $L = \displaystyle \lim_{x \to \infty} \frac{ax^p + bx^{p-2} + c}{dx^q + ex^{q-2} + k}$, *where* a, b, c, d, e *and* k *are constants, and* $p > 0, q > 0$.

Here, $\displaystyle \lim_{x \to \infty} \frac{ax^p + bx^{p-2} + c}{dx^q + ex^{q-2} + k}$

$$= \lim_{x \to \infty} \frac{x^p \left\{ a + b/x^2 + c/x^p \right\}}{x^q \left\{ d + e/x^2 + k/x^q \right\}}$$

Now, we have to consider here cases:

Case I : $\qquad\qquad p > q$

$$= \lim_{x \to \infty} \frac{x^{p-q} \left\{ a + \dfrac{b}{x^2} + \dfrac{c}{x^p} \right\}}{\left\{ d + \dfrac{e}{x^2} + \dfrac{k}{x^q} \right\}}$$

$$\{ \text{Since } p - q > 0 \Rightarrow x^{p-q} \to \infty \text{ as } x \to \infty \}$$

$$= \infty$$

Case II : $\qquad\qquad p = q$

$$= \lim_{x \to \infty} \frac{\left\{ a + \dfrac{b}{x^2} + \dfrac{c}{x^p} \right\}}{\left\{ d + \dfrac{e}{x^2} + \dfrac{k}{x^q} \right\}}$$

$$= \frac{a}{d}$$

Case III : $\quad p < q$

$$= \lim_{x \to \infty} \frac{\left\{ a + \dfrac{b}{x^2} + \dfrac{c}{x^p} \right\}}{x^{q-p} \left\{ a + \dfrac{e}{x^2} + \dfrac{k}{x^q} \right\}} \quad \{ \sin p - q < 0 \Rightarrow x^{p-q} \to 0 \text{ as } x \to \infty \}$$

$$= 0.$$

Hence, $\qquad \displaystyle \lim_{x \to \infty} \frac{ax^p + bx^{p-2} + c}{dx^q + ex^{q-2} + k} = \begin{cases} \infty, \text{ when } p > q \\ a/d, \text{when } p = q \\ 0, \text{ when } p < q. \end{cases}$

EXERCISES

1. Obtain

   (i) $\displaystyle \lim \frac{7x+1}{x+3}$ when x tends to $-3, +\infty, -\infty$,

   (ii) $\displaystyle \lim \frac{x^2-4}{x^2+7x+12}$ when x tends to $-4, -3, +\infty, -\infty$,

   (iii) $\displaystyle \lim \frac{x^2+1}{x^2-4}$ when x tends to $2, -2, \infty, -\infty$,

   (iv) $\displaystyle \lim \frac{x^2-4}{(x-2)^2(x+7)}$ when x tends to $2, -7, \infty, -\infty$,

2. Given that $f(x) = x^2 / [(x-1)(x-2)]$, show that

   (i) $\displaystyle \lim_{x \to (1-0)} f(x) = +\infty,$ $\qquad \displaystyle \lim_{x \to (1+0)} f(x) = -\infty,$

   (ii) $\displaystyle \lim_{x \to (2-0)} f(x) = -\infty,$ $\qquad \displaystyle \lim_{x \to (2-0)} f(x) = \infty,$

   (iii) $\displaystyle \lim_{x \to +\infty} f(x) = 1 = \lim_{x \to -\infty} f(x).$

3. Given that:

   (i) $f(x) = 1/x^2, g(x) = 1/x^4$ so that $\displaystyle \lim_{x \to 0} f(x) = \infty = \lim_{x \to 0} g(x)$.

      show that $\displaystyle \lim_{x \to 0} [f(x) - g(x)] = -\infty.$

   (ii) If $f(x) = \dfrac{1}{x^2} + 3, g(x) = \dfrac{1}{x^2} + 1$, show that

      $$\lim_{x \to 0} f(x) = \infty = \lim_{x \to 0} g(x)$$

      $$\lim_{x \to 0} [f(x) - g(x)] = 2$$

4. Give examples of pairs of functions f_1, g_1 and f_2, g_2 such that:

   (i) $\displaystyle \lim_{x \to \infty} f_1 = 0 = \lim_{x \to \infty} f_2; \quad \lim_{x \to \infty} g_1 = \infty = \lim_{x \to \infty} g_2$ but

      $$\lim_{x \to \infty} f_1 g_1 \neq \lim_{x \to \infty} f_2 g_2. \qquad\qquad (0, \infty \text{ form})$$

(ii) $\lim_{x \to c} f_1 = 0 = \lim_{x \to c} f_2; \quad \lim_{x \to c} g_1 = 0 = \lim_{x \to c} g_2$ but

$$\lim_{x \to c}(f_1 \div g_1) \neq \lim_{x \to c}(f_2 \div g_2).$$ (0/0 form)

(iii) $\lim_{x \to c} f_1 = \infty = \lim_{x \to c} f_2; \quad \lim_{x \to c} g_1 = \infty = \lim_{x \to c} g_2$ but

$$\lim_{x \to c}(f_1 \div g_1) \neq \lim_{x \to c}(f_2 \div g_2).$$ (∞/∞ form)

5. Find:

(i) $\lim_{x \to 1} \dfrac{x^n - 1}{x - 1}, n \in N,$ (ii) $\lim_{x \to 0} \dfrac{\sqrt{(1+x)} - 1}{x},$ (iii) $\lim_{x \to 3} \dfrac{\sqrt{(3x)} - 3}{\sqrt{(2x - 4)} - \sqrt{2}}.$

(In each case we have the indeterminate form 0/0)

6. Show that: $\lim_{x \to +\infty} \dfrac{\sqrt{(x^2 + 1)}}{x + 1} = 1; \quad \lim_{x \to -\infty} \dfrac{\sqrt{(x^2 + 1)}}{x + 1} = -1.$

7. Find: $\lim_{x \to 2}\left(\dfrac{1}{2 - x} - \dfrac{1}{4 - x^2} \right).$

3.5. Some useful limits.

We shall now consider the following two results:

(i) $\lim_{x \to 0} \dfrac{\sin x}{x}$ (ii) $\lim_{x \to \infty}(1 + x)^{1/x}$

The consideration of the limit on the right will lead to the introduction of an irrational number to be denoted by e. These two limits (i) and (ii) play a fundamental role in the derivation of the trigonometric and logarithmic functions as will be seen in the following chapter.

3.5.1. $\lim_{x \to 0} \dfrac{\sin x}{x} = 1$

We notice that $\lim_{x \to 0} \sin x = 0$ and $\lim_{x \to 0} x = 0$ so that $\sin x/x$ assumes the Indeterminate form 0/0 when x tends to 0.

We have shown on page 43, that when $x \in [0, \pi/2]$

$$\sin x < x < \tan x$$

$\Rightarrow \quad 1 < \dfrac{x}{\sin x} < \dfrac{1}{\cos x} \Rightarrow 1 > \dfrac{\sin x}{x} > \cos x.$

While we have derived this result for $x > 0$, the same also holds when $x < 0$ in that
$$\sin(-x)/(-x) = \sin x/x \text{ and } \cos(-x) = \cos x.$$

Now since $\lim_{x \to 0} \cos x = 1$ it following that

$$\lim_{x \to 0} \dfrac{\sin x}{x} = 1.$$

EXAMPLES

1. *Evaluate:* $\lim_{x \to 0} \dfrac{1 - \cos x}{x^2}.$

$$\lim_{x \to 0} \dfrac{1 - \cos x}{x^2} = \lim_{x \to 0} \dfrac{2 \sin^2 x/2}{x^2}$$

$$\lim_{x \to 0} \frac{1}{2} \left(\frac{\sin x/2}{x/2} \right)^2 = \frac{1}{2} (1)^2 = \frac{1}{2}.$$

2. *Evaluate* : $\lim\limits_{x \to y} \dfrac{\sin^2 x - \sin^2 y}{x^2 - y^2}$

$$\lim_{x \to y} \frac{\sin^2 x - \sin^2 y}{x^2 - y^2} = \lim_{x \to y} \frac{\sin(x + y) \sin(x - y)}{(x + y)(x - y)}$$

$$= \lim_{x \to y} \frac{\sin(x + y)}{(x + y)} \cdot \lim_{x \to y} \frac{\sin(x - y)}{(x - y)}$$

$$= \frac{\sin 2y}{2y} \times 1 \left[\because x \to y \Rightarrow (x - y) \to 0, \text{ but } x + y \to 2y \right]$$

$$= \frac{\sin 2y}{2y}.$$

3. *Evaluate*: $\lim\limits_{x \to 0} \dfrac{\sin\left(\pi \cos^2 x \right)}{x^2}$

$$\lim_{x \to 0} \frac{\sin\left(\pi \cos^2 x \right)}{x^2} = \lim_{x \to 0} \frac{\sin\left\{ \pi \left(1 - \sin^2 x \right) \right\}}{x^2}$$

$$= \lim_{x \to 0} \frac{\sin\left(\pi - \pi \sin^2 x \right)}{x^2}$$

$$= \lim_{x \to 0} \frac{\sin\left(\pi \sin^2 x \right)}{x^2}$$

$$= \lim_{x \to 0} \left\{ \frac{\sin\left(\pi \sin^2 x \right)}{\pi \sin^2 x} \times \frac{\pi}{1} \times \frac{\sin^2 x}{x^2} \right\}$$

$$= \lim_{x \to 0} \frac{\sin\left(\pi \sin^2 x \right)}{\pi \sin^2 x} \times \pi \times \lim_{x \to 0} \frac{\sin^2 x}{x^2}$$

$$= 1 \times \pi \times 1 = \pi$$

4. $\lim\limits_{x \to \pi/4} \dfrac{\sqrt{1 - \sqrt{\sin 2x}}}{\pi - 4x}$

$$\lim_{x \to \pi/4} \frac{\sqrt{1 - \sqrt{\sin 2x}}}{\pi - 4x} = \lim_{x \to \pi/4} \frac{\sqrt{1 - \sqrt{\sin 2x}}}{\pi - 4x} \cdot \frac{\sqrt{1 + \sqrt{\sin 2x}}}{\sqrt{1 + \sqrt{\sin 2x}}}$$

$$= \lim_{x \to \pi/4} \frac{\sqrt{1 - \sin 2x}}{\pi - 4x} \cdot \lim_{x \to \pi/4} \frac{1}{\sqrt{1 + \sqrt{\sin 2x}}}$$

$$= \lim_{x \to \pi/4} \frac{\sin\left(\dfrac{\pi}{4} - x \right)}{4 \left(\dfrac{\pi}{4} - x \right)} \cdot 1$$

which gives R H L at $x = \dfrac{\pi}{4} = -\dfrac{1}{4}$ and L H L at $x = \dfrac{\pi}{4} = \dfrac{1}{4}$

∴ Limit does not exists.

EXERCISES

1. Show that

    (i) $\lim\limits_{x \to 0} \dfrac{\sin a\,x}{\sin b\,x} = \dfrac{a}{b},\ (b \neq 0)$ (ii) $\lim\limits_{x \to 0} \dfrac{\tan x}{x} = 1$ (iii) $\lim\limits_{x \to 0} \dfrac{1 - \cos x}{x^2} = \dfrac{1}{2}$

    (iv) $\lim\limits_{x \to \pi/2} \dfrac{\cos x}{\pi/2 - x} = 1$ (v) $\lim\limits_{x \to (0+0)} \dfrac{x}{\sqrt{(1 - \cos x)}} = \dfrac{2}{\sqrt{2}}$

    (vi) $\lim\limits_{x \to (0-0)} \dfrac{x}{\sqrt{(1 - \cos x)}} = -\dfrac{2}{\sqrt{2}}$ (vii) $\lim\limits_{x \to 0} \dfrac{\sin x}{\tan x} = 1$ (viii) $\lim\limits_{x \to 0} \dfrac{\tan x - \sin x}{x^3} = \dfrac{1}{2}$

    (ix) $\lim\limits_{x \to 0} \dfrac{3 \sin^{-1} x}{4x} = \dfrac{3}{4}$ (x) $\lim\limits_{x \to 0} \dfrac{\sin(1 + x) - \sin(1 - x)}{x} = 2 \cos 1$

    (In each of the above cases we deal with indeterminate form 0/0)

2. Show that

    (i) $\lim\limits_{x \to 0} x \cot x = 1$ (ii) $\lim\limits_{x \to 1} (1 - x) \tan \dfrac{\pi x}{2} = \dfrac{2}{\pi}$ (iii) $\lim\limits_{x \to 0} x \sqrt{(\cot x)} = 0$

3. Show that $\lim\limits_{x \to 0} \dfrac{\operatorname{cosec}^2 x}{\cot^2 x} = 1$

4. Show that $\lim\limits_{x \to 0} (\operatorname{cosec} x - \cot x) = 0$

5. Examine whether the following functions are continuous at $x = 0$.

    (i) $f(x) = \begin{cases} \sin x\,, & \text{when } x \neq 0 \\ 1\,, & \text{when } x = 0 \end{cases}$

    (ii) $f(x) \neq \begin{cases} \sin^2 x\,, & \text{when } x \neq 0 \\ 1\,, & \text{when } x = 0 \end{cases}$

    (iii) $f(x) = \begin{cases} \dfrac{\tan^2 x}{3x}\,, & \text{when } x \neq 0 \\[2mm] \dfrac{2}{3}\,, & \text{when } x = 0 \end{cases}$

ANSWERS

5. (i) discontinuous,   (ii) discontinuous,   (iii) discontinous.

3.5.2. The number e. *To consider*

$$\lim \left(1 + \frac{1}{n}\right)^n$$

when *n tends to infinity through positive integral* **values**.

Step 1. By the Binomial Theorem for a positive integral index, we have

$$\left(1 + \frac{1}{n}\right)^n = 1 + n \cdot \frac{1}{n} + \frac{n(n-1)}{1 \cdot 2} \cdot \frac{1}{n^2} + \frac{n(n-1)(n-2)}{1 \cdot 2 \cdot 3} \cdot \frac{1}{n^3}$$

$$+ \ldots + \frac{n(n-1)(n-2)\ldots\ldots(n-n+1)}{1 \cdot 2 \cdot 3 \ldots\ldots n} \cdot \frac{1}{n^n}$$

$$= 1 + 1 + \frac{1}{1.2}\left(1 - \frac{1}{n}\right) + \frac{1}{1.2.3}\left(1 - \frac{1}{n}\right)\left(1 - \frac{2}{n}\right)$$

$$+ \ldots + \frac{1}{1.2 \ldots n}\left(1 - \frac{1}{n}\right)\left(1 - \frac{2}{n}\right) \ldots \ldots \left(1 - \frac{n-1}{n}\right).$$

The expression on the right is a sum of $(n + 1)$ positive terms. Changing n to $n + 1$, we get

$$\left(1 + \frac{1}{n+1}\right)^{n+1} = 1 + 1 + \frac{1}{1.2}\left(1 - \frac{1}{n+1}\right) + \ldots \ldots$$

$$+ \frac{1}{1.2 \ldots (n+1)}\left(1 - \frac{1}{n+1}\right)\left(1 - \frac{2}{n+1}\right) \ldots \ldots \left(1 - \frac{n}{n+1}\right).$$

The right hand side consists of the sum of $(n + 2)$ positive terms. Also

$$1 - \frac{1}{n} < 1 - \frac{1}{n+1}, \ 1 - \frac{2}{n} < 1 - \frac{2}{n+1}, \ 1 - \frac{3}{n} < 1 - \frac{3}{n+1}, \ldots \ldots$$

so that each term in the expansion of

$$\left(1 + \frac{1}{n+1}\right)^{n+1}$$

is greater than the corresponding term in the expansion of

$$\left(1 + \frac{1}{n}\right)^{n}.$$

Thus we conclude that $\left(1 + \dfrac{1}{n+1}\right)^{n+1} > \left(1 + \dfrac{1}{n}\right)^{n}$

for all positive integral values of n, *i.e.*, $(1 + 1/n)^n$ *increases as n increases.*

Step II. We have

$$\left(1 + \frac{1}{n}\right)^{n} = 1 + 1 + \frac{1}{1.2}\left(1 - \frac{1}{n}\right) + \frac{1}{1.2.3}\left(1 - \frac{1}{n}\right)\left(1 - \frac{2}{n}\right)$$

$$+ \ldots + \frac{1}{1.2.3 \ldots n}\left(1 - \frac{1}{n}\right)\left(1 - \frac{2}{n}\right) \ldots \ldots \left(1 - \frac{n-1}{n}\right), \ \ldots(i)$$

$$< 1 + 1 + \frac{1}{2} + \frac{1}{1.2.3} + \ldots + \frac{1}{1.2.3 \ldots n}$$

$$< 1 + 1 + \frac{1}{2} + \frac{1}{2.2} + \ldots + \frac{1}{2.2.2 \ldots (n-1) \text{ factors}}$$

$$= 1 + 1 + \frac{1}{2} + \frac{1}{2^2} + \ldots + \frac{1}{2^{n-1}}$$

$$= 1 + \frac{1 - (1/2)^n}{1 - 1/2} = 3 - \frac{1}{2^{n-1}} < 3 \ \forall \ n. \quad \ldots(ii)$$

Thus we see that $2 < (1 + 1/n)^n$ *increases as n takes up successively positive integral values* 1, 2, 3, *etc., and remains less than 3 for all values of n.*

Hence $(1 + 1/n)^n$ *approaches a finite limit as* $n \to \infty$.

From (i) *and* (ii) *we see that* $(1 + 1/n)^n$ *for every value of n so that the limit is a number which lies between* 2 *and* 3.

It can be proved that $\lim (1 + 1/n)^n$ is an *irrational number.* This number is denoted by e. While we are not giving here any method for computing the value of the number e upto any given number of decimal places, we only state that its value upto 10 significant decimal places is

$$e = 2.7182818284$$

as can be shown by advanced methods.

Note 1. The reader may be able to appreciate the underlying idea of the proof better if he calculates the value of

$$a_n = (1 + 1/n)^n$$

for a new successive value of n.

Thus for example $a_1 = 2$, a_2, $= 2.25$, $a_3 = 2.37$, $a_4 = 2.441$,.........

Note 2. The proof for the existence of the limit has been based on an intuitively obvious fact that an *increasing function which remains less than a fixed number tends to a limit.* The formal proof of this fact is beyond the scope of this book.

Cor. 1. $\lim (1 + 1/x)^x = e$ *when x tends to infinity taking all the real numbers as its values.* If x be any positive real number, then there exists a positive integer n such that

$$n \leq x < n + 1$$

$$\Rightarrow \qquad \frac{1}{n} \geq \frac{1}{x} > \frac{1}{n + 1}$$

$$\Rightarrow \qquad 1 + \frac{1}{n} \geq 1 + \frac{1}{x} > 1 + \frac{1}{n + 1}$$

$$\Rightarrow \qquad \left(1 + \frac{1}{n}\right)^{n+1} \geq \left(1 + \frac{1}{x}\right)^x > \left(1 + \frac{1}{n + 1}\right)^n$$

(*In case the base is greater than* 1, *raising a greater number to a greater power, does not alter the direction of the inequality*).

We thus have

$$\left(1 + \frac{1}{n}\right)\left(1 + \frac{1}{n}\right)^n \geq \left(1 + \frac{1}{x}\right)^x > \left(1 + \frac{1}{n + 1}\right)^{n+1} \bigg/ \left(1 + \frac{1}{n + 1}\right).$$

Let $x \to \infty$. Then n and $(n + 1) \to \infty$ through positive integral values. We know that

$$\lim_{n \to \infty}\left(1 + \frac{1}{n}\right) = 1 = \lim_{n \to \infty}\left(1 + \frac{1}{n + 1}\right)$$

$$\lim_{n \to \infty}\left(1 + \frac{1}{n}\right)^n = e = \lim_{n \to \infty}\left(1 + \frac{1}{n + 1}\right)^{n+1}$$

Therefore $\qquad\qquad \lim_{x \to \infty}\left(1 + \frac{1}{x}\right)^x = e.$

Cor. 2. $\qquad\qquad\qquad \lim_{x \to -\infty}\left(1 + \frac{1}{x}\right)^x = e.$

Let $x = -y$ so that $x \to -\infty \Rightarrow y \to +\infty.$

We have

$$\left(1 + \frac{1}{x}\right)^x = \left(1 - \frac{1}{y}\right)^{-y} = \left(\frac{y}{y-1}\right)^y$$

$$= \left(1 + \frac{1}{y-1}\right)^y = \left(1 + \frac{1}{y-1}\right)^{y-1}\left(1 + \frac{1}{y-1}\right).$$

$$\therefore \quad \lim_{x \to -\infty}\left(1 + \frac{1}{x}\right)^x = \lim_{y \to \infty}\left[\left(1 + \frac{1}{y-1}\right)^{y-1}\left(1 + \frac{1}{y-1}\right)\right]$$

$$= \lim_{y \to \infty}\left(1 + \frac{1}{y-1}\right)^{y-1} \lim_{y \to \infty}\left(1 + \frac{1}{y-1}\right)$$

$$= e . 1 = e.$$

Cor. 3. $\lim (1 + z)^{1/z} = e$ when $z \to 0$.

Let $z = 1/x$ so that $x \to +\infty$ or $-\infty$ according as $z \to 0$ through positive or negative values.

Now

$$\lim_{z \to (0+0)} (1 + z)^{1/z} = \lim_{x \to \infty}\left(1 + \frac{1}{x}\right)^x = e,$$

$$\lim_{z \to (0-0)} (1 + z)^{1/z} = \lim_{x \to -\infty}\left(1 + \frac{1}{x}\right)^x = e.$$

Thus

$$\lim_{z \to (0)} (1 + z)^{1/z} = e.$$

3.5.3. We shall now consider three limits which will be used for the determination of the power series expansion of e^x, $(1 + x)^m$, $\log(1 + x)$.

I. Lim x^n. *When n tends to infinity through positive integral values; x being any given real number.*

We have to consider several cases :

(*i*) Let $x > 1$. We write $x = 1 + h$ so that h is positive.

By the Binomial theorem for positive integral index, we have

$$x^n = (1 + h)^n = 1 + nh + \frac{n(n-1)}{2!}h^2 + \ldots\ldots\ldots\frac{n(n-1)\ldots 1}{n!}h^n,$$

where each term is positive. Thus $x^n > 1 + nh$.

Let G be a given positive number. Then we have $1 + nh > G$, if $n > (G-1)/h$.

If Δ be a positive integer greater than $(G-1) h$, then

$$x_n > G \, \forall \, n > \Delta$$

Thus, $x > 1 \Rightarrow \lim x^n = \infty$.

(*ii*) Let $x = 1$. Here $x^n = 1$ for all values of n and therefore, $x^n \to 1$ in this case.

(*iii*) Let $0 < 1 < x$ so that, x^n is positive. Let $x = 1/y$ so that $y > 1$ and $\lim_{n \to \infty} y^n = \infty$.

Let $\varepsilon > 0$ be given. As $y^n \to \infty$, there exists m such that $y^n > \dfrac{1}{\varepsilon} \, \forall \, n > m \Rightarrow x^n < \varepsilon \, \forall \, n > m$. As x^n is necessarily positive, it follows that

$$\lim x^n = 0 \qquad \text{if } 0 < x < 1.$$

Thus $0 < x < 1 \Rightarrow \lim x^n = 0$

(*iv*) Let $x = 0$. Here $x^n = 0$ for all n so that $x^n \to 0$.

(*v*) Let $-1 < x < 0$. We write $x = -a$ so that a is a positive number less than 1. We have $|x^n| = a^n$. But $a^n \to 0$. Hence $x^n \to 0$

Thus, $-1 < x < 0 \Rightarrow \lim x^n = 0$.

(*vi*) Let $x = -1$. Since x^n is alternatively -1 and 1, therefore, it neither tends to any finite limit nor to $\pm \infty$.

(*vii*) Let $x < -1$. Here, x^n which is alternatively negative and positive, takes values numerically greater than any assigned number, x^n does not tend to a limit.

Thus, we see that $\lim x^n$, *when* $n \to \infty$, *exists finitely if, and only if,* $-1 < x \le 1$. *Also, it is* 0 *if* $|x| < 1$ *and is* 1 *for* $x = 1$.

II. Lim $x^n / n\,!$, *when n tends to infinity through positive integral values and x is any given real number.*

Let x be a positive real number and let $m, m + 1$ be two consecutive integers between which x lies so that we have

$$m \le x < m + 1.$$

We write

$$\frac{x^n}{n!} = \frac{x}{1} \cdot \frac{x}{2} \cdot \frac{x}{3} \cdots \cdots \frac{x}{m} \cdot \frac{x}{m+1} \cdot \frac{x}{m+2} \cdots \cdots \frac{x}{n}.$$

Let $p = \dfrac{x}{1} \cdot \dfrac{x}{2} \cdots \cdots \cdots \dfrac{x}{m}$ so that p is a positive number independent of n.

Also, each of $\dfrac{x}{m+2}, \dfrac{x}{m+3}, \cdots \cdots \cdots, \dfrac{x}{n}$ is $< \dfrac{x}{m+1}$.

$$\therefore \quad 0 \frac{x^n}{n!} < p \left(\frac{x}{m+1} \right)^{n-m} = p \left(\frac{x}{m+1} \right)^{-m} \left(\frac{x}{m+1} \right)^n$$

$$= k \left(\frac{x}{m+1} \right)^n , \text{ say}$$

where k is a positive number independent of n.

Now, $0 < \dfrac{x}{m+1} < 1 \Rightarrow \lim\limits_{n \to \infty} \left(\dfrac{x}{m+1} \right)^n = 0.$

Thus, if x be a positive number

$$\lim \frac{x^n}{n!} = 0, \text{ when } n \to \infty$$

If x be a negative number, say $-\alpha$, so that α is positive, we have

$$\left| \frac{x^n}{n!} \right| = \left| \frac{(-1)^n \alpha^n}{n!} \right| = \frac{\alpha^n}{n!}.$$

Hence $\lim\limits_{n \to \infty} \dfrac{x^n}{n!} = 0 \ \forall \ x \in R.$

III. $\operatorname*{Lim}_{n\to\infty} u_n = 0$ *where* $u_n = \dfrac{m(m-1)\ldots\ldots(m-n+1)}{(n-1)!} x^n$, if $|x| < 1$.

Changing n to $n+1$, we get

$$u_{n+1} = \frac{m(m-1)\ldots\ldots(m-n)}{n!} x^{n+1}.$$

$$\frac{u_{n+1}}{u_n} = \frac{m-n}{n} x = -\left(1 - \frac{m}{n}\right) x.$$

Thus, $\displaystyle\lim_{n\to\infty}\left|\frac{u_{n+1}}{u_n}\right| = |x|.$

Let k be a positive number, such that $|x| < k < 1$. These exists a positive integer p such that

$$\left|\frac{u_{n+1}}{u_n}\right| < k \ \forall \ n \geq p. \text{ Thus}$$

$$|u_{p+1}| < k |u_p|,$$

$$|u_{p+2}| < k |u_{p+1}|,$$

$$\ldots\ldots\ldots\ldots\ldots\ldots$$

$$\ldots\ldots\ldots\ldots\ldots\ldots$$

$$|u_n| < k |u_{n-1}|.$$

Multiplying, we get

$$|u_n| < k^{n-p} |u_p| = k^n |u_p| / k^p.$$

Now, $|u_p| / k^p$ is a constant independent of n and $k^n \to 0$ as $n \to \infty$

Thus, $\displaystyle\lim_{n\to\infty} u_n = 0$ if $|x| < 1$.

3.5.4. Some miscellaneous Limits.

(*i*) Logarithmic Limits :

These limits are based on expansion of logarithmic series

$$\log(1+x) = x - \frac{x^2}{2} + \frac{x^3}{3} + \ldots\ldots \text{ to } \infty$$

where $-1 \leq x \leq 1$ and this expansion is true only if the base is e.

To evaluate the logarithmic limit we use $\displaystyle\lim_{x\to 0} \frac{\log(1+x)}{x} = 1$.

(*ii*) Exponential limits

These limits are based on the expansion

$$e^x = 1 + x + \frac{x^2}{2!} + \ldots\ldots \infty$$

To evaluate exponential limit we use the following result,

$$\lim_{x\to 0} \frac{e^x - 1}{x} = 1 \text{ and } \lim_{x\to 0} \frac{a^x - 1}{x} = \log_e a$$

To evaluate the exponential form 1 we should use the following results:

If $\quad \lim_{x \to a} f(x) = \lim_{x \to a} g(x) = 0$, then $\lim_{x \to a} \left[1 + f(x) \right]^{1/g(x)} = e^{\lim_{x \to a} \frac{f(x)}{g(x)}}$

or, when $\quad\quad \lim_{x \to a} f(x) = 1$ and $\lim_{x \to a} g(x) = \infty$,

then $\quad\quad \lim_{x \to a} \left[f(x) \right]^{g(x)} = \lim_{x \to a} \left[1 + f(x) - 1 \right]^{g(x)}$

$$= e^{\lim_{x \to a} (f(x) - 1) g(x)}$$

we should also remember, while evaluating exponential limit, that

$$\lim_{x \to 0} (1 + x)^{1/x} = e \; ; \quad \lim_{x \to \infty} \left(1 + \frac{1}{x} \right)^x = e \; ;$$

$$\lim_{x \to 0} (1 + \lambda x)^{\lambda x} = e^\lambda \; ; \quad \lim_{x \to \infty} \left(1 + \frac{\lambda}{x} \right)^x = e^\gamma$$

If we have to calculate the limit O° of form, then $\lim_{x \to a} f(x) \neq 1$ but $f(x)$ is positive in the neighbourhood of $x = a$. In this case we write

$$\{ f(x) \}^{g(x)} = e^{\log_e \{ f(x) \}^{g(x)}}$$

$\Rightarrow \quad\quad \lim_{x \to a} \left[f(x) \right]^{g(x)} = e^{\lim_{x \to a} g(x) \log_e f(x)}$

Here, it is to be noted that if $f(x)$ is not throughout positive in the neighbourhood of $x = a$, then $\lim_{x \to a} \left(f(x) \right)^{g(x)}$ will not exist, because in this case function will not be defined in the neighbourhood of $x = a$.

EXAMPLES

1. *Evaluate:* $\lim_{h \to 0} \dfrac{\log_e (1 + 2h) - 2 \log_e (1 + h)}{h^2}$

Given that

$$= \lim_{h \to 0} \frac{\left[(2h) - \dfrac{(2h)^2}{2} + \dfrac{(2h)^3}{3} \ \ldots\ldots \ \infty \right] - 2 \left(h - \dfrac{h^2}{2} + \dfrac{h^3}{3} \ \ldots\ldots \right)}{h^2}$$

$$= \lim_{h \to 0} \frac{h^2 (-1 + 2h - \ldots\ldots\ldots)}{h^2}$$

$$= -1.$$

2. *Evaluate:* $\lim_{x \to 0} \dfrac{(ab)^x - a^x - b^x + 1}{x^2}.$

Given limit

$$= \lim_{x \to 0} \frac{a^x (b^x - 1) - (b^x - 1)}{x^2}$$

$$= \lim_{x \to 0} \frac{(a^x - 1)}{x} \times \lim_{x \to 0} \frac{(b^x - 1)}{x}$$

$$= \log a \times \log b.$$

3. *Evaluate:* $\lim\limits_{x \to 0} \dfrac{e^{\tan x} - e^x}{\tan x - x}$

Given limit $\qquad = \lim\limits_{x \to 0} \dfrac{e^x \left\{ e^{\tan x - x} - 1 \right\}}{\tan x - x}$ $\qquad$ [As $x \to 0$, $\tan x - x \to 0$]

$\qquad\qquad\qquad = e^0 \times 1 = 1$

4. *Evaluate:* $\lim\limits_{x \to a} \left(2 - \dfrac{a}{x} \right)^{\tan \frac{\pi x}{2a}}$

Given limit $\qquad = \lim\limits_{x \to a} \left\{ 1 + \left(1 - \dfrac{a}{x} \right) \right\}^{\tan \frac{\pi x}{2a}}$

$\qquad\qquad\qquad = e^{\lim\limits_{x \to a} \left(1 - \frac{a}{x} \right) \tan \frac{\pi x}{2a}}$

$\qquad\qquad\qquad = e^{\lim\limits_{x \to a} \left(\frac{x - a}{x} \right) \tan \frac{\pi x}{2a}}$

Let $x - a = h$

$\Rightarrow \qquad e^{\lim\limits_{h \to 0} \left(\frac{h}{a + h} \right) \tan \frac{\pi(a + h)}{2a}}$

$\Rightarrow \qquad e^{\lim\limits_{h \to 0} \frac{h}{a + h} \tan \left(\frac{\pi}{2} + \frac{\pi h}{2a} \right)}$

$\Rightarrow \qquad e^{\lim\limits_{h \to 0} \frac{h}{a + h} \left\{ - \cot \left(\frac{\pi h}{2a} \right) \right\}}$

$\Rightarrow \qquad e^{\lim\limits_{h \to 0} \frac{-h}{(a + h) \tan(\pi h / 2a)} \cdot \frac{\pi}{2a} \cdot \frac{1}{\pi/2a}}$

$\Rightarrow \qquad e^{\lim\limits_{h \to 0} \frac{-2a}{\pi(a + h)} \cdot \frac{\pi h / 2a}{\tan(\pi h / 2a)}}$

$\Rightarrow \qquad e^{\frac{-2a}{\pi(a)}} \qquad \Rightarrow e^{\frac{-2}{\pi}}$.

5. *Evaluate:* $\lim\limits_{x \to 0} \left\{ \tan \left(\dfrac{\pi}{4} + x \right) \right\}^{\frac{1}{x}}$

The given limit $\qquad = \lim\limits_{x \to 0} \left\{ \dfrac{1 + \tan x}{1 - \tan x} \right\}^{\frac{1}{x}}$

$\qquad\qquad\qquad = \lim\limits_{x \to 0} \left\{ 1 + \dfrac{2 \tan x}{1 - \tan x} \right\}^{\frac{1}{x}}$

$\qquad\qquad\qquad = e^{\lim\limits_{x \to 0} \left(\frac{2 \tan x}{1 - \tan x} \right) \frac{1}{x}}$

$\qquad\qquad\qquad = e^{\lim\limits_{x \to 0} \frac{2 \tan x}{x} \cdot \frac{1}{1 - \tan x}}$

$\qquad\qquad\qquad = e^2.$

EXERCISES

Evaluate the following limits :

1. $\lim\limits_{x \to a} \dfrac{\log\{1 + (x - a)\}}{(x - a)}$

2. $\lim\limits_{x \to 0} \dfrac{\log(5 + x) - \log(5 - x)}{x}$

3. $\lim\limits_{x \to 0} \dfrac{a^x - b^x}{x}$

4. $\lim\limits_{x \to \infty} \left(1 + \dfrac{2}{x}\right)^x$

5. $\lim\limits_{x \to \infty} \left(\dfrac{x + 6}{x + 1}\right)^{x + 4}$

6. $\lim\limits_{x \to 0} \left(1 + \tan^2 \sqrt{x}\right)^{1/2x}$

7. $\lim\limits_{x \to 0} \left(\dfrac{1 + 5x^2}{1 + 3x^2}\right)^{1/x^2}$

8. $\lim\limits_{x \to \infty} \left(\dfrac{x - 3}{x + 2}\right)^x$

9. $\lim\limits_{x \to \pi/4} \left(2 - \tan x\right)^{\log \tan x}$

10. $\lim\limits_{x \to 1} \left(1 + \sin \pi x\right)^{\cot \pi x}$

ANSWERS

1. 1 ;

2. $2/5$;

3. $\log (a/b)$;

4. e^2 ;

5. e^5 ;

6. $e^{1/2}$;

7. e^2 ;

8. e^{-5} ;

9. 1 ;

10. e^{-1}

MISCELLANEOUS EXAMPLES

1. **Solve** $\lim\limits_{x \to 0} \left[\dfrac{\sin |x|}{|x|}\right]$, *where* $[.]$ *denotes greatest integer function.*

We have

$$\lim_{x \to 0} \left[\frac{\sin |x|}{|x|}\right] \qquad\qquad .....(i)$$

we know:, $\dfrac{\sin x}{x} \to 1$ as $x \to 0$

or $\qquad \dfrac{\sin |x|}{|x|} \to 1$ as $x \to 0$ from right or left

i.e., $\qquad$ at $x = 0 + h$ or $x = 0 - h$ and less than one,

$$\frac{\sin |x|}{|x|} = 0 \ \text{as} \ \frac{\sin |x|}{|x|} < 1.$$

$\Rightarrow$ from equation (i) whether we find R H L or L H L.

$$\lim_{x \to 0} \left[\frac{\sin |x|}{|x|}\right] = \lim_{x \to 0} 0 \Rightarrow 0.$$

$\therefore \qquad\qquad \lim\limits_{x \to 0} \left[\dfrac{\sin |x|}{|x|}\right],$ exists and is 0.

2. Solve $\lim\limits_{x\to 0} \dfrac{\sin[x]}{[x]}$, where $[x]$ denotes the greatest integer function.

We have to find $\qquad \lim\limits_{x\to 0} \dfrac{\sin[x]}{[x]}$ ....(i)

Now, $\qquad 0 < 0 + h < 1 \Rightarrow [0 + h] = 0$

and $\qquad -1 < 0 - h < 0 \Rightarrow [0 - h] = -1$

Now,

R H L at $\qquad x = 0$

$$= \lim\limits_{h\to 0} \frac{\sin[0 + h]}{[0 + h]}$$

$$= \lim\limits_{h\to 0} \frac{\sin 0}{0} \Rightarrow \text{limit does not exist.}$$

L.H.L. at $\qquad x = 0$

$$= \lim\limits_{h\to 0} \frac{\sin[0 - h]}{[0 - h]}$$

$$= \lim\limits_{h\to 0} \frac{\sin(-1)}{-1} = \sin 1$$

As $\text{RHL} \neq \text{LHL}$, hence limit does not exist.

3. If α and β be the roots of $ax^2 + bx + c = 0$; then solve:

$$\lim\limits_{x\to \alpha} \left(1 + ax^2 + bx + c\right)^{2/x - \alpha}$$

we have, given limit $= \lim\limits_{x\to \alpha} \left[1 + a(x - \alpha)(x - \beta)\right]^{\frac{2}{x - \alpha}}$

$$= e^{\lim\limits_{x\to \alpha} a(x - \alpha)(x - \beta)\frac{2}{x - \alpha}}$$

$$= e^{2a(\alpha - \beta)}$$

4. If $\lim\limits_{x\to 0} \dfrac{x^a \sin^b x}{\sin x^c}$, where $a, b, c \in R - \{0\}$, exist and has non-zero value. Then show that $a + b = c$.

We have,

$$\lim\limits_{x\to 0} \frac{x^a \sin^b x}{\sin x^c} = \lim\limits_{x\to 0} x^a \cdot \left(\frac{\sin x}{x}\right)^b \left(\frac{x^c}{\sin x^c}\right) \cdot x^{b - c}$$

$$= \lim\limits_{x\to 0} x^{a + b - c} \cdot \left(\frac{\sin x}{x}\right)^b \left(\frac{x^c}{\sin x^c}\right)$$

It will have non-zero value if and only if it is independent of x,

i.e., $a + b - c = 0$

$\Rightarrow$ $a + b = c.$

5. *Evaluate:* $\displaystyle \lim_{x \to \pi/2} \sqrt{\dfrac{\tan x - \sin\left\{\tan^{-1}(\tan x)\right\}}{\tan x + \cos^2(\tan x)}}$

we have, right hand limit at $x = \pi/2$

$$= \lim_{x \to +\pi/2} \sqrt{\dfrac{\tan x - \sin(x - \pi)}{\tan x + \cos^2(\tan x)}}$$

$$\{\because \tan^{-1}(\tan x) = x - \pi, \text{ when } x > \pi/2\}$$

$$= \lim_{x \to +\pi/2} \sqrt{\dfrac{1 + \dfrac{\sin x}{\tan x}}{1 + \dfrac{\cos^2(\tan x)}{\tan x}}}$$

$$= \sqrt{\dfrac{1 + 0}{1 + 0}} = 1.$$

L H L at $x = \pi/2$

$$= \lim_{x \to \frac{\pi^-}{2}} \sqrt{\dfrac{\tan x - \sin(x)}{\tan x + \cos^2(\tan x)}}$$

$$\left\{\because \tan^{-1}(\tan x) = x, \text{ when } x < \pi/2\right\}$$

$$= \lim_{x \to \frac{\pi^-}{2}} \sqrt{\dfrac{1 - \dfrac{\sin x}{\tan x}}{1 + \dfrac{\cos^2(\tan x)}{\tan x}}}$$

$$= \sqrt{\dfrac{1 + 0}{1 + 0}} = 1$$

$\therefore$ RHL = LHL, hence the given limit = 1.

6. *Evaluate* $\displaystyle \lim_{x \to 4} (\cos\alpha)^x - (\sin\alpha)^x - \cos 2\alpha, \ \alpha \in (0, \pi/2)$

Given limit $= \displaystyle \lim_{x \to 4} (\cos\alpha)^x - (\sin\alpha)^x - \left(\cos^2\alpha - \sin^2\alpha\right)\left(\cos^2\alpha + \sin^2\alpha\right)$

$$= \lim_{x \to 4} (\cos\alpha)^x - (\sin\alpha)^x - \cos^4\alpha + \sin^4\alpha$$

$$= \lim_{x \to 4} (\cos \alpha)^4 \left\{ (\cos \alpha)^{x-4} - 1 \right\} - (\sin \alpha)^4 \cdot \lim_{x \to 4} \left\{ (\sin \alpha)^{x-4} - 1 \right\}$$

$$= (\cos \alpha)^4 \cdot \log (\cos \alpha) - (\sin \alpha)^4 \cdot \log (\sin \alpha)$$

$$\left[\because \lim_{x \to 0} \frac{a^x - 1}{x} = \log a \right]$$

7. Evaluate: $\displaystyle \lim_{x \to 0} \left(\frac{1^x + 2^x + 3^x + \dots + n^x}{n} \right)^{a/x}$

Given limit

$$= \lim_{x \to 0} \left(1 + \frac{1^x + 2^x + \dots + n^x}{n} - 1 \right)$$

$$= e^{\displaystyle \lim_{x \to 0} \left(\frac{1^x + 2^x + \dots + n^x}{n} - 1 \right) \cdot \frac{n}{x}}$$

$$= e^{\displaystyle \lim_{x \to 0} \left\{ \frac{(1^x - 1)}{x} + \frac{(2^x - 1)}{x} + \dots + \frac{(n^x - 1)}{x} \right\} \frac{a}{n}}$$

$$= e^{\{ \log 1 + \log 2 + \log 3 + \dots + \log n \} \frac{a}{n}}$$

$$= e^{\left(\log (n!) \right)^{a/n}} = (n!)^{a/n}$$

8. Evaluate: $\displaystyle \lim_{n \to \infty} n^{-n^2} \left[(n+1) \left(n + \frac{1}{2} \right) \left(n + \frac{1}{2^2} \right) \dots \left(n + \frac{1}{2^{n-1}} \right) \right]^n$

Given limit

$$= \lim_{n \to \infty} \left[\frac{(n+1) \left(n + \frac{1}{2} \right) \dots \left(n + \frac{1}{2^{n-1}} \right)}{n^n} \right]^n$$

$$= \lim_{n \to \infty} \left(\frac{n+1}{n} \right)^n \left(\frac{n + \frac{1}{2}}{n} \right)^n \dots \left(\frac{n + \frac{1}{2^{n-1}}}{n} \right)^n$$

$$= \lim_{n \to \infty} \left(1 + \frac{1}{n} \right)^n \cdot \left(1 + \frac{1}{2n} \right)^{\frac{2n}{2}} \dots \left(1 + \frac{1}{2^{n-1} n} \right)^{\frac{2^{n-1} n}{2^{n-1}}}$$

$$= e^1 \cdot e^{1/2} \cdot e^{1/4} \dots e^{1/2^{n-1}} \qquad \left[\because \lim_{n \to \infty} \left(1 + \frac{1}{n} \right)^{an} = e^a \right]$$

$$= e^{(1 + 1/2 + 1/4 + \dots)}$$

$$= e^{\frac{1}{1 - \frac{1}{2}}} = e^2.$$

9. *If the rth term t_r of a series is given by* $t_r = \dfrac{r}{1 + r^2 + r^4}$, *find the value of* $\lim\limits_{n \to \infty} \sum\limits_{r=1}^{n} t_r$

we have

$$t_r = \frac{r}{1 + r^2 + r^4} = \frac{1}{2}\left(\frac{2r}{\left(1 + r^2 + r\right)\left(1 + r^2 - r\right)}\right)$$

$$= \frac{1}{2}\left(\frac{1}{\left(1 + r^2 + r\right)} - \frac{1}{\left(1 + r^2 - r\right)}\right)$$

Thus,

$$t_r = \frac{1}{2}\left(\frac{1}{1 + r(r-1)} - \frac{1}{1 + r(r+1)}\right)$$

$$\therefore \quad t_1 = \frac{1}{2}\left(1 - \frac{1}{3}\right)$$

$$t_2 = \frac{1}{2}\left(\frac{1}{3} - \frac{1}{7}\right)$$

$$t_3 = \frac{1}{2}\left(\frac{1}{7} - \frac{1}{13}\right)$$

$$t_n = \frac{1}{2}\left(\frac{1}{1 + n(n-1)} - \frac{1}{1 + n(n+1)}\right)$$

$$\therefore \quad = \lim_{n \to \infty} \sum_{r=1}^{n} t_r = \lim_{n \to \infty}\left(1 - \frac{1}{1 + n(n+1)}\right)$$

$$= \frac{1}{2}\left(1 - \frac{1}{\infty}\right) = \frac{1}{2}.$$

10. *If* $a = \min\{x^2 + 2x + 3, x \in R\}$ *and* $b = \lim\limits_{\theta \to 0} \dfrac{1 - \cos\theta}{\theta^2}$, *find the value of* $\sum\limits_{r=0}^{n} a^r . b^{n-r}$

we have $a = \min\{x^2 + 2x + 3, x \in R\}$

$$a = \min\{(x+1)^2 + 2\}$$

$\therefore$ a is minimum when $x = -1$, *i.e.*, $a = 2$.

$$b = \lim_{\theta \to 0} \frac{2 \sin^2 \theta/2}{(\theta/2)^2} \cdot \frac{2}{4} = \frac{1}{2}.$$

$$\therefore \quad \sum_{r=0}^{n} a^r b^{n-r} = \sum_{r=0}^{n} 2^r \cdot \left(\frac{1}{2}\right)^{n-r}$$

$$= \frac{1}{2^n}\left\{2^0 + 2^2 + 2^4 + \ldots + 2^{2n}\right\}$$

$$= \frac{1}{2^n} \cdot \frac{4^{n+1} - 1}{2^n \times 3}$$

$$\therefore \text{ Given limit } = \frac{4^{n+1} - 1}{2^n \times 3}.$$

EXERCISES

Evaluate following limits:

1. $\lim\limits_{x \to 0} \left[\dfrac{|\sin x|}{|x|} \right]$, where [.] denotes the greatest integer function.

2. $\lim\limits_{x \to 0} \dfrac{\sin [x]}{[x]}$.

3. $\lim\limits_{x \to \pi/4} [\sin x + \cos x]$

4. $\lim\limits_{x \to 0} \dfrac{\tan^{-1} x - \sin^{-1} x}{\sin^3 x}$

5. $\lim\limits_{x \to 0} \dfrac{e - (1+x)^{1/x}}{\tan x}$

6. $\lim\limits_{x \to 0} |x|^{\cos x}$

7. $\lim\limits_{n \to \infty} \dfrac{a^n}{n!}, a \in R^+$

8. $\lim\limits_{x \to a} \dfrac{x \sin a - a \sin x}{x - a}$

9. $\lim\limits_{x \to 0} \dfrac{e^x - e^{x \cos x}}{x + \sin x}$

10. $\lim\limits_{x \to y} \dfrac{x^y - y^x}{x^x - y^y}$

11. $\lim\limits_{n \to \infty} \sum\limits_{r=1}^{n} \cot^{-1} \left(\dfrac{r^3 - r + \frac{1}{r}}{2} \right)$

12. $\lim\limits_{x \to \pi} \dfrac{\sqrt{2 + \cos x} - 1}{(\pi - x)^2}$

13. $\lim\limits_{n \to \infty} \dfrac{1^3 + 2^3 + 3^3 + \ldots\ldots + n^3}{(n^2 + 1)^2}$

14. If $f(x) = \begin{cases} \dfrac{\sin [x]}{[x]}, & \text{for } [x] \neq 0 \\ 0, & \text{for } [x] = 0 \end{cases}$

where [x] denotes the greatest integer less than or equal to x, then find $\lim\limits_{x \to 0} f(x)$.

15. Evaluate: $\lim\limits_{x \to \pi/3} \dfrac{\sin\left(\dfrac{\pi}{3} - x \right)}{2\cos x - 1}$

16. Evaluate: $\lim\limits_{x \to 1} \left[\tan\left(\dfrac{\pi}{4} + \log x \right) \right]^{1/\log x}$

17. Evaluate: $\lim\limits_{x \to \infty} \dfrac{[1^2 x] + [2^2 x] + [3^2 x] + \ldots\ldots + [n^2 x]}{n^3}$

where [.] denotes greatest integral function less than or equal to x.

18. Find $\lim\limits_{x \to 0} \left[\sqrt{2 -- x} + \sqrt{1 + x} \right]$, where $a \in \left[0, \dfrac{1}{2} \right]$ and [.] denotes the greatest integer function.

19. Evaluate $\lim\limits_{x \to 0} \dfrac{e^{x^2} - \cos x}{x^2}$

20. If x is a real number is [0, 1]. Then find the value of

$$\lim\limits_{m \to \infty} \lim\limits_{n \to \infty} \left[1 + \cos^{2m} \left(n! \, \pi x \right) \right].$$

ANSWERS

1. 0 ;	**2.** limit does not exist ;	**3.** -2	**4.** $-\dfrac{1}{2}$;	**5.** $\dfrac{e}{2}$;
6. 1 ;	**7.** 0 ;	**8.** $\sin a - \cos a$;	**9.** 0 ;	**10.** $\dfrac{1 - \log y}{1 + \log y}$;
11. $\pi / 2$;	**12.** 1/4 ;	**13.** 1/4 ; **14.** limit does not exist ;		**15.** $1/\sqrt{3}$;
16. e^2 ;	**17.** $x/3$;	**18.** 2 ;	**19.** 3/2 ,	**20.** 2 when $x \in Q$, 1 when $x \notin Q$

3.6. Another form of the definition of continuity. Comparing the definitions of continuity and limit as given in §§ 3.1.1, 3.1.2, we see that

f is continuous at an interior point c if and only if.

$$\lim\limits_{x \to c} f(x) = f(c).$$

Thus the limit as x tends to c of a function f which is continuous at c is equal to the value $f(c)$ of the function for c.

3.6.1. Continuity of the sum, difference, product and quotient of two continuous functions.

Let f, g be two functions which are continuous for a, so that

$$\lim\limits_{x \to a} f(x) = f(a), \quad \lim\limits_{x \to a} g(x) = g(a).$$

From the theorems in §3.3, we see that

$$\lim\limits_{x \to a} (f + g)(x) = \lim\limits_{x \to a} \left[f(x) + g(x) \right] = \lim\limits_{x \to a} f(x) + \lim\limits_{x \to a} g(x)$$

$$= f(a) + g(a) = (f + g)(a)$$

so that $f + g$ is continuous at a.

The continuity of $f - g$ and $f g$ may similarly be proved.

for the quotient, we have

$$\lim\limits_{x \to a} \dfrac{f(x)}{g(x)} = \dfrac{\lim\limits_{x \to a} f(x)}{\lim\limits_{x \to a} g(x)}, \text{ when } \lim\limits_{x \to a} g(x) \neq 0$$

$$= \dfrac{f(a)}{g(a)} = \text{value of } \dfrac{f(x)}{g(x)}, \text{ for } x = a.$$

so that $f \div g$ is also continuous at a, provided that $\lim g(x) = g(a) \neq 0$, when $x \to a$.

Thus, *the sum, the difference, the product and the quotient of two continuous functions are also continuous* (with one obvious exception in the case of the quotient).

3.6.2. Continuity of elementary functions.

Consider the polynomial function

$$f(x) = a_0 x^n + a_1 x^{n-1} + \ldots + a_{n-1} x + a_n.$$

As already seen we have $f(x) = a_0 x^n + a_1 x^{n-1} + \ldots + a_{n-1} x + a_n.$ where $a_0 \ldots a_n$ denote constant functions and I the Identity function.

The result now follows from the preceding theorem and the fact that every constant function and the Identity function are continuous over **R**.

A rational function

$$g(x) = \frac{a_0 x^n + a_1 x^{n-1} + \ldots + a_n}{b_0 x^m + b_1 x^{m-1} + \ldots + b_m}$$

is continuous for every value of x except for those for which the denominator becomes zero.

We have seen that $y = \sin x$ is continuous for every value of x. It may be similarly shown that $y = \cos x$ is also continuous for every value of x.

Hence from § 3.6.1, we see that the functions

$$y = \tan x = \frac{\sin x}{\cos x}, \; y = \cot x = \frac{\cos x}{\sin x},$$

$$y = \sec x = \frac{1}{\cos x}, \; y = \cos ec \, x = \frac{1}{\sin x}$$

are continuous for every member of their domain *i.e.* for all those values of x for which they are defined. Thus the discontinuities of these four functions arise only when the denominators become zero and for such values of x, these functions themselves cease to be defined.

Thus the *domain of continuity* of each functions with function values $\sin x$, $\cos x$, $\tan x$, $\cot x$, $\sec x$, and $\cos ec \, x$ coincides with the corresponding *domain of definition.*

3.6.3. Continuity of the Inverse of a continuous invertible function.
We assume without proof that if f is an invertible continuous function and g denotes its inverse, then g is also continuous.

It now follows that the functions

$$y = \sin^{-1} x, \; y = \cos^{-1} x, \; y = \tan^{-1} x,$$

$$y = \cot^{-1} x, \; y = \sec^{-1} x, \; y = \cos ec^{-1} x,$$

are continuous in their domains.

Finally we *assume* that the exponential function $y = a^x$ is continuous in its domain **R**. It follows that its inverse $y = \log_a x$ is also continuous in its domain $[\, 0, \infty \,]$.

3.6.4. Continuity of the composite of two continuous functions.
We *assume* without proof that the composite $g \, of$ of two continuous functions f and g is continuous.

3.7. Some properties of continuous functions.

3.7.1. *If f is a continuous for c and $f(c) \neq 0$ then there exists an interval $[c - \delta, c + \delta]$ around c such that $f(x)$ has the sign of $f(c)$ for every value of x in this interval.*

Its truth is obvious if we remember that a continuous function does not undergo *sudden* changes so that if $f(c)$ is positive for the value c of x and also f is continuous at c, it cannot suddenly become negative or zero and must, therefore, remain positive for values of x in a certain neighbourhood of c.

In the *formal* manner, it may be proved as follows :

To given $\varepsilon > 0$, there corresponds $\delta > 0$ such that

$$f(c) - \varepsilon < f(x) < f(c) + \varepsilon \; \forall \; x \in \left[c - \delta, c + \delta \right].$$

Let $f(c) > 0$. In this case we *take* ε a positive number less than $f(c)$ so that $f(c) - \varepsilon$ and $f(c) + \varepsilon$ are both positive. Thus $f(x)$ which lies between two positive numbers is itself positive when x lies between $c - \delta$ and $c + \delta$.

Let $f(c) < 0$. Now $f(c) - \varepsilon$ is negative and $f(c) + \varepsilon$ will be negative if $\varepsilon < -f(c)$ Thus in this case if we *take* ε a positive number smaller than the positive number $- f(c)$, then we see that $f(x)$ which lies between the two negative numbers, $f(c) - \varepsilon$ and $f(c) + \varepsilon$ is itself negative when x lies between $c - \delta$ and $c + \delta$.

Hence the result.

Note: *This simple but important property of continuous functions will be used in the chapters of Maxima, Minima and Concavity.*

3.7.2 *If f is continuous in a closed interval $[a, b]$ and $f(a), f(b)$ are of opposite signs then $f(x)$ is zero for at least one $x \in [a, b]$.*

Its truth in intuitively obvious, for a continuous curve $y = f(x)$ going from a point on one side of *x-axis* to a point lying on the other cannot do so without crossing it Fig. 3.10.

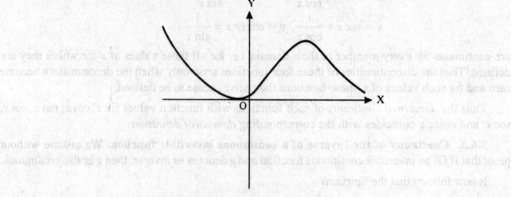

Fig. 3.10

Its formal proof, is however, beyond the scope of this book.

3.7.3. *If f is continuous in a **closed** interval $[a, b]$ then there exist points c and d in the interval $[a, b]$, where f assumed its greatest and least values M and m, i.e.,*

$$f(c) = M, f(d) = m.$$

The proof is beyond the scope of this book.

The theorem states that there is a value of a continuous function greater than every other of its values and also a value smaller than every other value.

A discontinuous function may *not* possess greatest or least value as we now illustrate. Consider the function defined as follows:

$$f(x) = \begin{cases} 1 - x, & \text{for } 0 < x \le 1 \\ \dfrac{1}{2}, & \text{for } x = 0 \end{cases}$$

Its graph consists of the point C (0, 1/2), and the line segment AB excluding the point B. The function possess no greatest value; 1 not being a value of the function. If we consider any value less than 1, howsoever near 1, it may be seen that there is a value of the function greater than that value.

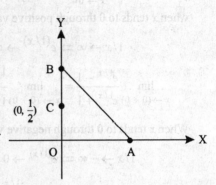

This is explained by the fact that the function is *not* continuous in the interval [0, 1] in that $x = 0$ is a point of discontinuity Fig. 3.11.

Note. This property will be required to prove Rolle's Theorem in Chap. 8.

Fig. 3.11

3.8. Discontinuity. As already defined a function is said to be discontinous at a point if it is not continuous at that point.

Note: *The question of continuity of a function at a point c does not arise if the function is not defined at that point.*

3.8.1. Types of Discontinuties.

1. Removable Discontinuity. f is said to have a removable discontinuity if $\lim\limits_{x \to c} f(x) \ne f(c)$.

This discontinuity can be removed by changing the value of the function at c.

2. Discontinuity of the first kind. f is said to have a discontinuity of the first kind at the point c if
$$\lim_{x \to c + 0} f(x) \ne \lim_{x \to c - 0} f(x)$$

We say that f has a discontinuity of the first kind from the left at c if $\lim\limits_{x \to c - 0} f(x) \ne f(c)$

Similarly f is said to have discontinuity of the first kind from the right at c if $\lim\limits_{x \to c + 0} f(x) \ne f(c)$

(Refer Examples: 5, 7)

3. Discontinuity of the second kind. f is said to have discontinuity of the second type at c if neither of $\lim\limits_{x \to c - 0} f(x)$ and $\lim\limits_{x \to c + 0} f(x)$ exists.

We say that f has a discontinuity of the second type from the left at c if $\lim\limits_{x \to c - 0} f(x)$ does not exist.

Similary f is said to have discontinuity of the second type from the right at c if $\lim\limits_{x \to c + 0} f(x)$ does not exist.   (Refer Example 8)

EXAMPLES

1. *Show that the function f is defined by $f(x) = (1 + 3x)^{1/x}$ when $x \ne 0, f(0) = e^3$ is continuous for $x = 0$.*

Now we have
$$\lim_{x \to 0} f(x) = \lim_{x \to 0}\left[(1 + 3x)^{1/3x}\right]^3 = e^3 = f(0).$$

Hence the result.

2. *Show that*

$$\lim_{x \to 0} \left[\left(e^{1/x} - 1 \right) \Big/ \left(e^{1/x} + 1 \right) \right] \textit{ does not exist.}$$

when x tends to 0 through positive values, we have

$$1/x \to \infty \Rightarrow e^{(1/x)} \to \infty \Rightarrow 1/e^{(1/x)} \to 0.$$

$$\lim_{x \to (0+0)} \frac{e^{1/x} - 1}{e^{1/x} + 1} = \lim_{x \to (0+0)} \frac{1 - \left(1/e^{1/x} \right)}{1 + \left(1/e^{1/x} \right)} = \frac{1 - 0}{1 + 0} = 1.$$

When x tends to 0 through negative values, we have

$$1/x \to -\infty \Rightarrow e^{(1/x)} \to 0$$

$$\Rightarrow \qquad \lim_{x \to (0-0)} \frac{e^{1/x} - 1}{e^{1/x} + 1} = \frac{0 - 1}{0 + 1} = -1.$$

Hence we see that

$$\lim_{x \to (0+0)} f(x) \neq \lim_{x \to (0-0)} f(x)$$

so that $\lim_{x \to 0} f(x)$ does not exist.

3. *Show that f defined as follows:*

$$f(x) = \frac{x - 1}{1 + e^{1/(x-1)}}, \, f(1) = 0$$

is continuous for x = 1.

It may be seen that

$$\lim_{x \to (1+0)} e^{1/(x-1)} = \infty, \quad \lim_{x \to (1-0)} e^{1/(x-1)} = 0$$

so that

$$\lim_{x \to (1+0)} f(x) = 0, \quad \lim_{x \to (1-0)} f(x) = 0.$$

Thus $\lim_{x \to 1} f(x) = 0 = f(0).$

Hence the result.

4. *Discuss the continuity for x = 0 of f defined by* $f(x) = 1 \Big/ \left(1 - e^{1/x} \right)$ *when x $\neq$ 0 and f(a) = 0.*

Now

$$1/x \to \infty \text{ or } -\infty \text{ according as } x \to (0+0) \text{ or } (0-0).$$

$$\Rightarrow \qquad e^{1/x} \to \infty \text{ or } 0 \text{ according as } x \to (0+0) \text{ or } (0-0).$$

Thus we have

$$\lim_{x \to (0+0)} f(x) = 0, \quad \lim_{x \to (0-0)} f(x) = 1.$$

Since these two limits are different, $\lim f(x)$ as $x \to 0$ does not exist. Thus f is discontinuous for x = 0 and the point of discontinuity is of the second kind.

Of course the function is continuous on the right and has a discontinuity of the second kind on the left of $x = 0$.

5. *Examine the continuity of the function defined by*

$$f(x) = \frac{|x - a|}{x - a}, \; x \neq a$$

$$= 1, x = a$$

at the point $x = a$.

We have

$$\lim_{x \to a + 0} f(x) = \lim_{x \to a + 0} \left(\frac{x - a}{x - a} \right) = 1$$

and

$$\lim_{x \to a - 0} f(x) = \lim_{x \to a - 0} \left(\frac{-(x - a)}{x - a} \right) = -1$$

$\Rightarrow$

$$\lim_{x \to a + 0} f(x) = f(a) \neq \lim_{x \to a - 0} f(x)$$

Therefore f has a discontinuity of the first kind from the left at $x = a$.

6. *Show that the function $f(x) = |x| + |x - 1| + |x - 2|$ is continuous at the points $x = 0, 1, 2$.*

We have

$$f(x) = \begin{cases} -x - (x - 1) - (x - 2) = -3x + 3 \; if \; x < 0 \\ x - (x - 1) - (x - 2) = -x + 3 \; if \; 0 \leq x < 1 \\ x + (x - 1) - (x - 2) = x + 1 \; if \; 1 \leq x < 2 \\ x + (x - 1) + (x - 2) = 3x - 3 \; if \; x \geq 2 \end{cases}$$

$$\lim_{x \to 0 + 0} f(x) = \lim_{x \to 0 + 0} (-x + 3) = 3$$

and

$$\lim_{x \to 0 - 0} f(x) = \lim_{x \to 0 - 0} (-3x + 3) = 3$$

$\therefore$

$$\lim_{x \to 0 + 0} f(x) = \lim_{x \to 0 - 0} f(x) = f(0)$$

$\Rightarrow$ f is continuous at $x = 0$

Again

$$\lim_{x \to 1 + 0} f(x) = \lim_{x \to 1 + 0} (x + 1) = 2$$

and

$$\lim_{x \to 1 - 0} f(x) = \lim_{x \to 1 - 0} (-x + 3) = 2$$

$\therefore$

$$\lim_{x \to 1 + 0} f(x) = \lim_{x \to 1 - 0} f(x) = f(2)$$

$\Rightarrow$ f is continuous at $x = 1$

Also

$$\lim_{x \to 2 + 0} f(x) = \lim_{x \to 2 + 0} (3x - 3) = 3$$

and

$$\lim_{x \to 2 - 0} f(x) = \lim_{x \to 2 - 0} (x + 1) = 3$$

so that

$$\lim_{x \to 2+0} f(x) = \lim_{x \to 2-0} f(x) = f(2)$$

∴ f is continuous at $x = 2$

7. *Examine the continuity of the function defined by*

$$f(x) = \begin{cases} -x^2, x \le 0 \\ 5x - 4, 0 < x < 2 \\ 4x^2 - 3x, 1 < x < 2 \\ 3x + 4, x \ge 2 \end{cases}$$

at the points $x = 0, 1, 2.$ *(Avadh 2001; Poorvanchal 2004)*

We have

$$\lim_{x \to 0+0} f(x) = \lim_{x \to 0+0} (5x - 4) = -4,$$

$$\lim_{x \to 0-0} f(x) = \lim_{x \to 0-0} \left(-x^2\right) = 0$$

so that

$$\lim_{x \to 0+0} f(x) \neq \lim_{x \to 0+0} f(x)$$

⇒ f is not continuous at $x = 0$

Again

$$\lim_{x \to 1+0} f(x) = \lim_{x \to 1-0} (4x^2 - 3x) = 1$$

and

$$\lim_{x \to 1-0} f(x) = \lim_{x \to 1-0} (5x - 4) = 1$$

Thus

$$\lim_{x \to 1+0} f(x) = \lim_{x \to 1-0} f(x) = 1 = f(1)$$

⇒ f is continuous at $x = 1$

Also

$$\lim_{x \to 2+0} f(x) = \lim_{x \to 2+0} (3x + 4) = 10$$

and

$$\lim_{x \to 2-0} f(x) = \lim_{x \to 2-0} \left(4x^2 - 3x\right) = 10$$

∴

$$\lim_{x \to 2+0} f(x) = \lim_{x \to 2-0} f(x) = f(2)$$

Thus f is continuous at $x = 2$

The function has discontinuity of the first kind from the right at the point $x = 0$ as

$$\lim_{x \to 0+0} f(x) \neq f(0) = \lim_{x \to 0-0} f(x)$$

8. *Show that the function f defined by* $f(x) = x - [x]$ *where* [x] *denotes the integral part of x is discontinuous for all integral values of x and continuous for all other.*

We have $$f(x) = \begin{cases} x - (\alpha - 1) \text{ for } \alpha - 1 < x < \alpha \\ 0 \text{ for } x = \alpha \\ x - \alpha \text{ for } \alpha < x < \alpha + 1 \end{cases}$$

where is α an integer.

$$f(\alpha + 0) = \lim_{x \to \alpha + 0} (x - \alpha) = 0$$

$$f(\alpha - 0) = \lim_{x \to \alpha - 0} (x - (\alpha - 1)) = 1$$

$$\therefore \qquad f(\alpha + 0) = f(\alpha) \neq f(\alpha - 0)$$

$\Rightarrow \qquad f$ is not continuous at $x = \alpha$

Thus $f(x)$ is discontinuous for all integral values of x and there is discontinuity of the second kind from the left at all integral values of x.

$f(x)$ is obviously continuous for all other values of x.

9. **If** $f(x) = \begin{cases} \left\{ \tan\left(\dfrac{\pi}{4} + x\right) \right\}^{1/x}, & x \neq 0 \\ k, & x = 0 \end{cases}$ *for what value of k, $f(x)$ is continuous at $x = 0$?*

$$\lim_{x \to 0} f(x) = \lim_{x \to 0} \left[\frac{1 + \tan x}{1 - \tan x}\right]^{1/x}$$

$$= \lim_{x \to 0} \left[1 + \frac{2 \tan x}{1 - \tan x}\right]^{1/x}$$

$$= e^{\lim_{x \to 0} \left(\frac{2 \tan x}{1 - \tan x}\right) \cdot \frac{1}{x}}$$

$$= e^2$$

$\therefore \qquad f(x)$ is continuous at $x = 0$, when

$$\lim_{x \to 0} f(x) = f(0), \ i.e., \ k = e^2.$$

10. *Discuss the continuity of the function*

$$f(x) = \lim_{n \to \infty} \frac{\log(2 + x) - x^{2n} \sin x}{1 + x^{2n}} \ at \ x = 1.$$

$$\lim_{n \to \infty} x^{2n} = \begin{cases} 0, \ if \ x^2 < 1 \\ \infty, \ if \ x^2 < 1. \end{cases}$$

$\therefore \quad$ for $x^2 > 1$

$$f(x) = \lim_{n \to \infty} \frac{\log(2 + x) - x^{2n} \sin x}{1 + x^{2n}}$$

$$\to \log(2 + x)$$

$\therefore \quad$ for $x^2 > 1$

$$f(x) = \lim_{x \to \infty} \frac{x^{\frac{1}{2n}} \log(2 + x) - \sin x}{1 + x^{\frac{1}{2n}}}$$

$$\to - \sin x$$

$\therefore \qquad \lim_{x \to 1 + 0} f(x) = -\sin 1 \quad$ and $\quad \lim_{x \to 1 - 0} f(x) = \log 3$

Also, $f(1) = \dfrac{1}{3}(\log 3 - \sin 1).$

$\therefore$ $\lim\limits_{x \to 1+0} f(x) \neq \lim\limits_{x \to 1-0} f(x) = f(1).$

Hence $f(x)$ is not continuous at $x = 1$

11. *Discuss the continuity of the function:*

$f(x) = [x] + [-x]$ *at integral values of x.*

(i) If x is an integer,

$$[x] = x \text{ and } [-x] = -x \Rightarrow f(x) = 0$$

(ii) if x is not an integer,

Let $x = n + f$ where n is an integer and $f \in (0, 1).$

$\Rightarrow$ $[x] = n$ and $[-x] = [-n-f]$

$= [(-n-1) + (1-f)] = (-n-1).$

$\therefore$ $f(x) = n + (-n-1) = -1.$

$\therefore$ $f(x) = \begin{cases} 0, \text{ if } x \text{ is an integer} \\ -1, \text{ if } x \text{ is not an int eger} \end{cases}$

At $x = a$, where a is an integer

$\lim\limits_{x \to a-0} f(x) = -1$

and $\lim\limits_{x \to a+0} f(x) = +1$ (As $a + 0$ and $a - 0$ are not integers)

But $f(a) = 0$ an a is an integer

Hence, $f(x)$ has a removable discontinuity at integral values of x.

12. *Discuss the continuity of* $f(x)$ *in* $[0, 2]$ *where*

$$f(x) = \begin{cases} [\cos \pi x], \; x \leq 1 \\ |(2x-3)| [x-2], \; x > 1, \end{cases}$$

where [.] *denotes the greatest integral function.*

First consider $x \in [0, 1]$

then $f(x) = [\cos \pi x]$ is discontinuous where $\cos \pi x \in I.$

In $[0, 1]$, $\cos \pi x$ is an integer at $x = 0, 1/2, 1.$

$\Rightarrow$ $x = 0, \dfrac{1}{2}$ and 1 may be the points at which $f(x)$ may be discontinuous.

Again, consider $x \in [1, 2]$

If we consider $f(x) = [x-2]|2x-3|$

then if $x \in (1, 2); [x-2] = -1$

and for $x = 2 ; [x-2] = 0$

Also, $|2x-3| = 0$ at $x = 3/2$

$\Rightarrow$ $x = 3/2$ and 2 may be the points at which $f(x)$ may be discontinuous.

Hence, the possible points of discontinuity may be $x = 0, 1/2, 1, 3/2, 2$.

Now,

$$f(x) = \begin{cases} 1; & x = 0 \\ 0; & 0 < x \le \dfrac{1}{2} \\ -1; & \dfrac{1}{2} < x \le 1 \\ -(3 - 2x); & 1 < x \le \dfrac{3}{2} \\ -(2x - 3); & 3/2 < x \le 2 \\ 0; & x = 2 \end{cases}$$

Now, we will consider the cases one by one.

$$\lim_{x \to 0 + 0} f(x) = 0 \text{ and } f(0) = 1$$

$\therefore$ $f(x)$ is discontinuous at $x = 0$.

$$\lim_{x \to \frac{1}{2} - 0} f(x) = \lim_{x \to \frac{1}{2} - 0} 0 = 0$$

$$\lim_{x \to \frac{1}{2} + 0} f(x) = \lim_{x \to \frac{1}{2} + 0} (-1) = -1$$

$\therefore$ $f(x)$ is discontinuous at $x = \dfrac{1}{2}$.

$$\lim_{x \to 1 - 0} f(x) = \lim_{x \to 1 - 0} (-1) = -1$$

$$\lim_{x \to 1 + 0} f(x) = \lim_{x \to 1 + 0} -(3 - 2x) = -1$$

and $\qquad f(1) = -1$

$\therefore$ $f(x)$ is continuous at $x = 1$.

$$\lim_{x \to \frac{3}{2} - 0} f(x) = \lim_{x \to \frac{3}{2} - 0} (2x - 3) = 0$$

$$\lim_{x \to \frac{3}{2} + 0} f(x) = \lim_{x \to \frac{3}{2} + 0} (3 - 2x) = 0$$

$$f(3/2) = 0$$

$\therefore$ $f(x)$ is continuous at $x = 3/2$.

$$\lim_{x \to 2 - 0} f(x) = \lim_{x \to 2 - 0} (3 - 2x) = -1$$

$$f(2) = 0$$

$f(x)$ is discontinuous at $x = 2$.

$\therefore$ $f(x)$ is continuous when

$$x \in [0, 2] - \{0, 1/2, 2\}.$$

13. *Given*

$$f(x) = \begin{cases} (\cos x - \sin x)^{\cosec x}, & -\dfrac{\pi}{2} < x < 0 \\ a, & x = 0 \\ \dfrac{e^{1/x} + e^{2/x} + e^{3/x}}{ae^{2/x} + be^{3/x}}, & 0 < x < \pi/2 \end{cases}$$

If $f(x)$ is continuous at $x = 0$, find a and b

$$\lim_{x \to (0+0)} f(x) = \lim_{h \to 0} \frac{e^{1/h} + e^{2/h} + e^{3/h}}{ae^{2/h} + be^{3/h}}$$

$$= \lim_{h \to 0} \frac{e^{\frac{1}{2h}} + e^{\frac{1}{h}} + 1}{e^{\frac{1}{h}} + b}$$

$$= \frac{1}{b}.$$

$$\lim_{x \to (0-0)} f(x) = \lim_{h \to 0} (\cos h + \sin h)^{-\cosec h}$$

$$= \lim_{h \to 0} \{1 + (\cos h + \sin h - 1)\}^{-1/\sin h}$$

$$= e^{\lim_{h \to 0}\{-2\sin h/2 + 2\sin h/2 \cos h/2\}\left\{-\dfrac{1}{2\sin h/2 \cos h/2}\right\}}$$

$$\Rightarrow \quad \lim_{e^{h} \to 0} \frac{\sin h/2 - \cos h/2}{\cos h/2} = e^{-1}$$

and $$f(0) = a$$

$$\therefore \quad a = e^{-1} = \frac{1}{b} \quad \Rightarrow \quad a = 1/e, \ b = e.$$

14. *Discuss the continuity of the function*

$$f(x) = \begin{cases} \dfrac{a^{2[x] + \{x\}} - 1}{2[x] + \{x\}}, & x \neq 0 \\ \log_e a, & x = 0 \end{cases}$$

at $x = 0$, where [.] denotes greatest integral part and { . } denotes fractional part of x.

As $[x] + \{x\} = x$ for any $x \in R$, the given function becomes:

$$f(x) = \begin{cases} \dfrac{a^{[x] + x} - 1}{[x] + x}, & x \neq 0 \\ \log_e a, & x = 0. \end{cases}$$

Now, $\lim\limits_{x \to 0 + 0} f(x) = \lim\limits_{h \to 0} \dfrac{a^{[h] + h} - 1}{[h] + h}$

$= \lim\limits_{h \to 0} \dfrac{a^h - 1}{h} \quad [0 < h < 1 \Rightarrow [h] = 0]$

$= \log a$

$\lim\limits_{x \to 0 - 0} f(x) = \lim\limits_{h \to 0} \dfrac{a^{-1-h} - 1}{-1 - h} \quad [\because -1 < 0 - h < 0 \Rightarrow [0 - h] = -1]$

$= \dfrac{a^{-1} - 1}{-1} = 1 - \dfrac{1}{a}.$

obviously $f(0 + 0) \neq f(0 - 0)$

Hence, $f(x)$ is not continuous at $x = 0$.

15. *Discuss the continuity of the function* $f(x) = \dfrac{|x + 2|}{\tan^{-1}(x + 2)}$.

f is continuous except possibly at $x = -2$.

$\lim\limits_{x \to -2 + 0} f(x) = \lim\limits_{x \to -2 + 0} \dfrac{(x + 2)}{\tan^{-1}(x + 2)} = 1$

$\lim\limits_{x \to -2 - 0} f(x) = \lim\limits_{x \to -2 - 0} \dfrac{-(x + 2)}{\tan^{-1}(x + 2)} = -1$

obviously, $f(x)$ is continuous at $x \in R - \{-2\}$.

EXERCISES

1. Examine the continuity of the following at $x = 0$:

(i) $f(x) = \begin{cases} (1 + x)^{1/x} & , \text{ when } x \neq 0 \\ 1 & , \text{ when } x = 0. \end{cases}$

(ii) $f(x) = \begin{cases} (1 + 2x)^{1/x} & , \text{ when } x \neq 0 \\ e^2 & , \text{ when } x = 0. \end{cases}$

(iii) $f(x) = \begin{cases} e^{-1/x^2} & , \text{ when } x \neq 0 \\ 1 & , \text{ when } x = 0. \end{cases}$

(iv) $f(x) = \begin{cases} \dfrac{e^{-1/x}}{1 + e^{1/x}} & , \text{ when } x \neq 0 \\ 1 & , \text{ when } x = 0. \end{cases}$

(v) $f(x) = \begin{cases} \dfrac{e^{x^2}}{e^{1/x^2 - 1}} & , \text{ when } x \neq 0 \\ 1 & , \text{ when } x = 0. \end{cases}$

2. Examine the continuity of the function.

$f(x) = \dfrac{2[x]}{3x - [x]}$ at $x = \dfrac{-1}{2}$ and $x = 1$ where $[x]$ denotes the greatest integer not greater than x.

3. Examine the limit of the function $f(x)$ as $x \to 2$, where

$$f(x) = \begin{cases} \dfrac{|x-2|}{x-2} & , \quad x \neq 2 \\ 0 & , \quad x = 2 \end{cases}$$

4. Discuss the continuity of the function f defined by

$$f(x) = |x-2| + |x-3|$$

at $x = 2$ and $x = 3$.

5. Prove that the function f defined as

$$f(x) = \begin{cases} x & , \quad x \leq 1 \\ 2 - x & , \quad 1 < x \leq 2 \\ -2 + 3x - x^2 & , \quad x > 2 \end{cases}$$

is continuous at $x = 1$ and $x = 2$

6. Examine the continuity of the function

$$f(x) = 2x - [x] + \sin \frac{1}{x}, x \neq 0$$

$$= 0 \text{ when } x = 0$$

where $[x]$ denotes the greatest integer not greater than x; at $x = 0$ and $x = 2$.

7. Examine the continuity, the function f defined by

$$f(x) = \begin{cases} 2x + 1, 0 < x < \dfrac{1}{2} \\ 1, x = \dfrac{1}{2} \\ 1 - x, \dfrac{1}{2} < x < 1 \\ 2x - 2, x \geq 1 \end{cases}$$

at the points $x = \dfrac{1}{2}$ and 1. In case of discontinuity discuss the nature of discontinuity.

8. A function f is defined on [0, 1] as follows:

$$f(x) = \frac{1}{2^n} \text{ when } \frac{1}{2^{n+1}} < x \leq \frac{1}{2^n} \ (n = 0, 1, 2, \dots\dots)$$

$$f(0) = 0$$

Show that f is discontinuous at the points $\dfrac{1}{2}, \left(\dfrac{1}{2}\right)^2, \left(\dfrac{1}{2}\right)^3, \dots\dots$ and examine the nature of discontinuity.

9. Investigate the points of continuity and discontinuity of the function f defined as follows:

$$f(x) = \begin{cases} \dfrac{x^2}{a} - a & \text{for } x \leq a \\ a - \dfrac{a^2}{x} & \text{for } x > a \end{cases}$$

(Avadh 99)

10. A function $f(x)$ is defined by

$$f(x) = \begin{cases} \dfrac{[x^2] - 1}{x^2 - 1} & , \quad \text{for } x^2 \neq 1 \\ 0 & , \quad \text{for } x^2 = 1 \end{cases}$$

Discuss the continuity of $f(x)$ is continuous at $x = 1$.

11. Let

$$f(x) = \begin{cases} \{1 + |\sin x|\}^{a/|\sin x|} & ; & -\pi/6 < x < 0 \\ b & ; & x = 0 \\ e^{\tan 2x/\tan 3x} & ; & 0 < x < \pi/6 \end{cases}$$

Determine a and b such that $f(x)$ is continuous at $x = 0$.

12. Find a and b so that the function:

$$f(x) = \begin{cases} x + a\sqrt{2} \sin x & ; & 0 \le x < \pi/4 \\ 2x \cot x + b & ; & \pi/4 \le x \le \pi/2 \\ a \cos 2x - b \sin x & ; & \pi/2 < x \le \pi \end{cases}$$

is continuous for $x \in [0, \pi]$?

13. Given $f(x) = \begin{cases} \dfrac{1 - \cos 4x}{x^2} & , & x < 0 \\ a & , & x = 0 \\ \dfrac{\sqrt{x}}{\sqrt{16 + \sqrt{x}} - 4} & , & x < 0 \end{cases}$

Determine the value of a if possible, so that the function is continuous at $x = 0$

14. If $f(x) = \dfrac{\sin 2x + A \sin x + B \cos x}{x^3}$ is continuous at $x = 0$, find the values of A, B and $f(0)$.

15. Discuss the continuity of the function $f(x) = [\,[x]\,] - [x - 1]$, where [.] denotes the greatest integral function.

16. If $f(x) = \lim\limits_{n \to \infty} \sum\limits_{r = 1}^{n} \dfrac{[2rx]}{n^2}$, where [.] denotes greatest integral function, discuss the continuity of $f(x)$.

17. Prove that $f(x) = \lim\limits_{n \to \infty} \left(\lim\limits_{m \to \infty} \cos^{2m}(n! \pi x) \right)$ is nowhere continuous.

ANSWERS

1. (*i*) discontinuous   (*ii*) continuous   (*iii*) discontinuous   (*iv*) discontinuous
    (*v*) discontinuous

2. Discontinuous at $x = \dfrac{1}{2}$, continuous at $x = -\dfrac{1}{2}$ **3.** Limit does not exist

4. Continuous at $x = 2, 3$;   **6.** Discontinuous

7. Discontinuous at $x = \dfrac{1}{2}$, continuous at $x = 1$,   **8.** Discontinuity of 2nd kind

9. Continuous at $x = a$.   **10.** Discontinuous;

11. $a = 2/3, b = e^{2/3}$;   **12.** $a = \pi/6, b = -\pi/12$;

13. $a = 8$;   **14.** $A = -2, B = 0, f(0) = -1$;

15. Continuous on R;   **16.** Continuous everywhere.

OBJECTIVE QUESTIONS

For each of the following questions four alternatives are given for the answer, only one of them is correct. Choose the correct alternative:

1. Limit of a function $f(x)$ when $x \to a$ exists if:

   (a) anyone of $f(a-0)$ and $f(a+0)$ exists   (b) both $f(a-0)$ and $f(a+0)$ exist

   (c) $f(a-0) = 0 \neq f(a+0)$ (d) $f(a-0) = f(a+0)$

2. A function $f(x)$ is said to be continuous at $x = a$ if:

   (a) $\lim\limits_{x \to a} f(x)$ exists   (b) $f(a)$ exists

   (c) $\lim\limits_{x \to a} f(x) = f(a)$ (d) $\lim\limits_{x \to a} f(x) \neq f(a)$

3. $\lim\limits_{x \to a} \sin(1/x)$:

   (a) exists   (b) is equal to zero

   (c) is equal to ∞ (d) does not exists

4. $\lim\limits_{x \to a} x\sin(1/x)$: is equal to

   (a) 1   (b) 0

   (c) ∞ (d) oscillatory.

5. A polynomial function in **R** :

   (a) is never continuous in **R**.   (b) may or may not be continuous in **R**

   (c) is always continuous in **R**   (d) is continuous in **R** except at $x = 0$

6. The function $f(x) = x\sin\dfrac{1}{x}$, $x \neq 0$ and $f(0) = 0$ is:

   (a) continuous at 0 only   (b) discontinuous at 0 only

   (c) continuous for all values of x (d) discontinuous for all values of x

7. The function $f(x) = \begin{cases} \cos x, & \text{for } x \geq 0 \\ -\cos x, & \text{for } x < 0 \end{cases}$ is:

   (a) continuous for all values of x

   (b) continuous for all values of x except at $x = \pi/2$

   (c) continuous for all values of x except at $x = 0$

   (d) discontinuous for all values of x

8. The function

$$f(x) = \begin{cases} 1 + x & \text{if } x \leq 2 \\ 5 - x & \text{if } x > 2 \end{cases}, \text{ is}$$

   (a) continuous for all values of x

   (b) continuous for all values of x except $x = 2$

   (c) discontinuous at $x = 0$

   (d) discontinuous at $x = 2$.

9. $\lim\limits_{x \to \infty} \left(1 + \dfrac{1}{x}\right)$ is equal to

   (a) 1   (b) 0

(c) e   (d) ∞

10. The function $f(x) = |x|$ is:
    (a) continuous for all x (b) discontinuous at $x = 0$ only
    (c) continuous at $x = 0$ only   (d) discontinuous for all x

11. Let $f(x) = \dfrac{\sin x}{x}$, Then the value of $f(0)$ so that $f(x)$ becomes continuous at $x = 0$ is:
    (a) 0   (b) 1
    (c) e   (d) ∞

12. If the function $f(x)$ is defined on R by:
$$f(x) = \begin{cases} 1, & \text{when } x \text{ is rational} \\ -1, & \text{when } x \text{ is irrational} \end{cases}, \text{ then}$$

    (a) $f(x)$ is continuous at $x = 1$ only   (b) $f(x)$ is continuous at $x = -1$ only
    (c) $f(x)$ is continuous at $x = \pm 1$ only   (d) $f(x)$ is not continuous at any point.

13. $\lim\limits_{x \to 0} \dfrac{a^x - 1}{x}$, where $a > 0$, is equal to
    (a) 0   (b) 1
    (c) e^a (d) $\log a$

14. Which two of the functions $\sin x$, e^x and $\log x$ are continuous on **R**?
    (a) $\sin x$ and e^x (b) $\sin x$ and $\log x$
    (c) e^x and $\log x$ (d) none of these

15. Which of the following is continuous at $x = 0$?
    (a) $f(x) = 1/x$ (b) $f(x) = |x|/x$
    (c) $f(x) = |x|$ (d) $x = x/|x|$

16. Given the function
$$f(x) = \begin{cases} |x|, & \text{if } x \neq 0 \\ 1, & \text{if } x = 0 \end{cases}$$

    which of the following is incorrect?
    (a) $f(x)$ is continuous at $x = 0$ (b) $f(x)$ is discontinuous at $x = 0$
    (c) $f(x)$ is continuous at $x = 1$ (d) $f(x)$ is continuous in $R - \{0\}$

17. If $f(x) = [x]$ be the greatest integer function then $\lim\limits_{x \to 1} f(x)$ is equal to:
    (a) 0   (b) 1
    (c) 2   (d) does not exist

18. If $f(x) = \dfrac{\sin[x]}{[x]}$, $[x] \neq 0 = 0$, $[x] = 0$.

    where $[x]$ denotes the greatest integer less then or equal to x, then $\lim\limits_{x \to 0} f(x)$ equals to:
    (a) 1   (b) 0
    (c) -1 (d) None of these

19. If function $f(x) = (x+1)^{\cot x}$ is continuous at $x = 0$, $f(0)$ must be defined as:

    (a $f(0) = 0$ (b) $f(0) = e$

    (c) $f(0) = 1/e$ (d)   None of these.

20. $f(x) = \begin{cases} -1; & x \le -1 \\ -x; & -1 \le x \le 1 \\ 1; & x > 1 \end{cases}$

    is continuous:

    (a)   at $x = 1$ but not at $x = -1$ (b)   at $x = -1$ but not at $x = 1$

    (c)   at both $x = 1$ and $x = -1$ (d)   at none of $x = 1$ and $x = -1$.

ANSWERS						
1. (d)	**2.** (c)	**3.** (d)	**4.** (d)	**5.** (c)	**6.** (c)	**7.** (c)
8. (a)	**9.** (a)	**10.** (c)	**11.** (b)	**12.** (d)	**13.** (d)	**14.** (a)
15. (c)	**16.** (a)	**17.** (d)	**18.** (d)	**19.** (b)	**20.** (d)	

CHAPTER 4

DIFFERENTIATION

4.1 Introduction

Rate of change : The subject of Differential Calculus had its origin mainly in the geometrical problem of the determination of the *gradient* of a curve at a point thereof resulting in the determination of the tangent at the point. This subject has also rendered possible the precise formulations of a large number of physical concepts such as *Velocity at an instant, Acceleration at an instant, Curvature at a point, Density at a point, Specific heat at any temperature*, etc. each of which appears as a **Local or Instantaneous Rate of change** as against the **Average Rate of Change.**

The fundamental idea of *Local* or *Instantaneous* rate of change pervading all these concepts underlines the introduction of the notion of **'Derivative'.**

The ideas underlying Differential Calculus were first conceived by **Newton** (1643–1727) and **Leibnitz** (1646–1716).

4.1.1. Derivability and derivative. We consider a function f with some domain D. Let 'c' be an interior point of D.

We take another member $c + h$ of the domain which lies to the right or left of c (*i.e.* $c + h > c$ or $c + h < c$) according as h is positive or negative. The values of the function corresponding to c and $c + h$ are $f(c)$ and $f(c + h)$ respectively. Now, h, is the change in x and $f(c + h) - f(c)$ is corresponding change in $f(x)$.

The expression $[f(c + h) - f(c)]/h$, which is the ratio of these two changes denotes the average rate of change of the dependent variable $f(x)$ as the independent variable x changes from c to $c + h$. *It is possible that this ratio tends to a limit as h tends to 0. The limit, if it exists, is called the Derivative of f at c and denoted as $f'(c)$. Also the function f, then, is said to be Derivable at c.*

Definition. *A function f is said to be derivable at c if,*

$$\lim_{h \to 0} \frac{f(c + h) - f(c)}{h}$$

exists and the limit is called the **Derivative** of the function for c or at c and is denoted by $f'(c)$.

The function f is said to be finitely derivable at c if the derivative at c is finite.

Right hand and Left hand derivatives.

$$\lim_{h \to (0 + 0)} \frac{f(c + h) - f(c)}{h}$$

if it exists, is called the *Right hand derivative* at c and is denoted by $f'(c + 0)$ or $Rf'(c)$ Similarly,

$$\lim_{h \to (0-0)} \frac{f(c + h) - f(c)}{h},$$

if it exists, is called the *Left hand derivative at c* and is denoted by $f'(c - 0)$ or $Lf'(c.)$

Clearly f is derivable at an interior point if f' (c + 0) and f '(c – 0) both exist and are equal.

EXAMPLES

1. *Show that the function* $f(x) = x^2$ *is derivable for* $x = 1$.

Now $f(x) = x^2 \Rightarrow f(1) = 1^2 = 1.$

To find the derivative for, 1, we change x from 1 to $1 + h$ so that the function value changes from 1 to $(1 + h)^2$. We have

$$\frac{f(1 + h) - f(1)}{h} = \frac{2h + h^2}{h} = 2 + h, \text{ [as } h \neq 0]$$

$\Rightarrow$ $$\lim_{h \to 0} \frac{f(1 + h) - f(1)}{h} = 2 \Rightarrow f'(1) = 2.$$

Hence f is derivable at 1 and the derivative $f'(1)$ is 2.

2. *Show that* $f(x) = |x|$ *is not derivable at* 0.

If will be shown that $\lim [f(0 + h) - f(0)]/h$ when h tends to 0 does not exist. We have

$$f(h) = h \text{ if } h > 0, f(h) = -h \text{ if } h < 0.$$

$\Rightarrow$ $$\frac{f(0 + h) - f(0)}{h} = \frac{f(h)}{h} = 1 \text{ or } -1.$$

According as h is positive or negative

Thus $f'(0 + 0) = 1$ and $f'(0 - 0) = 1.$

Now $f'(0 + 0)$ and $f'(0 - 0)$ being *different*, it follows that $f'(0)$ does not exist and as such the function f is not derivable at 0.

3. *If* $f(x) = x \sin (1/x)$ *when* $x \neq 0$ *and* $f(0) = 0$, *show that f is continuous but not derivable for* $x = 0$.
                                                                                        (Osmania 2004)

We have

$$f(x) - f(0) = x \sin \frac{1}{x} - 0 = x \sin \frac{1}{x}$$

$\Rightarrow$ $$|f(x) - f(0)| = \left| x \sin \frac{1}{x} \right| = |x| \left| \sin \frac{1}{x} \right| \leq |x|.$$

Let ε be a given positive number. We have

$$|f(x) - f(0)| < \varepsilon \text{ when } |x - 0| < \varepsilon$$

$\Rightarrow$ $$\lim_{x \to 0} f(x) = 0 = f(0)$$

$\Rightarrow$ f is continuous at 0.

Again $$\frac{f(x) - f(0)}{x - 0} = \frac{x \sin(1/x)}{x} = \sin(1/x)$$

and, as seen in Ex. 3, p. 64 lim sin $(1/x)$ does not exist when $x \to 0$, Thus f is not derivable at 0.

Ex. Find the derivatives for the given values of x if

(i) $f(x) = 2x^2 + 3x - 4$ for $x = 5/2$.

(ii) $f(x) = 1/x$ for $x = 5$.

(iii) $f(x) = \sqrt{x}$ for $x = 1$.

(iv) $f(x) = 1/\sqrt{x}$ for $x = 1$.

ANSWERS

(i) 13; (ii) $-1/25$; (iii) $1/2$; (iv) $-1/2$.

4.1.2. Derived Function. While in 4.11, we have defined the derivability of a function at a particular value, we now consider the derivability of a function in its domain.

Let f be a function with $[a, b]$ as its domain. We suppose that f is finitely derivable at every point of $[a, b]$ and also that right-hand and left-hand derivatives exist finitely at a function denoted by f' such that $f'(x)$ denotes the derivative of f for $x \in [a, b]$

The function f' is called the *Derived function* or simply the *Derivative* of the function f. The derived function f' is given as follows :

$$\frac{dy}{dx} = f'(x), \qquad x \in [a, b]$$

Also we say that f is derivable in $[a, b]$.

EXAMPLES

1. *Find the derivative of f if $f(x) = x^2$.*

The domain of the function f is the entire set $\mathbf{R}$ of real numbers. Let $x \in \mathbf{R}$.

We have when $h \neq 0$

$$\frac{f(x + h) - f(x)}{h} = \frac{(x + h)^2 - x^2}{h} = (2x + h).$$

so that $\lim\limits_{h \to 0} \dfrac{f(x + h) - f(x)}{h}$ exists and equals $2x$.

Thus $\qquad\qquad f'(x) = 2x \; \forall \, x \in \mathbf{R}.$

2. *Find the derivative of f if $f(x) = \sqrt{x}$.*

The domain of the functon f is the set of all non-negative real numbers *i.e.*, the interval $[0, \infty]$:

Let $x > 0$. We have, when $h \neq 0$

$$\frac{f(x + h) - f(h)}{h} = \frac{\sqrt{(x + h)} - \sqrt{x}}{h}$$

$$= \left[\frac{\sqrt{(x + h)} - \sqrt{x}}{h} \right] \left[\frac{\sqrt{(x + h)} + \sqrt{x}}{\sqrt{(x + h)} + \sqrt{x}} \right]$$

$$= \left(\frac{1}{\sqrt{(x + h)} + \sqrt{x}} \right)$$

so that $\lim\limits_{h \to 0} \dfrac{f(x + h) - f(x)}{h} = \dfrac{1}{2\sqrt{x}}; x > 0$

Thus for $x > 0$, $f(x) = \sqrt{x} \Rightarrow f'(x) = 1/2\sqrt{x}$.

We start afresh to examine the existence of the derivative at 0.

We have

$$\frac{f(0 + h) - f(0)}{h} = \frac{\sqrt{h}}{h} = \frac{1}{\sqrt{h}} \to \infty$$

when $h \to 0$ through positive values ; $\sqrt{h}$ being not defined for negative values of h. We have

$$f'(0 + 0) = \infty$$

Thus $f'(x) = 1/2\sqrt{x},\ \forall x \in [0, \infty]$.

Ex. Find the derived function f' if $f(x)$ is given as follows :

    (i) $1/(x^2 + 3)$ (ii) $1/\sqrt{x}$

    (iii) x^3 (iv) $ax^2 + bx + c$

ANSWERS	
(i) $-2x/(x^2 + x^3)$	(ii) $-1/2x^{3/2}$;
(iii) $3x^2$;	(iv) $(2ax + b)$

4.1.3. Another notation. Consider a derivable function f. We write $f(x) = y$. Let x be a member of the domain of f.

It is usual to denote the change in x by Δx in place of h and the resulting change $f(x + \Delta x) - f(x)$ in the value of the function by Δy so that we have

$$\frac{f(x + \Delta x) - f(x)}{\Delta x} = \frac{\Delta y}{\Delta x}.$$

Thus, we have

$$\lim_{\Delta x \to 0} \frac{f(x + \Delta x) - f(x)}{\Delta x} = \lim_{\Delta x \to 0} \frac{\Delta y}{\Delta x}$$

$$\Rightarrow \qquad\qquad f'(x) = \lim_{\Delta x \to 0} \frac{\Delta y}{\Delta x}.$$

It us usual to denote $\lim\limits_{\Delta x \to 0} \dfrac{\Delta y}{\Delta x}$ by $\dfrac{dy}{dx}$.

Thus $f(x) = y \Rightarrow f'(x) = dy/dx$.

The derivative at c, is, in terms of this notation, is denoted as

$$\left(\frac{dy}{dx}\right)_{x=c} \text{ so that we have } f'(c) = \left(\frac{dy}{dx}\right)_{x=c}$$

Ex. (i) Find $(dy/dx)_{x=0}$ when $y = 2x/(x^2 + 1)$.

    (ii) If $y = \sqrt{(x^2 + 1)}$, find (dy/dx) when $x = -1$

ANSWERS	
(i) 2 ;	(ii) $-1/\sqrt{2}$

4.1.4. Derivability implying continuity

Theorem. *If f is finitely derivable at c, then f is also continuous at c.*

Let f be finitely derivable at c so that the expression   *(Banglore 2005, Poorvanchal 2004)*

$$[f(c+h)-f(c)]/h$$

tends to a finite limit as h tends to 0. We write

$$f(c+h) - f(c) = \frac{f(c+h) - f(c)}{h} \times h$$

$$\lim_{h \to 0}[f(c+h) - f(c)] = \lim_{h \to 0}\left[\frac{f(c+h) - f(c)}{h} \times h\right]$$

$$= \lim_{h \to 0}\frac{f(c+h) - f(c)}{h} \times \lim_{h \to 0}(h)$$

$$= f'(c) \times 0 = 0$$

$\Rightarrow \qquad \lim_{h \to c} f(x) = f(c) \Rightarrow f$ is continuous at c.

Thus, we have seen that f is finitely derivable at $c \Rightarrow f$ is continuous at c.

Cor. *If f is finitely derivable for every point of its domain then it is continuous in the domain.*

Note 1. If f is derivable at c, then the expression

$$[f(c+h)-f(c)]/h$$

has the **Indeterminate form** 0/0 when $h \to 0$.

Note 2. *The converse of this theorem is not necessarily* true, *i.e.*, a function may be continuous for a value of x without being derivable for that value. For example the function f where

$$f(x) = |x|$$

is continuous but not derivable for 0.

EXAMPLES

1. *Show that $f(x) = x|x|$ is derivable at the origin.*

   We have $\qquad\qquad f(x) = x^2 \text{ if } x \geq 0$

   $$= -x^2 \text{ if } x < 0$$

   $$Lf'(0) = \lim_{x \to 0+0}\frac{f(x) - f(0)}{x - 0} = \lim_{x \to 0+0}\frac{(-x^2)}{x} = 0$$

   $$Rf'(0) = \lim_{x \to 0+0}\frac{f(x) - f(0)}{x - 0} = \lim_{x \to 0+0}\frac{x^2}{x} = 0$$

   $\therefore \qquad\qquad Lf'(0) = Rf'(0)$

   $\Rightarrow$ f is derivable at the origin.

2. *Examine the function f where*

   $$f(x) = \frac{x(e^{-1/x} - e^{1/x})}{e^{-1/x} + e^{1/x}}, x \neq 0$$

   $$= 0, x = 0.$$

   as regards continuity and derivability at the origin

We have $f(0+0) = \lim\limits_{x \to 0+0} x\left(\dfrac{e^{-1/x} - e^{1/x}}{e^{-1/x} + e^{1/x}}\right)$

$$= \lim\limits_{x \to 0+0} x\left(\dfrac{e^{-2/x} - 1}{e^{-2/x} + 1}\right)$$

$$= \dfrac{0(0-1)}{0+1} = 0$$

$$f(0-0) = \lim\limits_{x \to 0-0} x\left(\dfrac{e^{-1/x} - e^{1/x}}{e^{-1/x} + e^{1/x}}\right)$$

$$= \lim\limits_{x \to 0-0} x\left(\dfrac{1 - e^{2/x}}{1 + e^{2/x}}\right)$$

$$= \dfrac{0(1-0)}{1+0} = 0$$

Also $f(0) = 0$

$\therefore$ $f(0+0) = f(0-0) = f(0)$

$\Rightarrow$ f is continuous at $x = 0$

Again $f'(0) = \lim\limits_{x \to 0} \dfrac{f(x) - f(0)}{x - 0}$

$$= \lim\limits_{x \to 0} \dfrac{f(x)}{x}$$

$$= \lim\limits_{x \to 0} \left(\dfrac{e^{-1/x} - e^{1/x}}{e^{-1/x} + e^{1/x}}\right)$$

But this limit does not exist as

$$\lim\limits_{x \to 0+0} \dfrac{e^{-1/x} - e^{1/x}}{e^{-1/x} + e^{1/x}} = \lim\limits_{x \to 0+0} \dfrac{e^{-2/x} - 1}{e^{-2/x} + 1}$$

$$= \dfrac{0 - 1}{0 + 1} = -1$$

and

$$\lim\limits_{x \to 0-0} \dfrac{e^{-1/x} - e^{1/x}}{e^{-1/x} + e^{1/x}} = \lim\limits_{x \to 0-0} \dfrac{1 - e^{2/x}}{1 + e^{2/x}}$$

$$= \dfrac{0 - 1}{0 + 1} = 1$$

Hence f is not derivable at $x = 0$

3. *If a function f is defined by*

$$f(x) = \dfrac{xe^{1/x}}{1 + e^{1/x}}, x \neq 0$$

$$= 0, x = 0$$

show that f is continuous but not derivable at x = 0   *(Devi Ahilya 2001)*

We have $\qquad f(0+0) = \lim\limits_{x \to 0+0} \dfrac{xe^{1/x}}{1+e^{1/x}}$

$$= \lim\limits_{x \to 0+0} \dfrac{x}{e^{-1/x}+1}$$

$$= 0$$

$$f(0-0) = \lim\limits_{x \to 0-0} \dfrac{xe^{1/x}}{1+e^{1/x}}$$

$$= 0$$

Also $\qquad f(0) = 0$

$\therefore \qquad f(0+0) = f(0-0) = f(0)$

$\Rightarrow \qquad f$ is continuous at $x = 0$

Again $\qquad f'(0+0) = \lim\limits_{x \to 0+0} \dfrac{f(x) - f(0)}{x - 0}$

$$= \lim\limits_{x \to 0+0} \dfrac{\dfrac{xe^{1/x}}{1+e^{1/x}} - 0}{x}$$

$$= \lim\limits_{x \to 0+0} \dfrac{e^{1/x}}{1+e^{1/x}}$$

$$= \lim\limits_{x \to 0+0} \dfrac{1}{e^{-1/x}+1}$$

$$= 1$$

$$f'(0-0) = \lim\limits_{x \to 0-0} \dfrac{f(x) - f(0)}{x - 0}$$

$$= \lim\limits_{x \to 0-0} \dfrac{\dfrac{xe^{1/x}}{1+e^{1/x}} - 0}{x}$$

$$= \lim\limits_{x \to 0-0} \dfrac{e^{1/x}}{1+e^{1/x}}$$

$$= 0$$

Since $f'(0+0) \neq f'(0-0)$, the derivate of $f(x)$ at $x = 0$ does not exist.

4. *Show that the function f defined by*
$$f(x) = |x-2| + |x| + |x+2|$$
is not derivable at $x = -2, 0$ *and* 2.

We have

$$f(x) = \begin{cases} -(x-2) - x - (x+2) &= -3x, & x > -2 \\ -(x-2) - x + (x+2) &= -x+4, & -2 \leq x < 0 \\ -(x-2) + x + (x+2) &= x+4, & 0 \leq x < 2 \\ (x-2) + x + (x+2) &= 3x, & x \geq 2 \end{cases}$$

$$f'(-2+0) = \lim_{x \to -2+0} \frac{f(x) - f(-2)}{x+2}$$

$$= \lim_{x \to -2+0} \frac{-x+4-6}{x+2} = \lim_{x \to -2+0} \frac{-(x+2)}{x+2}$$

$$= -1$$

$$f'(-2-0) = \lim_{x \to -2-0} \frac{f(x) - f(2)}{x+2}$$

$$= \lim_{x \to -2-0} \frac{-3x-6}{x+2}$$

$$= -3$$

$\therefore \qquad f'(-2+0) \neq f'(-2-0)$

$\Rightarrow$ f is not derivable at $x = -2$

Again, $\qquad f'(0+0) = \lim_{x \to 0+0} \frac{f(x) - f(0)}{x-0}$

$$= \lim_{x \to +0} \frac{x+4-4}{x}$$

$$= 1$$

$$f'(0-0) = \lim_{x \to 0-0} \frac{f(x) - f(0)}{x-0}$$

$$= \lim_{x \to 0-0} \frac{-x+4-4}{x}$$

$$= -1$$

$\therefore \qquad f'(0+0) \neq f'(0-0)$

$\Rightarrow$ f is not derivable at $x = 0$

Also $\qquad f'(2+0) = \lim_{x \to 2+0} \frac{f(x) - f(2)}{x-2}$

$$= \lim_{x \to 2+0} \frac{3x-6}{x-2}$$

$$= 3$$

$$f'(2-0) = \lim_{x \to 2-0} \frac{f(x) - f(2)}{x-2}$$

$$= \lim_{x \to 2-0} \frac{x+4-6}{x-2}$$

$$= 1$$

$\therefore \qquad f'(2+0) \neq f'(2-0)$

$\Rightarrow$ f is not derivable at $x = 2$

5. *Discuss the derivability of the functon*

$$f(x) = \begin{cases} x & , & x < 1 \\ 2 - x & , & 1 \le x \le 2 \\ -2 + 3x - x^2, & , & x > 2 \end{cases}$$

at $\qquad x = 1, 2$

We have $\qquad f'(1 + 0) = \lim\limits_{x \to 1 + 0} \dfrac{f(x) - f(1)}{x - 1}$

$$= \lim\limits_{x \to 1 + 0} \dfrac{2 - x - 1}{x - 1}$$

$$= \lim\limits_{x \to 1 + 0} \dfrac{1 - x}{x - 1}$$

$$= -1$$

$$f'(1 - 0) = \lim\limits_{x \to 1 - 0} \dfrac{f(x) - f(1)}{x - 1}$$

$$= \lim\limits_{x \to 1 - 0} \dfrac{x - 1}{x - 1}$$

$$= 1$$

$\therefore \qquad f'(1 + 0) \ne f'(1 - 0)$

$\Rightarrow$ f is not derivable at $x = 1$

Again $\qquad f'(2 + 0) = \lim\limits_{x \to 2 + 0} \dfrac{f(x) - f(2)}{x - 2}$

$$= \lim\limits_{x \to 2 + 0} \dfrac{-2 + 3x - x^2 - 0}{x - 2}$$

$$= \lim\limits_{x \to 2 + 0} \dfrac{(x - 2)(1 - x)}{x - 2}$$

$$= -1$$

$$f'(2 - 0) = \lim\limits_{x \to 2 - 0} \dfrac{f(x) - f(2)}{x - 2}$$

$$= \lim\limits_{x \to 2 - 0} \dfrac{2 - x - 0}{x - 2}$$

$$= -1$$

$$f'(2 + 0) = f'(2 - 0)$$

$\Rightarrow$ f is derivable at $x = 2$.

6. *If* $f(x) = x^2 \sin(1/x)$ *when* $x \ne 0$ *and* $f(0)$, *show that* f *is derivable for every value of* x *but the derivative is not continuous for* $x = 0$. *(Calicut 2004)*

For $\qquad x \ne 0, \; f'(x) = 2x \sin \dfrac{1}{x} + x^2 \cos\left(\dfrac{1}{x}\right)\left(-\dfrac{1}{x^2}\right)$

$$= 2x \sin \dfrac{1}{x} - \cos \dfrac{1}{x}.$$

For $x = 0$, we have

$$\frac{f(x) - f(0)}{x - 0} = \frac{x^2 \sin \dfrac{1}{x}}{x} = x \sin \frac{1}{x} \Rightarrow f'(0) = 0.$$

Thus the function possesses a derivative for every value of x given by

$$f'(x) = 2x \sin \frac{1}{x} - \cos \frac{1}{x} \text{ when } x \neq 0, f'(0) = 0.$$

We now show that f' is not continuous for $x = 0$.

We write

$$\cos \frac{1}{x} = 2x \sin \frac{1}{x} - \left(2x \sin \frac{1}{x} - \cos \frac{1}{x} \right) \qquad \qquad \text{...(i)}$$

Here $\lim\limits_{x \to 0} \left(2x \sin \dfrac{1}{x} \right) = 0.$

In case $\lim\limits_{x \to 0} f'(x)$, i.e., $\lim\limits_{x \to 0} \left(2x \sin \dfrac{1}{x} - \cos \dfrac{1}{x} \right)$ had existed, it would follow from (i) that

$\lim\limits_{x \to 0} \left(\cos \dfrac{1}{x} \right)$ would also exist. But this is not the case. Hence $\lim\limits_{x \to 0} f'(x)$ does not exist so that f' is
not continuous for $x = 0$.

7. *Examine the continuity and derivability in the interval* $]-\infty, \infty[$ *of the function defined as
follows* :

$$f(x) = 1 \text{ in }]-\infty, 0[$$

$$f(x) = 1 + \sin x \text{ in } \left[0, \frac{1}{2}\pi \right[$$

$$f(x) = 2 + \left(x - \frac{1}{2}\pi \right)^2 \text{ in } \left[\frac{1}{2}\pi, \infty \right[.$$

The function f is derivable for every value of x except perhaps for $x = 0$ and $x = \pi/2$.

1. Firstly we consider $x = 0$.

Now $f(0) = 1 + \sin 0 = 1$

$$\lim\limits_{x \to (0-0)} f(x) = 1 \text{ and } \lim\limits_{x \to (0+0)} f(x) = \lim\limits_{x \to (0+0)} (1 + \sin x) = 1$$

$\therefore$ $\lim\limits_{x \to 0} f(x) = 1 = f(0).$

Hence f is continuous for $x = 0$.

Also $x < 0$

$\Rightarrow$ $\quad \dfrac{f(x) - f(0)}{x - 0} = \dfrac{1 - 1}{x.} = 0 \Rightarrow \lim\limits_{x \to (0-0)} \dfrac{f(x) - f(x)}{x - 0} = 0$

Again $x > 0 \Rightarrow \dfrac{f(x) - f(0)}{x - 0} = \dfrac{1 + \sin x - 1}{x - 0} = \dfrac{\sin x}{x}$

$$\Rightarrow \quad \lim_{x \to (0+0)} \frac{f(x) - f(0)}{x - 0} = \lim_{x \to (0+0)} \frac{\sin x}{x} = 1.$$

Thus
$$\lim_{x \to (0+0)} \frac{f(x) - f(0)}{x - 0} \neq \lim_{x \to (0-0)} \frac{f(x) - f(0)}{x - 0}.$$

Hence the function is *not* derivable for $x = 0$.

II. Now, we consider $x = \pi/2$. We have

$$f(\pi/2) = 2 + \left(\frac{1}{2}\pi - \frac{1}{2}\pi\right)^2 = 2$$

$$\lim_{x \to \left(\frac{1}{2}\pi - 0\right)} f(x) = \lim_{x \to \left(\frac{1}{2}\pi - 0\right)} (1 + \sin x) = 1 + 1 = 2,$$

$$\lim_{x \to \left(\frac{1}{2}\pi - 0\right)} f(x) = \lim_{x \to \left(\frac{1}{2}\pi - 0\right)} \left[2 + \left(x - \frac{1}{2}\pi\right)^2\right] = 2$$

$$\therefore \quad \lim_{x \to \pi/2} f(x) = 2 = f(\pi/2).$$

Hence, f is continuous for $x = \pi/2$.

Again, for $0 \le x < \frac{1}{2}\pi$, we have

$$\frac{f(x) - f\left(\frac{1}{2}\pi\right)}{x - \frac{1}{2}\pi} = \frac{(1 + \sin x) - 2}{x - \frac{1}{2}\pi} = \frac{1 - \sin x}{\frac{1}{2}\pi - x}.$$

Putting $\frac{1}{2}\pi - x = t$, we see that

$$\frac{1 - \sin x}{\frac{1}{2}\pi - x} = \frac{1 - \sin\left(\frac{1}{2}\pi - t\right)}{t}$$

$$= \frac{1 - \cot t}{t} = \frac{2\sin^2 \frac{1}{2}t}{t} = \sin\frac{1}{2}t \frac{\sin\frac{1}{2}t}{\frac{1}{2}t}$$

$$\Rightarrow \lim_{x \to \left(\frac{1}{2}\pi - 0\right)} \frac{1 - \sin x}{\frac{1}{2}\pi - x} = 0.$$

For $x > \frac{1}{2}\pi$

$$\frac{f(x) - f\left(\frac{1}{2}\pi\right)}{x - \frac{1}{2}\pi} = \frac{2 + \left(x - \frac{1}{2}\pi\right)^2 - 2}{x - \frac{1}{2}\pi} = x - \frac{1}{2}\pi$$

$$\Rightarrow \quad \lim_{x \to \left(\frac{1}{2}\pi + 0\right)} \frac{f(x) - f\left(\frac{1}{2}\pi\right)}{x - \frac{1}{2}\pi} = 0.$$

$\therefore \quad f'\left(\frac{1}{2}\pi\right)$ exists and is equal to 0.

8. If $f(x) = \begin{cases} x\, exp\left[-\left(\dfrac{1}{|x|} + \dfrac{1}{x}\right)\right], & x \neq 0 \\ 0, & x = 0 \end{cases}$ then test whether (i) $f(x)$ is continuous at $x = 0$ (ii)

$f(x)$ is differentiable at $x = 0$.

We have

$$f(x) = \begin{cases} x e^{-\left\{\frac{1}{x} + \frac{1}{x}\right\}}, & x > 0 \\ x e^{-\left\{\frac{1}{-x} + \frac{1}{x}\right\}}, & x < 0 \\ 0, & x = 0 \end{cases}$$

$$\Rightarrow \qquad f(x) = \begin{cases} x e^{-2/x}, & x > 0 \\ x, & x < 0 \\ 0, & x = 0 \end{cases}$$

We have $\qquad f(0 + 0) = \lim_{x \to 0+0} x e^{-2/x}$

$$= \lim_{h \to 0} \frac{h}{e^{2/h}} = 0$$

$$f(0 - 0) = \lim_{x \to 0-0} x = \lim_{h \to 0} (0 - h) = 0$$

and $\qquad\qquad f(0) = 0$

$\therefore \quad f(x)$ is continuous at $x = 0$.

Again, $\qquad f'(0 - 0) = \lim_{h \to 0} \dfrac{f(0 - h) - f(0)}{-h}, h > 0$

$$= \lim_{h \to 0} \frac{-h - 0}{-h} = 1$$

$$f'(0 + 0) = \lim_{h \to 0} \frac{f(0 + h) - f(0)}{h}$$

$$= \lim_{h \to 0} \frac{h e^{-2/h} - 0}{h} = e^{-\infty} = 0$$

$\Rightarrow \ f'(0 - 0) \neq f'(0 + 0)$

So, $f(x)$ is not differentiable at $x = 0$.

9. *If $f(x) = [n + p \sin x]$, $x \in (0, \pi)$, $n \in z$ and p is a prime number, where [.] denotes the greatest integer function, then find the number of points where $f(x)$ is not differentiable.*

We have,

$f(x) = [n + p \sin x]$ is not differentiable at those points where $n + p \sin x$ is integer. As p is a prime number

$\Rightarrow$ $n + p \sin x$ is an integer if $\sin x = 1, -1, r/p$

i.e. $x = \pi/2, -\pi/2, \sin^{-1} \dfrac{r}{p}, \pi - \sin^{-1} r/p$

But $x \neq -\dfrac{\pi}{2}, 0$.

$\therefore$ Function is not differentiable at

$$x = \frac{\pi}{2}, \quad \sin^{-1}\frac{r}{p}, \quad \pi - \sin^{-1}\frac{r}{p}.$$

where $0 < r \leq p - 1$.

So, the required number of points are

$$= 1 + 2(p - 1) = 2p - 1.$$

10. If $f(x) = \begin{cases} |1 - 4x^2|, & 0 \leq x < 1 \\ [x^2 - 2x], & 1 \leq x < 2, \end{cases}$ *Where [.] denotes greatest integral function. Discuss*

the differentiability of $f(x)$ in $[0, 2]$

As $1 \leq x < 2 \Rightarrow -1 \leq x^2 - 2x < 0$

$\Rightarrow$ $[x^2 - 2x] = -1$ when $1 \leq x < 2$.

$\therefore$ $f(x) = \begin{cases} 1 - 4x^2 & , \quad 0 \leq x < 1/2 \\ 4x^2 - 1 & , \quad 1/2 \leq x < 1 \\ -1 & , \quad 1 \leq x < 2. \end{cases}$

$\Rightarrow$ $f'(x) = \begin{cases} -8x & , \quad 0 \leq x < 1/2 \\ 8x & , \quad 1/2 \leq x < 1 \\ 0 & , \quad 1 \leq x < 2. \end{cases}$

$\Rightarrow$ $f'(x) = \begin{cases} -4 & x < 1/2 \\ 4 & x > 1/2 \end{cases}$ and

          $f'(x) = \begin{cases} 8, & x > 1 \\ 0, & x > 1 \end{cases}$

$\therefore$ $f'\left(\dfrac{1}{2} + 0\right) = 4, \qquad f'\left(\dfrac{1}{2} - 0\right) = -4$

and $f'(1 + 0) = 0, f'(1 - 0) = 8$

$\therefore$ $f(x)$ is differentiable, $\forall x \in [0, 2] - \{1/2, 1\}$.

EXERCISES

1. Show that if $f(x) = |x| + |x - 1|$, the function f is continuous for every value of x but not derivable for $x = 0$ and $x = 1$.

2. Construct a function which is continuous in $[1, 5]$ but not derivable at 2, 3, 4.

3. Show that $f(x) = |x|, x \in \mathbf{R}$ is continuous at $x = -1$ but not derivable at $x = -1$.

4. Determine whether f is continuous and has a derivative at the origin where
$$f(x) = \begin{cases} 2 + x & \text{if } x \geq 0 \\ 2 - x & \text{if } x < 0 \end{cases}$$

5. Discuss the continuity and derivability of the function
$$f(x) = \begin{cases} 1 + x & , & x \leq 0 \\ x & , & 0 < x < 1 \\ 2 - x & , & 1 \leq x \leq 2 \\ 3x - x^2 & , & x > 2 \end{cases}$$
   at $x = 0, 1$ and 2.

6. Show that the function defined by
$$f(x) = \begin{cases} \dfrac{x}{1 + e^{1/x}} & , & x \neq 0 \\ 0 & , & x = 0 \end{cases}$$
   is continuous at $x = 0$ but not derivable at that point.

7. Discuss the continuity and the derivable of the function f where
$$f(x) = \begin{cases} 0, & \text{when } x \text{ is irrational or zero,} \\ \dfrac{1}{q^3}, & \text{when } x = \dfrac{p}{q}, \text{ a fraction in the lowest terms.} \end{cases}$$

8. Find the derivative of f where
$$f(x) = \frac{\sin x^2}{x}, \text{ when } x \neq 0, \text{ and } f(0) = 0$$
   and show that the derivative is continuous at $x = 0$.

9. Show that the function
$$f(x) = x\left\{1 + \frac{1}{3}\sin(\log x^2)\right\} \text{ ; for } x \neq 0, f(0) = 0$$
   is everywhere continuous but has no derivative at $x = 0$.

10. Show that the function defined as
$$f(x) = x^m \sin\frac{1}{x}, x \neq 0, \ m \text{ is a } + ve \text{ integer}$$
   is continuous at $x = 0$. Examine its derivability. Determine m when f' is continuous at $x = 0$.

11. If $f(x) = x \tan^{-1}(1/x)$ when $x \neq 0$ and $f(0) = 0$, show that f is continuous but not derivable for $x = 0$.

12. A function f is defined as
$$f(x) = \begin{cases} \dfrac{1}{2}(b^2 - a^2) & \text{for } 0 \leq x \leq a \\ \dfrac{1}{2}b^2 - \dfrac{x^2}{6} - \dfrac{a^3}{3x} & \text{for } a < x \leq b \\ \dfrac{1}{3}\left(\dfrac{b^3 - a^3}{x}\right) & \text{for } x > b. \end{cases}$$

Prove that f and f' are continuous but f'' is discontinuous.

13. Examine for continuity at $x = a$ the function f where

$$f(x) = \begin{cases} \dfrac{x^2}{a} - a, & 0 < x < a \\ 0, & x = a \\ a - \dfrac{a^3}{x^2}, & a > x. \end{cases}$$

   Also examine if the function is derivable at a.

14. Is the function f where

$$f(x) = (x - a)\sin\frac{1}{x - a} \text{ for } x \neq a, f(a) = 0$$

continuous and derivable at $x = a$?

Give your answer with reasons.

15. Discuss the continuity of f in the neighbourhood of the origin when $f(x)$ is defined as follows :
   (i) $f(x) = x \log \sin x$ for $x \neq 0$ and $f(0) = 0$.
   (ii) $f(x) = e^{1/x}$ when $x \neq 0$ and $f(0) = 0$.

16. $f(x)$ is defined as being equal to $-x^2$, when $x \leq 0$, to $5x - 4$ when $0 < x \leq 1$, to $4x^2 - 3x$ when $1 < x < 2$ and to $3x + 4$ when $x \geq 2$. Discuss the existence of f' for $x = 0$, 1 and 2.

17. Show that the function $f(x) = 2|x - 2| + 5|x - 3| \,\forall\, x \in \mathbf{R}$ is not derivable at $x = 2$ and $x = 3$.

ANSWERS

4. f is continuous but not derivable at $x = 0$.

5. f is continuous at 0, 1 but not derivable. Discontinuous at 2.

7. continuous when x is irrational or zero and discontinuous for other values of x. Differentiable for no value.

8. $f'(0) = 0$.

13. continuous and derivable at $x = a$.

14. continuous but not derivable at $x = a$.

15. (i) continuous (ii) discontinuous.

16. discontinuous at 0; derivable at 1; continuous but not derivable at 2.

4.1.5. Geometrical interpretation of a derivative. *To show that $f'(c)$, is the tangent of the angle which the tangent line to the curve $y = f(x)$ at the point $P\,[c, f(c)]$ makes with x-axis.*

   We take two points $P\,[c, f(c)]$ and $Q\,[c + h, f(c + h)]$ on the curve $y = f(x)$.

   Draw the ordinates PL, QM and draw $PN \perp MQ$. We have

$$PN = LM = h$$

and $$NQ = MQ - LP$$

$$= f(c + h) - f(c)$$

$$\therefore \quad \tan \angle XRQ = \tan \angle NPQ$$

$$= \frac{NQ}{PN}$$

$$= \frac{f(c + h) - f(c)}{h} \qquad \qquad ...(i)$$

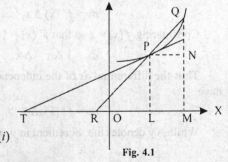

Fig. 4.1

Here, $\angle XRQ$ is the angle which the chord PQ of the curve makes with x-axis.

As h approcahes 0, the point Q moving along the curve approaches the point P, the chord PQ approaches the tangent line TP and $\angle XRQ$ approaches $\angle XTP$ which we denote by ψ. On taking limits, the equation (i) give

$$\tan \psi = f'(c).$$

Thus $f'(c)$ is the slope of the tangent to the curve $y = f(x)$ at the point $P[c, f(c)]$.

The slope of the tangent at a point of a curve is also known as the **Gradient** of the curve at the point.

Cor. *The equation of the tangent at the point $P[c, f(c)]$ of the curve $y = f(x)$ is*

$$Y - f(c) = f'(c)(X - c).$$

We shall in the following chapter 5 deal with detailed discussion about tangents to curves.

Note. The student should note that it is not necessary for every curve to have a tangent line at every point thereof. The existence of the tangent demands the existence of the derivative and we have seen in many examples that every function is not derivable for every value of x.

For example, the curve $y = |x|$ does not possess any tangent at $(0, 0)$. This fact may also be seen directly from the graph Fig. 2.5, p. 16.

EXERCISES

1. Find the slopes of the tangens to the parabola $y = x^2$ points $(2, 4)$ and $(-1, 1)$.

2. Show that the tangent to the hyperbola $y = 1/x$ at $(1, 1)$ makes an angle $3\pi/4$ with x-axis.

ANSWERS

1. $4, -2$.

4.1.6. Differentials. Differential co-efficient. Let a function f be derivable in $[a, b]$.

Let $x \in [1, b]$

We write $y = f(x)$.

Now when $\Delta x \to 0, \lim \dfrac{\Delta y}{\Delta x} = f'(x)$.

so that $\Delta y/\Delta x$ differs from $f'(x)$ by a variable which tends to zero when $x \to 0$.

We write $\dfrac{\Delta y}{\Delta x} = f'(x) + \alpha \Rightarrow \Delta y = f'(x)\,\Delta x + \alpha\,\Delta x.$

The change Δy in y consists of two parts, viz $f'(x)\,\Delta x$ and $\alpha\,\Delta x$. So far as contribution of these two parts to Δy is concerned that of $f'(x)\,\Delta x$ is more significant than that of $\alpha\,\Delta x$ in that $\alpha\,\Delta x$ is the product of two variables α and Δx each of which tends to 0. The part $f'\,\Delta x$ is called the **Differential** of y and is denoted by dy so that we have

$$dy = f'(x)\,\Delta x.$$

Considering $f(x) = x$ so that $f'(x) = 1$ we obtain

$$dx = 1\,\Delta x = \Delta x.$$

Thus the differential dx of the independent variable x is the same as the increment Δx. We thus have

$$dy = f'(x)\,dx.$$

While Δy denotes the increment in y, dy stands for the differential of y.

The derivative $f'(x)$ being the co-efficient of the differential dx is also known as **Differential co-efficient.**

The process of finding the derivative of a function is also called **Differentiation**.

■■■ EXAMPLES ■■■

1. *Find the differential dy and the increment Δy when $y = x^3$ for*

   (*i*) *arbitrary values of x and Δx.*

   (*ii*) $x = 10, \Delta x = 0.1$.

   We have $y + \Delta y = (x + \Delta x)^3$

   $$= x^3 + 3x^2 \Delta x + 3x (\Delta x)^2 + (\Delta x)^3$$

   $$\Delta y = 3x^2 \Delta x + 3x (\Delta x)^2 + (\Delta x)^3$$

   $$= 3x^2 \Delta x + [3x \Delta x + (\Delta x)^2] \Delta x.$$

   Here, $\alpha = 3x \Delta x + (\Delta x)^2$. We have

   $$dy = 3x^2 \, dx.$$

   For $x = 10$ and $\Delta x = 0.1$, we have

   $$3x^2 \, dx = 300 \,(0.1) = 30 = dy$$

   $$\Delta y = 30 + [30 \times 0.1 + (0.1)^2]\,(0.1)$$

   $$= 30 + .301 = 30.301.$$

   Thus considering dy in place of Δy, we have an error of .301.

2. Find dy in the following cases :

   (*i*) $y = x^2 + x^3$; for arbitrary x and dx.

   (*ii*) $y = 2x^2 - 2x$; $x = 1, dx = 0.01$.

   (*iii*) $y = \cos x$; $x = \pi/2, dx = 0.1$.

■■■ ANSWERS ■■■

2. (*i*) $(2x + 3x^2)\,dx$ (*ii*) .02   (*iii*) -0.1

Geometrical interpretation of the differential. From Fig. 4.2 it is clear that $\Delta x = PN$, $\Delta y = NQ$ and $dy = NR$, so that the differential of a function for given values of x and Δx is equal to the increment in the ordinate of the tangent to the curve $y = f(x)$ at $[x, f(x)]$.

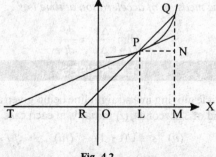

Fig. 4.2

4.1.7. Kinematical interpretation of a derivative. Every thinking person is aware of the concepts of *Velocity and Acceleration* of a moving point. The difficulty, however, arises in assigning *precise* measures to them. In practice, velocity at an instant is calculated by measuring the distance travelled in *some* short interval of time subsequent to the instant under consideration. This manner of calculating the velocity cannot clearly be considered precise, for different measuring agents may employ different intervals for the purpose. In fact this is only an approximate value of the actual velocity and some approximate value is all that we need in practice. The smaller the interval, the better is, of course, the approximation to the actual velocity.

In books not employing the method of Differential Calculus, velocity at an instant, is defined as the distance travelled in an *Infinitesimal* interval subsequent to the instant. Now there exists no such thing as an infinitesimal interval of time. We can take intervals of time as small as we like and, in fact, an interval with duration smaller than any other is conceivable. The definition, as it stands, is thus meaningless. A meaning can, however, be attached to the above definition by supposing that the words 'velocity' and 'infinitesimal' in it really stand for approximation to the 'velocity' and 'some short interval of time', respectively.

The precise meaning of the velocity of a moving particle at any instant can only be given by employing the notion of Derivative which we do in the following.

Expression for velocity. The motion of a particle along a straight line is analytically represented by an equation of the form

$$s = f(t).$$

Here 's' represents the distance of the particle measured from some fixed point O on the line at time t. Let P be the position of the particle at any time t. Let again Q be its position after some interval Δt and let $PQ = \Delta s$.

The ratio '$\Delta s/\Delta t$' is the average velocity over this interval and is an approximation to the actual velocity at P. We know intuitively that better approximations can be obtained by considering smaller values of Δt.

We are thus led to define the measure v of the velocity at time t as

$$\lim_{\Delta t \to 0} \frac{f(t + \Delta t) - f(t)}{\Delta t} = \lim_{\Delta t \to 0} \frac{\Delta s}{\Delta t} = \frac{ds}{dt}.$$

Expression for acceleration. Let v be the velocity at time 't' and let Δv be the velocity after some short interval of time Δt.

The ratio $\Delta v/\Delta t$ is the average acceleration during this interval Δt and is an approximation to the actual acceleration at time t.

The smaller values of Δt will correspond to better approximations for the acceleration at time t. We are thus led to *define the measure of acceleration at time t as*

$$\lim_{\Delta t \to 0} \frac{\Delta v}{\Delta t} = \frac{dv}{dt}.$$

EXERCISES

1. The motion of a particle moving in a straight line being given as follows, find the velocity and acceleration (*i*) at end of 3 seeconds, (*i*) initially, in each case :-

    (*i*) $s = t^2 + 2t + 3$ (*ii*) $s = 1/(t + 1)$ (*iii*) $s = \sqrt{(t + 1)}$.

2. A particle moves along a straight line such that 's' is a quadratic function of t; prove that its acceleration remains constant.

3. If $s = t^3 - 2t^2 + 3t - 4$, give the position, velocity and acceleration of the particle at the end of 0, 1, 2, seconds.

Note. The remaining part of this chapter is devoted to developing methods for the determination of the derivatives of the functions. Before, however, taking up some general theorems in § 4.3, we obtain the derivatives of a constant function and of the function $f(x) = x^n$ where n is a natural number.

4.2. Derivative of a constant function

We consider the constant function

$$y = c \; \forall \, x \in \mathbf{R}$$

where c is a given number.

We have

$$\frac{\Delta y}{\Delta x} = \frac{0}{\Delta x} = 0$$

$$\Rightarrow \qquad \frac{dy}{dx} = \lim \frac{\Delta y}{\Delta x} = 0 \; \forall \; x \in \mathbf{R}.$$

Thus, the derivative of a constant function is the zero function which is also a constant function.

Note. Looking at the *derivative as the rate of change*, this result appears almost intuitive in that the rate of change of anything which does not change is necessarily zero.

The result may also be geometrically inferred from the fact that the slope of the tangent at any point of the curve $y = c$, which is a straight line parallel to *x-axis*, is 0.

4.2.1. Derivative of $y = x^n$ where n is a positive integer.

The domain of the function $y = x^n$, $n \in \mathbf{N}$ is $\mathbf{R}$. Let x denote any given real number. Let Δy be the change in y corresponding to the change Δx in x.

We have

$$\frac{\Delta y}{\Delta x} = \frac{(x + \Delta x)^n \, (-x^n)}{\Delta x}, \; \Delta x \neq 0$$

$$= nx^{n-1} + n\,(n-1)\,x^{n-2}\,\Delta x + \ldots\ldots\ldots + (\Delta x)^{n-1}$$

$$\Rightarrow \qquad \frac{dy}{dx} = nx^{n-1}$$

$$\Rightarrow \qquad \frac{d(x^n)}{dx} = nx^{n-1}, \qquad \forall \, x \in \mathbf{R}, \; \text{and} \; \forall \, n \in \mathbf{N}.$$

It follows that when $n \in \mathbf{N}$, the function $y = x^n$ is derivable at every point of its domain $\mathbf{R}$ and its derived function is nx^{n-1}.

Ex. Give the derivatives of the functions

  (i) $y = x^3$.   *(ii)* $y = x^8$.

4.3. Some general theorems on Derivation

4.3.1. Derivaive of the sum or difference of two functions : Let u and v be two derivable functions of x.

We write

$$y = u + v \qquad \qquad \ldots(i)$$

Let $\Delta u, \Delta v, \Delta y$ be the respective changes in u, v, y corresponding to the change Δx in x so that we have

$$y + \Delta y = u + \Delta u + v + \Delta v \qquad \qquad \ldots(ii)$$

From (i) and (ii) we get

$$\frac{\Delta y}{\Delta x} = \frac{\Delta u}{\Delta x} + \frac{\Delta v}{\Delta x}$$

so that when $\Delta x \to 0$, we obtain

$$\lim \frac{\Delta y}{\Delta x} = \lim \left(\frac{\Delta u}{\Delta x} + \frac{\Delta v}{\Delta x} \right) = \lim \frac{\Delta u}{\Delta x} + \lim \frac{\Delta v}{\Delta x}.$$

$$\Rightarrow \qquad \frac{dy}{dx} = \frac{du}{dx} + \frac{dv}{dx}.$$

We may similarly prove that

$$\frac{d(u - v)}{dx} = \frac{du}{dx} - \frac{dv}{dx}.$$

The above results may also be seen to be true in the case of any finite number of functions.

Ex. Give the derivatives of the functions with the function values $x^4 + x^7,\ x^5 - x^3 + x,\ x^7 + x^4 - x^3$

ANSWERS

$$4x^3 + 7x^6,\ 5x^4 - 3x^2 + 1,\ 7x^6 + 4x^3 - 3x^2$$

4.3.2. Derivative of the product of two functions. Let u, v be two derivable functions of x.

We write

$$y = uv.$$

If $\Delta u, \Delta v, \Delta y$ be changes in u, v, y respectively corresponding to the change Δx in x, we have

$$y + \Delta y = (u + \Delta u)(v + \Delta v)$$

$$= uv + u\Delta v + v\Delta u + \Delta u.\,\Delta v$$

$$\Rightarrow \qquad \Delta y = u\Delta v + v\Delta u + \Delta u.\,\Delta v$$

$$\Rightarrow \qquad \frac{\Delta y}{\Delta x} = u\frac{\Delta v}{\Delta x} + v\frac{\Delta u}{\Delta x} + \Delta u.\frac{\Delta v}{\Delta x}.$$

Let $\Delta x \to 0$. Then Δu also $\to 0$, for u, being a derivable function, is continuous. We have

$$\frac{dy}{dx} = \lim_{\Delta x \to 0} \left[u\frac{\Delta v}{\Delta x} + v\frac{\Delta u}{\Delta x} + \Delta u.\frac{\Delta v}{\Delta x} \right]$$

$$= \lim \left(u\frac{\Delta v}{\Delta x} \right) + \lim \left(v\frac{\Delta u}{\Delta x} \right) + \lim \Delta u.\lim \frac{\Delta v}{\Delta x}$$

$$= u\frac{dv}{dx} + v\frac{du}{dx} + 0\frac{dv}{dx} = u\frac{dv}{dx} + v\frac{du}{dx}$$

$$\Rightarrow \qquad \frac{d\,(uv)}{dx} = u\frac{dv}{dx} + v\frac{du}{dx}.$$

Thus we have the theorem.

The product of two derivable fucntions is itself derivable and the derivative of the product of two functions = first function × derivative of the second + second function × derivative of the first.

Ex. Show that $\qquad y = u_1\, u_2\, u_3$

$$\Rightarrow \qquad \frac{dy}{dx} = u_1 u_2 \frac{du_3}{dx} + u_1 u_3 \frac{du_2}{dx} + u_2 u_3 \frac{du_1}{dx}.$$

The result can be easily extended to the product of any *finite* number of derivable functions.

4.3.3. Derivative of $y = cu$ where c is constant and u, is any derivable function of x.

Now

$$y = cu$$

$$\Rightarrow \qquad \frac{dy}{dx} = c\frac{du}{dx} + u\frac{dc}{dx} = \left(c\,\frac{du}{dx} + u\,.0 \right) = \frac{du}{dx}$$

$$\Rightarrow \qquad \frac{d\,(cu)}{dx} = c\frac{du}{dx}.$$

Ex. Give the derivatives of functions with function values

    (i) $(x^3 - 2x)\,(3x^2 + 4)$ (ii) $(x^2 - 1)^2$ (iii) $(x^3 + 3)\,(x^4 - 4)$.

ANSWERS

Ex.   (i) $15x^4 - 6x^2 - 8$;   (ii) $4x\,(x^2 - 1)$ (iii) $x^2\,(7x^4 + 12x - 12)$.

4.3.4. Derivative of a Quotient. Let u and v be two derivable functions of x. We suppose that v is not zero for the value of x under consideration.

We have

$$y = \frac{u}{v}$$

$$\Rightarrow \qquad y + \Delta y = \frac{u + \Delta u}{v + \Delta v}$$

$$\Rightarrow \qquad \Delta y = \frac{u + \Delta u}{v + \Delta v} - \frac{u}{v} = \frac{v\,\Delta u - u\,\Delta v}{v\,(v + \Delta v)}$$

$$\Rightarrow \qquad \frac{\Delta y}{\Delta x} = \frac{v\,.\dfrac{\Delta u}{\Delta x} - u\,.\dfrac{\Delta v}{\Delta x}}{v\,(v + \Delta v)}$$

Now, v, being a derivable functions, is continuous, so that

$$\Delta x \to 0 \Rightarrow \Delta v \to 0.$$

Thus,

$$\frac{d\left(\dfrac{u}{v}\right)}{dx} = \frac{dy}{dx} = \frac{v\dfrac{du}{dx} - u\dfrac{dv}{dx}}{v^2}$$

so that the derivative of the quotient of two functions.

$$= [(\text{derivative of Numer}) (\text{Denominator})$$
$$- (\text{Numer.}) \text{ Derivative of Denomi.}] \div \text{Square of Denominator}$$

Note. Being a derivable function v is continuous. Also we have supposed that the value of v is not zero for the value of x under consideration. There is therefore an interval around x such that no value of v is 0 for any point x in the interval. As such we suppose that $x + \Delta x$ belongs to the interval around x in question.

Cor. *Derivative of $y = x^m$ where $x \neq 0$ and m any non zero negative integer.*

Writing $\qquad m = -n$ so that $n \in \mathbf{N}$ and

$$y = x^m = 1/x^n, x \neq 0, \text{ we see that}$$

$$\frac{d(x^m)}{dx} = \frac{d(1/x^n)}{dx} = \frac{0.x^n - n.x^{n-1}}{x^{2n}} = nx^{-n-1} = mx^{m-1}.$$

In particular, we have

$$\frac{d(x^{-3})}{dx} = -3x^{-4}, \frac{d(x^{-7})}{dx} = -7x^{-8}.$$

Derivative of Polynomial function

Consider the polynomial function given by

$$y = a_0 x^n + a_1 x^{n+1} + \ldots\ldots\ldots + a_{n-1} x + a_n.$$

Employing the result of § 4.3, we see that $\forall\, x \in \mathbf{R}$

$$\frac{dy}{dx} = na_0 x^{n-1} + (n-1) a_1 x^{n-2} + \ldots\ldots\ldots + a_{n-1}$$

so that the derivative of a polynomial function of degree n is a polynomial function of degree $(n-1)$.

Derivative of a rational function. With the help of the general results obtained so far, we are in a position to find the derivative of every rational function. The domain of derivability of a rational function is the same as its domain of definition.

Example. *Differentiate $y = x/(1 + x)$.*

We have

$$\frac{dy}{dx} = \frac{(1+x)\dfrac{dx}{dx} - x\dfrac{d(1+x)}{dx}}{(1+x)^2} = \frac{(1+x).1 - x.1}{(1+x)^2} = \frac{1}{(1+x)^2}$$

EXERCISES

1. Find the derivatives of the following functions :

(i) $y = 2x^2 - 3x^3$ 　　　　　　(ii) $y = 5x^4 - 3x^3 + 1$ 　　　　　　(iii) $y = -x^7 + 2x^6 - x^3$

(iv) $y = (1 + 3x^2)(2x^3 - 1)$ 　　　(v) $y = (3x - 4)(x^2 - 6x + 7)$ 　(vi) $y = (2x - 1)(3x + 1)$

(vii) $y = \dfrac{x^2}{1 + x^3}$ 　　　　　　(viii) $y = \dfrac{2x - 4}{x^2 + 1}$ 　　　　　　(ix) $y = \dfrac{x - 1}{(x - 2)(x - 3)}$

(x) $y = \dfrac{x^3 + 1}{(x^2 - 1)(x^3 - 1)}$ 　　(xi) $y = \dfrac{3x + 4}{4x + 5}$ 　　　　　(xii) $y = \dfrac{x + 1}{(x + 2)^2}$

2. Given that

$$(i) \quad y = \frac{x^2 - 1}{(x-1)^2} \qquad\qquad (ii) \quad y = \frac{(x+4)^2}{(x-3)} \qquad\qquad (iii) \quad y = \frac{x^2 + 1}{x^2 - 3x + 2}$$

find dy/dx in each case.

Describe in each case the domain of the given function as also of the corresponding derived function.

3. Find the derivatives of the functions with the following function values :

$$(i) \quad \frac{x^2 + 1}{x^2 - x - 2} \qquad\qquad (ii) \quad \frac{(x-5)^2}{x^2 + 1} \qquad\qquad (iii) \quad \frac{x^4}{a^2 - x^2}$$

$$(iv) \quad x (x^2 - a^2)^2.$$

ANSWERS

1. (i) $4x - 9x^2$ (ii) $20x^3 - 9x^2$. (iii) $7x^6 + 12x^5 - 3x^2$.

   (iv) $6x (5x^3 + x - 1)$. (v) $9x^2 - 44x + 45$. (vi) $12x - 1$.

   (vii) $\dfrac{2x - x^4}{(1 + x^3)^2}$ $(viii)$ $\dfrac{-2x^2 + 8x + 2}{(x^2 + 1)^2}$

   (ix) $\dfrac{-x^2 + 2x + 1}{(x-2)^2 (x-3)^2}$. (x) $\dfrac{-2x^7 - 8x - 6x^2 + 6x^2 + 2x}{(x^2 - 1)^2 (x^3 - 1)^2}$.

   (xi) $-(4x - 5)^{-2}$. (xii) $-x (x + 2)^{-3}$.

2. (i) $\dfrac{-2}{(x-1)^2}$, $\mathbf{R} \sim \{1\}$. (ii) $\dfrac{(x+4)(x-10)}{(x-3)^2}$, $\mathbf{R} \sim \{3\}$.

   (iii) $\dfrac{-3x^2 + 2x + 3}{(x^2 - 3x + 2)^2}$, $\mathbf{R} \sim \{1, 2\}$.

3. (i) $\dfrac{-x^2 - 6x + 1}{(x^2 - x - 2)^2}$. (ii) $\dfrac{2(x-5)(5x+1)}{(x^2 + 1)^2}$.

   (iii) $\dfrac{2x^3 (2a^2 - x^2)}{(a^2 - x^2)}$. (iv) $(x^2 - a^2)(5x^2 - a^2)$.

4.3.5. Derivative of the inverse of an invertible function. Let $y = f(x)$ be a function derivable in its domain $[a, b]$. We suppose that f admits of an inverse function ϕ so that we have

$$y = f(x) \Leftrightarrow x = \phi(y).$$

We have to find a relation between $f' x$ and and $\phi'(y)$ for the corresponding values of x and y.

Let Δy be the change in y corresponding to the change Δx in x, as determined from $y = f(x)$. The change Δx in x corresponds to the change Δy in y as determined from $x = \phi(y)$. We suppose that $f'(x) \neq 0$. We have

$$1 = \frac{\Delta y}{\Delta x} \cdot \frac{\Delta x}{\Delta y} \Rightarrow \frac{\Delta x}{\Delta y} = 1 \div \left(\frac{\Delta y}{\Delta x} \right).$$

Let $\Delta x \to 0$.

$$\therefore \qquad \frac{dx}{dy} = \frac{1}{dy / dx} \Rightarrow \frac{dy}{dx} \cdot \frac{dx}{dy} = 1 \Rightarrow f'(x) \phi'(y) = 1.$$

Thus $\dfrac{dy}{dx}$ and $\dfrac{dx}{dy}$ are reciprocals of each other.

Ex. Verify the theorem for $y = x^3$ when $x = 2$.

Cor. Derivative of $y = x^{1/n}$ when $x > 0$ and $n \in \mathbf{N}$.

The function $y = x^{1/n}$ is the inverse of the function $y = x^n$ so that $y = x^{1/n} \Leftrightarrow x = y^n$.

Now $\qquad \dfrac{dy}{dx} \cdot \dfrac{dx}{dy} = 1 \Rightarrow \dfrac{dy}{dx} \cdot ny^{n-1} = 1$

$\Rightarrow \qquad \dfrac{dy}{dx} = \dfrac{1}{ny^{n-1}} = \dfrac{1}{n \left(x^{1/n} \right)^{n-1}} = \dfrac{1}{n} x^{1/n - 1}, x > 0.$

Thus $\qquad \dfrac{d(x^{1/n})}{dx} = \dfrac{1}{n} x^{1/n-1}, x > 0, n \in \mathbf{N}.$

4.3.6. Derivative of the function of a function. Let f and ϕ be two given derivable functions. We write

$$u = \phi(x), \qquad\qquad y = f(u).$$

Let the range of the function ϕ be a subset of the domain of the function f so that the composite $f \, 0 \, \phi$ has a meaning.

We have $\quad (f \, 0 \, \phi)(x) = f[\phi(x)].$

Let Δx be a change in x and Δu the corresponding change in u as determined from $u = \phi(x)$. Again correponding to the change Δu in u, let Δy be the change in y as determined from $y = f(u)$. We write

$$\frac{\Delta y}{\Delta x} = \frac{\Delta y}{\Delta u} \cdot \frac{\Delta u}{\Delta x}.$$

We suppose that there exists a neighbourhood of x such that for no $x + \Delta x$ in this neighbourhood $\Delta u = 0$. Now $\Delta x \to 0 \Rightarrow \Delta u \to 0$.

$$\lim_{\Delta x \to 0} \left(\frac{\Delta y}{\Delta x} \right) = \lim_{\Delta x \to 0} \left(\frac{\Delta y}{\Delta u} \cdot \frac{\Delta u}{\Delta x} \right)$$

$$= \lim_{\Delta x \to 0} \frac{\Delta y}{\Delta u} \cdot \lim_{\Delta x \to 0} \frac{\Delta u}{\Delta x}$$

$\Rightarrow \qquad \dfrac{dy}{dx} = \dfrac{dy}{du} \cdot \dfrac{du}{dx}.$

Thus $y = f(u)$, $u = \phi(x)$, so that $y = f(\phi(x))$

$\Rightarrow \qquad \dfrac{dy}{dx} = \dfrac{dy}{du} \cdot \dfrac{du}{dx} = f'(u) \, \phi'(x).$

The result is capable of generalisation to the case of the composites of more that two functions.

Note. We have shown that

$$[f(\phi(x))]' = f'(\phi(x)) \, \phi'(x)$$

we may also rewrite the result as follows

$$(f \, 0 \, \phi)' = (f' \, 0 \, \phi) \, \phi'.$$

Derivative of the function

$$y = x^a \quad x > 0 \text{ and } a \in \mathbf{Q}.$$

Let $\qquad a = p/q \,; q \in \mathbf{N}, p \in \mathbf{I}.$

We write

$$y = x^{p/q} = (x^{1/q})^p$$

and $\qquad\qquad u = x^{1/q}$ so that $y = u^p$.

Now $\qquad\qquad \dfrac{du}{dx} = \dfrac{1}{q} x^{1/q-1}, \dfrac{dy}{du} = pu^{p-1}$

$\therefore \qquad\qquad \dfrac{dy}{dx} = \dfrac{dy}{du} \cdot \dfrac{du}{dx} = \dfrac{p}{q} u^{p-1} \cdot x^{1/q-1}$

Thus $\qquad\qquad \dfrac{dx^a}{dx} = ax^{a-1} \; \forall \; a \in \mathbf{Q}; x > 0$

The present result is a generalisation of the corresponding results obtained earlier in §. (4.2 and 4.3.3 (cor.))

Ex. *Find the derivatives of the functions defined by the following expression :*

   (*i*) $\sqrt{(1 + x^2)}$, $\qquad\qquad$ (*ii*) $\sqrt{[(1 + x)/(1 - x)]}$

(*i*)   We write $u = 1 + x^2, \; y = \sqrt{u}$ so that $y = \sqrt{(1 + x^2)}$.

We have $\qquad\qquad \dfrac{du}{dx} = 2x, \dfrac{dy}{du} = \dfrac{1}{2} u^{-1/2} = \dfrac{1}{2}(1 + x^2)^{-1/2}$

Hence $\qquad\qquad \dfrac{dy}{dx} = \dfrac{dy}{du} \cdot \dfrac{du}{dx} = \dfrac{1}{2}(1 + x^2)^{-1/2} \, 2x = \dfrac{x}{\sqrt{(1 + x^2)}}.$

Or, directly without introducing u, we have

$$\frac{d\sqrt{(1 + x^2)}}{dx} = \frac{d\sqrt{(1 + x^2)}}{d(1 + x^2)} \cdot \frac{d(1 + x^2)}{dx}$$

$$= \frac{1}{2}(1 + x^2)^{-1/2} \cdot 2x = x/\sqrt{(1 + x^2)}.$$

(*ii*)   We have $\qquad y = \sqrt{[(1 + x)/(1 - x)]}.$

Let $\qquad\qquad u = \dfrac{1 + x}{1 - x}, y = u^{1/2}$

so that $\qquad\qquad y = \sqrt{\left(\dfrac{1 + x}{1 - x}\right)}.$

We have $\qquad\qquad \dfrac{du}{dx} = \dfrac{(1 - x)\dfrac{d(1 + x)}{dx} - (1 + x)\dfrac{d(1 - x)}{dx}}{(1 - x)^2}$

$$= \frac{(1 - x)1 - (1 + x)(-1)}{(1 - x)^2} = \frac{2}{(1 - x)^2}.$$

$$\frac{dy}{du} = \frac{1}{2} u^{-1/2} = \frac{1}{2}\left(\frac{1 + x}{1 - x}\right)^{-1/2}$$

Hence $\qquad\qquad \dfrac{dy}{dx} = \dfrac{dy}{du} \cdot \dfrac{du}{dx} = \dfrac{1}{2}\left(\dfrac{1 + x}{1 - x}\right)^{-1/2} \dfrac{2}{(1 - x)^2}$

$$= \frac{1}{(1+x)^{1/2}(1-x)^{3/2}}$$

or directly

$$\frac{d\sqrt{\left(\frac{1+x}{1-x}\right)}}{dx} = \frac{d\sqrt{\left(\frac{1+x}{1-x}\right)}}{d\left(\frac{1+x}{1-x}\right)} \cdot \frac{d\left(\frac{1+x}{1-x}\right)}{dx}$$

$$= \frac{1}{2}\left(\frac{1+x}{1-x}\right)^{-1/2} \frac{2}{(1-x)^2}$$

$$= \frac{1}{(1+x)^{1/2}(1-x)^{3/2}}.$$

Ex. Find the derivatives of the functions defined by the following expressions :

(i) $(ax+b)^n$

(ii) $1/(1+x^2)$

(iii) $\sqrt{(ax^2 + 2bx + c)}$

(iv) $\dfrac{\sqrt{(x^2+1)} - \sqrt{(x^2-1)}}{\sqrt{(x^2+1)} + \sqrt{(x^2-1)}}$

(v) $\sqrt{\left(\dfrac{a^2 - x^2}{a^2 + x^2}\right)}$

(vi) $\dfrac{1 - a\sqrt[3]{x^2}}{1 + a\sqrt[3]{x^2}}$

(vii) $\dfrac{x\sqrt{(x^2-4)}}{\sqrt{(x^2-1)}}$

(viii) $\sqrt{\left(\dfrac{1+x^2}{1-x^2}\right)}$

(ix) $\dfrac{2x^2 - 1}{x\sqrt{(1+x^2)}}$

ANSWERS

Ex. The derivatives are the function with the following function values :

(i) $an(ax+b)^{n-1}$. (ii) $-2x/(1+x^2)^2$.

(iii) $(ax+b)/\sqrt{(ax^2 + 2bx + c)}$.

(iv) $-2x/\sqrt{(x^4-1)}[x^2 + \sqrt{(x^4-1)}]$.

(v) $-\dfrac{2a^2 x}{(a^2 - x^2)^{1/2}(a^2 + x^2)^{3/2}}$.

(vi) $\dfrac{-4a}{3x^{1/3}(1 + ax^{2/3})^2}$

(vii) $\dfrac{x^4 - 2x^2 + 4}{(x^2 - 1)^{3/2}(x^2 - 4)^{1/2}}$.

(viii) $\dfrac{2x}{(1+x^2)^{1/2}(1-x^2)^{3/2}}$.

(ix) $\dfrac{4x^2 + 1}{x^2(1+x^2)^{3/2}}$.

4.4. Derivatives of Trigonometric functions :

Derivative of $y = \sin x, x \in \mathbf{R}$.

$$y = \sin x,$$

$$\Rightarrow \qquad \frac{\Delta y}{\Delta x} = \frac{\sin(x + \Delta x) - \sin x}{\Delta x}$$

$$= \frac{2 \cos \frac{1}{2}(2x + \Delta x) \sin \frac{1}{2} \Delta x}{\Delta x}$$

$$= \cos\left(x + \frac{1}{2}\Delta x\right) \frac{\sin \frac{1}{2}\Delta x}{\frac{1}{2}\Delta x}$$

$$\Rightarrow \qquad \frac{dy}{dx} = \lim_{\Delta x \to 0} \cos\left(x + \frac{1}{2}\Delta x\right) \cdot \lim_{\Delta x \to 0} \frac{\sin \frac{1}{2}\Delta x}{\frac{1}{2}\Delta x}$$

As cos x is a continuous function, we have

$$\lim \cos\left(x + \frac{1}{2}\Delta x\right) = \cos x \text{ when } \Delta x \to 0.$$

Also $\qquad \lim \dfrac{\sin \frac{1}{2}\Delta x}{\frac{1}{2}\Delta x} = 1 \text{ when } \Delta x \to 0.$

Thus, we have $\qquad \dfrac{dy}{dx} = \cos x$

$$\Rightarrow \qquad \frac{d(\sin x)}{dx} = \cos x.$$

Derivative of y = cos x, x ∈ R.

$$y = \cos x$$

$$\Rightarrow \qquad \frac{\Delta y}{\Delta x} = \frac{\cos(x + \Delta x) - \cos x}{\Delta x}$$

$$= \frac{-2 \sin \frac{1}{2}(2x + \Delta x)\sin \frac{1}{2}\Delta x}{\Delta x}$$

$$= -\sin\left(x + \frac{1}{2}\Delta x\right) \cdot \frac{\sin \frac{1}{2}\Delta x}{\frac{1}{2}\Delta x}$$

As sin x is a continuous function, we have

$$\lim \sin\left(x + \frac{1}{2}\Delta x\right) = \sin x \text{ when } x \to 0.$$

$$\therefore \qquad \frac{dy}{dx} = \lim_{\Delta x \to 0}\left[-\sin\left(x + \frac{1}{2}\Delta x\right)\right] \lim_{\Delta x \to 0} \frac{\sin \frac{1}{2}\Delta x}{\frac{1}{2}\Delta x}$$

$$= -\sin x . 1 = -\sin x.$$

Thus $\qquad \dfrac{d(\cos x)}{dx} = -\sin x.$

Ex. Find the derivatives of the functions defined by the following function values :

$\qquad$ (*i*) $\sin 2x$ $\qquad$ (*ii*) $\cos^3 x$ $\qquad$ (*iii*) $\sqrt{(\sin \sqrt{x})}$

(*i*) Let $\qquad y = \sin 2x$. We write

$\qquad u = 2x$ so that $y = \sin u$.

Now $\qquad \dfrac{dy}{dx} = \dfrac{dy}{du}.\dfrac{du}{dx} = \cos u.2 = 2\cos 2x$

or, *briefly*

$$\frac{d(\sin 2x)}{dx} = \frac{d(\sin 2x)}{d(2x)}.\frac{d(2x)}{dx} = \cos 2x.2 = 2\cos 2x.$$

(*ii*) Let $\qquad y = \cos^3 x = (\cos x)^3$. We write

$\qquad u = \cos x$ so that $y = u^3$.

Now $\qquad \dfrac{dy}{dx} = \dfrac{dy}{du}.\dfrac{du}{dx} = 3u^2(-\sin x) = -3\cos^2 x \sin x.$

or, *briefly*

$$\frac{d(\cos^3 x)}{dx} = \frac{d(\cos x)^3}{dx} = \frac{d(\cos x)^3}{d(\cos x)},\frac{d(\cos x)}{dx}$$

$$= 3(\cos x)^2 \times (-\sin x) = -3\cos^2 x \sin x$$

(*iii*) Let $\qquad y = \sqrt{(\sin \sqrt{x})}$. We write

$\qquad u = \sqrt{x}, \ v = \sin u$ so that $y = \sqrt{v},$

Now $\qquad \dfrac{dy}{dx} = \dfrac{dy}{dv},\dfrac{dv}{du}.\dfrac{du}{dx}$

$$= \frac{1}{2}v^{1/2}\cos u.\frac{1}{2}x^{-1/2} = \frac{1}{4}\frac{\cos\sqrt{x}}{\sqrt{(\sin\sqrt{x}}}.\frac{1}{\sqrt{x}}.$$

or, *briefly*

$$\frac{d\sqrt{(\sin\sqrt{x})}}{dx} = \frac{d\sqrt{(\sin\sqrt{x})}}{d\sin\sqrt{x}}.\frac{d\sin\sqrt{x}}{d\sqrt{x}}.\frac{d\sqrt{x}}{dx}$$

$$= \frac{1}{2}(\sin\sqrt{x})^{-1/2}.\cos\sqrt{x}.\frac{1}{2}x^{-1/2}.$$

$$= \frac{1}{4}.\frac{\cos\sqrt{x}}{\sqrt{x}\sqrt{(\sin\sqrt{x})}}.$$

EXERCISES

1. Find the slopes of the tangents to the curves (*i*) $y = \sin x$ at $(0, 0)$, $(\pi/2, 1)$ (*ii*) $y = \cos x$ at $(0, 1)$, $(\pi/2, 0)$. Also indicate the positions of the tangents on the corresponding curves.

2. Find the derivatives of the functions defined by the following function values :

$\qquad$ (*i*) $\sin^3 x$ $\qquad$ (*ii*) $\sqrt{(\cos 2x)}$ $\qquad$ (*iii*) $\sqrt[3]{\sin^2 x}$

(iv) $\dfrac{\sin x}{x}$ (v) $\dfrac{\sin^2 x}{1 + \cos x}$ (vi) $\cos\sqrt{(x^2 + 3x + 4)}$

(vii) $\sin^m x . \cos^n x.$ (viii) $\dfrac{\sin x}{1 + \cos x}$ (ix) $\sqrt{(x^2 + 1)} \sin^2 x.$

(x) $\dfrac{x \cos x}{\sqrt{(1 + x^2)}}$ (xi) $\dfrac{x - \cos x}{x + \cos x}$ (xii) $\sqrt{(1 + \sin^2 x)} \sqrt{x}.$

ANSWERS

1.   (i) $1, 0.$ (ii) $0, -1.$

2.   (i) $3 \sin^2 x \cos x.$ (ii) $\dfrac{-\sin 2x}{\sqrt{(\cos 2x)}}.$ (iii) $\dfrac{2}{3} . \dfrac{\cos x}{\sqrt[3]{\sin x}}.$

   (iv) $(x \cos x - \sin x)/x^2.$ (v) $\sin x.$

   (vi) $-\dfrac{1}{2}(2x + 3)(x^2 + 3x + 4)^{-1/2} \sin\sqrt{(x^2 + 3x + 4)}.$

   (vii) $(m \cos^2 x - n \sin^2 x) \sin^{m-1} x \cos^{n-1} x.$

   (viii) $\dfrac{1}{2} \sec^2 \dfrac{x}{2}.$ (ix) $\dfrac{\sin x}{\sqrt{(x^2 + 1)}} \{2(x^2 + 1)\cos x + \sin^2 x\}.$

   (x) $\dfrac{\cos x}{(1 + x^2)^{3/2}} - \dfrac{x \sin x}{\sqrt{(1 + x^2)}}.$ (xi) $\dfrac{\cos x + x \sin x}{(x + \cos x)^2}.$ (xii) $\dfrac{x \sin^2 x + \sin^2 x + 1}{2\sqrt{(1 + \sin^2 x)} . \sqrt{x}}.$

Derivative of $y = \tan x$.

$$y = \tan x = \frac{\sin x}{\cos x}, \quad x \in \mathbf{R} \sim \left\{(2n + 1)\frac{\pi}{2}; n \in \mathbf{I}\right\}$$

$$\Rightarrow \qquad \frac{dy}{dx} = \frac{\cos x . \dfrac{d(\sin x)}{dx} - \sin x . \dfrac{d(\cos x)}{dx}}{\cos^2 x}$$

$$= \frac{\cos x . \cos x + \sin x . \sin x}{\cos^2 x} = \frac{1}{\cos^2 x} = \sec^2 x.$$

$$\Rightarrow \qquad \frac{d(\tan x)}{dx} = \sec^2 x$$

for every x which belongs to the domain of $\tan x$.

We may similarly show that

$$\frac{d(\cot x)}{dx} = -\operatorname{cosec}^2 x \ \forall \ x \in \mathbf{R} \sim \{n\pi; n \in \mathbf{I}\}.$$

Its proof is left to the reader.

Ex. 1. Mark the positions of the tangents on the curves

   (i) $y = \tan x$ at $(0, 0)$ (ii) $y = \cot x$ at $(\pi/2, 0).$

2.   Find the derivatives of the functions defined by the following functions values :

   (i) $\tan x + \cot x$ (ii) $\sin x \tan 2x$ (iii) $x \tan x \cot 2x.$

(vi) $\dfrac{\tan x - \cot x}{\tan x + \cot x}$ (v) $\sqrt{\left(\dfrac{1-\tan x}{1+\tan x}\right)}$ (vi) $\sqrt{\left(\dfrac{1-\cos x}{1+\cos x}\right)}$

(vii) $\cot^2 3x$ $(viii)$ $\left(\tan\dfrac{x}{2}+\cot\dfrac{x}{2}\right)/x.$ (ix) $(\tan x + \cot x)$

ANSWERS

2. (i) $-4\,\mathrm{cosec}\,2x\,\cot 2x.$

(ii) $\cos x \tan 2x + 2 \sin x \sec^2 2x.$

(iii) $\dfrac{\sin 4x - 8x\sin x}{4\cos^2 x \sin 2x}$ (iv) $2\sin 2x.$

(v) $-\dfrac{1}{\sqrt{[(\cos 2x)]}\,(\cos x + \sin x)}.$ (vi) $\dfrac{1}{2}\sec^2\dfrac{x}{2}.$

(vii) $\dfrac{-6\cos 3x}{\sin^2 3x}.$ $(viii)$ $\dfrac{-2(\sin x + x\cos x)}{(x\sin x)^2}$

(ix) $\sec^2 x - \mathrm{cosec}^2\, x.$

Derivative of y = sec x.

We have

$$y = \frac{1}{\cos x}, x \in \mathbf{R}\sim\left\{(2n+1)\frac{\pi}{2}; n \in \mathbf{I}\right\}$$

so that $\cos x \neq 0.$

$$\Rightarrow \qquad \frac{dy}{dx} = \frac{\cos x.0 - 1(-\sin x)}{\cos^2 x} = \tan x \sec x.$$

Thus $\dfrac{d\,(\sec x)}{dx} = \tan x \sec x.$

for every x which belongs to the domain of the functions $\sec x.$

We may similarly show that

$$\frac{d\,(\mathrm{cosec}\,x)}{dx} = -\cot x\,\mathrm{cosec}\,x.$$

Its proof is left to the reader.

Ex. Find the derivatives of the functions defined by the following function values :

$(i,$ $\mathrm{cosec}^2 3x$ (ii) $\sqrt{[\sec(ax+b)]}$ (iii) $\tan x \sec^2 x$

(iv) $\sec\sqrt{(a+bx)}$ (v) $\sec(\mathrm{cosec}\,x)$ (vi) $\sec\sqrt{(\tan x)}.$

ANSWERS

Ex. (i) $-6\,\mathrm{cosec}^2\,3x\,\cot 3x.$

(ii) $\dfrac{1}{2}a\sqrt{[\sec(ax+b)]}\,\tan(ax+b).$

(iii) $\sec^4 x\,(1 + 2\tan x).$

(iv) $\left[\dfrac{1}{2}b\sec\sqrt{(a+bx)}\,\tan\sqrt{(a+bx)}\right]/\sqrt{(a+bx)}.$

(v) $- \sec (\operatorname{cosec} x) \tan (\operatorname{cosec} x) \cot x \operatorname{cosec} x.$

(vi) $\dfrac{\sec \sqrt{\tan x} \tan \sqrt{\tan x}}{\sqrt{2} \, \sqrt{(\sin 2x).\cos x}}$

4.5. Derivatives of inverse trigonometric functions.

The precise definitions of inverse trigonometrical functions as given in § 2.7, will have to be kept in mind to obtain their derivatives.

4.5.1. Derivative of $y = \sin^{-1} x$.

$$x \in [-1, 1], y \in [-\pi/2. \ \pi/2].$$

We have $y = \sin^{-1} x \Leftrightarrow x = \sin y.$

Now $x = \sin y \Rightarrow \dfrac{dx}{dy} = \cos y.$

so that $\dfrac{dy}{dx} = \dfrac{1}{\cos y}$ if $\cos y \neq 0$ *i.e.,* $y \neq \pm \dfrac{\pi}{2}$

$$= \pm \dfrac{1}{\sqrt{(1 - \sin^2 y)}} = \pm \dfrac{1}{\sqrt{(1 - x^2)}}$$

where the sign, plus or minus, is the same as that of $\cos y$, which is positive.

We note that

$$y \neq \pi/2 \Leftrightarrow x \neq \pm 1.$$

Hence, $\dfrac{d (\sin^{-1} x)}{dx} = \dfrac{1}{\sqrt{(1 - x^2)}}, \ \forall \ x \in \,]-1, \, 1[$

4.5.2. Derivative of $y = \cos^{-1} x$; $x \in [-1, 1], y \in [0, \pi]$

We have $y = \cos^{-1} x \Leftrightarrow x = \cos y$

Now $x = \cos y \Rightarrow \dfrac{dx}{dy} = - \sin y.$

$$\dfrac{dy}{dx} = - \dfrac{1}{\sin y} \text{ if } y \neq 0 \text{ } i.e., y \neq 0 \text{ } y = \pi$$

$$= \dfrac{-1}{\pm \sqrt{(1 - \cos^2 y)}} = \dfrac{-1}{\pm \sqrt{(1 - x^2)}}$$

where the sign, plus or minus, is the same as that of $\sin y$ which is necessarily positive.

Hence, $\dfrac{d (\cos^{-1} x)}{dx} = \dfrac{-1}{\sqrt{(1 - x^2)}}.$

$$x \in [-1, 1] \, ; \cos^{-1} x \in \,]0, \pi[$$

4.5.3. Derivative of $y = \tan^{-1} x$; $x \in \mathbf{R}, y \in]-\pi/2, \pi/2[$

We have $y = \tan^{-1} x \Leftrightarrow x = \tan y.$

Now $x = \tan y$

$\Rightarrow$ $\dfrac{dx}{dy} = \sec^2 y$ so that $dy/dx \neq 0$ for any y

$$\Rightarrow \qquad \frac{dy}{dx} = \frac{1}{\sec^2 y} = \frac{1}{1 + \tan^2 y} = \frac{1}{1 + x^2}.$$

Hence $\quad \dfrac{d(\tan^{-1} x)}{dx} = \dfrac{1}{1 + x^2}; \ x \in \mathbf{R}; \tan^{-1} x \in \left] -\dfrac{\pi}{2}, \dfrac{\pi}{2} \right[$

4.5.4. $\quad \dfrac{d(\cot^{-1} x)}{dx} = -\dfrac{1}{1 + x^2}; \ x \in \mathbf{R}; \cot^{-1} x \in \left]0, \pi\right[$

Its proof is left to reader.

4.5.5. Derivative of $y = \sec^{-1} x$;

$$x \in \mathbf{R} \sim \left]-1, 1\right[, \ y \in \left[0, \frac{\pi}{2}\right[\ \cup \ \left]\frac{\pi}{2}, \pi\right]$$

We have $\qquad y = \sec^{-1} x \Leftrightarrow x = \sec y.$

Thus $\qquad \dfrac{dy}{dx} = \sec y \tan y$

$\Rightarrow \qquad \dfrac{dy}{dx} = \dfrac{1}{\sec y \tan y}$; if any $y \neq 0 \ i.e., \ y \ 0, \ y \neq \pi.$

$$= \pm \frac{1}{\sec y \sqrt{\sec^2 y - 1)}} = \pm \frac{1}{x\sqrt{(x^2 - 1)}}$$

where the sign is the same as that of $\tan y$.

Now $\qquad x \in \left]1, \infty\right[\Rightarrow y \in \left]0, \dfrac{\pi}{2}\right[\Rightarrow \tan y > 0,$

$$x \in \left]\infty, -1\right[\Rightarrow y \in \left]\frac{\pi}{2}, \pi\right[\Rightarrow \tan y < 0.$$

Thus $\qquad x \in \left]1, \infty\right[\Rightarrow \dfrac{dy}{dx} = \dfrac{1}{x\sqrt{(x^2 - 1)}},$ and

$$x \in \left]\infty, -1\right[\Rightarrow \frac{dy}{dx} = -\frac{1}{x\sqrt{(x^2 - 1)}}.$$

If follows that

$$\frac{d(\sec^{-1} x)}{dx} = \frac{1}{|x|\sqrt{(x^2 - 1)}} \text{ for all admissible values of } x.$$

4.5.6. Derivative of $y = \mathrm{cosec}^{-1} x$;

$$x \in \mathbf{R} \sim \left]-1, 1\right[, y \in \left[-\frac{\pi}{2}, 0\right[\cup \left]0, \frac{\pi}{2}\right]$$

It may be shown that

$$\frac{d(\mathrm{cosec}^{-1}) x}{dx} = \frac{-1}{|x|\sqrt{(x^2 - 1)}}, \qquad \text{(for all admissible values of } x.)$$

EXERCISES

1. Find the derivatives of the following functions :

(i) $y = \sin^{-1}\sqrt{x}$,

(ii) $y = \sqrt{(\cot^{-1}\sqrt{x})}$,

(iii) $y = \tan^{-1}[(1+x)/(1-x)]$,

(iv) $y = \tan^{-1}(\cos\sqrt{x})$,

(v) $y = \sec^{-1}x^2$,

(vi) $y = \operatorname{cosec}^{-1}(x^{-1/2})$,

(vii) $y = \cos^{-1}\left(\dfrac{x - x^{-1}}{x + x^{-1}}\right)$,

(viii) $y = \dfrac{x\sin^{-1}x}{\sqrt{(1-x^2)}}$,

(ix) $y = \tan^{-1}\dfrac{4\sqrt{x}}{1-4x}$,

(x) $y = \cos^{-1}\left(\dfrac{a + b\cos x}{b + a\cos x}\right)$

2. Indicate the positions of the tangents to the curves :

(i) $y = \sin^{-1}x$ at $(0, 0)$,

(ii) $y = \cos^{-1}x$ at $(1, 0)$

ANSWERS

1. (i) $\dfrac{1}{2\sqrt{(x - x^2)}}$.

(ii) $\dfrac{-1}{4\sqrt{(1+x)}\sqrt{x}\sqrt{(\cot^{-1}\sqrt{x})}}$.

(iii) $\dfrac{1}{1 + x^2}$.

(iv) $-\dfrac{\sin x}{2\sqrt{x}}\cdot\dfrac{1}{1 + \cos^2\sqrt{x}}$.

(v) $\dfrac{2}{x\sqrt{(x^4 - 1)}}$.

(vi) $\dfrac{1}{2\sqrt{(x - x^2)}}$.

(vii) $-\dfrac{2}{1 + x^2}$.

(viii) $\dfrac{x\sqrt{(1-x^2)} + \sin^{-1}x}{(1-x^2)^{3/2}}$.

(ix) $\dfrac{2}{\sqrt{x}\,(1 + 4x)}$.

(x) $\dfrac{\sqrt{(b^2 - a^2)}}{b + a\cos x}$.

4.6. Derivative of $y = \log_a x$, $x \in [0, \infty]$, $a > 0$.

We have

$$\frac{\Delta y}{\Delta x} = \frac{\log_a(x + \Delta x) - \log_a x}{\Delta x}$$

$$= \frac{1}{\Delta x}\log_a\left(\frac{x + \Delta x}{x}\right)$$

$$= \frac{1}{x}\cdot\frac{x}{\Delta x}\log_a\left(1 + \frac{\Delta x}{x}\right)$$

$$\doteq \frac{1}{2}\log_a\left(1 + \frac{\Delta x}{x}\right)^{x/\Delta x}$$

$$\Rightarrow \quad \frac{d(\log_a x)}{dx} = \frac{1}{x}\log_a e, x \in\,]0, \infty[.$$

Cor. Let $a = e$. We have $y = \log_e x = \log x$

$$\Rightarrow \quad \frac{dy}{dx} = \frac{1}{x}\log_e e = \frac{1}{x}$$

$$\Rightarrow \quad \frac{d(\log x)}{dx} = \frac{1}{x}, x > 0.$$

Derivative of $y = a^x$, $x \in]-\infty, \infty[$, $a > 0$. We have

$$y = a^x \Leftrightarrow x = \log_a y.$$

Now $\qquad x = \log_a y \Rightarrow \dfrac{dx}{dy} = \dfrac{1}{y} \log_a e.$

Also $\qquad \dfrac{dy}{dx} \cdot \dfrac{dx}{dy} = 1 \Rightarrow \dfrac{dy}{dx} \cdot \dfrac{1}{y} \log_a e = 1$

$\Rightarrow \qquad \dfrac{dy}{dx} \cdot \dfrac{y}{\log_a e} = y \log_e a = a^x \log_e a.$

Cor, Let $a = e$ so that $y = e^x$. We obtain

$$\frac{de^x}{dx} = e^x \log_e e = e^x.$$

Ex. Find the derivative of the functions defined by the expressions :

(i) $\log \sin x$ (ii) $\cos (\log x)$

(iii) $e^{\sin x}$ (iv) $\log [\sin (\log x)]$

(v) $\log \sqrt{(x^2 + x + 1)}$ (vi) $\log \tan \left(\dfrac{1}{2}x + \dfrac{1}{4}\pi \right)$

(vii) $\log (\sec x + \tan x)$ (viii) $\dfrac{e^{2x}}{\log x}$

(ix) $\sqrt{(a^{\sqrt{x}})}$ (x) $\log_{10} (\sin^{-1} x^2)$

(xi) $\log[x + \sqrt{(x^2 + a^2)}]$ (xii) $\log (e^{mx} + e^{-mx})$

(xiii) $a^{2x} \sin^2 x$ (xiv) $e^{\sqrt[3]{ax}}$

(xv) $\log \dfrac{a + b \tan x}{a - b \tan x}$

ANSWERS

Ex. (i) $\cot x$,   (ii) $-\dfrac{\sin (\log x)}{x}$,   (iii) $e^{\sin x} \cos x$,

(iv) $\dfrac{\cot (\log x)}{x}$,   (v) $\dfrac{2x + 1}{2(x^2 + x + 1)}$,   (vi) $\sec x$,

(vii) $\sec x$,   (viii) $\dfrac{e^{2x} \cdot 2x \log x - e^{2x}}{x(\log x)^2}$,   (ix) $\dfrac{a^{\sqrt{x}} \log a}{4\sqrt{4}\sqrt{(a^{\sqrt{x}})}}$,

(x) $\dfrac{2x \cdot \log_{10} e}{\sqrt{(1 - x^4) \sin^{-1} x^2}}$,   (xi) $\dfrac{1}{\sqrt{(x^2 + a^2)}}$,   (xii) $\dfrac{m (e^{mx} - e^{-mx})}{e^{mx} + e^{-mx}}$,

(xiii) $a^2x (\sin^2 x . \log a^2 + \sin 2x)$, (xiv) $\dfrac{a}{3} \dfrac{e^{\sqrt[3]{ax}}}{\sqrt[3]{(a^2 x^2)}}$,   (xv) $\dfrac{2ab}{a^2 \cos^2 x - b^2 \sin^2 x}$.

4.7 Hyperbolic functions.

It is found useful to define $\sinh x$, $\cosh x$, $\tanh x$, $\coth x$, $\operatorname{sech} x$ and $\operatorname{cosech} x$ as follows :

$$\sinh x = \frac{e^x - e^{-x}}{2}, \qquad\qquad \cosh x = \frac{e^x + e^{-x}}{2},$$

$$\tanh x = \frac{\sinh x}{\cosh x} = \frac{e^x - e^{-x}}{e^x + e^{-x}}, \qquad \coth x = \frac{\cosh x}{\sinh x} = \frac{e^x + e^{-x}}{e^x - e^{-x}},$$

$$\text{sech } x = \frac{1}{\cosh x} = \frac{2}{e^x - e^{-x}}, \qquad \text{cosech } x = \frac{1}{\sinh x} = \frac{2}{e^x - e^{-x}}.$$

It may be seen that sinh x, cosh x, tanh x, sech x are defined $\forall\, x \in \mathbf{R}$ and coth x, cosech x for all non-zero values or $\mathbf{R}$.

The reader may prove directly that the following results hold good $\forall\, x \in \mathbf{R}$.

$$\sinh(-x) = -\sin h\, x, \qquad\qquad \cosh(-x) = \cosh^2 x,$$
$$\cosh^2 x - \sinh^2 x = 1, \qquad\qquad \cosh^2 x + \sinh^2 x = \cosh 2x$$
$$\sinh 2x = 2 \sinh x \cosh x,$$
$$\cosh(x \pm y) = \cosh x \cosh y \pm \sinh x \sinh y,$$
$$\sinh(x \pm y) = \sinh x \cosh y \pm \cosh x \sinh y.$$

It will be seen that sinh x, cosh x etc. have properties analogous to those of sin x, cos x.

The graphs of hyperbolic functions will be given in Chapter 7.

4.7.1. Derivatives of Hyperbolic Functions

Derivative of $y = \sinh x$; $x \in \mathbf{R}$

We have
$$y = \sinh x = \frac{e^x - e^{-x}}{2}$$

$$\Rightarrow \qquad \frac{dy}{dx} = \frac{d}{dx}\left[\frac{1}{2}(e^x - e^{-x})\right] = \frac{1}{2}(e^x + e^{-x}) = \cosh x$$

$$\Rightarrow \qquad \frac{d(\sinh x)}{dx} = \cosh x \,\forall\, x \in \mathbf{R}.$$

Derivative of $y = \cosh x$; $x \in \mathbf{R}$.

We have
$$y = \cosh x = \frac{e^x + e^{-x}}{2}$$

$$\Rightarrow \qquad \frac{dy}{dx} = \frac{e^x - e^{-x}}{2} = \sinh x$$

$$\Rightarrow \qquad \frac{d(\cosh x)}{dx} = \sinh x \,\forall\, x \in \mathbf{R}.$$

Derivative of $y = \tanh x$; $x \in \mathbf{R}$.

$$y = \tanh x = \frac{\sinh x}{\cosh x}; \cosh x \neq 0 \,\forall\, x \in \mathbf{R}$$

$$\Rightarrow \qquad \frac{dy}{dx} = \frac{\cosh x . \dfrac{d(\sin x)}{dx} - \sinh x . \dfrac{d(\cosh x)}{dx}}{\cosh^2 x}$$

$$= \frac{\cosh x . \cosh x - \sinh x . \sinh x .}{\cosh^2 x}$$

$$= \frac{\cosh^2 x - \sinh^2 x}{\cosh^2 x} = \frac{1}{\cosh^2 x} = \operatorname{sech}^2 x; \,\forall\, x \in \mathbf{R}$$

$$\frac{d(\tanh x)}{dx} = \operatorname{sech}^2 x.$$

It may be similarly shown that

$$\frac{d\,(\coth x)}{dx} = -\operatorname{cosech}^2 x.$$

Its proof is left to reader.

Derivative of $y = \operatorname{sech} x$ **;** $x \in \mathbf{R}$

$$y = \operatorname{sech} x = \frac{1}{\cosh x}.$$

$\Rightarrow \qquad \dfrac{dy}{dx} = \dfrac{\cosh x.0 - 1.\sinh x}{\cosh^2 x}$

$$= -\frac{\sinh x}{\cosh^2 x} = -\tanh x\,\operatorname{sech} x\;;\;\forall\,x \in \mathbf{R}.$$

$\Rightarrow \qquad = \dfrac{d\,(\operatorname{sech} x)}{dx} = -\tanh x\,\operatorname{sech} x$

It may be similary shown that

$$= \frac{d\,(\operatorname{cosech} x)}{dx} = -\coth x\,\operatorname{cosech} x.$$

Its proof is left to the reader.

4.7.2. Derivatives of inverse hyperbolic functions.

Derivative of $y = \sinh^{-1} x$**.**

Let $\qquad\qquad y = \sinh^{-1} x$ so that $x = \sinh y$.

$\therefore \qquad \dfrac{dx}{dy} = \cosh y.$

$\Rightarrow \qquad \dfrac{dy}{dx} = \dfrac{1}{\cosh y} = \pm\dfrac{1}{\sqrt{(1 + \sinh^2 y)}} = \pm\dfrac{1}{\sqrt{(1 + x^2)}},$

where the sign of the radical is the same as that of $\cosh y$ which we know, is always positive.

Hence, $\qquad \dfrac{d\,(\sinh^{-1} x)}{dx} = \dfrac{1}{\sqrt{(1 + x^2)}}.$

Derivative of $\cosh^{-1} x$**.**

Let $\qquad\qquad y = \cosh^{-1} x$ so that $x = \cosh y$.

$\Rightarrow \qquad \dfrac{dx}{dy} = \sinh y,$

$\Rightarrow \qquad \dfrac{dy}{dx} = \dfrac{1}{\sinh y} = \pm\dfrac{1}{\sqrt{(\cosh^2 y - 1)}} = \pm\dfrac{1}{\sqrt{(x^2 - 1)}}.$

where the sigh of the radical is the same as that of $\sinh y$.

Now, $\cosh^{-1} x$ i.e., y is always positive so that $\sinh y$ is positive.

Hence, $\dfrac{d\,(\cosh^{-1} x)}{dx} = \dfrac{1}{\sqrt{(x^2 - 1)}}.$

Derivative of tanh⁻¹ x. [| x | < 1].

Let $y = \tanh^{-1} x$ so that $x = \tanh y$.

$\Rightarrow \qquad \dfrac{dx}{dy} = \operatorname{sech}^2 y,$

$\Rightarrow \qquad \dfrac{dy}{dx} = \dfrac{1}{\operatorname{sech}^2 y} = \dfrac{1}{1 - \tanh^2 y} = \dfrac{1}{1 - x^2}.$

Thus $\dfrac{d\,(\tanh^{-1} x)}{dx} = \dfrac{1}{1 - x^2}.$

$\dfrac{d\,(\coth^{-1} x)}{dx} = -\dfrac{1}{x^2 - 1}$ [| x | > 1].

Its proof is left to the reader.

Derivative of sech⁻¹ x.

Let $y = \operatorname{sech}^{-1} x$ so that $x = \operatorname{sech} y$.

$\Rightarrow \qquad \dfrac{dx}{dy} = - \operatorname{sech} y \tanh y.$

$\Rightarrow \qquad \dfrac{dy}{dx} = - \dfrac{1}{\operatorname{sech} y \tanh y}$

$= \pm \dfrac{-1}{\operatorname{sech} y . \sqrt{(1 - \operatorname{sech}^2 y)}} = \pm \dfrac{-1}{x \sqrt{1 - x^2}}.$

where the sign of the radical is the same as that of tanh y.

But we know that sech⁻¹ x, *i.e.*, y is always positive, so that tanh y is always positive.

Hence, $\dfrac{d\,(\operatorname{sech}^{-1} x)}{dx} = -\dfrac{1}{x \sqrt{(1 - x^2)}}.$

Derivative of cosech⁻¹ x.

Let $y = \operatorname{cosech}^{-1} x$ so that $x = \operatorname{cosech} y$.

$\Rightarrow \qquad \dfrac{dx}{dy} = - \operatorname{cosech} y . \coth y.$

$\Rightarrow \qquad \dfrac{dy}{dx} = - \dfrac{1}{\operatorname{cosech} y . \coth y}$

$= \pm \dfrac{-1}{\operatorname{cosech} y \sqrt{(\operatorname{cosech}^2 y + 1)}} = \pm \dfrac{-1}{x \sqrt{(x^2 + 1)}},$

when the sign of the radical is the same as that of coth y.

Now, y, and therefore coth y is positive or negative according as x is positive or negative.

$\therefore \qquad \dfrac{dy}{dx} = \dfrac{-1}{x \sqrt{(x^2 + 1)}}$ if $x > 0$ and $= \dfrac{-1}{-x \sqrt{x^2 + 1}}$ if $x < 0$.

Thus $\dfrac{d\ (\operatorname{cosech}^{-1}\ x)}{dx} = \dfrac{-1}{|x|\ \sqrt{(x^2 + 1)}},$

for all values of x.

Ex. Find the derivatives of

(*i*) $\log (\cosh x)$ (*ii*) $e^{\sinh^2 x}$ (*iii*) $\tan x \,.\, \tanh x$.

ANSWERS

(*i*) $\tanh x$ (*ii*) $\sinh 2x\ e^{\sinh^2 x}$ (*iii*) $\sec^2 x \tanh x + \tan x \operatorname{sech}^2 x$

4.8. Derivation of parametircally defined functions.

Let $x = f(t),$ $y = g\,(t)$

be two derivable functions each defined in some interval. We also suppose that the function f is inveritble and ϕ denotes the inverse of f. Now f being derivable its inverse, ϕ is also derivable.

We have

$$x = f(t) \Leftrightarrow t = \phi\,(x).$$

Also $y = g\,(t),\quad t = \phi\,(x),\ \Rightarrow\ y = g\,[\phi\,(x)\,].$

We say that y is defined as a function of x defined with the help of the *parameter* t. Now we have

$$\frac{dy}{dx} = \frac{dy}{dt}\,.\,\frac{dt}{dx} = \frac{dy}{dt} \div \frac{dx}{dt},$$

where we suppose that $dx/\,dt \neq 0$ for the value of t in question.

Note. It will be seen in Chapter 7 that if $f'\,(t) = dx/\,dt$ is zero for only a *finite* number of values of t, then the domain of f can be divided into a finite number of intervals in each of which the function f is strictly increasing or decreasing and therefore, invertible. In the following, we shall suppose that this condition is always satisfied so that, we have $dy\,/\,dx = (dy/\,dt) \div (dx/\,dt)$ for all those values of t for which $dx/\,dt \neq 0$

EXAMPLES

1.   *Find $dy/\,dx$ when*

$$x = 2 \cos t - \cos 2t, \qquad y = 2 \sin t - \sin 2t.$$

We have,

$$\frac{dx}{dt} = -\,2 \sin t + 2 \sin 2t, \qquad \frac{dy}{dt} = 2 \cos t - 2 \cos 2t$$

Thus $\dfrac{dy}{dt} = \dfrac{dy}{dt} \div \dfrac{dx}{dt} = \dfrac{\cos t - \cos 2t}{\sin 2t - \sin t} = \tan \dfrac{3t}{2}$

for all those values of t for which $dx/\,dt \neq 0$.

The reader may give the set of values of t for which $dx/\,dt = 0$.

2.   *If $x = e^{-t^2}$ and $y = tan^{-1}\,(2t + 1),\,find\ \dfrac{dy}{dx}.$*

We have $\dfrac{dx}{dt} = e^{-t^2}\,(-\,2t)$

and $\dfrac{dy}{dt} = \dfrac{1}{1 + (2t + 1)^2}$ (2)

$$\therefore \quad \frac{dy}{dx} = \frac{dy/dt}{dx/dt} = \frac{\dfrac{2}{1+(4t^2+4t+1)}}{-2t/e^{t^2}}$$

$$= \frac{-e^{t^2}}{2t\,(2t^2+2t+1)}$$

EXERCISES

1. Find dy/dx, when

   (i) $x = a\,(\cos t + \sin t)$,   (ii) $x = 3\cos t - 2\cos^3 t$,

       $y = a\,(\sin t - t\cos t)$, $y = 3\sin t - 2\sin^3 t$,

  (iii) $x = a\cos^3 t$,   (iv) $x = a\,(t - \sin t)$,

       $y = a\sin^3 t$, $y = a\,(1 + \cos t)$.

2. Find dy/dx, when

   (i) $x = a\,\dfrac{1-t^2}{1+t^2}$, $y = b\,\dfrac{2t}{1+t^2}$.   (ii) $x = \dfrac{3at}{1+t^3}$, $y = \dfrac{3at^2}{1+t^2}$,

  (iii) $x = \dfrac{2at^2}{1+t^2}$, $y = \dfrac{2at^3}{1+t^2}$,

  (iv) $x = a\sqrt{\left(\dfrac{t^2-1}{t^2+1}\right)}$, $y = at\sqrt{\left(\dfrac{t^2-1}{t^2+1}\right)}$

ANSWERS

1. (i) $\tan t$,   (ii) $\cot t$,   (iii) $-\tan t$,   (iv) $-\tan t/2$.

2. (i) $\dfrac{t^2-1}{2at}$.   (ii) $\dfrac{t\,(2-t^3)}{1-2t^3}$,   (iii) $\dfrac{3}{2}\,t\,(1-t)^2$

  (iv) $\dfrac{2at}{\sqrt{(t^2-1)}\,(t^2+1)^{3/2}}$

4.9. Logarithmic differentiation :

In order to differentiation a function of the form $y = u^v$; where u, v are both variables, it i. round useful to take its logarithm and then differentiate. This process which is known as **Logarithmic differentiation** is also useful when the function to be differentiated is the puoduct of a number of factors. The following examples illustrate this process.

We use the process of logarithmic differentiation for differentiating the function $y = x^a$ where a is any real number. Now

$$y = x^a \Rightarrow \log y = a \log x$$

$$\Rightarrow \quad \frac{1}{y}\frac{dy}{dx} = \frac{a}{x} \Rightarrow \frac{dy}{dx} = \frac{ay}{x} = ax^{a-1}$$

EXAMPLES

1. *Find the derivative of* $y = x^{\sin x}$.

Now $y = x^{\sin x}$

$\Rightarrow$ $\log y = \log(x^{\sin x}) = \sin x \log x$.

Differentiating, we get

$$\frac{1}{y} \cdot \frac{dy}{dx} = \cos x \log x + \sin x \cdot \frac{1}{x}$$

$$\Rightarrow \qquad \frac{dy}{dx} = x^{\sin x} \left(\cos x \log x + \frac{\sin x}{x} \right).$$

2. *Differentiate* $y = [x^{\tan x} + (\sin x)^{\cos x}]$.

Let $\qquad u = x^{\tan x}, \ v = (\sin x)^{\cos x}$

$$\Rightarrow \qquad y = u + v \Rightarrow \frac{dy}{dx} = \frac{du}{dx} + \frac{dv}{dx}$$

By taking logarithmes, we may show that

$$\frac{du}{dx} = x^{\tan x} \left(\sec^2 x \log x + \frac{\tan x}{x} \right) \qquad \qquad \dots(i)$$

$$\frac{dv}{dx} = (\sin x)^{\cos x} \left(- \sin x \log \sin x + \frac{\cos^2 x}{\sin x} \right) \qquad \qquad \dots(ii)$$

Adding (*i*) and (*ii*), we obtain *dy/ dx*

3. *Differentiate* $\qquad y = \dfrac{x^{1/2} (1 - 2x)^{2/3}}{(2 - 3x)^{3/4} (3 - 4x)^{4/5}}$

Taking lograithms, we obtain

$$\log y = \frac{1}{2} \log x + \frac{2}{3} \log (1 - 2x) - \frac{3}{4} \log (2 - 3x) - \frac{4}{5} \log (3 - 4x)$$

Differentiating, we obtain

$$\frac{1}{y} \cdot \frac{dy}{dx} = \frac{1}{2} \cdot \frac{1}{x} + \frac{2}{3} \cdot \frac{-2}{1 - 2x} - \frac{3}{4} \cdot \frac{-3}{2 - 3x} - \frac{4}{5} \cdot \frac{-4}{3 - 4x}$$

$$= \frac{1}{2x} - \frac{4}{3 (1 - 2x)} + \frac{9}{4 (2 - 3x)} + \frac{16}{5 (3 - 4x)}$$

$$\Rightarrow \qquad \frac{dy}{dx} = y \left[\frac{1}{2x} - \frac{4}{3 (1 - 2x)} + \frac{9}{4 (2 - 3x)} + \frac{16}{5 (3 - 4x)} \right].$$

4. *If* $x^y \cdot y^x = 1$, *find* $\dfrac{dy}{dx}$

Taking logarithm

$$y \log x + x \log y = 0$$

Differentiating both sides, we get

$$y \cdot \frac{1}{x} + \log x \cdot \frac{dy}{dx} + 1 \cdot \log y + x \cdot \frac{1}{y} \frac{dy}{dx} = 0$$

$$\Rightarrow \qquad \left[\log x + \frac{x}{y} \right] \frac{dy}{dx} = - \left[\frac{y}{x} + \log y \right]$$

$$\Rightarrow \qquad \frac{dy}{dx} = - \frac{(y + x \log y)}{(x + y \log x)} \cdot \frac{y}{x}.$$

5. If $y^{\cot x} + (\tan^{-1})^y = 1$, find $\dfrac{dy}{dx}$.

Let
$$u = y^{\cot x} \text{ and } v = (\tan^{-1} x)^y, \text{ then}$$
$$u + v = 1 \qquad \qquad \qquad \qquad \text{...(i)}$$

Taking log on both sides for u and v, we obtain
$$\log u = \cot x \log y \text{ and } \log v = y \log (\tan^{-1} x)$$

On differentiating, we get
$$\frac{1}{u} \cdot \frac{du}{dx} = \cot x \cdot \frac{1}{y} \frac{dy}{dx} + \log y \cdot (- \operatorname{cosec}^2 x)$$

$$\Rightarrow \qquad \frac{du}{dx} = y^{\cot x} \left[\frac{\cot x}{y} \cdot \frac{dy}{dx} - (- \operatorname{cosec}^2 x) \log y \right]$$

Again,
$$\frac{1}{v} \cdot \frac{dv}{dx} = y \cdot \frac{1}{\tan^{-1} x} \cdot \frac{1}{1 + x^2} + \log (\tan^{-1} x) \frac{dy}{dx}$$

$$\Rightarrow \qquad \frac{dv}{dx} = (\tan^{-1} x)^y \left[\frac{y}{(1 + x^2) \tan^{-1} x} + \log (\tan^{-1} x) \frac{dy}{dx} \right]$$

Now, from (i)
$$\frac{du}{dx} + \frac{dv}{dx} = 0$$

$$\Rightarrow \qquad y^{\cot x} \left\{ \frac{\cot x}{y} \cdot \frac{dy}{dx} - (\operatorname{cosec}^2 x) \log y \right\}$$

$$+ (\tan^{-1} x)^y \left[\frac{y}{(1 + x^2) \tan^{-1} x} + \log (\tan^{-1} x) \cdot \frac{dy}{dx} \right] = 0$$

$$\Rightarrow \qquad \left\{ \frac{y^{\cot x} \cdot \cot x}{y} + (\tan^{-1} x)^y \cdot \log (\tan^{-1} x) \right\} \frac{dy}{dx}$$

$$= \left\{ y^{\cot x} \cdot (\operatorname{cosec}^2 x) \log y - \frac{(\tan^{-1} x)^y \cdot y}{(1 + x^2) \tan^{-1} x} \right\}$$

$$\therefore \qquad \frac{dy}{dx} = \frac{(y^{\cot x} \cdot \operatorname{cosec}^2 x \cdot \log y) - \left\{ \dfrac{(\tan^{-1} x)^{y-1} \cdot y}{1 + x^2} \right\}}{(y^{\cot x - 1} \cdot \cot x) + \left\{ (\tan^{-1} x)^y \cdot \log (\tan^{-1} x) \right\}}$$

EXERCISES

Find the derivatives of the functions with the following function values :

(i) $(\cos x)^{\log x}$

(ii) $(1 + x^{-1})^x$.

(iii) e^{x^x}

(iv) $x^{\sin x} + (\sin x)^x$

(v) $(\tan x)^{\cot x} + (\cot x)^{\tan x}$

(vi) $(\log x)^x + (\sin^{-1} x)^{\sin x}$.

(vii) $\dfrac{(1 - x)^{1/2} (2 - x^2)^{2/3}}{(3 - x^3)^{3/4} (4 - x^4)^{4/5}}$.

(viii) $\dfrac{x^3 \sqrt{(x^2 + 4)}}{\sqrt{(x^2 + 3)}}$

(ix) $\sin x . e^x . \log x . . x^x$

ANSWERS

Ex. (i) $(\cos x)^{\log x} [(\log \cos x)/x - \tan x \cdot \log x]$.

(ii) $(1 + x^{-1})^x [\log (1 + x^{-1}) - (1 + x)^{-1}]$.

(iii) $x^x \, e^{x^x} \log ex$.

(iv) $x^{\sin x} \left(\dfrac{\sin x}{x} + \cos x \log x \right) + (\sin x)^x \, (\log \sin x + x \cot x)$

(v) $(\tan x)^{\cot x} \operatorname{cosec}^2 x \log (e \cot x) - (\cot x)^{\tan x} \sec^2 x \log (e \tan x)$.

(vi) $(\log x)^x \left(\log \log x + \dfrac{1}{\log x} \right) + (\sin^{-1} x)^{\sin x} \left(\cos x \cdot \log \sin^{-1} x + \dfrac{\sin x}{\sqrt{(1 - x^2)} \sin^{-1} x} \right)$

(vii) $\dfrac{(1 - x)^{1/2} (2 - x^2)^{2/3}}{(3 - x^3)^{3/4} (4 - x^4)^{4/5}} \times \left(-\dfrac{1}{2 (1 - x)} - \dfrac{4x}{3 (2 - x^2)} + \dfrac{9x^2}{4 (3 - x^3)} + \dfrac{16x^3}{5 (4 - x^4)} \right)$

(viii) $(2x^4 + 15 \, x^2 + 36)/3 \, (x^2 + 3)^{2/3} (x^2 + 4)^{2/3}$

(ix) $\sin x \cdot e^x \cdot \log x \cdot x^x [\cot x + (x \log x)^{-1} + \log e^2 x]$.

4.10. Derivation of implicitly defined functions

Having already introduced the notion of implicitly defined function in chapter 2, we now illustrate the process of derivation of such functions without, even if it were possible, expressing explictly in terms of x. We shall also assume that in any given case the implicitly defined function is also derivable.

EXAMPLES

1.   *Find dy/dx when* $x^3 + y^3 = 3axy$.

Supposing that the given equation determines a function f, we see that

$$x^3 + [f(x)]^3 - 3ax [f(x)] = 0$$

for every value of x in the domain of the function f. We have here the zero function whose derivative is 0 for every admissible value of x. Using the different rules for derivation, we see that

$$3x^2 + 3[f(x)]^2 f'(x) - 3af(x) - 3ax f'(x) = 0$$

$\Leftrightarrow$ $\qquad x^2 - af(x) = \{ax - [f(x)]^2\} f'(x)$

$\Rightarrow$ $\qquad f'(x) = \dfrac{x^2 - af(x)}{ax - [f(x)]^2}$,

the result holding for all those values of x for which the denominator

$ax - [f(x)^2] \neq 0$. Replace $f(x)$ by y and $f'(x)$ by dy/dx, we obtain

While the above discussion is intended to describe the basic situation, in actual practice, however, we proceed as follows, supposing that y denotes an implicitly defined derivable function of x.

Making use of the various results on the derivation, we have

$$x^3 + y^3 = 3axy$$

$\Rightarrow$ $\qquad 3x^2 + 3y^2 \dfrac{dy}{dx} = 3a \left(y + x \dfrac{dy}{dx} \right)$

$$\Rightarrow \qquad \frac{dy}{dx} = \frac{x^2 - ay}{ax - y^2}.$$

the result holding for points (x, y) for which $ax - y^2 \neq 0$.

2. If $\sin y = x \sin(a + y)$, prove that $\dfrac{dy}{dx} = \dfrac{\sin^2(a + y)}{\sin a}$

We have $\qquad\qquad x = \dfrac{\sin y}{\sin(a + y)}$

$$\Rightarrow \qquad \frac{dx}{dy} = \frac{\cos y \sin(a + y) - \cos(a + y) \sin y}{\sin^2(a + y)}$$

$$= \frac{\sin(a + y - y)}{\sin^2(a + y)}$$

$$= \frac{\sin a}{\sin^2(a + y)}$$

$$\therefore \qquad \frac{dy}{dx} = \frac{\sin^2(a + y)}{\sin a}$$

3. If $x^y = e^{x-y}$, prove that $\dfrac{dy}{dx} = \dfrac{\log x}{(1 + \log x)^2}$

$$x^y = e^{x-y}$$

Taking log of both sides we get

$$y \log x = x - y$$

$$\Rightarrow \qquad\qquad y = \frac{x}{1 + \log x}$$

$$\Rightarrow \qquad\qquad \frac{dy}{dx} = \frac{1 + \log x - \frac{1}{x} \cdot x}{(1 + \log x)^2}$$

$$= \frac{\log x}{(1 + \log x)^2}$$

4. If $y\sqrt{1 - x^2} + x\sqrt{1 - y^2} = 1$, find $\dfrac{dy}{dx}$.

Let $\qquad\qquad x = \sin\theta,\ y = \sin\phi$, then

$$\sin\phi\,\sqrt{1 - \sin^2\theta} + \sin\theta\,\sqrt{1 - \sin^2\phi} = 1$$

$$\Rightarrow \qquad \sin(\theta + \phi) = 1$$

$$\Rightarrow \qquad \sin^{-1}x + \sin^{-1}y = \sin^{-1}1$$

Differentiating both sides,

$$\frac{1}{\sqrt{1 - x^2}} + \frac{1}{\sqrt{1 - y^2}} \cdot \frac{dy}{dx} = 0$$

$$\Rightarrow \qquad \frac{dy}{dx} = -\sqrt{\left(\frac{1-y^2}{1-x^2}\right)}$$

5. *Find* $\dfrac{dy}{dx}$, *where* $y = \left(\sqrt{x}\right)^{x^{x^{x \cdots to \infty}}}$

We have $\qquad y^2 = x^{x^{x^{x \cdots to \infty}}} \qquad \Rightarrow \qquad y^2 = x^{y^2}$

Taking logarithm, we get

$$2 \log y = y^2 \log x$$

On differentiation, we obtain

$$2 . \frac{1}{y} \frac{dy}{dx} = y^2 . \frac{1}{x} + 2 y . \frac{dy}{dx} . \log x$$

$$\Rightarrow \qquad \left[\frac{2}{y} - 2y \log x\right] \frac{dy}{dx} = \frac{y^2}{x}$$

$$\therefore \qquad \frac{dy}{dx} = \frac{y^3}{2x \, (1 - y^2 \, \log x)}$$

6. *If*

$$\cfrac{\sin x}{1 + \cfrac{\cos x}{1 + \cfrac{\sin x}{1 + \cos x \, \cdots \, \infty}}}$$

then prove that $\dfrac{dy}{dx} = \dfrac{(1 + y)\cos x + \sin x}{1 + 2y + \cos x - \sin x}$

Given function is

$$y = \frac{\sin x}{1 + \cfrac{\cos x}{1 + y}} = \frac{(1 + y)\sin x}{1 + y + \cos x}$$

or $\qquad y + y^2 + y \cos x = (1 + y) \sin x.$

On differentiation, we get

$$\frac{dy}{dx} + 2 y \frac{dy}{dx} + y(-\sin x) + \cos x . \frac{dy}{dx} = (1 + y)\cos x + \frac{dy}{dx} . \sin x$$

or $\qquad \dfrac{dy}{dx} (1 + 2y + \cos x - \sin x) = (1 + y) \cos x + y \sin x$

$$\therefore \qquad \frac{dy}{dx} = \frac{(1 + y)\cos x + y \sin x}{(1 + 2y + \cos x - \sin x)}$$

EXERCISES

1. Find $\dfrac{dy}{dx}$ if

 (i) $3x^2 - 2y^2 = 1$ for $(1, 1)$ and $(1, -1)$

 (ii) $x^4 + y^4 - a^2 xy = 0$ for (x, y)

2. If $x\sqrt{1 + y} + y\sqrt{1 + x} = 0$ prove that $\dfrac{dy}{dy} = \dfrac{-1}{(1 + x)^2}$

3. If $\dfrac{x}{x-y} = \log \dfrac{a}{x-y}$, prove that $\dfrac{dy}{dx} = 2 - \dfrac{x}{y}$

4. Find $\dfrac{dy}{dx}$ if y and x are related by the equation $x^2 \sin y = y^2 \sin x$.

5. Find $\dfrac{dy}{dx}$ if

    (i) $x^m y^m = (x+y)^{m+n}$ (ii) $y = \cos(x+y)$

    (iii) $x^{2/3} + y^{2/3} = a^{2/3}$ (iv) $y = x \log \dfrac{y}{a+bx}$

    (v) $x = y \log(xy)$ (vi) $x^v + y^x = c$

    (vii) $(\cos x)^y = (\sin y)^x$

6. Find $\dfrac{dy}{dx}$ when $x = e^{\tan^{-1}\left(\frac{y-x^2}{x^2}\right)}$

7. If $y = x \cos y + y \cos x$, find $\dfrac{dy}{dx}$

8. If $y = \sqrt{\sin x + \sqrt{\sin x + \sqrt{\sin x + \ldots \ldots \infty}}}$, find $\dfrac{dy}{dx}$.

9. If $y = \dfrac{x}{x + \dfrac{\sqrt[3]{x}}{x + \dfrac{\sqrt[3]{x}}{x + \sqrt[3]{x} + \ldots \ldots \infty}}}$ find $\dfrac{dy}{dx}$.

10. If $\sqrt{1-x^6} + \sqrt{1-y^6} = a^3(x^3 - y^3)$, find $\dfrac{dy}{dx}$

ANSWERS

1. (i) $\dfrac{3}{2}, -\dfrac{3}{2}$.   (ii) $(4x^3 - a^2y)/(a^2x - 4y^3)$;

4. $\dfrac{y^2 \cos x - 2x \sin y}{x^2 \cos y - 2y \sin x}$.

5. (i) $\dfrac{y}{x}$ (ii) $\dfrac{-\sin(x+y)}{1+\sin(x+y)}$ (iii) $-\left(\dfrac{y}{x}\right)^{1/3}$

    (iv) $\dfrac{y\{(a+bx)y - bx^2\}}{x(y-x)(a+bx)}$ (v) $\dfrac{y(x-y)}{x(x+y)}$ (vi) $\dfrac{(yx^{y-1} + x^y \log y)}{x^y \log x + xy^{x-1}}$

    (vii) $\dfrac{\log \sin y + y \tan x}{\log \cos x - x \cot y}$

6. $2x + x(2 \tan(\log x) + \sec^2(\log x))$

7. $\dfrac{\cos y - y \sin x}{1 + x \sin y - \cos x}$ 8. $\dfrac{\cos x}{2y - 1}$ 9. $\dfrac{\frac{5}{3}x^{2/3}(1-y)}{(x^{5/3} + 2y)}$

10. $\dfrac{x^2}{y^2}\sqrt{\dfrac{1-y^6}{1-x^6}}$

4.11. Preliminary transformation. In some cases, a preliminary transformation of the function to be differentiated facilitates the process of differentiation a good deal, as is illustrated by the following examples.

1. *Differentiate*

$$\sin^{-1}\frac{2x}{1+x^2}.$$

Putting $x = \tan\theta$, we have

$$y = \sin^{-1}\frac{2x}{1+x^2}$$

$$= \sin^{-1}\frac{2\tan\theta}{1+\tan^2\theta} = \sin^{-1}(\sin 2\theta) = 2\theta = 2\tan^{-1}x.$$

$$\therefore \quad \frac{dy}{dx} = \frac{2}{1+x^2}.$$

2. *Differentiate*

$$\tan^{-1}\frac{\sqrt{(1+x)}-\sqrt{(1-x)}}{\sqrt{(1+x)}+\sqrt{(1-x)}}$$

Putting $x = \cos\theta$, we have

$$\sqrt{(1+x)} = \sqrt{(1+\cos\theta)} = \sqrt{\left(2\cos^2\frac{\theta}{2}\right)} = \sqrt{2}\cos\frac{\theta}{2}.$$

$$\sqrt{(1-x)} = \sqrt{(1-\cos\theta)} = \sqrt{\left(2\sin^2\frac{\theta}{2}\right)} = \sqrt{2}\sin\frac{\theta}{2}.$$

$$\therefore \quad y = \tan^{-1}\frac{\cos\dfrac{\theta}{2}-\sin\dfrac{\theta}{2}}{\cos\dfrac{\theta}{2}+\sin\dfrac{\theta}{2}}$$

$$= \tan^{-1}\frac{1-\tan\dfrac{\theta}{2}}{1+\tan\dfrac{\theta}{2}}$$

$$= \tan^{-1}\left\{\tan\left(\frac{\pi}{4}-\frac{\theta}{2}\right)\right\}$$

$$= \frac{\pi}{4}-\frac{\theta}{2} = \frac{\pi}{4}-\frac{1}{2}\cos^{-1}x.$$

$$\therefore \quad \frac{dy}{dx} = \frac{1}{2\sqrt{(1-x^2)}}.$$

3. *Find the differential co-officient of* $\tan^{-1}\dfrac{2x}{1-x^2}$ *with respect to* $\sin^{-1}\dfrac{2x}{1+x^2}$.

Let $y = \tan^{-1}\dfrac{2x}{1-x^2}, \quad z = \sin^{-1}\dfrac{2x}{1+x^2}.$

Putting $x = \tan\theta$, we see that

$$y = \tan^{-1}\frac{2\tan\theta}{1-\tan^2\theta} = \tan^{-1}(\tan 2\theta) = 2\theta = 2\tan^{-1}x.$$

$$z = \tan^{-1}\frac{2\tan\theta}{1+\tan^2\theta} = \sin^{-1}(\sin 2\theta) = 2\theta = 2\tan^{-1}x.$$

$$\therefore \qquad \frac{dy}{dx} = \frac{2}{1+x^2}, \frac{dz}{dx} = \frac{2}{1+x^2}.$$

$$\therefore \qquad \frac{dy}{dz} = \frac{dy}{dx}\cdot\frac{dx}{dz} = 1.$$

Also otherwise, we have

$$y = z,$$

so that $\qquad \dfrac{dy}{dz} = 1.$

4. *If* $\sqrt{(1-x^2)} + \sqrt{(1-y^2)} = a(x-y)$, *prove that*

$$\frac{dy}{dx} = \frac{\sqrt{(1-y^2)}}{(1-x^2)}.$$

Putting $x = \sin\theta$ and $y = \sin\phi$, we have

$$\sqrt{(1-\sin^2\theta)} + \sqrt{(1-\sin^2\phi)} = a(\sin\theta - \sin\phi)$$

$$\Rightarrow \qquad \cos\theta + \cos\phi = a(\sin\theta - \sin\phi)$$

$$\Rightarrow \quad a = \frac{\cos\theta + \cos\phi}{\sin\theta - \sin\phi} = \frac{2\cos\frac{1}{2}(\theta-\phi)\cos\frac{1}{2}(\theta-\phi)}{2\cos\frac{1}{2}(\theta-\phi)\sin\frac{1}{2}(\theta-\phi)}$$

$$\therefore \qquad \cot\frac{1}{2}(\theta-\phi) = a$$

$$\Rightarrow \qquad \frac{1}{2}(\theta-\phi) = \cot^{-1}a$$

$$\theta - \phi = 2\cot^{-1}a.$$

$$\Rightarrow \quad \sin^{-1}x - \sin^{-1}y = 2\cot^{-1}a.$$

Differentiating, we get

$$\frac{1}{\sqrt{(1-x^2)}} - \frac{1}{\sqrt{(1-y^2)}}\frac{dy}{dx} = 0,$$

$$\Rightarrow \qquad \frac{dy}{dx} = \frac{\sqrt{(1-y^2)}}{\sqrt{(1-x^2)}}.$$

5. *Differentiate* $\cos^{-1}\left[\dfrac{3\cos x - 2\sin x}{\sqrt{13}}\right]$

w.r.t. $\qquad \sin^{-1}\left[\dfrac{5\sin x + 4\cos x}{\sqrt{41}}\right].$

Let $\qquad u = \cos^{-1}\left[\dfrac{3\cos x - 2\sin x}{\sqrt{13}}\right]$ and

$$v = \sin^{-1}\left[\frac{5\sin x + 4\cos x}{\sqrt{41}}\right]$$

Now, let $\quad \dfrac{3}{\sqrt{13}} = \cos\alpha \;$ and $\; \dfrac{2}{\sqrt{13}} = \sin\alpha$

$\therefore \quad u = \cos^{-1}(\cos\alpha\cos x - \sin\alpha\sin x) = x + \alpha \qquad \qquad …(i)$

$\Rightarrow \qquad\qquad \dfrac{du}{dx} = 1$

Again, let $\quad \dfrac{5}{\sqrt{41}} = \cos\beta \;$ and $\; \dfrac{4}{\sqrt{41}} = \sin\beta$

then $v = \sin^{-1}[\cos\beta\sin x + \sin\beta\cos x] = x + \beta$

$\therefore \qquad\qquad \dfrac{dv}{dx} = 1 \qquad\qquad\qquad\qquad\qquad\qquad …(ii)$

from (i) and (ii) $\quad \dfrac{du}{dv} = 1$.

EXERCISES

Find the derivatives of :

1. $y = \dfrac{x\cos^{-1}x}{\sqrt{(1-x^2)}}.$

2. $y = \dfrac{(3x^2 - 1)\sqrt{1+x^2}}{x^2}$

3. $y = \dfrac{x\sqrt{x^2 - 4a^2}}{\sqrt{(x^2 - a^2)}}.$

4. $y = \tan^{-1}\left(\dfrac{a}{x}\tan^{-1}\dfrac{x}{a}\right)$

5. $y = e^{(x)^x}.$

6. $y = \tan^{-1}\dfrac{x^{1/3} + a^{1/3}}{1 - a^{1/3}x^{1/3}}.$

7. $y = \tan^{-1}\dfrac{\cos x}{1 + \sin x}.$

8. $y = \tan^{-1}\left(\dfrac{1 - \cos x}{1 + \cos x}\right)^{1/2}$

9. $y = \dfrac{(1 - 2x)^{2/3}(1 + 3x)^{-3/4}}{(1 - 6x)^{5/6}(1 + 7x)^{-6/7}}.$

Obtain the derivatives of the functions given by the following function values :

10. $(x^x)^x.$

11. $x^{(x^x)}.$

12. $[1 - x^2]^{3/2}.\sin^{-1}x$

13. $\log\left[\tan\left(\dfrac{1}{2}x\right)\right].$

14. $\tan^{-1}\dfrac{\sqrt{(1+x^2)} + \sqrt{(1-x^2)}}{\sqrt{(1+x^2)} - \sqrt{(1-x^2)}}.$

15. $\sqrt{\left(\dfrac{1 + \tan x}{1 - \tan x}\right)}.$

16. $\sin^{-1}[2ax\sqrt{(1 - a^2x^2)}].$

17. $10^{\log\sin x}.$

18. $x\log x . \log(\log x).$

19. $\tan^{-1}\dfrac{a + b\cos x}{b + a\cos x}.$

20. $(\sin x)^{\cos^{-1}x}$

21. $e^{-ax^2}\sin(x\log x).$

22. $\sin^{-1}\left(\dfrac{e^{ax} - e^{-ax}}{e^{ax} + e^{-ax}}\right).$

23. $\tan^{-1}\dfrac{\sqrt{(1+x^2)} - 1}{x}.$

24. $\sin^{-1}\dfrac{x^2}{\sqrt{(x^4 + a^4)}}.$

25. $\log\left[e^x\left(\dfrac{x-2}{x+2}\right)^{3/4}\right].$

26. $\log\left(\dfrac{1 + \sqrt{x}}{1 - \sqrt{x}}\right).$

27. $\tan^{-1}\left(\dfrac{x\sin a}{1 - x\cos a}\right).$

28. $9x^4\sin(3x - 7)\log(1 - 5x)$

29. $e^{ax} \cos (b \tan^{-1} x)$ **30.** $\left(1 + \dfrac{1}{x}\right)^{x^2}$ **31.** $\cot x \coth x.$

32. $x\, a^x \sinh x.$ **33.** If $y = \tan^{-1} \dfrac{3a^4 x - x^3}{a(a^2 - 3x^2)}$, prove that $\dfrac{dy}{dx} = \dfrac{3a}{a^2 + x^2}$

34. Find $\dfrac{dy}{dx}$ when

(i) $x = \dfrac{\sin^3 t}{\sqrt{\cos 2t}}$, $y = \dfrac{\cos^3 t}{\sqrt{\cos 2t}}$ at $t = \pi/6$

(ii) $x = \sin t \sqrt{\cos 2t}$, $y = \cos t \sqrt{\cos 2t}$

(iii) $x = a\left(\cos t + \log \tan \dfrac{1}{2} t\right)$, $y = a \sin t$

35. Differentiate $\tan^{-1}\left[\left\{\sqrt{1 + x^2} - 1\right\}/x\right]$ with respect to $\tan^{-1} x$.

36. Differentiate $x^{\sin x}$ with respect to $(\sin x)^x$

37. Differentiate $\tan^{-1} \dfrac{\sqrt{1 + x^2} - \sqrt{1 - x^2}}{\sqrt{1 + x^2} + \sqrt{1 - x^2}}$ with respect to $\cos^{-1} x^2$.

38. Differentiate $(\log x)^{\tan x}$ with respect to $\sin (m \cos^{-1} x)$.

ANSWERS

1. $\dfrac{\cos^{-1} x - \sqrt{(1 - x^2)}}{(1 - x^2)^{3/2}}$

2. $\dfrac{3 - x^2}{x^4 \sqrt{(1 + x^2)}}$

3. $\dfrac{x^4 - 2a^2 x^2 - 4a^4}{(x^2 - a^2)^{3/2} (x^2 - 4a^2)^{1/2}}.$

4. $\dfrac{a^2}{a^2 + x^2 \left(\tan^{-1} \dfrac{x}{a}\right)^2} \left(\dfrac{1}{a} \tan^{-1} \dfrac{x}{a} + \dfrac{x}{a^2 + x^2}\right).$

5. $e x^x \cdot \log ex.$

6. $\dfrac{1}{x^{2/3} (1 + x^{2/3})}.$

7. $-\dfrac{1}{2}.$

8. $\dfrac{1}{2}.$

9. $\dfrac{(1 - 2x)^{2/3} (1 + 3x)^{-3/4}}{(1 - 6x)^{5/6} (1 + 7x)^{-6/7}} \times \left(\dfrac{5}{1 - 6x} + \dfrac{6}{1 + 7x} - \dfrac{4}{3(1 - 2x)} - \dfrac{9}{4(1 + 3x)}\right).$

10. $x^{(x^2 + 1)} \log (ex^2).$

11. $[1 + x \log x \log ex] x^{(x^x) + x - 1}$

12. $1 - x^2 - 3x\sqrt{(1 + x^2)} \cdot \sin^{-1} x.$

13. $\operatorname{cosech} x.$

14. $-\dfrac{x}{\sqrt{(1 - x^4)}}.$

15. $\dfrac{1}{(\cos x - \sin x)\sqrt{(\cos 2x)}}.$

16. $\dfrac{2a}{\sqrt{(1 - a^2 x^2)}}.$

17. $10^{\log \sin x} \cot x \cdot \log 10.$

18. $\log (e \log x) + \log x \cdot \log \log x.$

19. $\dfrac{(a^2 - b^2)\sin x}{(a^2 + b^2)(1 + \cos^2 x) + 4ab\cos x}.$

20. $(\sin x)^{\cos^{-1} x} \left(\cos^{-1} x \cdot \cot x - \dfrac{\log \sin x}{\sqrt{(1 - x^2)}} \right).$

21. $e^{-ax^2} [\cos (x \log x) \cdot \log ex - 2ax \sin (x \log x)].$

22. $a \operatorname{sech} ax.$

23. $\dfrac{1}{2(1 + x^2)}.$

24. $\dfrac{2a^2 x}{a^4 + x^4}.$

25. $\dfrac{x^2 - 1}{x^2 - 4}.$

26. $\dfrac{1}{\sqrt{x}\,(1 - x)}.$

27. $\dfrac{\sin a}{1 - 2x\cos a + x^2}.$

28. $9x^4 \sin (3x - 7) \log (1 - 5x) \times \left(\dfrac{4}{x} - \{\log(1 - 5x)\}^{-1} \dfrac{5}{(1 - 5x)} + 3\cot (3x - 7) \right).$

29. $e^{ax} [a \cos (b \tan^{-1} x) - b \sin (b \tan^{-1} x) (1 + x^2)^{-1}].$

30. $\left(1 + \dfrac{1}{x}\right)^{x^2} \left[2x \log\left(1 + \dfrac{1}{x}\right) - \dfrac{x}{1 + x} \right].$

31. $- \operatorname{cosec}^2 x \coth x - \operatorname{cosech}^2 x \cot x.$

32. $a^x \sinh x + xa^x \sinh x \log a + x\, a^x \cosh x$

34. (*i*) 0.   (*ii*) $- \tan 3t.$ (*iii*) $\tan t.$

35. $\dfrac{1}{2}$

36. $\dfrac{x^{\sin x - 1}(\sin x + x \cos x \log x)}{(\sin x)^{x - 1}(x \cos x + \sin x \log \sin x)}$

37. $- \dfrac{1}{2}$

38. $\dfrac{-\sqrt{(1 - x^2)}\,(\log x)^{\tan x}[\sec^2 x \log (\log x) + \tan x (x \log x)^{-1}]}{m \cos (m \cos^{-1} x)}$

4.12. Differentitation of a Determinant.

If a function be given in the form of a determienant. *i.e.*, if

$$y = \begin{vmatrix} p(x) & q(x) & r(x) \\ u(x) & v(x) & w(x) \\ \lambda(x) & \mu(x) & \upsilon(x) \end{vmatrix}$$

then $$\frac{dy}{dx} = \begin{vmatrix} p'(x) & q(x) & r(x) \\ u'(x) & v(x) & w(x) \\ \lambda'(x) & \mu(x) & \upsilon(x) \end{vmatrix} + \begin{vmatrix} p(x) & q'(x) & r(x) \\ u(x) & v'(x) & w(x) \\ \lambda(x) & \mu'(x) & \upsilon(x) \end{vmatrix} + \begin{vmatrix} p(x) & q(x) & r'(x) \\ u(x) & v(x) & w'(x) \\ \lambda(x) & \mu(x) & \upsilon'(x) \end{vmatrix},$$

The differentiation can also be done by row-wise.

Example : If $f(x) = \begin{vmatrix} x & x^2 & x^3 \\ 1 & 2x & 3x^2 \\ 0 & 2 & 6x \end{vmatrix}$, find $f'(x)$.

On differentiating, we get

$$f'(x) = \begin{vmatrix} \dfrac{d}{dx}(x) & x^2 & x^3 \\[2mm] \dfrac{d}{dx}(1) & 2x & 3x^2 \\[2mm] \dfrac{d}{dx}(0) & 2 & 6x \end{vmatrix} + \begin{vmatrix} x & \dfrac{d}{dx}(x^2) & x^3 \\[2mm] 1 & \dfrac{d}{dx}(2x) & 3x^2 \\[2mm] 0 & \dfrac{d}{dx}(2) & 6x \end{vmatrix} + \begin{vmatrix} x & x^2 & \dfrac{d}{dx}x^3 \\[2mm] 1 & 2x & \dfrac{d}{dx}3x^2 \\[2mm] 0 & 2 & \dfrac{d}{dx}6x \end{vmatrix}$$

$$= \begin{vmatrix} 1 & x^2 & x^3 \\ 0 & 2x & 3x^2 \\ 0 & 2 & 6x \end{vmatrix} + \begin{vmatrix} x & 2x & x^3 \\ 1 & 2 & 3x^2 \\ 0 & 0 & 6x \end{vmatrix} + \begin{vmatrix} x & x^2 & 3x^2 \\ 1 & 2x & 6x \\ 0 & 2 & 6 \end{vmatrix}$$

$$= 0 + 0 + 6\,(2x^2 - x^2) = 6x^2.$$

MISCELLANEOUS   EXAMPLES

1. *If* $f(x) = \sin\left\{\dfrac{\pi}{2}[x] - x^3\right\}, 1 < x < 2,$ *and* $[x]$ = *the greatest integer* $\leq x$, *then find*
$f'(\sqrt[3]{\pi/2}).$

We have, $1 < \sqrt[3]{\dfrac{\pi}{2}} < 2 \Rightarrow$ if $x = \sqrt[3]{\pi/2}, [x] = 1$

$\therefore \qquad f(x) = \sin\left\{\dfrac{\pi}{2} - x^3\right\} = \cos x^3$

$\qquad\qquad f'(x) = -\sin x^3 \cdot 3x^2$

$\therefore \qquad f'\left(\sqrt[3]{\dfrac{\pi}{2}}\right) = -3\left(\dfrac{\pi}{2}\right)^{2/3} \cdot \sin \pi/2 = -3\left(\dfrac{\pi}{2}\right)^{2/3}.$

2. *Let* $f(x)$ *be a polynomial function of second degree. If* $f(1) = f(-1)$ *and* p, q, r *are in A.P., then show that* $f'(p), f'(q), f'(r)$ *are also in A.P.*

Let $\qquad\qquad f(x) = a_1 x^2 + a_2 x + a_3$

Then $\qquad\quad f'(x) = 2a_1 x + a_2$

Now, $\qquad\quad f(1) = f(-1)$

$\Rightarrow \qquad a_1 + a_2 + a_3 = a_1 - a_2 + a_3$

$\Rightarrow \qquad\qquad a_2 = 0$

$\therefore \qquad\qquad f'(x) = 2a_1 x,$

and $\qquad\quad f'(p) = 2a_1 p, f'(q) = 2a_1 q, f'(r) = 2a_1 r.$

as p, q, r are in A.P., so $f'(p), f'(q), f'(r)$ will also be in A.P.

3. *If* $u = f(x^3), v = g(x^2), f'(x) = \cos x$ *and* $g'(x) = \sin x$, *then find* $\dfrac{du}{dv}$.

We have $\qquad \dfrac{du}{dx} = f'(x^3) \cdot \dfrac{d}{dx}(x^3)$

$\qquad\qquad\quad = \cos(x^3) \cdot 3x^2 = 3x^2 \cos x^3$

and
$$\frac{dv}{dx} = g'(x^2) \cdot \frac{d}{dx}(x^2)$$

$$= 2x \sin x^2$$

$$\therefore \quad \frac{du}{dv} = \frac{du/dx}{dv/dx} = \frac{3x^2 \cdot \cos x^3}{2x \cdot \sin x^2}$$

4. If $y = f\left(\dfrac{2x-1}{x^2+1}\right)$ and $f'(x) = \sin x^2$, find $\dfrac{dy}{dx}$.

We have
$$\frac{dy}{dx} = f'\left(\frac{2x-1}{x^2+1}\right) \cdot \frac{d}{dx}\left(\frac{2x-1}{x^2+1}\right)$$

$$= \sin\left(\frac{2x-1}{x^2+1}\right)^2 \cdot \left\{\frac{(x^2+1) \cdot 2 - (2x-1) \cdot 2x}{(x^2+1)^2}\right\}$$

$$= \sin\left(\frac{2x-1}{x^2+1}\right)^2 \cdot \frac{2(1+x-x^2)}{(x^2+1)^2}.$$

5. If $y = \tan^{-1}\left(\dfrac{a_1 x - \alpha}{a_1 \alpha + x}\right) + \tan^{-1}\left(\dfrac{a_2 - a_1}{1 + a_1 a_2}\right)$

$$+ \tan^{-1}\left(\frac{a_3 - a_2}{1 + a_3 a_2}\right) + \ldots\ldots + \tan^{-1}\left(\frac{a_n - a_{n-1}}{1 + a_n a_{n-1}}\right) - \tan^{-1}(a_n), \text{ then find } \frac{dy}{dx}.$$

We have $y = \tan^{-1}\left(\dfrac{a_1 - \dfrac{\alpha}{x}}{1 + \dfrac{a_1 \alpha}{x}}\right) + \{(\tan^{-1} a_2 - \tan^{-1} a_1) + (\tan^{-1} a_3 - \tan^{-1} a_2) + $

$$\ldots\ldots + (\tan^{-1} a_n - \tan^{-1} a_{n-1}) - \tan^{-1} a_n\}$$

$$= \tan^{-1} a_1 - \tan^{-1}\frac{\alpha}{x} - \tan^{-1} a_1$$

$$\therefore \quad y = -\tan^{-1}\left(\frac{\alpha}{x}\right)$$

$$\Rightarrow \quad \frac{dy}{dx} = -\frac{1}{1+(\alpha/x)^2} \cdot \left(\frac{-\alpha}{x^2}\right)$$

$$= \frac{\alpha}{x^2 + \alpha^2}.$$

6. If $f_n(x) = e^{f_{n-1}(x)}$ for all $n \in N$ and $f_0(x) = x$, then find $\dfrac{d}{dx}\{f_n(x)\}$.

We have,

$$\frac{d}{dx}\{f_n(x)\} = e^{f_{n-1}(x)} \cdot \frac{d}{dx}\{f_{n-1}(x)\}$$

$$= f_n(x) \cdot \frac{d}{dx}\{f_{n-1}(x)\} \qquad \ldots(i)$$

From (*i*), we have

$$\frac{d}{dx}\{f_{n-1}(x)\} = f_{n-1}(x).\frac{d}{dx}\{f_{n-2}(x)\}$$

$$\Rightarrow \qquad \frac{d}{dx}\{f_n(x)\} = f_n(x).f_{n-1}(x).\frac{d}{dx}\{f_{n-2}(x)\}$$

Similarly

$$\frac{d}{dx}\{f_n(x)\} = f_n(x).f_{n-1}(x).f_{n-2}(x)---f_2(x)f_1(x)\left\{\frac{d}{dx}f_0(x)\right\}$$

$$= f_n(x)f_{n-1}(x)f_{n-2}(x)---f_2(x)f_1(x),$$

7. *If for all x, y the function f is defined by :*

$$f(x) + f(y) + f(x).f(y) = 1 \text{ and } f(x) > 0. \text{ Then find } f'(x).$$

Putting $y = x$, we get

$$2f(x) + \{f(x)\}^2 = 1$$

On differentiating w. r. t. x,

$$2f'(x) + 2f(x)f'(x) = 0$$

or $\qquad\qquad 2f'(x)\{1+f(x)\} = 0$

$\Rightarrow \qquad\qquad f'(x) = 0; \text{ because } f(x) > 0$

9. *Let f(x) be a real function not identically zero such that* $f(x + y^{2n+1}) = f(x) + \{f(y)\}^{2n+1}$, $n \in$ *N and x, y are any real numbers and* $f'(0) \geq 0$.

Find the value of f(5) and f'(10).

Putting $\qquad\qquad x = 0, y = 0$, we get

$$f(0) = f(0) + \{f(0)\}^{2n+1} \Rightarrow f(0) = 0$$

$$f'(0) \geq 0 \text{ [Given]}$$

$$\Rightarrow \qquad\qquad \lim_{x \to 0}\frac{f(x) - f(0)}{x - 0} \geq 0$$

$$\Rightarrow \qquad\qquad \lim_{x \to 0}\frac{f(x)}{x} \geq 0$$

$$\therefore \qquad\qquad x > 0 \Rightarrow f(x) \geq 0.$$

Putting $x = 0, y = 1$ in the given relation $f(1) = f(0) + \{f(1)\}^{2n+1}$

$\Rightarrow \qquad\qquad f(1) = 0 \text{ or } 1.$

Now, putting $y = 1$, for all real x, we have $f(x + 1) = f(x) + [f(1)]^{2n+1}$

Case I : $\qquad$ If $f(1) = 0$

Then $\qquad\qquad f(x + 1) = f(x)$

$\Rightarrow \qquad\qquad f(1) = f(2) = f(3) = \ldots\ldots = 0$

It is not possible.

Case II : $\qquad$ If $f(1) = 1$

Then $\qquad\qquad f(x + 1) = f(x) + 1$

$$\Rightarrow \qquad f(2) = f(1) + 1 = 1 + 1 = 2$$
$$f(3) = f(2) + 1 = 2 + 1 = 3$$
$$f(4) = f(3) + 1 = 3 + 1 = 4$$

....

....

$$f(x) = x$$

and $\qquad f'(x) = 1$

$\therefore \qquad f'(10) = 1$ and $f(5) = 5$.

EXERCISES

1. If $5f(x) + 3f\left(\dfrac{1}{x}\right) = x + 2$ and $y = xf(x)$ then find $\dfrac{dy}{dx}$ at $x = 1$.

2. If $f'(x) = \sin x + \sin 4x . \cos x$, find $f'(2x^2 + \pi/2)$.

3. If $f'(x) = \sqrt{2x^2 - 1}$ and $y = f(x^2)$ then find $\dfrac{dy}{dx}$ at $x = 1$.

4. If $f(x) = (\alpha x + \beta) \cos x + (\gamma x + \delta) \sin x$ and $f'(x) = x \cos x$ is identity in x, find the values of α, β, γ and δ.

5. If $y = \tan^{-1} \dfrac{1}{x^2 + x + 1} + \tan^{-1} \dfrac{1}{x^2 + 3x + 3} + \tan^{-1} \dfrac{1}{x^2 + 5x + 7} + \dots$ to nth term,

   then show that $\dfrac{dy}{dx} = \dfrac{1}{(x + n)^2 + 1} - \dfrac{1}{x^2 + 1}$

6. If $f(\theta) = \cos \theta_1 . \cos \theta_2 . \cos \theta_3 \dots \cos \theta_n$, find the value of $\dfrac{f'(\theta)}{f(\theta)}$.

7. If $y = (1 + x)(1 + x^2)(1 + x^4) \dots (1 + x^{2^n})$, find $\dfrac{dy}{dx}$ at $x = 0$.

8. If $(\sin y)^{\frac{\sin \pi x}{2}} + \dfrac{\sqrt{3}}{2} \sec^{-1} 2x + 2^x \tan \{\log_e (x + 2)\} = 0$, find $\dfrac{dy}{dx}$ at $x = -1$

9. If $f_r(x)$, $g_r(x)$, $h_r(x)$; $r = 1, 2, 3$ are polynomials in x such that $f_r(a) = g_r(a) = h_r(a)$; $r = 1, 2, 3$ and

$$f(x) = \begin{vmatrix} f_1(x) & f_2(x) & f_3(x) \\ g_1(x) & g_2(x) & g_3(x) \\ h_1(x) & h_2(x) & h_3(x) \end{vmatrix},$$

   then find $f'(a)$.

10. If $y = \log \sqrt{\dfrac{x^2 + x + 1}{x^2 - x + 1}} + \dfrac{1}{2\sqrt{3}} \left\{ \tan^{-1} \dfrac{1 + 2x}{\sqrt{3}} + \tan^{-1} \dfrac{2x - 1}{\sqrt{3}} \right\}$

   prove that $\dfrac{dy}{dx} = \dfrac{1}{x^4 + x^2 + 1}$.

ANSWERS

1. $7/8$;

2. $(\cos 2x^2 - \sin 8 x^2 . \sin 2 x^2). 4x$

3. 2 ;

4. $\alpha = 0, \beta = 1, \gamma = 1, \delta = 0$;

6. $- \{\tan \theta_1 + \tan \theta_2 + \ldots + \tan \theta_n\}$

7. 1 ;

8. $\dfrac{3}{\pi\sqrt{\pi^2 - 3}}$

9. 0.

OBJECTIVE QUESTIONS

For each of the following questions four alternatives are given for the answer. Only one of them is correct. Choose the correct alternatives :

1. The value of the derivative of $|x - 1| + |x - 3|$ at $x = 2$ is :

   (a) -2 (b) 0 (c) 2 (d) None of these

2. At $x = 1$, the function

$$f(x) = \begin{cases} 2(x - 1), & 0 \le x \le 1 \\ (x - 1)(x + 1) & 1 < x \le 2 \end{cases} \text{ is}$$

   (a) Continuous and differentiable

   (b) Continuous but not differentiable

   (c) Not continuous but differentiable

   (d) Neither continuous nor differentiable

3. The derivative of $f(x) = |x|^3$ at $x = 0$ is :

   (a) -1 (b) not defined   (c) 0 (d) $1/2$

4. If $f'(x_0)$ exists, then $\lim\limits_{h \to 0} \dfrac{f(x_0 + h) - (x_0 - h)}{2h}$ is equal to

   (a) $\dfrac{1}{2} f'(x_0)$ (b $f'(x_0)$ (c) $2 f'(x_0)$ (d) None of these

5. If for a continuous function $f, f(0) = f(1) = 0, f'(1) = 2$ and $y(x) = f(e^x) e^{f(x)}$, then $y'(0)$ is equal to

   (a) 1 (b) 2 (c) 0 (d) None of these

6. If $y = \tan^{-1}\left(\dfrac{\sin x}{1 + \cos x}\right)$, then $\dfrac{dy}{dx} =$

   (a) $1/4$ (b) $1/2$ (c) $1 + \cos^2 x$ (d) $-1/4$

7. If $y = \tan^{-1}\left\{\dfrac{\cos x - \sin x}{\cos x + \sin x}\right\}$, then $\dfrac{dy}{dx} =$

   (a) 0 (b) 1 (c) -1 (d) None of these

8. The differential coefficient of $\log \tan x$ is

   (a) $2 \sec 2 x$ (b) $2 \operatorname{cosec} 2 x$ (c) $2 \sec^3 x$ (d) $2 \operatorname{cosec}^3 x$.

9. If $y = \tan^{-1}\left[\dfrac{\sqrt{1+x^2}+\sqrt{1-x^2}}{\sqrt{1+x^2}-\sqrt{1-x^2}}\right]$, then $\dfrac{dy}{dx}$ equals

(a) $\dfrac{1}{\sqrt{1-x^4}}$ (b) $-\dfrac{1}{\sqrt{1-x^4}}$ (c) $\dfrac{x}{\sqrt{1-x^4}}$ (d) $-\dfrac{x}{\sqrt{1-x^4}}$

10. The derivative of $\sin^{-1} x$ w.r.t. $\cos^{-1}\sqrt{(1-x^2)}$ is

(a) $1/\sqrt{1-x^2}$ (b) $\cos^{-1} x$ (c) 1 (d) None of these

11. If $x = a(1-\cos\theta)$, $y = a(\theta-\sin\theta)$, then $\left(\dfrac{dy}{dx}\right)_{\theta=\pi/2} =$

(a) -2 (b) -4 (c) 1 (d) -1

12. If $x = a\{\cos t + \log\tan(t(2))\}$, $y = a\sin t$, $dy/dx =$

(a) $\tan t$ (b) $\cos t$ (c) $\sec t$ (d) $\csc t$

13. If $\sin y = x\sin(a+y)$, then $\dfrac{dy}{dx} =$

(a) $\dfrac{\sin^2(a+y)}{\sin a}$ (b) $\sin(a+y)$ (c) $\sin^2(a+y)$ (d) $\dfrac{\sin(a+y)}{\sin a}$

14. The differential coefficient of $f(\log x)$ where $f(x) = \log x$ is :

(a) $x/\log x$ (b) $(\log x)/x$ (c) $(x\log x)^{-1}$ (d) None of these

15. If $y = \log|x|$, then $\dfrac{dy}{dx} =$

(a) $1/x$ (b) $-1/x$ (c) $\dfrac{1}{|x|}$ (d) None of these

16. If $f(x) = |\log x|$, then :

(a) $y'(1+0) = \dfrac{1}{x}$ (b) $y'(1-0) = \dfrac{1}{x}$ (c) $y'(1) = 1$ (d) $y'(0) = \infty$

17. If $\log_x(\log x)$, then $f'(x)$ at $x = e$ is :

(a) e (b) $\dfrac{1}{e}$ (c) $\dfrac{x}{y}$ (d) 0

18. If $x^p y^q = (x+y)^{p+q}$, then $\dfrac{dy}{dx}$ is equal to :

(a) $\dfrac{y}{x}$ (b) $\dfrac{py}{qx}$ (c) $\dfrac{x}{y}$ (d) $\dfrac{qy}{px}$

19. If $y = \log\left(\dfrac{1+\sqrt{x}}{1-\sqrt{x}}\right)$, then $\dfrac{dy}{dx} =$

(a) $\dfrac{1}{\sqrt{x}(1+x)}$ (b) $\dfrac{1}{\sqrt{x}(1-x)}$ (c) $\dfrac{\sqrt{x}}{(1+x)}$ (d) None of these

20. If $f(x) = \log_{x^2}(\log x)$ then $f'(x)$ at $x = e$ is :

(a) 0 (b) 1 (c) $1/e$ (d) $1/2e$

21. If $y = f\left(\dfrac{2x-1}{x^2+1}\right)$ and $f'(x) = \sin x^2$, then $\dfrac{dy}{dx}$ at $x = 0$ equals

(a) $\dfrac{1}{2}\sin 1$ (b) $\sin 1$ (c) $2\sin 1$ (d) None of these

22. If $2^x + 2^y = 2^{x+y}$, then the value of $\dfrac{dy}{dx}$ at $x = y = 1$ is :

(a) 1 (b) 2 (c) 0 (d) -1

23. If $y = \sqrt{\cos x + \sqrt{\cos x + \sqrt{\cos x + \,........\,\infty}}}$, then $\dfrac{dy}{dx} =$

(a) $\dfrac{\cos x}{1-2y}$ (b) $\dfrac{\sin x}{1-2y}$ (c) $\dfrac{-\sin x}{1-2y}$ (d) $\dfrac{-\cos x}{1-2y}$

24. If $\sin(x+y) = \log(x+y)$, then $\dfrac{dy}{dx} =$

(a) 2 (b) -2 (c) 1 (d) -1

25. If $y = e^{x + e^{x + \,......\,\infty}}$, then $\dfrac{dy}{dx} =$

(a) $\dfrac{1}{1-y}$ (b) $\dfrac{y}{1-y}$ (c) $\dfrac{2y}{1-y}$ (d) None of these

26. The value of $\dfrac{d}{dx}(x^x)$ is :

(a) $x \cdot x^{x-1}$ (b) $x^x \log e\, x$ (c) $x^x \log x$ (d) None of these

27. The derivative of $(\log x)^{\log x}$ at $x = e$ is equal to :

(a) 0 (b) 1 (c) e (d) $1/e$

28. If $y = 2^{2^x}$, then $\dfrac{dy}{dx} =$

(a) $y(\log_{10} 2)^2$ (b) $y(\log_e 2)^2$ (c) $y \cdot 2^x (\log_e 2)^2$ (d) $y \log_e 2$.

29. Let $f(x+y) = f(x)\, f(y)$ for all x and y. Suppose that $f(3) = 3$ and $f'(0) = 11$, then $f'(3)$ is equal to :

(a) 22 (b) 33 (c) 28 (d) None of these

30. If $x^y = e^{x-y}$, then $\dfrac{dy}{dx} =$

(a) $\dfrac{\log x}{(1+\log x)^2}$ (b) $\dfrac{x-y}{1+\log x}$ (c) $\dfrac{x-y}{(1+\log x)^2}$ (d) $\dfrac{y-x}{x(1+\log x)}$

ANSWERS

1. (b)	2. (a)	3. (c)	4. (b)	5. (b)	6. (b)	7. (c)		
8. (b)	9. (d)	10. (c)	11. (c)	12. (a)	13. (a)	14. (c)		
15. (a)	16. (a)	17. (b)	18. (a)	19. (b)	20. (d)	21. (c)		
22. (d)	23. (b)	24. (d)	25. (b)	26. (b)	27. (d)	28. (c)		
29. (b)	30. (a)							

CHAPTER 5

SUCCESSIVE DIFFERENTIATION

5.1 Higher order derivatives

The derivative f' of a derivable function f is itself a function. We suppose that it also possesses a derivative, which we denote by f'' and call it the *second derivative of f*. The *third derivative* of f which is the derivative of f'' is denoted by f''' and so on.

Thus, the successive derivatives of f are represented by the symbols,

$$f', f'', \ldots\ldots, f^n, \ldots\ldots$$

where each term is the derivates of the preceding one.

If we write

$$y = f(x),$$

the symbols

$$y_1, y_2, y_3, \ldots\ldots, y_n, \ldots\ldots$$

or

$$\frac{dy}{dx}, \frac{d^2y}{dx^2}, \frac{d^3y}{dx^3}, \ldots\ldots \frac{d^ny}{dx^n}, \ldots\ldots$$

are used to denote the successive derivatives of y.

The symbols

$$f^n(a), \left[\frac{d^ny}{dx^n}\right]_{x=a}, y_n(a)$$

denote the value of the *n*th derivative at *a*.

EXAMPLES

1. *If $x = a(\cos\theta + \theta \sin\theta)$, $y = a(\sin\theta - \theta \cos\theta)$, find d^2y/dx^2.*

We have $\quad dx/d\theta = a(-\sin\theta + \sin\theta + \theta \cos\theta) = a\theta \cos\theta$

$$dy/d\theta = a(\cos\theta - \cos\theta + \theta \sin\theta) = a\theta \sin\theta$$

$$\Rightarrow \qquad \frac{dy}{dx} = \frac{dy}{d\theta} \cdot \frac{d\theta}{dx} = \frac{dy}{d\theta} \div \frac{dx}{d\theta} = \tan\theta$$

$$\Rightarrow \qquad \frac{d^2y}{dx^2} = \sec^2\theta \frac{d\theta}{dx} = \frac{\sec^2\theta}{a\theta \cos\theta} = \frac{\sec^3\theta}{a\theta}.$$

[**Note.** The result holds good for all those values of θ for which $\dfrac{dx}{d\theta} \neq 0$.

What are these values ?]

166

2. *If $y = \sin(\sin x)$, prove that*

$$\frac{d^2 y}{dx^2} + \tan x \frac{dy}{dx} + y \cos^2 x = 0.$$

We have

$$\frac{dy}{dx} = \cos(\sin x) \cos x,$$

$$\frac{d^2 y}{dx^2} = -\sin(\sin x) \cos^2 x - \cos(\sin x) \sin x$$

$$\Rightarrow \qquad \frac{d^2 y}{dx^2} + y \cos^2 x = -\cos(\sin x) \cdot \cos x \cdot \frac{\sin x}{\cos x}$$

$$\Rightarrow \qquad \frac{d^2 y}{dx^2} + \tan x \frac{dy}{dx} + y \cos^2 x = 0.$$

The result holds good for all those values of x for which $\cos x \neq 0$.

3. *If $p^2 = a^2 \cos^2 \theta + b^2 \sin^2 \theta$, prove that*

$$p + \frac{d^2 p}{d\theta^2} = \frac{a^2 b^2}{p^3}$$

(Avadh 2000)

We have

$$p^2 = a^2 \cos^2 \theta + b^2 \sin^2 \theta \qquad \qquad ...(i)$$

Differentiating both sides of (i) w.r.t θ, we get

$$2p \frac{dp}{d\theta} = -2a^2 \cos \theta \sin \theta + 2b^2 \sin \theta \cos \theta$$

$$\Rightarrow \qquad p \frac{dp}{d\theta} = (b^2 - a^2) \sin \theta \cos \theta \qquad \qquad ...(ii)$$

Differentiating both sides of (ii) w.r.t. θ, we get

$$p \frac{d^2 p}{d\theta^2} + \left(\frac{dp}{d\theta}\right)^2 = (b^2 - a^2)(\cos^2 \theta - \sin^2 \theta)$$

$$= (b^2 \cos^2 \theta + a^2 \sin^2 \theta) - (a^2 \cos^2 \theta + b^2 \sin^2 \theta)$$

$$= (b^2 \cos^2 \theta + a^2 \sin^2 \theta) - p^2$$

$$\therefore \qquad p \frac{d^2 p}{d\theta^2} + p^2 = (b^2 \cos^2 \theta + a^2 \sin^2 \theta) - \left(\frac{dp}{d\theta}\right)^2$$

$$= (b^2 \cos^2 \theta + a^2 \sin^2 \theta) - \frac{(b^2 - a^2)^2 \sin^2 \theta \cos^2 \theta}{p^2} \qquad \text{(from (ii))}$$

$$= \frac{1}{p^2}\left[p^2 (b^2 \cos^2 \theta + a^2 \sin^2 \theta) - (b^2 - a^2)^2 \sin^2 \theta \cos^2 \theta\right]$$

$$= \frac{1}{p^2}\left[(a^2 \cos^2 \theta + b^2 \sin^2 \theta)(b^2 \cos^2 \theta + a^2 \sin^2 \theta) - (b^2 - a^2)^2 \sin^2 \theta \cos^2 \theta\right]$$

$$= \frac{1}{p^2}\left[a^2 b^2 \cos^4 \theta + a^2 b^2 \sin^4 \theta + 2a^2 b^2 \sin^2 \theta \cos^2 \theta\right]$$

$$= \frac{a^2 b^2}{p^2} (\cos^2 \theta + \sin^2 \theta)^2$$

$$= \frac{a^2 b^2}{p^2}$$

Dividing both sides by p, *we get*

$$\frac{d^2 p}{d\theta^2} + p = \frac{a^2 b^2}{p^3}$$

4. *If* $y = a \cos (\log x) + b \sin (\log x)$, *show that*

$$x^2 \frac{d^2 y}{dx^2} + x \frac{dy}{dx} + y = 0$$

We have $\qquad\qquad y = a \cos (\log x) + b \sin (\log x)$

$\Rightarrow \qquad\qquad \dfrac{dy}{dx} = [- a \sin (\log x)] \cdot \dfrac{1}{x} + [b \cos (\log x)] \cdot \dfrac{1}{x}$

$\Rightarrow \qquad\qquad x \dfrac{dy}{dx} = - a \sin (\log x) + b \cos (\log x)$

Differentiating again w.r.t. x, we get,

$$x \frac{d^2 y}{dx^2} + \frac{dy}{dx} = [- a \cos (\log x) \cdot \frac{1}{x} + [-b \sin (\log x)] \cdot \frac{1}{x}$$

$$= - \frac{1}{x} [a \cos (\log x) + b \sin (\log x)]$$

$$= - \frac{y}{x}$$

$\Rightarrow \qquad x^2 \dfrac{d^2 y}{dx^2} + x \dfrac{dy}{dx} + y = 0$

5. *Change the independent variable to* θ *in the equation*

$$\frac{d^2 y}{dx^2} + \frac{2x}{1 + x^2} \frac{dy}{dx} + \frac{y}{(1 + x^2)^2} = 0$$

by means of the transformantion $x = \tan \theta$.

We have $\qquad \dfrac{dy}{dx} = \dfrac{dy}{d\theta} \div \dfrac{dx}{d\theta} = \cos^2 \theta \dfrac{dy}{d\theta}$;

$$\frac{dx}{d\theta} = \sec^2 \theta \neq 0 \qquad \text{for any value of } \theta$$

Again $\qquad \dfrac{d^2 y}{dx^2} = - 2 \cos \theta \sin \theta \cdot \dfrac{d\theta}{dx} \cdot \dfrac{dy}{d\theta} + \cos^2 \theta \cdot \dfrac{d^2 y}{d\theta^2} \cdot \dfrac{d\theta}{dx}$

$$= - 2 \cos \theta \sin \theta \cos^2 \theta \cdot \frac{dy}{d\theta} + \cos^2 \theta \cdot \frac{d^2 y}{d\theta^2} \cdot \cos^2 \theta$$

$$= - 2 \sin \theta \cos^3 \theta \cdot \frac{dy}{d\theta} + \cos^4 \theta \cdot \frac{d^2 y}{d\theta^2}.$$

Substituting the values of x, dy/dx and d^2y/dx^2 in the given differential equation, we have

$$- 2 \sin \theta \cos^3 \theta \cdot \frac{dy}{d\theta} + \cos^4 \theta \cdot \frac{d^2 y}{d\theta^2} + \frac{2 \tan \theta}{1 + \tan^2 \theta} \cdot \cos^2 \theta \frac{dy}{d\theta} + \frac{y}{(1 + \tan^2 \theta)^2} = 0$$

$$\Rightarrow \qquad -2 \sin \theta \cos^3 \theta \, \frac{dy}{d\theta} + \cos^4 \theta . \frac{d^2 y}{d\theta} + 2 \sin \theta . \cos^3 \theta \, \frac{dy}{d\theta} + \cos^4 \theta . y = 0$$

$$\Rightarrow \qquad \frac{d^2 y}{d\theta^2} + y = 0.$$

6. *If* $x = f(t)$ *and* $y = f(t)$, *prove that* $\dfrac{d^2 y}{dx^2} = \dfrac{f_1 \phi_2 - f_2 \phi_1}{f_1^3}$, *where suffixes denote differentiation w.r.t. t?*

$$\frac{dy}{dx} = \frac{dy/dt}{dx/dt} = \frac{\phi_1 (t)}{f_1 (t)}$$

Again differentiating both sides w.r.t. t,

We have
$$\frac{d^2 y}{dx^2} = \frac{f_1 (t) . \phi_2 (t) - \phi_1 (t) . f_2 (t)}{[f_1 (t)]^2} . \frac{dt}{dx}$$

$$= \frac{f_1 (t) \phi_2 (t) - \phi_1 (t) . f_2 (t)}{[f_1 (t)]^2} . \frac{1}{f_1 (t)}$$

$$= \frac{f_1 \phi_2 - \phi_1 f_2}{f_1^3} .$$

7. *If* $\dfrac{x + b}{2} = a \tan^{-1} (a \log_e y)$, $a > 0$, *prove that* $yy'' - yy' \log y = (y')^2$.

We have $\qquad a \log y = \tan \left(\dfrac{x + b}{2a} \right)$

On differentiating w.r.t. x, we get
$$\frac{a}{y} . y' = \sec^2 \left(\frac{x + b}{2a} \right) . \frac{1}{2a}$$

$$\Rightarrow \qquad \frac{2a^2 \, y'}{y} = 1 + a \left(\log y \right)^2$$

Again differentiating both sides w.r.t. x,
$$\frac{2a^2 \, (yy'' - y' \, y')}{y^2} = 2a^2 \left(\log y \right) . \frac{1}{y} . y'$$

or $\qquad yy'' - (y')^2 = yy' \left(\log y \right)$

$\Rightarrow \qquad yy'' - yy' \left(\log y \right) = (y')^2.$

8. *If* $y = \left[\log \left(\dfrac{x + \sqrt{x^2 - a^2}}{a} \right) \right]^2 + k \log (x + \sqrt{x^2 - a^2})$,

then prove that $(x^2 - a^2) \dfrac{d^2 y}{dx^2} + x \dfrac{dy}{dx} = 2a.$

On differentiating w.r.t. x, we get
$$\frac{dy}{dx} = 2 \log \left(\frac{x + \sqrt{x^2 - a^2}}{a} \right) . \frac{a}{x + \sqrt{x^2 - a^2}} . \frac{(x + \sqrt{x^2 - a^2})}{\sqrt{x^2 - a^2}}$$

$$+ \frac{k}{(x + \sqrt{x^2 - a^2})} . \frac{(x + \sqrt{x^2 - a^2})}{\sqrt{x^2 - a^2}}$$

or $$\frac{dy}{dx} = \frac{2a}{\sqrt{x^2 - a^2}} \cdot \log\left(\frac{x + \sqrt{x^2 - a^2}}{a}\right) + \frac{k}{\sqrt{x^2 - a^2}}$$

or $$\sqrt{x^2 - a^2}\,\frac{dy}{dx} = 2a \log\left(\frac{x + \sqrt{x^2 - a^2}}{a}\right) + k$$

Again differentiating, we get

$$\sqrt{x^2 - a^2}\,\frac{d^2 y}{dx^2} + \frac{dy}{dx} \cdot \frac{x}{\sqrt{x^2 - a^2}} = \frac{2a}{x + \sqrt{x^2 - a^2}} \cdot \frac{x + \sqrt{x^2 - a^2}}{\sqrt{x^2 - a^2}}$$

or $$\sqrt{x^2 - a^2}\,\frac{d^2 y}{dx^2} + \frac{x}{\sqrt{x^2 - a^2}}\,\frac{dy}{dx} = \frac{2a}{\sqrt{x^2 - a^2}}$$

or $$(x^2 - a^2)\,\frac{d^2 y}{dx^2} + x\,\frac{dy}{dx} = 2a$$

9. If $f\left(\dfrac{x + y}{2}\right) = \dfrac{f(x) + f(y)}{2}$, $f'(0) = a$ and $f(0) = b$, then find $f''(x)$ where y is independent of x.

Since y is independent of x, hence $\dfrac{dy}{dx} = 0$.

On differentiating w.r.t. x, we get

$$f'\left(\frac{x + y}{2}\right) \cdot \frac{1}{2}\left(1 + \frac{dy}{dx}\right) = \frac{1}{2}\left\{f'(x) + f'(y) \cdot \frac{dy}{dx}\right\}$$

$$\Rightarrow \quad \frac{1}{2}\,f'\left(\frac{x + y}{2}\right) = \frac{1}{2}\,f'(x) \qquad\qquad ...(i)$$

Taking $x = 0$ and $y = x$ in (i), we get

$$f\left(\frac{0 + x}{2}\right) = f'(0)$$

$$\Rightarrow \qquad f'\left(\frac{x}{2}\right) = a$$

which shows $f\left(\dfrac{x}{2}\right) = a\left(\dfrac{x}{2}\right) + c$

$\therefore \qquad\qquad f(x) = ax + c.$

let $\qquad\qquad x = 0, \qquad\quad f(0) = b = c$

$\therefore \qquad\qquad f(x) = ax + b$

On differentiating $f'(x) = a$

and $\qquad\qquad f''(x) = 0$

EXERCISES

1. If $y = \dfrac{\log x}{x}$, show that $\dfrac{d^2 y}{dx^2} = \dfrac{2 \log x - 3}{x^3}$.

2. If $y = \log(\sin x)$, show that $y_3 = \dfrac{2 \cos x}{\sin^3 x}$.

3. Show that $y = x + \tan x$ satisfies the differential equation

$$\cos^2 x \, \frac{d^2 y}{dx} - 2y + 2x = 0.$$

4. If $y = [(a + bx)/(c + dx)]$, then $2y_1 y_3 = 3y_2^2$.

5. If $x = 2 \cos t - \cos 2t$ and $y = 2 \sin t - \sin 2t$, find the value of d^2y/dx^2 when $t = \dfrac{1}{2}\pi$.

6. If $x = a \sin 2\theta (1 + \cos 2\theta)$, $y = a \cos 2\theta (1 - \cos 2\theta)$, prove that

$$\frac{[1 + (dy/dx^2)]^{3/2}}{d^2 y/dx^2} = 4a \cos 3\theta.$$

7. If $y = (1/x)^x$, show that $y_2(1) = 0$.

8. If $y = x \log[(ax)^{-1} + a^{-1}]$, prove that $x(x+1)\dfrac{d^2 y}{dx^2} + x\dfrac{dy}{dx} = y - 1$.

9. If $x = \sin t$, $y = \sin pt$, prove that $(1 - x^2)\dfrac{d^2 y}{dx^2} - x\dfrac{dy}{dx} + p^2 y = 0$.

10. If $y^3 + 3ax^2 + x^3 = 0$. then prove that $\dfrac{d^2 y}{dx^2} + \dfrac{2a^2 x^2}{y^5} = 0$

11. If $ax^2 + 2hxy + by^2 = 1$, show that $\dfrac{d^2 y}{dx^2} = \dfrac{h^2 - ab}{(hx + by)^3}$

12. If $y = (\tan^{-1} x)^2$, prove that $(x^2 + 1)^2 \dfrac{d^2 y}{dx^2} + 2x(x^2 + 1)\dfrac{dy}{dx} - 2 = 0$

13. If $ay = \sin(x + y)$, prove that $y\left[1 + \left(\dfrac{dy}{dx}\right)\right]^3 = 0$

14. If $y = e^{-x}(A \cos x + B \sin x)$, prove that $\dfrac{d^2 y}{dx^2} + 2\dfrac{dy}{dx} + 2y = 0$

15. If $y = Ae^{px} + Be^{qx}$, show that $\dfrac{d^2 y}{dx^2} - (p + q)\dfrac{dy}{dx} + pqy = 0$

16. Change the independent variable to in the equation

$$\frac{d^2 y}{dx^2} + \cot x \frac{dy}{dx} + 4y \, \text{cosec}^2 x = 0$$

by means of the transformation $z = \log \tan \dfrac{x}{2}$.

17. If $x = \log \phi$, $y = \phi^2 - 1$. Find $\dfrac{d^2 y}{dx^2}$

18. If $y^3 - y = 2x$, then prove that $\dfrac{d^2y}{dx^2} = -\dfrac{24y}{(3y^2 - 1)^3}$.

19. If $2x = y^{1/5} + y^{-1/5}$ then express y as an explicit function of x and prove that

$$(x^2 - 1)\dfrac{d^2y}{dx^2} + x\dfrac{dy}{dx} = 25y.$$

20. If $f(x) = \begin{vmatrix} x^3 & \sin x & \cos x \\ 6 & -1 & 0 \\ a & a^2 & a^3 \end{vmatrix}$

Where a is constant, then prove that $\dfrac{d^3}{dx^3}[f(x)]$ at $x = 0$ is equal to 0.

ANSWERS

5. $-3/2$ 16. $\dfrac{d^2y}{dz^2} + 4y = 0$ 17. $4\phi^2$.

5.2. Calculation of the nth Derivative. Some standard results

5.2.1. Let
$$y = (ax + b)^m$$
$$y_1 = ma(ax + b)^{m-1}$$
$$y_2 = m(m-1)a^2(ax + b)^{m-2}$$
$$y_3 = m(m-1)(m-2)a^3(ax + b)^{m-3},$$

so that is general
$$y_n = m(m-1)(m-2)\ \text{.......}\ (m-n+1)a^n(ax+b)^{m-n}$$

In case, m is a positive integer, y_n can be written as

$$\dfrac{m!}{(m-n)!}a^n(ax+b)^{m-n},$$

so that, the mth derivative of $(ax + b)^m$ is a constant, viz., $m!\ a^m$ and the $(m+1)$th derivative is zero.

Cor.1. Putting $m = -1$, we get
$$y_n = (-1)(-2)\ \text{.......}\ (-n)a^n(ax+b)^{-1-n}]$$

$$\Rightarrow \quad \dfrac{d^n\left(\dfrac{1}{ax+b}\right)}{dx^n} = \dfrac{(-1)^n(n!)a^n}{(ax+b)^{n+1}}.$$

Cor.2. Let $\quad y = \log(ax + b)$ We have

$$y_1 = \dfrac{a}{ax+b} \Rightarrow y_n = \dfrac{(-1)^{n-1}(n-1)!}{(ax+b)^n}a^n$$

5.2.2. Let $\quad y = a^{mx}.$ We have

$$y_1 = ma^{mx}\log a \Rightarrow y_2 = m^2 a^{mx}(\log a)^2 \ ... \Rightarrow y_n = m^n a^{mx}(\log a)^n.$$

(Bhopal 2000, 2001 ; Jabalpur1994)

Cor. Putting e for a, we get

$$\dfrac{d^n(e^{mx})}{dx^n} = m^n e^{mx}.$$

5.2.3. Let $\quad y = \sin(ax + b)$. We have $\qquad\qquad$ *(Banglore 2004)*

$$y_1 = a\cos(ax + b) = a\sin\left(ax + b + \frac{1}{2}\pi\right)$$

$$y_2 = a^2\cos\left(ax + b + \frac{1}{2}\pi\right) = a^2\sin\left(ax + b + \frac{2}{2}\pi\right)$$

$$y_3 = a^3\cos\left(ax + b + \frac{2}{2}\pi\right) = a^3\sin\left(ax + b + \frac{3}{2}\pi\right)$$

$$\dotfill$$
$$\dotfill$$

Hence

$$\frac{d^n \sin(ax + b)}{dx^n} = a^n \sin\left(ax + b + \frac{n\pi}{2}\right).$$

5.2.4. Similarly :

$$\frac{d^n \cos(ax + b)}{dx^n} = a^n \cos\left(ax + b + \frac{n\pi}{2}\right).$$

5.2.5. Let $y = e^{ax} \sin(bx + c)$. We have $\qquad\qquad$ *(Jiwaji 2000, Gujrat 2005)*

$$y_1 = ae^{ax}\sin(bx + c) + e^{ax} b\cos(bx + c)$$
$$= e^{ax}[a\sin(bx + c) + b\cos(bx + c)].$$

In order to put y in a form which will enable us to make the required generalisation, we determine two constants r and ϕ such that

$$a = r\cos\phi, \; b = r\sin\phi$$

$\Rightarrow \qquad\qquad r = \sqrt{(a^2 + b^2)}, \qquad \phi = \tan^{-1}(b/a).$

Hence, we have

$$y_1 = re^{ax}\sin(bx + c + \phi).$$

Thus, y_1 arises on multiplying y by the constant r and increasing $bx + c$ by the constant ϕ.

Similary $\qquad\qquad y_2 = r^2 e^{ax} \sin(bx + c + 2\phi).$

Hence, in general

$$\frac{d^n [e^{ax} \sin(bx + c)]}{dx^n} = r^n e^{ax} \sin(bx + c\, n\, \phi)$$

where $\qquad\qquad r = \sqrt{(a^2 + b^2)}, \; \phi = \tan^{-1}(b/a).$

5.26 Similarly :

$$\frac{d^n [e^{ax} \cos(bx + c)]}{dx^n} = (a^2 + b^2)^{n/2} e^{ax} \cos\left(bx + c + n\tan^{-1}\frac{b}{a}\right). \qquad \textit{(Kanpur 2003)}$$

5.3. Determination of nth derivative of rational functions

Partial Fractions. In order to determine the nth derivative of a rational function, we have to decompose it into partial fractions. Sometimes it also becomes necessary to apply Demoivre's theoerm, which states that

$$(\cos\theta \pm i\sin\theta)^n = \cos n\theta \pm i\sin n\theta,$$

where n is any integer, positive or negative and $i = \sqrt{(-1)}$.

EXAMPLES

1. *Find the nth derivative of* $y = \dfrac{x^2}{(x+2)(2x+3)}$. *(MDU Rohtak 2000)*

Throwing $x^2/(x+2)(2x+3)$ into partial fractions, we obtain

$$\frac{x^2}{(x+2)(2x+3)} = \frac{1}{2}\left[1 - \frac{8}{x+2} + \frac{9}{2x+3}\right]$$

$$\therefore \quad \frac{d^n}{dx^n}\left[\frac{x^2}{(x+2)(2x+3)}\right] = -\frac{(-1)^n\,!\,8}{2\,(x+2)^{n+1}} + \frac{(-1)^n\,n!\,2^n\,.\,9}{2\,(2x+3)^{n+1}}$$

$$= \frac{(-1)^n n!}{2}\left[\frac{9\,.\,2^n}{(2x+3)^{n+1}} - \frac{8}{(x+2)^{n+1}}\right].$$

2. *Find the nth derivative of* $y = x/(x^2 + a^2)$.

We have $\qquad \dfrac{x}{x^2 + a^2} = \dfrac{x}{(x+ai)(x-ai)} = \dfrac{1}{2}\left[\dfrac{1}{x-ai} + \dfrac{1}{x+ai}\right]$.

$$\therefore \quad \frac{d^n}{dx^n}\left(\frac{x}{x^2+a^2}\right) = \frac{(-1)^n\,n!}{2}\left[\frac{1}{(x-ai)^{n+1}} + \frac{1}{(x+ai)^{n+1}}\right].$$

To render the result free from 'i' and express the same in real form, we determine two numbers r and θ, such that

$$x = r\cos\theta, \; a = r\sin\theta,$$

so that $\qquad\qquad r = \sqrt{(x^2 + a^2)},$

$$\theta = \tan^{-1}(a/x).$$

$$\therefore \quad \frac{1}{(x-ai)^{n+1}} = \frac{1}{r^{n+1}}\,.\,(\cos\theta - i\sin\theta)^{-(n+1)}$$

$$= \frac{1}{r^{n+1}}[\cos(n+1)\theta + i\sin(n+1)\theta]$$

and $\qquad \dfrac{1}{(x+ai)^{n+1}} = \dfrac{1}{r^{n+1}}(\cos\theta + i\sin\theta)^{-(n+1)}$

$$= \frac{1}{r^{n+1}}\{\cos(n+1)\theta - i\sin(n+1)\theta\}.$$

Hence. $\qquad \dfrac{d^n\,[x/(x^2+a^2)]}{dx^n} = \dfrac{(-1)^n\,n!\cos(n+1)\theta}{r^{n+1}}$

where $\qquad\qquad r = \sqrt{(x^2 + a^2)}, \qquad\qquad \theta = \tan^{-1}(a/x).$

3. *Show that the nth derivative of* $y = \dfrac{1}{1 + x + x^2 + x^3}$ *is*

$$\frac{1}{2}(-1)^n\,n!\,\sin^{n+1}\theta\left[\sin(n+1)\theta - \cos(n+1)\theta + (\sin\theta + \cos\theta)^{-n-1}\right]$$

where $\theta = \cot^{-1}x$

We have
$$y = \frac{1}{(1 + x)(1 + x^2)}$$

$$= \frac{1}{(1 + x)(x + i)(x - i)}$$

$$= \frac{1}{2}\left[\frac{1}{1 + x} + \frac{(i - 1)}{2} \cdot \frac{1}{x + i} - \frac{(i + 1)}{2} \cdot \frac{1}{x - i}\right]$$

$$\therefore \quad \frac{d^n y}{dx^n} = \frac{1}{2}(-1)^n \, n!\left[\frac{1}{(1 + x)^{n+1}} + \frac{(i - 1)}{2} \cdot \frac{1}{(x + i)^{n+1}} - \frac{(i + 1)}{2} \cdot \frac{1}{(x - i)^{n+1}}\right]$$

Put $x = r \cos \theta$, $\quad 1 = r \sin \theta$

so that $\quad r = \sqrt{x^2 + 1}$ and $\theta = \cot^{-1} x$

$$\frac{1}{(1 + x)^{n+1}} = \frac{1}{r^{n+1}}(\cos \theta + \sin \theta)^{-n-1},$$

$$\frac{1}{(x + i)^{n+1}} = \frac{1}{r^{n+1}}(\cos \theta + i \sin \theta)^{-(n+1)}$$

$$= \frac{1}{r^{n+1}}(\cos (n + 1)\theta - i \sin (n + 1)\theta)$$

and $\quad \dfrac{1}{(x - i)^{n+1}} = \dfrac{1}{r^{n+1}}(\cos \theta - i \sin \theta)^{-(n+1)}$

$$= \frac{1}{r^{n+1}}(\cos (n + 1)\theta + i \sin (n + 1)\theta)$$

$$\therefore \quad \frac{d^n y}{dx^n} = \frac{1}{2} \cdot \frac{(-1)^n \, n!}{r^{n+1}}\Big[(\cos \theta + \sin \theta)^{-n-1} + \frac{1}{2}\{(i - 1)(\cos (n + 1)\theta$$
$$- i \sin (n + 1)\theta - (i + 1)(\cos (n + 1)\theta + i \sin (n + 1)\theta)\}\Big]$$

$$= \frac{1}{2}\frac{(-1)^n \, n!}{r^{n+1}}\Big[(\cos \theta + \sin \theta)^{-n-1} + \sin (n + 1)\theta - \cos (n + 1)\theta\Big]$$

Now $\quad r^{n+1} = (x^2 + 1)^{(n+1)/2} = (\cot^2 \theta + 1)^{(n+1)/2} = \operatorname{cosec}^{n+1}\theta$

$$\Rightarrow \quad \frac{1}{r^{n+1}} = \sin^{n+1}\theta$$

$$\therefore \quad \frac{d^n y}{dx^n} = \frac{1}{2}(-1)^n \, n! \sin^{n+1}\theta\Big[(\cos \theta + \sin \theta)^{-n-1} + \sin (n + 1)\theta - \cos (n + 1)\theta\Big]$$

4. If $y = x \log \dfrac{x - 1}{x + 1}$, *prove that*

$$\frac{d^n y}{dx^n} = (-1)^n (n - 2)!\left[\frac{x - n}{(x + 1)^n} - \frac{x + n}{(x + 1)^n}\right]$$

We have $\quad y = x \log \dfrac{x - 1}{x + 1}$

$$\Rightarrow \quad \frac{dy}{dx} = \log \frac{x - 1}{x + 1} + x\left(\frac{x + 1}{x - 1}\right) \times \frac{2}{(x + 1)^2}$$

$$= \log \frac{x-1}{x+1} + \frac{2x}{(x-1)(x+1)}$$

$$= \log \frac{x-1}{x+1} + \frac{1}{x-1} + \frac{1}{x+1}$$

$$\Rightarrow \quad \frac{d^2 y}{dx^2} = \frac{2}{x^2 - 1} - \frac{1}{(x-1)^2} - \frac{1}{(x+1)^2}$$

$$= \frac{1}{x-1} - \frac{1}{x+1} - \frac{1}{(x-1)^2} - \frac{1}{(x+1)^2}$$

Differentiating $(n-2)$ times w.r.t. x, we get

$$\frac{d^n y}{dx^n} = (-1)^{n-2} \left[\frac{(n-2)!}{(x-1)^{n-1}} - \frac{(n-2)!}{(x+1)^{n-1}} - \frac{(n-1)!}{(x-1)^n} - \frac{(n-1)!}{(x+1)^n} \right]$$

$$= (-1)^{n-2} \left[\frac{(n+2)!}{(x+1)^n} \{(x-1) - (n-1)\} - \frac{(n-2)!}{(x+1)^n} \{(x+1) + (n-1)\} \right]$$

$$= (-1)^{n-2} (n-2)! \left[\frac{x-n}{(x-1)^n} - \frac{x+n}{(x+1)^n} \right]$$

$$= (-1)^n (n-2)! \left[\frac{x-n}{(x-1)^n} - \frac{x+n}{(x+1)^n} \right]$$

$$[\because (-1)^{n-2} = (-1)^n \cdot (-1)^{-2} = (-1)^n \cdot 1 = (-1)^n]$$

EXERCISES

1. Find the nth derivatives, given that

   (i) $y = \dfrac{x^4}{(x-1)(x-2)}$ (ii) $y = \dfrac{x+1}{x^2 - 4}$ *(Manipur 2000)*

   (iii) $y = \dfrac{4x}{(x-1)^2 (x+1)}$ (iv) $y = \dfrac{1}{(a^2 - x^2)}$ *(Sagar 97)*

2. Find the tenth and the nth derivatives of

$$y = \frac{x^2 + 4x + 1}{x^3 + 2x^2 - x - 2}.$$

3. Find the nth derivatives of the functions with the following function values :

   (i) $\dfrac{x}{1 + 3x + 2x^2}$ *(MDU Rohtak 1999)* (ii) $\dfrac{1}{x^4 - a^4}$

   (iii) $\dfrac{1}{x^2 + x + 1}$ (iv) $\dfrac{x}{x^2 + x + 1}$

4. Prove that the value of the nth derivative of $y = x^3 / (x^2 - 1)$ for $x = 0$ is zero, if n is even, and $-(n!)$ if n is odd and greater than 1.

5. Show that the nth derivatives of $y = \tan^{-1} x$ is

$$(-1)^{n-1} (n-1)! \sin n \left(\frac{1}{2} \pi - y \right) \sin^n \left(\frac{1}{2} \pi - y \right).$$
               (Gorakhpur 1999)

6. Find the nth derivative of

    (i) $y = \tan^{-1} \dfrac{1+x}{1-x}$ **(Sagar 2001)**   (ii) $y = \tan^{-1} \dfrac{x \sin \alpha}{1 - x \cos \alpha}$.   (iii) $y = \tan^{-1} \dfrac{2x}{1 - x^2}$

    (iv) $y = \sin^{-1} \dfrac{2x}{1 + x^2}$ (v) $y = \tan^{-1} \dfrac{\sqrt{1 + x^2} - 1}{x}$

7. If $y = x(x+1) \log(x+1)^3$, prove that

$$\frac{d^n y}{dx^n} = \frac{3(-1)^{n-1}(n-3)!(2x+n)}{(x+1)^{n-1}} \text{ provided that } n \geq 3.$$

ANSWERS

1.   (i) $(-1)^n \, n! \left[\dfrac{16}{(x-2)^{n+1}} - \dfrac{1}{(x-1)^{n+1}} \right], \; n > 2.$

    (ii) $\dfrac{(-1)^n \, n!}{4} \left[\dfrac{3}{(x-2)^{n+1}} - \dfrac{1}{(x+2)^{n+1}} \right].$

    (iii) $(-1)^n \, n! \left[\dfrac{1}{(x-1)^{n+1}} - \dfrac{1}{(x+1)^{n+1}} + \dfrac{2(n+1)}{(x-1)^{n+2}} \right].$

    (iv) $\dfrac{1}{2a} (-1)^n \, n! \left[\dfrac{1}{(x+a)^{n+1}} - \dfrac{1}{(x-a)^{n+1}} \right]$

2. $10! \left[\dfrac{1}{(x-1)^{11}} + \dfrac{1}{(x+1)^{11}} - \dfrac{1}{(x+2)^{11}} \right]$

     $(-1)^n \, n! \left[\dfrac{1}{(x-1)^{n+1}} - \dfrac{1}{(x+1)^{n+1}} - \dfrac{1}{(x+2)^{n+1}} \right].$

3.   (i) $(-1)^n \, n! \left[\dfrac{1}{(x+1)^{n+1}} - \dfrac{2^n}{(2x+1)^{n+1}} \right].$

    (ii) $\dfrac{(-1)^n \, n!}{4a^3} \left[\dfrac{1}{(x-a)^{n+1}} - \dfrac{1}{(x+a)^{n+1}} - \dfrac{2 \sin[(n+1)\cot^{-1}(x/a)]}{(a^2 + x^2)^{(n+1)/2}} \right].$

    (iii) $\dfrac{2(-1)^n \, n!}{\sqrt{3} \, r^{n+1}} \sin(n+1)\theta, \text{ where } r = \sqrt{(x^2 + x + 1)}, \; \theta = \cot^{-1}[(2x+1)/\sqrt{3}].$

6.   (i) $(-1)^{n-1}(n-1)! \sin^n \theta \sin n\theta$ where $\theta = \cot^{-1} x.$

    (ii) $(-1)^{n-1}(n-1)! \operatorname{cosec}^n \alpha \sin^n \theta \sin n\theta, \text{ where } \theta = \cot^{-1}[(x - \cos \alpha)/\sin x].$

    (iii) $2(-1)^{n-1}(n-1)! \sin^n \theta \sin n\theta$ where $\theta = \cot^{-1} x.$

    (iv) Same as (i)

    (v) $\dfrac{1}{2}(-1)^{n-1}(n-1)! \sin^n \theta \sin n\theta, \text{ where } \theta = \cot^{-1} x.$

5.4. The nth derivatives of the products of the powers of sines and cosines.

In order to find out the nth derivative of such a product, we express it as the sum of the sines and cosines of multiples of the independent variable Example 1 will illustrate the process.

EXAMPLES

1. *Find $d^n y / dx^n$ given that*

    (i) $y = \cos^4 x$ (ii) $y = e^{ax} \cos^2 x \sin x$

(i)   We have $\cos^2 x = \dfrac{1 + \cos 2x}{2}$

$\Rightarrow \qquad \cos^4 x = \left(\dfrac{1 + \cos 2x}{2}\right)^2$

$\qquad\qquad\qquad = \dfrac{1}{4} + \dfrac{\cos 2x}{2} + \dfrac{\cos^2 2x}{4}$

$\qquad\qquad\qquad = \dfrac{1}{4} + \dfrac{1}{2}\cos 2x + \dfrac{1}{8}(1 + \cos 4x) = \dfrac{3}{8} + \dfrac{1}{2}\cos 2x + \dfrac{1}{8}\cos 4x$

$\Rightarrow \qquad \dfrac{d^n (\cos 4x)}{dx^n} = \dfrac{1}{2} 2^n \cos\left(2x + \dfrac{n\pi}{2}\right) + \dfrac{1}{8} 4^n \cos\left(4x + \dfrac{n\pi}{2}\right).$

(ii) $\cos^2 x \sin x = \dfrac{1}{2}(1 + \cos 2x) \sin x$

$\qquad\qquad\qquad = \dfrac{1}{2}\sin x + \dfrac{1}{4} . 2 \sin x \cos 2x$

$\qquad\qquad\qquad = \dfrac{1}{2}\sin x + \dfrac{1}{4}(\sin 3x - \sin x)$

$\qquad\qquad\qquad = \dfrac{1}{4}\sin x + \dfrac{1}{4}\sin 3x.$

Hence

$\dfrac{d^n}{dx^n}(e^{ax}\cos^2 x \sin x) = \dfrac{1}{4}\dfrac{d^n}{dx^n}(e^{ax}\sin x) + \dfrac{1}{4}\dfrac{d^n}{dx^n}(e^{ax}\sin 3x)$

$\qquad\qquad\qquad = \dfrac{1}{4}(a^2 + 1)^{n/2} e^{ax} \sin\left(x + n\tan^{-1}\dfrac{1}{a}\right)$

$\qquad\qquad\qquad\qquad + \dfrac{1}{4}(a^2 + 9)^{n/2} e^{ax} \sin\left(3x + n\tan^{-1}\dfrac{3}{a}\right).$

2. *If $y = \sin ax + \cos ax$, prove that*

$$y_n = a^n \sqrt{\{1 + (-1)^n \sin 2ax\}}$$

(Awadh 2001, Kanpur 2001, Reewa 1996, 2000 ; Sagar 99)

Differentiating n times, we get

$$y_n = a^n \sin\left(ax + \dfrac{n\pi}{2}\right) + a^n \cos\left(ax + \dfrac{n\pi}{2}\right)$$

$$= a^n \sqrt{\left\{\sin\left(ax + \dfrac{n\pi}{2}\right) + \cos\left(ax + \dfrac{n\pi}{2}\right)\right\}^2}$$

$$= a^n \sqrt{\{1 + \sin(2ax + n\pi)\}}$$

$$= a^n \sqrt{1 + \sin 2ax \cos n\pi)}$$

$$= a^n \sqrt{(1 + (-1)^n \sin 2ax)}$$

EXERCISES

Find the nth derivatives of the functions with the following function values :

1. $\sin^3 x$ *(Calicut 2004)*
2. $\cos x \cos 2x \cos 3x$
3. $\sin 5x \sin 3x$
4. $\sin^2 x \cos^3 x$
5. $e^{ax} \cos^2 bx$
6. $e^x \sin^4 x$
7. $e^x \cos x \cos 2x$
8. $e^{2x} \cos x \sin^2 2x$

ANSWERS

1. $\dfrac{3}{4} \sin\left(x + \dfrac{1}{2} n\pi\right) - \left(\dfrac{3}{4}\right)^n \sin\left(3x + \dfrac{1}{2} n\pi\right).$

2. $\dfrac{1}{4}\left[2^n \cos(2x + \dfrac{1}{2} n\pi) + 4^n \cos(4x + \dfrac{1}{2} n\pi) + 6^n \cos\left(6x + \dfrac{1}{2} n\pi\right)\right].$

3. $\dfrac{1}{2}\left[2^n \cos\left(2x + \dfrac{n\pi}{2}\right) - 8^n \cos\left(8x + \dfrac{n\pi}{2}\right)\right]$

4. $\dfrac{1}{16}\left[2 \cos\left(x + \dfrac{1}{2} n\pi\right) - 3^n \cos\left(3x + \dfrac{1}{2} n\pi\right) - 5^n \cos\left(5x + \dfrac{1}{2} n\pi\right)\right].$

5. $\dfrac{e^{ax}}{2}\left[a^n + (a^2 + 4b^2)^{n/2} e^{ax} \cos\left(2bx + n \tan^{-1}\left(\dfrac{2b}{2}\right)\right)\right].$

6. $\dfrac{1}{8} e^x\left[3 - 4^{n/2} \cos(2x + n\tan^{-1} 2) - 17^{n/2} \cos\left(4x + n \tan^{-1} 4\right)\right]$

7. $\dfrac{e^x}{2}\left[(10)^{n/2} \cos(3x + n \tan^{-1} 3) + (2)^{n/2} \cos\cos\left(x + \dfrac{n\pi}{4}\right)\right]$

8. $\dfrac{1}{4} e^{2n}\left[2.5^{n/2} \cos(x + n \tan^{-1} /2) - 13^{n/2} \cos(3x + n \tan^{-1} 3/2) - 29^{n/2} \cos\left(5x + n \tan^{-1} 5/2\right)\right]$

5.5. Leibnitz's Theorem. The nth derivative of the product of two functions.

If u and v be two functions of x possessing derivatives of the nth order, then

$$(uv)_n = u_n + {}^nC_1 u_{n-1} v_1 + {}^nC_2 u_{n-2} v_2 + \ldots\ldots + \ldots\ldots + {}^nC_r u_{n-r} v_r + \ldots\ldots + {}^nC_n uv_n.$$

(Reewa 97, 99S; Bhopal 99, 2000; Vikram 99, 2001; Bilaspur 98, 2001; Bhuj 99 ; Jabalpur 99; Indore 2001, Avadh 98, 2002, 2005; Devi Ahilya 2001; Manipur 2000, 2002, Patna 2002, Gujrat 2005, GNOU 2004)

This theorem will be proved by *Mathematical induction.*

Step I. By direct differentiation, we have

$$(uv)_1 = u_1 v + uv_1,$$

and $\qquad (uv)_2 = u_2v + u_1v_1 + u_1v_1 + uv_2$

$$= u_2v + {}^2C_1 u_1v_1 + {}^2C_2 uv_2.$$

Thus the theorem is true for $n = 1, 2$.

Step II. We *assume* that the theorem is true for a particular value of n, say m, so that we have

$$(uv)_m = u_m v + {}^mC_1 u_{m-1} v_1 + {}^mC_2 u_{m-2} v_2 + \ldots\ldots$$

$$+ {}^mC_{r-1} u_{m-r+1} v_{r-1} + {}^mC_r u_{m-r} v_r + \ldots\ldots + {}^mC_m uv_m.$$

Differentiating both sides, we get

$$(uv)_{m+1} = u_{m+1} \, v + u_m \, v_1 + {}^mC_1 \, u_m \, v_1 + {}^mC_1 \, u_{m-1} \, v_2$$
$$+ \, {}^mC_2 \, u_{m-1} \, v_2 + {}^mC_2 \, u_{m-2} \, v_3 + \ldots + {}^mC_{r-1} \, u_{m-r+2} \, v_{r-1}$$
$$+ \, {}^mC_{r-1} \, u_{m-r+1} \, v_r + {}^mC_r \, u_{m-r+1} \, v_r + {}^mC_r \, u_{m-r} \, v_{r+1} + \ldots + \ldots + {}^mC_m \, uv_{m+1}$$
$$= u_{m+1} \, v + (1 + {}^mC_1) \, u_m \, v_1 + ({}^mC_1 + {}^mC_2) \, u_{m-1} \, v_2 + \ldots +$$
$$({}^mC_{r-1} + {}^mC_r) \, u_{m-r+1} \, v_r + \ldots + {}^mC_m \, uv_{m+1}.$$

We know that

$$\begin{aligned} {}^mC_{r-1} + {}^mC_r &= {}^{m+1}C_r, \\ 1 + {}^mC_1 = 1 + m &= {}^{m+1}C_1, \\ {}^mC_m = 1 &= {}^{m+1}C_{m+1}. \end{aligned}$$

$$\therefore \quad (uv)_{m+1} = u_{m+1} \, v + {}^{m+1}C_1 \, u_m \, v_1 + {}^{m+1}C_2 \, u_{m-1} \, v_2 + \ldots \ldots$$
$$+ \, {}^{m+1}C_r \, u_{m-r+1} \, v_r + \ldots + {}^{m+1}C_{m+1} \, uv_{m+1},$$

from which we see that if the theorem is true for any value m of n, then it is also true for the next higher value $m + 1$ or n.

Conclusion. In step I, we have seen that the theorem is true for $n = 2$. Therefore it must be true for $n = 2 + 1$, i.e., 3 and so for $n = 3 + 1$, i.e., 4, and so for every value of n.

EXAMPLES

1. *Find* $(x^2 \, e^x \cos x)_n$

To find the nth derivative, we look upon $e^x \cos x$ as the first factor and x^2 as the second.

$$\therefore \; (x^2 \, e^x \cos x)_n = (e^x \cos x)_n \, x^2 + {}^nC_1 \, (e^x \cos x)_{n-1} \cdot 2x + {}^nC_2 \, (e^x \cos x)_{n-2} \cdot 2$$
$$= 2^{n/2} \cdot e^x \cos (x + n \tan^{-1} 1) \cdot x^2 + n \, 2^{(n-1)/2} \, e^x \cos [x + (n-1) \cdot \tan^{-1} 1] \, 2x$$
$$+ \frac{n \, (n-1)}{2} \cdot 2^{(n-2)/2} \, e^x \cos [x + (n-2) \tan^{-1} 1] \, 2.$$
$$= 2^{(n-2)/2} \, e^x \left[2x^2 \cos \left(x + n \frac{\pi}{4} \right) \right] +$$
$$2^{3/2} \cdot nx \cdot \cos \left(x + \overline{n-1} \, \frac{\pi}{4} \right) + n \, (n-1) \cos \left(x + \overline{n-2} \, \frac{\pi}{4} \right) \bigg].$$

2. *If* $\cos^{-1} \left(\dfrac{y}{b} \right) = log \left(\dfrac{x}{n} \right)^n$, *prove that*

$$x^2 \, y_{n+2} + (2n + 1) \, xy_{n+1} + 2n^2 \, y_n = 0 \qquad \textbf{\textit{(Ravishankar 1999, Jabalpur 2000)}}$$

We have $\qquad \cos^{-1} \left(\dfrac{y}{b} \right) = n \log \dfrac{x}{n}$

Differentiating both sides w.r.t. x, we get

$$\frac{-y_1}{\sqrt{b^2 - y^2}} = \frac{n}{x}$$

$\Rightarrow \qquad\qquad\qquad x^2 \, y_1{}^2 = n^2 \, (b^2 - y^2)$

Differentiating again w.r.t. x, we get

$$2x^2 y_1 \, y_2 + 2xy_1{}^2 = -2n^2 \, yy_1$$

$\Rightarrow \qquad\qquad\qquad x^2 y_2 + xy_1 + n^2 y = 0$

Applying Leibnitz's theorem, we get

$$x^2 y_{n+2} + {}^nC_1 \cdot y_{n+1} \cdot 2x + {}^nC_2 \cdot y_n \cdot 2 + xy_{n+1} + {}^nC_1 y_n + n^2 y_n = 0$$

or $\quad x^2 y_{n+2} + (2n+1) xy_{n+1} + 2n^2 y_n = 0$

3. *If $y^{1/m} + y^{-1/m} = 2x$, prove that $(x^2 - 1) y_{n+2} + (2n+1) xy_{n+1} (n^2 - m^2) y_n = 0$*

(Indore 98 ; Reewa 98 ; Vikram 99 ; Bhopal 2000, 2001;
Avadh 99; Rohilkhand 2000; Kanpur 99)

We have $\qquad\qquad y^{1/m} + y^{-1/m} = 2x$

$\Rightarrow \qquad\qquad y^{2/m} - 2xy^{1/m} + 1 = 0$

$\therefore \qquad\qquad\qquad y^{1/m} = \dfrac{2x \pm \sqrt{4x^2 - 4}}{2}$

$$= x \pm \sqrt{x^2 - 1}$$

$\Rightarrow \qquad\qquad\qquad y = (x \pm \sqrt{x^2 - 1})^m$

$\therefore \qquad\qquad y_1 = m (x \pm \sqrt{x^2 - 1})^{m-1} \left(1 \pm \dfrac{x}{\sqrt{x^2 - 1}}\right)$

Squaring both sides, we get

$$(x^2 - 1) y_1^2 = m^2 y^2$$

Differentiating again w.r.t. x, we get

$$2 (x^2 - 1) y_1 y_2 + 2xy_1^2 = 2m^2 yy_1$$

$\Rightarrow \qquad (x^2 - 1) y_2 + xy_1 - m^2 y = 0$

Applying Leibnitz's Theorem, we get

$$(x^2 - 1) y_{n+2} + {}^nC_1 y_{n+1} \cdot 2x + {}^nC_2 y_n \cdot 2 + xy_{n+1} + {}^nC_1 y_n - m^2 y_n = 0$$

$\Rightarrow \quad (x^2 - 1) y_{n+2} + (2n+1) xy_{n+1} (n^2 - m^2) y_n = 0$

4. *If $y = \dfrac{\log x}{x}$, prove that*

$$y_n = \frac{(-1)^n \, n!}{x^{n+1}} \left[\log x - 1 - \frac{1}{2} - \frac{1}{3} - \ldots - \frac{1}{n} \right] \text{(Jabalpur 1995, Jiwaji 98)}$$

We have $\qquad\qquad y = \dfrac{1}{x} \log x$

Applyling Leibnitz's Theorem, we get

$$y_n = \frac{(-1)^n \, n!}{x^{n+1}} \log x + {}^nC_1 \frac{(-1)^{n-1} (n-1)!}{x^n} \frac{1}{x}$$

$$+ {}^nC_2 \frac{(-1)^{n-2} (n-2)!}{x^{n-1}} (-1) \cdot \frac{1}{x^2} + {}^nC_3 \frac{(-1)^{n-3} (n-3)!}{x^{n-2}} (-1) (-2) \cdot \frac{1}{x^3}$$

$$+ \ldots + \frac{1}{x} \frac{(-1)^{n-1} (n-1)!}{x^n}$$

$$= \frac{(-1)^n \, n!}{x^{n+1}} \log x + \frac{n (-1)^{n-1} (n-1)!}{x^{n+1}} + \frac{n (n-1)}{1.2} \frac{(-1)^{n-1} (n-2)!}{x^{n+1}}$$

$$+ \frac{n (n-1) (n-2)}{1.2.3} \frac{(-1)^{n-1} (n-3)}{x^{n+1}} \cdot 1 \cdot 2 + \ldots + \frac{(-1)^{n-1} (n-1)!}{x^{n+1}}$$

$$= \frac{(-1)^n \; n!}{x^{n+1}} \left[\log x - 1 - \frac{1}{2} - \frac{1}{3} - \ldots\ldots - \frac{1}{n} \right]$$

5. If $I_n = \dfrac{d^n}{dx^n} (x^n \log x)$, *prove that* $I_n = n I_{n-1} + (n-1)$ **(Sagar 1996, Calicut 2004)**

$$I_n = \frac{d^{n-1}}{dx^{n-1}} \left[\frac{d}{dx} (x^n \log x) \right]$$

$$= \frac{d^{n-1}}{dx^{n-1}} \left[n \, x^{n-1} \log x + x^{n-1} \right]$$

$$= n \frac{d^{n-1}}{dx^{n-1}} (x^{n-1} \log x) + \frac{d^{n-1}}{dx^{n-1}} (x^{n-1})$$

$$= n \, I_{n-1} + (n-1)!$$

6. If $x + y = 1$, *then prove that*

$$\frac{d^n}{dx^n} (x^n \, y^n) = n! \left[y^n - ({}^nC_1)^2 \, y^{n-1} . x \; ({}^nC_2)^2 \, y^{n-2} \, x^2 + \ldots\ldots (-1)^n \, x^n \right]$$

(Jabalpur 1995 ; Jiwaji 98)

$$x + y = 1 \; \Rightarrow \; y = 1 - x$$

Let $x^n = u$ and $y^n = v$, then

$$u_n = n!, \qquad u_{n-1} = n! \, x, \qquad u_{n-2} = \frac{n!}{2} \, x^2 \ldots\ldots$$

Then $\dfrac{d^n (x^n y^n)}{dx^n} = D^n \, (uv)$

$$= u_n \, v + {}^nC_1 \, u_{n-1} \, v_1 + {}^nC_2 \, u_{n-2} \, v_2 + \ldots\ldots\ldots + {}^nC_n \, v_n$$

$$= n! \, y^n + {}^nC_1 \, n! \, x \, (-{}^nC_1 \, y^{n-1}) + {}^nC_2 \left(\frac{n!}{2!} \, x^2 \right) {}^nC_2 \, 2! \, y^{n-2} + \ldots\ldots + x^n \, (-1)^n \, n!$$

$$= n! \, [y^n - ({}^nC_1)^2 \, y^{n-1} + ({}^nC_2)^2 \, y^{n-2} \, x^2 + \ldots\ldots + (-1)^n \, x^n]$$

7. *Prove that*

$$\frac{d^n}{dx^n} \left\{ \frac{\sin x}{x} \right\} = \frac{1}{x^{n+1}} \left[P \sin \left(x + \frac{1}{2} \, n\pi \right) + Q \cos \left(x + \frac{1}{2} \, n\pi \right) \right]$$

where $P = x^n - n \, (n-1) \, x^{n-2} + n \, (n-1) \, (n-2) \, (n-3) \, x^{n-4}$

and $Q = n \, x^{n-1} - n \, (n-1) \, (n-2) \, x^{n-3} + \ldots\ldots$

$$D^n (\sin x) = \sin \left(x + \frac{1}{2} \, n\pi \right)$$

$$D^{n-1} (\sin x) = \sin \left[x + \frac{1}{2} \, (n-1) \, \pi \right]$$

$$= - \cos \left(x + \frac{1}{2} \, n\pi \right)$$

$$D^{n-2} (\sin x) = \sin \left[x + \frac{1}{2} \, (n-2) \, \pi \right] = \sin \left(x + \frac{1}{2} \, n\pi \right) \text{ etc.}$$

By Leibnitz Theorem

$$D^n\left(\frac{\sin x}{x}\right) = D^n\,(\sin x)\cdot\frac{1}{x} + n\,D^{n-1}\,(\sin x).\,D\left(\frac{1}{x}\right) + \frac{n\,(n-1)}{2\,!}\,D^{n-2}\,(\sin x)\cdot D^2\left(\frac{1}{x}\right)$$

$$+ \frac{n\,(n-1)\,(n-2)}{3\,!}\,D^{n-3}\,(\sin x)\cdot D^3\left(\frac{1}{x}\right) + ...\; \sin\left(x + \frac{n\pi}{2}\right)\cdot\frac{1}{x} - n\left\{-\cos\left(x + \frac{n\pi}{2}\right)\right\}$$

$$+ \frac{n\,(n-1)}{2\,!}\left\{-\sin\left(x + \frac{n\pi}{2}\right)\right\}\left(\frac{2}{x^3}\right) + \frac{n\,(n-1)\,(n-2)}{3\,!}\left\{\cos\left(x + \frac{n\pi}{2}\right)\right\}\left(\frac{-6}{x^4}\right)$$

$$+ \frac{n\,(n-1)\,(n-2)\,(n-3)}{4\,!}\left\{\sin\left(x + \frac{n\pi}{2}\right)\right\}\left(\frac{24}{x^5}\right) + ...$$

$$= \frac{1}{x^{n+1}}\left[\,\{x^n - n\,(n-1)\,x^{n-2} + n\,(n-1)\,(n-2)\,(n-3)\,x^{n-4} + ...\}\sin\left(x + \frac{1}{2}\,n\pi\right)\right.$$

$$\left. + \{\,n\,x^{n-1}\,n\,(n-1)\,(n-2)\,x^{n-3} + ... \}\cos\left(x + \frac{1}{2}\,n\pi\right)\right]$$

$$= \frac{1}{x^{n+1}}\left[P\sin\left(x + \frac{n\pi}{2}\right) + Q\cos\left(x + \frac{n\pi}{2}\right)\right]$$

8. *By finding in two different ways the nth derivative of x^{2n}, prove that*

$$1 + \frac{n^2}{1^2} + \frac{n^2\,(n-1)^2}{1^2.\,2^2} + \frac{n^2\,(n-1)^2\,(n-2)^2}{1^2.\,2^2.\,3^2} + ... = \frac{(2\,n)\,!}{(n\,!)^2} \qquad \textbf{(Rohilkhand 1995)}$$

$$D^n\,(x^{2n}) = 2n\,(2n-1)\,(2n-2)\,.......\,(2n-n+1)\,(x)^{2n-n}$$

$$= 2n\,(2n-1)\,(2n-2)\,.....\,(n+1)\,x^n$$

$$= \frac{(2n)\,!}{n\,!}\,x^n \qquad\qquad ...(i)$$

Also, by Leibnitz's Theorem, we get

$$D^n\,(x^{2n}) = D^n\,(x^n\,.\,x^n)$$

$$= D^n\,(x^n)\,.\,x^n + {}^nC_1\,D^{n-1}\,(x^n)\,.\,D(x^n) + {}^nC_2\,D^{n-2}\,(x^n).D^2(x^n)$$

$$+ {}^nC_3\,D^{n-3}\,(x^n).\,D^3\,(x^n) + ...$$

$$= (n\,!\,)\,x^n + {}^nC_1\,(n\,!\,x)\,nx^{n-1} + {}^nC_2\left(\frac{n\,!\,x^2}{2\,!}\right)n\,(n-1)\,x^{n-2}$$

$$+ {}^nC_3\,\frac{n\,!\,x^3}{3\,!}\{n\,(n-1)\,(n-2)\,x^{n-3}\} + ...$$

$$= n\,!\,x^n\left[1 + \frac{n^2}{1^2} + \frac{x^2\,(n-1)^2}{1^2 - 2^2} + \frac{x^2\,(n-1)^2\,(n-2)^2}{1^2.\,2^2.\,3^2} + ...\right] \qquad ...(ii)$$

Equating the values of $D^n\,(x^{2n})$ as given by (i) and (ii) we get

$$x^n\,n\,!\left[1 + \frac{n^2}{1^2} + \frac{n^2\,(n-1)^2}{1^2.\,2^2} + \frac{n^2\,(n-1)^2\,(n-2)^2}{1^2.\,2^2.\,3^2} + ...\right] = \frac{(2n)!\,x^n}{n!}$$

or $\quad 1 + \dfrac{n^2}{1^2} + \dfrac{n^2\,(n-1)^2}{1^2.\,2^2} + \dfrac{n^2\,(n-1)^2\,(n-2)^2}{1^2.\,2^2.\,3^2} + ... = \dfrac{(2n)!}{(n!)^2}$

9. *Find the value of the nth derivative of* $y = e^{m\sin^{-1}x}$ *for* $x = 0$.

(Gorakhpur 2001, Vikram 97; Jabalpur 97)

Let $y = e^{m\sin^{-1}x}$ ...(i)

so that $y_1 = e^{m\sin^{-1}x} \cdot \dfrac{m}{\sqrt{(1-x^2)}}$ ...(ii)

$\Rightarrow$ $(1-x^2)\,y_1^{\,2} = m^2 y^2$

Differentiating, we get

$$(1-x^2)\,2y_1\,y_2 - 2xy_1^{\,2} = 2m^2\,yy_1$$

Dividing by $2y_1$, we obtain

$$(1-x^2)\,y_2 - xy_1 = m^2\,y$$ ...(iii)

Differentiating n times by Leibnitz's theorem, we get

$$(1-x^2)\,y_{n+2} - 2nxy_{n+1} - n\,(n-1)\,y_n - xy_{n+1} - ny_n = m^2 y_n$$
$$\Rightarrow \ (1-x^2)\,y_{n+2} - (2n+1)\,xy_{n+1} - (n^2+m^2)\,y_n = 0$$

Putting $x = 0$, we get

$$y_{n+2}\,(0) = (n^2+m^2)\,y_n\,(0)$$ ...(iv)

From (i), (ii) and (iii), we obtain

$y\,(0) = 1$, $y_1\,(0) = m$, $y_2\,(0) = m^2$.

Putting $n = 1, 2, 3, 4$ etc. in (iv) we get

$$y_3\,(0) = (1^2+m^2)\,y_1\,(0) = m\,(1^2+m^2)\;;$$
$$y_4\,(0) = (2^2+m^2)\,y_2\,(0) = m^2\,(2^2+m^2)\;;$$
$$y_5\,(0) = (3^2+m^2)\,y_3\,(0) = m\,(1^2+m^2)\,(3^2+m^2)\;;$$
$$y_6\,(0) = (4^2+m^2)\,y_4\,(0) = m^2\,(2^2+m^2)\,(4^2+m^2).$$

In general

$$y_n\,(0) = \begin{cases} m^2\,(2^2+m^2)\,(4^2+m^2)\ldots\ldots[(n-2)^2+m^2], & \text{when } n \text{ is even,} \\ m\,(1^2+m^2)\,(3^2+m^2)\ldots\ldots[(n-2)^2+m^2], & \text{when } n \text{ is odd.} \end{cases}$$

10. *Find* $y_n\,(0)$ *when* $y = \log(x + \sqrt{1+x^2})$

We have $y = \log(x + \sqrt{1+x^2})$ ...(i)

$\Rightarrow$ $y_1 = \dfrac{1}{x+\sqrt{1+x^2}} \cdot \left(1 + \dfrac{x}{\sqrt{1+x^2}}\right)$

              $= \dfrac{1}{\sqrt{1+x^2}}$ ...(ii)

$\Rightarrow$ $(1+x^2)y_1^{\,2} = 1$

Differentiating again w.r.t. x, we get

$$2y_1\,y_2\,(1+x^2) + 2xy_1^{\,2} = 0$$
$$\Rightarrow \qquad (1+x^2)\,y_2 + xy_1 = 0$$ ...(iii)

Applying Leibnitz's Theorem, we get

$$(1+x^2)\,y_{n+2} + {}^nC_1 y_{n+1} \cdot 2x + {}^nC_2\,y_n \cdot 2 + xy_{n+1} + {}^nC_1\,y_n = 0$$

$\Rightarrow$ $\qquad\qquad (1 + x^2) y_{n+2} + (2n+1) xy_{n+1} + n^2 y_n = 0$

Putting $x = 0$, we get

$$y_{n+2}(0) = - n^2 y_n(0) \qquad\qquad\qquad ...(iv)$$

From (i), (ii) and (iii), we get

$$y = (0) = 1, y_1(0) = 1, y_2(0) = 0$$

Putting $n = 1, 2, 3, 4$ etc. in (iv), we get

$$y_3(0) = - 1^2 y_1(0) = - 1^2.$$
$$y_4(0) = - 2^2 y_2(0) = 0$$
$$y_5(0) = - 3^2 y_3(0) = (-1)^2 \, 1^2.\, 3^2$$
$$y_6(0) = - 4^2 y_4(0) = 0$$

In general

$$y_n(0) = 0 \text{ if } n \text{ is even}$$
$$= (-1)^{(n-1)/2} \, 1^2.\, 3^2.\, 5^2 ...(n-2)^2, \text{ if } n \text{ is odd}.$$

11. *If $y = \cos(m \sin^{-1} x)$ show that*

$$(1 - x^2) y_{n+2} - (2n+1) xy_{n+1} + (m^2 - n^2) y_n = 0$$

and hence find $y_n(0)$. **(Kumaon 2002; Gorakhpur 99, 2000 Calicut 2004)**

We have $\qquad\qquad y = \cos(m \sin^{-1} x) \qquad\qquad\qquad ...(i)$

$$y_1 = - \sin(m \sin^{-1} x) \times \frac{m}{\sqrt{(1 - x^2)}} \qquad\qquad ...(ii)$$

$\Rightarrow \qquad\qquad (1 - x^2) y_1{}^2 = m^2 \sin^2(m \sin^{-1} x)$
$$= m^2 (1 - y^2)$$

Differentiating we get

$$(1 - x^2) \, 2y_1 y_2 - 2xy_1{}^2 = - 2m^2 \, yy_1$$

$\Rightarrow \qquad\qquad (1 - x^2) y_2 - xy_1 + m^2 y = 0 \qquad\qquad ...(iii)$

Applying Leibnitz's Theorem, we get

$(1 - x^2) y_{n+2} + {}^nC_1. - 2x. y_{n+1} + nC_2. - 2. y_n - xy_{n+1} - nC_1 y_n + m^2 y_n = 0$
$\Rightarrow \;\; (1 - x^2) y_{n+2} - (2n+1) xy_{n+1} + (m^2 - n^2) y_n = 0$

Putting $x = 0$, we get

$$y_{n+2}(0) = (n^2 - m^2) y_n(0) \qquad\qquad\qquad ...(iv)$$

From (i), (ii) and (iii) we have

$$y(0) = 1, y_1(0) = 0, y_2(0) = m^2 y(0) = m^2$$

Putting $n = 1, 2, 3, 4$, etc. in (iv) we get

$$y_3(0) = (1^2 - m^2) y_1(0) = 0$$
$$y_4(0) = (2^2 - m^2) y_2(0) = m^2 (2^2 - m^2)$$
$$y_5(0) = (3^2 - m^2) y_3(0) = 0$$
$$y_6(0) = (4^2 - m^2) y_4(0)$$
$$= m^2 (2^2 - m^2) (4^2 - m^2)$$

In general $\qquad y_n(0) = 0$, if n is odd

$$= m^2 (2^2 - m^2) (4^2 - m^2) .... ((n-2)^2 - m^2), \text{ if } n \text{ is even}.$$

EXERCISES

1. Find $d^n y/dx^n$ for the following functions :

    (i) $x^n e^x$,   (ii) $x^3 \cos x$,   **(Sagar 93, Gujrat 2005)**

    (iii) $e^{ax} [a^2 x^2 - 2nax + n (n + 1)]$,(iv) $e^x \log x$.

2. If $y = x^2 \sin x$, prove that

$$\frac{d^n y}{dx^n} = (x^2 - n^2 + n) \sin\left(x + \frac{n\pi}{2}\right) - nx \cos\left(x + \frac{n\pi}{2}\right).$$

3. If $f(x) = x^2 \tan x$, prove that

$$f^n (0) - {}^nC_2 f^{n-2} (0) + {}^nC_4 f^{n-4} (0) - \ldots\ldots\ldots = \sin\frac{n\pi}{2}.$$

[Write $f(x) \cos x = \sin x$ and apply Leibnitz theorem.]

4. Differentiate the following equations :

    (i) $(1 - x^2)\dfrac{d^2 y}{dx^2} - x\dfrac{dy}{dx} + a^2 y = 0.$

    (ii) $x^2\dfrac{d^2 y}{dx^2} - x\dfrac{dy}{dx} + (a^2 - m^2)y = 0.$

    n times with respect to x.

5. If $y = (\sin^{-1} x)^2$, prove that

$$(1 - x^2)\frac{d^2 y}{dx^2} - x\frac{dy}{dx} - 2 = 0.$$

    Differentiate the above equation n times with respect to x and show that
$$(1 - x^2) y_{n+2} - (2n + 1) xy_{n+1} - n^2 y_n = 0.$$

(Bilaspur 2000 ; Reewa 2000 ; Ravishankar 2000 ; Gorakhpur 2002, Calcutta 2005, Banglore 2005)

6. If $u = \tan^{-1} x$, prove that

$$(1 - x^2)\frac{d^2 u}{dx^2} - 2x\frac{du}{dx} = 0$$

    and hence determine that values of the derivatives of u when $x = 0$.   **(Avadh 2000)**

7. If $y = \sin (m \sin^{-1} x)$, show that
$$(1 - x^2) y_{n+2} = (2n + 1) xy_{n+1} (n^2 - m^2) y_n$$
    and find $y_n (0)$.

(Sagar 2000 ; Bilaspur 2001 ; Jabalpur 2001 ; Indore 99 ; Reewa 2000, 2001; Garhwal 2001; Kanpur 2002 ; Gorakhpur 2003 ; Avadh 2003; MDU Rohtak 99, Poorvanchal 2004)

8. Find $y_n (0)$, when $y = [x \sqrt{(1 + x^2)}]^m$.

(Kanpur 2006, Rohilkhand 2001, Vikram 2000 ; Indore 2001; Gorakhpur 2000, Devi Ahilya 2000)

9. If $e^{m\cos^{-1} x}$ show that
$$(1 - x^2) y_{n+2} - (2n + 1) xy_{n+1} - (n^2 + m^2) y_n = 0 \text{ and find } y_n (0).$$ **(Banglore 2004)**

10. If $y = \dfrac{\sin^{-1} x}{\sqrt{1 - x^2}}$, show that
$$(1 - x^2)y_{n+2} - (2n + 3) xy_{n+1} - (n + 1)^2 y_n = 0$$

11. If $y = a \cos (\log x) + b \sin (\log x)$, show that
$$x^2 y_{n+2} + (2n + 1) xy_n + 1 + (n^2 + 1) y_n = 0$$

(Ravishankar 1997, 99 ; Bhopal 98 ; Jabalpur 98 ; Indore 96, MDU Rohtak 2000)

12. If $y = (\sinh^{-1} x)^2$, prove that

$$(1 + x^2)y_{n+2} + 2xy_{n+1} \, n^2 \, y_n = 0$$

13. If $y = (x^2 - 1)^n$, prove that

$$(x^2 - 1) \, y_{n+2} + 2xy_{n+1} - n(n+1) \, y_n = 0 \qquad \textbf{\textit{(Kumaon 2003 ; Reewa 97; Avadh 96)}}$$

(**Hint.** $\log y = n \log (x^2 - 1) \Rightarrow \dot{y}_1 (x^2 - 1) = 2nxy$)

14. Prove that the nth dirrerential coefficient of $x^n (1 - x)^n$ is

$$n!(1 - x)^n \left[1 - \frac{n^2}{1^2} \frac{x}{1-x} + \frac{n^2 (n-1)^2}{1^2 . 2^2} \cdot \frac{x^2}{(1-x)^2} + \ldots \ldots \right]$$

15. If $y - e^{\tan^{-1} x}$ prove that

$$(1 + x^2) \, y_{n+1} + (2nx - 1) \, y_n + n(n-1) \, y_{(n-1)} = 0$$

$$\textbf{\textit{(Bilaspur 97; Indore 2000; MDU Rohtak 2001)}}$$

16. If $y = e^{a\sin^{-1} x}$ = prove that

$$(1 - x^2) \, y_{n+2} - (2n+1) \, x \, y_{n+1} - (n^2 + a^2) \, y_n = 0$$

$$\textbf{\textit{(Avadh 2002 ; Ravishankar 2001 ; Jabalpur 97, 99 ; Indore 95 ; Bhopal 95 ; Vikram 97 ; Reewa 99 ; Kumaon 2001 ; Bilaspur 2001 ; Rohilkhand 2003 ; Garhwal 2001; Gujrat 2005; Gauhati 2005)}}$$

17. If $y = \left[\log_e \{x + \sqrt{(1 + x^2)}\} \right]^2$ find $(y_n)_0$

$$\textbf{\textit{(Vikram 1996, Indore 2000, Avadh 99; Rohilkhand 2002, 2004)}}$$

18. If $y = \log \left[x + \sqrt{(x^2 + a^2)} \right]$, prove that $(a^2 + x^2) \, y_2 + xy_1 = 0$. Differentiate this differential

equation n times and prove that $\lim\limits_{x \to 0} \dfrac{y_{n+2}}{y_n} = -\dfrac{n^2}{a^2}$ $\qquad$ **_(Rohilkhand 2002)_**

ANSWERS

1. (i) $e^x \left[x^n + \dfrac{n^2}{1!} x^{n-1} + \dfrac{n^2 (n-1)^2}{2!} x^{n-2} + \ldots \dfrac{n^2 (n-1)^2 \ldots 1^2}{n!} \right]$

(ii) $x^3 \cos \left(x + \dfrac{1}{2} n\pi \right) + 3nx^2 \cos \left[x + \dfrac{1}{2}(n-1)\pi \right] + 3n(n-1)x \cos \left[x + \dfrac{1}{2}(n-2)\pi \right]$

$$+ n(n-1)(n-2) \cos \left[x + \dfrac{1}{2}(n-3)\pi \right].$$

(iii) $a^{n+2} \cdot e^{ax} \cdot x^2$.

(iv) $e^x [\log x + {}^n c_1 x^{-1} - {}^n c_2 x^{-2} + {}^n c_3 \, 2! \, x^{-3} - \ldots \ldots + (-1)^{n-1} \, {}^n c_n \, (n-1)! \cdot x^{-n}]$.

4. (i) $(1 - x^2) \, y_{n+2} - (2n+1) \, xy_{n+1} + (a^2 - n^2) \, y_n = 0$.

(ii) $x^2 y_{n+2} \, (2n+1) \, xy_{n+1} + (n^2 + a^2 - m^2) \, y_n = 0$.

6. $U_n(0) = 0$, if n is even ; and $(-1)^{(n-1)/2} (n-1)!$ if n is odd.

7. $y_n(0) = \begin{cases} 0, \text{ if } n \text{ is even} \\ m(1^2 - m^2)(3^2 - m^2) \ldots (n-2)^2 - m^2), \text{ if } n \text{ is odd} \end{cases}$

8. $y_{2n}(0) = m^2 (m^2 - 2^2) (m^2 - 4^2) \ldots [m^2 - (2n-2)^2]$;

$y_{2n+1}(0) = m (m^2 - 1^2)) (m^2 - 3^2) (m^2 - 5^2) \ldots [m^2 - (2n-1)^2]$.

9. $y_{2n+1} = -e^{m\pi/2}\, m\, (1^2 + m^2)\, (3^2 + m^2) \dots ((2n-1)^2 + m^2).$

$y_{2n} = e^{m\pi/2}\, m^2\, (2^2 + m^2)\, (4^2 + m^2) \dots ((2n-2)^2 + m^2).$

17. When n is odd 0, when n is even

$$(-1)^{\frac{n-2}{a}}\, 2^{n-1}\left[\left(\frac{n-2}{2}\right)!\right]^2$$

MISCELLANEOUS EXERCISES

1. Show that if $x(1-x)y_2 - (4-12x)y_1 - 36y = 0$, then

$x(1-x)y_{n+2} - [4 - n - (12-2n)x]\,y_{n+1} - (4-n)(9-n)\,y_n = 0.$

2. If y_n denotes the nth derivative of $e^{ax}\sin bx$ and $\theta = \tan^{-1}(b/a)$, prove that

$$y_n = (a\cos\theta)^n\, e^{ax}\sin(bx + n\theta).$$

Also show that $y_{y+1} - 2ay_n + (a^2 + b^2)y_{n-1} = 0.$

3. If $y = (a + bx)\cos kx + (c + dx)\sin kx$, prove that

$$\frac{d^4 y}{dx^4} + 2k^2\frac{d^2 y}{dx^2} + k^4 y = 0.$$

4. Find the third derivative of $y = \tan^{-1}\dfrac{x}{\sqrt{(1-x^2)}}$.

5. Prove that $\dfrac{d^4}{dx^4}\sqrt{(1+x^2)} = \dfrac{12x^2 - 3}{\sqrt{(1+x^2)^7}}.$

6. Find the value of the nth derivative of $y = \dfrac{x^2 - x}{(x^2 - 4)^2}$.

for $x = 0$.

7. If U_n denotes the nth derivative of $(Lx + M)/(x^2 - 2Bx + C)$, prove that

$$\frac{x^2 - 2Bx + C}{(n+1)(n+2)}U_{n+2} + \frac{2(x-B)}{n+1}U_{n+1} + U_n = 0.$$

8. If $y = x^2 e^x$, then

$$\frac{d^n y}{dx^n} = \frac{1}{2}n(n-1)\frac{d^2 y}{dx} - (n-2)\frac{dy}{dx} + \frac{1}{2}(n-1)(n-2).y. \qquad \text{(Bhopal 97)}$$

9. If $y = (\tan^{-1}x)^2$, then

$$(x^2 + 1)^2\frac{d^2 y}{dx^2} + 2x(x^2 + 1)\frac{dy}{dx} = 2. \qquad \text{(Manipur 2002)}$$

Deduce that

$$(x^2 + 1)^2\frac{d^{n+2}}{dx^{n+2}} + (4n+2)x(x^2+1)\frac{d^{n+1}y}{dx^{n+1}} + 2n^2(3x^2+1)\frac{d^n y}{dx^n}$$
$$+ 2n(n-1)(2n-1)x\frac{d^{n-1}y}{dx^{n-1}} + n(n-1)^2(n-2)\frac{d^{n-2}y}{dx^{n-2}} = 0.$$

10. If $y = e^{x^2/2}\cos x$, show that

$$y_{2n+2}(0) - 4ny_{2n}(0) + (2n-1)\,2ny_{2n-2}(0) = 0.$$

11. If $y = (1+x^2)^{m/2}\sin(m\tan^{-1}x)$, show that

$$y_{2n}(0) = 0 \text{ and } y_{2n+1}(0) = (-1)^n\, m\,(m-1)(m-2)\dots(m-2n).$$

ANSWERS

4. $(1 + 2x^2)/(1 - x^2)^{5/2}$

6. If n is even, $y_n(0) = 0$; if n is odd,

$$y_n(0) = n!(3n - 5)\frac{1}{2^{n+4}}.$$

OBJECTIVE QUESTIONS

Note : *For each of the following questions, four alternatives are given for the answer. Only one of them is correct. Choose the correct alternative.*

1. If $x = t - \sin t$, $y = 1 - \cos t$, value of $\dfrac{d^2 y}{dx^2}$ at $(\pi, 2)$ will be :

    (a) 0   (b) 1   (c) π (d) ∞

2. Value of $D^n (ax + b)^n$ is   *(Avadh 2002)*

    (a) $n a^n$ (b) $n! a^n$ (c) $n a b^n$ (d) $n! b^n$

3. Value of $D^n\left(\dfrac{1}{ax + b}\right)$ is

    (a) $\dfrac{n! a^n}{(ax + b)^n}$ (b) $\dfrac{(-1)^n n! a^n}{(ax + b)^n}$ (c) $\dfrac{(-1)^n n! a^n}{(ax + b)^{n+1}}$ (d) 0

4. Value of $D^n\{\sin (ax + b)\}$ is

    (a) $a^n \sin (ax + b + n\pi)$ (b) $b^n \sin (ax + \dfrac{1}{2} n\pi)$

    (c) $a^n \sin (ax + b + \dfrac{1}{2} n\pi)$ (d) $b^n \sin (ax + b + n\pi)$ *(Manipur 2000)*

5. Leibnitz's theorem is used to find the nth differential coefficient of :
    (a) trigonometric functions only
    (b) exponential function only
    (c) sum and difference of two functions
    (d) product of two functions   *(Rohilkhand 2002)*

6. $(r + 1)$ th term in the expression of $D^n (uv)$ is :

    (a) $^nC_{r+1} D^{n-r} u D^r v$ (b) $^{n+1}C_2 D^{n-r} u D^r v$

    (c) $^nC_r D^{n-r} u D^r v$ (d) $^nC_{r-1} D^{n-r} u D^r v$ *(Avadh 2005)*

7. If $y = \sin (m \sin^{-1} x)$, then
    (a) $(1 - x^2) y_2 - xy_1 + m^2 y = 0$ (b) $(1 - x^2) y_2 + xy_1 - m^2 y = 0$
    (c) $(1 - x^2) y_2 - xy_1 - m^2 y = 0$ (d) None of these

8. If $y = (\tan^{-1} x)^2$, then $(x^2 + 1)^2 y_2 + 2x (x^2 + 1)y_1$, =
    (a) 2   (b) -2 (c) $2y$ (d) $-2y$

9. If $y = [\log \{x + \sqrt{(x^2 + 1)}\}]^2$, then value of $(1 + x^2) y_2 + 2y_1$ is equal to

    (a) 1   (b) 2   (c) -1 (d) None of these

10. If $y^{1/m} - y^{-1/m} = 2x$, then $(x^2 - 1) y_2 + xy_1 =$
    (a) $m^2 y$ (b) $-m^2 y$ (c) $\pm m^2 y$ (d) None of these

11. If $y = b \cos \{n \log (x/n)\}$, then

    (a) $x^2 y_2 - xy_1 + x^2 y = 0$ (b) $x^2 y_2 + xy_1 - x^2 y = 0$

    (c) $x^2 y_2 - xy_1 - x^2 y = 0$ (d) $x^2 y_2 + xy_1 + x^2 y = 0$

12. Function $y = \left[x + \sqrt{x^2 + 1} \right]^k$ satisfies :

    (a) $(x^2 + 1) y' = k^2 y$ (b) $(1 + x^2) y'' + ky' - xy = 0$

    (c) $\sqrt{(x^2 + 1)}\, y' = ky$ (d) $(x^2 + 1) y'' + xy' - k^2 y = 0$

13. If $y = 1/(a^2 - x^2)$, then $y_n =$

    (a) $(-1)^n \dfrac{n!}{2a}[(x + a)^{-n} - (x - a)^{-n}]$ (b) $(-1)^n \dfrac{n!}{2a}[(x - a)^{-n} - (x + a)^{-n}]$

    (c) $(-1)^n \dfrac{n!}{2a}[(x - a)^{-n} + (x + a)^{-n}]$ (d) None of these

14. If $u = \sin nx + \cos nx$, then $u_r =$

    (a) $[1 + (-1)^r \sin 2nx]^{1/2}$ (b) $n^r [1 + \sin 2nx]^{1/2}$

    (c) $n^r[1 + (-1)^r \sin 2nx]^{1/2}$ (d) None of these

15. $D^n(\sin x \sin 3x) =$

    (a) $\dfrac{1}{2}\left[\cos\left(2x + \dfrac{n\pi}{2} \right) - \cos\left(4x + \dfrac{n\pi}{2} \right) \right]$

    (b) $\dfrac{1}{2}\left[\cos\left(4x + \dfrac{n\pi}{2} \right) - \cos\left(2x + \dfrac{n\pi}{2} \right) \right]$

    (c) $\dfrac{1}{2}\left[\cos\left(4x + \dfrac{n\pi}{2} \right) + \cos\left(2x + \dfrac{n\pi}{2} \right) \right]$

    (d) None of these

16. If y is a polynomial of degree n in x and first coefficient is 2, then $D^{n-1}(y) =$

    (a) $2\,(n!)$ (b) $2\,(n!)\,x$ (c) $2\,(n-1)!\,x$ (d) None of these

17. If $y = x^2 e^x$ and $y_n = {}^nC_2\, y_2 + ay_1 + {}^{(n-1)}C_2 y$, then $a =$

    (a) $n\,(n-1)$ (b) $n\,(n-2)$ (c) $-n\,(n-1)$ (d) $-n\,(n-2)$

18. If $y = x^{n-1} \log x$, then $xy_n =$

    (a) $n!$ (b) $(n-1)!$ (c) $(n-2)!$ (d) None of these

19. If $y = e^{\tan^{-1} x}$, then $(1 + x^2) y_{n+2} + [2(n+1)x - 1] y_{n+1} =$

    (a) $n^2 y_n$ (b) $n(n+1) y_n$ (c) $-n(n+1)y_n$ (d) $-n^2 y_n$

20. If $y = \sin^{-1} x$, then

    $(1 - x^2) y_{n+2} - (2n+1)xy_{n+1} =$

    (a) $n^2 y_n$ (b) $-n^2 y_n$ (c) $-(n^2 + 1)y_n$ (d) None of these

ANSWERS

1. (a)	2. (b)	3. (c)	4. (c)	5. (d)	6. (c)	7. (a)
8. (a)	9. (b)	10. (a)	11. (d)	12. (c)	13. (d)	14. (c)
15. (d)	16. (d)	17. (d)	18. (b)	19. (c)	20. (a)	

CHAPTER 6

EXPANSION OF FUNCTIONS

Introduction. Ordinary algebraic processes, taking limits, differentiation etc, though applicable to the sum of a finite number of terms, may break down for infinite series. In this chapter, we will expand functions in terms of infinite series by using the methods of differential calculus. The infinite series found below are to be regarded as formal expansions, which may not be true in exceptional cases.

6.1. Maclaurin's Theorem.

(Sagar 1997; Bilaspur 97, 2001; Jabalpur 2000 ; Ravishankar 2000 ; Avadh 2003, 2005; Gujrat 2005)

Let $f(x)$ be a funtion of x. Let this function is to be expanded in ascending powers of x and let the expansion be differentiable term by term any number of times.

Let
$$f(x) = A_0 + A_1 x + A_2 x^3 + A_3 x^3 + A_4 x^4 + \dots$$

Now by successive differentiation, we get
$$f'(x) = A_1 + 2 A_2 x + 3 A_3 x^2 + 4 A_4 x^3 + \dots$$
$$f''(x) = 2 \cdot 1 A_2 + 3 \cdot 2 A_3 x + 4 \cdot 3 A_4 x^2 + \dots$$
$$f'''(x) = 3 \cdot 2 \cdot 1 A_3 + 4 \cdot 3 \cdot 2 A_4 x + \dots$$
$$= \dots$$
$$f^n(x) = n(n-1)(n-2) \dots 2 \cdot 1 A_n + (n+1) n (n-1) \dots 3 \cdot 2 A_{n+1} \cdot x + \dots$$

Putting $x = 0$ in each term, we get
$$f(0) = A_0, f'(0) = A_1, f''(0) = 2 ! A_2$$
$$f''(0) = 3 ! A_3, \dots f^n(0) = n ! A_n \dots$$

with the help of these values of $A_0, A_1, A_2, A_3, \dots$ we have

$$f(x) = f(0) + x f'(0) + \frac{x^2}{2!} f''(0) + \dots + \frac{x^n}{n!} f^n(0)$$

This result is konwn as **Maclaurin's Theorem**.

EXAMPLES

1. *Expand cos x by Maclaurin's series.*   *(Kanpur 2006; Srivenkateshwara 2005)*

We have $y = \cos x,\ y_1 = -\sin x = \cos\left(\frac{\pi}{2} + x\right)$

$$y_2 = -\sin\left(\frac{\pi}{2} + x\right) = \cos\left(x + \frac{2\pi}{2}\right)$$

then $\qquad y_n = \cos\left(x + \dfrac{n\pi}{2}\right)$

Now put $x = 0$; then

$$(y)_0 = 1, (y_1)_0 = 0, (y_2)_0 = -1$$

$$(y_3)_0 = \cos\left(\frac{3\pi}{2}\right) = 0$$

In general $\quad (y_n)_0 = \cos\dfrac{n\pi}{2}$

Now, put $\qquad n = 4$; then $(y_4)_0 = \cos 2\pi = 1$

Hence $\qquad (y_n)_0 = \cos\dfrac{n\pi}{2} = 0$, if n is odd

$$= (-1)^{n/2} \text{ if } n \text{ is even.}$$

Then $\qquad f(x) = f(0) + x f'(0) + \dfrac{x^2}{2!} f''(0) + ...$

$$\cos x = 1 - \frac{x^2}{2!} + \frac{x^4}{4!} - \frac{x^6}{6!} + ... + (-1)^{n/2}\frac{x^n}{n!} + ...$$

when n is even.

2. *Obtain the expansion of* $\log \cosh x$ *in powers of* x *by Maclaurin's Theorem.*

We have $\qquad y = \log \cosh x$

$\qquad\qquad y_1 = \tanh x$

$\qquad\qquad y_2 = \operatorname{sech}^2 x = 1 - \tanh^2 x = 1 - y^2_1$

$\qquad\qquad y_3 = -2\, y_1 y_2$

$\qquad\qquad y_4 = -2\,(y_1 y_3 + y^2_2),$

$\qquad\qquad y_5 = -2\,(y_1 y_4 + y_2 y_3 + 2\, y_2 y_3)$

$\qquad\qquad y_6 = -2\,(y_1 y_5 + 4\, y_2 y_4 + 3\, y^2_3)$

Now putting $\quad x = 0$, we obtain

$$(y)_0 = 0, (y_1)_0 = 0, (y_2)_0 = 1, (y_3)_0 = 0,$$

$$(y_4)_0 = -2, (y_5)_0 = 0 \text{ and } (y_6)_0 = 16 \text{ etc.}$$

Substitute the values in the Maclaurin's series

$$y = (y)_0 + x\,(y_1)_0 + \frac{x^2}{2!}\,(y_2)_0 + ...$$

we obtain $\qquad y = \log \cosh x = \dfrac{x^2}{2} - \dfrac{x^4}{12} + \dfrac{x^6}{45} - ...$

3. *Expand* $\log(1 + x)$ *by Maclaurin's Theorem.* $\qquad$ **(*Sagar 1995, 2000 ; Indore 99*)**

We have $\qquad y = \log(1 + x) \;\Rightarrow\; (y)_0 = 0$

$\qquad\qquad y_1 = \dfrac{1}{1+x} \quad\Rightarrow\quad (y_1)_0 = 1$

$$y_2 = -\frac{1}{(1+x)^2} \qquad \Rightarrow \qquad (y_2)_0 = -1$$

$$y_3 = \frac{2}{(1+x)^3} \qquad \Rightarrow \qquad (y_3)_0 = 2$$

$$y_4 = -\frac{6}{(1+x)^4} \qquad \Rightarrow \qquad (y_4)_0 = -6$$

....

....

$$y_n = \frac{(-1)^{n-1}(n-1)!}{(1+x)^n} \qquad \Rightarrow \qquad (y_n)_0 = (-1)^{n-1}(n-1)!$$

$$\therefore \quad \log(1+x) = x - \frac{x^2}{2} + \frac{x^3}{3} - \frac{x^4}{4} + ... + \frac{(-1)^{n-1}(n-1)!\,x^n}{n!}$$

$$= x - \frac{x^2}{2} + \frac{x^3}{3} - \frac{x^4}{4} + ... \frac{(-1)^{n-1}x^n}{n} + ...$$

4. *Use Maclaurin's Theorem to find the expansion in ascending powers of x of* $\log(1+e^x)$ *to the terms containing* x^4. **(Garhwal 2002; Vikram 97 ; Sagar 98)**

We have $\qquad y = \log(1+e^x)$,

i.e., $\qquad e^y = 1 + e^x,$ $\qquad\qquad\qquad\qquad\qquad\qquad\qquad$ ...(i)

$$e^y \cdot y_1 = e^x \qquad\qquad\qquad\qquad\qquad\qquad ...(ii)$$

$$e^y \cdot y_2 + e^y \cdot y_1{}^2 = e^x \qquad\qquad\qquad\qquad ...(iii)$$

$$e^y \cdot y_3 + e^y\,y_1\,y_2 + e^y\,y_1{}^2 + e^y\cdot 2\,y_1\,y_2 = e^x \qquad ...(iv)$$

$$e^y\,y_1\,y_3 + e^y\,y_4 + 3\,e^y\,(y_2{}^2 + y_1\,y_3) + 3\,e^y\,y_1{}^2\,y_2 + e^y\cdot y_1{}^4 + e^y\cdot 3\,y_1{}^2\,y_2 = e^x \qquad ...(v)$$

Now put $x = 0$,

in (i), $\qquad (y)_0 = \log 2$

in (ii), $\qquad (y_1)_0 = \dfrac{1}{e^{\log 2}} = \dfrac{1}{2}$

in (iii), $\qquad 2\,(y_2)_0 + 2\cdot\dfrac{1}{4} = 1$

or $\qquad (y_2)_0 = \dfrac{1}{4},$

in (iv), $\qquad 2\cdot\dfrac{1}{2}\cdot(y_3)_0 + 2\cdot\dfrac{1}{2}\cdot\dfrac{1}{4} + 2\cdot\dfrac{1}{4} + 1\cdot 2\cdot\dfrac{1}{2}\cdot\dfrac{1}{4} = 1$

or $\qquad (y_3)_0 = 0$

in (v), $\qquad (y_4)_0 = -\dfrac{1}{6}$

Hence by Maclaurin's Theorem,

$$\log(1+e^x) = \log 2 + \frac{1}{2}x + \frac{1}{8}x^2 - \frac{1}{192}x^4 + ...$$

5. *Expand* $e^{a \sin^{-1} x}$ *in powers of x by Maclaurin's Theorem and hence obtain the value of* e^{θ}.

(*Kumaon 2003; Kuvempu, 2005*)

We have $\qquad y = e^{a \sin^{-1} x}, \quad y_1 = e^{a \sin^{-1} x} \cdot \dfrac{a}{\sqrt{(1 - x^2)}}$

or $\qquad (1 - x^2) y_1^2 = a^2 y^2$

Again differentiating

$$2 y_1 y_2 (1 - x^2) - 2 x y_1{}^2 = 2 a^2 y y_1$$

or $\qquad y_2 (1 - x^2) - x y_1 = a^2 y$

Differentiating n times more, we have

$$y_{n+2} - (2n + 1) y_{n+1} \cdot x - (n^2 + a^2) y_n = 0$$

Put $x = 0$, then

$$(y_{n+2})_0 = (n^2 + a^2) (y_n) 0$$

Now, $\qquad (y)_0 = 1, (y_1)_0 = a, (y_2)_0 = a^2$

$$(y_3)_0 = (1^2 + a^2) (y_1)_0 = (1^2 + a^2) \cdot a$$

$$(y_4)_0 = (2^2 + a^2) (y_2)_0 = (2^2 + a^2) \cdot a^2$$

Then by Maclaurin's Theorem

$$e^{a \sin^{-1} x} = 1 + a \cdot x + \frac{a^2}{2!} x^2 + \frac{x^3}{3!} (1^2 + a^2) a + \frac{x^4}{4!} (2^2 + a^2) a^2 + \dots$$

Now put $a \sin^{-1} x = \theta$, *i.e.*, $x = \sin \dfrac{\theta}{a}$ and put $a = 1$; then

$$e^{\theta} = 1 + \sin \theta + \frac{\sin^2 \theta}{2!} + 2 \cdot \frac{\sin^3 \theta}{3!} + \dots$$

6. *Show if* $y = \sin (m \sin^{-1} x)$, *then*

$$(1 - x^2) \frac{d^2 y}{d x^2} - x \frac{d y}{d x} + m^2 y = 0$$

Hence or otherwise expand sin m θ in powers of sin θ.

We have $\qquad y = \sin (m \sin^{-1} x),$

then $\qquad y_1 = \cos (m \sin^{-1} x) \cdot \dfrac{m}{\sqrt{(1 - x^2)}}$

or $\qquad (1 - x^2) y_1{}^2 = m^2 (1 - y^2)$

or $\qquad 2 y_1 y_2 (1 - x^2) + y_1{}^2 (- 2x) = - 2 m^2 y y_1$

or $\qquad (1 - x^2) y_2 - x y_1 + m^2 y = 0$

Differentiating n times, we have

$$y_{n+2} (1 - x^2) + n y_{n+1} (- 2x) + \frac{n (n-1)}{2!} y_n (-2) - x y_{n+1} - n y_n + m^2 y_n = 0$$

or $\qquad (1 - x^2) y_{n+2} - (2n + 1) x y_{n+1} + (m^2 - n^2) = 0.$

Now, put $x = 0$, then

$$(y_{n+2})_0 = (n^2 - m^2)(y_n)_0$$

Now, $(y)_0 = 0; (y_1)_0 = m, (y_2)_0 = 0$

$$(y_3)_0 = (1^2 - m^2). m ; (y_4) = 0$$

$$(y_5)_0 = (3^2 - m^2)(1^2 - m^2). m \text{ etc}$$

hence by Maclaurin's Theorem, we get

$$\sin(m \sin^{-1} x) = mx + m(1^2 - m^2)\frac{x^3}{3!} + m(3^2 - m^2).\frac{x^5}{5!} + \dots$$

Put $\sin^{-1} x = 0$; then $x = \sin \theta$

Hence $\sin m\theta = m \sin \theta + \dfrac{1}{3!} m(m^2 - 1)\sin^3 \theta + m(m^2 - 1^2)(m^2 - 3^2)\sin^5 \theta + \dots.$

7. *Expand $\log\{1 - \log(1-x)\}$ in powers of x by Maclaurin's Theorem as far as the term in x^3.*
By substituting $\dfrac{x}{1+x}$ for x deduce the expansion of $\log\{1 + \log(1+x)\}$ as far as the term x^3.

Let $y = \log\{1 - \log(1-x)\}$,   ...(i)

or $e^y = 1 - \log(1-x)$

Then $e^y \cdot y_1 = \dfrac{1}{(1-x)}$

or $(1-x)y_1 = e^{-y}$ ...(ii)

or $y_2(1-x) - y_1 = e^{-y}y_1,$ ...(iii)

$y_3(1-x) - 2y_2 = e^{-y}y_1^2 - e^{-y}y_2$ ...(iv)

Put $x = 0$

in (i) $(y)_0 = 0$

in (ii) $(y_1)_0 = 1$

in (iii) $(y_2)_0 = 0$

in (iv) $(y_3)_0 = 1,$ etc.

Hence $\{1 - \log(1-x)\} = x + \dfrac{x^3}{6} + \dots$

Now substituting $\dfrac{x}{1+x}$ for x, we obtain

$$\log\left\{1 - \log\left(1 - \frac{x}{1+x}\right)\right\} = \frac{x}{1+x} + \frac{1}{6}\left(\frac{x}{1+x}\right)^3 + \dots.$$

or $\log\{1 + \log(1+x)\} = x(1+x)^{-1} + \dfrac{1}{6}x^3(1+x)^{-3} + \dots$

$$= x(1 - x + x^2 + \dots) + \frac{1}{6}x^3(1 - 3x + \dots)$$

$$= x - x^2 + \frac{7}{6}x^3 + \dots$$

8. *Expand* $(\sin^{-1} x)^2$ *in ascending powers of* x.

Let $$y = f(x) = (\sin^{-1} x)^2, \qquad \qquad ...(i)$$

then $$y_1 = 2 \sin^{-1} x . \frac{1}{\sqrt{(1 - x^2)}} \qquad \qquad ...(ii)$$

or $(1 - x^2) y_1^2 = 4 y.$

Differentiating again

$$(1 - x^2). 2y_1 y_2 + y_1^2 (- 2 x) = 4 y_1$$

or $$(1 - x^2) y_2 = x y_1 + 2 \qquad \qquad ...(iii)$$

Differentiating n times, we get

$$(1 - x^2) \, y_{n+2} + n \, y_{n+1} (- 2x + \frac{n (n - 1)}{2 !} \, y_n (- 2) - x y_{n+1} - n y_n (1) = 0 \qquad ...(iv)$$

Putting $x = 0$ in (i), (ii), (iii) and (iv), we get

$$(y)_0 = 0; (y_1)_0 = 0 \, ; (y_2)_0 = 2 \text{ and } (y_{n+2})_0 = n^2 \, (y_n)_0$$

from which we get

$$(y_3)_0 = 1^2 . (y_1)_0 \, ; (y_4)_0 = 2^2 . (y_2)_0 = 2^2 . 2;$$
$$(y_5)_0 = 3^2 . (y_3)_0 = 0; (y_6)_0 = 4^2 (y_4)_0 = 4^2 . 2^2 . 2 \, ; \text{ etc.}$$

Hence by Maclaurin's Theorem, we get

$$(\sin^{-1} x)^2 = (y)_0 + x \, (y_1)_0 + \frac{x^2}{2 !} \, (y_2)_0 + \frac{x^3}{3 !} \, (y_3)_0 + ...$$

$$= \frac{x^2}{2 !} \, (2) + \frac{x^4}{4 !} \, (2^2 .2) + \frac{x^6}{6 !} \, (4^2 . 2^2) + ...$$

$$= \frac{2 x^2}{2 !} + 2 . 2^2 . \frac{x^4}{4 !} + 2 . 2^2 . 4^2 . \frac{x^6}{6 !} + ...$$

9. *If* $y = \dfrac{\sin^{-1} x}{\sqrt{(1 - x^2)}}$, *where* $- 1 < x < 1$ *and* $- \dfrac{\pi}{2} < \sin^{-1} x < \dfrac{\pi}{2}$, *prove that*

$$(1 - x^2) y_{n+1} - (2n + 1) x y_n - n^2 y_{n-1} = 0.$$

Assuming that y can be expanded in ascending powers of x in the form

$$a_0 + a_1 x + a_2 x^2 + ... + a_n x^n + ...$$

prove that $(n + 1) a_{n+1} = n a_{n-1}$, *and hence obtain the general term of the expansion.*

                                                                        (Garhwal 97)

We have $$y = \frac{\sin^{-1} x}{\sqrt{(1 - x^2)}}$$

or $$(1 - x^2) y^2 = (\sin^{-1} x)^2 \qquad \qquad ...(i)$$

Differentiating (i) with respect to x,

$$(1 - x^2) . 2 y y_1 - 2x y^2 = 2 \sin^{-1} x . \frac{1}{\sqrt{1 - x^2}} = 2y$$

or $$(1 - x^2) y_1 - x y - 1 = 0 \qquad \qquad ...(ii)$$

Differentiating n times by Leibnitz's Theorem we get

$$(1 - x^2) y_{n+1} - 2 \cdot {}^nC_1 x y_n - 2 \cdot {}^nC_2 y_{n-1} - (x y_n + {}^nC_1 y_{n-1}) = 0$$

or $\quad (1 - x^2) y_{n+1} - (2n + 1) x y_n - n^2 y_{n-1} = 0$ ...(iii)

Putting $x = 0$ in (i), (ii) and (iii), we have

$$(y)_0 = 0, (y_1)_0 = 1$$

and $\quad (y_{n+1})_0 = n^2 (y_{n-1})_0$ ...(iv)

Now by hypothesis,

$$y = a_0 + a_1 x + a_2 x^2 + \ldots + a_n x^n + \ldots$$ ...(v)

and from Maclaurin's Theorem

$$y = (y)_0 + x (y_1)_0 + \frac{x^2}{2!} (y_2)_0 + \ldots + \frac{x^n}{n!} (y_n)_0 + \ldots$$ ...(vi)

Equating the two values of y_1 we get

$$a_0 = (y)_0, a_1 = (y_1)_0 ; \ldots (y_{n-1})_0 = (n-1)! \, a_{n-1}$$

$$(y_n)_0 = n! \, a_n, (y_{n+1})_0 = (n+1)! \, a_{n+1} \text{ etc.}$$

By putting the values of $(y_{n+1})_0$ and $(y_{n-1})_0$ in (iv),

$$(n+1)! \, a_{n+1} = n^2 (n-1)! \, a_{n-1}$$

or $\quad (n+1) a_{n+1} = n a_{n-1}$

or $\quad a_{n+1} = \dfrac{n}{n+1} a_{n-1}$ ...(vii)

putting $\quad n = 1, 3, 5, \ldots (2m+1)$ in (vii),

$$a_2 = a_4 = a_6 = \ldots = a_{2m} = 0, \text{ as } a_0 = (y)_0 = 0.$$

putting $\quad n = 2, 4, 6, \ldots 2m$ in (vii),

$$3a_3 = 2a_1 = 2, \text{ as } a_1 = (y_1)_0 = 1$$

$$5a_5 = 4a_3,$$

$$7a_7 = 6a_5,$$

$$\ldots\ldots\ldots$$

$$(2m+1) a_{2m+1} = 2m \cdot a_{2m-1}$$

multiplying columnwise, we obtain

$$3 . 5 . 7 \ldots (2m+1) a_3 . a_5 . a_7 \ldots a_{2m+1} = (2 . 4 . 6 \ldots 2m) 1 . a_3 . a_5 \ldots a_{2m-1}$$

$$\therefore \qquad a_{2m+1} = \frac{2 . 4 . 6 \ldots 2m}{3 . 5 . 7 \ldots (2m+1)}$$

EXERCISES

Apply Maclaurin's Theorem to find the expansion of :

1. (a) $\sin x$;

                    (Kanpur 2001 ; Ravishankar 98, 99 S,

         Bilaspur 96; Manipur 99, 2001, 2002; Gujrat, 2005)

  (b) $\sec x$ *(MDU Rohtak 2000)*

  (c) $\sin^{-1} x$

  (d) $e^{\sin x}$

  (e) $\tan x$ *(Mysore 2004)*

2. (a) $e^{x \cos x}$

 (b) $\log_e (1 + \sin x)$

3. (a) $\log_e (1 + \tan x)$

  (b) $e^x \log_e (1 + x)$;

  (c) $e^{a \cos^{-1} x}$

4. (a) $e^x \sec x$ or $\dfrac{e^x}{\cos x}$ (*Kumaon 2000, 2002*)

  (b) a^x (*Sagar 98*)

5. Prove that

$$e^{ax} \cos b\, x = 1 + ax + \frac{a^2 - b^2}{2!} x^2 + \frac{a(a^2 - 3b^2)}{3!} x^3$$

$$+ \dots + \frac{(a^2 + b^2)^{n/2}}{n!} x^n \cos(n \tan^{-1} b/a) + \dots$$

6. Prove that

$$e^x \cos x = 1 + x - \frac{2x^3}{3!} - \frac{2^2 x^4}{4!} - \frac{2^2 x^5}{5!} + \frac{2^3 x^7}{7!} + \dots$$

7. Prove that

$$\log_e \sec x = \frac{1}{2} x^2 + \frac{1}{12} x^4 + \frac{1}{45} x^6 + \dots$$ (*Garhwal 98 ; Kumaon 99*)

8. Apply Maclaurin's Theorem to obtain the term upto x^4 in the expansion of

  $\log_e (1 + \sin^2 x)$ (*Jabalpur 1997 ; Sagar 2002*)

9. Prove that,

$$\tan^{-1} x = x - \frac{1}{3} x^3 + \frac{1}{5} x^5 - \dots$$ (*Manipur 2000*)

 Hence obtain the expansion of

$$\sin^{-1} \frac{2x^2}{1 + x^4}, \text{ giving the general term.}$$

10. By Maclaurin's Theorem, find three non-vanishing terms in the expansion of $\dfrac{e^x}{1 + e^x}$

11. Prove that for all finite values of x,

$$e^x \sin x = x + x^2 + \frac{2}{3!} x^3 - \frac{2^2}{5!} x^5 - \dots + \sin\left(\frac{n\pi}{4}\right) \frac{2^{n/2}}{n!} x^n + \dots$$

12. If $y^3 - 6xy - 8 = 0$, prove that

$$y = 2 + x - \frac{1}{2} \cdot \frac{x^3}{3!} + \frac{x^4}{4!} + \dots$$ (*Kumaon 96*)

13. Expand $\log_e \cos x$ in powers of x and verify the coefficient of x^4 algebraically by the use of well known expansions of $\cos x$ and $\log(1 - x)$.

14. If $y = e^{m \tan^{-1} x} = a_0 + a_1 x + a_2 x^2 + \ldots + a_n x^n + \ldots$

prove that $(n + 1) a_{n+1} + (n - 1) a_{n-1} = m a_n$

and $y = 1 + mx + \dfrac{m^2}{2!} x^2 + \dfrac{m(m^2 - 2)}{3!} x^3 + \dfrac{m^2(m^2 - 8)}{4!} x^4 + \ldots$

15. Prove by Maclaurin's Theorem, that

$$[x + \sqrt{1 + x^2}]^m = 1 + mx + \frac{m^2}{2!} \cdot x^2 + \frac{m(m^2 - 1^2)}{3!} x^3 + \frac{m^2(m^2 - 2^2)}{4!} x^4 + \ldots$$

and by using

$$a^m = 1 + m \log_e a + \frac{m^2}{2!} (\log_e a)^2 + \ldots \text{ deduce the expansions of}$$

$$\log_e [x + \sqrt{1 + x^2}] \text{ and } \left[\log_e \left\{ x + \sqrt{(1 + x^2)} \right\} \right]^2$$

16. If $y = \sqrt{(1 - x^2)} \sin^{-1} x$, prove that

(a) $y_3 (1 - x^2) - 3 y_2 x + 2 = 0$,

(b) $y_{n+3} (1 - x^2) - (2n + 3) x y_{n+2} - n(n + 2) y_{n+1} = 0$;

(c) $\sqrt{(1 - x^2)} \sin^{-1} x = x - \dfrac{x^3}{3} - \dfrac{2}{3} \dfrac{x^5}{5} - \dfrac{2.4}{3.5} \dfrac{x^7}{7} - \ldots$

(d) $\theta \cot \theta = 1 - \dfrac{\sin^2 \theta}{3} - \dfrac{2}{3} \dfrac{\sin^4 \theta}{5} - \dfrac{2.4}{3.5} \dfrac{\sin^6 \theta}{7} - \ldots$

(e) $\dfrac{\pi}{4} = 1 - \dfrac{1}{3}\left(\dfrac{1}{2}\right) - \dfrac{2}{3} \cdot \dfrac{1}{5}\left(\dfrac{1}{2}\right)^2 - \dfrac{2}{3} \cdot \dfrac{4}{5} \cdot \dfrac{1}{7}\left(\dfrac{1}{2}\right)^3 - \ldots$

ANSWERS

1. (a) $x - \dfrac{x^3}{3!} + \dfrac{x^5}{5!} - \ldots$ 　　　　　　　(b) $1 + \dfrac{x^2}{2!} + \dfrac{5x^4}{4!} + \dfrac{61 x^6}{6!} + \ldots$

(c) $x + 1^2 \cdot \dfrac{x^3}{3!} + 3^2 \cdot 1^2 \dfrac{x^5}{5!} + 5^2 \cdot 3^3 \cdot 1^2 \cdot \dfrac{x^7}{7!} + \ldots$ 　(d) $1 + x + \dfrac{x^2}{2} - \dfrac{x^4}{8} + \ldots$

(e) $x + \dfrac{2^3}{3} + \dfrac{x^5}{15} + \ldots$

2. (a) $1 + x + \dfrac{x^2}{2} - \dfrac{x^3}{3} - \dfrac{11 x^4}{24} - \dfrac{x^5}{5} + \ldots$ 　　(b) $x - \dfrac{x^2}{2} + \dfrac{x^3}{6} - \dfrac{x^4}{12} + \dfrac{x^5}{24} + \ldots$

3. (a) $x - \dfrac{1}{2} x^2 + \dfrac{2}{3} x^3 - \ldots$ 　　　　　　(b) $x + \dfrac{x^2}{2!} + \dfrac{2x^3}{3!} + \dfrac{9 x^5}{5!} + \ldots$

(c) $e^{a\pi/2} \left\{ 1 - ax + \dfrac{a^2 x^2}{2!} - a(1 + a^2) \dfrac{x^3}{3!} + (2^2 + a^2) \dfrac{a^2 x^4}{4!} + \ldots \right\}$

4. (a) $1 + x + x^2 + \dfrac{2x^3}{3!} + ...$ (b) $1 + x \log a + \dfrac{x^2}{2!} (\log a)^2 + ... + \dfrac{x^n}{n!} (\log a)^n$

8. $x^2 - \dfrac{5x^4}{6} + ...$

9. $2 \left[x^2 - \dfrac{x^6}{3} + \dfrac{x^{10}}{5} - ... + (-1)^{n-1} \dfrac{x^{4n-2}}{2n-1} + ... \right]$

10. $\dfrac{1}{2} + \dfrac{1}{4} x - \dfrac{1}{8} \cdot \dfrac{x^3}{3!} + ...$

13. $-\dfrac{x^2}{2!} - 2 \dfrac{x^4}{4!} - 16 \dfrac{x^6}{6!} - ...$

15. $x - \dfrac{1^2}{3!} x^3 + ..., \; 2 \left\{ \dfrac{x^2}{2!} - \dfrac{2^2}{4!} x^4 + ... \right\}$

6.2 Taylor's Theorem   *(Ravishankar 2000 ; Bhuj 99; Poorvanchal 2004)*

Let $f(x + h)$ be a function of h (x being independent of h) which can be expanded in powers of h and the expansion be differentiable any number of times, then

$$f(x + h) = f(x) + h f'(x) + \frac{h^2}{2!} f'(x) + ... + \frac{h^n}{n!} f^n(x). + ...$$

Let $f(x + h) = A_0 + A_1 h + A_2 h^2 + A_3 h^3 + A_4 h^4 + ...$ where $A_0, A_1, A_2, ...$ are functions of x alone which are to be determined.

Now by successive differentiation with respect to h, we get

$$f(x + h) = A_1 + 2 A_2 h + 3 A_3 h^2 + 4 A_4 h^3 + ...$$
$$f''(x + h) = 2.1 A_2 + 3.2 A_3 h + 4.3 A_4 h^2 + ...$$
$$f'''(x + h) = 3.2.1 A_3 + 4.3.2 A_4 h + ... \text{ etc.}$$

Putting $h = 0$

$$f(x) = A_0, f'(x) = A_1 f''(x) = 2! A_2, f'''(x) = 3! A_3, ...$$

Hence,

$$f(x + h) = f(x) + h f'(x) + \frac{h^2}{2!} f''(x) + \frac{h^3}{3!} f'''(x) + ... \frac{h^n}{n!} f^n(x) + ...$$

Other forms :

Putting $x = a$ in Taylor's Theorem.

$$f(a + h) = f(a) + h f'(a) + \frac{h^2}{2!} f''(a) + ... \frac{h^n}{n!} f^n(a) + ... \quad ...(i)$$

Now, putting $a = 0$ and $h = x$ in (i), we get

$$f(x) = f(0) + x f'(0) + \frac{x^2}{2!} f''(0) + ... \frac{x^n}{n!} f^n(0) + ... \qquad \textit{(Vikram 1996)}$$

this is nothing but Maclaurin's Theorem.

Putting $h = x - a$ in (i), we get the expansion of $f(x)$ in powers of $(x - a)$ which is given below :

$$f(x) = f(a) + (x - a) f'(a) + \frac{(x - a)^2}{2!} f''(a) + \frac{(x - a)^3}{3!}$$

$$f'''(a) + \ldots + \frac{(x - a)^n}{n!} f^n(a) + \ldots$$

EXAMPLES

1. *Expand* $\log (x + a)$ *in powers of* x *by Taylor's Theorem.*

                              (Bhopal 1998, 2001 ; Sagar 98 ; Reewa 2000 ; Indore 2002; Avodh 2005)

Here $f(x + a) = \log (x + a)$

$\therefore$ $f(a) = \log a$

therefore $f'(a) = \dfrac{1}{a} ; f''(a) = -\dfrac{1}{a^2} ; f'''(a) = \dfrac{2}{a^3} \quad \ldots f^n(a) = \dfrac{(-1)^{n-1}(n-1)!}{a^n}.$

Then by Taylor's Theorem

$$\log (x + a) = \log a + \frac{x}{a} - \frac{x^2}{2a^2} + \frac{x^3}{3a^3} - \ldots$$

2. *Expand* $\log_e \cos (x + h)$ *in powers of* h *by Taylor's theorem.*

Let $f(x + h) = \log_e \cos (x + h)$

so that $f(x) = \log_e \cos x$

then $f'(x) = -\tan x$

                $f''(x) = -\sec^2 x$

                $f'''(x) = -2 \sec^2 x \tan x$ etc.

Hence, $\log_e \cos (x + h) = \log_e \cos x + h(-\tan x) + \dfrac{h^2}{2!}(-\sec^2 x) + \dfrac{h^3}{3!}(-2 \sec^2 x \tan x) + \ldots$

$$= \log_e \cos x - h \tan x - \frac{h^2}{2} \sec^2 x - \frac{h^3}{3} \sec^2 x \tan x + \ldots$$

3. *Expand* $\log \sin x$ *in powers of* $(x - 2)$.

                             (Garhwal, 96, 2001 ; Kumaon 2001 ; Bhopal 96; Poorvanchal 2004)

We have $f(x) = \log \sin x$

If can be adjusted as

         $f(x - 2 + 2) = \log \sin (x - 2 + 2)$

It is to be expanded in powers of $(x - 2)$

therefore, $f(2) = \log \sin 2,$

            $f'(2) = \cot 2, f''(2) = -\text{cosec}^2 2$

            $f'''(2) = +2 \text{ cosec}^2 2 \cot 2, \ldots$ etc.

Hence by Taylor's Theorem

$$f(x) = f(2) + (x - 2) f'(2) + \frac{(x - 2)}{2!} f''(2) + \ldots$$

$$\log \sin x = \log \sin 2 + (x - 2) \cot 2 + \frac{(x - 2)^2}{2!}(-\text{cosec}^2 2) + \ldots$$

4. *Expand* $2x^3 + 7x^2 + x - 6$ *in powers of* $(x-2)$.

(*Kumaon 97 ; Jiwaji 99, Reewa 2000 ; Jabalpur 2001; Vikram 98, Bilaspur 2000*)

We have

$$f(x) = 2x^3 + 7x^2 + x - 6.$$

By Taylor's Theorem , we have

$$f(x) = f(2) + (x-2) f'(2) + \frac{(x-2)^2}{2!} f''(2)\; \frac{(x-2)^3}{3!} f'''(2) + \dots$$

Now, $f(2) = 2 \cdot 2^3 + 7 \cdot 2^2 + 2 - 6 = 40$

Again $f'(x) = 6x^2 + 14x + 1, f'(2) = 53.$

$$f''(x) = 12x + 14, f''(2) = 38$$

and $f'''(x) = 12$

Hence $f(x) = 40 + (x-2) \cdot 58 + \dfrac{(x-2)^2}{2!} \cdot 88 + \dfrac{(x-2)^3}{3!} \cdot 12$

$$= 40 + 58(x-2) + 44(x-2)^2 + 2(x-2)^3$$

5. *Expand* $\sin x$ *in powers of* $\left(x - \dfrac{\pi}{2} \right)$.

(*Gorakhpur 2001; Bilaspur 95, 99 ; Ravishankar 98 ; Bhopal 97 ; Vikram 97 ; Indore 2000 ; Jabalpur 98, 99 ; Jiwaji 2000, 2001 ; Sagar 2002*)

Let $f(x) = \sin x$ then $f\left(\dfrac{\pi}{2} \right) = 1$

$$f'(x) = \cos x \qquad\qquad f'\left(\dfrac{\pi}{2} \right) = 0$$

$$f''(x) = -\sin x \qquad\qquad f''\left(\dfrac{\pi}{2} \right) = -1$$

$$f'''(x) = -\cos x \qquad\qquad f'''\left(\dfrac{\pi}{2} \right) = 0$$

$$f^{iv}(x) = \sin x \qquad\qquad f^{iv}\left(\dfrac{\pi}{2} \right) = 1, \text{ etc}$$

Hence by Taylor's Theorem,

$$\sin x = f\left(\frac{\pi}{2} \right) + \left(x - \frac{\pi}{2} \right) f'\left(\frac{\pi}{2} \right) + \frac{\left(x - \dfrac{\pi}{2} \right)}{2!} f''\left(\frac{\pi}{2} \right) + \frac{\left(x - \dfrac{\pi}{2} \right)^3}{3!} f'''\left(\frac{\pi}{2} \right) + \dots$$

$$= 1 - \frac{1}{2!} \left(x - \frac{\pi}{2} \right)^2 + \frac{1}{4!} \left(x - \frac{\pi}{2} \right)^4 - \dots$$

6. *Use Taylor's theorem to prove that*

$$\tan^{-1}(x + h) = \tan^{-1} x + h \sin x \cdot \frac{\sin z}{1} - (h \sin z)^2 \cdot \frac{\sin 2z}{2} + (h \sin z)^3 \frac{\sin 3z}{3} \dots$$

where $z = \cot^{-1} x$.

(*Gorakhpur 2003, Kumaon 2000, 2004; Reewa 97 ; Jabalpur 97; Indore 99 ; Bilaspur 2001*)

Let $\qquad f(x+h) = \tan^{-1}(x+h)$

$\therefore \qquad f(x) = \tan^{-1} x$

then $\qquad f^n(x) = (-1)^{n-1}(n-1)! \sin^n z \sin nz$

where $\qquad z = \tan^{-1} \dfrac{1}{x} = \cot^{-1} x$

$\therefore \qquad f'(x) = \sin z \sin z$

$\qquad f''(x) = -(1!) \sin^2 z \sin 2z,$

$\qquad f'''(x) = (2!) \sin^3 x \sin 3z,$ etc.

$\therefore \quad \tan^{-1}(x+h) = \tan^{-1} x + \dfrac{h}{1!} \sin z \cdot \sin z$

$$- \dfrac{h^2}{2!}(1!) \sin^2 z \sin 2z + \dfrac{h^3}{3!}(2!) \sin^3 z \sin 3z - \dots$$

$$= \tan^{-1} x + h \sin z \cdot \dfrac{\sin z}{1} - (h \sin z)^2 \cdot \dfrac{\sin 2z}{2} + (h \sin z)^3 \cdot \dfrac{\sin 3z}{3} - \dots$$

7. Prove that

$$f\left(\dfrac{x^2}{1+x}\right) = f(x) - \dfrac{x}{1+x} f'(x) + \dfrac{x^2}{(1+x)^2} \dfrac{f''(x)}{2!} + \dots$$

(Rohilkhand 2000 ; Garhwal 97)

We can write

$$f\left(\dfrac{x^2}{1+x}\right) = \left(x - \dfrac{x}{1+x}\right)$$

Also, by Taylor's Theorem

$$f(x+h) = f(x) + h f'(x) + \dfrac{h^2}{2!} f''(x) + \dots$$

Putting $h = -\left(\dfrac{x}{1+x}\right)$ in this expansion, we get

$$f\left(x - \dfrac{x}{1+x}\right) = f(x) - \dfrac{x}{1+x} f'(x) + \dfrac{x^2}{(1+x)^2} \cdot \dfrac{1}{2!} f''(x) + \dots$$

or $\quad f\left(\dfrac{x^2}{1+x}\right) = f(x) - \left(\dfrac{x}{1+x}\right) f'(x) + \dfrac{x^2}{(1+x)^2} \dfrac{f''(x)}{2!} + \dots$

EXERCISES

1. Prove that $\log_e \sin(x+h) = \log_e \sin x + h \cot x - \dfrac{h^2}{2} \operatorname{cosec}^2 x$

$$+ \dfrac{h^3}{3} \operatorname{cosec}^2 x \cot x + \dots$$

2. Apply Taylor's theorem to prove ; $e^{x+h} = e^x + he^x + \dfrac{h^2}{2!} e^x + \dots$

3. Prove that $\dfrac{1}{x+h} = \dfrac{1}{x} - \dfrac{h}{x^2} + \dfrac{h^2}{x^3} - \dfrac{h^3}{x4} + ...$ (*Bhopal 1998*)

4. Expand $\sin^{-1}(x+h)$ in powers of x as far as the terms in x^3.   (*Garhwal 2003*)

5. Expand $\tan^{-1} x$ in powers of $\left(x - \dfrac{\pi}{4} \right)$.   (*Sagar 1997 ; Indore 97 ; Bhopal 2002 ;*

Ravishankar 98, 99 S, 2001 ; Vikram 99)

6. Expand $\sin\left(\dfrac{\pi}{4} + \theta \right)$ in powers of θ.   (*Rohilkhand 2001 ; Bhopal 96*)

7. Prove that

$$f(mx) = f(x) + (m-1)\,x\,f'(x) + \dfrac{1}{2!}(m-1)^2\,x^2.f''(x)$$

$$+ \dfrac{1}{3!}(m-1)^3\,x^3\,f'''(x) + ...$$ (*Jiwaji 98*)

ANSWERS

4. $\sin^{-1} h + x(1-h^2)^{-1/2} + \dfrac{x^2}{2!}\,h\,(1-h^2)^{-3/2} + \dfrac{x^3}{3!}\left\{ (1-h^2)^{-5/2}\,(1+2h^2) \right\} + ...$

5. $\tan^{-1}\dfrac{\pi}{4} + \dfrac{\left(x - \dfrac{\pi}{4} \right)}{\left(1 + \dfrac{1}{16}\pi^2 \right)} - \dfrac{\pi\left(x - \dfrac{\pi}{4} \right)^2}{4\left(1 + \dfrac{1}{16}\pi^2 \right)^2} + ...$

6. $\left(\dfrac{1}{\sqrt{2}} \right)\left(1 + \theta - \dfrac{\theta^2}{2!} - \dfrac{\theta^3}{3!} + \dfrac{\theta^4}{4!} + \dfrac{\theta^5}{5!} + ... \right)$

OBJECTIVE QUESTIONS

Note : *For each of the following questions four alternatives are given for the answer, only one of them is correct. Choose the correct alternative.*

1. The $(n+1)^{th}$ term in Maclaurin's series is :

(a) $\dfrac{x^n}{n}\,f^{(n)}(a)$ (b) $\dfrac{x^n}{n!}\,f^{(n)}(a)$

(c) $\dfrac{x^n}{n!}\,f^{(n)}(0)$ (d) $f^{(n)}(0)$

2. The $(n+1)^{th}$ term in Maclaurin's series of $\tan x$ is :

(a) $\dfrac{x^n}{n!}$ (b) $\dfrac{x^{2n+1}}{2n+1}$

(c) $\dfrac{(-1)^n\,x^n}{n}$ (d) None of these

3. $f(x) = \log x$ can be expanded in powers of $x - 1$ by using :

    (a) Maclaurin's theorem

    (b) Taylor's theorem

    (c) Leibnitz theorem

    (d) Gregory's theorem

4. The nth term in the expansion of $f(a + h)$ is :

    (a) $\dfrac{1}{n!} f^n(a)$

    (b) $\dfrac{h}{n!} f^n(a)$

    (c) $\dfrac{h^n}{n!} f^n(a)$

    (d) None of these

5. If $e^{a \sin^{-1} x} = a_0 + a_1 x + a^2 x^2 + ...$, then $a_2 =$

    (a) 1

    (b) a

    (c) $\dfrac{a^2}{2!}$

    (d) $\dfrac{(1^2 + a^2) a}{3!}$

6. If $\cos x = a_0 + a_1 x + a_2 x^2 + ...$, then value of $a_3 =$

    (a) 0

    (b) -1

    (c) 1

    (d) $-\dfrac{1}{3!}$

ANSWERS

1. (c)	2. (d)	3. (b)	4. (c)	5. (c)	6. (a).

CHAPTER

7

TANGENTS AND NORMALS

7.1 Introduction.

This chapter will be devoted to some elementary applications to Geometry involving only first order derivatives.

7.2 Equations of the tangent and normal.

7.2.1. Cartesian Equations. As shown in Chapter 4 the angel ψ which the tangent at any point (x, y) on the curve $y = f(x)$ makes with x-axis, is given by

$$\tan \psi = \frac{dy}{dx} = f'(x).$$

Thus, *the equation of the tangent at a point* (x, y) *on the curve* $y = f(x)$ *is*

$$Y - y = f'(x)(X - x)$$

where (X, Y) is an arbitray point on the tangent.

The normal at a point being the line through the point and perpendicular to the tangent at the point, *the equation of the normal at* (x, y) *to the curve* $y = f(x)$ *is*

$$Y - y = -[1/f'(x)](X - x) \Leftrightarrow (X - x) + f'(x)(Y - y) = 0, f'(x) \neq 0.$$

Ex. What happens if $f'(x) = 0$.

7.2.2. Parametric Cartesian Equations. At any point 't' of the curve $x = f(t), y = F(t)$, where $\frac{dx}{dt} = f'(t) \neq 0$, we have

$$\frac{dy}{dx} = \frac{dy}{dt} \Big/ \frac{dx}{dt} = \frac{F'(t)}{f'(t)}.$$

Thus the equations of the tangent and the normal at the point 't' of the curve $x = f(t), y = F(t)$ are

$$[X - f(t)] F'(t) - [Y - F'(t)] f'(t) = 0$$

and $\quad [X - f(t)] f'(t) + [Y - F'(t)] F(t) = 0$

EXAMPLES

1. *Find the equations of the tangent and the normal at any point* (x, y) *of the curve* $y = c \cosh(x/c)$.

Also show that the length of the perpendicular from the foot of the ordinate on any tangent to the curve is constant.

The equations of the tangent and the normal to $y = c \cosh \dfrac{x}{c}$ at the point (x, y) are

$$Y - c \cosh \frac{x}{c} = (X - x) \sinh \frac{x}{c},$$

$$\left(Y - c \cosh \frac{x}{c} \right) \sinh \frac{x}{c} + X - x = 0.$$

respectively.

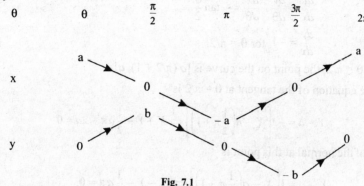

Fig. 7.1

Also the length of the prependicualr from the foot $(x, 0)$ of the ordinate of (x, y) to the tangent

$$X \sinh \frac{x}{c} - Y + \left(c \cosh \frac{x}{c} - x \sinh \frac{x}{c} \right) = 0$$

is $\dfrac{x \sinh x/c - 0 + (c \cosh x/c - x \sinh x/c)}{\sqrt{(\sinh^2 x/c + 1)}} = \dfrac{c \cosh x/c}{\cosh x/c} = c,$

which is free of x, y and is, therefore, constant.

2.   *Find the equations of the tangent and the normal at any point (x, y) of the curve*

$$\frac{x^m}{a^m} + \frac{y^m}{b^m} = 1$$

We have $\dfrac{x^m}{a^m} + \dfrac{y^m}{b^m} - 1 = 0$

Differentiating $\dfrac{m x^{m-1}}{a^m} + \dfrac{m y^{m-1}}{b^m} \dfrac{dy}{dx} = 0$

$\Rightarrow$ $\dfrac{dy}{dx} = -\dfrac{b^m x^{m-1}}{a^m y^{m-1}}$

Therefore, the equation of the tangent at (x, y) is

$$Y - y = -\frac{b^m x^{m-1}}{a^m y^{m-1}} (X - x)$$

or $\dfrac{X x^{m-1}}{a^m} + \dfrac{Y y^{m-1}}{b^m} = \dfrac{x^m}{a^m} + \dfrac{y^m}{b^m} = 1,$

for (x, y) lies on the curve.

Also, the equation of the normal at (x, y) is

$$Y - y = \frac{a^m y^{m-1}}{b^m x^{m-1}} (X - x),$$

or $\dfrac{X - x}{b^m x^{m-1}} = \dfrac{Y - y}{a^m y^{m-1}}.$

3. *Find the equations of the tangent and normal at* $\theta = \pi/2$ *to the curve* $x = a\,(\theta + \sin\theta)$, $y = a\,(1 + \cos\theta)$.

We have
$$\frac{dx}{d\theta} = a(1 + \cos\theta) = 2a\cos^2\frac{\theta}{2},$$

$$\frac{dy}{d\theta} = -a\sin\theta = -2a\sin\frac{\theta}{2}\cos\frac{\theta}{2}$$

$\Rightarrow$
$$\frac{dy}{dx} = \frac{dy}{d\theta} \Big/ \frac{dx}{d\theta} = -\tan\frac{\theta}{2}$$

$\Rightarrow$
$$\frac{dy}{dx} = -1 \text{ for } \theta = \pi/2.$$

Also, for $\theta = \pi/2$, the point on the curve is $[a\,(\pi/2 + 1), a]$.

Hence the equation of the tangent at $\theta = \pi/2$ is

$$Y - a = -1\left[X - a\left(\frac{1}{2}\pi + 1\right)\right] \Leftrightarrow X + Y - \frac{1}{2}a\pi - 2a = 0$$

and that of the normal at this point is

$$Y - a = 1\left[X - a\left(\frac{1}{2}\pi + 1\right)\right] \Leftrightarrow X - Y - \frac{1}{2}a\pi = 0.$$

4. *Show that the length of the portion of the tangent to the curve* $x = a\cos^3\theta$, $y = a\sin^3\theta$ *intercepted between the co-ordinate axes is constant.*

We have
$$\frac{dx}{d\theta} = -3a\cos^2\theta\,\sin\theta, \frac{dy}{d\theta} = 3a\sin^2\theta\cos\theta$$

$\Rightarrow$
$$\frac{dx}{dx} = \frac{dy}{d\theta} \Big/ \frac{dx}{d\theta} = -\tan\theta.$$

The tangent at the point 'θ' viz.

$$y - a\sin^3\theta = -\tan\theta\,(x - a\cos^3\theta)$$

$\Rightarrow$
$$x\sin\theta + y\cos\theta - a\sin\theta\cos\theta = 0$$

meets the co-ordinate axes at the points $(a\cos\theta, 0)$, $(0, a\sin\theta)$, Thus the length of the portion of the tangent intercepted between the co-ordinate axes is

$$\sqrt{(a^2\cos^2\theta + a^2\sin^2\theta)} = a, \text{ which is constant.}$$

5. *Prove that the equation of the normal to the Astroid*

$$x^{2/3} + y^{2/3} = a^{2/3}$$

may be written in the form

$$x\sin\phi - y\cos\phi + a\cos 2\phi = 0.$$

Differentiating

$$x^{2/3} + y^{2/3} = a^{2/3}, \qquad\qquad\qquad ...(i)$$

we get

$$\frac{2}{3}x^{-1/3} + \frac{2}{.3}y^{-1/3}\frac{dy}{dx} = 0$$

or
$$\frac{dy}{dx} = -\frac{y^{1/3}}{x^{1/3}}.$$

Therefore the slope of the normal at any point $(x, y) = x^{1/3}/y^{1/3}$

But the slope of the given line $= \tan \phi$.

We write $\qquad x^{1/3}/y^{1/3} = \tan \phi$ $\qquad$ ...(ii)

Equations (i) and (ii) have now to be solved to find x and y.

We get

$$y^{2/3} \tan^2 \phi + y^{2/3} = a^{2/3},$$

or $\qquad y^{2/3} = a^{2/3} \cos^2 \phi, \ i.e., \ y = a \cos^3 \phi.$

Substituting this value of y in (ii), we get

$$x^{1/3} = a^{1/3} \sin \phi, \ i.e., \ x = a \sin^3 \phi.$$

Thus $\tan \phi$ the slope of the normal at $(a \sin^3 \phi, a \cos^3 \phi,)$. The equation of the normal at this point is

$$y - a\cos^3 \phi = \frac{\sin\phi}{\cos\phi}(x - a\sin^3\phi),$$

or $\qquad y \cos \phi - a \cos^4 \phi = x \sin \phi - a \sin^4 \phi,$

i.e., $\qquad x \sin \phi - y \cos \phi + a (\cos^2 \phi - \sin^2 \phi)(\cos^2 \phi + \sin^2 \phi) = 0,$

i.e., $\qquad x \sin \phi - y \cos \phi + a \cos 2 \phi = 0,$

which is the required equation.

6. *Find the condition for the line*

$$x \cos \theta + y \sin \theta = p, \qquad ...(i)$$

to touch the curve

$$x^m/a^m + y^m/b^m = 1. \qquad ...(ii)$$

In Ex. 3, (ii) it was shown that the equation of the tangent is

$$\frac{X x^{m-1}}{a^m} + \frac{Y y^{m-1}}{b^m} = 1; \qquad ...(iii)$$

where X, Y are the current co-ordinates.

We re-write the equation (i) in the form

$$X \cos \theta + Y \sin \theta - p = 0, \qquad ...(iv)$$

taking X, Y as current co-ordinates instead of x, y.

The equations (iii) and (iv) represent the same line.

$$\Rightarrow \qquad \frac{x^{m-1}}{a^m \cos \theta} = \frac{y^{m-1}}{b^m \sin \theta} = \frac{1}{p},$$

$$\Rightarrow \qquad x = \left(\frac{a^m \cos\theta}{p}\right)^{1/(m-1)}, y = \left(\frac{b^m \sin\theta}{p}\right)^{1/(m-1)}$$

The point (x, y) lies on the given curve. Therefore

$$\Rightarrow \qquad \frac{1}{a^m}\left(\frac{a^m \cos\theta}{p}\right)^{m/(m-1)} + \frac{1}{b^m}\left(\frac{b^m \sin\theta}{p}\right)^{m/(m-1)} = 1,$$

$\Rightarrow$ $(a \cos \theta)^{m/(m-1)} + (b \sin \theta)^{m/(m-1)} = p^{m/(m-1)},$

which is the required condition.

7. *Show that the length of the portion of the tangent to the astroid*

$$x^{2/3} + y^{2/3} = a^{2/3}$$

intercepted between the co-ordinate axes is constant.

The equation of the tangent at *any* point (x, y) (Ex. 5) is

$$Y - y = -\frac{y^{1/3}}{x^{1/3}}(X - x),$$

The tangent meets X-axis where $X = 0$

$\therefore$ $y = \frac{-y^{1/3}}{x^{1/3}}(X - x)$

$\Rightarrow$ $X = x^{1/3} a^{2/3}$

Again the tangent meets Y-axis where $Y = 0$

$\therefore$ $Y - y = \frac{-y^{1/3}}{x^{1/3}}(-x)$

$\Rightarrow$ $Y = y^{1/3} a^{2/3}$

$\therefore$ Length of tangent intercepted between co-ordinate axes

$$= \sqrt{x^{2/3} a^{4/3} + y^{2/3} a^{4/3}}$$

$$= \sqrt{a^{2/3} a^{4/3}}$$

$$= \sqrt{a^2} = a.$$

8. *Prove that, the portion of the tangent to the curve*

$$\frac{x + \sqrt{a^2 - y^2}}{a} = \log\frac{a + \sqrt{a^2 - y^2}}{y}$$

intercepted between the point of contact and the x-axis is constant.

We have $\dfrac{x + \sqrt{a^2 - y^2}}{a} = \log\dfrac{a + \sqrt{a^2 - y^2}}{y}$

Differentiating w.r.t. x, we get,

$$\frac{1}{a}\left[1 - \frac{y}{\sqrt{a^2 - y^2}}\frac{dy}{dx}\right] = \frac{y}{a + \sqrt{a^2 - y^2}} \times \frac{\dfrac{-y^2}{\sqrt{a^2 - y^2}} - (a + \sqrt{a^2 - y^2})}{y^2}\cdot\frac{dy}{dx}$$

$$= -\frac{a}{y\sqrt{a^2 - y^2}}\frac{dy}{dx}$$

$\Rightarrow$ $\left(\dfrac{y}{a\sqrt{a^2 - y^2}} - \dfrac{a}{y\sqrt{a^2 - y^2}}\right)\dfrac{dy}{dx} = \dfrac{1}{a}$

$$\Rightarrow \qquad \frac{dy}{dx} = \frac{-y}{\sqrt{a^2 - y^2}}$$

Equation of the tangent at the point (x, y) is

$$Y - y = \frac{-y}{\sqrt{a^2 - y^2}} (X - x)$$

This meets x-axis where $Y = 0$ so that

$$X = x + \sqrt{a^2 - y^2}$$

∴ the portion of the tangent intercepted between the point of contact (x, y) and the point $(x + \sqrt{a^2 - y^2}, 0)$ is $\sqrt{a^2 - y^2 + y^2} = a =$ constant.

9. *Tangents are drawn from the origin to the curve $y = \sin x$. Prove that their point of contact lies on $x^2 y^2 = x^2 - y^2$*

$$y = \sin x \qquad\qquad\qquad ...(i)$$

$$\Rightarrow \qquad \frac{dy}{dx} = \cos x$$

Equation of the tangent at the point (x, y) is
$$Y - y = \cos x \, (X - x)$$

Since it passes through the origin, therefore

$$- y = - x \cos x$$

$$\Rightarrow \qquad \frac{y}{x} = \cos x \qquad\qquad\qquad ...(ii)$$

The point of contact lies on the locus given by these equations (i) and (ii) and that is

$$y^2 + \frac{y^2}{x^2} = 1$$
$$\Rightarrow \qquad x^2 y^2 = x^2 - y^2.$$

10. *If $ax + bx = 1$ is a normal to the parabola $y^2 = 4\lambda x$, then prove that $\lambda a^3 + 2a \lambda b^2 = b^2$.*

Let the normal is drawn at the point (x, y) then

$$y_1^2 = 4\lambda x_1$$

from the given curve, we have

$$\left(\frac{dy}{dx}\right)_{(x_1, y_1)} = \frac{2\lambda}{y_1}$$

Hence, the equation of normal is

$$y - y_1 = -\frac{y_1}{2\lambda} (x - x_1)$$

$$\Rightarrow \qquad y - y_1 = -\frac{y_1}{2\lambda}\left(x - \frac{y_1^2}{4\lambda}\right)$$

$$\Rightarrow \qquad 4\lambda y_1 x + 8 \lambda^2 y = y_1^3 + 8\lambda^2 y_1$$

Comparing this equation of normal with given equation

$$ax + by = 1$$

we get $\qquad \dfrac{4\lambda\, y_1}{a} = \dfrac{8\lambda^2}{b} = \dfrac{y_1^{\,3} + 8\lambda^2\, y_1}{1}$

$\qquad\qquad (i) \qquad (ii) \qquad (iii)$

From (i) and (ii), we get

$$y_1 = \frac{2a\lambda}{b}$$

From (i) and (iii), we get

$$\frac{4\lambda}{a} = y_1^{\,2} + 8\lambda^2$$

From (ii) and (iii), we get

$$\frac{4\lambda}{a} = \frac{4a^2\lambda^2}{b^2} + 8\lambda^2$$

$\Rightarrow \qquad\qquad 4\lambda\, b^2 = 4a^3\,\lambda^2 + 8a\,\lambda^2\, b^2$

$\Rightarrow \qquad\qquad b^2 = a^3\,\lambda + 2a\,\lambda\, b^2.$

11. *For the curve $xy = c^2$, prove that the portion of the tangent intercepted between the coordinate axes is bisected at the point of contact.*

Let the point at which tangent is drawn be (x_1, y_1) on the curve. Thus

$$x_1\, y_1 = c^2$$

On differentitation given curve, we have

$$\frac{dy}{dx} = -\frac{c^2}{x^2}$$

$\Rightarrow \qquad \left(\dfrac{dy}{dx}\right)_{(x_1,\, y_1)} = -\dfrac{c^2}{x_1^{\,2}} = -\dfrac{y_1}{x_1}$

Thus, the equation of tangent is

$$y - y_1 = -\frac{y_1}{x_1}(x - x_1)$$

$\Rightarrow \qquad\qquad y\,x_1 - x_1 y_1 = -x\,y_1 + x_1\, y_1$

$\Rightarrow \qquad\qquad \dfrac{x}{2x_1} + \dfrac{y}{2y_1} = -1$

obviously, the tangent live cuts x-and y-axes at $(2x_1, 0)$ and $(0, 2y_1)$ and the middle point of the line joining these points is (x_1, y_1).

12. *If the tangent at (x_0, y_0) to the curve $x^3 + y^3 = a^3$ meets the curve again at (x_1, y_1) then prove that $\dfrac{x_1}{x_0} + \dfrac{y_1}{y_0} = 1$.*

Differentiating the equation of curve

$$3x^2 + 3y^2 \cdot \frac{dy}{dx} = 0 \qquad \Rightarrow \qquad \frac{dy}{dx} = -\frac{x^2}{y^2}$$

∴ Equation of tangent as (x_0, y_0) is

$$y - y_0 = -\frac{x_0^2}{y_0^2}(x - x_0)$$

$\Rightarrow \qquad x\,x_0^2 + y\,y_0^2 = x_0^3 + y_0^3$

$\Rightarrow \qquad x\,x_0^2 + y\,y_0^2 = a^3$

This passes through (x_1, y_1)

$\Rightarrow \qquad x_1\,x_0^2 + y_1\,y_0^2 = a^3$

Also $\qquad x_1^3 + y_1^3 = a^3 = x_0^3 + y_0^3$ {As tangent passes through (x_1, y_1)}

$\Rightarrow \qquad x_1^3 - x_0^3 = y_0^3 - y_1^3$

$\Rightarrow \quad (x_1 - x_0)(x_1^2 + x_1 x_0 + x_0^2) = (y_0 - y_1)(y_1^2 + y_0 y_1 + y_0^2)$

$\Rightarrow \qquad \dfrac{y_0 - y_1}{x_0 - x_1} = -\dfrac{(x_1^2 + x_0 x_1 + x_0^2)}{(y_1^2 + y_0 y_1 + y_0^2)}$

Slope of the line joining (x_0, y_0) and (x_1, y_1)

$$= \frac{y_0 - y_1}{x_0 - x_1} = -\frac{x_0^2}{y_0^2}$$

From these two relations, we have

$$\frac{x_0^2}{y_0^2} = \frac{x_0^2 + x_1^2 + x_0 x_1}{y_0^2 + y_1^2 + y_0 y_1}$$

$\Rightarrow \qquad x_0^2 y_1^2 - x_1^2 y_0^2 + x_0 y_0 (x_0 y_1 - x_1 y_0) = 0$

$\Rightarrow \qquad (x_0 y_1 - x_1 y_0)(x_0 y_1 + x_1 y_0 - x_0 y_0) = 0$

$\Rightarrow \qquad \qquad x_0 y_1 + x_1 y_0 - x_0 y_0 = 0 \qquad \qquad [\because x_0 y_1 \neq x_1 y_0]$

$\Rightarrow \qquad \dfrac{y_1}{y_0} + \dfrac{x_1}{x_0} = 1.$

13. *Find the point as the curve $3x^2 - 4y^2 = 72$ which is nearest to the line $3x + 2y + 1 = 0$*

Slope of the given line is $(-3/2)$. To find the nearest point we will find a point on the curve at which the tangent is parallel to the given line.

Differentiating the given curve, we have

$$\frac{dy}{dx} = \frac{3x}{4y}$$

∴ $\qquad \left(\dfrac{dy}{dx}\right)_{(x_1, y_1)} = \dfrac{3x_1}{4y_1} = -\dfrac{3}{2}$

Also the point (x_1, y_1) lies on the curve, hence

$\qquad 3x_1^2 - 4y_1^2 = 72$

Solving these, we get

$$y_1^2 = 9 \Rightarrow \quad y_1 = \pm 3$$

∴ Required points are $(-6, 3)$ and $(6, -3)$ now, distance of $(-6, 3)$ from the given line

$$= \frac{|-18 + 6 + 1|}{\sqrt{3}} = \frac{11}{\sqrt{13}}$$

and distance of $(6, -3)$ from the given line

$$= \frac{|-18 - 6 + 1|}{\sqrt{13}} = \frac{13}{\sqrt{13}}$$

$\therefore$ $(-6, 3)$ is the required point.

14. *If the tangent at any point $(4m^2, 8m^3)$ of $x^3 = y^2$ is a normal to the curve at the point where it meets the curve again, then find the value of m.*

$$y^2 = x^3 \qquad \qquad ...(i)$$

$$\Rightarrow \qquad \frac{dy}{dx} = \frac{3x^2}{2y}$$

$\therefore$ Slope at $(4m^2, 9m^3) = 3m$

$\therefore$ Equation of tangent at $(4m^2, 8m^3)$ is

$$y - 8m^3 = 3m\,(x - 4m^2)$$

$$\Rightarrow \qquad \qquad y = 3m\,x - 4m^3 \qquad \qquad ...(ii)$$

Solving (i) and (ii), we get

$$x = 4m^2,\ m^2$$

$\therefore$ $A\,(4m^2, 9m^3)$ and $B\,(m^2, -m^3)$ are the points of intersection of (i) and (ii).

Now, Slope of tangent at $B = -\dfrac{3}{2}m$

$\Rightarrow$ Slope of normal at $B = \dfrac{2}{3m}$

Since tangent at A is normal at B, hence

$$\frac{2}{3m} = 3m \quad \Rightarrow \quad m^2 = \frac{2}{3}$$

$$\Rightarrow \qquad \qquad m = \sqrt{\frac{2}{3}}.$$

15. *Three normals are drawn from the point $(c, 0)$ to the curve $y^2 = x$, show that c must be greater than 1/2. One normal is always the x-axis. Find c for which the other two normals are perpendicular to each other.*

Differentiating equation of the curve, we have

$$\frac{dy}{dx} = \frac{1}{2y}$$

Let any point on the curve, is $\left(\dfrac{m^2}{4}, \dfrac{m}{2}\right)$

$$\left(\frac{dy}{dx}\right)_{\left(\frac{m^2}{4}, \frac{m}{2}\right)} = \frac{1}{m}$$

So, equation of the normal at $\left(\dfrac{m^2}{4}, \dfrac{m}{2}\right)$ is

$$y = mx - \frac{m}{2} - \frac{m^3}{4}$$

This passes through $(c, 0)$, then

$$0 = mc - \frac{1}{2}m - \frac{1}{4}m^3$$

$$\Rightarrow \qquad m = 0, \frac{m^2}{4} - c + \frac{1}{2} = 0$$

$$\Rightarrow \qquad m = 0, m = \pm 2\sqrt{(c - 1/2)}$$

For $m = 0$, normal is x-axis.

For other two values, it will be real if

$$c > 1/2.$$

For the other two normals to be perpendicular to each other, we must have

$$\left(2\sqrt{c - \frac{1}{2}}\right)\left(-2\sqrt{c - \frac{1}{2}}\right) = -1$$

$$\Rightarrow \qquad c = 3/4.$$

16. *Tangent at a point P_1, (other than $(0, 0)$) on the curve $y = x^3$ meets the curve again at P_2. The tangent at P_2 meets the curve at P_3, and so on, show that the abscissae of $P_1, P_2, P_3 \ldots P_n$, form a G.P.*

Let $P_1 (x_1, y_1)$ be a point an the curve

$$y = x^3 \qquad \qquad \qquad \ldots(i)$$

$$\therefore \qquad y_1 = x_1^3 \qquad \qquad \qquad \ldots(ii)$$

Now, $\qquad \dfrac{dy}{dx} = 3x^2$

$\therefore$ Slope of tangent at $P_1 = m_1 = 3x_1^2$

Equation of the tangent at $P_1 (x_1, y_1)$ is

$$y - x_1^3 = 3x_1^2 (x - x_1)$$

$$\Rightarrow \qquad y = 3x_1^2 x - 2x_1^3 \qquad \qquad \ldots(iii)$$

Solving (i) and (ii) we get

$$x^3 - 3x_1^2 x + 2x_1^3 = 0$$

$$\Rightarrow \quad (x - x_1)^2 (x + 2x_1) = 0$$

$$\therefore \qquad \qquad x = x_1 \text{ or } x = -2x_1$$

$$\therefore \qquad \qquad x_2 = -2x_1, y_2 = x_2^3 = -8x_1^3$$

$$\therefore \qquad \qquad P_2 (x_2, y_2) = (-2x_1, -8x_1^3)$$

Slope of the tangent at $P_2 = 3x_2^2 = 12x_1^2$

$\therefore$ Equation of tangent at P_2 is

$$y - x_2^3 = 3x_2^2 (x - x_2) \qquad \qquad \ldots(iv)$$

Solving (i) and (iv) poin $P_3 (x_3, y_3)$ will be $(-2x_2, -8x_2^3) = (4x_1, 64x_1^3)$ and so on.

$\therefore$ Abscissae of $P_1, P_2, P_3, \ldots$ are given by $x_1, -2x_1, 4x_1, -8x_1, \ldots$, which is G.P. with comman ratio -2.

17. *Show that there is no cubic curve for which the tangent lines at two distinct points coincide.*

Let
$$y = ax^3 + bx^2 + cx + d = 0 \ (a \neq 0)$$
be a cubic curve.

Let us assume that (x_1, y_1) and (x_2, y_2), $(x_1 < x_2)$ are two district points on the curve at which tangents coincide.

Then by mean value thereon, there exists x_3 $(x_1 < x_3 < x_2)$ such that
$$\frac{y_2 - y_1}{x_2 - x_1} = y'(x_3).$$

Since, x_1, x_2, x_3 are solutions of the equation $3ax^2 + 2bx + c = m$.

But it is a quadratic can not have more that two roots. Therefore, no such cubic is possible.

EXERCISES

1. Find the tangents and normals to the given curves at the given points :

(i) $y^2 = 4ax$ at $(a, -2a)$.

(ii) $x^2/a^2 + y^2/b^2 = 1$ at (x', y').

(iii) $xy = c^2$ at $(cp, c/p)$.

(iv) $(x^2 + y^2) x - ay^2 = 0$ at $x = a/2$.

(v) $(x^2 + y^2) x - a (a^2 - y^2) = 0$ for $x = -3a/5$.

(vi) $x^2 (x - y) + a^2 (x + y) = 0$ at $(0, 0)$

(vii) $x = 2a \cos \theta - a \cos 2\theta$, $y = 2a \sin \theta - a \sin 2\theta$ at $\theta = \pi/2$.

(viii) $c^2 (x^2 + y^2) = x^2 y^2$ at $(c/\cos \theta, c/\sin \theta)$.

(ix) $x = \dfrac{2at^2}{1 + t^2}$, $y = \dfrac{2at^3}{1 + t^2}$ for $t = \dfrac{1}{2}$.

2. Find the equation of the tangent to the curve $c^2 (x^2 + y^2) = x^2 y^2$ in the form $x \cos^3 \theta + y \sin^3 \theta = c$.

3. Prove that the equation of the tangent at any point $(4m^2, 8m^3)$ of the semi-cubical parabola $x^3 - y^2 = 0$ is $y = 3mx - 4m^3$ and show that it meets the curve again at $(m^2, -m^3)$, where it is normal if $9m^2 = 2$.

4. Find where the tangent is parallel to x-axis and where it is parallel to y-axis for the following curves :

(i) $x^3 + y^3 = a^3$. (ii) $x^3 + y^3 = 3axy$. (iii) $25x^2 + 12xy + 4y^2 = 1$.

5. Tangent at any point of the curve $x = a \cos^3 \theta$, $y = b \sin^3 \theta$ meets the co-ordinates axes in A and B; show that the locus of the point with (OA, OB) as co-ordinates is $x^2/a^2 + y^2/b^2 = 1$.

6. Find the equations of the tangent and the normal to the curve $y (x - 2) (x - 3) - x + 7 = 0$ at the point where it meets x-axis.

7. Show that the tangent at any point (x, y) on the curve $y^m = ax^{m-1} + x^m$, makes intercepts
$$\frac{ax}{(m - 1)a + mx} \text{ and } \frac{ay}{m(a + x)}$$
on the co-ordinates axes.

8. Prove that the sum of intercepts on the co-ordinates axes of any tangent to $\sqrt{x} + \sqrt{y} = \sqrt{a}$ is constant.

9. Show that the normal at any point of the curve

$x = a \cos\theta + a\,\theta \sin\theta, \; y = a \sin\theta - a\,\theta \cos\theta$

is at a constant distance from the origin.

10. Show that the length of the portion of the normal to the curve
$x = a\,(4 \cos^3\theta - 3 \cos\theta), \; y = a\,(4 \sin^3\theta - 9 \sin\theta)$ intercepted between the co-ordinate axes in constant.

11. Show that the tangent and the normal at any point of the curve $x = ae^\theta\,(\sin\theta - \cos\theta)$, $y = ae^\theta\,(\sin\theta + \cos\theta)$ are equidistant from the origin.

12. Show that the distance from the origin of the normal at any point of the curve

$$x = ae^\theta\left(\sin\frac{\theta}{2} + 2\cos\frac{\theta}{2}\right), \; y = ae^\theta\left(\cos\frac{\theta}{2} - 2\sin\frac{\theta}{2}\right)$$

is twice the distance of the tangent at the point from the origin.

13. Show that the tangent at the point of the curve

$x = a\,(t + \sin t \cos t), \; y = a\,(1 + \sin t)^2$ makes angle $\dfrac{1}{4}(\pi + 2t)$ with x-axis.

14. The tangent at any point on the curve $x^3 + y^3 = 2a^3$ cuts off lengths p and q on the co-ordinate axes; show that $p^{-3/2} + q^{-3/2} = 2^{-1/2}\,a^{-3/2}$.

15. The tangent at any point P of the curve $y = x^2 - x^3$ meets it again at Q. Find the co-ordinates of the mid-point of PQ and show that its locus is

$$y = 1 - 9x + 28x^2 - 28x^3.$$

16. Show that the line

$$x \cos\theta + y \sin\theta - p = 0$$

will touch the curve

$x^m\,y^n - a^{m+n} = 0,$

if $p^{m+n}\,m^m\,n^n = (m+n)^{m+n}\,a^{m+n}\cos^m\theta\,\sin^n\theta.$

17. If $p = x \cos\theta + y \sin\theta$, touches the curve

$$\left(\frac{x}{a}\right)^{\frac{n}{n-1}} + \left(\frac{y}{b}\right)^{\frac{n}{n-1}} = 1$$

prove that $p^n = (a \cos\theta)^n + (b \sin\theta)^n$

18. Prove that the curve $\left(\dfrac{x}{a}\right)^n + \left(\dfrac{y}{b}\right)^n = 2$ touches the straight line $\dfrac{x}{a} + \dfrac{y}{b} = 2$ at the point (a, b), whatever be the value of n.

19. Find the point on the curve $y - e^{xy} + x = 0$, at which we have vertical tangent.

20. If the line $ax + by + c = 0$ is normal to the curve $xy + 5 = 0$, then show that a and b have same sign.

21. Find the equation of normal to the curve $x + y = x^y$, whose it cuts x-axis.

22. Find the equation of normal to the curve $y = (1 + x)^y + \sin^{-1}(\sin^2 x)$ at $x = 0$.

23. Find the equations of tangents to the ellipse $3x^2 + y^2 + x + 2y = 0$ which are pespendicular to the line $4x - 2y = 1$.

24. Find the equation of tangent and normal to the curve $y^2 (x + a) = x^2 (3a - x)$ at the point where $x = a$.

25. Find all tangents to the curve $y = \cos(x + y), -2\pi \leq x \leq 2\pi$ that are parallel to the line $x + 2y = 0$

26. Find the point on the curve $\dfrac{x^n}{a^n} + \dfrac{y^n}{b^n} = 2$, so that it touches the line $\dfrac{x}{a} + \dfrac{y}{b} = 2$

27. Prove that the portion of the tangent to the curve $x^m y^n = a^{m+n}$ intercepted between the axes is divided in the ration $m : n$ at the point of contact.

28. Prove that the equation of the tangent at a point on the curve $x = a\left\{\dfrac{\phi(t)}{f(t)}\right\}$ and $y = a\left\{\dfrac{\psi(t)}{f(t)}\right\}$ may be written in the form

$$\begin{vmatrix} x & y & a \\ \phi(t) & \psi(t) & f(t) \\ \phi'(t) & \psi'(t) & f'(t) \end{vmatrix} = 0$$

29. If p_1 and p_2 be the perpendiculars from the origin on the tangent and normal respectively at any point (x, y) on curve, then show that

$p_1 = |x \sin \phi - y \cos \phi|, p_2 = |x \cos \phi + y \sin \phi|$, where $\tan \phi = \dfrac{dy}{dx}$. If in the above case curve be $x^{2/3} + y^{2/3} = a^{2/3}$, then show that $4p_1^2 + p_2^2 = a^2$.

30. If the tangent at (a, b) to the curve $x^3 + y^3 = c^3$ meets the curve again in (a_1, b_1) then prove that

$$\dfrac{a_1}{a} + \dfrac{b_1}{b} + 1 = 0.$$

31. Show that the angle between the tangents at any point P an the line joining P to the origin O is the same at all points of the curve $\log(x^2 + y^2) = c \tan^{-1}(y/x)$, where c is constant.

32. The tangent to the curve $y = x - x^3$ at a point P meets the curve again at Q. Prove that one point of trisection of PQ lies on the y-axis. Find the locus of the other point of trisection.

ANSWERS

1. (i) $x + y + a = 0, x - y = 3a$.

   (ii) $b^2 x' x + a^2 y' y = 3a$.

   (iii) $x + p^2 y = 2cp ; p^3 x - py = c (p^4 - 1)$.

   (iv) $4x \pm 2y - a = 0 ; 2x \pm 4y = 3a$.

   (v) $31x \pm 8y + 9a = 0 ; 8x \pm 31y + 42a = 0$.

   (vi) $x + y = 0 ; x - y = 0$.

   (vii) $x + y = 3a ; x - y + a = 0$.

   (viii) $x \cos^3 \theta + y \sin^3 \theta = c$;

   $x \sin^3 \theta - y \cos^3 \theta + 2c \cot 2\theta = 0$.

   (ix) $13x - 16y = 2a ; 16x + 13y = 9a$.

4. (i) $(0, a) ; (a, 0)$

   (ii) $(\sqrt[3]{2}\,a, \sqrt[3]{4}\,a) ; (\sqrt[3]{4}\,a, \sqrt[3]{2}\,a)$

   (iii) $\left(\pm\dfrac{3}{20}, \pm\dfrac{5}{8}\right), \left(\pm\dfrac{1}{4}, \pm\dfrac{3}{8}\right)$.

6. $x - 20y = 7, 20x + y = 140$.

15. $\left(\dfrac{1 - 2x_1}{2}, \dfrac{2x_1 - 7x_1^2 + 7x_1^3}{2}\right)$.

19. $(1, 0)$;	**21.** $y = x - 1$;	
22. $x + y = 1$;	**23.** $x + 2y = 0$, $3x + 6y + 13 = 0$	
24. $x + 2y + 3a = 0$, $2x - y - 3a = 0$;		
25. $x + 2y = \dfrac{\pi}{2}$, $x + 2y = -\dfrac{3\pi}{2}$		
26. (a, b)	**32.** $y = x - 5x^3$.	

7.3. Angle of intersection of two curves

Def. *The angle of intersection of two curves at a point of intersection is the angle between the tangents to the curves at the point.*

EXAMPLES

1. *Find the angle of intersection of the parabolas $y^2 = 4ax$ and $x^2 = 4by$, at their point of intersection other that the origin.*

Now $(0, 0)$ and $(4a^{1/3} b^{2/3}, 4a^{2/3} b^{1/3})$ are the two points of intersection of the parabolas.

For $\qquad\qquad y^2 = 4ax$,

$\left(\dfrac{dy}{dx}\right)$ at the point $(4a^{1/3} b^{2/3}, 4a^{2/3} b^{1/3})$ is $a^{1/3}/2b^{1/3}$

For $\qquad\qquad x^2 = 4by$,

$\left(\dfrac{dy}{dx}\right)$ at $(4a^{1/3} b^{2/3}, 4a^{2/3} b^{1/3})$ is $2a^{1/3}/b^{1/3}$.

Thus, if m, m' be the slopes of the tangents to the two curves, we have $m = a^{1/3}/2b^{1/3}$, $m' = 2a^{1/3}/b^{1/3}$ so that the required angle

$$= \tan^{-1} \frac{m' - m}{1 + m'm} = \tan^{-1} \frac{\dfrac{2a^{1/3}}{b^{1/3}} - \dfrac{a^{1/3}}{2b^{1/3}}}{1 + \dfrac{2a^{1/3}}{b^{1/3}} \cdot \dfrac{a^{1/3}}{2b^{1/3}}} = \tan^{-1} \frac{3a^{1/3} b^{1/3}}{2(a^{2/3} + b^{2/3})}.$$

2. *Find the condition for the curves*

$$ax^2 + by^2 = 1, \ a_1x^2 + b_1y^2 = 1$$

to intersect orthogonally.   *(Avadh 2001)*

If (x', y') be a point of intersection of the two curves,

we have $\qquad x'^2 = (b_1 - b)/(ab_1 - a_1b)$, $y'^2 = (a - a_1)/(ab_1 - a_1b)$.

For $\qquad ax^2 + by^2 = 1$, we get $\left(\dfrac{dy}{dx}\right)_{(x',y')} = -\dfrac{ax'}{by'}$,

and for $\qquad a_1x^2 + b_1y^2 = 1$, we have $\left(\dfrac{dy}{dx}\right)_{(x',y')} = -\dfrac{a_1x'}{b_1y'}$

For orthogonal intersection, we have

$$\frac{-ax'}{by'} \times \frac{-a_1x'}{b_1y'} = -1$$

$\Rightarrow \qquad aa_1 x'^2 + bb_1 y'^2 = 0$

$\Rightarrow \qquad \dfrac{aa_1 (b_1 - b)}{ab_1 - a_1 b} + \dfrac{bb_1 (a - a_1)}{ab_1 - a_1 b} = 0$

$\Rightarrow \qquad \dfrac{b_1 - b}{bb_1} + \dfrac{a - a_1}{aa_1} = 0 \qquad \Rightarrow \qquad \dfrac{1}{b} - \dfrac{1}{b_1} = \dfrac{1}{a} - \dfrac{1}{a_1};$

as the required condition.

3. *Find the acute angle between the curve $y = |x^2 - 1|$ and $y = |x^2 - 3|$ at their points of intersection when $x > 0$.*

For the intersection of the given curves

$$|x^2 - 1| = |x^2 - 3| \Rightarrow (x^2 - 1)^2 = (x^2 - 3)^2$$

$\Rightarrow \qquad [(x^2 - 1) - (x^2 - 3)]\,[(x^2 - 1) + (x^2 - 3)] = 0$

$\Rightarrow \qquad 4\,[x^2 - 2] = 0 \Rightarrow \quad x = \pm\sqrt{2}$

Neglecting $\qquad x = -\sqrt{2}$ as $x > 0$.

Now, $\qquad\qquad y = |x^2 - 1| = (x^2 - 1)$ in the neighbourhood of

$\qquad\qquad\qquad x = \sqrt{2}$

and $\qquad\qquad y = -(x^2 - 3)$ in the neighbourhood of $x = \sqrt{2}$

$\Rightarrow \qquad\qquad \left(\dfrac{dy}{dx}\right)_{C_1} = 2x = 2\sqrt{2}$

and $\qquad\qquad \left(\dfrac{dy}{dx}\right)_{C_2} = -2x = -2\sqrt{2}$

Then $\qquad\qquad \tan\theta = \left|\dfrac{2\sqrt{2} - (-2\sqrt{2})}{1 + 2\sqrt{2}\,(-2\sqrt{2})}\right|$

$$= \left|\dfrac{4\sqrt{2}}{-7}\right| = \dfrac{4\sqrt{2}}{7}$$

$\therefore \qquad\qquad \theta = \tan^{-1}\left(\dfrac{4\sqrt{2}}{7}\right).$

4. *Find the angle of intersection of curves, $y = [\,|\sin x| + |\cos x|\,]$ and $x^2 + y^2 = 5$ whose $[.]$ denotes greatest integer function.*

We know that

$$1 \le |\sin x| + |\cos x| \le 2$$

For all real values of

$$y = [\,|\sin x| + |\cos x|\,] = 1$$

Let P and Q be the points or intersection of given curves.

Clearly the given curves meet at points where $y = 1$ so, we get

$$x^2 + 1 = 5 \Rightarrow x = 2$$

Now, $\qquad P\,(2, 1)$ and $Q\,(-2, 1)$

we have $\qquad x^2 + y^2 = 5$

$$\Rightarrow \qquad 2x + 2y\frac{dy}{dx} = 0 \qquad \Rightarrow \qquad \frac{dy}{dx} = -\frac{x}{y}$$

$$\left(\frac{dy}{dx}\right)_{(2,1)} = -2 \quad \text{and} \quad \left(\frac{dy}{dx}\right)_{(-2,1)} = 2$$

Clearly, the slope of line $y = 1$ is zero and the slope of the tangents at P and Q are -2 and 2 respectively.

Thus, the angle of intersection is $\tan^{-1}(2)$.

EXERCISES

1. Find the angle between the pairs of curves.

   (i) $x^2 - y^2 = a^2, x^2 + y^2 = a^2\sqrt{2}$.   (ii) $y^2 = ax, x^3 + y^3 = 3axy$.

   (iii) $y = \sin x, y = \cos x$ (iv) $2y^2 = x^3, y^2 = 32x$

2. Prove that the curves $x^3 + 2xy^2 - 10a^2y + 3a^3 = 0$ and $y^3 + 2xy^2 - 5a^2x - a^3 = 0$, intersection at an angle $\tan^{-1}(88/73)$ at the point $(3a - 2a)$.

3. Prove that the curves $x^2 + 2xy - y^2 + 2ax = 0$ and $3y^3 - 2a^2x - 4a^2y + a^3 = 0$ intersect at an angle $\tan^{-1}\left(\dfrac{9}{8}\right)$ at the point $(a, -a)$.

4. Find the condition for the line $y = mx$ to cut at right angles the conic $ax^2 + 2hxy + by^2 = 1$. Hence find the directions of the axes of the conic.

5. Show that the curves $\dfrac{x^2}{a} + \dfrac{y^2}{b} = 1$ and $\dfrac{x^2}{a_1} + \dfrac{y^2}{b_1} = 1$ will cut orthogonally if $a - b = a_1 - b_1$

6. Find the angle of intersectin of $y = a^x$ and $y = b^x$.

7. Find the condition that the curves $ax^2 + by^2 = 1$ and $a'x^2 + b'y^2 = 1$ intersect orthogonally and hence show that the curves

$$\frac{x^2}{a^2 + \lambda} + \frac{y^2}{b^2 + \lambda} = 1 \quad \text{and} \quad \frac{x^2}{a^2 + \mu} + \frac{y^2}{b^2 + \mu} = 1$$

intersect orthoganally.

ANSWERS

1. (i) $\pi/4$.   (ii) $\tan^{-1}\sqrt[3]{16}$.   (iii) $\tan^{-1}2\sqrt{2}$ (iv) $\dfrac{\pi}{2}$ and $\tan^{-1}\dfrac{1}{2}$

4. $m^2h + m(a - b) - h = 0$. The roots of this quadratic equation in, m, are the slopes of the two axes.

6. $\tan\theta = \dfrac{\log a/b}{1 + \log a \log b}$

7.4. Length of the tangent, normal, sub-tangent and sub-normal at any point of a curve.

Let the tangent and the normal at any point (x, y) of the curve meet the x-axis at T and G respectively. Draw the ordinate PM.

Then *the lines TM, MG are called the sub-tangent and sub-normal respectively.*

Then tangents *PT*, *PG* are sometimes referred to as the lengths of the tangent and the normal respectively.

Clearly

$$\angle MPG = \psi.$$

Also

$$\tan \psi = \frac{dy}{dx}.$$

From the figure, we have

(i) **Tangent** $= TP = MP \operatorname{cosec} \psi = y \sqrt{(1 + \cot^2 \psi)}$ Fig. 7.2

$$= y \sqrt{\left[1 + \left(\frac{dx}{dy}\right)^2\right]}.$$

(ii) **Sub-tangent** $= TM = MP \cot \psi = y \dfrac{dx}{dy}.$

(iii) **Normal** $= GP = MP \sec \psi = y \sqrt{(1 + \tan^2 \psi)}$

$$= y \sqrt{\left[1 + \left(\frac{dy}{dx}\right)^2\right]}.$$

(iv) **Sub-normal** $= MG = MP \tan \psi = y \dfrac{dy}{dx}.$

EXAMPLES

1. *Show that in the curve* $y = be^{x/a}$, *the sub-tangent is of constant length and the subnormal varies as the square of ordinate?*

Let us consider a point (x_1, y_1) on the curve.

Differentiating the equation of curve w.r.t. x,

$$\frac{dy}{dx} = be^{x/a} \cdot \frac{1}{a}$$

$$\therefore \qquad \left(\frac{dy}{dx}\right)_{(x_1, y_1)} = \frac{b}{a} e^{x_1/a}$$

Length of sub-tangent $= y_1 \left(\dfrac{dx}{dy}\right)_{(x_1, y_1)}$

$$= y_1 \cdot \frac{a}{be^{x/a}}$$

$$= a$$

Length of sub-normal $= y_1 \left(\dfrac{dy}{dx}\right)_{(x_1, y_1)}$

$$= be^{x_1/a} \cdot \frac{be^{x_1/a}}{a}$$

$$= \frac{1}{a} y^2.$$

2. *Show that in the case of the curve* $\beta y^2 = (x + \alpha)^3$, *the square of the sub-tangent varies as the sub-normal.*

Differentating the equation of curve

$$2\beta y.\frac{dy}{dx} = 3(x + \alpha)^2$$

$\Rightarrow$

$$\frac{dy}{dx} = \frac{3(x + \alpha)^2}{2\beta y}$$

$\therefore$ Sub-tangent

$$= \frac{y}{(dy/dx)} = \frac{2\beta y^2}{3(x + \alpha)^2}$$

$$= \frac{2(x + \alpha)^3}{3(x + \alpha)^2} = \frac{2}{3}(x + \alpha).$$

Sub-normal

$$= y.\frac{dy}{dx} = y.\frac{3(x + \alpha)}{2\beta y}$$

$$= \frac{3(x + \alpha)^2}{2\beta}$$

$\therefore$

$$\frac{(\text{Sub-tangent})^2}{(\text{Sub-normal})} = \frac{\frac{4}{9}(x + \alpha)^2}{\frac{3}{2}\frac{(x + \alpha)^2}{\beta}}$$

$$= \frac{8\beta}{27} = \text{constant}$$

3. *If the relation between sub-normal SN and sub-tangent ST at any point S on the curve* $by^2 = (x + a)^3$ *is* $\lambda (SN) = \mu (ST)^2$, *then find the value of* $\dfrac{\lambda}{\mu}$.

Differentiating the equation of the given curve, we get

$$2by\frac{dy}{dx} = 3(x + a)^2$$

$\Rightarrow$

$$\frac{dy}{dx} = \frac{3}{2}.\frac{(x + a)^2}{by}$$

$\therefore$ Length of sub-tangent $ST = \dfrac{y}{dy/dx}$

$$= \frac{2by^2}{3(x + a)^2}$$

Length of sub-normal $SN = y\dfrac{dy}{dx} = \dfrac{3}{2}\dfrac{(x + a)^2}{b}$

$\therefore$

$$\frac{\lambda}{\mu} = \frac{(ST)^2}{SN} = \frac{(2by^2)^2.2b}{\{3(x + a)^2\}^2.3(x + a)^2}$$

$$= \frac{8b}{27}.$$

EXERCISES

1. Prove that the sub-normal at any point of the parabola

$$y^2 = 4ax$$

   is constant.

2. Show that the sub-normal at any point of the curve

$$y^2 x^2 = a^2 (x^2 - a^2)$$

   varies inversely as the cube of its abscissa.

3. Find the length of the tangent, length of the normal, sub-tangent and sub-normal at the point θ, on the four cusped hypocyloid

$$x = a \cos^3 \theta, \, y = a \sin^3 \theta.$$

4. Show that the sub-tangent at any point of the curve

$$x^m y^n = a^{m+n}$$

   varies as the abscissa.

5. Show that for the curve

$$x = a + b \log[b + \sqrt{(b^2 - y^2)}] - \sqrt{(b^2 - y^2)}$$

   sum of the sub-normal and sub-tangent is constant.

6. Show that for the curve

$$y = hn^{-1/2} e^{-h^2 n^2},$$

   the product of the abscissa and the sub-tangent is constant.

7. Show that the sub-tangent at any point of the exponential curve, $y = ae^{x/b}$ is constant.

8. For the catenary

$$y = c \cosh (x/c),$$

   prove that the length of the normal is y^2/c.

9. Prove that the sum of the tangent and sub-tangent at any point of the curve

$$e^{y/a} = x^2 - a^2.$$

   varies as the product of the corresponding co-ordinates.

10. Show that in the curve

$$y = a \log (x^2 - a^2).$$

    the sum of the tangent and the sub-tangent varies as the product of the co-ordinates of the point. [Compare Ex. 9 above].

11. Prove that in the ellipse $x^2/a^2 + y^2/b^2 = 1$, the length of the normal varies inversely as the perpendicular from the origin on the tangent.

12. In the curve $x^{m+n} = a^{m-n} y^{2n}$, prove the the mth power of the sub-tangent varies as the nth power of the sub-normal.

ANSWERS

3. $a \sin^2 \theta$, $a \tan \theta \sin^2 \theta$, $a \sin^2 \theta \cos \theta$, $a \sin^3 \theta \tan \theta$.

7.5 Pedal equations or p, r equations.

Def. *A relation between the distance, r, of any point on the curve from the origin (or pole), and the length of the perpendicular, p, from the origin (or pole) to the tangent at the point is called pedal equation of the curve.*

7.5.1. *To determine the pedal equation of a curve whose Cartesian equation is given.*

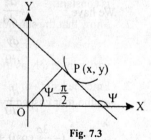

Fig. 7.3

Let the equation of the curve be

$$y = f(x). \qquad \qquad ...(i)$$

Equation of the tangent at any point (x, y) is

$$Y - y = f'(x)(X - x)$$

or $\quad Xf'(x) - Y + [y - xf'(x)] = 0. \qquad ...(ii)$

If, p, be the length of the perpendicular from $(0, 0)$ to this tangent, we have

$$p = \frac{y - x f'(x)}{\sqrt{[1 + f'^2(x)]}}.$$

Also $\quad r^2 = x^2 + y^2. \qquad \qquad ...(iii)$

Eliminating x, y between (i), (ii) and (iii),

we obtain the required pedal equation of the curve (i).

EXAMPLES

1. *Find the pedal equation of the parabola*

$$y^2 = 4a(x + a) \quad \textbf{(Kanpur 2001; Guwahati 2005; Bangalore 2006)}$$

Tangent at (x, y) is

$$Y - y = \frac{2a}{y}.(X - x)$$

or $\qquad 2a - yY + (y^2 - 2ax) = 0. \qquad \qquad ...(i)$

The length, p, of the perpendicular from the origin to the tangent (i) is given by

$$p = \frac{y^2 - 2ax}{\sqrt{(4a^2 + y^2)}}$$

$$= \frac{4a(x + 1) - 2ax}{\sqrt{[(4a^2 + 4a(x + y^2)]}}$$

$$= \frac{2a(x + 2a)}{\sqrt{[4a(x + 2a)]}} = \sqrt{[a(x + 2a)]}. \qquad ...(ii)$$

Also $\qquad r^2 = x^2 + y^2 = x^2 + 4a(x + a) = (x + 2a)^2. \qquad ...(iii)$

From (ii) and (iii), we obtain

$$p^2 = ar,$$

which is the required pedal equations.

2. *Find the pedal equation of the curve*

$$2x = 3a \cos \theta + a \cos 3\theta$$
$$2y = 3a \sin \theta - a \sin 3\theta.$$

We have
$$\frac{dx}{d\theta} = -\frac{3a}{2}(\sin \theta + \sin 3\theta)$$

$$\frac{dy}{d\theta} = \frac{3a}{2}(\cos \theta - \cos 3\theta)$$

$$\therefore \quad \frac{dy}{d\theta} = \frac{dy/d\theta}{dx/d\theta} = \frac{\dfrac{3a}{2}(\cos \theta - \cos 3\theta)}{-3a/2(\sin \theta + \sin \theta)} = -\tan \theta$$

$$\therefore \quad p = \frac{(-\dfrac{3a}{2}\cos \theta + \dfrac{a}{2}\cos 3\theta)\tan \theta - \dfrac{3a}{2}\sin \theta + \dfrac{a}{2}\sin 3\theta}{\sqrt{(1 + \tan^2 \theta)}}$$

$$= \frac{\dfrac{-3a}{2}\dfrac{(\sin \theta \cos \theta + \cos \theta \sin \theta)}{\cos \theta} - \dfrac{a}{2}\dfrac{(\cos 3\theta \sin \theta - \sin 3\theta \cos \theta)}{\cos \theta}}{\sec \theta}$$

$$= -a \sin 2\theta \quad \Rightarrow \quad p^2 = a^2 \sin^2 2\theta$$

Now,
$$r^2 = x^2 + y^2 = (9a^2 + a^2 + 6a^2 \cos 4\theta)/4$$
$$= 10a^2 + 6a^2 - 12a^2 \sin^2 2\theta$$
$$= 16a^2 - 12a^2 \cdot \frac{p^2}{a^2}$$

$$\therefore \quad r^2 = 4a^2 - 3p^2 \text{ is the pedal equation.}$$

EXERCISES

1. Show that the pedal equation of ellipse
$$x^2/a^2 + y^2/b^2 = 1,$$

is
$$\frac{1}{p^2} = \frac{1}{a^2} + \frac{1}{b^2} - \frac{r^2}{a^2 b^2}.$$
                                                                            (Calcutta 2005)

2. Show that the pedal equation of the astroid
$$x = a \cos^3 \theta, \; y = a \sin^3 \theta, \qquad \text{(Bangalore 2005)}$$

is
$$r^2 = a^2 - 3p^2.$$

3. Show that the pedal equation of the curve
$$x = 2a \cos \theta - a \cos 2\theta, \; y = 2a \sin \theta - a \sin 2\theta,$$

is
$$9(r^2 - a^2) = 8p^2.$$

4. Show that the pedal equation of the curve
$$x = ae^\theta (\sin \theta - \cos \theta), \; y = ae^\theta (\sin \theta + \cos \theta),$$

is
$$r = \sqrt{2}p.$$

5. Show that the pedal equation of the curve
$$x = a(3 \cos \theta - \cos^3 \theta), \; y = a(3 \sin \theta - \sin^3 \theta),$$

is
$$3p^2 (7a^2 - r^2) = (10a^2 - r^2)^2.$$

6. Show that the pedal equation of the curve
$$c^2 (x^2 + y^2) = x^2 y^2.$$
    is
$$1/p^2 + 3/r^2 = 1/c^2.$$

POLAR CO-ORDINATES

7.6 Angle between radius vector and tangent.

Consider the point $P(r, \theta)$ on the curve $r = f(\theta)$. We transform to cartesian co-ordinates and obtain

$$x = r \cos \theta = f(\theta) \cos \theta, \quad y = r \sin \theta = f(\theta) \sin \theta$$

Considering $x = f(\theta) \cos \theta$, $y = f(\theta) \sin \theta$ as cartesian paraetric equations of the given curve with (0) as parameter, we obtain

$$\tan \psi = \frac{dy}{dx} = \frac{dy}{d\theta} \div \frac{dx}{d\theta} = \frac{f'(\theta)\sin\theta + f(\theta)\cos\theta}{f'(\theta)\cos\theta - f(\theta)\sin\theta}$$

Dividing the numerator and denminator by $f'(\theta) \cos \theta$, we obtain

$$\tan \psi = \frac{\tan\theta + f(\theta)/f'(\theta)}{1 - [f(\theta)/f'(\theta)]\tan\theta} \qquad \qquad ...(i)$$

Denoting the angle between the radius vector and the tangent by ϕ, we have

$$\psi = \theta + \phi \qquad \qquad ...(ii)$$

Comparing (i) and (ii), we see that

Cor. Angle of intersection of two curves. If ϕ_1, ϕ_2, be the angles between the common radius vector and the tangents to the two given curves at a point of intersection, then the angle of intersection of the curves is

$$|\phi_1 - \phi_2|.$$

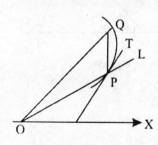

Note. Precise meaning of ϕ. For a point p of the curve,

$$r = f(\theta),$$

ϕ is precisely defined to be the angle through which the positive direction of the radius vector (i.e., the direction of the radius vector produced) has to rotate to coincide with the direction of the tangent in which θ increases. The direction in which θ increases is taken as the positive direction of the tangent.

Fig. 7.4

EXAMPLES

1. *Show that the radius vector is inclined at a constant angle to the tangent at any point on the equiangular spiral*

$$r = ae^{b\theta}. \qquad \qquad ...(i) \quad (Kuvempu \ 2005)$$

    Differentiting (i) w.r.t. θ, we get

$$\frac{dr}{d\theta} = b.ae^{b\theta} = br \qquad \Rightarrow \qquad r\frac{d\theta}{dr} = \frac{1}{b}$$

$\Rightarrow \qquad \qquad \tan \phi = \frac{1}{b},$

$\Rightarrow \qquad \qquad \phi = \tan^{-1}\frac{1}{b},$

which is a constant. This property of equiangular spiral justifies the adjective '*Equiangular*'.

2. *Find the angle of intersection of the cardioides*

$$r = a \, (1 + \cos \theta), \, r = b \, (1 - \cos \theta).$$ *(Bangalore 2005)*

Let $P \, (r_1, \theta_1)$ be a point of intersection. Let ϕ_1, ϕ_2 be the angles which OP makes with the two tangents to the two curves at P.

For the curve $r = a \, (1 + \cos \theta)$ we have

$$\frac{dr}{d\theta} = -\, a \sin \theta$$

$$\therefore \qquad r \frac{d\theta}{dr} = -\frac{a(1 + \cos\theta)}{a\sin\theta} = \frac{-2\cos^2\theta/2}{2\sin\theta/2.\cos\theta/2} = -\cot\frac{\theta}{2}$$

$$\Rightarrow \qquad \tan\phi = \tan\left(\frac{\pi}{2} + \frac{\theta}{2}\right),$$

$$\Rightarrow \qquad \phi = \frac{\pi}{2} + \frac{\theta}{2}$$

Hence $$\phi_1 = \frac{\pi}{2} + \frac{\theta_1}{2}.$$

3. *Prove that the normal at any point (r, θ) of the curve $r^n = a^n \cos n\, \theta$ makes an angle $(n + 1)\, \theta$ with the initial line.*

The given curve is $r^n = a^n \cos n\, \theta$

Taking log of both the sides, we obtain

$$n \log r = n \log a + \log \cos n\, \theta$$

Differentiating,

$$\frac{n}{r}\frac{dr}{d\theta} = -\, n \tan n\theta$$

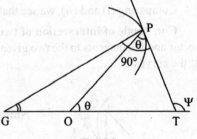

Fig. 7.5

Hence $$\tan\phi = r\frac{d\theta}{dr} = -\, n\cot n\theta = \tan\left(\frac{\pi}{2} + n\theta\right)$$

Therefore $$\phi = \frac{\pi}{2} + n\theta.$$

Let α be the angle which the normal makes with the initial line.

Then $$\psi = 90^0 + \alpha = \frac{\pi}{2} + \alpha$$

and $$\psi = \theta + \phi = \theta + \frac{1}{2}\pi + n\theta = \frac{1}{2}\pi + (n + 1)\theta$$

Then $$\frac{\pi}{2} + (n + 1)\theta = \frac{\pi}{2} + \alpha \qquad \Rightarrow \qquad \alpha = (n + 1)\,\theta$$

4. *Find the angle ϕ for the curve $a\theta = (r^2 - a^2)^{1/2} - a \cos^{-1}(a/r)$.*

Differentiating the given equation *w.r.t.* θ,

we get $$a = \frac{1}{2}(r^2 - a^2)^{-1/2}.2r\frac{dr}{d\theta} + \frac{a}{\sqrt{1 - (a/r)^2}}\left(-\frac{a}{r^2}\right)\frac{dr}{d\theta}$$

or $\qquad a\dfrac{d\theta}{dr} = \dfrac{r}{\sqrt{r^2 - a^2}} - \dfrac{a^2}{r\sqrt{r^2 - a^2}}$

$\qquad\qquad\qquad = \dfrac{\sqrt{r^2 - a^2}}{r}$

$\therefore \qquad \dfrac{r\,d\theta}{dr} = \dfrac{\sqrt{r^2 - a^2}}{a} \qquad\Rightarrow\qquad \tan\phi = \dfrac{\sqrt{r^2 - a^2}}{a}$

or $\qquad \cos\phi = \dfrac{1}{\sqrt{1 + (r^2 - a^2)/a^2}} = \dfrac{a}{r}$

$\therefore \qquad \phi = \cos^{-1}(a/r).$

5. *Show that the curve $r^n = a^n \sec(n\theta + \alpha)$ and $r^n = b^n \sec(n\theta + \beta)$ intersect at an angle which is independent of a and b.*

Taking logarithm of both sides of first curve,

we get $\qquad n\log r = n\log a + \log\sec(n\theta + \alpha)$

Differentiating

$$\dfrac{n}{r}\dfrac{dr}{d\theta} = n\tan(n\theta + \alpha)$$

$\Rightarrow \qquad \tan\phi_1 = r\left(\dfrac{d\theta}{dr}\right) = \cot(n\theta + \alpha)$

$\Rightarrow \qquad \phi_1 = \dfrac{1}{2}\pi - (n\theta + \alpha)$

Similarly for second curve, we get

$$\phi_2 = \dfrac{1}{2}\pi - (n\theta + \beta)$$

Angle of intersection of curves $= \phi_1 - \phi_2$

$$= \left\{\dfrac{1}{2}\pi - (n\theta + \alpha)\right\} - \left\{\dfrac{1}{2}\pi - (n\theta + \beta)\right\}$$

$$= \beta - \alpha, \text{ which is indenpendent of } a \text{ and } b.$$

6. *If $u = 1/r$, prove that*

$$\dfrac{d\phi}{d\theta} = \dfrac{u(d^2 u/d\theta^2) - (du/d\theta)^2}{u^2 + (du/d\theta)}$$

We have $\qquad \tan\phi = r\dfrac{d\theta}{dr}$

and $\qquad u = \dfrac{1}{r} \quad\Rightarrow\quad r = \dfrac{1}{u} \quad\Rightarrow\quad \dfrac{dr}{d\theta} = -\dfrac{1}{u^2}\dfrac{du}{d\theta}$

$$\therefore \qquad \tan\phi = \frac{1}{u}\left[-u^2\frac{d\theta}{du}\right] = -u\frac{d\theta}{du}$$

or $\qquad \tan\phi = \dfrac{-u}{du/d\theta}$

or $\qquad \phi = -\tan^{-1}\left[u/\left(\dfrac{du}{d\theta}\right)\right]$

$$\therefore \qquad \frac{d\phi}{d\theta} = -\frac{1}{1+\left[u/\left(\dfrac{du}{d\theta}\right)\right]^2}\times\frac{d}{d\theta}\left[\frac{u}{du/d\theta}\right]$$

$$= -\frac{(du/d\theta)^2}{(du/d\theta)^2+u^2}\times\frac{(du/d\theta)(du/d\theta)-u(d^2u/d\theta^2)}{(du/d\theta)^2}$$

$$= \frac{u(d^2u/d\theta^2)-(du/d\theta)^2}{u^2+(du/d\theta)^2}$$

EXERCISES

1. Find ϕ for the curves

   (i) $r = a(1-\cos\theta)$. (Cardioide)

   (ii) $r^m = a^m\cos m\theta$.

   (iii) $2a/r = 1+\cos\theta$. (Parabola)

   (iv) $r^m = a^m(\cos m\theta + \sin m\theta)$.

2. Show that the two curves

   $r^2 = a^2\cos 2\theta$ and $r = a(1+\cos\theta)$

   intersect at an anlge $3\sin^{-1}(3/4)^{1/4}$.

3. Show that the curves $r^m = a^m\cos m\theta$, $r^m = a^m\sin m\theta$ cut each other orthogonally.

   (Garhwal 2000; Guwahati 2005)

4. Find the angle between the curves

   (i) $r = a\theta$, $r\theta = a$;   **(Kanpur 2002)**

   (ii) $r = a\theta/(1+\theta)$, $r = a/(1+\theta)^2$;

   (iii) $r = a\,\text{cosec}^2(\theta/2)$, $r = b\sec^2(\theta/2)$;   **(Kuvempu 2005)**

   (iv) $r = a\log\theta$, $r = a/\log\theta$;

   (v) $r^2\sin 2\theta = 4$, $r^2 = 16\sin 2\theta$;

   (vi) $r = ae^\theta$, $re^\theta = b$.

5. Show that in the case of the curve $r = a(\sec\theta \pm \tan\theta)$, if a radius vector OPP' be drawn cutting the curve in P and P' and if the tangents at P, P' meet in T, then $TP = TP'$.

6. Show that the tangents to the cardioide

   $$r = a(1+\cos\theta)$$

   at the points whose vectorial angles are $\pi/3$ and $2\pi/3$ are respectively parallel and perpenducular to the initial line.

7. The tangents at two points P, Q lying on the same side of the initial line of the cardioide $r = a(1+\cos\theta)$, are prependicular to each other; show that the line PQ subtends an angle $\pi/3$ at the pole.

8. Show that the tangents drawn at the extermities of any chord of the cardioide $r = a(1 + \cos\theta)$ which passes through the pole are perpendicular to each other.

9. If two tangents to the cardioide $r = a(1 + \cos\theta)$ are parallel, show that the line joining their points of contact subtend an angle $2\pi/3$ at the pole.

10. Show that $\tan\phi = \left(x\dfrac{dy}{dx} - y\right) \bigg/ \left(y\dfrac{dy}{dx} + x\right)$.

ANSWERS

1. (i) $\theta/2$. (ii) $\pi/2 + m\theta$. (iii) $\pi/2 - \theta/2$. (iv) $\pi/4 + m\theta$.

4. (i) $\pi/2$. (ii) $\tan^{-1} 3$. (iii) $\pi/2$. (iv) $\tan^{-1}[2e/(e^2 - 1)]$.

   (v) $2\pi/3$. (vi) $\pi/2$.

7.7. Length of the perpendicular from pole to the tangent.

Let $OY = p$ be the perpendicular from the pole to the tangent to the curve $r = f(\theta)$ at $P(r, \theta)$.

From the $\triangle OPY$, we get

$$\frac{p}{r} = \sin\phi$$

or $\qquad p = r\sin\phi,$ $\hfill$ *(Bangalore 2006)*

Now $\qquad \tan\phi = r\dfrac{d\theta}{dr} = r \div \dfrac{dr}{d\theta}$

$\Rightarrow \qquad \sin\phi = \pm\dfrac{r}{\sqrt{[r^2 + (dr/d\theta)^2]}}.$

$\Rightarrow \qquad p = \pm\dfrac{r^2}{\sqrt{[r^2 + (dr/d\theta)^2]}}.$

which we can rewrite as

$$\frac{1}{p^2} = \frac{r^2 + (dr/d\theta)^2}{r^4} = \frac{1}{r^2} + \frac{1}{r^4}\left(\frac{dr}{d\theta}\right)^2. \qquad ...(i) \qquad \textit{(Kanpur 2003)}$$

Yet, another form of p will be obtained, if we write

$$r = 1/u,$$

so that

Substituting these values in (i), we get

$$\frac{1}{p^2} = u^2 + \left(\frac{du}{d\theta}\right)^2. \qquad\qquad \textit{(Rohilkhand 2002)}$$

7.8. Length of polar sub-tangent and polar sub-normal.

Let the line through the pole O, drawn perpendicular to the radius vector OP, meet the tangent and normal at P in points T and G respectively.

Then, *OT, OG are respectively called the polar sub-tangent and polar sub-normal at P.*

From the $\triangle OPT$, we get

$$\frac{OT}{OP} = \tan\phi$$

$\Rightarrow$ $\qquad$ $OT = OP \tan \phi = r^2 \dfrac{d\theta}{dr}$.

Hence polar sub-tangent

$$\mathbf{r^2 \dfrac{d\theta}{dr}} = -\dfrac{d\theta}{du} \text{ where } u = \dfrac{1}{r}.$$

From the $\triangle OPG$, we get

$$\dfrac{OG}{OP} = \cot\phi,$$

$\Rightarrow$ $\qquad$ $OG = OP \cot\phi$

$$= r.\dfrac{1}{r}.\dfrac{dr}{d\theta} = \dfrac{dr}{d\theta}.$$

Hence, **polar sub-normal** $= \dfrac{dr}{d\theta} = -\dfrac{1}{u^2}.\dfrac{du}{d\theta}$ where $u = \dfrac{1}{r}$

Note. The lengths PT and PG are called the lengths of the polar tangents and polar at P. We have

$$PT = OP \sec\phi = r\sqrt{[1 + \tan^2\phi]} = r\sqrt{\left[1 + r^2\left(\dfrac{d\theta}{dr}\right)^2\right]}$$

$$PG = OP \operatorname{cosec}\phi = r\sqrt{[1 + \cot^2\phi]} = \sqrt{\left[r^2 + \left(\dfrac{dr}{d\theta}\right)^2\right]}.$$

7.9. Pedal equations.

To obtain the pedal equation of a curve whose polar equation is given.

Let

$\qquad\qquad r = f(\theta)$ be the given curve. $\qquad\qquad\qquad\qquad\qquad$ ...(i)

We have $\qquad\qquad \dfrac{1}{p^2} = \dfrac{1}{r^2} + \dfrac{1}{r^4}\left(\dfrac{dr}{d\theta}\right)^2$. $\qquad\qquad\qquad\qquad$ ...(ii)

Eliminating θ between (i) and (ii), we obtain the required pedal equation of the curve.

The pedal equation is somtimes more conveniently obtained by eliminating θ and ϕ between (i)

$$\tan\phi = r\dfrac{d\theta}{dr},$$

and $\qquad\qquad p = r \sin\phi.$

EXAMPLES

1. *If $r = a(1 + \cos\theta)$, find the polar sub-tangent, polar sub-normal and the length of polar tangent and polar normal when $\theta = \tan^{-1}(3/4)$.*

We have $\qquad\qquad r = a(1 + \cos\theta);$ $\qquad\qquad \therefore \dfrac{dr}{d\theta} = (-a\sin\theta)$

Now, at $\qquad\qquad \theta = \tan^{-1}\dfrac{3}{4}$ or $\sin^{-1}\dfrac{3}{5}$ or $\cos^{-1}\dfrac{4}{5}$

$$r = a\left(1 + \frac{4}{5}\right) = \frac{9}{5}a$$

and

$$\frac{dr}{d\theta} = -\frac{3a}{5}$$

Then, polar sub-tangent $= r^2\frac{d\theta}{dr}$

$$= \left(\frac{9}{5}a\right)^2\left(-\frac{5}{3a}\right) = -\frac{27}{5}a.$$

Polar sub-normal

$$= \frac{dr}{d\theta} = -\frac{3a}{5}$$

Length of polar tangent

$$= r\sqrt{1 + \left(r\frac{d\theta}{dr}\right)^2}$$

$$= \frac{9}{5}a\sqrt{1 + \left\{\frac{9}{5}a\left(-\frac{5}{3a}\right)\right\}^2}$$

$$= \frac{9}{5}a\sqrt{10}$$

Length of polar normal

$$= \sqrt{\left\{r^2 + \left(\frac{dr}{d\theta}\right)^2\right\}}$$

$$= \sqrt{\left(\frac{9}{5}a\right)^2 + \left(\frac{-3a}{5}\right)^2} = \frac{3}{5}a\sqrt{(10)}.$$

2. *Find the perpendicular length from the pole on the tangent to the curve*
$$r\,(\theta - 1) = a\,\theta^2.$$

The given curve is $\quad r = \dfrac{a\theta^2}{\theta - 1}$

Then

$$\frac{dr}{d\theta} = \frac{2a\theta(\theta - 1) - a\theta^2}{(\theta - 1)^2}$$

$$= \frac{a\theta(\theta - 2)}{(\theta - 1)^2}$$

Now,

$$\frac{1}{p^2} = \frac{1}{r^2} + \frac{1}{r^4}\left(\frac{dr}{d\theta}\right)^2$$

$$= \frac{1}{a^2\theta^4/(\theta - 1)^2} + \frac{1}{a^4\theta^8/(\theta - 1)^4}\left[\frac{a^2\theta^2(\theta - 2)^2}{(\theta - 1)^4}\right].$$

$$= \frac{1}{a^2}\left[\frac{1}{a^2} - \frac{2}{\theta^3} + \frac{2}{\theta^4} - \frac{4}{\theta^5} + \frac{4}{\theta^6}\right]$$

3. *Find the pedal equation of the curve* $r^m = a^m \cos m\theta$.   (**Bangalore 2004**)

The given curve is

$$r^m = a^m \cos m\theta$$

Then $m \log r = \log a + \log \cos m\theta$

and $\dfrac{m}{r} \dfrac{dr}{d\theta} = -m \tan m\theta$

Hence $\tan \phi = \dfrac{1}{r} \dfrac{d\theta}{dr} = -\cot m\theta$

$$= \tan\left(\dfrac{\pi}{2} + m\theta\right)$$

$\therefore$ $\phi = \dfrac{\pi}{2} + m\theta.$

But $p = r \sin \phi = r \sin\left(\dfrac{\pi}{2} + m\theta\right)$

or $p = r \cos m\theta = r \dfrac{r^m}{a^m}$

Then $a^m p = r^{m+1},$

which is the required pedal equation.

4. *Prove that the locus of the extremity of the polar sub-normal of the curve* $r = f(\theta)$ *is the curve* $r = f'(\theta - \pi/2)$.

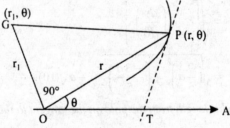

Fig. 7.6

Let P be a point on the curve. Let the line drawn through O perpendicular to the radius vector OP meets the normal at P in G. Let be (r_1, θ_1)

Then $r_1 = OG = $ polar sub-normal

$$= \dfrac{dr}{d\theta} = f'(\theta)$$

Also $\theta_1 = \angle GOA = 90° + \theta$ or $\theta = \theta_1 - \dfrac{\pi}{2}.$

Hence $r_1 = f'(\theta_1 - \pi/2)$

$\therefore$ Locus of $G(r_1, \theta_1)$ is $r = f'(\theta - \pi/2)$

5. *Prove that the locus of the extremity of the polar sub-tangent of the curve* $\dfrac{1}{r} = f(\theta)$ *is*

$$\dfrac{1}{r} + f'\left(\theta + \dfrac{\pi}{2}\right) = 0.$$

Let $P(r, \theta)$ be any point on the curve OP is the radius vector and PT the tangent to the curve at P. Draw a line OT at right angles to the radius vector OP meeting the tangent in T. Then T is extermity of the polar sub-tangent. Let T be (r_1, θ_1) and we are to find the locus of $T(r_1, \theta_1)$.

The equation of curve is

$$\frac{1}{r} = f(\theta) \quad ...(i)$$

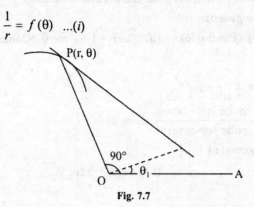

Fig. 7.7

Differentiating

$$-\frac{1}{r^2}\frac{dr}{d\theta} = f'(\theta) \qquad\qquad\qquad\qquad ...(ii)$$

$\therefore$ Polar sub-tangent $OT = r_1 = r^2 \dfrac{d\theta}{dr}$

$$= -\frac{1}{f'(\theta)}$$

or $\dfrac{1}{r_1} + f'(\theta) = 0$

Also from the figure it is evident that

$$\angle POA = \theta \text{ so we have}$$

$$\theta_1 = \angle POA - 90° = \theta - \frac{\pi}{2}$$

or $\qquad\qquad \theta = \dfrac{\pi}{2} + \theta_1$

Hence from (iii), we get

$$\frac{1}{r_1} + f'\left(\frac{\pi}{2} + \theta_1\right) = 0$$

Hence locus of T is

$$\frac{1}{r} + f'\left(\frac{\pi}{2} + \theta\right) = 0.$$

EXERCISES

1. Show that for the curve

$$\theta = \cos^{-1}\frac{r}{k} - \sqrt{\left[\frac{k^2 - r^2}{r^2}\right]},$$

the length of polar tangent is constant

2. Prove that for the curve
$$r^{-1} = A\,\theta + B,$$
the polar sub-tangent is constant.

3. Show that for the spiral $r = a\,\theta$, the polar sub-normal is constant.   (*Kanpur 2001*)

4. Find the polar sub-tangent of
   (*i*) $r = a\,(1 + \cos\theta)$. (Cardioide)   (*ii*) $2a/r = 1 + e\cos\theta$. (Conic)
   (*iii*) $r = a\,\theta^2/(\theta - 1)$.

5. Show that for the curve : $r = ae^{b\theta^2}$.
$$\frac{\text{polar sub-normal}}{\text{polar sub-tangent}}\ \text{varies as}\ \theta^2,$$

6. Find the polar sub-normal of
   (*i*) $r = a/\theta$;   (*ii*) $r = a + b\cos\theta$.

7. Find p for the curve
$$r = ae^{\theta}/(1 + \theta)^2.$$

8. Find the pedal equations of
   (*i*) $r = a/\theta$.   (*ii*) $r = ae^{\theta\cot\alpha}$
   (*iii*) $l/r = 1 + e\cos\theta$.   (*iv*) $r = a\,\theta$.
   (*v*) $r = a\,(1 + \cos\theta)$. (*Kovempu 2005, Calcutta 2005*)   (*vi*) $r = a\sin m\,\theta$.
   (*vii*) $r\left(1 - \sin\dfrac{1}{2}\theta\right)^2 = a.$ (*viii*) $r = a + b\cos\theta$.
   (*ix*) $r^m = a^m\sin m\,\theta + b^m\cos m\,\theta$.

9. Show that the pedal equation of the spiral $r = a\operatorname{sech} n\,\theta$ is of the form
$$\frac{1}{p^2} = \frac{A}{r^2} + B.$$

10. For the curve $r = \sin^2\dfrac{\theta}{2}$, prove that $\phi = \dfrac{\theta}{2}$, $p^2 = r^3$, polar sub-tangent $= \sin^2\dfrac{\theta}{2}\tan\dfrac{\theta}{2}$.

ANSWERS

4. (*i*) $2a\cos^3\dfrac{1}{2}\theta\cosec\dfrac{1}{2}\theta.$ (*ii*) $\dfrac{2a}{e\sin\theta}$ (*iii*) $\dfrac{a\theta^3}{\theta - 2}$

6. (*i*) $-a/\theta^2.$ (*ii*) $-b\sin\theta.$

7. $p = ae/(1 + \theta)\sqrt{(2 + 2\theta^2)}.$

8. (*i*) $1/p^2 = 1/r^2 + 1/a^2.$ (*ii*) $p = r\sin\alpha$

   (*iii*) $\dfrac{1}{p^2} = \dfrac{(e^2 - 1)}{l^2} + \dfrac{2}{lr}.$ (*iv*) $1/p^2 = 1/r^2 + a^2/r^4.$

   (*v*) $r^3 = 2ap^2$ (*vi*) $r^4 = p^2\,[a^2\,m^2 + (1 - m^2)\,r^2].$
   (*vii*) $ar^3 = 4p^4.$ (*viii*) $r^4 = (b^2 - a^2 + 2ar)p^3.$

   (*ix*) $r^{m+1} = p\sqrt{(a^{2m} + b^{2m})}$

7.10. Derivative of Arcs

We are familar with the determination of the circumference of a circle by finding the limit of sum of the sides of a inscribed polygon when the number of sides is indefinitely increased and length of each side approaches to zero.

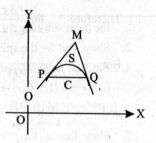

Thus the length of the circumference of circle is defined as the limit of such a sum of straight line segments.

Similarly the lengths of arcs of various curves may be determined.

We consider an arc AB, which consists of several arcs.

Fig. 7.8

i.e., $\qquad A\, P_1, P_1, P_2, P_2\, P_3, ..., P_{n-1}\, B$

Hence the arc $AB = \text{arc}\, AP_1 + \text{arc}\, P_1\, P_2 + ... + \text{arc}\, P_{n-1}\, B$ and $P_1, P_2, P_3, ..., P_{n-1}$ are points of division. Then we increase indefinitely the number of points of division; thus the consecutive points coincide along the curve and then the sum of chords approaches a limit. This limit is defined as the length of arc AB.

Let us again consider arc PQ and chord $PQ = c$.

Chord $PQ < \text{arc}\, PQ < PM + MQ$ where PM and QM are any two lines enclosing the curve.

Then $\qquad 1 < \dfrac{\text{arc}\, PQ}{\text{chord}\, PQ} < \dfrac{PM}{PQ} + \dfrac{MQ}{PQ}$

If QM and PM are at right angles, then

$$1 < \frac{\text{arc}\, PQ}{\text{chord}\, PQ} < \sin MQP + \cos MQP$$

If $Q \to P$, the chord PQ tends to the tangent PM and the angle $MQP = 0$

Hence $\qquad \displaystyle\lim_{Q \to P} \frac{\text{arc}\, PQ}{\text{chord}\, PQ} = 1.$

7.11. Differential Coefficient of Length of Arc

Let us consider the curve to be $y = f(x)$. A is some fixed point upon the curve and P anyother point on the curve whose coordinates are (x, y).

Let s be the length of arc, AP, *i.e.*, the distance along the curve from A to P and δs be the increase in the length of arc AP, *i.e.*, δy, when x and y are increased by δx and y. Let Q be the point whose coordinates are $x + \delta x$, $y + \delta y$ and arc PQ be δs.

Now $\qquad \dfrac{\delta s}{\delta x} = \dfrac{\delta s}{\delta x} \cdot \dfrac{\text{chord}\, PQ}{\text{arc}\, \delta x}$

$\qquad\qquad = \dfrac{\delta s}{\text{arc}\, PQ} \cdot \dfrac{\text{chord}\, PQ}{\delta x}$

Now chord $\quad PQ = \sqrt{(PR^2 + RQ)^2}$

$\qquad\qquad\qquad = \sqrt{\{(\delta x)^2 + (\delta y)^2\}}$

Hence $\qquad \dfrac{\delta s}{\delta x} = \dfrac{\delta s}{\text{arc}\, PQ} \cdot \dfrac{\sqrt{\{(\delta x)^2 + (\delta y)^2\}}}{\delta x}$

$\qquad\qquad = \dfrac{\delta s}{\text{arc}\, PQ} \sqrt{1 + \left(\dfrac{\delta y}{\delta x}\right)^2}$

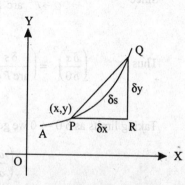

Fig. 7.9

Therefore $\quad\lim\limits_{\delta x \to 0} \dfrac{\delta s}{\delta x} = \lim\limits_{\delta x \to 0} \dfrac{\delta x}{\text{arc } PQ}\, \lim\limits_{\delta x \to 0} \sqrt{1 + \left(\dfrac{\delta y}{\delta x}\right)^2}$

But $\qquad \dfrac{\delta s}{\text{arc } PQ} = 1$ when $\delta x \to 0$

and $\qquad \lim\limits_{\delta x \to 0} \dfrac{\delta y}{\delta x} = \dfrac{dy}{dx}$

Then $\qquad \dfrac{ds}{dx} = \sqrt{\left\{1 + \left(\dfrac{dy}{dx}\right)^2\right\}}$

Similarly we can prove that

$$\dfrac{ds}{dy} = \sqrt{\left\{1 + \left(\dfrac{dx}{dy}\right)^2\right\}}$$

7.12. Derivative of Arc Length (Polar Formula)

Let the curve be $r = f(\theta)$ and P any point (r, θ) and Q another point on the curve whose coordinates are $(r + \delta r, \theta + \delta\theta)$ which is a nearby point of the curve. We denote the arc PQ by δs. PR is perpendicular to OQ. In right triangle PQR, (chord $PQ)^2 = PR^2 + RQ^2$

and $\qquad RP = OP \sin POQ$

$\qquad\qquad = r \sin \delta\theta$

and $\qquad RQ = OQ - OR = r + \delta r - r \cos \delta\theta$

$\qquad\qquad = \delta r + r(1 - \cos \delta\theta)$

Then (chord $PQ)^2 = r^2 \sin^2 \delta\theta + [\delta r + r(1 - \cos \delta\theta)]^2$

$$= r^2 \sin^2 \delta\theta + \left[\delta r + 2r \sin^2 \dfrac{\delta\theta}{2}\right]^2$$

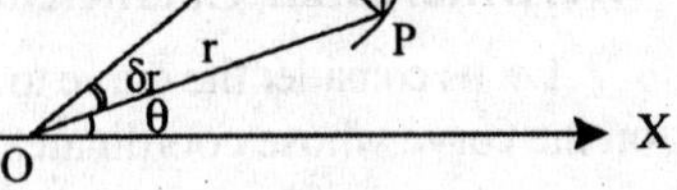

Now $\qquad \left(\dfrac{\delta s}{\delta\theta}\right)^2 = \left(\dfrac{\delta s}{\text{arc } PQ}\right)^2 \left(\dfrac{\text{chord } PQ}{\delta\theta}\right)^2$

Since $\qquad \lim\limits_{Q \to P} \dfrac{\text{chord } PQ}{\text{arc } PQ} = 1.$

Thus $\qquad \left(\dfrac{\delta s}{\delta\theta}\right)^2 = \left(\dfrac{\delta s}{\text{arc } PQ}\right)^2 \left[r^2 \left(\dfrac{\sin \delta\theta}{\delta\theta}\right)^2 + \left(\dfrac{\delta r}{\delta\theta} + \dfrac{2r \sin^2 \dfrac{\delta\theta}{2}}{\delta\theta}\right)^2\right]$

Taking limits as $\delta\theta \to 0$ we get

$$\left(\dfrac{ds}{d\theta}\right)^2 = 1\left[r^2 + \left(\dfrac{dr}{d\theta}\right)^2\right]$$

Fig. 7.10

$$\therefore \qquad \frac{ds}{d\theta} = \sqrt{\left\{ r^2 + \left(\frac{dr}{d\theta} \right)^2 \right\}}$$

Similarly we can prove that

$$\frac{ds}{dr} = \sqrt{\left\{ 1 + r^2 \left(\frac{d\theta}{dr} \right)^2 \right\}}$$

7.13. Other Formulae

(*i*) Also, $\dfrac{dy}{dx} = \tan \psi$ where ψ is the angle of the tangent with the *x*-axis.

Then $\qquad \dfrac{ds}{dx} = \sqrt{(1 + \tan^2 \psi)} = \pm \sec \psi.$

Therefore $\qquad \dfrac{dx}{ds} = \cos \psi$

(*ii*) $\qquad \dfrac{dy}{ds} = \dfrac{dy}{dx} \cdot \dfrac{dx}{ds} = \tan \psi \cdot \cos \psi$

$\therefore \qquad \dfrac{dy}{ds} = \sin \psi.$

(*iii*) If *x* and *y* are given in terms of a parameter *t*

$$x = f_1(t), y = f_2(t),$$

then $\qquad \dfrac{ds}{dt} = \dfrac{ds}{dx} \cdot \dfrac{dx}{dt}$

$$= \sqrt{\left\{ 1 + \left(\frac{dy}{dx} \right)^2 \right\}} \cdot \frac{dx}{dt}$$

$$= \sqrt{\left(\frac{dx}{dt} \right)^2 + \left(\frac{dy}{dx} \cdot \frac{dx}{dt} \right)^2}$$

$$= \sqrt{\left\{ \left(\frac{dx}{dt} \right)^2 + \left(\frac{dy}{dt} \right)^2 \right\}}$$

$\therefore \qquad \dfrac{ds}{dt} = \sqrt{\left\{ \left(\dfrac{dx}{dt} \right)^2 + \left(\dfrac{dy}{dt} \right)^2 \right\}}$

(*iv*) Also $\qquad \tan \phi = r \dfrac{d\theta}{dr}$

Hence $\qquad \cos \phi = \dfrac{1}{\sec \phi} = \dfrac{1}{\sqrt{\left\{ 1 + r^2 \left(\dfrac{d\theta}{dr} \right)^2 \right\}}}$

$$= \frac{(dr/d\theta)}{\left\{\left(\frac{dr}{d\theta}\right)^2 + r^2\right\}^{1/2}}$$

$$= \frac{dr/d\theta}{ds/d\theta} = \frac{dr}{ds}$$

Thus $\dfrac{dr}{ds} = \cos\phi.$

and $\sin\phi = \tan\phi\cos\phi = r\dfrac{d\theta}{dr}\cdot\dfrac{dr}{ds}$

$\therefore$ $r\dfrac{d\theta}{ds} = \sin\phi$

Corollary.   We have proved that

$$\sin\phi = r\frac{d\theta}{ds}$$

$\therefore$ $p = r\sin\phi = r^2\dfrac{d\theta}{ds}$

$\therefore$ $\dfrac{ds}{d\theta} = \dfrac{r^2}{p}$

Again $\cos\phi = \dfrac{dr}{ds}.$

or $\sqrt{(1 - \sin^2\phi)} = \dfrac{dr}{ds}$

$$\sqrt{\left(1 - \frac{p^2}{r^2}\right)} = \frac{dr}{ds}$$

i.e., $\dfrac{ds}{dr} = \dfrac{r}{\sqrt{(r^2 - p^2)}}$

EXAMPLES

1. *Find* $\dfrac{ds}{dx}$ *for the curve* $y = a\cos h\dfrac{x}{a}.$ *(Kuvempu 2005)*

The given curve is

$$y = a\cosh\frac{x}{a}$$

$\therefore$ $\dfrac{dy}{dx} = \sinh\dfrac{x}{a}$

Thus $\dfrac{ds}{dx} = \sqrt{1 + \sinh^2\left(\dfrac{x}{a}\right)} = \cosh\dfrac{x}{a}$

2. *Calculate* $\dfrac{ds}{dx}$ *and* $\dfrac{ds}{dy}$ *for the curve* $y = a \log \sec \dfrac{x}{a}$ *and prove that* $x = a\,\psi$. *Also prove*

that $\dfrac{d^2 x}{ds^2} = \dfrac{1}{2a} \sin \dfrac{2x}{a}.$

The given curve is

$$y = a \log \sec \frac{x}{a}$$

Then $\qquad \dfrac{dy}{dx} = \dfrac{a}{\sec \dfrac{x}{a}} . \sec \dfrac{x}{a} \tan \dfrac{x}{a} . \dfrac{1}{a}$

$$= \tan \frac{x}{a}$$

Hence $\qquad \tan \psi = \tan \dfrac{x}{a}$; *i.e.* $x = a\,\psi$ $\qquad\qquad$ ...(*i*)

Now $\qquad \dfrac{ds}{dx} = \sqrt{\left\{ 1 + \left(\dfrac{dy}{dx} \right)^2 \right\}}$

$$= \sqrt{\left(1 + \tan^2 \frac{x}{a} \right)} = \sec \frac{x}{a}$$

and $\qquad \dfrac{ds}{dy} = \sqrt{\left\{ 1 + \left(\dfrac{dx}{dy} \right)^2 \right\}} = \sqrt{\left(1 + \cot^2 \frac{x}{a} \right)}$

$$= \operatorname{cosec} \frac{x}{a}$$

Now, again

$$\frac{dx}{ds} = \cos \frac{x}{a}$$

then $\qquad \dfrac{d^2 x}{ds^2} = - \sin \dfrac{x}{a} . \dfrac{1}{a} \dfrac{dx}{ds}$

$$= - \sin \frac{x}{a} . \frac{1}{a} \cos \frac{x}{a}$$

$$= - \frac{1}{2a} \sin \frac{2x}{a}$$

3. *In the curve* $r^m = a^m \cos m\theta$

prove that $\qquad \dfrac{ds}{d\theta} = a (\sec m\theta)^{(m-1/m)}$

and $\qquad a^{2m} \dfrac{d^2 r}{ds^2} + m r^{2m-1} = 0$

The given curve is

$$r^m = a^m \cos m\,\theta \qquad \dots(i)$$

Taking logarithms of both sides and then differentiating with respect to θ, we get

$$\frac{m}{r}\frac{dr}{d\theta} = -m \tan m\theta$$

or

$$\frac{dr}{d\theta} = -r \tan m\theta \qquad \dots(ii)$$

$\therefore$

$$\frac{ds}{d\theta} = \sqrt{r^2 + \left(\frac{dr}{d\theta}\right)^2}$$

$$= \sqrt{r^2 + r^2 \tan^2 m\theta}$$

$$= r \sec m\,\theta$$

$$= \frac{a(\cos m\,\theta)^{1/m}}{\cos m\,\theta}$$

$$= a\,(\cos m\,\theta)^{(1-m)/m}$$

$$= a \sec^{(m-1)/m} m\,\theta$$

We have already proved that

$$\frac{ds}{d\theta} = r \sec m\theta = \frac{r}{\cos m\theta}$$

and

$$\frac{dr}{d\theta} = -r \tan m\theta$$

$\therefore$

$$\frac{dr}{ds} = \frac{dr}{d\theta}\cdot\frac{d\theta}{ds} = -\frac{r \tan m\theta}{a^m\, r^{1-m}}$$

$$= -a^{-m}\, r^m \tan m\,\theta$$

$$\frac{d^2 r}{ds} = \frac{d}{ds}\left(\frac{dr}{ds}\right) = \frac{d}{dr}\left(\frac{dr}{ds}\right)\frac{dr}{ds}$$

$$= -a^{-m}\, r^m \tan m\theta\cdot\frac{d}{dr}\left\{-a^{-m}\, r^m \tan m\theta\right\}$$

$$= -a^{-2m}\, r^m \tan m\theta\left[m r^{m-1} \tan m\theta + m r^m \sec^2 m\theta\,\frac{d\theta}{dr}\right]$$

$\Rightarrow$

$$= ma^{-2m}\, r^m + \tan\theta\left[r^{m-1} \tan m\theta - r^m \sec^2 m\theta\cdot\frac{1}{r}\cot m\theta\right]$$

$$= m\,a^{-2m}\, r^{2\,m-1}\,[\tan^2 m\,\theta - \tan m\,\theta \sec^2 m\,\theta \cot m\,\theta]$$

$$= ma^{-2m}\, r^{2\,m-1}\,[\tan^2 m\,\theta - \sec^2 m\,\theta]$$

$$= ma^{-2m}\, r^{2\,m-1}\,(-1)$$

$\therefore$

$$\frac{d^2 r}{ds^2} + m\,a^{-2m}\, r^{2m-1} = 0$$

or $$a^{2m}\frac{d^2r}{ds^2} + m\,r^{2m-1} = 0$$

4. *Prove that for any curve*

$$\sin^2\phi\,\frac{d\phi}{d\theta} + r\,\frac{d^2r}{ds^2} = 0$$

We know that

$$\cos\phi = \frac{dr}{ds}$$

then $-\sin\phi.\dfrac{d\phi}{ds} = \dfrac{d^2r}{ds^2}$, differentiating w. r. t. s.

or $$-\sin\phi.\frac{d\phi}{d\theta}.\frac{d\theta}{ds} = \frac{d^2r}{ds^2}$$

or $$-\sin\phi.\frac{d\phi}{d\theta}.\frac{\sin\phi}{r} = \frac{d^2r}{ds^2}$$ $\qquad$ (since $r\dfrac{d\theta}{ds} = \sin\phi$)

Hence $$r\frac{d^2r}{ds^2} + \sin^2\phi\,\frac{d\phi}{d\theta} = 0$$

EXERCISES

1. Find $\dfrac{ds}{dx}$ for the following curves :

    (*i*) $\;3\,a\,y^2 = x\,(x-a)^2$; $\qquad\qquad$ (*ii*) $y = \log_e \dfrac{e^x - 1}{e^x + 1}$.

2. For the parabola $y^2 = 4\,a\,x$, prove that

$$\frac{ds}{dx} = \sqrt{\left(\frac{a+x}{x}\right)}.$$

3. Calculate $\dfrac{ds}{dt}$ for the following curves :

    (*i*) $\;x = t^2, y = t - 1$;

    (*ii*) $\;x = a\sec t, y = b\tan t$;

    (*iii*) $x = a\,(\cos t + t\sin t), y = a\,(\sin t - t\cos t)$;

    (*iv*) $x\sin t + y\cos t = f'\,(t), x\cos - y\sin t = f''\,(t)$;

    (*v*) $\;x = e^t\sin t, y = e^t\cos t$.

4. Find $\dfrac{ds}{dy}$ for the curve $\{4\,(x^2 + y^2) - a^2\}^3 = 27\,a^4\,y^2$.

5. Calculate $\dfrac{ds}{d\theta}$ for the following curves :

    (*i*) $\;r = \log_e \sin 3\theta$ $\qquad$ (*ii*) $\;r = a\,(1 + \cos\theta)$; $\qquad$ (*iii*) $r^2 = a^2\cos 2\theta$;

    (*iv*) $\;r = a\,e^{\theta\cot\alpha}$; $\qquad\qquad$ (*v*) $\;r = a\,(\theta - 1)$; $\qquad\qquad$ (*vi*) $r = \dfrac{a}{\left(\theta^2 - 1\right)}$

6. Prove that for the ellipse $\dfrac{x^2}{a^2} + \dfrac{y^2}{b^2} = 1$, if $x = a\cos\phi$

$$\frac{ds}{d\phi} = a\sqrt{\left(1 - e^2\cos^2\phi\right)}.$$

7. Prove that in the curve $r\theta = a$

$$\frac{ds}{dr} = \frac{\sqrt{\left(r^2 + a^2\right)}}{r}$$

ANSWERS

1. (i) $\dfrac{3x + a}{2\sqrt{3}\,ax}$;　　　　　　(ii) $\dfrac{e^{2x} + 1}{e^{2x} - 1}$

2. (i) $\sqrt{\left(1 + 4r^2\right)}$;　　　　(ii) $(a^2\sin^2 t + b^2)^{1/2}\sec^2 t$;　　　(iii) $a\,t$;

   (iv) $f'(t) + f''(t)$　　　　(v) $\sqrt{2}\,e^t$.

4.　$a^{2/3}\,(a^{2/3} + 2y^{2/3})/4xy^{1/3}$;

5. (i) $\sqrt{\left\{(\log\sin 3\theta)^2 + 9\cot^2 3\theta\right\}}$;

  (ii) $2a\cos\dfrac{\theta}{2}$;　　　　　(iii) $a\sqrt{(\sec 2\theta)}$;　　　(iv) $a\,\mathrm{cosec}\,\alpha\,e^{\theta\cot\alpha}$;

  (v) $a\,(\theta^2 + 1)$;　　　　　(vi) $a\,(\theta^2 + 1)/(\theta^2 - 1)^2$

OBJECTIVE QUESTIONS

Note : *For each of the following questions, four alternatives are given for the answer. Only one of them is correct. Choose the correct allernative.*

1. Angle ϕ between the tangent and radius vector is given by:

  (a) $\tan\phi = \dfrac{1}{r}\dfrac{d\theta}{dr}$　　　　　(b) $\tan\phi = \dfrac{1}{r}\dfrac{dr}{d\theta}$

  (c) $\tan\phi = r\dfrac{dr}{d\theta}$　　　　　(d) $\tan\phi = r\dfrac{d\theta}{dr}$

2. For the equiangular spiral $r = ae^{\theta\cot\alpha}$, the angle ϕ between the tangent and the radius vector at any point on the curve is:

  (a) $\dfrac{1}{2}\alpha$　　　　　　　(b) α

  (c) 2α　　　　　　　(d) $\theta + \alpha$　　　　　*(Avadh 2003)*

3. The angle of intersection of two curves is defined as the angle between their:

  (a) normals　　　　　　(b) radius vectors

  (c) tangents　　　　　　(d) none of these

4. If ϕ_1 and ϕ_2 be the angles which the tangets at P (the point of intersection of two curves) to the two curves make with the radius vector OP, then the angle between the curves is given by:

(a) $\tan^{-1}\left(\dfrac{\tan\phi_1 + \tan\phi_2}{1 - \tan\phi_1 \tan\phi_2}\right)$

(b) $\tan^{-1}\left(\dfrac{\tan\phi_1 \sim \tan\phi_2}{1 - \tan\phi_1 \tan\phi_2}\right)$

(c) $\tan^{-1}\left(\dfrac{\tan\phi_1 \sim \tan\phi_2}{1 + \tan\phi_1 \tan\phi_2}\right)$

(d) $\tan^{-1}\left(\dfrac{\tan\phi_1 + \tan\phi_2}{1 + \tan\phi_1 \tan\phi_2}\right)$

5. Polar sub-tangent is equal to :

(a) $\dfrac{d\theta}{dr}$

(b) $r\dfrac{d\theta}{dr}$

(c) $r^2\dfrac{d\theta}{dr}$

(d) $\dfrac{1}{r}\dfrac{d\theta}{dr}$

(Garhwal 2002 ; Awadh 2002 ; Kanpur 2001)

6. Polar sub-normal is equal to :

(a) $\dfrac{dr}{d\theta}$

(b) $\dfrac{d\theta}{dr}$

(c) $r\dfrac{dr}{d\theta}$

(d) $r\dfrac{d\theta}{dr}$ *(Awadh 2003)*

7. Which of the following formulae is incorrect?

(a) $\dfrac{dx}{ds} = \sin\psi$

(d) $\dfrac{dy}{ds} = \sin\psi$

(c) $\dfrac{dx}{ds} = \cos\psi$

(d) $\dfrac{dy}{ds} = \tan\psi$

8. For the curve $r = a\,(1 - \cos\theta)$, the angle at which the radius vector cuts the curve is :

(a) θ

(b) 2θ

(c) $\theta/2$

(d) $\dfrac{\pi}{2} - \dfrac{\theta}{2}$

9. Polar sub-tangent for the curve $r = a\theta$ is :

(a) r^2/a

(b) a

(c) a/r^2

(d) $1/a$

10. Polar sub-tangent for the curve $r = a\theta$ is :

(a) r^2/a

(b) a

(c) a/r^2

(d) $1/a$

11. Polar sub-tangent for the curve $l/r = 1 + e\cos\theta$ is:

(a) $l/e\cos\theta$

(b) $l/e\sin\theta$

(c) $\dfrac{e\cos\theta}{l}$

(d) $\dfrac{e\sin\theta}{l}$

12. Polar sub-tangent for the curve $r = a\,(1 + \cos\theta)$ is :

(a) $2a\cos^2\dfrac{\theta}{2}\cot\dfrac{\theta}{2}$

(b) $2a\cos^2\dfrac{\theta}{2}\tan\dfrac{\theta}{2}$

(c) $2a\sin^2\dfrac{\theta}{2}\tan\dfrac{\theta}{2}$

(d) $2a\sin^2\dfrac{\theta}{2}\cot\dfrac{\theta}{2}$

13. Polar sub-tangent for the curve $r = a(1 - \cos\theta)$ is :

   (a) $2a\cos^2\dfrac{\theta}{2}\cot\dfrac{\theta}{2}$

   (b) $2a\cos^2\dfrac{\theta}{2}\tan\dfrac{\theta}{2}$

   (c) $2a\sin^2\dfrac{\theta}{2}\tan\dfrac{\theta}{2}$

   (d) $2a\sin^2\dfrac{\theta}{2}\cot\dfrac{0}{2}$

14. Angle of intersection of the curves $r = a(1 + \cos\theta),\ r = b(1 - \cos\theta)$

   (a) π

   (b) $\pi/2$

   (c) 0

   (d) $-\pi/2$ 　　　　　　　　　 **(Garhwal 2001)**

15. Angle of intersection of the curves $r = a(1 + \sin\theta)$ and $r = a(1 - \sin\theta)$ is:

   (a) 0

   (b) $\pi/2$

   (c) π

   (d) $-\pi/2$

16. Angle of intersection of the curves $r = a\cos\theta,\ 2r = a$ is:

   (a) $\pi/2$

   (b) $\pi/4$

   (c) $\pi/3$

   (d) 0

17. For the cardioides $r = a(1 - \cos\theta)$:

   (a) $\phi = \dfrac{1}{2}\theta$

   (b) $\phi = \theta$

   (c) $\phi = \dfrac{1}{3}\theta$

   (d) $\phi = \pi/2$

18. Pedal equation of the curve $r^n = a^n \sin n\theta$ is:

   (a) $p = r$

   (b) $p = r\sin\theta$

   (c) $p = r\sin n\theta$

   (d) $p = r\cos n\theta$

19. Pedal equation of the curve $r = a\,e^{\theta\cot\alpha}$ is :

   (a) $p = r\sin\alpha$

   (b) $p = r\cos\alpha$

   (c) $p = r$

   (d) $p = r\sin 2\alpha$

20. For any curve, we have

   (a) $\dfrac{ds}{d\theta} = \dfrac{r}{\sqrt{r^2 - p^2}}$

   (b) $\dfrac{ds}{d\theta} = \dfrac{r^2}{p}$

   (c) $\dfrac{ds}{d\theta} = \text{const.}$

   (d) $\dfrac{ds}{d\theta} = 0$

ANSWERS					
1. (d)	2. (b)	3. (c)	4. (c)	5. (c)	6. (a)
7. (a)	8. (c)	9. (a)	10. (b)	11. (b)	12. (a)
13. (c)	14. (b)	15. (b)	16. (c)	17. (a)	18. (c)
19. (a)	20. (b)				

CHAPTER
8
MEAN VALUE THEOREMS

Introduction

By now, the student must have learnt to distinguish between theorems applicable to a class of functions and those concerning some particular functions like $f(x) = \sin x$, $f(x) = \log x$ etc. The theorems applicable to a class of functions are known as *General theorems*.

Of these *Rolle's theorem* is the most fundamental.

8.1 Rolle's Theorem.

If a function f is (i) continuous in a closed interval [a, b], (ii) derivable in the open interval] a, b [and (iii) $f(a) = f(b)$, then there exists at least one value 'c' $\in$] a, b [such that $f'(c) = 0$.

(Kumaon 2002, 2004; Calicut 2004; Garhwal 2004, 2005; Mysore 2004; Calcutta 2005)

By virtue of the continuity of the function in the closed interval [a, b] it has a greatest value M and a least value m in the interval, (3.7.3 page 117) so that there are two numbers c and d such that

$$f(c) = M, \qquad f(d) = m.$$

Now either $\qquad M = m$ ....(i)

or $\qquad M \neq m$ ....(ii)

When the greatest value coincides with the least value as in case (i), the function reduces to a constant so that the derivative is equal to 0 for every value of x and therefore the theorem is true in this case.

When M and m are unequal, as in case (ii), at least one of them must be different from the equal values $f(a)$, $f(b)$. Let $M = f(c)$ be different from them. The number c being different from a and b, lies within the interval [a, b] and as such belongs to the open interval] a, b [is, in particular, derivable at c, so that

$$\lim \frac{f(c + h) - f(c)}{h} \text{ when } h \to 0$$

exists and is the same when $h \to 0$ through positive values.

Also $f(c)$ is the greatest value of the function, we have

$$f(c + h) \leq f(c)$$

whatever positive or negative values h has. Thus

$$\begin{cases} \dfrac{f(c + h) - f(c)}{h} \leq 0 \text{ for } h > 0. & ....(iii) \\[2mm] \dfrac{f(c + h) - f(c)}{h} \geq 0 \text{ for } h < 0. & ....(iv) \end{cases}$$

Let $h \to 0$ through positive values. From (iii), we get
$$f'(c) \le 0 \qquad \qquad \qquad \qquad \qquad \qquad \dots(v)$$
Let $h \to 0$ through negative values. From (iv), we get
$$f'(c) \ge 0 \qquad \qquad \qquad \qquad \qquad \qquad ..(vi)$$
The relations (v) and (vi) will both be true if, and only if
$$f'(c) = 0.$$
The same conclusion would be similarly reached if, it is the least value which differs from $f(a)$ and $f(b)$.

Hence the theorem.

Note 1. When geometrically interpreted, the conclusion of the theorem states that the ordinates of the end points A, B being equal, there is a point on the curve the tangent at which is parallel to the chord AB.   **(Gorakhpur 2001)**

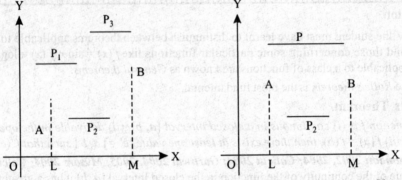

Fig. 8.1

Note.2. The conclusion of Rolle's theorem may not hold good for a function which does not satisfy any of its conditions. To illustrate this remark, we consider the function given by $y = f(x) = |x|$ in the interval $[-1, 1]$.

The function is continuous in $[-1, 1]$ and $f(1) = f(-1)$. As for the derivative, f being *not* derivable at 0, the function is *not* derivable at 0, the function is *not* derivable in $]-1, 1[$. Thus not all the 3 conditions of the Rolle's theorem are satisfied. Also the conclusion is not valid in that $f'(x)$ vanishes for *no* value of x.

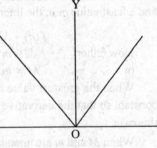

Fig. 8.2

EXAMPLES

1. *Verify Rolle's theorem for*

   (i) $f(x) = x^2$ in $[-1, 1]$,   **(Avadh 2002; Mysore 2004)**

   (ii) $f(x) = x(x + 3) e^{-x/2}$ in $[-3, 0]$.

   (i) Let $f(x) = x^2$, $x \in [-1, 1]$ so that $f(1) = 1 = f(-1)$.

   Also the function is derivable in $[-1, 1]$.

   The conditions of the theorem being satisfied, the derivative $f'(x)$ must vanish for at least one value of $x \in]-1, 1[$.

   Directly, we see that the derivative vanishes for $x = 0$ which belongs to $]-1, 1[$. Hence the verification.

   (ii) Let $f(x) = x(x + 3) e^{-x/2}$, $x \in [-3, 0]$ so that

$$f(-3) = 0 = f(0).$$

Also the function is derivable in the interval $[-3, 0]$. We have

$$f'(x) = (2x + 3) e^{-x/2} + x (x + 3) e^{-x/2} \left(-\frac{1}{2} \right)$$

$$= \frac{(-x^2 + x + 6)}{2} e^{-x/2}$$

Now $$f'(x) = 0 \Leftrightarrow -x^2 + x + 6 = 0.$$

The equation $-x^2 + x + 6 = 0$. is satisfied by $x = -2, 3$. Of these two values of x for which $f'(x)$ is zero, -2 belongs to the open interval $]-3, 0[$ under consideration. Hence the verification.

2. *Prove that if $a_0, a_1, a_2, ..., a_n$ are real number such that*

$$\frac{a_0}{n + 1} + \frac{a_1}{n} + ... + \frac{a_{n-1}}{2} + a_n = 0,$$

then there exists at least one number x between 0 and 1 such that

$$a_0 x^n + a_1 x^{n-1} + a_2 x^{n-2} + ... + a_n = 0$$

Consider the function f defined as

$$f(x) = \frac{a_0}{n + 1} x^{n+1} + \frac{a_1}{n} x^n + ... + \frac{a_{n-1}}{2} x^2 + a_n x, \; x \in [0,1]$$

f being a polynomial satisfies the following conditions :

   (i) f is continuous in $[0, 1]$

   (ii) f is derivable in $]0, 1[$

   (iii)   Since $f(0) = 0$ and $f(1) = 0$ by hypothesis,

$\therefore$ $$f(0) = f(1)$$

Hence $\exists$ some $x \in \,] 0, 1 \, [$ such that $f'(x) = 0$

i.e., $$\frac{a_0}{n + 1} (n + 1) x^n + \frac{a_1}{n} n x^{n-1} + ... + \frac{a_{n-1}}{2} . 2x + a_n = 0$$

i.e., $$a_0 x^n + a_1 x^{n-1} + ... + a_{n-1} x + a_n = 0$$

3.   *Discuss applicability of Rolle's Theorem to the function $f(x) = |x|$ in $[-1, 1]$.*

Here $$f(1) = |1| = 1 \text{ and } f(-1) = |-1| = 1$$

$\therefore$ $$f(1) = f(-1) = 1.$$

Also we observe that the function $f(x)$ is continuous for all values of x in $(-1, 1)$ except perhaps at $x = 0$.

We find $$f(0) = |0| = 0$$

Also, $$\lim_{h \to 0} f(0 + h) = \lim_{h \to 0} |0 + h| = |0| = 0$$

$$\lim_{h \to 0} f(0 - h) = \lim_{h \to 0} |0 - h| = |0| = 0$$

$\therefore$ $$\lim_{h \to 0} f(0 + h) = \lim_{h \to 0} (0 - h) = f(0).$$

Hence the function is continuous at $x = 0$ and thus the function $f(x)$ is continuous at every point in $[-1, 1]$.

Also,
$$Rf'(0) = \lim_{h \to 0} \frac{f(0+h) - f(0)}{h}$$

$$= \lim_{h \to 0} \frac{|0+h| - |0|}{h}$$

$$= \lim_{h \to 0} \frac{h-0}{h} = 1$$

and
$$Lf'(0) \lim_{h \to 0} \frac{f(0-h) - f(0)}{-h}$$

$$= \lim_{h \to 0} \frac{|0-h| - f(0)}{-h}$$

$$= \lim_{h \to 0} \frac{h-0}{-h} = -1.$$

$\therefore$ We find that $Rf'(0) \neq Lf'(0)$ and so $f'(0)$ does not exist. Hence the function $f(x)$ is not differentiable in the entire open interval $]-1, 1[$ and so Rolle's Theorem is not applicable to the given function $f(x)$ in $[-1, 1]$

4. *Discuss the applicablility of Rolle's Theorem to the function*

$$f(x) = x^2 + 1, \qquad \text{when } 0 \leq x \leq 1$$
$$= 3 - x, \qquad \text{when } 1 < x \leq 2.$$

Solution. Here $f(0) = 0^2 + 1 = 1$ and $f(2) = 3 - 2 = 1$

$\therefore$ $\qquad\qquad f(0) = 1 = f(2).$

Also we observe that the function $f(x)$ is continuous for all x in the range $(0, 2)$ except perhaps at $x = 1$.

Also $\qquad\qquad f(1) = 1^2 + 1 = 2$

Again $\qquad f(1 + 0) = \lim_{h \to 0} [3 - (1 + h)]$

$$= \lim_{h \to 0} (2 - h) = 2;$$

and $\qquad f(1 - 0) = \lim_{h \to 0} (1 - h)^2 + 1$

$$= \lim_{h \to 0} (2 - 2h + h^2) = 2$$

Hence $f(1 - 0) = f(1) = f(1 + 0)$ and so the function $f(x)$ is continuous at $x = 1$, and thus continuous in the whole interval $(0, 2)$.

Again $\qquad\qquad f'(x) = 2x, \qquad \text{when } 0 \leq x \leq 1$

$$= -1 \qquad \text{when } 1 < x \leq 2.$$

$\therefore$ $f(x)$ is differentiable in the interval $(0, 2)$ except perhaps at $x = 1$.

Now $\qquad\qquad Rf'(1) = \lim_{h \to 0} \frac{f(1+h) - f(1)}{h}$

$$= \lim_{h \to 0} \frac{\{3 - (1+h)\} - 2}{h}$$

$$= \lim_{h \to 0} \frac{2 - h - 2}{h} = \lim_{h \to 0} (-1) = -1$$

And
$$Lf'(1) = \lim_{h \to 0} \frac{f(1 - h) - f(1)}{h}$$

$$= \lim_{h \to 0} \frac{[(1 + h)^2 + 1] - 2}{-h}$$

$$= \lim_{h \to 0} \frac{2h - h^2}{h} = = \lim_{h \to 0} (2 - h) = 2.$$

∴ We find that $Rf'(1) \neq Lf'(1)$ and so $f'(1)$ does not exist. Hence the function $f(x)$ is not differentiable in the entire range $(0, 2)$ and therefore Rolle's theorem is not applicable to the given function $f(x)$ in $(0, 2)$.

5. *The function $f(x)$ is defined in $[0, 1]$ as follows*

$$f(x) = 1 \quad for \ 0 < x < \frac{1}{2}$$

$$= 2 \ for \ \frac{1}{2} \leq x \leq 1.$$

Show that $f(x)$ satisfies none of the conditions of Rolle's Theorem, yet $f'(x) = 0$ for many points in $[0, 1]$.

Solution. Here we find that

$$f\left(\frac{1}{2} - 0\right) = 1 \ whereas \ f\left(\frac{1}{2} + 0\right) = 2$$

i.e. $f\left(\frac{1}{2} - 0\right) \neq f\left(\frac{1}{2} + 0\right)$, though $f\left(\frac{1}{2} + 0\right) = 2 = f\left(\frac{1}{2}\right)$ so the function $f(x)$ is discontinuous at $x = \frac{1}{2}$.

Also the continuity being a necessary condition for the existence of a finite derivative, the function $f'(x)$ does not exist for every point in $0 \leq x \leq 1$.

Also $\quad f(0) = 1$ and $f(1) = 2$ (given)

So $\quad f(0) \leq f(1)$.

Hence all the three conditions of Rolle's Theorem are not satisfied by $f(x)$ in $[0, 1]$.

Here $f(x)$ being a function free from x (*i.e.*, constant) in $[0, 1]$, there is possibility of the value of $f'(x)$ to be zero at many points in $[0, 1]$.

6. *Verify Rolle's theorem, for the function*:

$$f(x) = (x - a)^m (x - b)^n; \ m, n \ being \ positive \ integers; \ x \in [a, b].$$

Let $\quad f(x) = (x - a)^m (x - b)^n$

f being a polynomial function satisfies the following conditions:

(*i*) f is continuous in $[a, b]$

(*ii*) f is derivable in $]a, b[$

(*iii*) $f(a) = 0 = f(b)$

Hence there exists some $c \in] a, b [$ such that

$$f'(c) = 0$$

$\Rightarrow \qquad m (c - a)^{m-1} (c - b)^n + n (c - a)^m (c - b)^{n-1} = 0$

$\Rightarrow \qquad (c - a)^{m-1} (c - b)^{n-1} (mc - mb + nc - na) = 0$

$\Rightarrow \qquad c = \dfrac{mb + na}{m + n}$

7. If $f(x)$, $\phi(x)$, $\psi(x)$ have derivatives when $a \le x \le b$, show that there is a value ξ of x lying between a and b such that

$$\begin{vmatrix} f(a) & \phi(a) & \psi(a) \\ f(b) & \phi(b) & \psi(b) \\ f'(\xi) & \phi'(\xi) & \psi'(\xi) \end{vmatrix} = 0$$

Consider the function $F(x)$, where

$$F(x) = \begin{vmatrix} f(a) & \phi(a) & \psi(a) \\ f(b) & \phi(b) & \psi(b) \\ f(x) & \phi(x) & \psi(x) \end{vmatrix}.$$

We put $x = a$ in the determinant; we observe that 1st row and third row become identical; thus $F(a) = 0$.

Similarly, $F(b) = 0$.

(i) $F(a) = F(b) = 0.$

(ii) $F(x)$ is continuous in $a \le x \le b$, since $f(x)$, $\phi(x)$, $\psi(x)$ have derivatives in $a \le x \le b$.

(iii)   It is derivable, $a < x < b$.

All the conditions of Rolle' theorem are satisfied; then there exists a point $x = \xi$, where $F'(\xi) = 0$.

Thus
$$\begin{vmatrix} f(a) & \phi(a) & \psi(a) \\ f(b) & \phi(b) & \psi(b) \\ f'(\xi) & \phi'(\xi) & \psi'(\xi) \end{vmatrix} = 0.$$

8. Let $f(x) = (x - a)(x - b)(x - c)$, $a < b < c$, show that $f'(x) = 0$ has two roots one belonging to $] a, b [$ and other belonging to $] b, c [$.

Here, $f(x)$ being a polynomial is continuous and differentiable for all real values of x.

We also have $f(a) = f(b) = f(c) = 0.$

Then by Rolle's theorem, $f'(x) = 0$ would have at least one root in $] a, b[$ and at least one root in $] b, c [$.

But $f'(x) = 0$ is a polynomial of degree two, hence $f'(x) = 0$ cannot have more than two roots.

Hence exactly one root of $f'(x) = 0$ would lie in $] a, b [$ and exactly one in $] b, c [$.

EXERCISES

1. Verify Rolle's Theorem, for the function:
    (i) $f(x) = \log [(x^2 + ab)/(a + b) x]$, $x \in [a, b]$
    (ii) $f(x) = 2 + (x - 1)^{2/3}$, $x \in [0, 2]$ *(Osmania 2004)*
2. Verify Rolle's Theorem for the function
    (i) $f(x) = \sin x / e^x$; $x \in [0, \pi]$
    (ii) $f(x) = e^x (\sin x - \cos x)$; $x \in [\pi / 4, 5\pi / 4]$.

3. By considering the function $f(x) = (x - 2) \log x$, show that equation $x \log x = 2 - x$ is satisfied by at least one value of x lying between 1 and 2.

4. Show that between any two roots of $e^x \cos x = 1$, there exists at least one root of $e^x \sin x - 1 = 0$.

5. Show that there is no real number k for which the equation $x^2 - 3x + k = 0$ has two distinct roots in $[0, 1]$.

6. If f and F are continuous in $[a, b]$ and derivable in $] a, b [$ with $F'(x) \neq 0 \forall x \in] a, b [$, prove that there exists $c \in] a, b [$ such that

$$\frac{f'(c)}{F'(c)} = \frac{f(b) - f(a)}{F(b) - F(a)}$$

[**Hint:** Consider the function

$\varphi(x) = f(x) [F(b) - F(a)] - [f(x) - f(a)] [F(x) - F(a)]$

7. Show that there is no real number k, for which the equation $x^3 - 3x + k = 0$ has two distinct roots in $[0, 1]$.

8. If f', g' are continuous and differentiable on $[a, b]$, then show that $a < c < b$.

$$\frac{f'(b) - f(a) - (b - a) F'(a)}{g(b) - g(a) - (b - a) g'(a)} = \frac{f''(c)}{g''(c)}$$

[**Hint:** Apply Rolle's theorem to the function

$\phi(x) = f(x) + (b - x) f'(x) + A (g(x) + (b - x) g'(x))$]

9. Verify Rolle's Theorem for the function $f(x) = \sin(1/x)$ in the interval $(0, \dfrac{3\pi}{4})$.

10. Any of the three conditions of Rolle's Theorem is not necessary for $f'(x)$ to vanish at some points in (a, b). Illustrate this by considering the function $f(x) = \sin(1/x)$ in $(-1, 1)$.

8.2. Lagrange's Mean Value Theorem

If a function f is (i) continuous in a closed interval [a, b] and (ii) derivable in the open interval $] a, b [$, *then there exists at least one value* $c \in] a, b [$ *such that*

$$\frac{\mathbf{f(b) - f(a)}}{\mathbf{b - a}} = \mathbf{f'(c)}.$$

(Kanpur 2001; Kumaon 1999, 2001, 2003, 2005; Avadh 2001; Manipur 1999, 2000; Srivenkateshwara 2005, Bangalore 2005, Kuvempu 2005)

To prove the theorem, we define a new function φ involving f and designed so as to satisfy the conditions of Rolle's theorem. Let

$$\varphi(x) = f(x) + Ax, x \in [a, b]$$

where A is a constant to be determined such that

$$\varphi(a) = \phi(b).$$

Thus, $\qquad f(a) + Aa = f(b) + Ab \Rightarrow A = -\dfrac{f(b) - f(a)}{b - a}$

Now, the function f is derivable in $] a, b [$. Also x is derivable and A, is a constant. Therefore, ϕ is derivable in $] a, b [$.

Thus, φ satisfies all the conditions of Rolle's theorem. There is, therefore, at least one value 'c' $\in] a, b [$ such that $\varphi'(c) = 0$, so that we have

$$0 = \varphi'(c) = f'(c) + A \Rightarrow f'(c) = -A$$

$$\Rightarrow \qquad \frac{f(b) - f(a)}{b - a} = f'(c). \qquad\qquad\qquad ...(i)$$

Another form of Statement of Lagrange's Mean Value Theorem. *If a function f is continuous in a closed interval* $[a, a+h]$ *and derivable in the open interval* $]a, a+h[$, *then there exists at least one number* 'θ' $\in]0, 1[$ *such that*

$$f(a+h) = f(a) + hf'(a+\theta h).$$

We write $b - a = h$ so that h denotes the length of the interval $[a, b]$ which may now be rewritten as $[a, a+h]$. The number, c' which lies between a and $a+h$ is greater than a and less than $a+h$ so that we may write $c = a + \theta h$, where θ is some number between 0 and 1. Thus the equation (i) becomes

$$\frac{f(a+h) - f(a)}{h} = f'(a+\theta h)$$

$\Rightarrow$ $\qquad$ $f(a+h) = f(a) + hf'(a+\theta h)$ $\qquad\qquad\qquad\qquad\qquad$ $[0 < \theta < 1]$

Geometrically interpreted, the conclusion of the theorem is that there is a point P on the curve, the tangent at which is parallel to the chord AB joining the extermities of the curve.

Let P be the point $[c, f(c)]$ on the curve such that

$$\frac{f(b) - f(a)}{b - a} = f'(c)$$

The slope of the chord $AB = \dfrac{f(b) - f(a)}{b - a}$ and that of the tangent at P $[c, f(c)]$ is $f'(c)$.

These being equal, by (i), it follows that there exists a point P on the curve, tangent at which is parallel to the chord AB.

Note 1. Now $[f(b) - f(a)]$ is the change in the function f as x changes from a to b so that $[f(b) - f(a)]/(b-a)$ is the *average rate of change* of the function over the interval $[a, b]$. Also $f'(c)$ is the *actual* rate of change of the

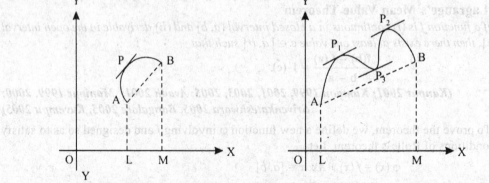
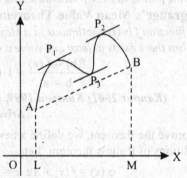

Fig. 8.3

function for $x = c$. Thus, the theorem states that the average rate of change of a function over an interval is also the actual rate of change of the function at some point of the interval. In particular, for instance, the average velocity of a particle over an interval of time is equal to the velocity at some instant belonging to the interval.

This interpretation of the theorem justifies the name 'Mean Value' for the theorem.

Note 2. If we draw some curves satisfying the conditions of the theorem, we will realise that the theorem, as stated in the geometrical form, is almost self-evident.

EXAMPLE

1. *If* $f(x) = (x-1)(x-2)(x-3); x \in [0, 4]$, *find c.* $\qquad\qquad$ **(Gorakhpur 2001)**

$\quad$ We have $\qquad$ $f(4) = 6,$ $\qquad$ $f(0) = -6,$

$$\Rightarrow \qquad \frac{f(4) - f(0)}{4 - 0} = \frac{12}{4} = 3$$

Also, $f'(x) = (x-2)(x-3) + (x-3)(x-1) + (x-1)(x-2)$

$$= 3x^2 - 12x + 11.$$

We have $\qquad \dfrac{f(4) - f(0)}{4 - 0} = f'(c)$

$$\Rightarrow \qquad 3 = 3c^2 - 12c + 11 \Rightarrow c = \frac{6 + 2\sqrt{3}}{8}, \frac{6 - 2\sqrt{3}}{8}$$

Taking $\sqrt{3} = 1.732........$ we may see that both these values of 'c' belong to the open interval] 0, 4 [, so that we have two numbers c in this case.

2. *Prove that for any quadratic function $px^2 + qx + r$, the value of θ in Lagrange's theorem is always $\dfrac{1}{2}$ whatever p, q, r, a , h may be.*

Let $\qquad f(x) = px^2 + qx + r, \quad x \in [a, a+h]$

Since f is a polynomial, it satisfies the conditions of Lagrange's Mean Value Theorem and hence there exists θ $(0 < \theta < 1)$ satisfying

$$f(a+h) - f(a) = hf'(a + \theta h)$$

$$\Rightarrow p(a+h)^2 + q(a+h) + r - pa^2 - qa - r = h[2p(a+\theta h) + q]$$

$$\Rightarrow 2pah + ph^2 + qh = 2pah + 2p\theta h^2 + qh$$

$$\Rightarrow \qquad \theta = \frac{1}{2}$$

3. *Let f be defined and continuous in $[a-h, a+h]$ and derivable in $]a-h, a+h[$. Prove that there is a real number θ between 0 and 1 such that*

$$f(a+h) - f(a-h) = h[f'(a+\theta h) + f'(a-\theta h)]$$

Define a function

$$\varphi(x) = f(a+hx) - f(a-hx), x \in [0, 1]$$

As x varies in $[0, 1]$, $a+hx$ varies in $[a, a+h]$ and $a-hx$ varies in $[a-h, a]$. Thus φ is continuous in $[0, 1]$ and derivable in $] 0, 1[$. Hence by Lagrange's Mean Value Theorem there exists θ $(0 < \theta < 1)$ such that

$$\frac{\varphi(1) - \varphi(0)}{1 - 0} = \varphi'(\theta)$$

$$\Rightarrow \qquad f(a+h) - f(a-h) = h[f'(a+\theta h) + f'(a-\theta h)]$$

where $\qquad \varphi(1) = f(a+h) - f(a-h)$

$$\varphi(0) = 0$$

4. *If a function f is such that its derivative f' is continuous in $[a, b]$ and derivable in $]a, b[$ then prove that there exists a number c $(a < c < b)$ such that*

$$f(b) = f(a) + (b-a)f'(a) + \frac{1}{2}(b-a)^2 f''(c)$$

Consider a function φ such that

$$\varphi(x) = f(b) - f(x) - (b-x)f'(x) - (b-x)^2 A$$

where A is a constant to be determined such that $\varphi\,(a) = \varphi\,(b)$.

φ is continuous in $[a, b]$ and derivable in $]\,a, b\,[$. Thus φ satisfies all the conditions of Rolle's Theorem where A is given by

$$f(b) - f(a) - (b-a)f'(a) - (b-a)^2 A = 0 \qquad\qquad ...(i)$$

$\therefore\qquad$ there exists $c \in\,]\,a, b\,[$ such that $\varphi'(c) = 0$

We have $\qquad\quad \varphi'\,(x) = -f'(x) + f'(x) - (b-x)f''(x) + 2\,(b-x)\,A$

$$= -(b-x)f''\,(x) + 2\,(b-x)\,A$$

$$\varphi'(c) = 0 \Rightarrow -(b-c)f''(c) + 2\,(b-c)\,A = 0$$

$\Rightarrow\qquad\qquad (b-c)\,(2A - f''\,(c)) = 0$

$\Rightarrow\qquad\qquad\quad A = \dfrac{1}{2}f''(c) \qquad\qquad \{b \ne c\}$

From (i), we have

$$f(b) = f(a) + (b-a)f'(a) + \frac{1}{2}(b-a)^2\,f''(c)$$

5. *Assuming f'' to be continuous in $[a, b]$ show that*

$$f(c) - \frac{b-c}{b-a}\,f(a) - \frac{c-a}{b-a}\,f(b) = \frac{1}{2}\,(c-a)\,(c-b)f''(\xi)$$

where c and ξ both lie in $[a, b]$

We define a function $F(x)\ \forall\, x \in [a, b]$ as follows:

$$F(x) = (b-a)f(x) - (b-x)f(a) - (x-a)f(b) - (x-a)\,(x-b)\,(b-a)\,A$$

where A is a constant to be determined such that $F(c) = 0$

i.e. $(b-a)f(c) - (b-c)f(a) - (c-a)f(b) - (c-a)\,(c-b)\,(b-a)\,A = 0 ...(i)$

Also $\qquad\qquad F(a) = 0 = F(b)$

The function $F(x)$ is continuous in $[a, c]$ and $[c, b]$ and derivable in $]\,a, c\,[$ and $]\,c, b[$ respectively. Thus F satisfies all the conditions of Rolle's theorem in $[a, c]$ and $[c, b]$.

There exists $\xi_1 \in\,]\,a, c\,[$ and $\xi_2 \in\,]\,c, b\,[$ such that

$$F'\,(\xi_1) = 0 \text{ and } F'\,(\xi_2) = 0$$

$$F'\,(x) = (b-a)f'(x) + f(a) - f(b) - (b-a)\,A\,[2x - (a+b)]$$

F' is continuous in $[a, b]$ and derivable in $]a, b[$, hence it is continuous in $[\xi_1, \xi_2]$ and derivable in $]\,\xi_1, \xi_2\,[\ \{a < \xi_1 < c < \xi_2 < b\}$ and

$$F'\,(\xi_1) = 0 = F'\,(\xi_2)$$

$\therefore$ F' satisfies all the conditions of Rolle's Theorem in $[\xi_1, \xi_2]$.

There exist some $\xi \in\,]\,\xi_1, \xi_2\,[$ such that

$$F''(\xi) = 0$$

$\Rightarrow\quad (b-a)f''\,(\xi) - 2A\,(b-a) = 0$

$\Rightarrow\qquad (b-a)\,(f''(\xi) - 2A) = 0$

$$\Rightarrow \qquad A = \frac{1}{2}f''(\xi)$$

Substituting this value in (i), we get:

$$(b-a)f(c)-(b-c)f(a)-(c-a)f(b)-\frac{1}{2}(c-a)(c-b)(b-a)f''(\xi)=0$$

$$\Rightarrow \quad f(c)-\frac{(b-c)}{b-a}f(a)-\frac{c-a}{b-a}f(b)=\frac{1}{2}(c-a)(c-b)f''(\xi)$$

EXERCISES

1. Verify the mean value theorem for

   (i) $f(x) = \log x$ in $[1, e]$,   *(Kanpur 2003; Manipur 2002)*

   (ii) $f(x) = x^3$ in $[a, b]$,   *(Kanpur 2001)*

   (iii) $f(x) = lx^2 + mx + n$ in $[a, b]$.

2. Find 'c' of the mean value theorem, if

   $$f(x) = x\,(x-1)\,(x-2); \; x \in \left[0, \frac{1}{2}\right]$$

3. Find 'c' so that $f'(c) = [f(b) - f(a)]/(b - a)$ in the following cases :

   (i) $f(x) = x^2 - 3x - 1; a = -11/7, b = 13/7,$

   (ii) $f(x) = \sqrt{(x^2 - 4)}; \; a = 2, b = 3,$

   (iii) $f(x) = e^x; a = 0, b = 1.$

4. Exaplain the failure of the theorem in the interval $[-1, 1]$, when
   $$f(x) = 1/x, (x \neq 0); f(0) = 0.$$

5. Deduce Lagrange's Mean Value Theorem from Rolle's Theorem by considering the derivable function $\varphi(x) = f(x) - f(a) - A(x - a)$ where A is a constant to be determined such that $\varphi(b) = 0$.

6. A twice differentiable function f is such that $f(a) = f(b) = 0$ and $f(c) > 0$ for $a < c < b$. Prove that there is at least one value φ between a and b for which $f''(\varphi) = 0$

7. If f', F are continuous and differentiable, then show that for $a < c < b$
   $$\frac{f(b) - f(a) - (b - a)\,f'(a)}{F(b) - F(a) - (b - a)\,F'(a)} = \frac{f''(c)}{F''(c)}$$

ANSWERS

2. $c = (6 - \sqrt{21})/6$

3.   (i) $\dfrac{1}{7}$ (ii) $\sqrt{5}$ (iii) $\log(e - 1)$

8.3. Meaning of the Sign of Derivative.

We consider a function f which satisfies the conditions of the Lagrange's mean value theorem in $[a, b]$. Let x_1, x_2 be any two points of $[a, b]$ such that $x_1 < x_2$.

Applying the mean value theorem to the interval $[x_1, x_2]$, we see that there exists $\xi \in \,]\,x_1, x_2\,[$ such that

$$f(x_2) - f(x_1) = (x_2 - x_1) f'(\xi).$$ ...(i)

8.3.1. *Let* $f'(x) = 0 \,\forall\, x \in \,]\, a, b \,[.$

From (i), we get $f(x_2) = f(x_1).$

As x_1, x_2 are any two members of $[a, b]$, it follows that every two values of the function are equal and as such f is a constant function is the zero function.

This is the converse of the theorem, "Derivative of a constant function is the zero function."

Cor. If two functions f and F have the same derivative $\forall\, x \in \,]\, a, b \,[$, then they differ only by a constant.

Now $\varphi = f - F$

$\Rightarrow$ $\varphi'(x) = f'(x) - F'(x) = 0 \,\forall\, x \in \,]\, a, b \,[$

$\Rightarrow$ φ is a constant function.

8.3.2. *Let* $f'(x) > 0 \,\forall\, x \in \,]\, a, b \,[.$

From (i) we get

$$f(x_2) - f(x_1) > 0 \Rightarrow f(x_2) > f(x_1)$$

for, $x_2 - x_1$ and $f'(\xi)$ are both positive.

Thus, $f'(x) > 0 \,\forall\, x \in \,]\, a, b \,[\, \Rightarrow f$ is a strictly increasing function in $[a, b]$.

Cor. The function f is strictly increasing in $[a, b]$ if

$$f'(x) > 0 \,\forall\, x \in \,]\, a, b \,[$$

except for a finite number of points where it is zero.

We consider the case of the derivative becoming 0 at *only* one point $\alpha \in \,]\, a, b \,[.$

As $f'(x) > 0 \,\forall\, x \in \,]\, a, \alpha \,[$ and $\forall\, x \in \,]\, \alpha, b \,[$ the function f is strictly increasing in $[a, \alpha]$ and $[\alpha, b]$ and as such also in $[a, b]$.

The result may now be extended to the case when the derivative is 0 at any finite number of points.

8.3.3. *Let* $f'(x) < 0 \,\forall\, x \in \,]\, a, b \,[.$

From (i), we get

$$f(x_2) - f(x_1) < 0 \Rightarrow f(x_2) < f(x_1)$$

for $x_2 - x_1$ is positive and $f'(\xi)$ is negative.

Thus $f'(x) < 0 \,\forall\, x \in \,]\, a, b \,[\, \Rightarrow f$ is a strictly decreasing function in $[a, b]$.

Cor. The function f is strictly decreasing in $[a, b]$ if $f'(x) < 0 \,\forall\, x \in \,]\, a, b \,[$ *except* for a finite number of points where the derivative is zero.

Note. The results obtained in $\xi\, 8.3.2$ and $\xi\, 8.3.3$ will be seen to play a very important part in the chapters on Maxima and Minima, Concavity, points of Inflexion and Curve tracing.

8.4. Graphs of hyperbolic functions

Graph of y = sin h x. We have $\forall\, x \in \mathbf{R}$

$$\frac{d(\sinh x)}{dx} = \cosh x\,.$$

Now $\forall\, x \in \,]-\infty, \infty\,[,\ e^x > 0$ and $e^{-x} > 0$

so that $\cosh x = \dfrac{1}{2}(e^x + e^{-x}) > 0$

Thus sinh x is strictly increasing in $]-\infty, \infty[$. Also $\sinh 0 = 0$ and

$\sinh x < 0 \ \forall x \in \ ]-\infty, 0\ [$ and $> \ \forall 0 \in \ ]0, \infty[$.

Finally

$$\lim_{x \to +\infty} \sinh x = \lim_{x \to +\infty} \frac{e^x - e^{-x}}{2} = +\infty$$

$$\lim_{x \to -\infty} \sinh x = \lim_{x \to -\infty} \frac{e^x - e^{-x}}{2} = -\infty$$

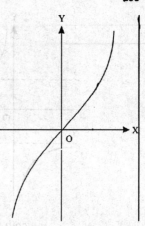

Fig. 8.4

It follows that the graph of $y = \sinh x$ is as given in Fig. 8.4.

Graph of y = cosh x. We have $\forall \ x \in R$

$$\frac{d(\cosh x)}{dx} = \sinh x.$$

Now $\sinh x < 0 \ \forall x \in \ ]-\infty, 0\ [$ and $> 0 \ \forall x \in \ ]0, \infty[$.

Thus $\cosh x$ is strictly decreasing in $[-\infty, 0]$ and strictly increasing in $[0, \infty]$.

Also $\lim_{x \to -\infty} \cosh x = \infty = \lim_{x \to \infty} \cosh x.$

Finally $\cosh 0 = 1$. The graph of $y = \cosh x$ is given in Fig. 8.5.

Graph of y = tanh x. We have $\forall \ x \in R$

$$\frac{d(\tanh x)}{dx} = \mathrm{sech}^2 x$$

Thus $\tanh x$ is strictly increasing in $]-\infty, \infty[$.

Also $\tanh 0 = 0$ and

$$\lim_{x \to \infty} \tanh x = \lim_{x \to \infty} \frac{e^x - e^{-x}}{e^x + e^{-x}}$$

$$= \lim_{x \to +\infty} \frac{1 - e^{-2x}}{1 + e^{-2x}} = 1$$

$$\lim_{x \to -\infty} \tanh x = \lim_{x \to -\infty} \frac{e^x - e^{-x}}{e^x + e^{-x}}$$

$$= \lim_{x \to -\infty} \frac{e^{2x} - 1}{e^{2x} + 1} = -1$$

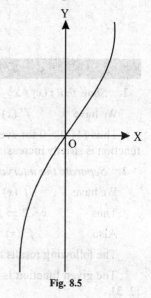

Fig. 8.5

The graph of $y = \tanh x$ is as given in Fig. 8.6.

The line $y = x$ is tangent to $y = \tanh x$ at $(0, 0)$. Also $y = 1$ and $y = -1$ are the two asymptotes of the curve $y = \tanh x$.

Graphs of y = coth x, y = sech x, y = cosech x.

We geve below the graphs without giving the details which are left to the reader.

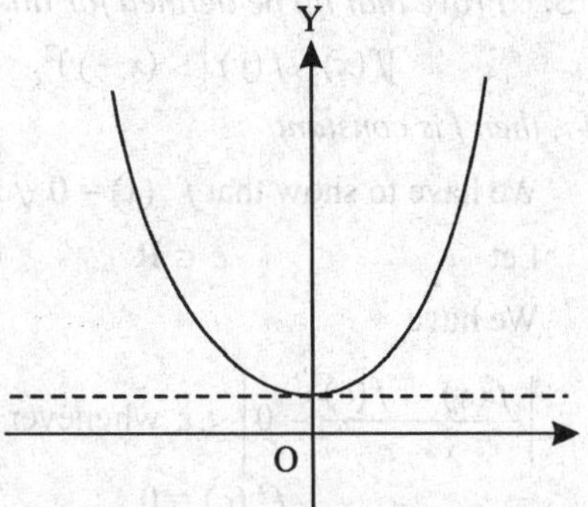

Fig. 8.6

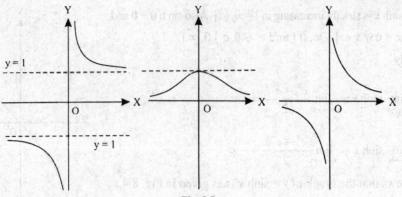

Fig. 8.7

$y = \coth x$ $y = \operatorname{sech} x$ $y = \operatorname{cosech} x$

Each of these curves possesses asymptotes. Also the asymptotes of

$$y = \coth x \text{ are } x = 0, y = 1 \text{ and } y = -1;$$
$$y = \operatorname{sech} x \text{ is } y = 0,$$
$$y = \operatorname{cosech} x \text{ are } y = 0 \text{ and } x = 0.$$

EXAMPLES

1. *Show that* $f(x) = x^3 - 3x^2 + 3x + 2$ *is strictly increasing in every interval.*

We have $f'(x) = 3x^2 - 6x + 3 = 3(x-1)^2.$

Thus $f'(x) > 0$ for every value of x except one point *viz.* 1; where it is zero. Hence the given function is strictly increasing in every interval.

2. *Separate the intervals in which* $f(x) = 2x^3 - 15x^2 + 36x + 1$ *is increasing or decreasing.*

We have $f'(x) = 6x^2 - 30x + 36 = 6(x-2)(x-3).$

Thus $x < 2 \Rightarrow f'(x) > 0;\ 2 < x < 3 \Rightarrow f'(x) < 0;\ x > 3 \Rightarrow f'(x) > 0$

Also $f'(x) = 0$ for $x = 2$ and 3.

The following results now follow.

The given function is strictly increasing in $]-\infty, 2\,[$, and $[3, \infty\,[$; and strictly decreasing in $[2, 3]$.

3. *Prove that if f be defined for all real x such that*

$$|f(x) - f(y)| < (x - y)^2$$

then f is constant.

We have to show that $f'(x) = 0 \; \forall \; x \in \mathbf{R}.$

Let $c \in \mathbf{R}.$

We have

$$\left| \frac{f(x) - f(c)}{x - c} - 0 \right| < \varepsilon \text{ whenever } |x - c| < \varepsilon \qquad \text{(using given condition)}$$

$\Rightarrow$ $f'(c) = 0$ (Taking $\delta = \varepsilon$)

$\Rightarrow$ f is a constant function.

4. *Find the interval in which the function* $f(x) = \sin(\log_e x) - \cos(\log_e x)$ *is strictly increases.*

Here domain is $x > 0$ as $\log_e x$ exists when $x > 0.$

$$f'(x) = \frac{\cos(\log_e x) + \sin(\log_e x)}{x}$$

$$= \frac{\sqrt{2}\left\{\sin\dfrac{\pi}{4}\cos(\log_e x) + \cos\dfrac{\pi}{4}\sin(\log_e x)\right\}}{x}$$

$$= \frac{\sqrt{2}\sin\left(\dfrac{\pi}{4} + \log_e x\right)}{x}$$

$\therefore f'(x)$ strictly increases when $f'(x) \geq 0$,

i.e., $\quad \sin\left(\dfrac{\pi}{4} + \log x\right) \geq 0$

$\Rightarrow \qquad\qquad 2n\pi \leq \dfrac{\pi}{4} + \log x \leq (2n+1)\pi$

$\Rightarrow \qquad\qquad 2n\pi - \dfrac{\pi}{4} \leq \log x \leq 2n\pi + \pi - \dfrac{\pi}{4}$

$\Rightarrow \qquad\qquad e^{2n\pi - \pi/4} \leq x \leq e^{2n\pi + \frac{3\pi}{4}}$

$\therefore \quad f(x)$ is strictly increasing when

$$x \in \left[e^{2n\pi - \frac{\pi}{4}}, e^{2n\pi + \frac{3\pi}{4}}\right]$$

5. Let $g(x) = f(x) + f(1-x)$ *and* $f''(x) > 0$, $\forall x \in (0, 1)$. *Find the intervals of increase and decrease of* $g(x)$.

We have $\quad g(x) = f(x) + f(1-x)$

Then $\qquad\qquad g'(x) = f'(x) - f'(1-x)$

We are given that $f''(x) > 0$, $\forall x \in (0, 1)$ it clearly means that $f'(x)$ would be increasing on $(0, 1)$ Hence two cases arise:

Case I. $x > 1 - x$ and $f'(x)$ is increasing.

$\Rightarrow \qquad\qquad f'(x) > f'(1-x)$, $\forall\, x > \dfrac{1}{2}$

or $\qquad\qquad g'(x) = f'(x) - f'(1-x) > 0$, $\forall\, \dfrac{1}{2} < x < 1$

$\therefore \quad g(x)$ is increasing in $\left(\dfrac{1}{2}, 1\right)$.

Case II. $x < 1 - x$ and $f'(x)$ is increasing.

$\Rightarrow \qquad\qquad f'(x) < f'(1-x)$, $\forall x < \dfrac{1}{2}$

or $\qquad\qquad g'(x) = f'(x) - f'(1-x) < 0$, $\forall\, 0 < x < \dfrac{1}{2}$

$\therefore \quad g(x)$ is decreasing in $\left(0, \dfrac{1}{2}\right)$

6. *Show that*

(a) $x/(1 + x) < \log(1 + x) < x \ \forall x > 0$

(b) $x - \dfrac{x^2}{2} < \log(1 + x) < x - \dfrac{x^2}{2(1 + x)} \ \forall x > 0$

(a) Now $f(x) = \log(1 + x) - \dfrac{x}{(1 + x)}$, $x > 0$

$\Rightarrow \qquad f'(x) = \dfrac{1}{1 + x} - \dfrac{1}{(1 + x)^2} = \dfrac{x}{(1 + x)^2}$

$\Rightarrow \qquad f'(x) > 0 \ \forall x > 0$ and is 0 for $x = 0$

$\Rightarrow f$ is strictly increasing in $[0, \infty[$,

Also $\qquad f(0) = 0$.

It follows that $f(x) > f(0) = 0 \ \forall x > 0$

$\Rightarrow \qquad \log(1 + x) > \dfrac{x}{1 + x}$ ...(i)

Again $\qquad F(x) = x - \log(1 + x)$, $x > 0$

$\Rightarrow \qquad F'(x) = 1 - \dfrac{1}{1 + x} = \dfrac{x}{1 + x}$.

$\Rightarrow \qquad F'(x) > 0 \ \forall x > 0$ and is 0 for $x = 0$

$\Rightarrow F(x)$ is strictly increasing in the interval $[0, \infty[$

Also $\qquad F(0) = 0$

Thus $\quad F(x) > F(0) = 0 \Rightarrow x > \log(1 + x) \ \forall x > 0$ ...(ii)

From (i) and (ii) we get the required result.

(b) Consider $f(x) = \log(1 + x) - x + \dfrac{x^2}{2}$

$\qquad f'(x) = \dfrac{1}{1 + x} - 1 + x$

$\qquad = \dfrac{x^2}{1 + x} > 0 \ \forall x > 0$

$\Rightarrow f(x)$ is an increasing function for $x > 0$

$\Rightarrow \qquad f(x) > f(0) = 0$

$\Rightarrow \qquad \log(1 + x) > x - \dfrac{x^2}{2}$ for $x > 0$ ...(i)

Similarly by considering the function

$$F(x) = x - \dfrac{x^2}{2(1 + x)} - \log(1 + x)$$

we can show that

$$\log(1 + x) < x - \dfrac{x^2}{2(1 + x)}$$

From (*i*) and (*ii*) we get the result.

7. *Using Lagrange's mean value theorem show that*

$$\frac{x}{1+x} < \log(1+x) < x \quad x > 0$$

Hence show that $0 < [\log(1+x)]^{-1} - x^{-1} < 1 \,\forall\, x > 0$

Consider $f(x) = \log(1+x)$ in $[0, x]$

$f(x)$ satisfies the condition of Lagrange's mean value theorem in $[0, x]$, thus there exists θ ($0 < \theta < 1$) such that

$$\frac{f(x) - f(0)}{x - 0} = f'(\theta x)$$

$\Rightarrow$ $\dfrac{1}{x} \log(1+x) = \dfrac{1}{1 + \theta x}$

Here $0 < \theta < 1$ and $x > 0$

Now $0 < \theta < 1, \quad x > 0 \Rightarrow \theta x < x$

$\Rightarrow$ $1 + \theta x < 1 + x$

$\Rightarrow$ $\dfrac{1}{1 + \theta x} > \dfrac{1}{1 + x}$

$\Rightarrow$ $\dfrac{x}{1 + \theta x} > \dfrac{x}{1 + x}$ ...(*i*)

Again $0 < \theta < 1, x > 0$

$\Rightarrow$ $\theta x > 0$

$\Rightarrow$ $1 + \theta x > 1$

$\Rightarrow$ $\dfrac{1}{1 + \theta x} < 1 \Rightarrow \dfrac{x}{1 + \theta x} < x$ ...(*ii*)

(*i*) and (*ii*) give

$$\frac{x}{1+x} < \log(1+x) < x$$

$\Rightarrow$ $\dfrac{1+x}{x} > [\log(1+x)]^{-1} > \dfrac{1}{x}$

$\Rightarrow$ $\dfrac{1}{x} + 1 > [\log(1+x)]^{-1} > \dfrac{1}{x}$

$\Rightarrow$ $1 > [\log(1+x)]^{-1} - x^{-1} > 0$

8. *Show that* $\dfrac{v-u}{1+v^2} < \tan^{-1} v - \tan^{-1} u < \dfrac{v-u}{1+v^2}$, $0 < u < v$ *and deduce that*

$\dfrac{\pi}{4} + \dfrac{3}{25} < \tan^{-1} \dfrac{4}{3} < \dfrac{\pi}{4} + \dfrac{1}{6}$ **(Gujrat 2005, Nagpur 2005)**

Let $f(x) = \tan^{-1} x, u < x < v$

$\Rightarrow$ $f'(x) = \dfrac{1}{1 + x^2}$

By Lagrange's Mean Value Theorem, there exists $c \in\,] u, v\, [$ such that

$$\frac{f(v) - f(u)}{v - u} = f'(c)$$

$$\Rightarrow \quad \frac{\tan^{-1} v - \tan^{-1} u}{v - u} = \frac{1}{1 + c^2}$$

$$u < c \Rightarrow u^2 < c^2 \Rightarrow 1 + u^2 < 1 + c^2$$

$$\Rightarrow \quad \frac{1}{1 + u^2} > \frac{1}{1 + c^2}$$

$$\therefore \quad \frac{\tan^{-1} v - \tan^{-1} u}{v - u} < \frac{1}{1 + u^2}$$

$$\Rightarrow \quad \tan^{-1} v - \tan^{-1} u < \frac{v - u}{1 + u^2} \qquad \qquad \text{...(}i\text{)}$$

Again $\qquad\qquad c < v \Rightarrow c^2 < v^2 \Rightarrow 1 + c^2 < 1 + v^2$

$$\Rightarrow \quad \frac{1}{1 + c^2} > \frac{1}{1 + v^2}$$

$$\Rightarrow \quad \frac{\tan^{-1} v - \tan^{-1} u}{v - u} > \frac{1}{1 + v^2}$$

$$\Rightarrow \quad \tan^{-1} v - \tan^{-1} u > \frac{v - u}{1 + v^2} \qquad \qquad \text{....(}ii\text{)}$$

From (*i*) and (*ii*), we have

$$\frac{v - u}{1 + v^2} < \tan^{-1} v - \tan^{-1} u < \frac{v - u}{1 + u^2}$$

Let $\qquad\qquad v = \frac{4}{3}, u = 1$

$$\therefore \qquad \frac{3}{25} < \tan^{-1} \frac{4}{3} - \tan^{-1} 1 < \frac{1}{6}$$

$$\Rightarrow \qquad \frac{3}{25} + \frac{\pi}{4} - < \tan^{-1} \frac{4}{3} < \frac{1}{6} + \frac{\pi}{4}$$

9. *Show that* $\dfrac{\tan x}{x} > \dfrac{x}{\sin x}$ *for* $0 < x < \pi/2$

We have to show that

$$\frac{\tan x \sin x - x^2}{x \sin x} > 0 \text{ for } 0 < x < \pi/2$$

Since $x \sin x > 0$ for $0 < x < \dfrac{\pi}{2}$; it is enough to show that

$$\tan x \sin x - x^2 > 0, \ 0 < x < \pi/2$$

Let $\qquad\qquad f(x) = \tan x \sin x - x^2, \ 0 < x < \pi/2$

$$f'(x) = \sin x \sec^2 x + \sin x - 2x$$

$$f''(x) = \cos x \sec^2 x + \sin x . 2 \sec^2 x \tan x + \cos x - 2$$

$$= \sec x + \cos x - 2 + 2 \sin x \tan x \sec^2 x$$

$$= \left(\sqrt{\sec x} - \sqrt{\cos x}\right)^2 + 2\sin x \tan x \sec^2 x > 0 \text{ for } 0 < x < \pi/2$$

$\therefore f'$ is strictly increasing in $[0, \pi/2]$. Also $f'(0) = 0$

$\Rightarrow \qquad\qquad f'(x) > 0$ for $0 < x < \pi/2$

$\Rightarrow f$ is strictly increasing in $[0, \pi/2]$. Also $f(0) = 0$

$\Rightarrow \qquad\qquad f(x) > 0$ for $0 < x < \pi/2$

$\Rightarrow \qquad \tan x \sin x - x^2 > 0$ for $0 < x < \pi/2$

$\therefore \qquad \dfrac{\tan x \sin x - x^2}{x \sin x} > 0,\ 0 < x < \pi/2$

$\Rightarrow \qquad \dfrac{\tan x}{x} > \dfrac{x}{\sin x},\ 0 < x < \pi/2$

10. *Show that* $\dfrac{2}{\pi} < \dfrac{\sin x}{x} < 1,\ 0 < x < \pi/2$

Let $\qquad\qquad f(x) = \begin{cases} \dfrac{\sin x}{x}, & x \neq 0 \\ 1, & x = 0 \end{cases}$

f is continuous in $[0, \pi/2]$ and derivable in $]0, \pi/2[$

$$f'(x) = \frac{x \cos x - \sin x}{x^2}$$

Let $\qquad\qquad F(x) = x \cos x - \sin x,\ x \in [0, \pi/2]$

$\qquad\qquad\qquad F'(x) = \cos x - x \sin x - \cos x$

$\qquad\qquad\qquad\qquad = -x \sin x < 0,\ x \in]0, \pi/2[$

$\Rightarrow \quad F$ is strictly decreasing in $[0, \pi/2]$

$\Rightarrow \qquad \cdot F(x) < F(0) = 0,\ x \in [0, \pi/2]$

$\Rightarrow \qquad\qquad f'(x) < 0,\ x \in]0, \dfrac{\pi}{2}]$

$\Rightarrow \quad f$ is strictly decreasing in $[0, \pi/2]$

$\Rightarrow \qquad f(0) > f(x) > f(\pi/2)$ for $0 < x < \pi/2$.

$\Rightarrow \qquad\qquad 1 > \dfrac{\sin x}{x} > \dfrac{1}{\pi/2}$

$\Rightarrow \qquad\qquad \dfrac{2}{\pi} < \dfrac{\sin x}{x} < 1$ for $0 < x < \pi/2$

EXERCISES

1. Show that the functions
   (i) $f(x) = \sin hx$ and $f(x) = \tan hx$ are strictly increasing in $]-\infty, \infty[$.
   (ii) $f(x) = \cosh x$ is strictly decreasing in $]-\infty, 0[$ and strictly increasing in the $[0, \infty[$.

2. Examine the functions $f(x) = \coth x$, $\sech x$, $\cosech x$ in respect of monotonicity.

3. Separate the intervals in which the polynomials.
   (i) $-x^3 + 7x^2 - 8x - 18$ \qquad (ii) $x^3 + 8x^2 + 5x - 2$
   (iii) $(x-2)^2(x+1)$ \qquad\qquad (iv) $x^4 - 13x^2 + 36$

(v) $-2x^4 + 3x^2 - 1$ (vi) $(4 - x^2)^2$

are increasing or decreasing.

4. Find the greatest and least values of the function
$$f(x) = x^3 - 9x^2 + 24x \text{ in } [0, 6].$$

5. Separate the intervals in which the rational function

$f(x) = (x^2 + x + 1)/(x^2 - x + 1)$ is increasing or decreasing.

6. Determine the intervals in which the function

$f(x) = (x^4 + 6x^3 + 17x^2 + 32x + 32) e^{-x}$ is increasing or decreasing.

7. Show that

(i) $x/\sin x$ increases in the interval $]0, \pi/2[$.

(ii) $x/\tan x$ decreases in the interval $]0, \pi/2[$.

(iii) the equation $\tan x - x = 6$ has one and only one root in the interval $] -\pi/2, \pi/2[$.

(iv) $\tan^{-1} x > \dfrac{x}{1 + \dfrac{1}{3}x^2}$ if $x \in] 0, \infty [$.

8. Show that $f(x) = x - \sin x$ is an increasing function throughout every interval. Determine for what values of a, $ax - \sin x$ is an increasing function.

9. Show that $x^{-1} \log (1 + x)$ decreases as x increases from 0 to ∞.

10. Show that $\forall x > 0$

$$x - \frac{x^2}{2} + \frac{x^3}{3(1 + x)} < \log(1 + x) < x - \frac{x^2}{2} + \frac{x^3}{3}$$

11. Show that

(i) $1 + x < e^x < 1 + xe^x \ \forall x \geq 0$

(ii) e^{-x} lies between $1 - x$ and $1 - x + \dfrac{x^2}{2}, \ \forall \ x \in \mathbf{R}.$

(iii) $\sin x$ lies between $x - \dfrac{x^3}{6}$ and $x - \dfrac{x^3}{6} + \dfrac{x^5}{120}, \forall x \in \mathbf{R}.$

12. Let $f(x) = x^3 + ax^2 + bx + 5 \sin^2 x$ be an increasing function on the set R. Then show that $a^2 - 3b + 15 = 0.$

13. Find the set of all values of a for which

$$f(x) = \left(\frac{\sqrt{a + 4}}{1 - a} - 1\right) x^5 - 3x + \log 5 \text{ monotonically decreses for all } x.$$

14. Let $g(x) = 2f(\dfrac{x}{2}) + f(2 - x)$ and $f''(x) < 0 \ \forall x \in (0, 2)$. Find the intervals of increase and decrease of $g(x).$

15. Show that

$$x \in] 0, 1 [\Rightarrow x < -\log (1 - x) < x (1 - x)^{-1}$$

$$x \in] 0, 1 [\Rightarrow 2x < \log \frac{1 + x}{1 - x} < 2x\left(1 + \frac{1}{3} \cdot \frac{x^2}{1 - x^2}\right)$$

$$x \in \,] \, 1, \infty \, [\Rightarrow x - 1 > \log x > (x - 1) x^{-1}$$

$$x \in \,] \, 1, \infty [\Rightarrow x^2 - 1 > 2x \log x > 4 \, (x - 1) - 2 \log x$$

16. The derivative of a function f is positive for every value of x in an interval $[c - h, c]$, and negative for every value of x in $[c, c + h]$; show that $f(c)$ is the greatest value of the function in the interval $[c - h, c + h]$.

ANSWERS

3. (i) Decreasing in $] - \infty, \infty [$.

(ii) Increasing in $] + \infty, -5]$ and $\left[-\dfrac{1}{3}, + \infty \right[$; Decreasing in $\left[-5, -\dfrac{1}{3} \right]$.

(iii) Increasing in $[- \infty, 0]$ and $[2, \infty \, [$; Decreasing in $[0, 2]$.

(iv) Decreasing in $] - \infty, -\sqrt{3}/2 \,]$ and $[0, \sqrt{3}/2 \, [$.

Increasing in $[-\sqrt{3}/2 , 0]$ and $[\sqrt{3}/2, \infty \, [$.

(v) Increasing in $] - \infty, -\sqrt{3}/2 \,]$ and $[0, \sqrt{3}/2 \,]$.

Increasing in $[-\sqrt{3}/2 , 0]$ and $[\sqrt{3}/2, \infty \, [$.

(vi) Decreasing in $] - \infty, -2]$ and $[0, 2]$.

Increasing in $[-2, 0]$ and $[2, \infty . [$

4. $36, 0$.

5. Increasing in $[-1, 1]$; decreasing in $] - \infty, -1]$ and $[1, \infty [$.

6. Increasing in $[-2, -1]$ and $[0, 1]$;

Decreasing in $] - \infty], -2], [-1, 0]$ and $[1, \infty]$.

13. $a \in \left(-4, \dfrac{3 - \sqrt{21}}{2} \right) \cup (1, \infty)$

14. Increasing in $\left(0, \dfrac{4}{3} \right)$, decreasing in $\left(\dfrac{4}{3}, 2 \right)$

8.5. Cauchy's mean value theorem.

If two functions $f(x)$ and $F(x)$ are derivable in closed interval $[a, b]$ and $F'(x) \neq 0$ for any value of x in $\lfloor a, b \rfloor$, then there exists at least one value 'c' of x belonging to the open interval $] \, a, \, b[$ such that **(Kumaon 2000, Gorakhpur 2002; Manipur 2002; Gujrat 2005)**

$$\frac{f(b) - f(a)}{F(b) - F(a)} = \frac{f'(c)}{F'(c)}$$

Firstly, we note that $[F(b) - F(a)] \neq 0$; for if it were 0, then $F(x)$ would satisfy the conditions of the Rolle's theorem and its derivative would therefore vanish for at least one value of x and the hypothesis that $F'(x)$ is never 0 would be contradicted.

Now, we define a new function $\phi(x)$ invloving $f(x)$ and $F(x)$ and designed so as to satisfy the conditions of Rolle's theorem.

Let $\phi(x) = f(x) + AF(x)$

where A is a constant to be determined such that

$$\phi(a) = \phi(b).$$

Thus
$$f(a) + AF(a) = f(b) + AF(b)$$

$$\Rightarrow \qquad A = -\frac{f(b) - f(a)}{F(b) - F(a)}, \text{ for } [F(b) - F(a)] \neq 0.$$

Now, $f(x)$ and $F(x)$ are derivable in $[a, b]$. Also A is a constant. Therefore, $\phi(x)$ is derivable in $[a, b]$ and its derivative is

$$f'(x) + AF'(x).$$

Thus, $\phi(x)$ satisfies the conditions of Rolle's theorem. There is, therefore at least one value'c' of x belonging to the open interval $]a, b[$ such that

$$\phi'(c) = 0$$

$$\Rightarrow \qquad f'(c) + AF'(c) = 0$$

$$\Rightarrow \qquad f'(c) = -AF'(c),$$

$$\Rightarrow \qquad \frac{f'(c)}{F'(c)} = -A = \frac{f(b) - f(a)}{F(b) - F(a)}, \text{ as } F'(c) \neq 0.$$

Hence the theorem.

Another form of statement of Cauchy's mean value theorem. If two functions $f(x)$ and $F(x)$ derivable in a closed interval $[a, a + h]$ and $F'(x) \neq 0$ for any x in $[a, a + h]$ then there exists at least one number θ belonging to the open interval $]0, 1[$ such that

$$\frac{f(a + h) - f(a)}{F(a + h) - F(a)} = \frac{f'(a + \theta h)}{F'(a + \theta h)} \qquad\qquad (0 < \theta < 1)$$

The equivalence of the two statements can be easily seen as in the case of Lagrange's mean value theorem.

Note. Taking $F(x) = x$, we may easily see that Lagrange's theorem is only a particular case of Cauchy's.

EXAMPLES

1. If in the Cauchy's Mean Value Theorem, $f(x) = e^x$ and $F(x) = e^{-x}$, show that c is arithmetic mean between a and b.

$$\frac{f(b) - f(a)}{F(b) - F(a)} = \frac{e^b - e^a}{e^{-b} - e^{-a}}$$

$$= -e^a e^b$$

$$= -e^{a+b}$$

$$\frac{f'(x)}{F'(x)} = \frac{e^x}{-e^{-x}}$$

$$\Rightarrow \qquad \frac{f'(c)}{F'(c)} = -e^{2c}$$

$$\therefore \qquad -e^{a+b} = -e^{2c}$$

$$\Rightarrow \qquad c = \frac{a + b}{2}$$

2. *Show that* $\dfrac{\sin \alpha - \sin \beta}{\cos \beta - \cos \alpha} = \cot \theta,\ 0 < \alpha < \theta < \beta < \pi/2$

Let $\qquad\qquad f(x) = \sin x,\ F(x) = \cos x, x \in [\alpha, \beta].$

f and F are continuous in $[\alpha, \beta]$ and derivable in $]\,\alpha, \beta[$. Thus there exists $\theta \in\ ]\,\alpha, \beta\,[$ such that

$$\frac{\sin \beta - \sin \alpha}{\cos \beta - \cos \alpha} = -\frac{\cos \theta}{\sin \theta}$$

$\Rightarrow \qquad \dfrac{\sin \alpha - \sin \beta}{\cos \beta - \cos \alpha} = \cot \theta,\ 0 < \alpha < \theta < \beta < \pi/2$

3. *Use Cauchy's Mean Value theorem to evaluate*

$$\lim_{x \to 1} \left[\frac{\cos \dfrac{1}{2}\pi x}{\log (1/x)} \right]$$

From Cauchy's Mean Value theorem, we have

$$\frac{f(b) - f(a)}{\phi(b) - \phi(a)} = \frac{f'(c)}{\phi'(c)} \quad \text{where } a < c < b \qquad\qquad\qquad ...(i)$$

Let $\qquad\qquad f(x) = \cos\left(\dfrac{1}{2}\pi x\right);\ \ \phi(x) = \log x$

$$a = x,\ b = 1.$$

Then from (i)

$$\frac{\cos\left(\dfrac{\pi}{2}\right) - \cos\left(\dfrac{\pi}{2} - x\right)}{\log 1 - \log x} = \frac{-\dfrac{1}{2}\pi \sin\left(\dfrac{1}{2}\pi c\right)}{1/c}$$

where $\qquad\qquad a < c < b,\ i.e.,\ x < c < 1.$

Taking limits as $x \to 1$ which inplies that $c \to 1$, we have

$$\lim_{x \to 1}\left\{ \frac{-\cos\left(\dfrac{1}{2}\pi x\right)}{\log (1/x)} \right\} = \lim_{c \to 1}\left\{ \frac{-\dfrac{1}{2}\pi \sin\left(\dfrac{1}{2}\pi c\right)}{1/c} \right\}$$

$\Rightarrow \qquad -\lim\limits_{x \to 1}\left\{ \dfrac{\cos\left(\dfrac{1}{2}\pi x\right)}{\log (1/x)} \right\} = \dfrac{1}{2}\pi,\ \text{as } \sin\left(\dfrac{1}{2}\pi c\right) \to 1 \text{ as } c \to 1$

$\Rightarrow \qquad \lim\limits_{x \to 1}\left\{ \dfrac{\cos\left(\dfrac{1}{2}\pi x\right)}{\log (1/x)} \right\} = \dfrac{1}{2}\pi$

EXERCISES

1. Verify Cauchy's Mean Value theorem for the functions x^2 and x^4 in the interval $[a, b]$; a, b being positive.

2. If in the Cauchy's mean value theorem, we write for $f(x)$, $F(x)$; (i) x^2, x; (ii) $\sin x$, $\cos x$; show that in each case 'c' is the arithmetic mean beween a and b.

3. If, in the Cauchy's mean value theorem, we write for $f(x)$ and $F(x)$, $\sqrt{x}$ and $\dfrac{1}{\sqrt{x}}$ reprectively

   then, c, is the geometric mean between a and b, $\dfrac{1}{x^2}$ and $\dfrac{1}{x}$ if we write and then, c is the

   harmonic mean between a and b. It is understood that a, and b are both positive.

   (Kuvempu 2005)

4. Let the function f be continuous in [a, b] and derivable in] a, b [. Show that there exists a number $c \in$] a, b [such that

$$2c\,[f(a) - f(b)] = f'(c)\,[a^2 - b^2].$$

8.6. Higher derivatives.

Let f be derivable *i.e.*, f' exist in a certain neighbourhood of c. This implies that f is defined and continous in a neighbourhood of c. If the funciton f' has derivative at c, then this derivative is called the seocnd derivative of f at c, and is denoted by $f''(c)$. In this case f' is necessarily continuous at c.

In general, if $f^{n-1}(x)$ exists in a neighbourhood of c, then derivative of f^{n-1} at c, in case it exists, is called the nth derivative of f at c and is denoted by $f^n(c)$.

8.6.1. Generalised Mean Value Theorem : Talylor's theorem. *If a function f is such that*

(i)   *the $(n-1)$th derivative f^{n-1} is continuous in* [a, $a + h$],

(ii) *the nth derivative f^n exists in*] a, $a + h$ [,

and (iii) *p is a given positive integer,*

then there exists at least one number, θ, between 0 and 1 such that

(Manipur 1999; Osmania 2004)

$$f(a + h) = f(a) + hf'(a) + \frac{h^2}{2!}\,f''(a) + \ldots + \frac{h^{n-1}}{(n-1)!}\,f^{n-1}(a)$$

$$+ \frac{h^n(1-\theta)^{n-p}}{(n-1)!\,p}\,f^n(a + \theta h). \qquad \ldots(i)$$

The condition (i) implies the continuity of

$$f, f', f'', \ldots\ldots, f^{n-2} \text{ in } [a, a + h].$$

Let a function φ be defined by

$$\varphi(x) = f(x) + (a + h - x)\,f'(x) + \frac{(a + h - x)^2}{2!}\,f''(x) + \ldots$$

$$+ \frac{(a + h - x)^{n-1}}{(n-1)!}\,f^{n-1}(x) + A\,(a + h - x)^p$$

where A is constant to be determined such that

$$\varphi(a) = \varphi(a + h).$$

Thus A is given by

The function is continuous in $[a, a+h]$, derivable in $]a, a+h[$ and $\varphi(a) = \varphi(a+h)$. Hence, by Rolle's theorm, there exist at least one number, θ, between 0 and 1 such that

$$\varphi'(a + \theta h) = 0.$$

But

$$\varphi'(x) = \frac{(a + h - x)^{n-1}}{(n-1)!} f^n(x) - pA(a + h - x)^{p-1}.$$

$$\therefore 0 = \varphi'(a + \theta h) = \frac{h^{n-1}(1-\theta)^{n-1}}{(n-1)!} f^n(a + \theta h) - pA(1-\theta)^{p-1} h^{p-1}$$

$$\Rightarrow A = \frac{h^{n-p}(1-\theta)^{n-p}}{p.(n-1)!} \cdot f^n(a + \theta h), \text{ for } (1-\theta) \neq 0 \text{ and } h \neq 0.$$

Substituting the value of A in (ii), we get the required result (i).

(i) *Remainder after n terms.* Then term

$$R_n = \frac{h^n(1-\theta)^{n-p}}{p(n-1)!} f^n(a + \theta h),$$

is known as the Taylor's remainder R_n after n terms and is due to *Schlomilch and Roche.*

   (ii) Putting $p = 1$, we obtain

$$R_n = \frac{h^{n-1}(1-\theta)^{n-1}}{(n-1)!} f^n(a + \theta h), \qquad \text{(MDU Rohtak 2000)}$$

which form of remainder is due to *Cauchy.*

   (iii) Putting $p = n$, we obtain

$$R_n = \frac{h^n}{n!} f^n(a + \theta h), \qquad \text{(MDU Rohtak 1998, 99)}$$

which is due to Lagrange.

   Cor. 1. Let x be a point of the interval $[a, a+h]$. Let f satisfy the conditions of Taylor's theorem in $[a, a+h]$ so that it satisfies the conditions for $[a, x]$ also.

Changing $a + h$ to x *i.e.,* h to $x - a$, in (i), we obtain

$$f(x) = f(a) + (x-a) f'(a) + \frac{(x-a)^2}{2!} f^n(a) + \frac{(x-a)^3}{3!} f'''(a) + \dots +$$

$$\frac{(x-a)^{n-1}}{(n-1)!} f^{n-1}(a) + \frac{(x-a)^n (1-\theta)^{n-p}}{p.(n-1)!} f^n[a + \theta(x-a)], 0 < \theta < 1.$$

This result holds $\forall x \in [a, a+h]$. Of course, θ may be different for different points x.

   Cor. 2. Maclaurin's theorem. Putting $a = 0$, we see that if x $[0, h]$, then

$$f(x) = f(0) + xf'(0) + \frac{x^2}{2!} f''(0) \frac{x^3}{3!} f'''(0) + \dots +$$

$$\frac{x^{n-1}}{(n-1)!} f^{n-1}(0) + \frac{x^n(1-\theta)^{n-p}}{p(n-1)!} f^n(\theta x)$$

which holds when

(i) f^{n-1} is continuous in $[0, h]$, and (ii) f^n exists in$] 0, h[$.

Putting $p = 1$ and $p = n$, respectively in the Schlomilch form of remainder

$$\frac{x^n (1 - \theta)^{n-p} f^n (\theta x)}{p \cdot (n-1)!}$$

we see that Cauchy's and Lagrange's forms are respectively

$$\frac{x^n (1 - \theta)^{n-1}}{(n-1)!} f^n (\theta x) \text{ and } \frac{x^n}{n!} f^n (\theta x).$$ **(MDU Rohtak 2001)**

EXERCISES

1. Show that $\forall x \in \mathbf{R}$

(i) $e^x = 1 + x + \dfrac{x^2}{\underline{2}} + ... + \dfrac{x^n}{\underline{n-1}} + \dfrac{x^n}{\underline{n}} e^{\theta x}.$

(ii) $\sin x = x - \dfrac{x^3}{\underline{3}} + \dfrac{x^5}{\underline{5}} ... + (-1)^{n-1} \dfrac{x^{2n-1}}{\underline{2n-1}} + (-1)^n \dfrac{x^{2n}}{\underline{2n}} \sin \theta x.$

(iii) $\cos x = 1 - \dfrac{x^2}{\underline{2}} + \dfrac{x^4}{\underline{4}} ... + (-1)^n \dfrac{x^{2n}}{\underline{2n}} + (-1)^{n+1} \dfrac{x^{2n+1}}{\underline{2n+1}} \sin \theta x.$

2. Find by Maclaurin's theorem, the first four terms and the remainder after n terms of the Maclaurin's expression of $e^{ax} \cos bx$ in terms of the ascending powers of x.

ANSWERS

2. $e^{ax} \cos bx = 1 + ax + \dfrac{a^2 - b^2}{2!} x^2 + \dfrac{a(a^2 - 3b^2)}{3!} x^3 + ...$

$$...... + \frac{x^n}{n!} (a^2 + b^2)^{n/2} e^{a\theta x} \cos \left(b\theta x + n \tan^{-1} \frac{b}{a} \right)$$

8.6.2. Failure of Taylor's Series.

The expression of a function $f(x)$ by Taylor's Theorem will fail for those values of x for which

(i) $f(x)$ or any of its differential coefficient becomes infinite.

(ii) $f(x)$ or any of its differential coefficient is discontinuous, and

(iii) $\lim\limits_{n \to \infty} R_n \neq 0$, i.e., $\lim\limits_{n \to \infty} \dfrac{h^n}{n!} f^n (x + \theta h) \neq 0$

i.e., the series $\sum\limits_{h = n}^{\infty} \dfrac{h^n}{n!} f^n (x)$ is not convergent.

8.6.3. Failure of Maclaurin's Series

The expansion of a function $f(x)$ by Maclaurin's Theorem is not valid for those values of x for which

(i) $f(0)$ or anyone of f'(0), $f''(0), ... , f'''(0)$ is not finite,

(ii) $f(x)$ or any of its derivatives is discontinuous as x passes through zero.

(iii) $\lim\limits_{h \to \infty} R_n \neq 0$, i.e., $\lim\limits_{h \to \infty} \dfrac{x^n}{n!} f^n(\theta x) \neq 0,$

i.e., the series $\displaystyle\sum_{n=0}^{\infty} \frac{x^n}{n!} f^n(0)$ in not convergent.

Maclaurin's Power series for a given Function.

Let a function f possess continuous derivatives of all orders in the interval $[0, x]$ so that we have

$$f(x) = f(0) + x f'(0) + ... + \frac{x^{n-1}}{\underline{|n-1}} f^{n-1}(0) + R_n$$

where R_n denotes Lagrange's or Taylor's form of remainder as obtained above.

Thus we have

$$f(x) = f(0) + x f'(0) + ... + \frac{x^n}{\underline{|n}} f^n(0) + ... \qquad ...(i)$$

valid for those of the values of n for which

$$\lim R_n = 0 \text{ as } n \to \infty$$

The series (*i*) is known as Maclaurin's infinite series for the expansison of $f(x)$ as **power series**.

We shall now obtain the Maclaurin's expansion of

$$e^x, \sin x, \cos x, (1 + x)^m, \log(1 + x)$$

and in the process also obtain the values of x for which the corresponding expressions are vaild.

Expansion of e^x.

We write $\qquad f(x) = e^x$.

We have $\qquad f^n(x) = e^x \; \forall \; x \in \mathbf{R}.$

Denoting by R_n, the Lagrange's form of remainder after n terms, we have

$$R_n = \frac{x^n}{n!} f^n(\theta x) = \frac{x^n}{n!} e^{\theta x}, \; 0 < \theta < 1.$$

Now $\qquad x > 0 \Rightarrow \theta x < x \Rightarrow e^{\theta x} < e^x$

$$x < 0 \Rightarrow e^{\theta x} < 1.$$

Assuming that for all x $\qquad$ (Refer appendix I)

$$\lim_{n \to \infty} \frac{x^n}{n!} = 0,$$

we see that $R_n \to 0$ as $n \to \infty \; \forall x \in \mathbf{R}.$

Thus, we obtain the expansion

$$e^x = 1 + x + \frac{x^2}{2!} + \ldots\ldots + \frac{x^n}{n!} + \ldots\ldots$$

valid $\forall \; x \in \mathbf{R}.$

Expansion of $\sin x$.

We write $\qquad f(x) = \sin x$ so that

$$f^n(x) = \sin\left(x + \frac{1}{2} n\pi\right) \forall \; x \in \mathbf{R}.$$

Denoting by R_n the Lagrange's form of remainder, we have

$$R_n = \frac{x^n}{n!} f^n(\theta x) = \frac{x^n}{n!} \sin\left(\theta x + \frac{n\pi}{2}\right)$$

so that
$$|R_n| = \left| \frac{x^n}{n!} \right| \left| \sin\left(\theta x + \frac{n\pi}{2}\right) \right| \le \frac{x^n}{n!}$$

$\Rightarrow$ $R_n \to 0$ as $n \to \infty \ \forall \ x \in \mathbf{R}$.

It follows that $\forall \ x \in \mathbf{R}$.

$$\sin x = x - \frac{x^3}{3!} + \frac{x^5}{5!} - \ldots\ldots$$

Expansion of cos x. We may show that

$$\cos x = 1 - \frac{x^2}{2!} + \frac{x^4}{4!} - \frac{x^6}{6!} + \ldots\ldots \ \forall \ x \in \mathbf{R}.$$

Expansion of $(1+x)^m$.

We write $f(x) = (1+x)^m$.

Now, m being any real number, $(1+x)^m$ possess continuous derivatives of every order when $(1+x) > 0$. *i.e.*, when $x > -1$.

Also, $f^n(x) = m(m-1)(m-2) \ldots\ldots (m-n+1)(1+x)^{m-n}$.

Considering Cauchy's form of remainder, we have

$$R_n = \frac{x^n}{(n-1)!}(1-\theta)^{n-1} f^n(\theta x)$$

$$= \frac{x^n}{(n-1)!}(1-\theta)^{n-1} m(m-1) \ldots (m-n+1)(1+\theta x)^{m-n}$$

$$= x^n \cdot \frac{m(m-1)\ldots(m-n+1)}{(n-1)!}\left(\frac{1-\theta}{1+\theta x}\right)^{n-1}(1+\theta x)^{m-1}.$$

Let $|x| < 1 \Leftrightarrow -1 < x < 1.$

Now $-1 < x \Rightarrow -\theta < \theta x \Rightarrow 1 - \theta < 1 + \theta x$

$\Rightarrow$ $(1-\theta)/(1+\theta x) < 1.$

Thus, we have $0 < \left(\dfrac{1-\theta}{1+\theta x}\right)^{n-1} < 1.$

Let $(m-1)$ be positive. We have

$$0 < 1 + \theta x < 1 + 1 = 2 \Rightarrow 0 < (1+\theta x)^{m-1} < 2^m.$$

Let $(m-1)$ be negative. We have

$$\theta x \ge -|x| \Rightarrow 1 + \theta x \ge 1 - |x|$$

$\Rightarrow$ $(1+\theta x)^{m-1} \le (1-|x|)^{m-1}.$

We know that

$$\lim_{n\to\infty} \frac{m(m-1)\ldots(m-n+1)}{(n-1)!} x^n = 0$$

we see that $R_n \to 0$ as $n \to \infty$ if $|x| < 1.$

Making substitutions, we get

$$(1 + x)^m = 1 + mx + \frac{m(m-1)}{2!}x^2 + \frac{m(m-1)(m-2)}{3!}x^3 + \dots$$

when $\qquad -1 < x < 1.$

Note. If m is a positive integer, it may be seen that we get a finite series expansion of $(1 + x)^m$ valid $\forall\, x \in \mathbf{R}$.

Expansion of log $(1 + x)$.

Let $\qquad\qquad f(x) = \log(1 + x).$

We know that $\log(1 + x)$ possesses derivatives of every order when $(1 + x) > 0$, *i.e.*, when $x > -1.$

Also, $\qquad\qquad f^n(x) = \dfrac{(-1)^{n-1}(n-1)!}{(1+x)^n}$

Taking the Cauchy's form of remainder, we have

$$R_n = \frac{x^n}{(n-1)!}(1-\theta)^{n-1} f^n(\theta x)$$

$$= \frac{x^n}{(n-1)!}(1-\theta)^{n-1}\frac{(-1)^{n-1}(n-1)!}{(1+\theta x)^n}$$

$$= (-1)^{n-1}x^n \frac{1}{(1+\theta x)}\left(\frac{1-\theta}{1+\theta x}\right)^{n-1} \qquad\qquad \textbf{\textit{(Manipur, 2001)}}$$

Let $|x| < 1$. We have

$$-1 < x < 1 \Rightarrow -\theta < \theta x < \theta \Rightarrow 1 - \theta < 1 + \theta x < 1 + \theta.$$

Thus, we have

$$0 < \frac{1-\theta}{1+\theta x} < 1 \Rightarrow \left(\frac{1-\theta}{1+\theta x}\right)^{n-1} < 1.$$

As we have,

$$\theta x > -|x| \Rightarrow 1 + \theta x > 1 - |x| \Rightarrow \frac{1}{1+\theta x} < \frac{1}{1-|x|}.$$

Thus, $\forall\, n$

$$|R_n| < |x|^n \frac{1}{1-|x|} \to 0 \text{ as } n \to \infty.$$

It follows that when $|x| < 1$,

$$\log(1 + x) = x - \frac{x^2}{2} + \frac{x^3}{3} - \frac{x^4}{4} + \dots + \frac{(-1)^{n-1}}{n-1}x^{n-1} + \dots$$

By adopting Lagrange's form of remainder, we may show that the infinite series expansion is valid for $x = 1$ also.

8.7 Formal Expansions of Functions

We have seen that in order to find out if a given functions can be expanded as a infinite Maclaurin series, it is necessary to examine the behaviour of R_n as n tends to infinity. To put down R_n, we require to obtain the general expression for the nth derivative of the function, so that we fail to apply Maclaurin's theoram to expand in a power series a function for which a general expression for the nth derivative cannot be determined. Other more powerful methods have, accordingly, been discovered to obtain such expansions whenever they are possible. But to deal with these methods is *not* within the scope of this book.

Formal expressions of a function as a power series may, however, be obtained by *assuming* that it can be so expanded, *i.e.*, R_n does tend to 0 as n tends to infintiy. Thus we have the result :

If $f(x)$ **can** *be expanded as an infinite maclaurin's series, then*

$$f(x) = f(0) + xf'(0) + \frac{x^2}{2!}f''(0) + \ldots\ldots \qquad \ldots(i)$$

Such an investigation will not give any information as to the range of values of x for which the expansion is valid.

To obtain the expression of a function, on the assumption that it is possible, we have only to calculate the values of its derivatives for $x = 0$ and substitute them in (i).

EXAMPLES

1.   *Can the function $f(x)$ defined by $f(x) = e^{1/x}$ for $x \neq 0$ and $f(0) = 0$ be expanded in ascending powers of x by Maclaurin's Theorem ?*

$$\lim_{x \to 0} f(x) = \lim_{x \to 0} e^{1/x}$$

$$= \lim_{x \to 0}\left[1 + \frac{1}{x} + \frac{1}{2x^2} + \ldots\right] = \infty, \ i.e., \text{not finite}$$

Hence the given function $f(x)$ is discontinuous at $x = 0$.

Hence the function $f(x)$ can not be expanded in ascending power of x by Maclaurin's Theorem.

2.   *Can the function $f(x) = \sqrt{x}$ be expanded in ascending powers of x by Maclaurin's Theorem?*

We have $f(x) = \sqrt{x}$

$$f'(x) = \frac{1}{2}x^{-1/2}, \ f''(x) = -\frac{1}{4}x^{-3/2}, \ldots$$

$\therefore$ $f'(0), f''(0), f'''(0)$ etc. are all infinite and hence $\sqrt{x}$ can not be expanded in ascending powers of x by Maclaurin's Theorem.

3.   *Prove that*

$$\sin ax = ax - \frac{a^3 x^3}{3!} + \frac{a^5 x^5}{5!} \ldots + \frac{a^{n-1} x^{n-1}}{(n-1)!}\sin\left(\frac{n-1}{2}\pi\right)$$

$$+ \frac{a^n x^n}{n!}\sin\left(a\theta x + \frac{n\pi}{2}\right).$$

Let $f(x) = \sin ax$

then $f'(x) = a\cos ax = a\sin\left(ax + \frac{\pi}{2}\right)$

$$f''(x) = a^2 \cos\left(ax + \frac{\pi}{2}\right) = a^2 \sin\left(ax + 2 \cdot \frac{\pi}{2}\right)$$

... ...
... ...

$$f^{n-1}(x) = a^{n-1} \sin\left(ax + \frac{n-1}{2}\pi\right)$$

and $$f^n(x) = a^n \sin\left(ax + \frac{n\pi}{2}\right)$$

so that $$f^n(\theta x) = a_n \sin\left(a\theta x + \frac{n\pi}{2}\right)$$

∴ $$f(0) = \sin 0 = 0,$$

$$f'(0) = a\sin\frac{\pi}{2} = a, \ f''(0) = a^2 \sin \pi = 0,$$

$$f'''(0) = a^3 \sin\frac{3\pi}{2} = -a^3, f^{\mathrm{iv}}(0) = a^4 \sin 2\pi = 0,$$

$$\dots f^{n-1}(0) = a^{n-1} \sin\left(\frac{n-1}{2}\right)\pi$$

Putting these values in

$$f(x) = f(0) + xf'(0) + \frac{x^2}{2!}f''(0) + \dots + \frac{x^{n-1}}{(n-1)!}f^{n-1}(0) + \frac{x^n}{n!}f^n(\theta x),$$

We get

$$\sin ax = 0 + xa + 0 - \frac{x^3}{3!}a^3 + 0 + \frac{x^5}{5!}a^5 + \dots + \frac{x^{n-1}}{(n-1)!}a^{n-1}\sin\frac{n-1}{2}\pi$$

$$+ \frac{x^n}{n!}a^n \sin\left(a\theta x + \frac{n\pi}{2}\right)$$

or $$\sin ax = ax - \frac{a^3 x^3}{3!} + \frac{a^5 x^5}{5!} \dots + \frac{x^{n-1}a^{n-1}}{(n-1)!}\sin\left(\frac{n-1}{2}\pi\right) + \frac{a^n x^n}{n!}\sin\left(a\theta x + \frac{n\pi}{2}\right).$$

4. *A function f (x) is defined as*

$$f(x) = e^{-1/x^2}, \ x \neq 0$$

and $$f(x) = 0, \ x = 0.$$

Can it be expanded in ascending powers of x by Maclaurin's Theorem.

Given $$f(x) = e^{-1/x^2}, \ x \neq 0 \qquad \dots (i)$$

∴ $$R f'(0) = \lim_{h \to 0}\left[\frac{f(0+h) - f(0)}{h}\right]$$

$$= \lim_{h \to 0}\left[\frac{e^{-1/h^2} - 0}{h}\right]$$

$$= \lim_{\theta \to \infty}\left(\frac{e^{-\theta^2}}{1/\theta}\right), \quad \text{where } \theta = 1/h.$$

$$= \lim_{\theta \to \infty}\left[\frac{\theta}{e^{\theta^2}}\right] = \lim_{\theta \to \infty}\left[\frac{1}{2\theta\, e^{\theta^2}}\right]$$

$$= 0. \qquad\qquad\qquad \text{[By } L' \text{ Hospital's Rule]}$$

Also, $\qquad L f'(0) = \lim_{h \to 0}\left[\dfrac{f(0-h) - f(0)}{-h}\right]$

$$= \lim_{h \to 0}\left[\frac{e^{1/h^2} - 0}{-h}\right] = 0 \qquad \text{as before}$$

$\therefore \qquad\qquad\qquad f'(0) = 0$

Also from (i) we get

$$f'(x) = \frac{2}{x^3} e^{-1/x^2}, \; x \neq 0$$

$\therefore \qquad \lim_{x \to 0}[f'(x)] = \lim_{x \to 0}\dfrac{2}{x^3} e^{-1/x^2}$

$$= \lim_{t \to \infty}[2t^3 e^{-t^2}], \; \text{where } t = \frac{1}{x}$$

$$= \lim_{t \to \infty}\left[\frac{2t^3}{e^{t^2}}\right] = 0 = f'(0)$$

Hence $f'(x)$ is continuous at $x = 0$.

If we find the higher derivatives of $f(x)$ for $x \neq 0$ we shall get e^{-1/x^2} multiplied by a polynomial

in $\left(\dfrac{1}{x}\right)$. Hence all the higher derivatives of $f(x)$ will be zero at $x = 0$.

Therefore the function $f(x)$ possesses continuous derivative for every value of x.

$\therefore \quad$ By Maclaurin's Theorem, we get

$$f(x) = x \cdot f'(0) + \frac{x^2}{2!} f''(0) + \ldots + \frac{x^{n-1}}{(n-1)!} + f^{n-1}(0) + R_n$$

i.e., $\quad e^{-1/x^2} = 0 + x \cdot 0 + \dfrac{x^2}{2!} \cdot 0 + \ldots + \dfrac{x^{n-1}}{(n-1)!} \cdot 0 + R_n$

i.e., $\qquad\qquad R_n = e^{-1/x^2}$

$\therefore \quad \lim_{n \to \infty} R_n \neq 0$ as e^{-1/x^2} does not vanish as $n \to \infty$.

Hence the given function $f(x)$ can not be expanded in ascending powers of x by Maclaurin's Theorem.

5. If $f(x) = f(0) + xf'(0) + \dfrac{x^2}{2!} f''(\theta x)$, *find the value of θ as x tends to 1, $f(x)$ being* $(1-x)^{5/2}$. $\qquad\qquad$ *(Manipur 2001)*

We have $\qquad f(0) = 1, \; f'(0) = -\dfrac{5}{2}, \; f''(\theta x) = \dfrac{15}{4}(1 - \theta x)^{1/2}$

Hence substituting these values in (i), we get

$$(1-x)^{5/2} = 1 - \frac{5}{2}x + \frac{x^2}{2!} \times \frac{15}{4}(1 - \theta x)^{1/2}$$

Therefore as $x \to 1$, we get

$$0 = 1 - \frac{5}{2} + \frac{1}{2!} \cdot \frac{15}{4}(1 - \theta)^{1/2}$$

or $\quad (1 - \theta)^{1/2} = \frac{4}{5} \quad$ or $\quad 1 - \theta = \frac{16}{25}$

i.e., $\qquad\qquad \theta = \frac{9}{25}.$

EXERCISES

1. Can the functions

    (i) $f(x) = \tan^{-1}\sqrt{(x/a)};$ (ii) $f(x) = \log x$

    be expanded in ascending powers of x by Maclaurin's Theorem.

2. If $f''(a), f'''(a), \dots, f^{n-1}(a)$ are zero but $f^n(x)$ is continuous non-zero at $x = a$, then show that

    $$\lim_{h \to 0} (\theta_n - 1) = \frac{1}{n},$$

    where $f(a + h) = f(a) + hf'(a) + \dots + \dfrac{h^{n-1}}{(n-1)!} f^{n-1}(x + \theta_{n-1} h).$

3. If $f(x + h) = f(x) + hf'(x) + \dfrac{h^2}{2!} f''(x + \theta h),$ find the value of θ as $x \to a$, $f(x)$ being $(x - a)^{5/2}$.

ANSWERS

1. (i) No, (ii) No;   3. $\theta = \dfrac{64}{225}.$

OBJECTIVE QUESTIONS

Note : *For each of the following questions four alternatives are given for the answer. Only one of them is correct. Choose the correct alternative.*

1. "If f is continuous in $[a, b]$, differentiable in (a, b) and $f(a) = f(b)$, then there exist at least one point $c \in (a, b)$ such that $f'(c) = 0$." This result is known as

    (a) Lagrange's theorem   (b) Euler's theorem

    (c) Rolle's theorem   (d) Cauchy's theorem

2. To which of the following, Rolle's theorem can be applied :

    (a) $f(x) = \tan x$ in $[0, \pi]$ (b) $f(x) = \cos(1/x)$ in $[-1, 1]$

    (c) $f(x) = x^2$ in $[2, 3]$ (d) $f(x) = x(x + 3)e^{-x/2}$ in $[-3, 0]$

3. The 'c' of Rolle's theorem for the function $f(x) = \sin x$ in $[0, \pi]$ is

    (a) 0   (b) $\dfrac{1}{6}\pi$

    (c) $\dfrac{1}{3}\pi$ (d) $\dfrac{1}{2}\pi$

4. If f is continuous in $[a, b]$ and differentiable in (a, b), then there exists at least one point c in (a, b) such that $f'(c)$ is equal to

   (a) $\dfrac{f(b) + f(a)}{b + a}$

   (b) $\dfrac{f(b) - f(a)}{b + a}$

   (c) $\dfrac{f(b) - f(a)}{b - a}$

   (d) $\dfrac{f(b) + f(a)}{b - a}$

5. Lagrange's mean value theorem can be proved for a function $f(x)$ by applying Rolle's theorem to the function

   (a) $\phi(x) = f(x) + kx^2$

   (b) $\phi(x) = f(x) - kx^2$

   (c) $\phi(x) = f(x) + kx$

   (d) $\phi(x) = \{f(x)\}^2 + kx^2$

6. The function $f(x) = x(x+3)e^{-x/2}$ satisfy all conditions of Rolle's theorem in the interval $[-3, 0]$. Then the value of c :

   (a) 0

   (b) 1

   (c) 2

   (d) -2

7. The value of c of the mean value theorem, if $f(x) = x(x-1)(x-2)$; $a = 0$, $b = \dfrac{1}{2}$ is :

   (a) 1

   (b) $(1 + \sqrt{(21)}/6)$

   (c) $(1 - \sqrt{(21)}/6)$

   (d) none of these.

8. The value of c of the mean value theorem, if $f(x) = 2x^2 + 3x + 4$ in $[1, 2]$ is :

   (a) 1

   (b) 2

   (c) 3/2

   (d) 2/3   (Garwhal 2005)

9. The function $f(x) = 2x^3 + x^2 + x^2 - 4x - 2$ satisfy all conditons of Rolle's Theorem at the interval $[-\sqrt{2}, \sqrt{2}]$. Then the value of c is :

   (a) -1

   (b) $-2/3$

   (c) 1

   (d) 3/2

10. Let $f(x) = (x-4)(x-5)(x-6)(x-7)$; then

   (a) $f'(x) = 0$ has four roots

   (b) three roots of $f'(x) = 0$ lie in $(4, 5) \cup (5, 6) \cup (6, 7)$

   (c) the equation $f'(x) = 0$ has only one real root.

   (d) three roots of $f'(x) = 0$ lie in $(3, 4) \cup (4, 5) \cup (5, 6)$.

11. Lagrange's form of remainder after n terms in Tayler's development of the function e^x in a finite form in the interval $[a, a+h]$ is

   (a) $\dfrac{h^n}{n!} e^{a+\theta h}$

   (b) $\dfrac{h^n}{n!} e^{\theta h}$

   (c) $\dfrac{h^n(1-\theta)}{n!} e^{a+\theta h}$

   (d) $\dfrac{h^n(1+\theta)^n}{n!} e^{a+\theta h}$ (Manipur 2002)

ANSWERS					
1. (c)	2. (d)	3. (d)	4. (c)	5. (c)	6. (d)
7. (c)	8. (c)	9. (a)	10. (b)	11. (b)	

CHAPTER 9

MAXIMA AND MINIMA

9.1 In this chapter we shall be concerned with the application of Differential Calculus to the determination of the values of a function which are greatest or least in their immediate neighbourhoods technically known as relatively greatest and least or maximum and minimum values.

It will be assumed that given function possesses continuous derivatives in appropriate intervals of every order that come in question.

Maximum value of a function. We say that $f(c)$ is a maximum value of a function f, if it is the greatest of all its values for values of x in some neighbourhood of c.

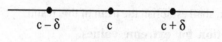

$$c - \delta \qquad c \qquad c + \delta$$

Fig. 9.1

Thus $f(c)$ is a maximum value, if there exists some interval $]c - \delta, c + \delta[$ around 'c' such that $f(c) > f(x)$ for all values of x, other than c, lying in this interval.

This is equivalent to saying that $f(c)$ is a maximum value of f if there exists $\delta > 0$ such that

$$f(c) > f(c + h) \Leftrightarrow f(c + h) - f(c) < 0 \text{ for } 0 < |h| < \delta.$$

Minimum value of function. $f(c)$ is said to be a minimum value of f, if it is the least of all its values for values of x in some neighbourhood of c.

Thus $f(c)$ is a minimum value of f, if there exists some interval $]c - \delta, c + \delta[$ around 'c' such that $f(c) < f(x)$ for all values of x, other than c, lying in this interval.

This is equivalent to saying that $f(c)$ is a minimum value of f if there exists a positive δ such that

$$f(c) < f(c + h) \Leftrightarrow f(c + h) - f(c) > 0 \text{ for } 0 < |h| < \delta.$$

The maximum and minimum values of a function are also known as *Relatively* greatest and least values of the function in that these are the greatest and least values of the function relatively to some neighbourhoods of the points in question.

Note 1. The term *extreme* value is used both for a maximum as well as for a minimum value.

Note 2. While ascertaining whether a value $f(c)$ is an extreme value of f or not, we compare $f(c)$ with the values of f for values of x in any neighbourhood of c, so that the values of the function outside the neighbourhood do not come into question.

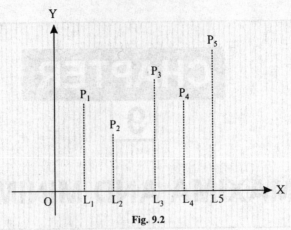

Fig. 9.2

Thus, a maximum (minimum) values of a function may not be the greatest (least) of all the values of the function in a finite interval. In fact a function can have several maximum and minimum values and a minimum value may even be greater than a maximum value.

A glance at the adjoining graph shows that the ordinates of points P_1, P_3, P_5 are the maximum and the ordinates of the points P_2, P_4 are the minimum values of the corresponding function f and the ordinate of P_4 which is a minimum is greater than the ordinate of P_1 which is a maximum.

Also the reader may perhaps guess that the tangents at the points P_1, P_2, P_3, P_4, P_5 are parallel to x-axis so that if c_1, c_2, c_3, c_4, c_5 are the abscissae of these points, then each of $f'(c_1), f'(c_2), f'(c_3), f'(c_4), f'(c_5)$ is zero.

In the next section, we shall establish the truth of this result.

9.2. A necessary condition for extreme values.

A necessary condition for $f(c)$ to be an extreme value of f is that $f'(c) = 0$ so that

$$f(c) \text{ is an extreme value} \Rightarrow f'(c) = 0.$$

Here we assume that f is derivable at c.

Let $f(c)$ be a maximum value of f.

There exists an interval $]c - \delta, c + \delta[$, around c, such that, if $c + h$ is a number other than c, belonging to this interval, we have

$$f(c + h) < f(c).$$

Here, h may be positive or negative. Thus

$$h > 0 \Rightarrow \frac{f(c + h) - f(c)}{h} < 0 \qquad \qquad ...(i)$$

$$h < 0 \Rightarrow \frac{f(c + h) - f(c)}{h} > 0 \qquad \qquad ...(ii)$$

From (*i*) and (*ii*), we have

$$\lim_{h \to (0+0)} \frac{f(c + h) - f(c)}{h} \leq 0 \text{ and } \lim_{h \to (0-0)} \frac{f(c + h) - f(c)}{h} \geq 0 \qquad ...(iii)$$

The relations (*iii*) will simultaneously be true, if and only if

$$f'(c) = 0.$$

It can similarly be shown that $f'(c) = 0$, if $f(c)$ is a minimum value of f.

Note 1. The vanishing of $f'(c)$ is only a necessary but not a sufficient condition for $f(c)$ to be an extreme value as we now show.

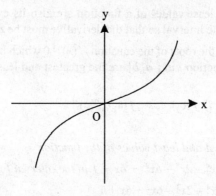

Fig. 9.3

Consider $f(x) = x^3$ for $x = 0$.

Now $\qquad x > 0 \Rightarrow f(x) > 0 = f(0)$; $x < 0 \Rightarrow f(x) < 0 = f(0)$.

Thus, $f(0)$ is not an extreme value even though $f'(0) = 0$.

Note 2. $f(0) = 0$ is a minimum value of $f(x) = |x|$ even though f is not derivable at 0.

Example. *Prove that the function f defined by*

$$f(x) = 3|x| + 4|x-1| \quad \forall x \in R$$

has a minimum value 3 at x = 1.

We have $\qquad f(x) = \begin{cases} 4 - 7x & \text{if } x < 0 \\ 4 & \text{if } x = 0 \\ 4 - x & \text{if } 0 < x < 1 \\ 3 & \text{if } x = 1 \\ 7x - 4 & \text{if } x > 1 \end{cases}$

From above it is clear that f has minimum value 3 at $x = 1$.

Also

$$Lf'(1) = -1 \quad \text{and} \quad Rf'(1) = 7$$

Show that function is not derivable at $x = 1$.

Ex. 1. Prove that the function f defined by

$$f(x) = 2|x - 2| + 5|x - 3| \quad \forall x \in R$$

has a minimum value 2 at $x = 3$.

Stationary Values : A function f is said to be stationary for 'c' and $f(c)$ a stationary value of f if $f'(c) = 0$. The rate of change of a function is zero at a stationary point. While a maximum or minimum value of a stationary value may neither be a maximum nor a minimum value.

Cor. Greatest and least values of function in an interval. *The greatest and least values of a function f in an interval* $[a, b]$ *are* $f(a)$ *or* $f(b)$ *or are given by the values of* x *for which* $f'(x) = 0$.

The greatest and the least values of a function are also its extreme values in case they are attained at points within the interval so that the derivative must be zero at corresponding point.

If $c_1, c_2, \ldots\ldots\ldots, c_k$ be the roots of the equation $f'(x) = 0$ which belog to $]a, b[$, then the greatest and least values of the function f in $[a, b]$ are the greatest and least members respectively of the finite set

$$\{f(a), f(c_1), f(c_2), \ldots\ldots, f(c_k), f(b)\}$$

Hence the result.

Ex. 1. *Find the greatest and least values of the function*

$$f(x) = 3x^4 - 2x^3 - 6x^2 + 6x + 1 \text{ in the interval } [0, 2].$$

We have $\qquad f(x) = 3x^4 - 2x^3 - 6x^2 + 6x + 1$

$\Rightarrow \qquad\qquad f'(x) = 12x^3 - 6x^2 - 12x + 6$

$\qquad\qquad\qquad = 6(x - 1)(x + 1)(2x - 1)$

$\Rightarrow \qquad\qquad f'(x) = 0 \text{ for } x = 1, -1, \dfrac{1}{2}.$

The number -1 does not belong to the interval $[0, 2]$ and is not, therefore, to be considered. Now

$$f(1) = 2, \ f\left(\frac{1}{2}\right) = \frac{39}{16}, \ f(0) = 1, \ f(2) = 21.$$

Thus, the least value is 1 and the greatest value is 21.

EXERCISES

1. Find the greatest and least values of the function

    $x^4 - 4x^3 - 2x^2 + 12x + 1$ in the interval $[-2, 5]$.

2. Find the greatest and least values of the function $2x^3 - 15x^2 + 36x + 1$ in the interval $[2, 3]$.

    Also find the greatest and least values in the interval $[0, 4]$.

ANSWERS

1. $136, -8$ $\qquad\qquad$ 2. $29, 28; 33, 1.$

Def. Change of Sign. A function is said to change sign from positive to negative as x passes through a number c, if there exists some left-handed neighbourhood $]c - h, c[$ of c for every point of which the function is positive and also there exists some right handed neighbourhood $]c, c + h[$ of c for every point of which the function is negative.

A similar meaning with obvious alterations can be assigned to the statement.

A function changes sign from negative to positive as x passes through 'c'.

It is clear that if a continuous function $f(x)$ changes sign as x passes through c' we must have $f(c) = 0$.

Ex. 1. Show that the function

$$f(x) = (x + 2)(x - 1)^2(2x - 1)(x - 3)$$

changes sign from positive to negative as x passes through 1/2 and from negative to positive as x passes through -2 or 3; also show that it does not change sign as x passes through 1.

Ex.2. Show that the function $f(x) = (2x + 3)(x + 4)(x - 2)(x - 1)^3$ changes sign from positive to negative as x passes through -4 and 1 and from negative to positive as x passes through $-3/2$ and 2.

9.3. Sufficient condition for extreme values

Theorem. $f(c)$ *is an extreme value of f if and only if $f'(x)$ changes sign as x passes through c.*

The extreme value $f(c)$ is a maximum value if the sign changes from positive to negative and minimum value in the contrary case.

Case I. *Let $f'(x)$ changes sign from positive to negative as x passes through c.*

In some left-handed neighbourhood $]c - \delta, c[$ of c, $f'(x)$ is positive and so f is strictly increasing implying that $f(c)$ is the greatest of all the values of f in this neighbourhood.

In some right-handed neighbourhood $]c, c + \delta[$ of c, $f'(x)$ is negative and so f is strictly decreasing implying that $f(c)$ is the greatest of all the values of f in this neighbourhood.

Thus, $f(c)$ is the greatest of all the values of $f(x)$ when $x \in]c - \delta, c + \delta[$, *i.e.*, is the greatest of all the values of f in a certain complete neighbourhood of c and so, by definition, $f(c)$ is a maximum value of f.

Case II. *Let $f'(x)$ change sign from negative to positive as x passes through c.* It may be shown that $f(c)$ is the least of all the values of f in a certain complete neighbourhood of c and so, by definition, $f(c)$ is a minimum value of f.

Case III. If $f'(x)$ does not change sign, *i.e.*, has the same sign in a certain complete neighbourhood of c, then $f(x)$ is either strictly increasing or decreasing throughout this neighbourhood implying that $f(c)$ is *not* an extreme value of f.

Note. Geometrically interpreted, the theorem states that the tangent to a curve at every point in a certain left-handed neighbourhood of the point P whose ordinate is a maximum (minimum) makes an acute angle (obtuse angle) and the tangent at any point in a certain right-handed neighbourhood of P makes an obtuse angle (or acute angle) with x-axis. In case, the tangent on *either* side of P makes an acute angle (or obtuse angle), the ordinate of P is neither a maximum nor a minimum.

EXAMPLES

Ex. 1. *Examine the polynomial function given by*
$$10x^6 - 24x^5 + 15x^4 - 40x^3 + 108$$
for maximum and minimum values.

We have $\qquad f(x) = 10x^6 - 24x^5 + 15x^4 - 40x^3 + 108$

$\Rightarrow \qquad f'(x) = 60x^5 - 120x^4 + 60x^3 - 120x^2$

$\qquad\qquad\qquad = 60x^2(x^3 - 2x^2 + x - 2)$

$\qquad\qquad\qquad = 60x^2(x^2 + 1)(x - 2)$

$\Rightarrow \qquad f'(x) = 0$ for $x = 0$ and $x = 2$.

Thus, we expect extreme values of the function for $x = 0$ and 2 only. Now,

$\qquad x < 0 \Rightarrow f'(x) < 0, 0 < x < 2 \Rightarrow f'(x) < 0,$

$\qquad x > 2 \Rightarrow f'(x) > 0.$

Here $f'(x)$ does not change sign as x passes through 0 so that $f'(0)$ is neither a maximum nor a minimum value even through $f'(0) = 0$.

Also since $f'(x)$ changes sign from negative to positive as x passes through 2, therefore, $f(2) = -100$ is a minimum values.

The function has only extreme value, *viz.*, that at 2.

Ex. 2. *If* $f'(x) = (x-a)^{2n}(x-b)^{2m+1}$ *where m and n are positive integers is the derivative of a function f. Then show that x = b gives a minimum but x = a is neither a maximum nor a minimum.*

We have $\quad f'(x) = (x-a)^{2n}(x-b)^{2m+1}$

Now $\quad\quad f'(x) = 0 \Rightarrow x = a, b$

For $x < b$, $(x-a)^{2n}(x-b)^{2m+1} < 0$ and

for $x > b$, $(x-a)^{2n}(x-b)^{2m+1} > 0$

Thus f' changes sign from negative to positive as x passes through b as such has minimum at b. Again since $2n$ is an even integer $(x-a)^{2n}(x-b)^{2m+1}$ does not change sign as x passes through a. This function has neither maximum nor minimum for $x = a$.

EXERCISES

1. Find the extreme values of
$$5x^6 + 18x^5 + 15x^4 - 10.$$

2. Show that the maximum and minimum values of
$$(x+1)(x+4)(x-1)(x-4)$$
are -9 and $-1/9$ respectively.

3. Show that $9x^5 + 30x^4 + 35x^3 + 15x^2 + 1$

   is maximum when $x = -2/3$ and minimum when $x = 0$.

   Also find the greatest and least values in the intervals $[-2/3, 0]$ and $[-2, 2]$.

4. Show that $x^5 - 5x^4 + 5x^2 - 1$

   has a maximum value when $x = 1$ a minimum value when $x = 3$ and neither when $x = 0$.

   (Kumaon 2003)

ANSWERS

1. max. -8 ; min. $-10, -26,$　　3.　$55/27, 1$; $1109, -27.$

9.4. Use of second order derivatives

The derivatives of the first order only have so far been employed for determining and distinguishing between the extreme values of a function. As shown in the present article, the same thing can sometimes be done more conveniently by employing derivatives of second and higher orders.

All along the discussion, it will be assumed that f possesses continuous derivatives of second order, that come in question, in some neighbourhood of the point c.

Theorem 1. $f(c)$ *is a maximum value of the function f if*
$$f'(c) = 0 \text{ and } f''(c) > 0.$$

As $f''(c)$ is positive, there exists an interval $[c-\delta, c+\delta]$, around, c for every point x of which the second derivative $f''(x)$ is positive. This implies that $f'(x)$ is strictly increasing in $[c-\delta, c+\delta]$. Also since $f'(c) = 0$, it follows that

$$f'(x) < 0 \;\forall\, x \in [c-\delta, c[\text{ and } f(x) > 0 \;\forall\, x \in\,]c, c+\delta].$$

Thus, f is strictly decreasing in $[c-\delta, c]$ and strictly increasing in $[c, c+\delta]$ so that $f(c)$ is a minimum value of $f(x)$.

Theorem 2. $f(c)$ is a maximum value of the function f, if

$$f'(c) = 0 \text{ and } f''(c) < 0.$$

As $f''(c)$ is negative, there exist an interval $[c - \delta, c + \delta]$ around c for every point x of which the second derivative is negative. Thus, $f'(x)$ is strictly decreasing in $[c - \delta, c + \delta]$. Also, $f'(c) = 0$. It follows that

$$f'(x) > 0 \ \forall \ x \in [c - \delta, c \ [, f'(x) < 0 \ \forall \ x \in \ ] \ c, c + \delta].$$

Thus f is strictly increasing in $[c - \delta, c]$ and strictly decreasing in $[c, c + \delta]$ so that $f(c)$ is a maximum value of $f(x)$.

Note. The theorem, as stated above, ceases to be helpful if for some $c, f'(c)$ and $f''(c)$ are both zero. To prove for this deficiency, we need to consider higher order derivatives. It will be shown in Ch. 14 that if each of

$$f'(c), f''(c), \dots\dots\dots, f^{n-1}(c)$$

is 0 and $f^n(c) \neq 0$, then

(i) $f(c)$ is not an extreme value if n is odd,

(ii) $f(c)$ is an extreme value if n is even. Also $f(c)$ is a maximum or minimum value according as

$$f^n(c) > 0 \text{ or } f^n(c) > 0.$$

EXAMPLES

1. Find the maximum and minimum values of the polynomial function f given by

$$f(x) = 8x^5 - 15x^4 + 10x^2.$$

We have $\quad f'(x) = 40x^4 - 60x^3 + 20x$

$$= 20x(2x^3 - 3x^2 + 1) = 20x(x - 1)^2(2x + 1)$$

$\Rightarrow \qquad f'(x) = 0$ for $x = 0, 1, -1/2$.

Again $\quad f''(x) = 160x^3 - 180x^2 + 20 = 20(8x^3 - 9x^2 + 1)$

Now $\qquad f'(0) = 0$ and $f''(0) = 20 > 0$

$\Rightarrow \qquad f(0) = 0$ is a minimum value.

$\qquad f(-1/2) = 0$ and $f''(-1/2) = -45 < 0$

$\Rightarrow \qquad f(-1/2) = 21/61$ is a maximum value.

While $\qquad f'(1) = 0$ and $f''(1) = 0$, we see that

as x passes through, $1, f'(x)$ does not change sign.

Thus $f(1)$ is neither a maximum nor a minimum value.

2. Investigate for maximum and minimum values the function given by $y = \sin x + \cos 2x$.

(Garhwal 1997, G.N.D.U. Amritsar 2004)

We have $\qquad \dfrac{dy}{dx} = \cos x - 2 \sin 2x = \cos x - 4 \sin x \cos x$

$\Rightarrow \qquad \dfrac{dy}{dx} = 0$ when $\cos x = 0$ or $\sin x = 1/4$.

We consider values of x between 0 and 2π only, for the given function is periodic with period 2π. Now

$$\cos x = 0 \Rightarrow x = \pi/2 \text{ and } 3\pi/2 \text{ and}$$

$$\sin x = 1/4 \Rightarrow x = \sin^{-1} 1/4 \text{ and } \pi - \sin^{-1} 1/4,$$

$\sin^{-1} 1/4$ lying between 0 and $\pi/2$.

Now $\qquad \dfrac{d^2y}{dx^2} = -\sin x - 4\cos 2x$

$x = \dfrac{\pi}{2} \Rightarrow \dfrac{d^2y}{dx^2} = 3 > 0$

$x = 3\pi/2 \Rightarrow \dfrac{d^2y}{dx^2} = 5 > 0$

$x = \sin^{-1} 1/4 \Rightarrow \dfrac{d^2y}{dx^2} = -\sin x - 4(1 - 2\sin^2 x) = -15/4 < 0$

$x = \pi - \sin^{-1} 1/4 \Rightarrow \dfrac{d^2y}{dx^2} = -15/4 < 0$

Therefore y is maximum for $x = \sin^{-1} 1/4$, $\pi - \sin^{-1} 1/4$ and is a minimum for $x = \pi/2$, $3\pi/2$.

Putting these values of x in $\sin x + \cos 2x$ we see that 9/8, 9/8 are its two maximum values and $0, -2$ are its two minimum values.

3. *Show that the maximum value of* $(1/x)^x$ *is* $(e)^{1/e}$.

Let $\qquad y = (1/x)^x$

$\Rightarrow \qquad \log y = -x \log x$

$\Rightarrow \qquad \dfrac{1}{y}\dfrac{dy}{dx} = -(1 + \log x)$

$\Rightarrow \qquad \dfrac{dy}{dx} = -(1 + \log x)(1/x)^x$

$\qquad \dfrac{dy}{dx} = 0 \Rightarrow \log x + 1 = 0 \Rightarrow x = e^{-1}$

Again $\qquad \dfrac{d^2y}{dx^2} = -\dfrac{1}{x}\left(\dfrac{1}{x}\right)^x + (1 + \log x)^2 (1/x)^x$

$\qquad$ at $x = e^{-1}$, $\dfrac{d^2y}{dx^2} = -e(e)^{1/e} < 0$

$\Rightarrow$ y has maximum for $x = e^{-1}$ and the maximum value is $(e)^{1/e}$.

4. *Show that the function* f *defined by*

$$f(x) = x^p (1-x)^q \qquad\qquad \forall\, x \in R$$

where p, q *are positive integers has a maximum value for* $x = \dfrac{p}{p+q}$ *for all* p, q.

We have $f(x) = x^p (1-x)^q$

$\Rightarrow \qquad f'(x) = px^{p-1}(1-x)^q - qx^p(1-x)^{q-1}$

$\qquad\qquad = x^{p-1}(1-x)^{q-1}(p - x(p+q))$

$\qquad f'(x) = 0 \Rightarrow x = 0, 1, p/p + q$

Again

$$f''(x) = (p-1) x^{p-2} (1-x)^{q-1} [p - x(p+q)]$$
$$- (q-1) x^{p-1} (1-x)^{q-2} [p - x(p+q)] - (p+q) x^{p-1} (1-x)^{q-1}$$

$$f''(p/p+q) = -(p+q) \left(\frac{p}{p+q}\right)^{p-1} \left(\frac{q}{p+q}\right)^{q-1}$$

< 0 where p and q are integers.

Thus the function has a max. value at $x = p/p + q$ for all integers p and q and the max. value is

$$p^p q^q / (p+q)^{p+q}.$$

5. *Show that the function f defined by*

$$f(x) = |x|^p |x-1|^q \qquad \forall \, x \in \mathbf{R}$$

has a maximum value $p^p q^q / (p+q)^{p+q}$, p, q being positive.

We have

$$f(x) = \begin{cases} (-1)^p \, x^p \, (1-x)^q & , & x < 0 \\ 0 & , & x = 0 \\ x^p \, (1-x)^q & , & 0 < x < 1 \\ 1 & , & x = 1 \\ x^p \, (x-1)^q & , & x > 1 \end{cases}$$

Consider $f(x) = x^p (1-x)^q , 0 < x < 1$

By previous example, f has a maximum value at

$$x = \frac{p}{p+q}, \, 0 < \frac{p}{p+q} < 1$$

The max. value is $\dfrac{p^p q^q}{(p+q)^{p+q}}$.

6. *Find the maxima and minima of the function*

$$f(x) = \sin x + \frac{1}{2} \sin 2x + \frac{1}{3} \sin 3x \, \forall \, x \in [0, \pi].$$

We have $f(x) = \sin x + \dfrac{1}{2} \sin 2x + \dfrac{1}{3} \sin 3x$

$\Rightarrow$ $\qquad f'(x) = \cos x + \cos 2x + \cos 3x$

$\qquad\qquad\quad = \cos 2x + 2 \cos 2x \cos x$

$\qquad\qquad\quad = \cos 2x (1 + 2 \cos x)$

$f'(x) = 0 \Rightarrow x = \dfrac{\pi}{4}, \dfrac{2\pi}{4}, \dfrac{2\pi}{3}, \dfrac{3\pi}{4}$

$f''(x) = -\sin x - 2 \sin 2x - 3 \sin 3x$

$f''\left(\dfrac{\pi}{4}\right) = \dfrac{-4}{\sqrt{2}} - 2 < 0 \Rightarrow f$ has a max. at $\pi/4$.

$f''\left(\dfrac{2\pi}{3}\right) = \dfrac{\sqrt{3}}{2} > 0 \Rightarrow f$ has a min. at $\dfrac{2\pi}{3}$

$$f''\left(\frac{3\pi}{4}\right) = 2 - 2\sqrt{2} < 0 \Rightarrow f \text{ has a max. at } 3\pi/4.$$

7. *If A denotes the arithmetic mean of the real numbers* $a_1, a_2, \ldots, a_n$*, show that* $\displaystyle\sum_{i=1}^{n}(x - a_i)^2$

has a minimum at A.

Let $\qquad f(x) = \displaystyle\sum_{i=1}^{n}(x - a_i)^2$

$\Rightarrow \qquad f'(x) = \displaystyle\sum_{i=1}^{n}2(x - a_i)$

$\qquad\qquad\qquad = 2nx - (a_1 + a_2 \ldots\ldots + a_n)$

$\qquad\qquad\qquad = 2nx - 2nA$ where $A = \dfrac{a_1 + a_2 + \ldots + a_n}{n}$

$\qquad\qquad f'(x) = 0 \Rightarrow x = A.$

Again $\qquad f''(x) = 2n > 0$

Thus the given sum has a minimum for $x = A$.

8. *Investigate the points for maxima and minima, the function*

$$f(x) = \int_{1}^{x}\left\{2(t-1)(t-2)^3 + 3(t-1)^2(t-2)^2\right\}dt.$$

$\qquad f'(x) = 2(x-1)(x-2)^3 + 3(x-1)^2(x-2)^2 \qquad\qquad\qquad$ [By Leibnitz-rule]

$\qquad\qquad = (x-2)^2(x-1)(5x-7)$

$\therefore \qquad\qquad f'(x) = 0$ for $\qquad x = 1, 2, 7/5.$

Now, $\qquad f''(x) = (x-2)^2(10x-12) + (5x^2 - 12x + 7). 2(x-2)$

Substituting $\quad x = 1, 2, 7/5$, we get

$\qquad\qquad f''(1) = -2 < 0 \qquad \therefore \qquad\qquad$ maximum when $x = 1$

$\qquad\qquad f''(2) = 0$

$\qquad f''(7/5) > 0 \qquad\qquad\qquad \therefore \qquad$ minimum when $x = 7/5.$

9. *If* $a > b > 0$ *and* $f(\theta) = \dfrac{(a^2 - b^2)\cos\theta}{a - b\sin\theta}$*, then find the maximum value of* $f(\theta)$.

We have $\qquad f(\theta) = \dfrac{a^2 - b^2}{a\sec\theta - b\tan\theta} = \dfrac{a^2 - b^2}{\phi(\theta)},$

where $\qquad \phi(\theta) = a\sec\theta - b\tan\theta$

$\therefore \quad f(\theta)$ is maximum and minimum as $\phi(\theta)$ is minimum and maximum respectively. Now,

$\qquad\qquad \phi'(\theta) = a\sec\theta\tan\theta - b\sec^2\theta$

For maximum and minimum

$\qquad\qquad \sec\theta(a\tan\theta - b\sec\theta) = 0$

$\Rightarrow \qquad\qquad \sin\theta = \dfrac{b}{a} \qquad\qquad\qquad\qquad \ldots(i) \qquad\qquad$ [as $\sec\theta \neq 0$]

$\phi''(\theta) = \sec\theta\tan\theta(a\tan\theta - b\sec\theta) + (a\sec^2\theta - b\sec\theta\tan\theta)\sec\theta$

$$= \frac{a + a \sin^2 \theta - 2b \sin \theta}{\cos^3 \theta}$$

Putting $\quad \sin \theta = \dfrac{b}{a}$

$$\phi''(b/a) = \frac{a + a \cdot \dfrac{b^2}{a^2} - 2b \cdot \dfrac{b}{a}}{\cos^3 \theta}$$

$$= \frac{a^2 - b^2}{a \cos^3 \theta} > 0 \quad \left\{ \because a > b > 0 \text{ and } \sin \theta = \frac{b}{a} \text{ is } + ve \right\}$$

$\therefore \quad \phi(\theta)$ is minimum when $\sin \theta = \dfrac{b}{a}$.

$\therefore \quad f(\theta)$ is maximum when $\sin \theta = \dfrac{b}{a}$.

$$\therefore \quad [f(\theta)]_{max} = \frac{(a^2 - b^2) \cdot \dfrac{\sqrt{(a^2 - b^2)}}{a}}{a - b \cdot \left(\dfrac{b}{a} \right)}$$

$$= \sqrt{(a^2 - b^2)}.$$

EXERCISES

1. Find the extreme values of
$$f(x) = 5x^6 + 18x^5 + 15x^4 - 10.$$

2. Show that the maximum and minimum values of the function defined by $(x + 1)(x + 4) / (x - 1)(x - 4)$ are -9 and $-1/9$ respectively.

3. Find the greatest and least values of the function
$$f(x) = x^4 - 8x^3 + 22x^2 - 24x + 1 \text{ in } [0, 2], [2, 4]$$

4. Investigate the maximum and minimum values of the function given by
   (i) $2x^3 - 15x^2 + 36x + 10$ (ii) $3x^4 - 4x^3 + 5$.

5. Find the values of x for which the function defined by
$$x^6 - 6ax^5 + 9a^2 x^4 + a^6 \text{ has minimum values } (a > 0).$$

6. Determine the values of x for which the function given by
$$12x^5 - 45x^4 + 40x^3 + 6$$
   attains (i) a maximum value (ii) a minimum value.

7. Find the maximum value of the function
$$f(x) = (x - 1)(x - 2)(x - 3).$$

8. Find the maxima and minima as well as the greatest and the least values of the function $f(x) = x^3 - 12x^2 + 45x$ in the interval $[0, 7]$.

9. Find for what values of x the following expression is a maximum or minimum :
$$2x^3 - 21x^2 + 36x - 20.$$

10. Find the extreme values of the expression $x^3 / (x^4 + 1)$.

11. Determine the value for which $\dfrac{x}{1 + x \tan x}$ has a maximum value.

(*Kumaon* 1997)

12. Find the maxima and minima of the radii vector of the curve

$$\dfrac{c^4}{r^2} = \dfrac{a^2}{\sin^2 \theta} + \dfrac{b^2}{\cos^2 \theta}.$$

(*Garhwal* 1999)

13. Show that the expression $(x + 1)^2 / (x + 3)^3$ has a maximum value 2/27 and a minimum value 0.

14. Show that $f(x) = x^5 - 5x^4 + 5x^3 - 1$ has a maximum value when $x = 1$, a minimum value when $x = 3$ and neither when $x = 0$.

15. Show that x^x is a minimum for $x = e^{-1}$.

16. Show that the maximum value of $(x)^{1/x}$ is $e^{1/e}$. (*Calicut 2004*)

17. Find the maximum value of $(\log x)/x$ in $]\,0, \infty\,[$. (*Bangalore 2005*)

18. Find the extreme value of $a^{x+1} - a^x - x$, $(a > 1)$.

19. Find the maximum and minimum values of $x + \sin 2x$ in $[0, 2\,\pi]$.

20. Find the values of x for which $\sin x - x \cos x$ is a maximum or a minimum.

21. Find in $[-\pi, \pi]$ the maximum and minimum values of

$$x - \sin 2x + \dfrac{1}{3} \sin 3x.$$

22. Show that $\sin x\,(1 + \cos x)$ is maximum when $x = \dfrac{1}{3}\,\pi$. (*Kumaon* 1998)

23. Discuss the maxima and minima in the interval $]\,0, \pi\,[$ of

(*i*) $\sin x + \dfrac{1}{2} \sin 2x + \dfrac{1}{3} \sin 3x,$

(*ii*) $\cos x + \dfrac{1}{2} \cos 2x + \dfrac{1}{3} \cos 3x.$

Also, obtain their greatest and the least values in the given interval.

24. Find the minimum and maximum values of

(*i*) $\sin x \cos 2x$ (*ii*) $a \sec x + b \csc x,\ (0 < a < b).$

(*iii*) $\sin x \cos^2 x$ (*iv*) $e^x \cos(x - a).$

ANSWERS

1. Max. -8 ; min. $-10, -26$

3. Greatest is 0 and the least is -8 in $[0, 2]$; the greatest is 1 and the least is -8 in $[2, 4]$.

4. (*i*) Max. 38; min. 37   (*ii*) min. 4.

5. Max. for $x = 2a$ and min. for $x = 0$ and $x = 3a$.

6. Max. for $x = 1$, min. for $x = 2$.   7. $2\sqrt{3}/9$.

8. Max. 54, min. 50; least 0, greatest 70.

9. Max. for $x = 1$; min. for $x = 6$.   10. Max. for $x = \sqrt[4]{3}$; min. for $x = -\sqrt[4]{3}$;

11. $x = \cos x$;   12. $r_{\max} = \dfrac{c^2}{(a + b)^2}$

17. Min. value : log $[(ae - e) \log a]/ \log a$.

18. Max. values : $(2\pi + 3\sqrt{3})/6$; $(8\pi + 3\sqrt{3})/6$.

    Min. values : $(4\pi - 3\sqrt{3})/6$; $(10\pi - 3\sqrt{3})/6$.

19. e^{-1}.

20. Max. for $x = n\pi$ where n is an odd positive or even negative integer, and minimum for $x = n\pi$ where n is an even positive or odd negative integer.

21. Min. values : $-\dfrac{1}{6}(5\pi + 3\sqrt{3} + 2), \dfrac{1}{6}(2\pi - 3\sqrt{3} + 2)$

    Max. value; $\dfrac{1}{6}(-\pi + 3\sqrt{3} - 2), \dfrac{1}{6}(5\pi + 3\sqrt{3} + 2)$.

23. (i) Min. value $\dfrac{\sqrt{3}}{4}$; max. values $\dfrac{4\sqrt{2} \pm 3}{6}$.

    Least value 0; greatest value $\dfrac{4\sqrt{2} + 3}{6}$.

    (ii) Min. values : $-\dfrac{1}{2}, -\dfrac{5}{6}$; max. values : $\dfrac{11}{6}, -\dfrac{5}{12}$.

    Least value : $-\dfrac{5}{6}$; greatest value : $\dfrac{11}{6}$.

24. (i) Min. $-1, -2/3\sqrt{6}$; max. $1, 2/3\sqrt{6}$.

    (ii) Min $(a^{2/3} + b^{2/3})^{3/2}$; max. $-(a^{2/3} + b^{2/3})^{3/2}$.

    (iii) Min. $0, -2/3\sqrt{3}$; max. $0, 2/3\sqrt{3}$.

    (iv) where n is any integer.

9.5. Application to Problems.

In the following, we shall apply the theory of maxima and minima to solve problems involving the use of the same. It will be seen that, in general, we shall not need to find the second derivative and complete decision would be made at the stage of the first derivative only when we have obtained the stationary values. In this connection, it will be found useful to determine the limits between which the independent variable lies *i.e.*, the domain of the independent values. Suppose that these limits are a, b.

If y is 0 for $x = a$ as well as for $x = b$ and positive otherwise and has only *one* stationary value, then the stationary value is necessarily the maximum and the greatest.

If $y \to \infty$ as $x \to a$ and $x \to b$ and has only *one* stationary value, then the stationary value is necessarily the minimum and the least.

In connection with the problems concerning spheres, cones and cylinders, the following results would be often needed :

1. *Sphere of Radius r.*

$$\text{Volume} = \frac{4}{3}\pi r^3, \qquad \text{Surface} = 4\pi r^2.$$

2. *Circular cylinder of height h and radius of base r.*

$$\text{Volume} = \pi r^2 h, \qquad \text{Curved surface} = 2\pi rh.$$

The area of each plane face $= \pi r^2$.

3. *Circular Cone of height, h and radius of base, r.*

Semi-vertical angle $= \tan^{-1}(r/h)$, Slant height $= \sqrt{(r^2 + h^2)}$.

Volume $= \dfrac{1}{3}\,\pi r h,$, Curved surface $= \pi r\,\sqrt{(r^2 + h^2)}$.

Here cones and cylinders are always supposed to be right circular.

EXAMPLES

1. *Show that the height of an open cylinder of given surface and greatest volume is equal to the radus of its base.*   **(G.N.D.U. Amritsar 2004)**

Let r be the radius of the circular base; h, the height; S the surface and V, the volume of the open cylinder so that

$$S = \pi r^2 + 2\,\pi r h, \qquad\qquad\qquad ...(i)$$
$$V = \pi r^2 h. \qquad\qquad\qquad ...(ii)$$

Here, as given, S is a constant and V a variable. Also, h, r are variables. Substituting the values of h, as obtained from (i) in (ii), we get

$$V = \pi r^2 \left(\frac{S - \pi r^2}{2\,\pi r}\right) = \frac{Sr - \pi r^2}{2}, \qquad\qquad ...(iii)$$

which gives V in terms of one variable r.

As V must be necessarily non-negative, we have

$$Sr - \pi r^3 \geq 0 \Rightarrow \pi r^3 \leq Sr \Rightarrow r \leq \sqrt{(S/\pi)}$$

Also r is non-negative.

Thus r varies in the interval, $[0, \sqrt{(S/\pi)}]$.

Now $\qquad \dfrac{dV}{dr} = \dfrac{S - 3\,\pi r^2}{2},$

so that $dV/dr = 0$ only when $r = \sqrt{(S/3\pi)}$: negative value of r being inadmissible. Thus V has only *one stationary* value.

Now $V = 0$ for the end points $r = 0$ and $\sqrt{(S/\pi)}$ and positive for every other admissible value of r. Hence V is greatest for

$$r = \sqrt{(S/3\,\pi)}.$$

Substituting this value of r in (i), we get

$$h = \frac{S - \pi r^2}{2\,\pi r} = \frac{S - \pi\,(S/3\,\pi)}{2\,\pi\,\sqrt{(S/3\,\pi)}}$$

$$= \frac{2S}{3} \cdot \frac{1}{2\,\pi}\sqrt{\left(\frac{3\,\pi}{S}\right)} = \sqrt{\left(\frac{S}{3\,\pi}\right)}.$$

Hence $h = r$ for the cylinder of greatest volume and given surface.

2. *Show that the radius of right circular cylinder of greatest curved surface which can be inscribed in a given cone is half that of the cone.*

Let r be the radius OA of the base and h, the height OV of the given cone.

We inscribe in it a cylinder, the radius of whose base is $OP = x$, as shown in the figure. We note that x may take up any value between 0 and r so that $x \in [0, r]$.

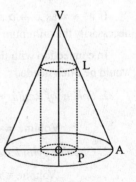

Fig. 9.4

To determine the height PL, of this cylinder, we have

$$\frac{PL}{OV} = \frac{PA}{OA}$$

$\Rightarrow$ $\qquad \dfrac{PL}{h} = \dfrac{r-x}{r}$

$\Rightarrow$ $\qquad PL = \dfrac{h(r-x)}{r}.$

If S be the curved surface of the cylinder,

we have $\qquad S = 2\pi . OP . PL = \dfrac{2\pi xh(r-x)}{r} = \dfrac{2\pi h}{r}(rx - x^2),$

$\Rightarrow$ $\qquad \dfrac{dS}{dx} = \dfrac{2\pi h}{r}(r - 2x) = 0$ for $x = r/2.$

Thus S has only one stationary value.

Now S is 0 for $x = 0$ as well as for $x = r$ and is positive for values of x lying between 0 and r. Therefore S is greatest for $x = r/2$.

3. *Find the surface of the right circular cylinder of greatest surface which can be inscribed in a sphere of radius r.*

We construct a cylinder as shown in the figure 9.5. OA is the radius of the base and CB is the height of this cylinder.

Let $\angle AOB = \theta$, so that θ lies between 0 and π.

We have $\qquad \dfrac{OA}{OB} = \cos \theta$

$\Rightarrow$ $\qquad OA = OB \cos \theta = r \cos \theta.$

Also $\qquad \dfrac{AB}{OB} = \sin \theta$

$\Rightarrow$ $\qquad AB = OB \sin \theta = r \sin \theta.$

If S be the surface, we have

$\qquad S = 2\pi . OA^2 + 2\pi . OA.BC = 2\pi r^2 (\cos^2 \theta + \sin 2\theta)$ ...(i)

$\Rightarrow$ $\qquad dS/d\theta = 2\pi r^2 (-2 \cos \theta \sin \theta + 2 \cos 2\theta)$

$\qquad\qquad\qquad = 2\pi r^2 (2 \cos 2\theta - \sin 2\theta)$

$\Rightarrow$ $\qquad dS/d\theta = 0$ for

$2 \cos 2\theta - \sin 2\theta = 0 \Leftrightarrow \tan 2\theta = 2$ ...(ii)

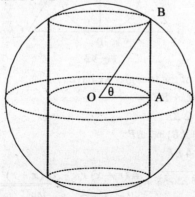

Fig. 9.5

The reader may established that the equation $\tan 2\theta = 2$ admits of *only* one value of $\theta \in [0, \pi/2]$ as its solution.

Let $\theta_1 \in [0, \pi/2]$ be the root of $\tan 2\theta = 2$.

Now $\tan 2\theta_1 = 2 \Rightarrow \sin 2\theta_1 = 2/\sqrt{5}$ and $\cos 2\theta_1 = 1/\sqrt{5}$.

From (*i*) we see that

$$\theta = 0 \Rightarrow S = 2\pi r^2$$

$$\theta = \frac{\pi}{2} \Rightarrow S = 0$$

$$\theta = \theta_1 \Rightarrow S = 2\pi r^2 \left(\frac{1 + \cos 2\theta_1}{2} + \sin 2\theta_1 \right)$$

$$= \frac{\pi r^2 (5 + 5\sqrt{5})}{5}$$

which is greater than $3\pi r^2$.

Hence $\dfrac{\pi r^2 (5 + \sqrt{5})}{5}$ is the required greatest surface.

4. *Prove that the least perimeter of an isosceles triangle in which a circle of radius r can be inscribed is $6r\sqrt{3}$.*

We take one vertex A of the triangle at a distance x from the centre O that $OA = x$. Surely $x > r$. Let AO meet the circle at P. The two tangents from A and the tangent at P determine an isosceles triangle ABC circumscribing the given circle. We have $OL = r$.

$\therefore$ $AL = \sqrt{(OA^2 - OL^2)} = \sqrt{(x^2 - r^2)}.$

Also $BP = AP \tan \angle BAP = AP \tan \angle LAO$

$$= AP \cdot \frac{OL}{AL} = (r + x)\frac{r}{\sqrt{(x^2 - r^2)}}.$$

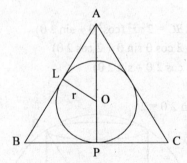

Fig. 9.6

If, p, denotes the perimeter of the triangle, we have

$$p = AB + AC + BC$$

$$= 2AB + 2BP$$

$$= 2(AL + LB) + 2BP$$

$$= 2AL + 4BP \qquad\qquad\qquad \text{(for } BL = BP)$$

$$= 2\sqrt{(x^2 - r^2)} + \frac{4(r + x)r}{\sqrt{(x^2 - r^2)}} = 2\frac{(x + r)^2}{\sqrt{(x^2 - r^2)}} \qquad\qquad ...(i)$$

$$\Rightarrow \qquad \frac{dp}{dx} = 2 \, \frac{2\,(x+r)\,\sqrt{(x^2-r^2)} - x\,(x+r^2)\,(x^2-r^2)^{-1/2}}{(x^2-r^2)^{3/2}}$$

$$= 2 \, \frac{(x+r)^2\,(x-2r)}{(x^2-r^2)^{3/2}}$$

$\Rightarrow \;\; dp/dx = 0$ for $x = 2r$; negative value, $-r$, of x being inadmissible.

Now, x may take up any value $> r$ only so that it varies in the interval $]\,r,\,\infty\,[$.

From (i), we see that $x \to r \Rightarrow p \to \infty$. Also $x \to \infty \Rightarrow p \to \infty$.

Hence p is least for $x = 2r$. Putting this value of x in (i), we see that the least value of p is $6r\sqrt{3}$.

5. *A cone is circumscribed to a sphere of radius r; show that when the volume of the cone is least its altitude is 4r and its semi-vertical angle is* $\sin^{-1} \dfrac{1}{3}$.

We take the vertex A of the cone at a distance x from the centre O of the sphere. (See Fig. 9.6 page 299).

By drawing tangent lines from A, as shown in the figure, we construct the cone circumscribing the sphere,

Let the semi-vertical angle BAP of the cone be θ.

Now, if v be the volume of the cone, we have

$$v = \frac{1}{3}\,\pi \cdot \mathrm{BP}^2 \cdot \mathrm{AP}.$$

which will now be expressed in terms of x.

We have $\qquad \mathrm{AP} = \mathrm{AO} + \mathrm{OP} = r + x.$

Also $\qquad \sin\theta = \dfrac{\mathrm{OL}}{\mathrm{OA}} = \dfrac{r}{x} \Rightarrow \tan\theta = \dfrac{r}{\sqrt{(x^2-r^2)}}$

Again $\qquad \dfrac{\mathrm{BP}}{\mathrm{AP}} = \tan\theta$

$\Rightarrow \qquad \mathrm{BP} = \mathrm{AP}\tan\theta = (r+x)\cdot\dfrac{r}{\sqrt{(x^2-r^2)}}$

Thus $\qquad v = \pi\,\dfrac{(r+x)^2\,r^2}{3\,(x^2-r^2)}\,(x+r) = \dfrac{\pi\,r^2\,(x+r)^2}{3\,(x-r)}.$

$\Rightarrow \qquad \dfrac{dv}{dx} = \dfrac{\pi\,r^2(x+r)\,(x-3r)}{3\,(x-r)^2}$

$\Rightarrow \;\; dv/dx$ is 0 for $x = 3r$.

Here x can take up any value $\geq r$.

Also $x \to r \Rightarrow v \to \infty$ and $x \to \infty \Rightarrow v \to \infty$.

Thus v is least for $x = 3r$.

Hence, for least volume, the altitude of the cone $= \mathrm{AP} = r + 3r = 4r$

and the semi-vertical angle $\theta = \sin^{-1}\dfrac{r}{x} = \sin^{-1}\dfrac{r}{3r} = \sin^{-1}\dfrac{1}{3}$.

6. *Normal is drawn at a variable point P of an ellipse*

$$x^2/a^2 + y^2/b^2 = 1 \; ;$$

find the maximum distance of the normal from the centre of the ellipse.

We take any point P ($a \cos \theta$, $b \sin \theta$) on the ellipse; θ being the eccentric angle of the point.

Because of the symmetry of the ellipse about the two coordinate axes, it is enough to consider only those of the values of θ which lie between 0 and $\pi/2$.

The equation of the tangent at P is $\dfrac{x \cos \theta}{a} + \dfrac{y \sin \theta}{b} = 1.$

Hence the slope of the normal at $P = a \sin \theta / b \cos \theta$.

Therefore equation of the normal at P is $ax \sin \theta - by \cos \theta = (a^2 - b^2) \sin \theta \cos \theta$.

If, 'P' be its perpendicular distance from the centre (0, 0), we obtain

$$p = \frac{(a^2 - b^2) \sin \theta \cos \theta}{\sqrt{(a^2 \sin^2 \theta + b^2 \cos^2 \theta)}}. \qquad \qquad ...(i)$$

$$\Rightarrow \qquad \frac{dp}{d\theta} = (a^2 - b^2) \frac{b^2 \cos^4 \theta - a^2 \sin^4 \theta}{(a^2 \sin^2 \theta + b^2 \cos^2 \theta)^{3/2}}$$

$$\Rightarrow \qquad \frac{dp}{d\theta} = 0 \text{ for } \tan^4 \theta = \frac{b^2}{a^2} \text{ i.e., for } \tan \theta = \pm \sqrt{\frac{b}{a}}$$

Now, $p = 0$ when $\theta = 0$ or $\pi/2$ and p is positive when θ lies between 0 and $\pi/2$. Therefore p is maximum when $\tan \theta = \sqrt{(b/a)}$. Substituting this value in (i), we see that the maximum value of p is $a - b$.

7. *Assuming that the petrol burnt (per hour) in driving a motor boat varies as the cube of its velocity. Show that the most economical speed when going against a current of c miles per hour is* $\dfrac{3}{2}$ *c miles per hour.*

Let v miles per hour be the velocity of the boat so that $v - c$ miles per hour is its velocity relative to water going against the current.

Therefore the time required to cover a distance of d miles

$$= d/(v - c) \text{ hours.}$$

The petrol burnt per hour = kv^3, where k is a constant. Thus the total amount, y, of petrol burnt is given by

$$y = k \cdot \frac{v^2 d}{v - c} = kd \frac{v^3}{v - c}$$

$$\Rightarrow \qquad \frac{dy}{dv} = kd \frac{3v^2 (v - c) - 1 . v^3}{(v - c)^2}$$

$$\Rightarrow \qquad \frac{dy}{dv} = 0 \text{ for } v = 0 \text{ and } \frac{3}{2} c, \text{ of these } v = 0 \text{ is inadmissible.}$$

Also $\qquad v \to c \quad \Rightarrow y \to \infty, v \to \infty \Rightarrow y \to \infty.$

Thus $v = \dfrac{3}{2} c$ gives the least value of c.

8. *A lane runs at right angles to a road 'a' feet wide. Find how many feet wide the lane must be if it is just possible to carry a pole 'b' feet long (b > a) from the road into the lane, keeping it horizontal.*

Let AB be the pole and Q be the angle it makes with the road.

Then

$$AC = a \operatorname{cosec} \theta.$$

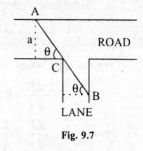

Hence $BC = AB - AC$

$$= b - a \operatorname{cosec} \theta$$

and $BD = BC \cos \theta$ (length of lane)

$$= (b - a \operatorname{cosec} \theta) \cos \theta$$

$$= b \cos \theta - a \cot \theta$$

Fig. 9.7

For maxima or minima,

$$\frac{d}{d\theta}(BD) = 0, \ i.e., -b \sin \theta + a \operatorname{cosec}^2 \theta = 0$$

or $\sin^3 \theta = \dfrac{a}{b}; \ \sin \theta = \left(\dfrac{a}{b}\right)^{1/3}$

Hence the required length $= b \cos \theta - a \cot \theta$

$$= b \sqrt{1 - \sin^2 \theta} - a \frac{\sqrt{1 - \sin^2 \theta}}{\sin \theta}$$

$$= b \sqrt{\frac{b^{2/3} - a^{2/3}}{b^{2/3}}} - a \cdot \sqrt{\frac{b^{2/3} - a^{2/3}}{a^{2/3}}}$$

$$= (b^{2/3} - a^{2/3}) \sqrt{(b^{2/3} - a^{2/3})}$$

$$= (b^{2/3} - a^{2/3})^{3/2}$$

9. *A Person being in a boat 'a' miles from the nearest point of the beach wishes to reach as quickly as possible a point 'b' miles from that point along the shore. The ratio of his rate of walking to his rate of rowing is sec α.*

Prove that they should land at a distance (b – a cot α) from the place to be reached.

Let B is the boat and $BC =$ breadth of the river as 'a' miles.

Let the person lands, at a distance x from C, $i.e., CP = x,$

$$PA = b - x \text{ and } BP$$

$$= \sqrt{(a^2 + x^2)}$$

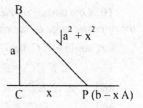

$$\frac{\text{Rate of walking}}{\text{Rate of rowing}} = \sec \alpha,$$

$i.e., \ \dfrac{W}{R} = \sec \alpha,$ where $W =$ rate of walking and $R =$ rate of rowing.

The person walked $= b - x$

$$\therefore \qquad t_1 = \frac{b - x}{R \sec \alpha}$$

and the boat rowed $\sqrt{(a^2 + x^2)}$

$$\therefore \qquad t_2 = \frac{\sqrt{(a^2 + x^2)}}{R}$$

$\therefore$ Total time $T = \dfrac{b - x}{R \sec \alpha} + \dfrac{\sqrt{(a^2 + x^2)}}{R}$

or $\qquad T = \dfrac{1}{R}[(b - x) \cos \alpha + \sqrt{(a^2 + x^2)}]$

Time is least, when

$$\frac{dT}{dx} = 0$$

Now, $\qquad \dfrac{dT}{dx} = \dfrac{1}{R}\left[- \cos \alpha + \dfrac{1}{2}(a^2 + x^2)^{-1/2} \cdot 2x \right]$

$$= \frac{1}{R}\left[- \cos \alpha + \frac{x}{\sqrt{(a^2 + x^2)}} \right]$$

$$\cos \alpha = \frac{x}{\sqrt{(a^2 + x^2)}}$$

or $\qquad a^2 + x^2 = x^2 \sec^2 \alpha$

or $\qquad a^2 = x^2(\sec^2 \alpha - 1) = x^2 \tan^2 \alpha$

$\therefore \qquad x = a \cot \alpha$

Now, $\qquad \dfrac{d^2 T}{dx^2} = \dfrac{1}{R}\left[\dfrac{\sqrt{(a^2 + x^2)} - x(a^2 + x^2)^{-1/2} \, x}{a^2 + x^2} \right]$

$$= \frac{1}{R}\left[\frac{a^2 + x^2 - x^2}{(a^2 + x^2)^{3/2}} \right]$$

$$= \frac{a^2}{R(a^2 + x^2)^{3/2}}$$

Hence $\dfrac{d^2 T}{dx^2}$ is positive at $x = a \cot \alpha$. Therefore, for $x = a \cot \alpha$, time is minimum.

Hence the required distance $= b - x = b - a \cot \alpha$

10. *One corner of a long rectangular sheet of paper of width 1 foot is folded over so as to reach the opposite edge of the sheet. Find the minimum length of the crease.*

Let angle FEC' be θ. FC' is the edge which has fallen at points F and C'. Then

$$\angle FEC = \theta$$

and $\qquad \angle C'ED = \pi - 2\theta$

and $\qquad EC' = EC = l \cos \theta$

where $\qquad l = \text{length of } ED.$

and then $\qquad ED = EC' \cos C'ED$

$$= l \cos \theta \cos (\pi - 2\theta)$$

$$= - l \cos \theta \cos 2\theta$$

Now $\qquad AB = 1' \text{ given.}$

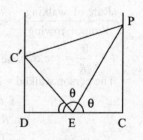

Fig. 9.8

$\therefore \qquad\qquad DC = 1'$

Hence, $\qquad\qquad DE + EC = 1$

or $\qquad\qquad -l \cos \theta \cos 2\theta + l \cos \theta = 1$

$\therefore \qquad\qquad \dfrac{1}{l} = \cos \theta \, (1 - \cos 2\theta)$

$-\dfrac{1}{l^2} \dfrac{dl}{d\theta} = (-\sin \theta)(1 - \cos 2\theta) + \cos \theta \, (\sin 2\theta) \cdot 2$

For maxima or minima, $\dfrac{dl}{d\theta} = 0$

$\therefore \qquad \sin \theta \, (1 - \cos 2\theta) - 2 \sin 2\theta \cos \theta = 0$

or $\qquad \sin \theta - \sin \theta \cos 2\theta - 2 \sin 2\theta \cos \theta = 0$

or $\qquad \sin \theta - \sin \theta + 2 \sin^3 \theta - 4 \sin \theta \cos^2 \theta = 0$

or $\qquad\qquad 2 \sin \theta \, (\sin^2 \theta - 2 \cos^2 \theta) = 0$

Either $\qquad \sin \theta = 0 \qquad$ or $\qquad \tan \theta = \pm \sqrt{2}$

Then it can be easily shown that $\dfrac{d^2 l}{d\theta^2}$ is negative for $\tan \theta = \sqrt{2}$, i.e., $\dfrac{1}{l}$ is maximum or l is minimum.

Hence minimum value of $l = \dfrac{1}{\cos \theta \, (1 - \cos 2\theta)} = \dfrac{1}{2 \sin^2 \theta \cos \theta}$

$\qquad\qquad\qquad = \dfrac{3\sqrt{3}}{4}$ ft.

11. *The range R of a shell (in empty space), fired with an initial velocity v_0 from a gun inclined to the horizontal at an angle ϕ, is determined by the formula*

$$R = \dfrac{v_0^2 \sin 2\phi}{g}$$

where g is the acceleration due to gravity. Determine the angle ϕ for which the range will be maximum for a given initial velocity v_0.

The quantity **R** is a function of the variable angle ϕ. We have to test this function for a maximum in the interval $0 \le \phi \le \dfrac{\pi}{2}$,

$$\dfrac{dR}{d\phi} = \dfrac{2 v_0^2 \cos 2\phi}{g}$$

For maxima or minima $\qquad \dfrac{dR}{d\phi} = 0$

i.e. $\dfrac{2 v_0^2 \cos 2\phi}{g} = 0 \qquad$ or $\qquad \cos 2\phi = 0$

$\therefore \qquad\qquad \phi = \dfrac{\pi}{4}.$

Also, $\qquad \dfrac{d^2 R}{d\phi^2} = -\dfrac{4 v_0^2 \sin 2\phi}{g}$

At $\phi = \dfrac{\pi}{4}, \dfrac{d^2 R}{d\phi^2} = -\dfrac{4 v_0^2}{g} = $ negative

Hence, for the value $\phi = \dfrac{\pi}{4}$ the function R has a maximum.

Then $R_{max} = \dfrac{v_0^2}{g}$

12. *Tangents are drawn to the ellipse* $\dfrac{x^2}{a^2} + \dfrac{y^2}{b^2} = 1$ *and the circle* $x^2 + y^2 = a^2$ *at the points where a common ordinate cuts them. Show that if* θ *be the greatest inclination of the tangents, then*

$$\tan \theta = \dfrac{a - b}{2 \sqrt{ab}}.$$

Let the common ordinate cut the ellipse and the circle in Q and P respectively. If the eccentric angle of Q be ϕ, then from the properties of the ellipse $\angle POM = \phi$. Thus the coordinates of Q and P are respectively $(a \cos \phi, b \sin \phi)$ and $(a \cos \phi, a \sin \phi)$.

Tangent to the given ellipse at θ is

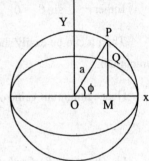

Fig. 9.9

$\dfrac{x}{a} \cos \phi + \dfrac{y}{b} \sin \phi = 1$

Hence gradient of this tangent $= -\dfrac{b}{a} \cot \phi = m_1$ (say)

Also the equation of the tangent to the given circle at P is

   $x \cos \phi + y \sin \phi = a$

$\therefore$ Its gradient $= - \cot \phi = m_2$ (say)

If θ be the angle between these tangents, then we have

$$\tan \theta = \dfrac{-\dfrac{b}{a} \cot \phi + \cot \phi}{1 + \dfrac{b}{a} \cot^2 \phi} = \dfrac{a - b}{a \tan \phi + b \cot \phi} \qquad \qquad ...(i)$$

Now θ will be maximum when $\tan \theta$ is maximum; *i.e.* when $a \tan \phi + b \cot \phi$ is minimum. Now, let

$z = a \tan \phi + b \cot \phi$

$\therefore$ $\dfrac{dz}{d\phi} = a \sec^2 \phi - b \ \mathrm{cosec}^2 \phi$

and $\dfrac{d^2 z}{d\phi^2} = 2a \sec^2 \phi \tan \phi + 2b \ \mathrm{cosec}^2 \phi \cot \phi$

For a maximum or a minimum z,

$\dfrac{dz}{d\phi} = 0$.

This gives $\tan^2 \phi = \dfrac{b}{a}$

i.e. $\tan \phi = \pm \sqrt{\left(\dfrac{b}{a}\right)}$

when $\quad \tan \phi = \sqrt{\dfrac{b}{a}}, \dfrac{d^2 z}{d\phi^2} = $ positive,

because for this ϕ, all the trigonometrical ratios are positive.

When $\quad \tan \phi = \sqrt{\left(\dfrac{b}{a}\right)}$, then $\dfrac{d^2 z}{d\phi^2} = $ negative

Putting this value in (i) we have

θ is maximum when $\tan \phi = -\sqrt{\left(\dfrac{b}{a}\right)}$,

Than $\tan \phi = -\dfrac{a-b}{a\sqrt{\dfrac{b}{a}} + b\sqrt{\dfrac{a}{b}}} = \dfrac{a-b}{2\sqrt{ab}}$

13. *A window of fixed perimeter (including the base of the arc) is in the form of a rectangle surmounted by a semi-circle. The semi-circular portion is fitted with coloured glass while the rectangular portion is fitted with clear glass. The clear glass transmits three times as much light per square meter as the coloured glass does. What is the ratio of the sides of the rectangle so that the window transmits the maximum light ?*

Let $2b$ be the diameter of the circular portion and a be the lengths of the other side of the rectangle.

∴ Total perimeter $= 2a + 4b + \pi b = \lambda$ (say)   ...(1)

Let the light transmission rate of the coloured glass be L per square meter and Q be the total amount of transmitted light. Then

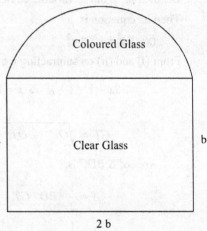

Fig. 9.10

$$Q = 2ab \cdot 3L + \frac{1}{2} \pi b^2 \cdot L$$

$$= \frac{L}{2} [\pi b^2 + 6b (\lambda - 4b - \pi b)]$$

$$= \frac{L}{2} [6\lambda b - 24b^2 - 5\pi b^2]$$

∴ $\quad \dfrac{dQ}{db} = \dfrac{L}{2} [6\lambda - 48b - 10\pi b] = 0$

⇒ $\quad b = \dfrac{6\lambda}{48 + 10\pi}$ ...(2)

and $\quad \dfrac{d^2 Q}{db^2} = \dfrac{L}{2} [-48 - 10\pi] < 0$

Thus, Q is maximum and from (1) and (2)

$(48 + 10\pi) b = 6\lambda = 6 [2a + 4b + \pi b]$

∴ The ratio $= \dfrac{2b}{a} = \dfrac{6}{6 + \pi}$.

14. *The circle* $x^2 + y^2 = 1$ *cuts the x-axis at A and B. Another circle with centre at B and variable radius intersects the first circle at C above the x-axis and the line segment AB at D. Find the maximum area of the triangle BDC.*

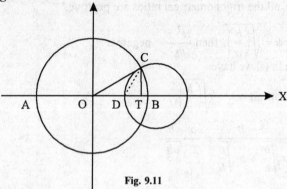

Fig. 9.11

The centre of the circle

$$x^2 + y^2 = 1 \qquad \qquad ...(i)$$

is $(0, 0)$ and coordinates of B are $(1, 0)$.

Let the radius of the variable circle be r.

Then its equation is

$$(x-1)^2 + y^2 = r^2 \qquad \qquad ...(ii)$$

From (i) and (ii) on subtracting we get

$$2x - 1 = 1 - r^2 \Rightarrow x = 1 - \frac{r^2}{2} = OT$$

$$\therefore \qquad CT = \sqrt{OC^2 - OT^2} = \sqrt{1 - \left(1 - \frac{r^2}{2}\right)^2}$$

$\therefore$ Area of $\Delta\, BDC$ is,

$$A = \frac{1}{2} \cdot BD \cdot CT$$

$$\Rightarrow \qquad A^2 = \frac{1}{4} r^2 \left(r^2 - \frac{r^4}{4}\right)$$

$$= \frac{1}{16} (4r^4 - r^6)$$

$$\frac{d(A^2)}{dr} = \frac{1}{16} (16r^3 - 6r^5) = 0 \qquad \qquad \text{[For maxima, minima]}$$

$$\Rightarrow \qquad r = 2\sqrt{\frac{2}{3}}.$$

Also, $\qquad \dfrac{d^2(A^2)}{dr^2} = \dfrac{1}{16}(48\, r^2 - 30\, r^4) = -\dfrac{16}{3} < 0 \qquad \qquad \text{[For } r = 2\sqrt{\dfrac{2}{3}} \text{]}$

Hence, area is maximum at $r = 2\sqrt{\dfrac{2}{3}}$ and $A_{max} = \dfrac{4}{3\sqrt{3}}$ sq. units.

15. *Narendra has x children by his first wife. Pramila has (x + 1) children by her first husband. They marry and have children of their own. The whole family has 24 children. Assuming the two* *children of same parents do not fight, then find the maximum possible number of fights that can take place.*

Since the whole family has 24 children, those of Narendra and Pramila are

$$24 - x - (x + 1) = 23 - 2x$$

$$F = \text{Total number of fights}$$

= (number of fights when a Narendra's child fights a Pramila's child) + (number of fights when a Narendra's child fights a Narendra-Pramila child) + (number of fights when a Pramila's child fights a Narendra-Pramila child)

$$= x(x + 1) + x(23 - 2x) + (x + 1)(23 - 2x)$$

$$= 23 + 45x - 3x^2$$

For maximum

$$\frac{dF}{dx} = 45 - 6x = 0 \Rightarrow x = 7.5$$

But in this case fractional value is not possible.

The nearest integral values are $x = 7$ and $x = 8$.

In either case the total number of fights

$$= 23 + 45 \times 7 - 3 \times (7)^2 = 191.$$

16. *From point A located on a highway a man has to get by bus to his office B located in a lane at a distance l from the highway in the least possible time. At what distance from D should the bus leave the highway when the bus moves in the lane n times slower than on highway?*

Let $AD = s$ and $CD = x$, where C is a point where the bus leaves the highway.

Let the speed of the bus is v on highway.

$\therefore$ Total time taken

$$t = \frac{AC}{v} + \frac{BC}{(v/n)}$$

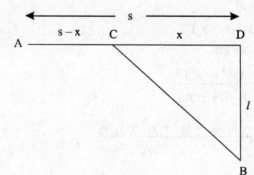

Fig. 9.12

$$= \frac{s - x}{v} + \frac{n\sqrt{l^2 + x^2}}{v}$$

For extreme, differentiating w.r.t. x, we have

$$\frac{dt}{dx} = \frac{1}{v}\left[1 + \frac{n}{2\sqrt{l^2 + x^2}}(2x)\right] = 0$$

$$\Rightarrow \qquad n^2 x^2 = l^2 + x^2 \Rightarrow x = \frac{l}{\sqrt{(n^2 - 1)}}.$$

Thus t is minimum when $x = \dfrac{l}{\sqrt{(n^2 - 1)}}$.

17. *Two men are walking on a path* $x^3 + y^3 = a^3$ *when the first man arrives at a point* (x_1', y_1), *he finds the second man in the direction of his own instantaneous motion. If the coordinates of the second man are* (x_2, y_2) *then show that*

$$\frac{x_2}{x_1} + \frac{y_2}{y_1} + 1 = 0.$$

Since (x_1, y_1) and (x_2, y_2) lie on the curve.

$$x_1^3 + y_1^3 = a^3 \qquad\qquad\qquad ...(i)$$
and $\qquad x_2^3 + y_2^3 = a^3 \qquad\qquad\qquad ...(ii)$

Subtracting

$$(x_2^3 - x_1^3) = -(y_2^3 - y_1^3) \qquad\qquad\qquad ...(iii)$$

Differentiating $x^3 + y^3 = a^3$ w.r.t. x., we get

$$3x^2 + 3y^2 \frac{dy}{dx} = 0$$

Slope of tangent at $(x_1, y_1) = -\dfrac{x_1^2}{y_1^2}$

$\therefore$ Equation of tangent at (x_1, y_1) :

$$y - y_1 = -\frac{x_1^2}{y_1^2}(x - x_1)$$

It passes through (x_2, y_2)

$$\therefore \qquad y_2 - y_1 = -\frac{x_1^2}{y_1^2}(x_2 - x_1)$$

or $\quad x_1^2 (x_2 - x_1) = -y_1^2 (y_2 - y_1) \qquad\qquad\qquad ...(iv)$

Dividing (iii) and (iv), we get

$$\frac{x_2^3 - x_1^3}{x_1^2 (x_2 - x_1)} = \frac{-(y_2^3 - y_1^3)}{-y_1^2 (y_2 - y_1)}$$

$$\Rightarrow \qquad \frac{x_2^2 + x_1^2 + x_1 x_2}{x_1^2} = \frac{y_2^2 + y_1^2 + y_1 y_2}{y_1^2}$$

$$\Rightarrow \qquad \left(\frac{x_2}{x_1}\right)^2 - \left(\frac{y_2}{y_1}\right)^2 = \left(\frac{y_2}{y_1}\right) - \left(\frac{x_2}{x_1}\right)$$

$$\Rightarrow \left(\frac{x_2}{x_1} - \frac{y_2}{y_1}\right)\left(\frac{x_2}{x_1} + \frac{y_2}{y_1} + 1\right) = 0$$

i.e.,   either $\dfrac{x_2}{x_1} = \dfrac{y_2}{y_1}$ or $\dfrac{x_2}{x_1} + \dfrac{y_2}{y_1} + 1 = 0$

But, $\dfrac{x_2}{x_1} \neq \dfrac{y_2}{y_1}$

$\therefore$ $\dfrac{x_2}{x_1} + \dfrac{y_2}{y_1} + 1 = 0.$

EXERCISES

1. Divide a number 15 into two parts such that the square of one multiplied with the cube of the other is a maximum. **(G.N.D.U Amritsar 2004)**

2. Show that of all rectangles of given area, the square has the smallest perimeter.

3. Find the rectangle of greatest perimeter which can be inscribed in a circle of radius a. **(Banglore 2005)**

4. If 40 square feet of sheet metal are to be used in the construction of an open tank with a square base, find the dimensions so that the capacity is greatest possible.

5. A figure consists of a semi-circle with a rectangle on its diameter. Given that the perimeter of the figure is 20 feet, find its dimensions in order that its area may be maximum.

6. A, B are fixed points with co-ordinates $(0, a)$ and $(0, b)$ and P is a variable point $(x, 0)$ referred to rectangular axes; prove that $x^2 = ab$ when the angle APB is a maximum.

7. A given quantity of metal is to be cast into a half cylinder, *i.e.*, with a rectangular base and semi-circular ends. Show that in order that the total surface area may be minimum the ratio of the length of the cylinder to the diameter of its circular ends is $\pi/(\pi + 2)$.

8. The sum of the surface of a cube and a sphere is given; show that when the sum of their volumes is least, the diameter of the sphere is equal to the edge of the cube.

9. The strength of a beam varies as the product of its breadth and the square of its depth. Find the dimensions of the strongest beam that can be cut from a circular log of wood of radius a units.

10. The amount of fuel consumed per hour by a certain steamer varies as the cube of its speed. When the speed is 15 miles per hour, the fuel consumed is $4\dfrac{1}{2}$ tons of coal per hour at Rs. 4 per ton. The other expenses total Rs. 100 per hour. Find the most economical speed and the cost of a voyage of 1980 miles.

11. Show that the semi-vertical angle of the cone of maximum volume and of given slant height is $\tan^{-1}\sqrt{2}$.

12. Show that the right circular cylinder of the given surface and maximum volume is such that its height is equal to the diameter of its base.

13. Show that the height of a closed cylinder of given volume and least surface is equal to its diameter.

14. Given the total surface of the right circular cone, show that when the volume of the cone is maximum, then the semi-vertical angle will be $\sin^{-1} 1/3$.

15. Show that the right cone of least curved surface and given volume has an altitude equal to $\sqrt{2}$ times the radius of its base. **(Kumaon 1995)**

16. Show that the curved surface of a right circular cylinder of greatest curved surface which can be inscribed in a sphere in one half of that of the sphere.

17. A cone is inscribed in a sphere of radius r, prove that its volume as well as its curved surface is greatest when its altitude is $4r/3$.

18. Find the volume of the greatest cylinder that can be inscribed in a cone of height h and semi-vertical angle α.   *(Garhwal 2003)*

19. Prove that the area of the triangle formed by the tangent at any point of the ellipse $x^2/a^2 + y^2/b^2 = 1$ and its axes is a minimum for the point
$$(a/\sqrt{2}, b/\sqrt{2}).$$

20. Find the area of the greatest isosceles triangle that can be inscribed in a given ellipse, the triangle having its vertex coincident with one extremity of the major axis.

21. A perpendicular is let fall from the centre to a tangent to an ellipse. Find the greatest volume of the intercept between the point of contact and the foot of the perpendicular.

22. A tangent to an ellipse meets the axes in P and Q : show that the least value of PQ is equal to the sum of the semi-axes of the ellipse and also that PQ is divided at the point of contact in the ratio of its semi-axes.

23. N is the foot of the perpendicular drawn from the centre O on to the tangent at a variable point P on the ellipse $x^2/a^2 + y^2/b^2 = 1$ $(a > b)$. Prove that the maximum area of the triangle OPN is $(a^2 - b^2)/4$.

24. One corner of a rectangular sheet of the paper, width one foot, is folded over so as just reach the opposite edge of the sheet; find the minimum length of the crease.

25. A grocer requires cylindrical vessels of thin metal with lids, each to contain exactly a given volume V. Show that if he wishes to be as economical as possible in metal, the radius r of the base is given by $2\,\pi r^3 = V$.

    If, for other reasons, it is impracticable to use vessels in which the diameter exceeds, three-fourths of the height, what should be the radius of the base of each vessel ?

26. A tree trunk, l feet long is in the shape of a frustum of a cone the radii of its ends being a and b feet $(a > b)$. It is required to cut from it a beam of uniform square section. Prove that the beam of the greatest volume that can be cut is $al/3\,(a - b)$ feet long.

27. Find the volume of the greatest right circular cone that can be described by the revolution about a side of a right-angled triangle of hypotenuse 1 foot.

28. A rectangular sheet of metal has four equal square portions removed at the corners, and the sides are then turned up so as to form an open rectangular box. Show that when volume contained in the box is a maximum, the depth will be
$$\frac{1}{4}\left[(a + b) - (a^2 - ab + b^2)^{1/2}\right],$$
where a, b are the sides of the original rectangle.

29. The parcel post regulations restrict parcels to be such that the length plus the girth must not exceed 6 feet and the length must not exceed $3\frac{1}{2}$ feet. Determine the parcels of greatest volume that can be sent up by post if the form of the parcel be a right circular cylinder. Will the result be affected if the greatest length permitted were only $1\frac{3}{4}$ feet ?

30. Show that the maximum rectangle inscribed in a circle is a square.

31. Find the point on the curve $4x^2 + a^2y^2 = 4a^2$; $4 < a^2 < 8$ that is farthest from the point $(0, -2)$.

32. A despatch rider is in open country at a distance of 6 km from the nearest point P of a straight road. He wishes to proceed as quickly as possible to a point Q on the road 20 km from P. If his maximum speed across country is 40 km/hr, than at what distance from P, he should strike the road?

33. LL' be latus-rectum of the parabola $y^2 = 4ax$ and PP' is a double ordinate between the vertex and the latus-rectum. Show that the area of trapezium $PP'L'L$ is maximum when the distance of PP' from the vertex is $a/9$.

34. A small disc A slides with zero initial velocity from the top of a smooth hill of height H having a horizontal portion. Determine the ratio h/H so that R is maximum.

35. Two corridors of widths a and b are at right angles. Show that the length of the longest pipe that can be passed round the corner horizontally is $(a^{2/3} + b^{2/3})^{3/2}$.

ANSWERS

1. 9, 6.

3. Square with side $\sqrt{2}a$.

4. $\sqrt{(40/3)}, \sqrt{(40/3)}, \frac{1}{2}\sqrt{(40/3)}$.

5. Diameter of the semi-circle $= 40/(\pi + 4)$.

   Height of the rectangle $= 20/(\pi + 4)$.

9. Breadth $\dfrac{\sqrt{4}}{3}a$, depth $\dfrac{\sqrt{8}}{3}a$.

10. $15\left(\dfrac{8}{3}\right)^{2/3}$ miles per hour, $19800\left(\dfrac{3}{5}\right)^{2/3}$ rupees.

18. $\dfrac{4}{27}\pi h^3 \tan^2\alpha$.   20. $3\sqrt{3}\,ab/4$.   21. $a - b$.

24. $3\sqrt{3}/4$.   25. $(3v/8\pi)^{1/3}$.   27. $3\sqrt{3}\,\pi/27$.

29. Length 2 ft., girth 4 ft.; yes, length should now be $1\dfrac{3}{4}$ ft. and girth $4\dfrac{1}{4}$ ft.

31. $(0, 2)$ 32. 8 km   34. $1/2$.

Use of third and higher order derivatives for the determination of maximum and minimum values.

In, § 9.4, we showed that if $f'(c) = 0$ and $f''(c) \neq 0$, then

$\qquad f''(c) > 0 \Rightarrow f(c)$ is a minimum value of $f(x)$,

$\qquad f''(c) < 0 \Rightarrow f(c)$ is a maximum value of $f(x)$.

This result does not refer to what happens if $f''(c) = 0$. We now make use of the Higher Mean Value Theorem to obtain a generalisation of this result. As shown in the following article the same thing can sometimes be done more conveniently by employing derivation of the second and higher orders. It will be assumed that f possesses continuous derivatives of every order that appear in the discussion in some neighbourhood of c. We suppose that $c + h$ belongs to this neighbourhood.

We suppose that

$$0 = f''(c) = f'''(c) = \ldots\ldots = f^{n-1}(c) \text{ and } f^n(c) \neq 0.$$

We now have

$$f(c + h) = f(c) + hf'(c) + \frac{h^2}{2!}f''(c) + \ldots$$

$$+ \ldots\ldots + \frac{h^{n-1}}{(n-1)!}f^{n-1}(c) + \frac{h^n}{n!}f^n(c + \theta h)$$

so that under the conditions as stated, we have

$$f(c + h) - f(c) = \frac{h^n}{n!} f^n(c + \theta h), \quad 0 < \theta < 1.$$

Let $f^n(c) > 0$. There exists a neighbourhood $]\,c - \delta, c + \delta\,[$ of c for every point x of which $f^n(x) > 0$. It follows that if n is even, then for every point $c + h$ of this neighbourhood

$$f(c + h) - f(c) > 0$$

and as such $f(c)$ is a minimum value of the function.

Also if n is odd,

$$f(c + h) - f(c) < 0 \qquad \text{when} \quad c + h \in [\,c - \delta, c\,[$$

and $\qquad f(c + h) - f(c) > 0 \qquad \text{when} \quad c + h \in]\,c, c + \delta\,[$

and it follows that in this case $f(c)$ is not an extreme value of the function.

Let $f^n(c) < 0$. As above it may be shown that $f(c)$ is a maximum value if n is even and not an extreme value if n is odd.

EXAMPLE

1. *Find the maximum and minimum values of the function*

$$f(x) = 8x^5 - 15x^4 + 10x^3.$$

We have $\quad f'(x) = 40x^4 - 60x^3 + 30x^2$

$$= 20x\,(x - 1)^2\,(2x + 1)$$

$$f'(x) = 0 \text{ for } x = 0, 1, -1/2$$

We have already discussed the cases when $x = 0, -1/2$

for $\qquad\qquad x = 1, f''(1) = 0.$

$$f'''(x) = 20\,(16x^2 - 18x)$$

$$f'''(1) \neq 0$$

∴ For $x = 1$ function has neither maximum or minimum.

9.6 Maxima and minima of functions of two variables

Let $f(x, y)$ be a function of two independent variables x, y such that it is continuous and finite for all values of x and y in the neighbourhood of their values a and b (say) respectively. Then the value of $f(a, b)$ is called maximum or minimum value of $f(x, y)$ according as $f(a + h, b + k)$, $<$ or $> f(a, b)$ whatever be the relative values of h and k, positive or negative, provided h and k are finite and sufficienlty small.

9.6.1 Condition for the existence of Maxima or Minima

We know by Taylor's expansion in two variables, that

$$f(x + h, y + k) = f(x, y) + \left(h\frac{\partial f}{\partial x} + k\frac{\partial f}{\partial y} \right)$$

$$+ \frac{1}{2!}\left(h^2\frac{\partial^2 f}{\partial x^2} + 2hk\frac{\partial^2 f}{\partial x\partial y} + k^2\frac{\partial^2 f}{\partial y^2} \right) + \dots.$$

or $\quad f(x + h, y + k) - f(x, y)$

$$= \left(h\frac{\partial f}{\partial x} + k\frac{\partial f}{\partial y} \right) + \text{terms of second and higher orders} \qquad \dots(i)$$

Now, from the definition of maxima or minima, it is evident that there shall be a maximum or a minimum for these values of x and y for which the expression $f(x+h, y+k) - f(x, y)$ for all finite by sufficiently small positive or negative values of h and k is of invariable sign.

Now if h and k sufficiently small, then the sign of the right hand of (i) will depend on the first degree terms in h, k and therefore the sign of left hand side of (i) will also depend on the first degree terms in h, k on the right hand side of (i). Thus if the sign of both h and k be changed then the sign of the left hand side of (i) will also be changed. Hence the necessary condition for a maximum or minimum values is

$$h \frac{\partial f}{\partial x} + k \frac{\partial f}{\partial y} = 0 \qquad \qquad ...(ii)$$

As (ii) is true for all values of h and k, so we have

$$\frac{\partial f}{\partial x} = 0, \frac{\partial f}{\partial y} = 0 \qquad ...(iii) \qquad \qquad \textbf{\textit{(Sagar 2000; Jabalpur 2002)}}$$

as the necessary condition for a maximum or a minimum value of $f(x, y)$.

Note that the above conditions are not sufficient for the existence of a maximum or a minimum value of $f(x, y)$.

9.6.2 Stationary and extreme Points

A point (a, b) is called a *stationary point* if all the first order partial derivatives of the function $f(x, y)$ vanish at that point. A *stationary point*, if it is a maximum or a minimum is known as an *extreme point* and the value of the function at an *extreme point* is known as an *extreme value*.

From above we conclude that a stationary point may be a maximum or a minimum or neither of these two.

9.6.3 Sign of quadratic expressions : Algebraic Lemma

We know that

$$ax^2 + 2hxy + by^2 = \frac{1}{a}[a^2 x^2 + 2ahxy + aby^2]$$

$$= \frac{1}{a}[(ax + by)^2 + (ab - h^2) y^2] \qquad \qquad ...(i)$$

If $ab - h^2 > 0$, then from (i) the sign of expression $ax^2 + 2hxy + by^2$ will be the same as that of a.

If $ab - h^2 > 0$, then nothing can be said about the sign of expression in the square bracket on the right hand of (i) and hence that of the quadratic expression $ax^2 + 2hxy + by^2$.

9.6.4 Lagrange's condition for maximum and minimum values of a function of two variables.

(Gorakhpur 2002)

If r, s, t denote the values of $\frac{\partial^2 f}{\partial x^2}, \frac{\partial^2 f}{\partial x \partial y}, \frac{\partial^2 f}{\partial y^2}$ when $x = a, y = b$ then supposing that the

necessary conditions for the maximum and minimum are satisfied, i.e., $\partial f / \partial x = 0, \partial f / \partial y = 0$, when $x = a, y = b$, we can write

$$f(a + h, b + k) - f(a, b) = \frac{1}{2!}[r h^2 + 2 shk + t k^2] + R \qquad \qquad ...(i)$$

Where R consists of terms of higher orders of h and k.

Now if h and k be sufficiently small, then sign of the right hand side of (i) depend on the second degree terms on the right hand side of (i) and therefore the sign of the left hand side of (i) depends on the second degree terms on the right hand side of (i). If for all positive values of h, k this sign is positive then we have a minimum but if it is negative then we have a maximum at (a, b).

Now, by Algebraic Lemma of § 9.6.3 the second degree terms $rh^2 + 2hsk + tk^2$ shall always be positive if $rt - s^2$ is positive and r is positive.

Hence **Lagrange's Condition for minimum is**

$$rt - s^2 > 0 \text{ and } r > 0$$

Also the second degree term $rh^2 + 2hsk + tk^2$ in (i) shall always be negative if $rt - s^2$ is positive and r is negative.

Hence **Lagrange's Condition for maximum is**

$$rt - s^2 > 0 \text{ and } r < 0$$

But if $rt - s^2 < 0$; then there is **neither a maximum nor a minimum.**

If $rt - s^2 = 0$, then from (i) we have the quadratic expression $rh^2 + 2hsk + tk^2$

$= \dfrac{1}{r}(rh + ks)^2$ and therefore of the same sign as r. Hence in this case the maximum or minimum value of the function will depend on the sign of r.

If $rt - s^2 = 0$ and $\dfrac{h}{k} = -\dfrac{s}{r} = \alpha$ (say), then second degree terms of right hand side of (i) vanish and we shall have to consider higher degree terms on the right hand side of (i). Now the third degree terms must collectively vanish when $\dfrac{h}{k} = \alpha$, otherwise by changing the sign of both h and k, the sign of the expression $f(a + h, b + k) - f(a, b)$ can be changed and in this case the fourth degree terms must collectively be of the same sign as r and t when $h/k = \alpha$.

EXAMPLES

1. *Discuss the maximum or minimum values of u given by*
   $$u = x^3 y^2 (1 - x - y).$$ *(Gorakhpur 2000, 2002 ; Kumaon 2000, 2002 ; Sagar 1997 ; Bhopal 96, Jiwaji 2000, 2002 ; Jabalpur 96 ; Bilaspur 98 ; Indore 2001)*

We have
$$\frac{\partial u}{\partial x} = 3 x^2 y^2 (1 - x - y) + x^3 y^2 (-1)$$

$$= 3 x^2 y^2 - 4 x^3 y^2 - 3 x^2 y^3,$$

$$\frac{\partial u}{\partial y} = 2 x^3 y (1 - x - y) + x^3 y^3 (-1)$$

$$= 2 x^3 y - 2 x^4 y - 3 x^3 y^2$$

$$r \equiv \frac{\partial^2 u}{\partial x^2} = 6xy^2 - 12x^2 y^2 - 6x y^3,$$

$$t = \frac{\partial^2 u}{\partial y^2} = 2x^3 - 2x^4 - 6x^3 y$$

and
$$s = \frac{\partial^2 u}{\partial x \partial y} = 6x^2 y - 8x^3 y - 9x^2 y^2$$

Now for maximum or minimum we must have
$$\frac{\partial u}{\partial x} = 0 \text{ and } \frac{\partial u}{\partial y} = 0$$

from $\dfrac{\partial u}{\partial x} = 0$, we get

$$3x^2 y^2 - 4 x^3 y^2 - 3 x^2 y^3 = 0$$ ...(*i*)

or $\qquad x^2 y^2 (3 - 4x - 3y) = 0,$

Hence we get $x = 0, y = 0, 4x + 3y = 3$

Also from $\dfrac{\partial u}{\partial y} = 0$, we get

$$2 x^3 y - 2 x^4 y - 3 x^3 y^2 = 0$$

or $\qquad x^3 y (2 - 2 x - 3 y) = 0$

Hence we get $x = 0, y = 0, 2x + 3 y = 2$

Solving $\qquad 4 x + 3 y = 3$ and $2 x + 3 y = 2$

we get $\qquad x = \dfrac{1}{2}, y = \dfrac{1}{3}$

Hence the solutions are

$$x = 0, y = 0, \; x = \dfrac{1}{2}, y = \dfrac{1}{3}, x = \dfrac{1}{3}, y = 0 \text{ and } x = 0 \;\; y = \dfrac{1}{3}.$$

when $\qquad x = \dfrac{1}{2}, y = \dfrac{1}{3}$ we have

$$r = 6 \left(\dfrac{1}{2}\right)\left(\dfrac{1}{3}\right)^3 - 12 \left(\dfrac{1}{2}\right)^2 \left(\dfrac{1}{3}\right)^2 - 6 \left(\dfrac{1}{2}\right)\left(\dfrac{1}{3}\right)^3 = -\dfrac{1}{4}$$

$$s = 6 \left(\dfrac{1}{2}\right)^2 \left(\dfrac{1}{3}\right) - 8 \left(\dfrac{1}{2}\right)^3 \left(\dfrac{1}{3}\right) - 9 \left(\dfrac{1}{2}\right)^2 \left(\dfrac{1}{3}\right)^3 = \dfrac{-1}{12}$$

$$t = 2 \left(\dfrac{1}{2}\right)^3 - 2 \left(\dfrac{1}{3}\right)^3 - 6 \left(\dfrac{1}{2}\right)^3 \left(\dfrac{1}{3}\right) = -2/27$$

From these we have $rt - s^2 > 0$ and $r < 0$, so there is maximum at $x = \dfrac{1}{2}, y = \dfrac{1}{3}$.

Similarly we can discuss other solutions.

2. *Show that minimum value of*

$$u = x y + (a^3 / x) + (a^3 / y) \text{ is } 3 \, a^2.$$

(*Indore* **1996** ; *Jabalpur* **2002** ; *Kanpur* **2001**; *Kumaon* **2004**)

We have $\qquad \dfrac{\partial u}{\partial x} = y - \dfrac{a^3}{x^2} ; \dfrac{\partial u}{\partial y} = x - \dfrac{a^3}{y^2},$

$$r = \dfrac{\partial^2 u}{\partial x^2} = \dfrac{2a^3}{x^3} ; \quad s = \dfrac{\partial^2 u}{\partial x \, \partial y} = 1$$

$$t = \dfrac{\partial^2 u}{\partial y^2} = \dfrac{2a^3}{y^3}.$$

Now for maximum or minimum we must have

$$\frac{\partial u}{\partial x} = 0, \frac{\partial u}{\partial y} = 0$$

So from $\quad\dfrac{\partial u}{\partial x} = 0$

we get $\quad y - \dfrac{a^3}{x^2} = 0$

or $\quad\quad x^2 y = a^3$ ...(i)

and from $\quad\dfrac{\partial u}{\partial y} = 0$, we have $x - \dfrac{a^3}{y^2} = 0$

or $\quad\quad x y^2 = a^3$ ...(ii)

Solving (i) and (ii) we find that

$$x^2 y = x y^2 \text{ or } x y (x - y) = 0$$

or $\quad\quad x = 0, y = 0 \quad\text{and}\quad x = y$

From (i) and (ii) we see that $x = 0$ and $y = 0$ do not hold as it gives $a = 0$, which is inadmissible.

Hence we have $x = y$ and from (i) we get $x = a$ and therefore we have

$$x = y = a.$$

At $x = y = a$, we have

$$r = 2\frac{a^3}{a^3} = 2, s = 1, t = 2$$

$\therefore\quad\quad r t - s^2 = (2)(2) - 1^2 = 3 > 0.$

Also $\quad\quad r = 2 > 0.$

Hence there is a minima at $x = y = a$.

Hence the minimum value of $u = a + a\dfrac{a^3}{a} + \dfrac{a^3}{a} = a^2 + a^2 + a^2 = 3a^2.$

3. *Discuss the maximum and minimum values of*

$$2 \sin \frac{1}{2} (x + y) \cos \frac{1}{2} (x - y) + \cos (x + y).$$

or

$$u = \sin x + \sin y + \sin (x + y)$$

(Bilaspur 2001; Vikram 2001; Jabalpur 98 ; Bhopal 1998, 2000; Indore 2000 ; Kanpur 2002; MDU Rohtak 2001)

Let $\quad\quad u = 2 \sin \dfrac{1}{2} (x + y) \cos \dfrac{1}{2} (x - y) + \cos (x + y)$

$$= \sin x + \sin y + \cos (x + y)$$

$\therefore\quad\dfrac{\partial u}{\partial x} = \cos x - \sin (x + y); \dfrac{\partial u}{\partial y} = \cos y - \sin (x + y);$

$$r = \frac{\partial^2 u}{\partial x^2} = -\sin x - \cos (x + y); s = \frac{\partial^2 u}{\partial x \partial y} = -\cos (x + y);$$

$$t = \frac{\partial^2 u}{\partial y^2} = -\sin y - \cos(x+y)$$

For maximum or minimum we must have $\frac{\partial u}{\partial x} = 0$, $\frac{\partial u}{\partial y} = 0$.

From $\qquad \frac{\partial u}{\partial x} = 0$, $\qquad$ we get $\cos x - \sin(x+y) = 0$ $\qquad\qquad$ ...(i)

From $\qquad \frac{\partial u}{\partial y} = 0$, $\qquad$ we get $\cos y - \sin(x+y) = 0$ $\qquad\qquad$ ...(ii)

Solving (i) and (ii) we get

$$\cos x = \cos y$$

This gives $x = 2 n \pi \pm y$, where n is any interger. In particular $x = y$.

When $x = y$, from (i)

we get $\qquad \cos x - \sin 2x = 0$ $\qquad$ or $\qquad \cos x (1 - 2\sin x) = 0$.

This gives $\quad \cos x = 0$, $\sin x = \frac{1}{2}$.

If $\sin x = \frac{1}{2}$, then $x = n\pi + (-1)^n \cdot \frac{\pi}{6} = y$

And for these values of x and y we have

$$r = -\frac{1}{2} - \cos\left[2 n \pi + (-1)^n \cdot \frac{\pi}{3}\right] < 0$$

Similarly, $t < 0$ and $s < 0$ and $r > s$, $t > s$.

$\therefore \qquad r t - s^2 > 0$, $r < 0$.

Hence there is a maximum when $x = n\pi + (-1)^n \cdot \frac{1}{6}\pi = y$.

If $\cos x = 0$, then $x = 2n\pi \pm \frac{1}{2}\pi = y$

From here we get $x = y = \pm\frac{1}{2}\pi, \frac{3}{2}\pi, \frac{5}{2}\pi$ etc.

If $\qquad x = \frac{\pi}{2} = y$ then $r = -1 + 1 = 0$

$$s = 1, t = 0.$$

$\therefore \quad r t - s^2 < 0$. Hence there is neither maximum nor minimum at $x = \frac{\pi}{2} = y$.

If $\qquad x = -\frac{\pi}{2} = y$, then $r = 1 + 1 = 2 = t$, $s = 1$

$\therefore \qquad r t - s^2 = (2 \times 2) - 1 = 3 > 0$. Also $r > 0$.

Hence there is a minimum at $x = -\frac{\pi}{2} = y$. In a similar way we may discuss other values too.

4. *Find the maximum value of u where*

$$u = \frac{xyz}{(a+x)(x+y)(y+z)(z+b)}.$$

We have $\log u = \log x + \log y + \log z - \log(a+x) - \log(x+y) - \log(y+z) - \log(z+b)$

Hence, $\dfrac{1}{u}\dfrac{\partial u}{\partial x} = \dfrac{1}{x} - \dfrac{1}{a+x} - \dfrac{1}{x+y} = \dfrac{ay - x^2}{x(a+x)(x+y)}$

or $\dfrac{\partial u}{\partial x} = \dfrac{(ay - x^2)u}{x(a+x)(x+y)}$

Similarly $\dfrac{\partial u}{\partial y} = \dfrac{(xz - y^2)u}{y(x+y)(y+z)}$

and $\dfrac{\partial u}{\partial z} = \dfrac{(by - z^2)u}{z(y+z)(z+b)}$

Hence for maxima or minima of u, we have

$$ay - x^2 = 0,\ xz - y^2 = 0 \text{ and } by - z^2 = 0$$

From the above equation, it follows that a, x, y, z, b are in geometrical progression. Let r be the common ratio, then

$$ar^4 = b \text{ or } r = \left(\dfrac{b}{a}\right)^{1/4}$$

Also $x = ar,\ y = ar^2,\ z = ar^3$

Substituting these values, we get

$$u = \dfrac{ar \cdot ar^2 \cdot ar^3}{a(1+r) \cdot ar(1+r) \cdot ar^2(1+r) \cdot ar^3(a+r)}$$

$$= \dfrac{1}{a(1+r)^4} = \dfrac{1}{a\left[1 + \left(\dfrac{b}{a}\right)^{1/4}\right]^4}$$

$$= \dfrac{1}{(a^{1/4} + b^{1/4})^4}.$$

To decide whether this value of u is a maximum or minimum, we have

$$\dfrac{\partial^2 u}{\partial x^2} = -\dfrac{2xu}{x(a+x)(x+y)} + (ay - x^2)\dfrac{\partial}{\partial x}\left(\dfrac{u}{x(a+x)(x+y)}\right)$$

Therefore $A = -\dfrac{2u}{a^3 r(1+r)^2}$,   since $x = ar,\ y = ar^2,\ z = ar^3$

or, $A = -ve$ quantity

Similarly $B, C, \ldots$ can be found and we shall finally arrive at a result that this value of u is a maximum.

5. *Find a point within a triangle such that the sum of the square of its distances from the three vertices is a minimum.*   **(Ravishankar 1995; Rewa 98, Garhwal 2005)**

Let $(x_r, y_r),\ r = 1, 2, 3$, be the vertices of the triangle and (x, y) be any point inside the triangle.

Let $u = \displaystyle\sum_{r=1}^{3}[(x - x_r)^2 + (y - y_r)^2]$

For maximum or minimum of u, we have

$$\frac{\partial u}{\partial x} = \Sigma\, 2\,(x - x_r) = 0$$

i.e., $(x - x_1) + (x - x_2) + (x - x_3) = 0$

or $x = \dfrac{x_1 + x_2 + x_3}{3}$

Similarly $\dfrac{\partial u}{\partial y} = 0$ will give

$$y = \frac{y_1 + y_2 + y_3}{3}$$

Now $r = \dfrac{\partial^2 u}{\partial x^2} = 6, \quad s = \dfrac{\partial^2 u}{\partial y\,\partial x} = 0,$

$t = \dfrac{\partial^2 u}{\partial y^2} = 6,$ so that $rt - s^2 = 36 = +\,ve.$ Also r is $+\,ve.$

Hence u is minimum when

$$x = \frac{x_1 + x_2 + x_3}{3}, \; y = \frac{y_1 + y_2 + y_3}{3}$$

Thus the required point is $\left(\dfrac{x_1 + x_2 + x_3}{3},\, \dfrac{y_1 + y_2 + y_3}{3}\right).$

6. *Show that the point such that the sum of the squares of its distances from n given points shall be minimum, is the centre of the mean position of given points.*

Let the n given points be $(a_1, b_1, c_1), (a_2, b_2, c_2), (a_3, b_3, c_3),...$ etc. and let (x, y, z) be the coordinates of the required point. Then

$$u = \Sigma\,[\,(x - a_1)^2 + (y - b_1)^2 + (z - c_1)^2\,]$$

where the summation extends to each of the n points.

For a maximum or minimum of u, we have

$$\frac{\partial u}{\partial x} = 2\,\Sigma\,(x - a_1) = 2nx - 2\,\Sigma\,a_1 = 0,$$

$$\frac{\partial u}{\partial y} = 2\,\Sigma\,(y - b_1) = 2ny - 2\,\Sigma\,b_1 = 0,$$

$$\frac{\partial u}{\partial z} = 2\,\Sigma\,(z - c_1) = 2nz - 2\,\Sigma\,c_1 = 0,$$

hence $x = \dfrac{\Sigma\,a_1}{n}, \; y = \dfrac{\Sigma\,b_1}{n}, \; z = \dfrac{\Sigma\,c_1}{n}.$

Now, $A = \dfrac{\partial^2 u}{\partial x^2} = 2n.$

Since A is positive here, u is a minimum when the point (x, y, z) is the centre of the mean position of the given points. As we can easily show that

$$\begin{vmatrix} A & H \\ H & B \end{vmatrix}, \begin{vmatrix} A & H & G \\ H & B & F \\ G & F & C \end{vmatrix} \text{ are } +\,ve.$$

7. *Find the maximum value of*

$$(ax + by + cz)\, e^{-\alpha^2 x^2 - \beta^2 y - \gamma z^2}.$$

Let $$u = (ax + by + cz)\, e^{-\alpha^2 x^2 - \beta^2 y^2 - \gamma^2 z^2}$$

then $$\log u = \log (ax + by + cz) - (\alpha^2 x^2 + \beta^2 y^2 + \gamma^2 z^2)$$

Differentiating partially w.r.t. x, we get

$$\frac{1}{u}\frac{\partial u}{\partial x} = \frac{a}{ax + by + cz} - 2\alpha^2 x = 0$$

Similarly

$$\frac{1}{u}\frac{\partial u}{\partial y} = \frac{b}{ax + by + cz} - 2\beta^2 y = 0$$

and $$\frac{1}{u}\frac{\partial u}{\partial z} = \frac{c}{ax + by + cz} - 2\gamma^2 \beta = 0.$$

These give

$$x\,(ax + by + cz) = \frac{a}{2\alpha^2} \qquad\qquad\qquad ...(1)$$

$$y\,(ax + by + cz) = \frac{b}{2\beta^2} \qquad\qquad\qquad ...(2)$$

$$z\,(ax + by + cz) = \frac{c}{2\gamma^2} \qquad\qquad\qquad ...(3)$$

Multiplying (1), (2), (3) by a, b, c and adding, we get

$$(ax + by + cz)^2 = \frac{1}{2}\left(\frac{a^2}{\alpha^2} + \frac{b^2}{\beta^2} + \frac{c^2}{\gamma^2}\right)$$

or $$(ax + by + cz) = \sqrt{\left\{\frac{1}{2}\left(\frac{a^2}{\alpha^2} + \frac{b^2}{\beta^2} + \frac{c^2}{\gamma^2}\right)\right\}}$$

$$= A \text{ (say)}.$$

Then $$x = \frac{a}{2\alpha^2 A},\; y = \frac{b}{2\beta^2 A},\; z = \frac{c}{2\gamma^2 A}$$

Again, $$\frac{1}{u}\frac{\partial^2 u}{\partial x^2} - \frac{1}{u^2}\left(\frac{\partial u}{\partial x}\right)^2 = -\frac{a^2}{(ax + by + cz)^2} - 2\alpha^2$$

or $$\frac{\partial^2 u}{\partial x^2} = -u\left[\frac{a^2}{(ax + by + cz)^2} + 2\alpha^2\right]$$

since $$\frac{\partial u}{\partial \alpha} = 0$$

Hence for these values of x, y, z, u will be maximum.

Maximum value of u is given by

$$= \sqrt{\left\{\frac{1}{2}\left(\frac{a^2}{\alpha^2} + \frac{b^2}{\beta^2} + \frac{c^2}{\gamma^2}\right)\right\}}\; e^{-\frac{1}{4A^2}\left(\frac{a2}{\alpha^2} + \frac{b^2}{\beta^2} + \frac{c^2}{\gamma^2}\right)}$$

$$= \sqrt{\left\{\frac{1}{2}\left(\frac{a^2}{\alpha^2} + \frac{b^2}{\beta^2} + \frac{c^2}{\gamma^2}\right)\right\}} \; e^{-1/2}$$

$$= \sqrt{\left\{\frac{1}{2e}\left(\frac{a^2}{\alpha^2} + \frac{b^2}{\beta^2} + \frac{c^2}{\gamma^2}\right)\right\}}$$

EXERCISES

1. Discuss the maximum or minimum values of u, when

   (i) $u = x^3 + y^3 - 3\,a\,x\,y$.　　　(ii) $u = ax^2y^2 - x^4y^2 - x^3y^3$.

   (*Sagar* 1998; *Ravishankar* 2003; *Kumaon* 99; *Indore* 96, 98, 2002;
   MDU Rohtak 2000; *Garhwal* 2004)

2. Discuss the maximum values of u, where
   $$u = 2a^2 xy - 3ax^2y - ay^3 + x^3y + xy^3.$$

3. Find the maxima or minima of u, when $u = 2\,xy - 3x^2y - y^3 + x^3y + xy^3$.

4. Discuss the maxima or minima of $u = x^4 + y^4 - 2\,x^2 + 4xy - 2y^2$.

5. Examine for maximum and minimum values of the function
   $$z = x^2 - 3\,xy + y^2 + 2x.$$

6. Discuss the maximum and minimum values of $x^4 + 2\,x^2y - x^2 + 3\,y^2$.

7. Find the maximum and minimum values of the function $z = \sin x \sin y \sin (x + y)$.

8. Show that the distance of any point (x, y, z) on the plane $2x + 3y - z = 12$, from the origin is given by
   $$l = \sqrt{x^2 + y^2 + (2x + 3y - 12)^2}\,.$$

   Hence find a point on the plane that is nearest to the origin.

9. Find a point within a triangle such that the sum of the distances from the angular points may be maximum.

10. Find the maximum and minimum ordinates of the curve
    $$u = ax\, y^2z^3 - x^3y^2z^3 - xy^3z^3 - xy^3z^3 - xy^2z^4.$$

11. Show that the maxima and minima of the fraction
    $$\frac{ax^2 + by^2 + c + 2hxy + 2gx + 2fy}{a'x^2 + b'y^2 + c' + 2h'xy + 2g'x + 2f'y}$$

    are given by the roots of the equation
    $$\begin{vmatrix} a - a'u & h - h'u & g - g'u \\ h - h'u & b - b'u & f - f'u \\ g - g'u & f - f'u & c - c'u \end{vmatrix} = 0.$$

12. *ABCD* is a quadrilateral having no re-entrant angle and P is a point in its plane. Find the position of P for which the sum of its distances from the vertices is a minimum.

13. If x, y, z are angles of a tirangle, then find the maximum value of $\sin x \sin y \sin z$.

    (*Bhopal* 1998; *Indore* 2001)

14. Find three positive numbers whose sum is 30 and whose product is maximum. (*Jiwaji* 2003)

ANSWERS

1. (*i*) No maxima or minima at $(0, 0)$, at (a, a) maximum or minimum according as $a < 0$ or $a > 0$.

   (*ii*) u is maximum when $x = a/2$, $y = a/3$.

2. Maximum at $\left(\dfrac{a}{2}, \dfrac{a}{2}\right), \left(\dfrac{3}{2}a, -\dfrac{a}{2}\right)$; minima at $\left(\dfrac{a}{2}, -\dfrac{a}{2}\right), \left(\dfrac{3}{2}a, \dfrac{a}{2}\right)$, no maxima or minima at $(a, a), (a, -a)$.

3. Maximum at $\left(\dfrac{1}{2}, \dfrac{1}{2}\right), \left(\dfrac{3}{2}, -\dfrac{1}{2}\right)$; minimum at $\left(\dfrac{1}{2}, -\dfrac{1}{2}\right), \left(\dfrac{3}{2}, \dfrac{1}{2}\right)$; no mixima or minima at $(1, 1), (1, -1)$.

4. Minima when $x = \pm\sqrt{2}$, $y = \mp\sqrt{2}$.

5. Neither maxima nor minima at $x = \dfrac{4}{5}$, $y = \dfrac{6}{5}$.

6. Max. when $x = \dfrac{\sqrt{3}}{2}$, $y = -\dfrac{1}{4}$ and min. when $x = \dfrac{-\sqrt{3}}{2}$, $y = \dfrac{-1}{4}$

7. Max. at $x = y = \dfrac{\pi}{3}$ and min., at $x = y = 0$.

8. $\left(\dfrac{12}{7}, \dfrac{18}{7}, -\dfrac{6}{7}\right)$

9. Point is such that each side subtends an angle of $120°$ there.

10. Max. value $= 108\, a^7/7^7$

12. P is the intersection of the diagonals of the quadrilateral.

14. $10, 10, 10$

9.7 Lagrange's Method of Undetermined Multipliers (for several independent variables)

Let $u = \phi\,(x_1, x_2, x_3, \dots x_n)$ be a function of n variables $x_1, x_2, x_3, \dots, x_n$.

Let these variables be connected by following r equation

$$\left.\begin{array}{l} f_1\,(x_1, x_2, \dots, x_n) = 0 \\ f_2\,(x_1, x_2, \dots, x_n) = 0 \\ \quad\dots \quad\quad \dots \quad\quad \dots \quad\quad \dots \\ f_r\,(x_1, x_2, \dots\ x_n) = 0 \end{array}\right\} \qquad \dots(i)$$

Thus there are only $n - r$ independent variables out of these n variables.

For maxima or minima of u, we have

$$du = \frac{\partial u}{\partial x_1}\,dx_1 + \frac{\partial u}{\partial x_2}\,dx_2 + \dots + \frac{\partial u}{\partial x_n}\,dx_n = 0 \qquad \dots(ii)$$

Also $\qquad df_1 = \dfrac{\partial f_1}{\partial x_1}\,dx_1 + \dfrac{\partial f_1}{\partial x_2}\,dx_2 + \dots + \dfrac{\partial f_1}{\partial x_n}\,dx_n = 0 \qquad \dots(iii)$

$$df_2 = \frac{\partial f_2}{\partial x_1} dx_1 + \frac{\partial f_2}{\partial x_2} dx_2 + \dots + \frac{\partial f_2}{\partial x_n} dx_n = 0 \qquad \dots(iv)$$

. . . .

. . . .

$$df_r = \frac{\partial f_r}{\partial x_1} dx_1 + \frac{\partial f_r}{\partial x_2} d x_2 + \dots + \frac{\partial f_r}{\partial x_n} d x_n = 0 \qquad \dots(v)$$

Multiplying equations $(ii), (iii), (iv) \dots (v)$ by $1, \lambda_1, \lambda_2, \dots \lambda_r$ respectively and adding, we get

$$A_1 \, dx_1 + A_2 d x_2 + \dots + A_n dx_n = 0 \qquad \dots(vi)$$

Where $\quad A_i = \dfrac{\partial u}{\partial x_i} + \lambda_1 \dfrac{\partial f_1}{\partial x_i} + \lambda_2 \dfrac{\partial f_2}{\partial x_i} + \dots + \lambda_r \dfrac{\partial f_r}{\partial x_i}$ and $i = 1, 2, \dots n$.

Now, we have at our choice the r quantities $\lambda_1, \lambda_2, \dots, \lambda_r$ and we choose them in such a way that

$$A_1 = 0 = A_2 = A_3 = \dots = A_r \qquad \dots(vii)$$

Then the equation (vi) is reduced to

$$A_{r+1} \, d x_{r+1} + A_{r+2} dx_{r+2} + \dots + A_n \, d x_n = 0 \qquad \dots(viii)$$

Let us now suppose that out of n variables the following $(n - r)$ variables are independent.

$$x_{r+1}, x_{r+2}, \dots, x_n.$$

Then as $dx_{r+1}, dx_{r+2}, \dots, dx_n$ are independent, so their coefficients are separately zero and we have

$$A_{r+1} = A_{r+2} = \dots = A_n = 0 \qquad \dots(ix)$$

$\therefore$ From (vii) and (ix), we have

$$A_1 = 0 = A_2 = A_3 = \dots = A_r = A_{r+1} = \dots A_n \qquad \dots(x)$$

Also from (i) we have

$$f_1 = 0 = f_2 = \dots = f_r \qquad \dots(xi)$$

From (x) and (xi) we get $(n + r)$ equations which determine the r multipliers $\lambda_1, \lambda_2, \dots, \lambda_r$ and get the possible values of u.

EXAMPLES

1. *Find the maxima and minima of* $x^2 + y^2 + z^2$ *subject to the conditions*
    $ax^2 + by^2 + cz^2 = 1$ *and* $lx + my + nz = 0$.   *(Kanpur 96; Vikram 1998; Indore 96,*
    2000; Sagar 99, 2000 ; Bhopal 96)

Let $\qquad u = x^2 + y^2 + z^2$ $\qquad\qquad\qquad\qquad$ ...(i)

Given $\qquad ax^2 + by^2 + cz^2 = 1$ $\qquad\qquad\qquad$ ...(ii)

and $\qquad lx + my + nz = 0$ $\qquad\qquad\qquad\qquad$ ...(iii)

From (i) we get

$$d u = 2 x \, d x + 2 y \, dy + 2 z \, dz = 0 \qquad \dots(iv)$$

From (ii) and (iii) we have

$$2 \, a x \, dx + 2 \, b y \, dy + 2 \, c z \, dz = 0 \qquad \dots(v)$$

and $\qquad l \, dx + m \, dy + n \, dz = 0$ $\qquad\qquad\qquad$ ...(vi)

Multiplying $(iv), (v)$ and (vi) by $1, \lambda_1, \lambda_2$ and adding we get

$$(x \, d x + y \, d y + z \, d z) + \lambda_1 (a x \, d x + b y \, dy + cz \, dz) + \lambda_2 (ldx + m \, d y + n \, dz) = 0$$

or $(x + a\lambda_1 x + l\lambda_2)\, dx + (y + b\lambda_1 y + m\lambda_2)\, dy + (z + c\lambda_1 z + n\lambda_2)\, dz = 0$

Equating to zero the coefficients of $dx, dy,$ and dz we get

$$\left. \begin{array}{l} x + a\lambda_1 x + l\lambda_2 = 0 \\ y + b\lambda_1 y + m\lambda_2 = 0 \\ z + c\lambda_1 z + n\lambda_2 = 0 \end{array} \right\} \qquad \dots(viii)$$

Multiplying these by x, y, z and adding, we get

$$(x^2 + y^2 + z^2) + \lambda_1 (ax^2 + by^2 + cz^2) + \lambda_2 (lx + my + nz) = 0$$

or $u + \lambda_1 (1) + \lambda_2 (0) = 0$, from $(i), (ii)$ and (iii)

or $u + \lambda_1 = 0$ or $\lambda_1 = -u$

$\therefore$ From $(viii)$ we have

$$x - a\,x\,u + l\lambda_2 = 0 \qquad y - b\,y\,u + m\lambda_2 = 0,$$
$$z - c\,z\,u + n\lambda_2 = 0$$

or $x(1 - au) = -l\lambda_2, \qquad y(1 - bu) = -m\lambda_2,$
$$z(1 - cu) = -n\lambda_2.$$

or $$x = \frac{l\lambda_2}{au - 1}, \; y = \frac{m\lambda_2}{bu - 1}, \; z = \frac{n\lambda_2}{cu - 1}$$

Substituting these values in (iii), we get

$$\frac{l^2}{(au - 1)} + \frac{m^2}{(bu - 1)} + \frac{n^2}{(cu - 1)} = 0$$

which gives the maxima and minima of u, *i.e.*, $x^2 + y^2 + z^2$.

2.   *Discuss the maxima and minima of the function*

$$u = \sin x \sin y \sin z,$$

where x, y, z are the angles of a triangle.   *(**Kanpur** 94 ; **Bhopal** 1998)*

Since x, y, z are the angles of a triangle, hence

$$x + y + z = \pi \qquad \dots(i)$$

Given $u = \sin x \sin y \sin z$ $\dots(ii)$

For max. or min. We have from (ii)

$du = \cos x \sin y \sin z\, dx + \sin x \cos y \sin z\, dy + \sin x \sin y \cos z\, dz = 0$ $\dots(iii)$

Also from $(i)\; dx + dy + dz = 0$ $\dots(iv)$

Multiplying (iii) by 1 and (iv) by λ and adding, we get

$(\cos x \sin y \sin z + \lambda)\, dx + (\sin x \cos y \sin z + \lambda)\, dy + (\sin x \sin y \cos z + \lambda)\, dz = 0$

Equating the coefficients of dx, dy, dz to zero, we get

$\cos x \sin y \sin z + \lambda = 0$, $\sin x \cos y \sin z + \lambda = 0$

$\sin x \sin y \cos z + \lambda = 0$ $\dots(v)$

$\therefore$ $\cos x \sin y \sin z = \sin x \cos y \sin z$

$$= \sin x \sin y \cos z = -\lambda$$

or $\cot x = \cot y = \cot z$, dividing each term by $\sin x \sin y \sin z$

or $x = y = z = \dfrac{1}{3}\pi,$ from (i)

Let x and y be independent and z be dependent, as the relation (i) exists to find r, s and t.

From (ii) $\qquad \dfrac{\partial u}{\partial x} = \cos x \sin y \sin z + \sin x \sin y \cos z \dfrac{\partial z}{\partial x}$

But from (i) we get

$$1 + \dfrac{\partial z}{\partial x} = 0, \ i.e., \ \dfrac{\partial z}{\partial x} = -1$$

$\therefore \qquad \dfrac{\partial u}{\partial x} = \cos x \sin y \sin z - \sin x \sin y \cos z$

and $\qquad \dfrac{\partial^2 u}{\partial x^2} = \left(-\sin x \sin y \sin z + \cos x \sin y \cos z . \dfrac{\partial z}{\partial x} \right)$

$$- \left(\cos x \sin y \cos z - \sin x \sin y \sin z . \dfrac{\partial z}{\partial x} \right)$$

$$= (-\sin x \sin y \sin z - \cos x \sin y \cos z)$$

$$- (\cos x \sin y \cos z + \sin x \sin y \sin z)$$

i.e., $\qquad r = -2 (\sin x \sin y \sin z + \cos x \sin y \cos z)$

and $\qquad \dfrac{\partial^2 u}{\partial y \partial x} = \cos x \cos y \sin z + \cos x \sin y \cos z \dfrac{\partial z}{\partial y}$

$$- \sin x \cos y \cos z + \sin x \sin y \sin z \dfrac{\partial z}{\partial y},$$

$$\text{where } \dfrac{\partial z}{\partial y} = -1 \text{ as before}$$

or $\qquad s = \cos x \cos y \sin z - \cos x \sin y \cos z - \sin x \cos y \cos z - \sin x \sin y \sin z$

Putting $x = y = z = \dfrac{\pi}{3}$, we get

$$r = -2 \left(\dfrac{1}{2}\sqrt{3} . \dfrac{1}{2}\sqrt{3} . \dfrac{1}{2}\sqrt{3} + \dfrac{1}{2} . \dfrac{1}{2}\sqrt{3} . \dfrac{1}{2} \right) = -\sqrt{3}$$

$$s = \dfrac{1}{2} . \dfrac{1}{2} . \dfrac{1}{2}\sqrt{3} - \dfrac{1}{2} . \dfrac{1}{2}\sqrt{3} . \dfrac{1}{2} - \dfrac{1}{2}\sqrt{3} . \dfrac{1}{2} . \dfrac{1}{2} - \dfrac{1}{2}\sqrt{3} . \dfrac{1}{2}\sqrt{3} . \dfrac{1}{2}\sqrt{3}$$

$$= -\dfrac{1}{2}\sqrt{3} .$$

By symmetry $\quad t = \dfrac{\partial^2 u}{\partial y^2} = -\sqrt{3}, \ \text{at } x = y = z = \dfrac{\pi}{3}$

$\therefore \qquad rt - s^2 = (-\sqrt{3})(-\sqrt{3}) - \left(-\dfrac{1}{2}\sqrt{3} \right)^2$

$$= 3 - \dfrac{3}{4} = \dfrac{9}{4} = +ve$$

Also, $\qquad r = -\sqrt{3} = -ve$

Hence there is maximum at

$$x = y = z = \dfrac{\pi}{3}.$$

3. *Show that the maximum and minimum of radii vectors of the section of the surface*

$$(x^2 + y^2 + z^2)^2 = \frac{x^2}{a^2} + \frac{y^2}{b^2} + \frac{z^2}{c^2}. \text{ by the plane } \lambda x + \mu y + vz = 0$$

are given by the equation

$$\frac{a^2 \lambda^2}{1 - a^2 r^2} + \frac{b^2 \mu^2}{1 - b^2 r^2} + \frac{c^2 v^2}{1 - c^2 r^2} = 0.$$

We have to find the maximum value of radius vector r, where

$$r^2 = x^2 + y^2 + z^2.$$

It is given that

$$\frac{x^2}{a^2} + \frac{y^2}{b^2} + \frac{z^2}{c^2} = r^4$$

and $\qquad \lambda x + \mu y + vz = 0.$

Differentiating these equations and putting the condition $dr = 0$, we have

$$x\, dx + y\, dy + z\, dz = 0 \qquad\qquad\qquad\qquad\qquad ...(1)$$

$$\frac{x}{a^2}\, dx + \frac{y}{b^2}\, dy + \frac{z}{c^2}\, dz = 0 \qquad\qquad\qquad\qquad ...(2)$$

and $\qquad \lambda\, dx + \mu\, dy + v\, dz = 0 \qquad\qquad\qquad\qquad ...(3)$

Multiplying (1), (2), (3) by $1, \lambda_1, \lambda_2$ respectively and adding and then equating to zero the coefficients of dx, dy, dz we get

$$x + \frac{x}{a^2}\lambda_1 + \lambda \lambda_2 = 0 \qquad\qquad\qquad\qquad\qquad ...(4)$$

$$y + \frac{y}{b^2}\lambda_1 + \mu \lambda_2 = 0 \qquad\qquad\qquad\qquad\qquad ...(5)$$

$$z + \frac{z}{c^2}\lambda_1 + v \lambda_2 = 0 \qquad\qquad\qquad\qquad\qquad ...(6)$$

Multiplying (4), (5), (6) by x, y, z respectively and adding, we get

$$r^2 + r^4 \lambda_1 = 0 \text{ or } \lambda_1 = -\frac{1}{r^2}$$

Therefore from (4),

$$x\left(1 - \frac{1}{a^2 r^2}\right) + \lambda \lambda_2 = 0$$

or $\qquad\qquad\qquad\qquad\qquad x = \dfrac{a^2 r^2 \lambda \lambda_2}{1 - a^2 r^2}$

Similarly, $y = \dfrac{b^2 r^2 \mu \lambda_2}{1 - b^2 r^2}$ and $z = \dfrac{c^2 r^2 v \lambda_2}{1 - c^2 r^2}$

Substituting the values of x, y, z in $\lambda x + \mu y + vz = 0$, we get

$$\frac{a^2 \lambda^2}{1 - a^2 r^2} + \frac{b^2 v^2}{1 - b^2 r^2} + \frac{c^2 v^2}{1 - c^2 r^2} = 0$$

4. *Find the maximum and minimum values of* $\dfrac{x^2}{a^4} + \dfrac{y^2}{b^4} + \dfrac{z^2}{c^4}$, *when* $lx + my + nz = 0$ *and*

$\dfrac{x^2}{a^2} + \dfrac{y^2}{b^2} + \dfrac{z^2}{c^2} = 1$. *Interpret the result geometrically.*

Let $u = \dfrac{x^2}{a^4} + \dfrac{y^2}{b^4} + \dfrac{z^2}{c^4}$; then

$$du = \frac{2x}{a^4} dx + \frac{2y}{b^4} dy + \frac{2z}{c^4} dz = 0 \qquad \qquad …(1)$$

$$l\,dx + m\,dy + n\,dz = 0 \qquad \qquad …(2)$$

$$\frac{2x}{a^2} dx + \frac{2y}{b^2} dy + \frac{2z}{c^2} dz = 0 \qquad \qquad …(3)$$

Multiplying (1), (2) and (3) by $1, \lambda_1, \lambda_2$ respectively and adding, and then equating the coefficients of dx, dy and dz to zero, we get

$$\frac{x}{a^4} + \lambda_1 l + \lambda_2 \frac{x}{z^2} = 0, \qquad \qquad …(4)$$

$$\frac{y}{b^4} + \lambda_1 m + \lambda_2 \cdot \frac{y}{b^2} = 0 \qquad \qquad …(5)$$

and $\qquad \dfrac{z}{c^4} + \lambda_1 n + \lambda_2 \cdot \dfrac{z}{c^2} = 0 \qquad \qquad …(6)$

Multiplying (4) by x, (5) by y, and (6) by z and adding, we get

$$u + \lambda_2 = 0 \text{ or } \lambda_2 = -u,$$

$$\therefore \qquad \frac{x}{a^4} + \lambda_1 l - \frac{x}{a^2} \cdot u = 0 \text{ etc.}$$

Hence $x = -\dfrac{\lambda_1 l a^4}{1 - a^2 u}$

Similarly, $y = \dfrac{-x_1 m b^4}{1 - b^2 u}$ and $z = \dfrac{-\lambda_1 n c^4}{1 - c^2 u}$

Substituting these values of x, y, z in $lx + my + nz = 0$, we get

$$\frac{l^2 a^4}{1 - a^2 u} + \frac{m^2 b^4}{1 - b^2 u} + \frac{n^2 c^4}{1 - c^2 u} = 0 \qquad \qquad …(7)$$

The equation (7) gives the required maximum or minimum values of u.

Geometrical Interpretation. The tangent to the ellipsoid $\dfrac{x^2}{a^2} + \dfrac{y^2}{b^2} + \dfrac{z^2}{c^2} = 1$ at (x', y', z') is

$\dfrac{xx'}{a^2} + \dfrac{yy'}{b^2} + \dfrac{zz'}{c^2} = 1$. Perpendicular distance to the tangent plane from origin, *i.e.*, p is given by

$$p^2 = \frac{1}{\dfrac{x'^2}{a^4} + \dfrac{y'^2}{b^4} + \dfrac{z'^2}{c^4}}.$$

If the point (x', y', z') lies on the given plane $lx + my + nz = 0$, the given problem consists of finding out the maximum or minimum value of the perpendicular distance from the origin to the tangent planes to the ellipsoid at the points common to the plane $lx + my + nz = 0$ and the ellipsoid.

5. *Prove that of all rectangular parallelopiped of the same volume, the cube has the least surface.*   (*Jiwaji* **1998, 2000**)

Let x, y, z be the dimensions of the parallelopiped. Let V denotes its volume and S its surface. Then

$$V = xyz \qquad\qquad ..(1)$$

and

$$S = 2xy + 2yz + 2zx \qquad\qquad ...(2)$$

For maxima and minima of S, we have

$$dS = 2(y+z)\,dx + 2(z+x)\,dy + 2(x+y)\,dz$$

i.e., $(y+z)\,dx + (z+x)\,dy + (x+y)\,dz = 0 \qquad\qquad ...(3)$

Since V is constant, hence

$$yz\,dx + zx\,dy + xy\,dz = 0 \qquad\qquad ...(4)$$

Multiplying eqn. (3) by 1 and eqn. (4) by λ and adding and then equating the coefficients of dx, dy, dz to zero, we get

$$(y+z) + \lambda yz = 0 \qquad\qquad ...(5)$$

$$(z+x) + \lambda zx = 0 \qquad\qquad ...(6)$$

$$(x+y) + \lambda xy = 0 \qquad\qquad ...(7)$$

Solving (5), (6) and (7), we get

$$-\lambda = \frac{1}{y} + \frac{1}{z} = \frac{1}{z} + \frac{1}{x} = \frac{1}{x} + \frac{1}{y} \quad \text{or} \quad \frac{1}{y} - \frac{1}{x} = 0 \text{ or } x = y$$

Similarly, $y = z$

Since $x = y = z = V^{1/3}$, from (1)

Let us regard x and y as independent of each other. Then

$$\frac{\partial S}{\partial x} = 2y + 2y\frac{\partial z}{\partial x} + 2z + 2x\frac{\partial z}{\partial x}$$

Also from (1) $yz + xy\dfrac{\partial z}{\partial x} = 0$ *i.e.*, $\dfrac{\partial z}{\partial x} = -\dfrac{z}{x}$.

$$\frac{\delta S}{\delta y} = 2y - \frac{2yz}{x} + 2z - 2z = 2y - \frac{2yz}{x}$$

and

$$\frac{\delta^2 S}{\delta y^2} = \frac{2yz}{x^2} - \frac{2y}{x}\frac{\partial z}{\partial x} = \frac{2yz}{x^2} + \frac{2yz}{x^2} = 4 \text{ at } x = y = z.$$

Similarly $\dfrac{\delta^2 S}{\delta y^2} = 4$ at $x = y = z$.

and

$$\frac{\delta^2 S}{\delta y\,\delta x} = 2 - \frac{2y}{x} - \frac{2y}{x}\frac{\partial z}{\partial x}$$

$$= 2 - \frac{2z}{x} + \frac{2yz}{x^2}$$

$$= 2 - 2 + 2 = 2 \text{ at } x = y = z.$$

Hence $rt - s^2 = 4 \times 4 - 2^2 = 12 > 0$

Since r and $rt - s^2$ are both $+$ ve, S is least when $x = y = z$, *i.e.*, **parallelopiped of least surface is a cube.**

6. *Determine the greatest quadrilateral which can be formed with the four given sides* α, β, γ, δ *taken in this order.*

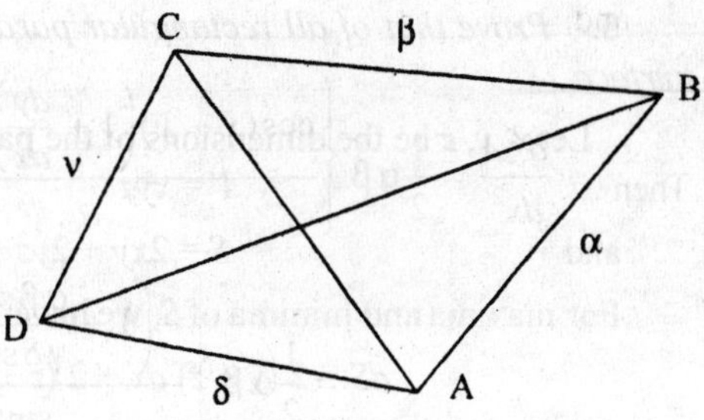

Let *x* denotes the angle between α and β. *y* the angle between γ and δ. Then if *A* denotes the area of the figure, we have

$$A = \frac{1}{2}(\alpha\beta\sin x + \gamma\delta\sin y) \qquad ...(1)$$

Also, if we draw a diagonal from the intersection of β and γ to the intersection of α and δ, (*AC*), we have two different values of the length of this diagonal

$$\alpha^2 + \beta^2 - 2\alpha\beta\cos x = \gamma^2 + \delta^2 - 2\gamma\delta\cos y,$$

i.e., $2\alpha\beta\cos x - 2\gamma\delta\cos y + \gamma^2 + \delta^2 - \alpha^2 - \beta^2 = 0$ ...(2)

Now for maxima or minima of *A*, we have

$$dA = \frac{1}{2}(\alpha\beta\cos x\, dx + \gamma\delta\cos y\, dy) = 0$$

i.e., $\alpha\beta\cos x\ dx + \gamma\delta\cos y\, dy = 0$ ...(3)

Also from (2),

$$-2\alpha\beta\sin x\, dx + 2\gamma\delta\sin y\, dy = 0 \qquad ...(4)$$

Multiplying (3) by 1 and (4) by λ and adding and then equating the coefficients of *dx, dy* to zero, we have

$$\alpha\beta\cos x + \lambda\,\alpha\beta\sin x = 0$$

and $\gamma\delta\cos y - \lambda\gamma\delta\sin y = 0$

These give $-\lambda = \cot x = -\cot y$

∴ Either $\cot x = \cot(-y)$ or $\cot(\pi - y)$

i.e., either $x = -y$, which is not possible in the present case.

or $x = \pi - y$, *i.e.,* $x + y = \pi$ ...(5)

This is the only applicable solution.

In this case only one variable, say *x* is to be regarded as independent.

∴ $$\frac{dA}{dx} = \frac{1}{2}\left(\alpha\beta\cos x + \gamma\delta\cos y\,\frac{dy}{dx}\right)$$

Also from (2),

$$-2\alpha\beta\sin x + 2\gamma\delta\sin y\,\frac{dy}{dx} = 0$$

Hence $\dfrac{dA}{dx} = \dfrac{1}{2}\alpha\beta\left(\cos x + \dfrac{\sin x\cos y}{\sin y}\right)$

$$= \frac{1}{2}\alpha\beta\,\frac{\sin(x + y)}{\sin y}$$

Fig. 9.13

and

$$\frac{d^2 A}{dx^2} = \frac{1}{2}\alpha\beta\left\{\frac{\cos(x+y)\left(1+\dfrac{dy}{dx}\right)\sin y - \sin(x+y)\cos y\, \dfrac{dy}{dx}}{\sin^2 y}\right\}$$

$$= \frac{1}{2}\alpha\beta\left\{\frac{-\left(1+\dfrac{\alpha\beta\sin x}{\gamma\delta\sin y}\right)\sin y}{\sin^2 y}\right\}, \quad \text{since } x+y = \pi$$

$$= -\frac{\alpha\beta(\gamma\delta + \alpha\beta)}{2\gamma\delta\sin y}, \quad \text{since } \sin x = \sin(\pi - y) = \sin y.$$

$$= a - ve \text{ quantity since } \sin y \text{ is } +ve, y \text{ being } < \pi.$$

Hence the greatest quadrilateral is such that it can be inscribed in a circle.

7. *Divide a number n into three parts x, y, z such that ayz + bzx + cxy shall have a maximum or minimum, and determine which it is.*

Let $\qquad u = ayz + bzx + cxy$ ...(1)

where $\qquad x + y + z = n$ ...(2)

On differentiating we have

$du = (bz + cy)\, dx + (cx + az)\, dy + (ay + bx)\, dz = 0 \text{ and } dx + dy + dz = 0.$

Multiplying (1) by 1 and (2) by λ, adding and equating the coefficients of dx, dy and dz to zero, we get ...(3)

$$2(ayz + bzx + cxy) + (x + y + z)\lambda = 0$$

or $\qquad 2u + n\lambda = 0$ from (1) and (2)

Thus equations (3) may be written as

$$0 \cdot x + cy + bz - \frac{2u}{n} = 0$$

$$cx + 0 \cdot y + az - \frac{2u}{n} = 0$$

$$bx + ay + 0 \cdot z - \frac{2u}{n} = 0$$

and $\qquad x + y + z - n = 0,$

Eliminating x, y, z from these equations, we get

$$\begin{vmatrix} 0 & c & b & \dfrac{2u}{n} \\ c & 0 & a & \dfrac{2u}{n} \\ b & a & 0 & \dfrac{2u}{n} \\ 1 & 1 & 1 & n \end{vmatrix} = 0 \qquad \qquad \text{...(4)}$$

This gives maximum or minimum value of u. To see whether it is maximum or minimum, we have

$$\frac{\partial u}{\partial x} = bz + cy + (ay + bx)\frac{\partial z}{\partial x},$$

where z is regarded as a function of x and y.

$$= bz + cy - (ay + bx), \text{ from (1)}$$

and

$$\frac{\partial^2 u}{\partial x^2} = -b + b\frac{\partial z}{\partial x} = -2b = r$$

$$\frac{\partial^2 u}{\partial x \partial y} = (c - a) + b\frac{\partial z}{\partial y} = (c - a - b) = s,$$

$$\frac{\partial u}{\partial y} = (az + cx) + (ay + bx)\frac{\partial z}{\partial y}$$

$$= ax + cx - ay - bx,$$

$$\frac{\partial^2 u}{\partial y^2} = a + a\frac{\partial z}{\partial y} = -2a = t$$

Now,

$$rt - s^2 = 4\,ab - (c - a - b)^2$$

$$= -(a^2 + b^2 + c^2 - 2ab - 2bc - 2ca)$$

$$= a\,(b + c - a) + b\,(c + a - b) + c\,(a + b - c)$$

which is +ve if a, b, c are the sides of a triangle.

Hence a maximum value is given by (4), where a, b, c are such that a triangle could be constructed with these as sides.

8. *Find a maximum value of $(x + 1)(y + 1)(z + 1)$ where $a^x b^y c^z = A$. Interpret the result geometrically.*

Let $u = (x + 1)(y + 1)(z + 1)$ ..(1)

and $x \log a + y \log b + z \log c = \log A$ ...(2)

$\therefore \quad du = (y + 1)(z + 1)\,dx + (z + 1)(x + 1)\,dy + (x + 1)(y + 1)\,dz = 0$

and $dx \log a + dy \log b + dz \log c = 0$.

Hence

$$\left.\begin{array}{l} (y + 1)(z + 1) + \lambda \log a = 0 \\ (z + 1)(x + 1) + \lambda \log b = 0 \\ (x + 1)(y + 1) + \lambda \log c = 0 \end{array}\right\}$$ ...(3)

Multiplying the equations (3) by $(x + 1)$, $(y + 1)$, $(z + 1)$ respectively and adding, we get

$3u + \lambda\,[\log a + \log b + \log c + x \log a + y \log b + z \log c] = 0$

or $\quad 3u + \lambda\,[\log a + \log b + \log c + \log A] = 0$

or

$$\lambda = -\frac{3u}{\log(A\,abc)}$$ ...(4)

Also writing (3) as $(y + 1)(z + 1) = -\lambda \log a$ etc. and multiplying, we get

$$(x + 1)^2 (y + 1)^2 (z + 1)^2 = -\lambda^3 \log a \log b \log c$$

or

$$u^2 = -\frac{27 u^3}{[\log (Aabc)]^3} \log a \log b \log c$$

or

$$u = \frac{[\log (Aabc)]^3}{27 \log a \log b \log c}$$

$$= \frac{[\log (Aabc)]^3}{\log a^3 \log b^3 \log c^3} \qquad \qquad ...(5)$$

Now

$$\frac{\partial u}{\partial x} = (y + 1)(z + 1) + (x + 1)(y + 1)\frac{\partial z}{\partial x}$$

where

$$\log a + \log c \frac{\partial z}{\partial x} = 0, \text{ from (2)}$$

Hence

$$\frac{\partial u}{\partial x} = (y + 1)(z + 1) - (x + 1)(y + 1)\frac{\log a}{\log c},$$

$$\frac{\partial^2 u}{\partial x^2} = (y + 1)\frac{\partial z}{\partial x} - (y + 1)\frac{\log a}{\log c}$$

$$= -2(y + 1)\frac{\log a}{\log c} = -ve \text{ if } y \text{ is } + ve.$$

Hence if x, y, z are +ve, a maximum value is given by (5).

Geometrical Interpretation. $a^x b^y c^z = A$ denotes the surface and $(x + 1)(y + 1)(z + 1)$ will denote the volume of a rectangular parallelopiped, the end point of one of whose diagonals are $(-1, -1, -1)$ and (x, y, z) and whose edges are parallel to the coordinate axes. Hence the problem of finding out the maximum volume of such a solid when the point (x, y, z) moves on the surface $a^x b^y c^z = A$.

9. *Prove that if* $x + y + z = 1$, $ayz + bzx + cxy$ *has an extreme value equal to*

$$\frac{abc}{(2bc + 2ca + 2ab + a^2 - b^2 - c^2)}.$$

Prove also that if a, b, c *are all positive and* c *lies between* $a + b - 2\sqrt{(ab)}$ *and* $a + b + 2\sqrt{(ab)}$ *this value is true maximum and that if* a, b, c *are all negative and* c *lies between* $a + b \pm 2\sqrt{(ab)}$, *it is true minimum.*

Let

$$u = ayz + bzx + cxy \qquad \qquad ...(1)$$

and

$$x + y + z = 1 \qquad \qquad ...(2)$$

on differentiating, we get

$$du = (bz + cy) dx + (cx + az) dy + (ay + bx) dz = 0$$

and

$$dx + dy + dz = 0.$$

$$\therefore \quad bz + cy + \lambda = 0, \ cx + az + \lambda = 0, \ ay + bx + \lambda = 0 \qquad \qquad ...(3)$$

Hence

$$-\lambda = bz + cy = cx + az = ay + bx$$

These give

$$z = x + \left(\frac{a - c}{b}\right) y = y + \left(\frac{b - c}{a}\right) x$$

or $\dfrac{x}{a(b+c-a)} = \dfrac{y}{b(c+a-b)} = \dfrac{z}{(a+b-c)}$

$$= \dfrac{x+y+z}{2\sum bc - \sum a^2} = \dfrac{1}{2\sum bc - \sum a^2}$$

Hence, $u = ayz + bxz + cxy = \dfrac{abc}{(2bc + 2ca + 2ab - a^2 - b^2 - c^2)}$

Now, $\dfrac{\partial u}{\partial x} = (bz + cy) + (ay + bx)\dfrac{\partial z}{\partial x}$, regarding y as constant,

$$= bz + cy - ay - bx, \text{ from (2)}$$

$\therefore$ $\dfrac{\partial^2 u}{\partial x^2} = b\dfrac{\partial z}{\partial x} - b = -2b = r.$

$\dfrac{\partial^2 u}{\partial y \, \partial x} = b\dfrac{\partial z}{\partial x} + c - a = c - a - b = s$

Similarly, $\dfrac{\partial^2 u}{\partial y^2} = -2a = t,$

$\therefore$ $rt - s^2 = 4ab - (c - a - b)^2$

$$= \left\{ 2\sqrt{(ab)} + c - a - b \right\}\left\{ 2\sqrt{(ab)} - c + a + b \right\}$$

Hence $rt - s^2$ will be positive when $c > a + b - 2\sqrt{(ab)}$ and $< a + b + 2\sqrt{(ab)}$, whether a, b, are all $+ve$ or all negative. But when a, b, c are all $+ve$, r and t are negative. Therefore, we have a true maximum in this case and when a, b, c are all $-ve$, r and t are $+ve$, so that u is a true minimum in this case.

10. *If $F(a) = \mu \neq 0$, $F'(a) \neq 0$ and x, y, z satisfy the relation*

$$F(x) F(y) F(z) = \mu^3.$$

Prove that the function

$$\phi(x) + \phi(y) + \phi(z)$$

has a maximum when $x = y = z = a$, provided that

$$\phi'(a) \left\{ \dfrac{F''(a)}{F'(a)} - \dfrac{F'(a)}{F(a)} \right\} > \phi''(a).$$

Let $u = \phi(x) + \phi(y) + \phi(z)$ ...(i)

$F(x) F(y) F(z) = \mu^3$ ...(ii)

$\Rightarrow$ $\log F(x) + \log F(y) + \log F(z) = 3 \log \mu$ ...(iii)

For max or min., we have

$$du = \phi'(x)\, dx + \phi'(y)\, dy + \phi'(z)\, dz = 0 \qquad \text{...(iv)}$$

And from (iii)

$$\dfrac{F'(x)}{F(x)}\, dx + \dfrac{F'(y)}{F(y)}\, dy + \dfrac{F'(z)}{F(z)}\, dz = 0 \qquad \text{...(v)}$$

Multiplying (v) by λ and adding to (iv), we get

$$\phi'(x) + \lambda\, \dfrac{F'(x)}{F(x)} = 0, \text{ etc.}$$

For the function to be maximum at (a, a, a), we must necessarily have

$$\phi'(a) + \lambda \frac{F(a)}{F'(a)} = 0$$

i.e. $\lambda = -\dfrac{\phi'(a) \, F(a)}{F'(a)} = 0$, since $F'(a) \neq 0$.

and $F(a) \neq 0$.

Now, regard z as a function of x and y so that x, y are independent.

$$\therefore \quad \frac{\partial u}{\partial x} = \phi'(x) + \phi'(z) \frac{\partial z}{\partial x}.$$

From (*iii*), $\dfrac{F'(x)}{F(x)} + \dfrac{F'(z)}{F(z)} \cdot \dfrac{\partial z}{\partial x} = 0$

i.e., $\dfrac{\partial z}{\partial x} = -1$ at (a, a, a)

$$\therefore \qquad \frac{\partial u}{\partial x} = \phi'(x) - \frac{F'(x) \, F(z)}{F(x) \, F'(z)} \, \phi'(z)$$

$$\frac{\partial^2 u}{\partial x^2} = \phi''(x) - \frac{F'(x) \, F(z)}{F(x) \, F'(z)} \, \phi''(z) \frac{\partial z}{\partial x}$$

$$- \phi'(z) \left[\frac{\{F''(x) \, F(z) + F'(x) \, F'(z) \frac{\partial z}{\partial x}\} \, F(x) \, F'(z)}{[F(x) \, F'(z)]^2} \right]$$
$$\qquad\qquad\qquad \frac{- \{F'(x) \, F'(z) + F(x) \, F''(z) \frac{\partial z}{\partial x}\} \, F'(x) \, F(z)}{[F(x) \, F'(z)]^2}$$

$$= \phi''(a) + \phi''(a) - \phi'(a) \left[\frac{2F''(a) \, F(a) - 2\{F'(a)\}^2}{F(a) \, F'(a)} \right]$$

$$= 2 \left[\phi''(a) - \phi'(a) \left\{ \frac{F''(a)}{F(a)} \right\} - \frac{F'(a)}{F(a)} \right]$$

which is $-ve$ if $\phi'(a) \left\{ \dfrac{F''(a)}{F'(a)} - \dfrac{F'(a)}{F(a)} \right\} > \phi''(a)$

Hence u is maximum under the given condition.

EXERCISES

1. Determine the maxima and minima of $x^2 + y^2 + z^2$ when $ax^2 + by^2 + cz^2 = 1$.

    (Jiwaji 1999; Indore 96; Bilaspur 97, 2000)

2. Find the minimum value of $x^2 + y^2 + z^2$, given that $x + by + cz = p$.

    (Sagar 1998; Bilaspur 98; Jiwaji 2000; Ravishankar 96, Rewa 2001; Bhopal 2001; Indore 95, 96, 98, 2000; Jabalpur 2000)

3. Determine the minimum value of $x^2 + y^2 + z^2$ subject to the condition $x + 2y - 4z = 5$.

4. Find the minimum value of $x^2 + y^2 + z^2$ when $yz + zx + xy = 3a^2$.

5. In a plane triangle find the maximum value of $u = \cos A \cos B \cos C$.

6. If $u = x^2 + y^2 + z^2$, where

    $ax^2 + by^2 + cz^2 + 2 fyz + 2 gzx + 2 hxy = 1$,

    find the maximum values of u.

7. If two variables x and y are connected by the relation $ax^2 + by^2 = ab$, show that the maximum and minimum values of the functions $x^2 + y^2 + xy$ will be the values of u given by the equation
$$4(u - a)(u - b) = ab.$$

8. Find the maxima and minima of $x^2 + y^2 + z^2$ subject to the conditions : $ax^2 + by^2 + cz^2 + 2fyz + 2gzx + 2hxy = 1$ and $lx + my + nz = 0$.

9. Find the maximum and minimum values of $u = a^2x^2 + b^2y^2 + c^2z^2$, where $x^2 + y^2 + z^2 = 1$ and $lx + my + nz = 0$.   (*Sagar* 1997; *Vikram* 2000)

10. Find the maximum and minimum values of $x^2 + y^2 + z^2$ subject to the conditions $x + y + z = 1$ and $xyz + 1 = 0$.

11. Show that the point within a triangle for which the sum of the squares of its perpendicular distance from the sides is least is the centre of the cosine-circle.

12. Find a point such that sum of squares of its distances from four faces of a given tetrahedron shall be a minimum. Find its value.

13. Find maximum value of $f(x, y, z) = x^a y^b z^c$ subject to the condition $x + y + z = 1$ where a, b, c are positive numbers.   (*Kanpur* 96)

14. Find the volume of the greatest rectangular parallelopiped inscribed in the ellipsoid whose equation is $\dfrac{x^2}{a^2} + \dfrac{y^2}{b^2} + \dfrac{z^2}{c^2} = 1$.

15. In a plane triangle, find the maximum value of $u = \cos A \cos B \cos C$.
                              (*Rewa* 1999; *Ravishankar* 2000; *Bhopal* 97; *Bilaspur* 2002)

16. Find the maximum and minimum values of $ax^2 + by^2 + cz^2 + 2hxy + 2gzx + 2fyz = 0$, subject to the conditions
$$lx + my + nz = 0 \text{ and } x^2 + y^2 + z^2 = 1.$$

17. Find the triangular pyramid of given base and altitude which has the least surface.

18. Determine the maximum value of OP, O being the origin of coordinates where P describes the curve
$$x^2 + y^2 + 2z^2 = 5, x + 2y + z = 5.$$

19. Find the shortest distance between the points $A(x_1, y_1, z_1)$ and $B(x_2, y_2, z_2)$ if A lies on the plane $x + y + z = 2a$ and B lies on the ellipsoid $\dfrac{x^2}{a^2} + \dfrac{y^2}{b^2} + \dfrac{z^2}{c^2} = 1$.

ANSWERS

1. $u = 1/a, 1/b, 1/c$;

2. $\dfrac{p^2}{a^2 + b^2 + c^2}$;

3. $\dfrac{25}{21}$;

4. $3a^2$

5. $A = B = C = \dfrac{\pi}{3}$.

6. Root of the eqn.
$$\begin{vmatrix} a - \dfrac{1}{u} & h & g \\[2mm] h & b - \dfrac{1}{u} & f \\[2mm] g & f & 1 - \dfrac{1}{u} \end{vmatrix} = 0$$

9. $\dfrac{l^2}{u-a^2}+\dfrac{m^2}{u-c^2}+\dfrac{n^2}{u-c^2}=0$ 10.   Min. value = 3.

12. $u=\dfrac{gV^2}{\sum\Delta_i^2}$ 14. $\dfrac{8\,abc}{3\sqrt{3}}$;   15. $\dfrac{1}{8}$

16. Max. or Min. value of u are given by

$$\begin{vmatrix} a-u & h & g & l \\ h & b-u & f & m \\ g & f & c-u & n \\ l & m & n & 0 \end{vmatrix}=0$$

17. When the side faces are equally inclined to the base.

18. $\sqrt{5}$

19. $\dfrac{1}{\sqrt{5}}\left[\sqrt{(a^2+b^2+c^2)}-2a\right]$

OBJECTIVE QUESTIONS

For each of the following questions, four alternatives are given for the answer. Only one of them is correct. Choose the correct alternative.

1. The necessary condition for a function $f(x)$ to have a maxima at $x=c$ is that :
   (a) $f'(c)>0$ (b) $f'(c)=0$
   (c) $f'(c)<0$ (d) None of these

2. A function $f(x)$ has maximum value at $x=c$ if :
   (a) $f'(c)=0$ and $f''(c)>0$ (b) $f'(c)=0$ and $f''(c)<0$
   (c) $f'(c)=0$ and $f''(c)\neq0$ (d) $f'(c)\neq0$ and $f''(c)<0$

3. The maximum value of $\sin x+\cos x$ is :
   (a) 2   (b) $\sqrt{2}$
   (c) 1   (d) $1+\sqrt{2}$

4. If $f(x)=|x|$, then :
   (a) $f'(0)=0$ (b) $f(x)$ is maximum at $x=0$
   (c) $f(x)$ is minimum at $x=0$ (d) None of these

5. If $y=a\log x+bx^2+x$ has its extremum values at $x=-1$ and $x=2$; then :
   (a) $a=2,\,b=-1$ (b) $a=2,\,b=-\dfrac{1}{2}$
   (c) $a=-2,\,b=\dfrac{1}{2}$ (d) None of these

6. The function $f(x)=x^3-6x^2+24x+4$ has :
   (a) a maximum value at $x=2$
   (b) a minimum value at $x=2$
   (c) a maximum value at $x=4$ and a minimum at $x=6$
   (d) neither maximum nor minimum at any point

7. The necessary conditions for a function $f(x,y)$ to have maximum or minimum value at a point (a,b) are that :

(a) $\dfrac{\partial f}{\partial x} = \dfrac{\partial f}{\partial y}$ at (a, b)

(b) $\dfrac{\partial f}{\partial x} \ne \dfrac{\partial f}{\partial y}$ at (a, b)

(c) $\dfrac{\partial f}{\partial x} = 0 = \dfrac{\partial f}{\partial y}$ at (a, b)

(d) $\dfrac{\partial^2 f}{\partial x^2} = 0 = \dfrac{\partial^2 f}{\partial y^2}$ at (a, b)

8. The maximum value of $f(x) = \dfrac{1}{3} x^3 - 2x^2 + 3x + 1$ is :

(a) 3/7

(b) 7/3

(c) 1

(d) 7   *(Garhwal 2001)*

9. $\sin x\,(1 + \cos x)$ is a maximum at :

(a) $x = \pi/3$

(b) $x = \pi$

(c) $x = \pi/2$

(d) $x = 0$

10. Maximum value of $(\log x)/x$ is :

(a) e

(b) $1/e$

(c) 0

(d) 1

11. Function $u = \sin x \sin y \sin (x + y)$ at $x = y = \dfrac{\pi}{3}$ is

(a) Maximum

(b) Minimum

(c) Neither Maxima nor minima

(d) None of these   *(Kanpur 2001)*

12. The minimum value of $xy + \dfrac{a^3}{x} + \dfrac{a^3}{y}$ at $x = a, y = a$ is :

(a) a^2

(b) $2a^2$

(c) $3a^2$

(d) $4a^2$ *(Kanpur 2002)*

13. The minimum value of $a \tan^2 x + b \cot^2 x$ equals the maximum value of $a \sin^2 \theta + b \cos^2 \theta$, where $a > b > 0$, then :

(a) $a = b$

(b) $a = 2b$

(c) $a = 3b$

(d) $a = 4b$

14. The maximum value of $\cos (\cos (\sin x))$ is

(a) $\cos (\cos 1)$

(b) $\cos 1$

(c) 1

(d) 0

15. Minimum value of $4x^2 + 4x \mid \sin \theta \mid - \cos^2 \theta$ is :

(a) -2

(b) -1

(c) $-\dfrac{1}{2}$

(d) 0

16. For any real θ, the maximum value of $\cos^2 (\cos \theta) + \sin^2 (\sin \theta)$ is :

(a) 1

(b) $1 + \sin^2 1$

(c) $1 + \cos^2 1$

(d) does not exist

ANSWERS					
1. (b)	2. (b)	3. (b)	4. (d)	5. (c)	6. (d)
7. (c)	8. (b)	9. (a)	10. (b)	11. (c)	12. (c)
13. (d)	14. (a)	15. (b)	16. (b)		

CHAPTER 10

INDETERMINATE FORMS

10.1 Indeterminate Forms

We know that when $x \to a$

$$\lim \frac{f(x)}{F(x)} = \frac{\lim f(x)}{\lim F(x)} \quad \text{if } \lim F(x) \neq 0,$$

so that this result fails to give any information regarding the limit of a fraction whose denominator tends to zero as its limit.

Now, suppose, that the denominator tends to 0 as $x \to a$.

The numerator may or may not tend to zero.

Suppose, that the numerator does *not* tend to zero. Then $f(x) / F(x)$ cannot tend to a finite limit. For, if possible, let it tend to a finite limit, say l. We have in this case,

$$f(x) = \frac{f(x)}{F(x)} F(x)$$

$$\Rightarrow \qquad \lim f(x) = \lim \left[\frac{f(x)}{F(x)} F(x) \right]$$

$$= \lim \frac{f(x)}{F(x)} \lim F(x) = l.0 = 0.$$

Thus, we have a contradiction.

Three types of different behaviours are possible in this case in that the fraction may tend to $+\infty$ or $-\infty$ or that the limit may not exist as illustrated below.

The limit of the denominator in each of the following three cases is 0 when $x \to 0$:

(i) $\lim (1/x^2) = +\infty$ (ii) $\lim (1/-x^2) = -\infty$

(iii) $\lim (1/x)$ does not exist.

The case when the limit of the numerator is also zero is more important and interesting. A general method of determining the limit of such a fraction will be given in this chapter. As already stated in Chapter 3, we say that a *fraction whose numerator and denominator both tend to zero as x tends to a, assumes the* **indeterminate form 0/0 for $x = a$.**

The determination of the derivative dy/dx is itself equivalent to finding the limit of a fraction $\delta y/\delta x$ which assumes the indeterminate form 0/0.

Other cases of limits which are reducible to this form will also be considered in this chapter.

In what follows it will always be assumed that the functions f and F possess continuous derivatives of every order that come in question in a certain interval enclosing $x = a$. We shall be making use of the Cauchy's mean value theorem (Refer ξ 8.5) in this chapter.

10.2 The Indeterminate Form : 0/0   *(Gujrat 2004)*

If $\lim\limits_{x \to a} F(x) = 0$, $\lim\limits_{x \to a} f(x) = 0$, *then*

$$\lim_{x \to a} \frac{f'(x)}{F'(x)} = l \; \Rightarrow \; \lim_{x \to a} \frac{f(x)}{F(x)} = l.$$

Since the values of the functions f and F at a are not relevant to either the hypothesis or the conclusion of the theorem, we suppose that $f(a) = 0$, $F(a) = 0$ and thus render f and F continuous at a.

Also, since we are given that

$$\lim_{x \to a} \frac{f'(x)}{F'(x)} = l.$$

It follows that there exists a neighbourhood of a for every point x of which $f'(x) / F'(x)$ is defined so that $F'(x)$ is not zero. We have by Cauchy's mean value theorem,

$$\frac{f(a+h)}{F(a+h)} = \frac{f(a+h) - f(a)}{F(a+h) - F(a)}$$

$$= \frac{f'(a+\theta h)}{F'(a+\theta h)} \qquad\qquad 0 < \theta < 1$$

Since $\dfrac{f'(a+\theta h)}{F'(a+\theta h)} \to l$ as $h \to 0$, it follows that

$$\lim_{h \to 0} \frac{f(a+h)}{F(a+h)} = l \; \Leftrightarrow \; \lim_{h \to 0} \frac{f(x)}{F(x)} = l.$$

Generalisation. If lim $\lim\limits_{x \to a} F'(x) = 0$ and $\lim\limits_{x \to a} f'(x) = 0$, *then*

$$\lim_{x \to a} \frac{F''(x)}{f''(x)} = l \Rightarrow \lim_{x \to a} \frac{F'(x)}{f'(x)} = l$$

and so on.

EXAMPLES

1. *Determine* $lim \dfrac{e^x - e^{-x} - 2\log(1+x)}{x \sin x}$, *where* $x \to 0$.

                                                                      (G.N.D.U. Amritsar, 2004)

$$\begin{cases} f(x) = e^x - e^{-x} - 2\log(1+x) \\ F(x) = x \sin x \end{cases} \Rightarrow \begin{cases} \lim\limits_{x \to 0} f(x) = f(0) = 0, \\ \lim\limits_{x \to 0} F(x) = F(0) = 0. \end{cases}$$

$$\begin{cases} f'(x) = e^x + e^{-x} - \dfrac{2}{1+x} \\ F'(x) = x \cos x + \sin x \end{cases} \Rightarrow \begin{cases} \lim\limits_{x \to 0} f'(x) = f'(0) = 0, \\ \lim\limits_{x \to 0} F'(x) = F'(0) = 0. \end{cases}$$

$$\begin{cases} f''(x) = e^x - e^{-x} + \dfrac{2}{(1+x)^2}, \\ F''(x) = -x \sin x + 2 + \cos x, \end{cases} \Rightarrow \begin{cases} \lim\limits_{x \to 0} f''(x) = f''(0) = 2, \\ \lim\limits_{x \to 0} F''(x) = F''(0) = 2. \end{cases}$$

$$\therefore \lim_{x \to 0} \frac{e^x - e^{-x} - 2 \log(1+x)}{x \sin x} = \frac{2}{2} = 1.$$

The process may be conveniently exhibited as follows:-

$$\lim_{x \to 0} \frac{e^x - e^{-x} - 2 \log(1+x)}{x \sin x} = \lim_{x \to 0} \frac{e^x - e^{-x} - 2/(1+x)}{x \cos x + \sin x} = \lim_{x \to 0} \frac{e^x - e^{-x} + 2/(1+x)^2}{-x \sin x + 2 \cos x} = \frac{2}{2} = 1.$$

2. *Find the values of a and b in order that*

$$\lim_{x \to 0} \frac{x(1 + a \cos x) - b \sin x}{x^3}$$

may be equal to 1.   (*Rohilkhand* 2000, *Avadh* 99)

The function is of the form (0/0) for all values of *a* and *b* when $x \to 0$.

$$\therefore \lim_{x \to 0} \frac{x(1 + a \cos x) - b \sin x}{x^3} = \lim_{x \to 0} \frac{1 + a \cos x - ax \sin x - b \cos x}{3x^2}$$

The denominator being 0 for $x = 0$, the fraction will not tend to a finite limit unless the numerator is also zero for $x = 0$. This requires

$$1 + a - b = 0 \qquad \qquad \dots(i)$$

Again, supposing this relation is satisfied, we have

$$\lim_{x \to 0} \frac{1 + a \cos x - ax \sin x - b \cos x}{3x^2} = \lim_{x \to 0} \frac{-2a \sin x - ax \cos x + b \sin x}{6x}$$

$$= \lim_{x \to 0} \frac{-3a \cos x + ax \sin x + b \cos x}{6} = \frac{b - 3a}{6}$$

As given $\dfrac{b - 3a}{6} = 1 \Rightarrow b - 3a = 6.$ $\dots(ii)$

From (*i*) and (*ii*), we have

$$a = -\frac{5}{2}, \quad b = -\frac{3}{2}.$$

EXERCISES

1. Determine the limits of the following :

(i) $\dfrac{1 + \log x - x}{1 - 2x + x^2}$ $(x \to 1)$.

(ii) $\dfrac{\cos^2 \pi x}{e^{2x} - 2ex} \left(x \to \dfrac{1}{2} \right)$.

(iii) $\dfrac{\sinh x - x}{\sin x - x \cos x} (x \to 0)$.

(iv) $\dfrac{xe^x - \log(x+1)}{\cosh x - \cos x}, (x \to 0)$.

(v) $\dfrac{a^x - 1 - x \log_e a}{x^2}$ $(x \to 0)$ (*Garhwal* 96)

(vi) $\dfrac{\log(1 - x^2)}{\log \cos x}$ $(x \to 0)$ (*Kumaon* 1996, 99)

(vii) $\dfrac{e^{ax} - e^{-ax}}{\log(1 + bx)}$, $(x \to 0)$.

(viii) $\dfrac{e^{x} - e^{\sin x}}{x - \sin x}$ $(x \to 0)$ **(Patna 2003)**

2. Evaluate the following :

(i) $\displaystyle\lim_{x \to 0} \dfrac{xe^{x} - \log(1 + x)}{x^2}$.

(ii) $\displaystyle\lim_{x \to 0} \dfrac{x \cos x - \log(1 + x)}{x^2}$ **(MDU Rohtak 2000)**

(iii) $\displaystyle\lim_{x \to 0} \dfrac{e^{x} \sin x - x - x^2}{x^2 + x \log(1 - x)}$

(iv) $\displaystyle\lim_{x \to 0} \dfrac{x - \log(1 + x)}{x^2}$

(v) $\dfrac{\sin x - x + \dfrac{x^3}{6}}{x^5}$ $(x \to 0)$ **(Agra 2002; Kanpur 2002, Kumaon 2005)**

(vi) $\dfrac{a^{x} - b^{x}}{x}(x \to 0)$ **(Rohilkhand 2001)**

(vii) $\dfrac{\tan x - x}{x^2 \sin x}(x \to 0)$ **(Manipur 2002)**

(viii) $\dfrac{e^{x} - e^{-x} - 2\sin x}{x^3}(x \to 0)$ **(Calicut 2004)**

3. If the limit of $\dfrac{\sin 2x + a \sin x}{x^3}$ **(G.N.D.U. Amritsar 2004)**

as x tends to zero, be finite, find the value of a and the limit.

ANSWERS							
1.	(i) $-\dfrac{1}{2}$	(ii) $\dfrac{\pi^2}{2e}$.		(iii) $\dfrac{1}{2}$		(iv) $\dfrac{3}{2}$
	(v) $\log_e \dfrac{a^2}{2}$	(vi) 2.		(vii) $\dfrac{2a}{b}$		(viii) 1
2.	(i) $\dfrac{2}{3}$.	(ii) $\dfrac{1}{2}$.		(iii) $-\dfrac{2}{3}$.		(iv) $\dfrac{1}{2}$.
	(v) $\dfrac{1}{120}$	(vi) $\log_e(a/b)$	(vii) $1/3$		(viii) $\dfrac{2}{3}$
3.	$a = 2$; limit is -1.					

10.2.1. Preliminary transformation. Sometimes a preliminary transformation involving the use of known results on limits, such as

$$\lim_{x \to 0} \dfrac{\sin x}{x} = 1, \ \lim_{x \to 0} \dfrac{\tan x}{x} = 1,$$

simplifies the process a good deal. These limits may also be used to shorten the process at an intermediate stage.

EXAMPLES

1. *Find* $\lim \dfrac{1 + sin\, x - cos\, x + log\,(1 - x)}{x\, tan^2 x}$, $(x \to 0)$.   *(Kuvempu 2005)*

The inconvenience of continuously differentiating the denominator, which involves $\tan^2 x$ as a factor, may be partially avoided as follows. We write

$$\frac{1+\sin x-\cos x+\log(1-x)}{x\tan^2 x} = \frac{1+\sin x-\cos x+\log(1-x)}{x^3} \cdot \left(\frac{x}{\tan x}\right)^2$$

so that $\lim\limits_{x\to 0} \dfrac{1+\sin x-\cos x+\log(1-x)}{x\tan^2 x} = \lim\limits_{x\to 0} \dfrac{1+\sin x-\cos x+\log(1-x)}{x^3} \lim\limits_{x\to 0}\left(\dfrac{x}{\tan x}\right)^2$

$$= \lim_{x\to 0} \frac{1+\sin x-\cos x+\log(1-x)}{x^3}.1$$

$$= \lim_{x\to 0} \frac{1+\sin x-\cos x+\log(1-x)}{x^3}$$

To evaluate the limit on the R.H.S., we notice that the numerator and denominator both become 0 for $x = 0$.

$\therefore \quad \lim\limits_{x\to 0} \dfrac{1+\sin x-\cos x+\log(1-x)}{x^3} = \lim\limits_{x\to 0} \dfrac{\cos x-\sin x-[1/(1-x)]}{3x^2}$

$$= \lim_{x\to 0} \frac{-\sin x+\cos x-[1/(1-x)^2]}{6x}$$

$$= \lim_{x\to 0} \frac{-\cos x-\sin x-[2/(1-x)^3]}{6} = -\frac{3}{6} = -\frac{1}{2}$$

2. *Evaluate:* $\lim\limits_{x\to 0} \dfrac{\cosh x - \cos x}{x\, \sin x}$.

We have $\dfrac{\cosh x-\cos x}{x\sin x} = \dfrac{\cosh x-\cos x}{x^2} \cdot \dfrac{x}{\sin x}$

It may be shown that

$$\lim_{x\to 0} \frac{\cosh x-\cos x}{x^2} = 1.$$

So it follows that

$$\lim_{x\to 0} \frac{\cosh x-\cos x}{x\sin x} = 1.$$

3. *Evaluate* $\lim\limits_{x\to 0} \dfrac{(1+x)^{1/x} - e + \dfrac{1}{2}ex}{x^2}$.

Since $\lim\limits_{x\to 0}(1+x)^{1/x} = e$, this is a 0/0 form.

Let $y = (1+x)^{1/x}$

$\Rightarrow \log y = \dfrac{1}{x}\log(1+x)$

$$= \frac{1}{x}\left(x - \frac{x^2}{2} + \frac{x^3}{3} - \frac{x^4}{4} + \ldots\ldots\right)$$

$$= 1 - \frac{x}{2} + \frac{x^2}{3} - \frac{x^3}{4} + \ldots\ldots$$

$$\Rightarrow \qquad y = e^{1 - \frac{x}{2} + \frac{x^2}{3} - \frac{x^3}{4} + \ldots}$$

$$= e \cdot e^{-\frac{x}{2} + \frac{x^2}{3} - \frac{x^3}{4} + \ldots}$$

$$= e\left[1 + \left(-\frac{x}{2} + \frac{x^2}{3} - \frac{x^3}{4} + \ldots\ldots\right) + \frac{1}{2!}\left(-\frac{x}{2} + \ldots\ldots\right)^2 + \ldots\ldots\right]$$

$$= e\left[1 - \frac{x}{2} + \frac{11}{24}x^2 - \ldots\ldots\right]$$

$$\therefore \quad \lim_{x \to 0} \frac{(1+x)^{1/x} - e + \frac{1}{2}ex}{x^2} = \lim_{x \to 0} \frac{e\left[1 - \frac{x}{2} + \frac{11}{24}x^2 - \ldots\ldots\right] - e + \frac{ex}{2}}{x^2}$$

$$= \lim_{x \to 0}\left(\frac{11e}{24} + \text{terms containing } x\right)$$

$$= \frac{11e}{24}.$$

EXERCISES

1. Determine the limits of the following :—

   (i) $\dfrac{x^2 + 2\cos x - 2}{x\sin^3 x}, (x \to 0)$

   (ii) $\dfrac{\sin x - \log(e^x \cos x)}{x\sin x}, (x \to 0)$

   (iii) $\dfrac{x - \log(1+x)}{1 - \cos x}, (x \to 0)$

   (iv) $\log(1-x)\cot\dfrac{\pi}{2}x, (x \to 0)$

   (v) $\dfrac{\sin 2x + 2\sin^2 x - 2\sin x}{\cos x - \cos^2 x}, (x \to 0)$

   (*Calicut* 2004)

   (vi) $\dfrac{\tan x - x}{x^2 \tan x}, (x \to 0)$

   (vii) $\dfrac{\tan^2 x - x^2}{x^2 \tan^2 x}, (x \to 0).$

ANSWERS

1.   (i) $\dfrac{1}{12}$,   (ii) $-\dfrac{1}{2}$,   (iii) 1,   (iv) $-\dfrac{2}{\pi}$,   (v) 4,   (vi) $\dfrac{1}{3}$,   (vii) $\dfrac{2}{3}$

10.3 The Indeterminate From ∞/∞.

If $\lim_{x \to a} f(x) = \infty = \lim_{x \to a} F(x)$, *then*

$$\lim_{x \to a} \frac{f'(x)}{F'(x)} = l \Rightarrow \lim_{x \to a} \frac{f(x)}{F(x)} = l.$$

Suppose $x > a$

Let c and x be two numbers greater than a, where $a < x < c$.

By Cauchy's mean value theorem, there exists $\xi \in \,]x, c\,[$ such that

$$\frac{f(x) - f(c)}{F(x) - F(c)} = \frac{f(c) - f(x)}{F(c) - F(x)} = \frac{f'(\xi)}{F''(\xi)}$$

$$\Rightarrow \qquad \frac{f(x)}{F(x)}\left[\frac{1-f(c)/f(x)}{1-F(c)/F(x)}\right]=\frac{f'(\xi)}{F'(\xi)}$$

$$\Rightarrow \qquad \frac{f(x)}{F(x)}=\frac{1-F(c)/F(x)}{1-f(c)/f(x)}\cdot\frac{f'(\xi)}{F'(\xi)} \qquad\qquad\qquad ...(i)$$

Let $\varepsilon > 0$ be a given number.

As $\lim\limits_{x\to 0}\dfrac{f'(y)}{F'(y)}=l$, there exists a right handed neighbourhood $] a, a + \delta_1 [$ of a such that $\forall\ y \in$ $] a, a + \delta_1 [$,

$$\left|\frac{f'(y)}{F'(y)}-l\right|<\varepsilon$$

$$\Rightarrow \qquad l-\varepsilon<\frac{f'(y)}{F'(y)}<l+\varepsilon.$$

If c and x are chosen so as to belong to this neighbourhood, then ξ also belongs to the same neighbourhood and as such

$$l-\varepsilon<\frac{f'(\xi)}{F'(\xi)}<l+\varepsilon \qquad\qquad\qquad ...(ii)$$

Keeping c fixed, we see that

$$\lim_{x\to a}\frac{1-F(c)/F(x)}{1-f(c)/f(x)}=1$$

so that there exists a neighbourhood $] a, a + \delta_2 [$ of a such that $\forall\ x$ in this neighbourhood

$$\left|\frac{1-F(c)/F(x)}{1-f(c)/f(x)}-1\right|<\varepsilon$$

$$\Rightarrow \qquad 1-\varepsilon<\frac{1-F(c)/F(x)}{1-f(c)/f(x)}<1+\varepsilon \qquad\qquad\qquad ...(iii)$$

Let δ be the smaller of the two positive numbers δ_1 and δ

From (i), (ii) and (iii), it follows that $\forall\ x \in\ ] a, a + \delta [$,

We have

$$(l-\varepsilon)(1-\varepsilon)<\frac{f(x)}{F(x)}<(l+\varepsilon)(1+\varepsilon)$$

Thus we see that to an arbitrarily assigned $\varepsilon > 0$, there corresponds a neighbourhood of a such that $\forall x$ in this neighbourhood

$$(l-\varepsilon)(1-\varepsilon)<\frac{f(x)}{F(x)}<(l+\varepsilon)(1+\varepsilon).$$

It follows that

$$\lim_{x\to a+0}\frac{f(x)}{F(x)}=l=\lim_{x\to a+0}\frac{f'(x)}{F'(x)}.$$

We may similarly consider x tending to a from the left. Thus

$$\lim_{x\to a}\frac{f'(x)}{F'(x)}=l\Rightarrow\lim_{x\to a}\frac{f(x)}{F(x)}=l.$$

EXAMPLES

1. *Determine* $\lim \dfrac{\log(x-a)}{\log(e^x - e^a)}$ *as* $x \to a$.

Here the numerator and the denominator both tend to ∞ as x tends to a.

$$\therefore \quad \lim_{x \to a} \frac{\log(x-a)}{\log(e^x - e^a)} = \lim_{x \to a} \left(\frac{1}{x-a} \right) \Big/ \left(\frac{e^x}{e^x - e^a} \right)$$

$$= \lim_{x \to a} \frac{e^x - e^a}{e^x (x-a)} \quad \left(\frac{0}{0} \right)$$

$$= \lim_{x \to a} \frac{e^x}{e^x (x-a) + e^x} = \frac{e^a}{e^a} = 1 .$$

2. *Determine* $\lim \dfrac{\tan 3x}{\tan x}$ *as* $x \to \pi/2$.

We have
$$\lim_{x \to \pi/2} \frac{\tan 3x}{\tan x} = \lim_{x \to \pi/2} \frac{3 \sec^2 3x}{\sec^2 x}$$

$$= \lim_{x \to \pi/2} \frac{3 \cos^2 x}{\cos^2 3x} \qquad \left(\frac{0}{0} \text{ form} \right)$$

$$= \lim_{x \to \pi/2} \frac{\sin 2x}{\sin 6x}$$

$$= \lim_{x \to \pi/2} \frac{2 \cos 2x}{6 \cos 6x}$$

$$= \frac{1}{3} .$$

3. *Discuss the continuity of f at the origin when*

$$f(x) = x \log \sin x \text{ for } x = 0$$

and $\qquad f(0) = 0$

$$\lim_{x \to 0} f(x) = \lim_{x \to 0} x \log \sin x$$

$$= \lim_{x \to 0} \frac{\log \sin x}{1/x} \qquad \left(\frac{\infty}{\infty} \right)$$

$$= \lim_{x \to 0} \frac{\cot x}{-1/x^2}$$

$$= \lim_{x \to 0} - x . \frac{x}{\tan x}$$

$$= -\lim_{x \to 0} x . \lim_{x \to 0} \frac{x}{\tan x}$$

$$= 0$$

$$= f(0).$$

$\therefore \quad f$ is continuous at $x = 0$.

EXERCISES

1. Determine the following limits:

   (i) $\dfrac{\log \sin x}{\cot x}, (x \to 0)$ **(Mysore 2004)**

   (ii) $\dfrac{\tan x}{\tan 3x}, \left(x \to \dfrac{\pi}{2}\right)$

   (iii) $\dfrac{\log \tan x}{\log x}, (x \to 0)$

   (iv) $\dfrac{\log \tan 2x}{\log \tan x}, (x \to 0)$

   (v) $\log_{\tan x} \tan 2x, (x \to 0)$ **(Gujrat 2004)**

   $$\left[\textbf{Hint :} \ \log_{\tan x} \tan 2x = \dfrac{\log \tan 2x}{\log \tan x} \right]$$

   (vi) $\log (1 - x) \cot (x\,\pi/2), (x \to 1)$.

ANSWERS

1. (i) 0   (ii) 3   (iii) 1   (iv) 1   (v) 1   (vi) 0

10.4 The Indeterminate Form 0. ∞

We consider

$$\lim_{x \to a} [f(x).F(x)]$$

when $\lim_{x \to a} f(x) = 0, \ \lim_{x \to a} F(x) = \infty$.

We write

$$f(x).F(x) = \frac{f(x)}{1/F(x)} = \frac{F(x)}{1/f(x)}$$

so that these new forms are of the type 0/0 and ∞/∞ respectively and the limit can, therefore, be obtained by § 10.2 or by § 10.3.

EXAMPLE

1. *Determine lim(x log x) as x → 0.*

We write $x \log x = \dfrac{\log x}{1/x}$

$$\therefore \ \lim_{x \to 0} (x \log x) = \lim_{x \to 0} \frac{\log x}{1/x}, \left(\frac{\infty}{\infty}\right)$$

$$= \lim_{x \to 0} \left(\frac{1}{x} \div \frac{-1}{x^2}\right) = \lim_{x \to 0}(-x) = 0.$$

The reader may see that by writing

$$x \log x = \frac{x}{(1/\log x)}$$

which is of the form (0/0) and employing the corresponding result of § 10.2 would not be of any avail.

Note. Since log x is defined for positive values of x only we see that $\lim\limits_{x \to 0} (x \log x)$ really means $\lim\limits_{x \to (0 + 0)} (x \log x)$.

EXERCISES

1. Determine the following limits :–

   (i) $x \log \tan x, (x \to 0)$ *(Gujrat 2005)*

   (ii) $x \tan (\pi/2 - x), (x \to 0)$ (iii) $(a-x) \tan \dfrac{\pi x}{2a}, (x \to a)$

ANSWERS

1.   (i) 0   (ii)   1   (iii) $2a/\pi$.

10.5 The Indeterminate Form $\infty - \infty$

We consider

$$\lim_{x \to a} [f(x) - F(x)]$$

when $\lim\limits_{x \to a} f(x) = \infty = \lim\limits_{x \to a} F(x).$

We write

$$f(x) - F(x) = \left[\frac{1}{F(x)} - \frac{1}{f(x)} \right] \div \frac{1}{f(x)\,F(x)},$$

so that the numerator and denominator both tend to 0 as x tends to a. The limit may now be determined with the help of § 10.2.

Note. In order to evaluate the limit of an expression which assumes the form $\infty - \infty$, it is necessary to express the same in such a manner that it assumes the form 0/0 or ∞/∞.

EXAMPLES

1.   *Determine* $\lim \left\{ \dfrac{1}{x-2} - \dfrac{1}{\log(x-1)} \right\}$ *as* $x \to 2$

We write

$$\frac{1}{x-2} - \frac{1}{\log(x-1)} = \frac{\log(x-1) - (x-2)}{(x-2)\log(x-1)}$$

and see that the new form is of the type 0/0 when $x \to 2$. We have

$$\lim_{x \to 2} \left[\frac{1}{x-2} - \frac{1}{\log(x-1)} \right] = \log_{x \to 2} \frac{\log(x-1) - (x-2)}{(x-2)\log(x-2)}$$

The numerator and the denominator are both 0 for $x = 2$. On using the method of § 10.2 we may show that the required limit is $-\dfrac{1}{2}$.

2.   *Determine* $\lim \left(\dfrac{1}{x^2} - \dfrac{1}{\sin^2 x} \right)$ *as* $x \to 0$.

(Garhwal 2000; G.N.D.U. Amritsar 2004; Gujrat 2005)

We have $\lim\limits_{x \to 0} \left(\dfrac{1}{x^2} - \dfrac{1}{\sin^2 x} \right)$ $(\infty - \infty)$

$$= \lim_{x \to 0} \left(\frac{\sin^2 x - x^2}{x^2 \sin^2 x} \right)$$ $\left(\dfrac{0}{0} \right)$

$$= \lim_{x \to 0} \left(\frac{\sin^2 x - x^2}{x^4} \right) \left(\frac{x}{\sin x} \right)^2$$

$$= \lim_{x \to 0} \left(\frac{\sin^2 x - x^2}{x^4} \right) = \lim_{x \to 0} \frac{\sin 2x - 2x}{4x^3}$$

$$= \lim_{x \to 0} \frac{2\cos 2x - 2}{12x^2} = \lim_{x \to 0} \frac{\cos 2x - 1}{6x^2}$$

$$= \lim_{x \to 0} \frac{-2\sin^2 x}{6x^2} = \frac{-1}{3} \lim_{x \to 0} \left(\frac{\sin x}{x} \right)^2$$

$$= \frac{-1}{3}$$

EXERCISES

1. Determine the following limits :–

(i) $\left(\dfrac{1}{x} - \dfrac{1}{e^x - 1} \right), (x \to 0)$

(ii) $\left(\dfrac{a}{x} - \cot \dfrac{x}{a} \right), (x \to 0)$ $\qquad \left[\text{Write } \cot \dfrac{x}{a} = \dfrac{\cos (x/a)}{\sin (x/a)} \right]$

(iii) $\left(\dfrac{1}{x^2} - \cot^2 x \right), (x \to 0)$ **(Kumaon 97; Garhwal 98; G.N.D.U. Amritsar, 2004)**

(iv) $(\sec x - \tan x), \left(x \to \dfrac{\pi}{2} \right)$

(v) $\left(\dfrac{1}{x} - \dfrac{1}{\sin x} \right) (x \to 0).$

ANSWERS

1. (i) $\dfrac{1}{2}$ (ii) 0 (iii) $\dfrac{2}{3}$ (iv) 0 (v) 0

10.6. The Indeterminate Forms, $0°, 1^\infty, \infty°$.

We consider

$$\lim \left[f(x)^{F(x)} \right] \text{ as } x \to a$$

when

(i) $\lim f(x) = 0; \lim F(x) = 0.$

(ii) $\lim f(x) = 1; \lim F(x) = \infty.$

(iii) $\lim f(x) = \infty; \lim F(x) = 0.$

We write

$$y = [f(x)]^{F(x)}$$

$\Longrightarrow \qquad \log y = F(x) . \log f(x).$

In each of the three cases, we see that the right hand side assumes the indeterminate form $0 . \infty$ and its limit may, therefore, be determined by the method given in § 10.5.

Let $\qquad \lim\limits_{x \to a} [F(x) . \log f(x)] = l.$

$\therefore \qquad \lim \log y = l,$

$\Rightarrow \qquad \log \lim y = l \Rightarrow \lim y = e^l$

$\Rightarrow \qquad \lim [f(x)^{F(x)}] = e^l.$

Since we talk of $\log f(x)$, it is necessary that for values of x near a, $f(x)$ is positive.

EXAMPLES

1. *Determine* $\lim (x - a)^{x-a}$ *as* $x \to a$. $(0^0 form)$

We have $\qquad y = (x - a)^{x-a}.$

$\Rightarrow \qquad \log y = (x - a) \log (x - a) = \dfrac{\log (x - a)}{1/(x - a)} , \left(\dfrac{\infty}{\infty} \right)$

$\Rightarrow \qquad \lim\limits_{x \to a} \log y = \lim\limits_{x \to a} \dfrac{\log (x - a)}{1/(x - a)}$

$\qquad\qquad = \lim\limits_{x \to a} \left(\dfrac{1}{x - a} \div \dfrac{1}{(x - a)^2} \right) = \lim\limits_{x \to a} -(x - a) = 0$

$\Rightarrow \qquad \log \lim y = 0 \Rightarrow \lim y = e^0 = 1.$

Thus, $\lim (x - a)^{x-a} = 1$ when $x \to a.$

Note. Here it is understood that x tends to a through values greater than a , for otherwise the base $(x - a)$ would be negative and $(x - a)^{x-a}$ would have no meaning.

2. *Determine* $\lim (\cos x)^{1/x^2}$ *as* $x \to 0.$ $\qquad\qquad\qquad$ **(MDU Rohtak 2001)**

We have $\qquad y = (\cos x)^{1/x^2}$

$\Rightarrow \qquad \log y = \dfrac{\log \cos x}{x^2} \qquad\qquad\qquad\qquad\qquad\qquad \left(\dfrac{0}{0} \right)$

$\Rightarrow \qquad \lim\limits_{x \to 0} \log y = \lim\limits_{x \to 0} \dfrac{\log \cos x}{x^2}$

$\qquad\qquad = \lim\limits_{x \to 0} \dfrac{- \tan x}{2x} = -\dfrac{1}{2}$

$\Rightarrow \qquad \log (\lim\limits_{x \to 0} y) = -\dfrac{1}{2}$

$\Rightarrow \qquad \lim\limits_{x \to 0} y = e^{-1/2} \Rightarrow \lim\limits_{x \to 0} (\cos x)^{1/x^2} = e^{-1/2}$

3. *Evaluate:* $\lim\limits_{x \to 0} (\cos x)^{\cot x}$ $\qquad\qquad\qquad\qquad$ **(Kumaon 2001; Avadh 2003)**

We have $\qquad y = \lim\limits_{x \to 0} (\cos x)^{\cot x}$

$\Rightarrow \qquad \log y = \lim\limits_{x \to 0} \cot x \log \cos x$

$\qquad\qquad = \lim\limits_{x \to 0} \dfrac{\log \cos x}{\tan x} \qquad\qquad\qquad\qquad\qquad \left(\dfrac{0}{0} \right)$

$$= \lim_{x \to 0} (-\sin x \cos x) = 0$$

$$\therefore \qquad y = e^0 = 1.$$

4. Find $\lim_{x \to 0} \left(\dfrac{\tan x}{x} \right)^{1/x}$.   *(Garhwal 2003; Patna 2003)*

We have $y = \lim_{x \to 0} \left(\dfrac{\tan x}{x} \right)^{1/x}$

$\Rightarrow \qquad \log_e y = \lim_{x \to 0} \dfrac{1}{x} \log \left(\dfrac{\tan x}{x} \right)$

$$= \lim_{x \to 0} \dfrac{\log \tan x - \log x}{x}$$

$$= \lim_{x \to 0} \left[\dfrac{\sec^2 x}{\tan x} - \dfrac{1}{x} \right] \qquad\qquad\qquad (\infty - \infty)$$

$$= \lim_{x \to 0} \left[\dfrac{x - \sin x \cos x}{x \sin x \cos x} \right]$$

$$= \lim_{x \to 0} \left[\dfrac{2x - \sin 2x}{x \sin 2x} \right] \qquad\qquad\qquad \left(\dfrac{0}{0} \right)$$

$$= \lim_{x \to 0} \left[\dfrac{2 - 2\cos 2x}{\sin 2x + 2x \cos 2x} \right] \qquad\qquad \left(\dfrac{0}{0} \right)$$

$$= \lim_{x \to 0} \left[\dfrac{4 \sin 2x}{4 \cos 2x - 4x \sin 2x} \right]$$

$$= 0$$

$\Rightarrow \qquad y = e^0 = 1.$

5. Determine $\lim \left(2 - \dfrac{x}{a} \right)^{\tan \frac{\pi x}{2a}}$ as $x \to a$.   *(Rohilkhand 2002; Mysore 2004)*

Let $y = \left(2 - \dfrac{x}{a} \right)^{\tan \frac{\pi x}{2a}}$ (1^∞)

$$\log y = \tan \dfrac{\pi x}{2a} \log \left(2 - \dfrac{x}{a} \right)$$

$\Rightarrow \qquad \lim_{x \to a} \log y = \lim_{x \to a} \tan \dfrac{\pi x}{2a} \log \left(2 - \dfrac{x}{a} \right)$

$$= \lim_{x \to a} \dfrac{\log \left(2 - \dfrac{x}{a} \right)}{\cot \dfrac{\pi x}{2a}} \qquad\qquad \left(\dfrac{0}{0} \right)$$

$$= \lim_{x \to a} \dfrac{-\dfrac{1}{a} \cdot \dfrac{1}{2 - \dfrac{x}{a}}}{-\dfrac{\pi}{2a} \operatorname{cosec}^2 \dfrac{\pi x}{2a}}$$

$$= \frac{2}{\pi}$$

$$\Rightarrow \qquad \log \lim_{x \to a} y = \frac{2}{\pi}$$

$$\Rightarrow \qquad \lim_{x \to a} y = e^{2/\pi}$$

6. *Determine* $\lim (\cot x)^{1/\log x}, x \to 0$.

Let $$y = (\cot x)^{1/\log x}.$$

$$\log y = \frac{1}{\log x} \log (\cot x)$$

$$\Rightarrow \qquad \lim_{x \to 0} \log y = \lim_{x \to 0} \frac{\log \cot x}{\log x}$$

$$= \lim_{x \to 0} \frac{\dfrac{-\operatorname{cosec}^2 x}{\cot x}}{1/x} = \lim_{x \to 0} -\frac{x}{\sin x} \cdot \frac{1}{\cos x} = -1$$

$$\Rightarrow \qquad \log \lim_{x \to 0} y = -1 \Rightarrow \lim_{x \to 0} y = e^{-1} = 1/e.$$

7. *Determine* $\lim \left(\dfrac{\pi}{2} - x \right)^{\tan x}$ *as* $x \to \left(\dfrac{\pi}{2} - 0 \right)$.

Let $$y = \left(\frac{\pi}{2} - x \right)^{\tan x}$$

$$\log y = \tan x \log \left(\frac{\pi}{2} - x \right)$$

$$\lim_{x \to (\pi/2 - 0)} \log y = \lim_{x \to (\pi/2 - 0)} \tan x \log \left(\frac{\pi}{2} - x \right)$$

$$= \lim_{x \to \left(\frac{\pi}{2} - 0 \right)} \frac{\log \left(\dfrac{\pi}{2} - x \right)}{\cot x}$$

$$= \lim_{x \to \left(\frac{\pi}{2} - 0 \right)} \frac{-\sin^2 x}{- \left(\dfrac{\pi}{2} - x \right)}$$

$$= \lim_{x \to \left(\frac{\pi}{2} - 0 \right)} \frac{-2 \sin x \cos x}{1}$$

$$= 0$$

$$\log \lim_{x \to (\pi/2 - 0)} y = 0$$

$$\therefore \qquad \lim_{x \to (\pi/2 - 0)} y = e^0 = 1$$

EXERCISES

1. Determine the following limits :–

(i) x^x, $(x \to 0)$

(ii) $x^{(1-x^4)^{-1}}$, $(x \to 1)$

(iii) $(\cot x)^{\sin 2x}$, $(x \to 0)$

(iv) $(\sin x)^{\tan x}$, $(x \to \pi/2)$

(*Calicut 2004, Bangalore 2005, Calcutta 2005*)

(v) $\left(\dfrac{2x+1}{x+1} \right)^{x-1}$ $(x \to 0)$

(vi) $(1 + \sin x)^{\cot x}$, $(x \to 0)$

(vii) $\left(\dfrac{\sinh x}{x} \right)^{x^{-2}}$, $(x \to 0)$

(*Patna 2002*)

(viii) $\dfrac{1}{x^{x-1}}$, $(x \to 1)$

(ix) $\left(\dfrac{\tan x}{x} \right)^{1/x^2}$, $(x \to 0)$

(*Garhwal 2001; Kumaon 2002, G.N.D.U. Amritsar 2004*)

(x) $(\tan x)^{\tan 2x}$, $(x \to \pi/4)$

(*Gujrat 2004*)

(xi) $(\sec x)^{\cot x}$, $(x \to \dfrac{\pi}{2} + 0)$

(xii) $(\cot x)^{\sin x}$, $(x \to 0)$

(xiii) $\lim\limits_{x \to 0} \left(\dfrac{\tan x}{x} \right)^{1/x^3}$

(*Kanpur 2000; Avadh 2001*)

ANSWERS

1. (i) 1   (ii) e^{-1} (iii) 1   (iv) 1   (v) e (vi) e

(vii) $e^{1/6}$ (viii) e (ix) $e^{1/3}$ (x) $1/e$ (xi) 1   (xii) 1

(xiii) ∞.

MISCELLANEOUS EXERCISES

Determine the following limits :–

1. $\dfrac{e^x - e^{-x} - x}{x^2 \sin x}$, $(x \to 0)$

(*Manipur 2000*)

2. $\dfrac{\log x}{x^3}$, $(x \to \infty)$

3. $\dfrac{1 + x \cos x - \cosh x - \log(1+x)}{\tan x - x}$, $(x \to 0)$

4. $\dfrac{\log(1+x) \log(1-x) - \log(1-x^2)}{x^4}$, $(x \to 0)$

5. $\left(\dfrac{x-1}{2x^2} + \dfrac{e^{-x}}{2x \sinh x} \right)$, $(x \to 0)$

6. $\dfrac{1 - x + \log x}{1 - \sqrt{(2x - x^2)}}$, $(x \to 1)$

7. $(2x \tan x - \pi \sec x)$, $(x \to \pi/2)$

8. $\left(\dfrac{\sin x}{x} \right)^{1/x^2}$, $(x \to 0)$

(*Kumaon 2003; MDU Rohtak 1999; Manipur 2002*)

9. $\dfrac{(x - 4x^4)^{1/2} - (x/4)^{1/3}}{1 - (8x^3)^{1/4}}$, $(x \to 1/2)$.

10. $\log \dfrac{(1 + x \sin x)}{\cos x - 1}$, $(x \to 0)$

11. $(\cos ax)^{b/x^2}$, $(x \to 0)$

12. $(\sin x)^{\tan^2 x}$, $(x \to \pi/2)$

13. $(1 - x^2)^{1/\log(1-x)}, (x \to 1)$

14. $\left[\dfrac{2(\cosh x - 1)}{x^2} \right]^{1/x^2}, (x \to 0)$

15. $\left[\sin^2 \dfrac{\pi}{2 - ax} \right]^{\sec^2 \frac{\pi}{2 - bx}}, (x \to 0)$

16. $\dfrac{a^{\sin x} - a}{\log \sin x}, (x \to \pi/2)$

17. $\dfrac{x^x - x}{x - 1 - \log x}, (x \to 1)$

   (MDU Rohtak 2001)

18. $(\sec x)^{\cot x}, (x \to \pi/2)$

19. $(2 - x)^{\tan \pi x/2} (x \to 1)$

20. $\dfrac{1 - 4\sin^2 (\pi x/6)}{1 - x^2}, (x \to 1)$

21. $\dfrac{a^b - b^a}{a^a - b^b}, (a \to b)$

   (MDU Rohtak 1999)

22. $\dfrac{(1 + x)^{1/x} - e}{x}, (x \to 0)$

   (Agra 1998; Kumaon 2004; Avadh 99, 2000)

23. $\dfrac{\log_{\sec(\frac{1}{2}x)} \cos x}{\log_{\sec x} \cos \frac{1}{2} x}, (x \to 0)$

24. $\dfrac{a^x - x^a}{x^x - a^a}, (x \to 0)$

   (Rohilkhand 2003)

25. Find $\lim \dfrac{1 - ae^{-x} - be^{-2x} - ce^{-3x}}{1 - ae^x - be^{2x} - ce^{3x}}, (x \to 0)$

   when

   (i) $a = 3, b = -5, c = 4$;   (ii) $a = 3, b = -4, c = 2$;

   (iii) $a = 3, b = -3, c = 1$.

26. Find $\lim \dfrac{3x \log (\sin x / x)^2 + x^3}{(x - \sin x)(1 - \cos x)}, (x \to 0)$.

27. Find $\lim \dfrac{(1 + x) e^{-x} - (1 - x)e^x}{x(e^x - e^{-x}) - 2x^2 e^{-x}}, (x \to 0)$.

28. Obtain

   (i) $\lim\limits_{x \to 0} \left(\dfrac{1}{x} \right)^{\tan x}$ (ii) $\lim\limits_{x \to 0} \dfrac{e^{ax} - e^{-ax}}{\log (1 + bx)}$ (iii) $\lim\limits_{x \to 0} \dfrac{x^2 \sin 1/x}{\tan x}$.

29. Give the limits of the following when $x \to \infty$

   $\dfrac{\log x}{x^3}, \dfrac{\log x}{x^2}, \dfrac{\log x}{x}, \dfrac{\log x}{\sqrt{x}}$

   and of the following when $x \to (0 + 0)$

   $\sqrt{x} \log x, x \log x, x^2 \log x, x^3 \log x$.

30. If $f(x) = e^{-1/x^2}, x \neq 0$

   $f(0) = 0$,

   show that the derivative of every order of f vanishes for $x = 0$, *i.e.*,

   $f''(0) = 0, \forall n$.

31. Find the values of a, b and c such that

$$\lim_{x \to 0} \frac{x(a + b - \cos x) - c \sin x}{x^5} = 1.$$

(*Agra 2001, Garhwal 1999, Kanpur 2001*)

ANSWERS

1. $\dfrac{1}{3}$ **2.** 0 **3.** $-\dfrac{5}{2}$. **4.** $\dfrac{1}{12}$ **5.** $\dfrac{1}{6}$ **6.** -1.

7. -2. **8.** $e^{-1/6}$ **9.** $\dfrac{8}{9}$. **10.** -2 **11.** $e^{-a^2 b/2}$. **12.** $e^{-1/2}$.

13. e. **14.** $e^{1/12}$. **15.** e^{-a^2/b^2} **16.** $a \log a$. **17.** 2. **18.** 1.

19. $e^{2/\pi}$. **20.** $\sqrt{3}\,\pi/6$. **21.** $\log (e/b) \log be$. **22.** $-e/2$.

23. $11e/24$. **24.** $\dfrac{\log(a/e)}{\log ae}$ **25.** (*i*) 1. (*ii*) -1. (*iii*) -1.

26. $\dfrac{2}{5}$. **27.** $\dfrac{1}{8}$. **28.** (*i*) 1. (*ii*) $2a/b$. (*iii*) 0.

29. $0, 0, 0, 0;\ 0, 0, 0, 0$. **30.** Continuous at 0.

31. $a = 120,\ b = 60,\ c = 180$

OBJECTIVE QUESTIONS

For each of the following questions, four alternatives are given for the answer. Only one of them is correct. Choose the correct alternative.

1. Which of the following is not an indeterminate form?

(*a*) $\infty + \infty$ (*b*) $\infty - \infty$

(*c*) ∞/∞ (*d*) $0 \times \infty$ (*Agra 2001, Kanpur 2001*)

2. The formula of L' Hospital's rule is :

(*a*) $\lim\limits_{x \to a} \dfrac{f(x)}{g(x)} = \lim\limits_{x \to a} \left[\dfrac{f(x)}{g(x)} \right]$ (*b*) $\lim\limits_{x \to a} \dfrac{f(x)}{g(x)} = \lim\limits_{x \to a} \dfrac{f'(x)}{g'(x)}$

(*c*) $\lim\limits_{x \to a} \dfrac{f(x)}{g(x)} = \dfrac{f(a)}{g(a)}$ (*d*) $\lim\limits_{x \to a} \dfrac{f(x)}{g(x)} = \dfrac{f'(a)}{g'(a)}$

3. Consider (*i*) $\lim\limits_{x \to 0} \dfrac{\sin x}{x} = 1$, (*ii*) $\lim\limits_{x \to 0} \dfrac{\tan x}{x} = 1$, then :

(*a*) Both (*i*), (*ii*) are true (*b*) (*i*) is false, (*ii*) is true

(*c*) (*i*) is true, (*ii*) is false (*d*) Both (*i*), (*ii*) are false.

4. $\lim\limits_{x \to 0} \dfrac{a^x - b^x}{x}$ is equal to :

(*a*) 0 (*b*) ∞ (*c*) $\log (a/b)$ (*d*) $\log (a - b)$

(*Agra 2001, Kanpur 2001*)

5. $\lim\limits_{x \to 0} x \log x$ is equal to :

    (a) 0   (b) 1   (c) ∞ (d) -1

6. The value of $\lim\limits_{x \to 1} \dfrac{\log x}{x - 1}$ is :

    (a) -1 (b) ∞ (c) 1   (d) 0

                                                    (Garhwal 2001, Avadh 2005)

7. The value of $\lim\limits_{x \to 0} \dfrac{(1 + x)^n - 1}{x}$ is :

    (a) 1   (b) n (c) n^n (d) ∞

8. The value of $\lim\limits_{x \to 0} \dfrac{1 - \cos x}{3x^2}$ is :

    (a) 3   (b) $\dfrac{1}{3}$ (c) $\dfrac{1}{6}$ (d) $\dfrac{1}{9}$

                                                    (Garhwal 2002)

9. Value of $\lim\limits_{x \to 0} \dfrac{e^{ax} - e^{-ax}}{\log (1 + bx)}$ is :

    (a) $\dfrac{2a}{b}$ (b) $\dfrac{b}{2a}$ (c) $\dfrac{e^a - e^{-a}}{\log (1 + b)}$ (d) 0

10. Value of $\lim\limits_{x \to 0} \dfrac{\log (1 + kx^2)}{1 - \cos x}$ is :

    (a) 1   (b) -1 (c) 0   (d) ∞

11. $\lim\limits_{x \to 0} \dfrac{\cos hx - \cos x}{x \sin x}$ is :

    (a) 0   (b) -1 (c) 0   (d) ∞

12. $\lim\limits_{\theta \to 0} \dfrac{\sin \theta - \theta \cos \theta}{\sin \theta - \theta}$ is :

    (a) 2   (b) -2 (c) 0   (d) ∞

13. $\lim\limits_{x \to a} \left(\dfrac{\log x}{x} \right)$ is :

    (a) 1   (b) 0   (c) $\log (a/b)$ (d) $\log (b/a)$

14. $\lim\limits_{x \to 0} \dfrac{\log x}{\cot x}$ is :

    (a) 0   (b) 1   (c) -1 (d) ∞

15. $\lim\limits_{x \to 0} \dfrac{\log \sin x}{\log x}$ is :

    (a) 0   (b) 1   (c) 1   (d) ∞

16. The value of $\lim\limits_{x \to 0} \dfrac{\log x^2}{\cot x^2}$ is :

   (a) ∞ (b) 1   (c) 0   (d) -1

17. Value of $\lim\limits_{x \to 0} \dfrac{\log \tan x}{\log x}$ is :

   (a) 0   (b) 1   (c) -1 (d) ∞

18. Value of $\lim\limits_{x \to 0} \dfrac{3x + 4}{\sqrt{(2x^2 + 5)}}$ is :

   (a) $\dfrac{4}{\sqrt{5}}$ (b) $\dfrac{3}{\sqrt{5}}$ (c) $\dfrac{3}{5}$ (d) $\dfrac{4}{\sqrt{2}}$

19. $\lim\limits_{x \to \pi/2} \dfrac{\tan 5x}{\tan x}$ is :

   (a) $\dfrac{1}{5}$ (b) 5   (c) ∞ (d) 0

20. $\lim\limits_{x \to \infty} \left\{ x \tan \left(\dfrac{1}{x} \right) \right\}$ is :

   (a) 0   (b) ∞ (c) 1   (d) -1

21. $\lim\limits_{x \to 1} \left[(1 - x) \tan \dfrac{\pi x}{2} \right]$ is :

   (a) 0   (b) 1   (c) $\dfrac{\pi}{2}$ (d) $\dfrac{2}{\pi}$

22. $\lim\limits_{x \to \infty} (a^{1/x} - 1) x$ is:

   (a) 1   (b) ∞ (c) $\log a$ (d) $\log (1/a)$

23. Value of $\lim\limits_{x \to \pi/2} (\sin x) \tan x$ is :

   (a) 0   (b) ∞ (c) 0   (d) -1

24. Value of $\lim\limits_{x \to 0} 2^x \sin \left(\dfrac{a}{2^x} \right)$ is :

   (a) ∞ (b) $-\infty$ (c) a (d) $a/2$

25. Value of $\lim\limits_{x \to \pi/2} (\sec x - \tan x)$ is :

   (a) ∞ (b) 0   (c) $-\infty$ (d) 1

26. $\lim\limits_{x \to 0} (\operatorname{cosec} x - \cot x)$ is :

   (a) 0   (b) 1   (c) -1 (d) ∞

27. Value of $\lim\limits_{x \to 0} (1 + x)^{1/x}$ is :

   (a) 1   (b) -1 (c) e (d) $1/e$

                                                                         (Agra 2003; Kanpur 2003)

28. Value of $\lim\limits_{x \to 0} (\cos x)^{\cot^2 x}$ is:

(a) e　　　　　　(b) $\dfrac{1}{e}$　　　　　　(c) $\dfrac{1}{\sqrt{e}}$　　　　　　(d) $\sqrt{e}$

29. $\lim\limits_{x \to 0} x^x$ is :

(a) 0　　　　　　(b) 1　　　　　　(c) e　　　　　　(d) 1/e

30. Value of $\lim\limits_{x \to \infty} \left(\dfrac{1}{x}\right)^{1/x}$ is :

(a) 0　　　　　　(b) 1　　　　　　(c) e　　　　　　(d) 1/e

ANSWERS					
1. (a)	2. (b)	3. (a)	4. (c)	5. (a)	6. (c)
7. (b)	8. (c)	9. (a)	10. (d)	11. (a)	12. (b)
13. (b)	14. (a)	15. (c)	16. (c)	17. (b)	18. (b)
19. (a)	20. (c)	21. (d)	22. (c)	23. (a)	24. (c)
25. (b)	26. (a)	27. (c)	28. (a)	29. (b)	30. (b)

CHAPTER 11

PARTIAL DIFFERENTIATION

11.1. Introduction

The notion of continuity, limit and differentiation in relation to real valued functions of two real variables, will be briefly explained in this chapter. A few theorems of elementary character will also be given.

The subject of functions of two variables is capable of extension to functions of any number n of variables, but the treatment of the subject in this generalised form is not within the scope of this book. Only a few examples dealing with functions of three or more variables may be given.

11.2. Functions of two variables.

As in the case of functions of a single variable, we introduce the notion of functions of two variables by considering some examples.

(*i*) The relation

$$z = \sqrt{(1 - x^2 - y^2)}, \qquad \qquad ...(1)$$

between x, y, z, determines a value of z corresponding to every pair of numbers x, y, which are such that $x^2 + y^2 \leq 1$.

Denoting a pair of numbers x, y, geometrically by a point on a plane as explained in § 2.2, we see that the points (x, y) for which $x^2 + y^2 \leq 1$ lie on or within the circle whose centre is at the origin and radius is 1.

The region determined by the point (x, y) is called the *domain* of the point (x, y).

We say that the relation (1) associates a real number z to each pair (x, y) such that $x^2 + y^2 \leq 1$. We say that we have a function whose domain is the set

$$\{(x, y) : x^2 + y^2 \leq 1\}. \text{ We may write}$$

$$z = f(x, y) = \sqrt{(1 - x^2 - y^2)} : (x, y) \in \{(x, y) : x^2 + y^2 \leq 1\}$$

(*i*) Consider the relation

$$z = \sqrt{[(a - x)(x - b)]} + \sqrt{[(c - y)(y - d)]} \qquad \qquad ...(2)$$

where $a < b; c < d$.

Now $(a - x)(x - b)$ is non-negative if $a \leq x \leq b$ and $(c - y)(y - d)$ is non-negative if $c \leq y \leq d$.

The points (x, y) for which $a \leq x \leq b, c \leq y \leq d$ determine a rectangular domain bounded by the lines

$$x = a, x = b; y = c, y = d.$$

We see that the relation (2) associates a number z to every point (x, y) which is a member of the set

$$\{(x, y) : a \le x \le b \, ; c \le y \le d\}.$$

In general we say that a real valued function f of two variables is defined if to every member (x, y) of a given set D of ordered pairs of real numbers there is associated a real number.

The set D is called the domain of the function f.

(ii) The relation

$$z = e^{-x^2 - y^2}$$

determines a function of two variables (x, y); the domain of the function being the whole plane *i.e.*, the set of all the ordered pairs of real numbers.

We may also write this as follows :

$$z = f(x, y) = e^{-x^2 - y^2}$$

Ex. Give the domains of the following functions :

(i) $f(x, y) = 1/[\log x + \log y]$

(ii) $f(x, y) = \log (x + y)$

(iii) $f(x, y) = (x^y + y^x)$.

ANSWERS

(i) $\{(x, y) : 0 < x, 0 < y\}$

(ii) $\{(x, y) : 0 < x + y\}$

(iii) $\{(x, y) : 0 < x, 0 < y\}$

11.3. Neighbourhood of a point (a, b)

Let δ be any positive number. The points (x, y) such that

$$a - \delta \le x \le a + \delta, \, b - \delta \le y \le b + \delta$$

determine a square bounded by the lines

$$x = a - \delta, x = a + \delta \, ; y = b - \delta, y = b + \delta.$$

Its centre is at the point (a, b). This square is called a *Neighbourhood of the point (a, b)*. For every value of δ, we will get neighbourhood.

Thus the set

$$\{(x, y) : a - \delta \le x \le a + \delta, \, b - \delta \le y \le b + \delta\}$$

is a neighbourhood of the point (a, b).

11.4. Continuity of a function of two variables. Continuity at a point.

Let (a, b) be a point of the domain of   *(Manipur 2002)*

$$z = f(x, y)$$

As in the case of functions of one variable, we say that f is continuous at (a, b), if for points (x, y) near (a, b), the value $f(x, y)$ of the function is near $f(a, b)$ *i.e.*, $f(x, y)$ can be made as near f (a, b) as we like by taking the points (x, y) sufficiently near (a, b).

Formally, we say that a function f of two variables is continuous at a point (a, b), if corresponding to any pre-assigned positive number ε, there exists a positive number δ such that

$$|f(x, y) - f(a, b)| < \varepsilon$$

for all members of the neighbourhood

$$\{(x, y) : a - \delta \le x \le a + \delta, b - \delta \le y \le b + \delta\} \text{ of } (a, b).$$

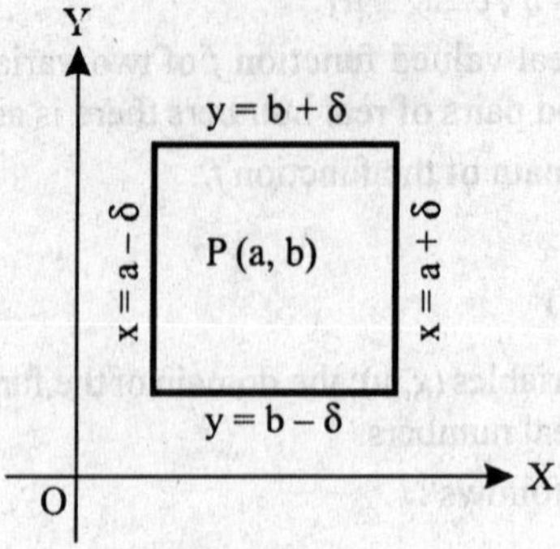

Fig. 11.1

Thus for continuity at (a, b), there exists a square bounded by the lines

$$x = a - \delta, x = a + \delta; y = b - \delta, y = b + \delta$$

such that, for any point (x, y) of this square, $f(x, y)$ lies between

$$f(a, b) - \varepsilon \text{ and } f(a, b) + \varepsilon$$

where ε is any positive number, however small.

11.4.1. Continuity of a function : *A function f is said to be continuous if it is continuous at every point of its domain.*

11.4.2. Special case. Let *f* be continuous at (a, b) and ε be any positive number, however small. Then there exists a square bounded by the lines

$$x = a - \delta, x = a + \delta; y = b - \delta, y = b + \delta.$$

such that for points (x, y) of the square, the numerical value of the difference between $f(x, y)$ and $f(a, b)$ is less than ε.

In particular, if we consider points of this square lying on the line $y = b$, we see that for values of $x \in [a - \delta, a + \delta]$ the numerical value of the difference between $f(x, b)$ and $f(a, b)$ is less than ε. Also such a choice of δ is possible for every positive number ε. This is equivalent to saying that $f(x) = f(x, b)$ is a continuous function of a single variable x for $x = a$.

It may be similarly shown that $f(y) = f(a, y)$ is a continuous function of a single variable y for $y = b$.

Thus we have shown that a *continuous function of two variables is also a continuous function of each variable separately.*

11.5. Limit of a function of two variables.

$$\lim f(x, y) = l \text{ as } (x, y) \to (a, b).$$

A function f is said to tend to the limit l, as (x, y) tends to (a, b), if, corresponding to any pre-assigned positive number ε, there exists a positive number δ such that

$$|f(x, y) - l| < \varepsilon$$

for all points (x, y), other than possibly (a, b), such that

$$a - \delta \le x \le a + \delta; b - \delta \le y \le b + \delta.$$

i.e., for all members of the set

$$\{(x, y) : x \in [a - \delta, a + \delta], y \in [b - \delta, b + \delta]\} \sim \{(a, b)\}.$$

This means that corresponding to every positive number ε there exists a neighbourhood of (a, b) such that for every point (x, y) of this neighbourhood, other than possibly (a, b), $f(x, y)$ lies between $1 - \varepsilon$ and $1 + \varepsilon$.

11.5.1. Limit of a continuous function. Comparing the definitions of limit and continuity as given above in § 11.4, and § 11.5, we see that f *is continuous at* (a, b), *if and only if*

$$\lim f(x, y) = f(a, b) \text{ as } (x, y) \to (a, b).$$

i.e., the limit of the function = value of the function.

The same thing may also be expressed by saying that for continuity at (a, b), we have

$$\lim f(a + h, b + k) = f(a, b)$$

as $(h, k) \to (0, 0)$.

11.6. Partial derivatives.

Consider $\quad z = f(x, y)$.

Then $\quad \lim \dfrac{f(a + h, b) - f(a, b)}{h}, h \to 0,$

if it exists, is said to be the partial derivative of f w. r. t. x at (**a, b**) *and is denoted by*

$$\left(\frac{\partial z}{\partial x} \right)_{(\mathbf{a,\,b})} \text{ or } \mathbf{f}_x\,(\mathbf{a, b}).$$

It will thus be seen that the partial derivative of f w. r. t. x at (a, b) is the derivative of the function $f(x) = f(x, b)$ of a single real variable at $x = a$.

Again

$$\lim \frac{f(a, b + k) - f(a, b)}{k} \text{ as } k \to 0$$

if it exists, is called the partial derivative of $f(x, y)$ w. r. to y at (a, b) *and is denoted by*

$$\left[\frac{\partial z}{\partial y} \right]_{(\mathbf{a,\,b})} \text{ or } \mathbf{f}_y\,(\mathbf{a, b})$$

so that the partial derivative *w. r. t. y* at (a, b) is the derivative of the function $f(y) = f(a, y)$ of a single real variable at $y = b$.

If f possesses a partial derivative *w. r.* to x at every point of its domain, then there arises a new function $(a, b) \to f_x\,(a, b)$ and denoted by f_x of two variables whose domain is the same as that of the given function f.

Thus we have

$$f_x\,(x, y) = \lim \frac{f(x + h, y) - f(x, y)}{h} \text{ as } h \to 0.$$

11.6.1. Partial derivatives of higher orders. We can form partial derivatives of the first order partial derivatives just as we formed those of f.

Thus we have

$$\frac{\partial}{\partial x} \left(\frac{\partial z}{\partial x} \right), \frac{\partial}{\partial y} \left(\frac{\partial z}{\partial x} \right) \text{ or equivalently } f_{xx}, f_{yx}$$

called the second order partial derivatives of f. Now f_{xx}, may also be denoted by f_{x^2}.

We write
$$\frac{\partial}{\partial x}\left(\frac{\partial z}{\partial x}\right)=\frac{\partial^2 z}{\partial x^2}, \quad \frac{\partial}{\partial y}\left(\frac{\partial z}{\partial x}\right)=\frac{\partial^2 z}{\partial y\,\partial x}$$

Similarly the second order partial derivatives
$$\frac{\partial}{\partial x}\left(\frac{\partial z}{\partial y}\right), \quad \frac{\partial}{\partial y}\left(\frac{\partial z}{\partial y}\right)$$

are respectively denoted by
$$\frac{\partial^2 z}{\partial x\,\partial y}, \quad \frac{\partial^2 z}{\partial y^2} \text{ or } f_{xy}, f_{y^2}.$$

Thus, there are *four* second order partial derivatives of z at any point (a, b).

The partial derivatives $\partial^2 z/\partial y\,\partial x$ and $\partial^2 z/\partial x\,\partial y$ are distinguished by the *order* in which z is *successively* differentiated w. r. to x and y. But, it will be seen that, in *general*, they are equal. The proof is given in ξ 11.10.1.

11.7. Geometrical representation of a function of two variables

We take a pair of perpendicular lines OX and OY. Through the point O, we draw a line $Z'\,OZ$ perpendicular to the XY plane and call it z-axis.

Any one of the two sides of z-axis may be assigned a positive sense. Length, z, will be measured parallel to this axis.

The three co-ordinate axes, taken in pairs determine three planes, viz., XY, YZ and ZX which are taken as the *co-ordinate planes*.

Let f be a function defined in some domain in the XY plane.

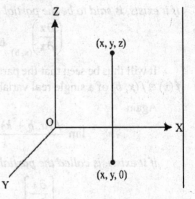

Fig. 11.2

To each point (x, y) of this domain, there corresponds a value of z. Through the point (x, y) we draw the line perpendicular to the XY plane equal in length to z, so that we arrive at another point P denoted as (x, y, z), lying on one or the other side of the plane according as z is positive or negative.

Thus to each point of the domain of the function f in the XY plane there corresponds a point P. The set of the points P determines a surface which is said to represent the function geometrically.

11.7.1. Geometrical interpretation of partial derivatives of the first order.

Consider $z = f(x, y)$ ...(i)

We have seen that the functional equation (i) represents a *surface* geometrically. We now seek the geometrical interpretation of the partial derivatives
$$\left[\frac{\partial z}{\partial x}\right]_{(a,\,b)} \text{ and } \left[\frac{\partial z}{\partial y}\right]_{(a,\,b)}.$$

The point $P\,[(a, b), f(a, b)]$ on the surface corresponds to the point (a, b) of the domain of the function.

If a variable point, starting from P, changes its position on the surface such that y remains constantly equal to b, then it is clear that the locus of the point is the curve of intersection of the surface and the plane $y = b$.

On this curve x and z vary according to the relation

$$z = f(x, b).$$

Also, $\left(\dfrac{\partial z}{\partial x}\right)_{(a, b)}$ is the ordinary derivative of $f(x, b)$.

w.r. to x for $x = a$.

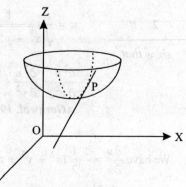

Fig. 11.3

Hence we see that $\left[\dfrac{\partial z}{\partial x}\right]_{(a, b)}$ denotes the tangent of the angle which the tangent to the curve, in which the plane $y = b$ parallel to the ZX plane cuts the surface at $P\,[(a, b), f(a, b)]$, makes with x-axis.

Similarly, it may be seen that $\left[\dfrac{\partial z}{\partial y}\right]_{(a, b)}$ is the tangent of the angle which the tangent at $P\,[(a, b),$ $f(a, b)]$ to the curve of intersection of the surface and the plane $x = a$ parallel to the ZY plane, makes with y-axis.

EXAMPLES

1. If $u = x^2 \tan^{-1} \dfrac{y}{x} - y^2 \tan^{-1} \dfrac{x}{y}$; $xy \neq 0$

prove that

$$\frac{\partial^2 u}{\partial x \partial y} = \frac{x^2 - y^2}{x^2 + y^2}.$$

(*Ravishankar, 1997, 2002; Jabalpur, 1998, 1999; Bhopal, 1994, 1995*)

We have

$$\frac{\partial u}{\partial y} = x^2 \frac{1}{1 + \dfrac{y^2}{x^2}} \cdot \frac{1}{x} - 2y \tan^{-1} \frac{x}{y} + y^2 \frac{1}{1 + \dfrac{x^2}{y^2}} \cdot \frac{x}{y^2}$$

$$= \frac{x^3}{x^2 + y^2} + \frac{xy^2}{x^2 + y^2} - 2y \tan^{-1} \frac{x}{y}$$

$$= x - 2y \tan^{-1} \frac{x}{y}.$$

$$\frac{\partial^2 u}{\partial x \partial y} = \frac{\partial}{\partial x}\left(\frac{\partial u}{\partial y}\right)$$

$$= 1 - 2y \frac{1}{1 + \dfrac{x^2}{y^2}} \cdot \frac{1}{y} = 1 - \frac{2y^2}{y^2 + x^2} = \frac{x^2 - y^2}{x^2 + y^2}.$$

2. If $\qquad u = \dfrac{1}{\sqrt{(x^2 + y^2 + z^2)}}; \ x^2 + y^2 + z^2 \neq 0,$

show that

$$\frac{\partial^2 u}{\partial x^2} + \frac{\partial^2 u}{\partial y^2} + \frac{\partial^2 u}{\partial z^2} = 0.$$

*(**Garhwal, 1998; Sagar, 1996, 1997, 2001; Vikram, 1998; Ravishankar, 1996; Jabalpur, 1994; Indore, 2000; Manipur 1999; Banglore 2005**)*

We have $\dfrac{\partial u}{\partial x} = -\dfrac{1}{2}(x^2 + y^2 + z^2)^{-3/2} \ 2x = -x(x^2 + y^2 + z^2)^{-3/2},$

$$\frac{\partial^2 u}{\partial x^2} = -(x^2 + y^2 + z^2)^{-3/2} + 3x^2(x^2 + y^2 + z^2)^{-5/2}.$$

Similarly or by symmetry

$$\frac{\partial^2 u}{\partial y^2} = -(x^2 + y^2 + z^2)^{-3/2} + 3y^2(x^2 + y^2 + z^2)^{-5/2}.$$

$$\frac{\partial^2 u}{\partial z^2} = -(x^2 + y^2 + z^2)^{-3/2} + 3z^2(x^2 + y^2 + z^2)^{-5/2}.$$

Adding, we obtain the result as given.

3. If $u = e^{xyz}$, *show that*

$$\frac{\partial^3 u}{\partial x \partial y \partial z} = (1 + 3xyz + x^2 y^2 z^2)\, e^{xyz}. \quad \textbf{(Kumaon 2001; Gorakhpur 2003; Avadh 2003)}$$

We have $\qquad u = e^{xyz},$

then $\qquad \dfrac{\partial u}{\partial x} = e^{xyz} \cdot yz \qquad\qquad\qquad\qquad$ (treating y and z constant)

Again $\qquad \dfrac{\partial^2 u}{\partial y \partial x} = \dfrac{\partial}{\partial y}\left(\dfrac{\partial u}{\partial x}\right)$

$$= \frac{\partial}{\partial y}\left(yz\, e^{xyz}\right)$$

$$= z\left[ye^{xyz} \cdot xz + e^{xyz}\right]$$

Then $\qquad \dfrac{\partial^2 u}{\partial x \partial y} = xyz^2 e^{xyz} + z e^{xyz}$

and $\qquad \dfrac{\partial}{\partial z}\left(\dfrac{\partial^2 u}{\partial x \partial y}\right) = \dfrac{\partial}{\partial z}\left[xyz^2 e^{xyz} + z e^{xyz}\right]$

$$= xy\left[z^2 e^{xyz} \cdot xy + e^{xyz} \cdot 2z\right] + \left[z e^{xyz} \cdot zy + e^{xyz}\right]$$

$$= e^{xyz}(1 + 3xyz + x^2 y^2 z^2)$$

$\therefore \qquad\qquad \dfrac{\partial^3 u}{\partial x \partial y \partial z} = \left(1 + 3xyz + x^2 y^2 z^2\right) e^{xyz}$

4. *If* $u = \log(x^3 + y^3 + z^3 - 3xyz)$ *show that*

(a) $\left(\dfrac{\partial}{\partial x} + \dfrac{\partial}{\partial y} + \dfrac{\partial}{\partial z} \right)^2 u = -\dfrac{9}{(x+y+z)^2}$

(Indore, 2002; Jabalpur, 1999; Bilaspur, 1996, 2000; Kanpur, 2000; Kumaon 2004; Avadh 2002)

and (b) $\dfrac{\partial^2 u}{\partial x^2} + \dfrac{\partial^2 u}{\partial y^2} + \dfrac{\partial^2 u}{\partial z^2} = \dfrac{-3}{(x+y+z)^2}.$

(Ravishankar, 1998; Indore, 2001; Vikram, 1999; Jabalpur 2000)

We have

(a) $\qquad u = \log(x^3 + y^3 + z^3 - 3xyz)$

$\therefore \qquad \dfrac{\partial u}{\partial x} = \dfrac{3x^2 - 3xyz}{x^3 + y^3 + z^3 - 3xyz}$ ...(i)

(treating y and z constant)

Similarly,

$\dfrac{\partial u}{\partial y} = \dfrac{3y^2 - 3xz}{x^3 + y^3 + z^3 - 3xyz}$ ...(ii)

(treating x and z constant)

and $\qquad \dfrac{\partial u}{\partial z} = \dfrac{3z^2 - 3xy}{x^3 + y^3 + z^3 - 3xyz}$ ...(iii)

(treating x and y constant)

$\therefore \qquad \dfrac{\partial u}{\partial x} + \dfrac{\partial u}{\partial y} + \dfrac{\partial u}{\partial z} = \dfrac{(x^2 + y^2 + z^2 - xy - yz - zx)}{x^3 + y^3 + z^3 - 3xyz}$

$= \dfrac{3}{x+y+z}$

Now, $\left(\dfrac{\partial}{\partial x} + \dfrac{\partial}{\partial y} + \dfrac{\partial}{\partial z} \right)^2 u = \left(\dfrac{\partial}{\partial x} + \dfrac{\partial}{\partial y} + \dfrac{\partial}{\partial z} \right) \left(\dfrac{\partial u}{\partial x} + \dfrac{\partial u}{\partial y} + \dfrac{\partial u}{\partial z} \right)$

$= \left(\dfrac{\partial}{\partial x} + \dfrac{\partial}{\partial y} + \dfrac{\partial}{\partial z} \right) \left(\dfrac{3}{x+y+z} \right)$

$= \dfrac{\partial}{\partial x} \left(\dfrac{3}{x+y+z} \right) + \dfrac{\partial}{\partial y} \left(\dfrac{3}{x+y+z} \right) + \dfrac{\partial}{\partial z} \left(\dfrac{3}{x+y+z} \right)$

$= -\dfrac{3}{(x+y+z)^2} - \dfrac{3}{(x+y+z)^2} - \dfrac{3}{(x+y+z)^2}$

$= -\dfrac{9}{(x+y+z)^2}$

(*b*) Differentiating (*i*) partially, we get

$$\frac{\partial^2 u}{\partial x^2} = \frac{6x\,(x^3 + y^3 + z^3 - 3xyz) - (3x^2 - 3yz)\,(3x^2 - 3yz)}{(x^3 + y^3 + z^3 - 3xyz)^2}$$

$$= -\frac{3\,(x^4 - 2xy^3 - 2xz^3 + 3y^2z^2)}{(x^3 + y^3 + z^3 - 3xyz)^2}$$

Similarly,

$$\frac{\partial^2 u}{\partial y^2} = -\frac{3\,(y^4 - 2yz^3 - 2yx^3 + 3z^2x^2)}{(x^3 + y^3 + z^3 - 3xyz)^2}$$

and

$$\frac{\partial^2 u}{\partial z^2} = -\frac{3\,(z^4 - 2zx^3 - 2xy^3 + 3x^2y^2)}{(x^3 + y^3 + z^3 - 3xyz)^2}$$

On adding, we will have

$$\frac{\partial^2 u}{\partial x^2} + \frac{\partial^2 u}{\partial y^2} + \frac{\partial^2 u}{\partial z^2}$$

$$= -\frac{3\,(x^2 + y^2 + z^2 - xy - yz - zx)^2}{(x^3 + y^3 + z^3 - 3xyz)^2}$$

$$= -\frac{3}{(x + y + z)^2}$$

5. *If $x^x\,y^y\,z^z = c$ show that at $x = y = z$,*

$$\frac{\partial^2 z}{\partial x \partial y} = -\,(x \log e x)^{-1}.$$ (*Kumaon, 1998; Rohilkhand, 1999*)

Given that $x^x\,y^y\,z^z = c$

Taking log of both the sides, we have $x \log x + y \log y + z \log z = \log c$.

Differentiating partially *w. r. t. x*, we get

$$x \cdot \frac{1}{x} + \log x + \left(z \cdot \frac{1}{z} + \log z\right)\frac{\partial z}{\partial x} = 0$$

i.e.,

$$\frac{\partial z}{\partial x} = -\frac{(1 + \log x)}{(1 + \log z)}$$

Similarly, we have

$$\frac{\partial z}{\partial y} = -\frac{(1 + \log y)}{(1 + \log z)}$$

$$\therefore \quad \frac{\partial^2 z}{\partial x \partial y} = \frac{(1 + \log x)}{(1 + \log z)^2} \cdot \frac{1}{z}\,\frac{\partial z}{\partial y}$$

$$= -\frac{(1 + \log x)}{(1 + \log z)^2} \cdot \frac{1}{z} \cdot \frac{(1 + \log y)}{(1 + \log z)}$$

$$= -\frac{(1 + \log x)^2}{(1 + \log x)^3} \cdot \frac{1}{x} \qquad (\text{since } x = y = z)$$

$$= -\frac{1}{x\,(1 + \log x)} = -\frac{1}{x \log ex}$$

$$= -(x \log ex)^{-1}.$$

6. If $u = \log (\tan x + \tan y + \tan z)$, *prove that*

$$(\sin 2x)\,\frac{\partial u}{\partial x} + (\sin 2y)\,\frac{\partial u}{\partial y} + (\sin 2z)\,\frac{\partial u}{\partial z} = 2.$$

Given $\qquad u = \log (\tan x + \tan y + \tan z)$

$$\therefore \qquad \frac{\partial u}{\partial x} = \frac{1}{(\tan x + \tan y + \tan z)} \sec^2 x$$

or $\quad (\sin 2x)\,\dfrac{\partial u}{\partial x} = \dfrac{\sec^2 x \sin 2x}{\tan x + \tan y + \tan z}$

or $\quad (\sin 2x)\,\dfrac{\partial u}{\partial x} = \dfrac{2 \tan x}{\tan x + \tan y + \tan z}$ ...(i)

Similarly $\quad (\sin 2y)\,\dfrac{\partial u}{\partial y} = \dfrac{2 \tan y}{\tan x + \tan y + \tan z}$ ...(ii)

and $\quad (\sin 2z)\,\dfrac{\partial u}{\partial z} = \dfrac{2 \tan z}{\tan x + \tan y + \tan z}$ ...(iii)

Adding (i), (ii) and (iii) we get

$$(\sin 2x)\,\frac{\partial u}{\partial x} + (\sin 2y)\,\frac{\partial u}{\partial y} + (\sin 2z)\,\frac{\partial u}{\partial z} = \frac{2 \tan x + 2 \tan y + 2 \tan z}{\tan x + \tan y + \tan z} = 2.$$

7. If $u = 3 (lx + my + nz)^2 - (x^2 + y^2 + z^2)$ *and* $l^2 + m^2 + n^2 = 1$,

show that $\qquad \dfrac{\partial^2 u}{\partial x^2} + \dfrac{\partial^2 u}{\partial y^2} + \dfrac{\partial^2 u}{\partial z^2} = 0.$

Given $u = 3 (lx + my + nz)^2 - (x^2 + y^2 + z^2)$.

$$\therefore \qquad \frac{\partial u}{\partial x} = 6l (lx + my + nz) - 2x \qquad \therefore \qquad \frac{\partial^2 u}{\partial x^2} = 6l^2 - 2.$$

Similarly, $\quad \dfrac{\partial^2 u}{\partial y^2} = 6m^2 - 2$ and $\dfrac{\partial^2 u}{\partial z^2} = 6n^2 - 2.$

$$\therefore \quad \frac{\partial^2 u}{\partial x^2} + \frac{\partial^2 u}{\partial y^2} + \frac{\partial^2 u}{\partial z^2} = (6l^2 - 2) + (6m^2 - 2) + (6n^2 - 2)$$

$$= 6 (l^2 + m^2 + n^2) - 6$$

$$= 6 (1) - 6, \qquad \therefore\ l^2 + m^2 + n^2 = 1 \text{ (given)}$$

$$= 0.$$

8. If $v = At^{-1/2} e^{-x^2/4a^2 t}$, prove that $\dfrac{\partial v}{\partial t} = a^2 \dfrac{\partial^2 v}{\partial x^2}$.

We have $v = At^{-1/2} e^{-x^2/4a^2 t}$

$$\Rightarrow \qquad \frac{\partial v}{\partial x} = At^{-1/2} \cdot e^{-x^2/4a^2 t} \left(\frac{-2x}{4a^2 t} \right)$$

$$= - \frac{x}{2a^2 t} v$$

$$\Rightarrow \qquad \frac{\partial^2 v}{\partial x^2} = - \frac{1}{2a^2 t} \left[v + x \frac{\partial v}{\partial x} \right]$$

$$= - \frac{1}{2a^2 t} \left[v + x \left(\frac{-vx}{2a^2 t} \right) \right]$$

$$= \frac{v}{4a^4 t^2} \left(- 2a^2 t + x^2 \right)$$

Again,

$$\frac{\partial v}{\partial t} = At^{-1/2} \cdot e^{-x^2/4a^2 t} \left(\frac{x^2}{4a^2 t^2} \right) - A \cdot \frac{1}{2} t^{-3/2} e^{-x^2/4a^2 t}$$

$$= At^{-1/2} \cdot e^{-x^2/4a^2 t} \left[\frac{x^2}{4a^2 t^2} - \frac{1}{2t} \right]$$

$$= \frac{v}{4a^2 t} \left(x^2 - 2a^2 t \right)$$

Clearly, $\dfrac{\partial v}{\partial t} = a^2 \dfrac{\partial^2 v}{\partial x^2}$.

EXERCISES

1. Find the first order partial derivatives of

  (i) $\tan^{-1}(x+y)$.   (ii) $e^{ax} \sin by$.   (iii) $\log(x^2 + y^2)$.

2. Find the second order partial derivatives of

  (i) e^{x-y} (ii) e^{xy} (iii) $\tan(\tan^{-1} x + \tan^{-1} y)$.

3. Verify that $\dfrac{\partial^2 u}{\partial x \partial y} = \dfrac{\partial^2 u}{\partial y \partial x}$, when u is

  (i) $\sin^{-1} \dfrac{x}{y}$.   (ii) $\dfrac{xy}{\sqrt{(1 + x^2 + y^2)}}$.   (iii) $\log(y \sin x + x \sin y)$.

4. Find the value of $\dfrac{1}{a^2} \dfrac{\partial^2 z}{\partial x^2} + \dfrac{1}{b^2} \dfrac{\partial^2 z}{\partial y^2}$,

  when $a^2 x^2 + b^2 y^2 - c^2 z^2 = 0$.

5. If $z = \tan(y + ax) + (y - ax)^{3/2}$, find the value of

$$\frac{\partial^2 z}{\partial x^2} - a^2 \frac{\partial^2 z}{\partial y^2}.$$

6. If $z = \tan^{-1}(y/x)$, verify that

$$\frac{\partial^2 z}{\partial x^2} + \frac{\partial^2 z}{\partial y^2} = 0.$$

7. If $z(x+y) = x^2 + y^2$, show that

$$\left(\frac{\partial z}{\partial x} - \frac{\partial z}{\partial y}\right)^2 = 4\left(1 - \frac{\partial z}{\partial x} - \frac{\partial z}{\partial y}\right). \qquad \textbf{\textit{(Kumaon, 2000)}}$$

8. If $z = 3xy - y^3 + (y^2 - 2x)^{3/2}$, verify that

$$\frac{\partial^2 z}{\partial x \partial y} = \frac{\partial^2 z}{\partial y \partial x} \text{ and } \frac{\partial^2 z}{\partial x^2}\frac{\partial^2 z}{\partial y^2} = \left(\frac{\partial^2 z}{\partial x \partial y}\right)^2.$$

9. If $z = f(x + ay) + \varphi(x - ay)$, prove that

$$\frac{\partial^2 z}{\partial y^2} = a^2 \frac{\partial^2 z}{\partial x^2}. \qquad \textbf{\textit{(Rewa, 2001; Jiwaji, 2002)}}$$

10. If $z = (x + y) + (x + y)\,\varphi(y/x)$, prove that

$$x\left(\frac{\partial^2 z}{\partial x^2} - \frac{\partial^2 x}{\partial y \partial x}\right) = y\left(\frac{\partial^2 z}{\partial y^2} - \frac{\partial^2 z}{\partial x \partial y}\right).$$

11. If $u = f(ax^2 + 2hxy + by^2)$, $v = \varphi(ax^2 + 2hxy + by^2)$, prove that

$$\frac{\partial}{\partial y}\left(u\,\frac{\partial v}{\partial x}\right) = \frac{\partial}{\partial x}\left(u\,\frac{\partial v}{\partial y}\right).$$

12. If $\theta = t^n\, e^{-r^2/4t}$, find the value of n which will make

$$\frac{1}{r^2}\frac{\partial}{\partial r}\left(r^2\,\frac{\partial \theta}{\partial r}\right) = \frac{\partial \theta}{\partial t}.$$

13. If $u = f(r)$ where $r = \sqrt{(x^2 + y^2)}$, prove that

$$\frac{\partial^2 u}{\partial x^2} + \frac{\partial^2 u}{\partial y^2} = f''(r) + \frac{1}{r}\,f'(r).$$

$$\textbf{\textit{(Vikram, 1997; Sagar, 1997; Gorakhpur, 2000)}}$$

14. If $u = \log(x^2 + y^2 + z^2)$, prove that

$$x\,\frac{\partial^2 u}{\partial y \partial z} = y\,\frac{\partial^2 u}{\partial z \partial x} = z\,\frac{\partial^2 u}{\partial x \partial y}. \qquad \textbf{\textit{(Kumaon, 2002)}}$$

15. If $\dfrac{x^2}{a^2 + u} + \dfrac{y^2}{b^2 + u} + \dfrac{z}{c^2 + u} = 1$, prove that

$$u_x^2 + u_y^2 + u_z^2 = 2(x\,u_x + y\,u_y + z\,y_z). \qquad \textbf{\textit{(Garhwal 2004)}}$$

16. If $\tan u = \dfrac{\cos x}{\sin hy}$ and $\tan hv = \dfrac{\sin x}{\cos hy}$ prove that

$$\frac{\partial u}{\partial x} = \frac{\partial v}{\partial y} \text{ and } \frac{\partial u}{\partial y} = -\frac{\partial v}{\partial x}. \qquad \textbf{\textit{(Garhwal, 1999)}}$$

17. If $u = \tan^{-1} \dfrac{xy}{\sqrt{(1 + x^2 + y^2)}}$, show that

$$\frac{\partial^2 u}{\partial x \partial y} = \frac{1}{(1 + x^2 + y^2)^{3/2}}.$$

(Vikram, 2000; Ravishankar, 2003; Bilaspur, 2000; Rewa, 2001)

18. If $u = x^2 (y - z) + y^2 (z - x) + z^2 (x - y)$, prove that

$$\frac{\partial u}{\partial x} + \frac{\partial u}{\partial y} + \frac{\partial u}{\partial z} = 0.$$

19. If $u = \log_e \sqrt{(x^2 + y^2 + z^2)}$, prove that

$$(x^2 + y^2 + z^2) \left(\frac{\partial^2 u}{\partial x^2} + \frac{\partial^2 u}{\partial y^2} + \frac{\partial^2 u}{\partial z^2} \right) = 1. \qquad \textbf{(Ravishankar, 1998)}$$

20. If $u = e^x (x \cos y - y \sin y)$, show that

$$\frac{\partial^2 u}{\partial x^2} + \frac{\partial^2 u}{\partial y^2} = 0.$$

21. If $u = (1 - 2xy + y^2)^{-1/2}$, prove that

$$\frac{\partial}{\partial x} \left\{ (1 - x^2) \frac{\partial u}{\partial x} \right\} + \frac{\partial}{\partial y} \left(y^2 \frac{\partial u}{\partial y} \right) = 0. \qquad \textbf{(Bilashpur, 2000; Jiwaji, 1995)}$$

22. If $a^2 x^2 + b^2 y^2 - c^2 z^2 = 0$, show that

$$\frac{\partial^2 z}{\partial x^2} + \frac{\partial^2 z}{\partial y^2} = \frac{a^2 b^2}{c^4 z^3} (x^2 + y^2).$$

23. If $u = \begin{vmatrix} x^2 & y^2 & z^2 \\ x & y & z \\ 1 & 1 & 1 \end{vmatrix}$, show that

$$u_x + u_y + u_z = 0. \qquad \textbf{(Bilaspur, 1995; Ravishankar, 2002)}$$

24. If $u = x + y + z$, $v = x^2 + y^2 + z^2$, $w = x^3 + y^3 + z^3 - 3 x y z$, prove that

$$\begin{vmatrix} u_x & u_y & u_z \\ v_x & v_y & v_z \\ w_x & w_y & w_z \end{vmatrix} = 0.$$

25. If $u = \sin^{-1} \dfrac{x}{y} + \tan^{-1} \dfrac{y}{x}$, show that

$$x \frac{\partial u}{\partial x} + y \frac{\partial u}{\partial y} = 0. \qquad \textbf{(Indore, 1995; Jiwaji, 1995, 2003; Sagar, 1998)}$$

ANSWERS

1. (i) $\dfrac{1}{1 + (x + y)^2} , \dfrac{1}{1 + (x + y)^2}.$ (ii) $ae^{ax} \sin by,\ be^{ax} \cos by.$

(iii) $\dfrac{2x}{(x^2 + y^2)} , \dfrac{2y}{(x^2 + y^2)}.$

2. (i) $e^{x-y}, -e^{x-y}, -e^{x-y}, e^{x-y}.$

(ii) $ye^{(x^y)} x^{y-2} [yx^y + y - 1].$

   $e^{(x^y)} x^{y-1} [1 + (1 + x^y) \log x^y].$

   $e^{(x^y)} x^{y-1} [1 + (1 + x^y) \log x^y].$

   $e^{(x^y)} x^y (1 + x^y) (\log x)^2.$

(iii) $\dfrac{2y (1 + y^2)}{(1 - xy)^3}, \dfrac{2 (x + y)}{(1 - xy)^3}, \dfrac{2 (x + y)}{(1 - xy)^3}, \dfrac{2x (1 + x^2)}{(1 - xy)^3}.$

4. $1/c^2 z.$ **5.** 0.   **12.** $-3/2.$

11.8. Homogeneous Functions   *(Manipur 2000)*

Consider the function

$$f(x, y) = a_0 x^n + a_1 x^{n-1} y + a_2 x^{n-2} y^2 + \ldots + a_{n-1} xy^{n-1} + a_n y^n \quad \ldots (i)$$

We see that the expression, $f(x, y)$ is a polynomial in (x, y) such that the degree of each of the terms is the same *viz.*, n.

Because of this f is called a homogeneous function of degree n.

This definition of *homogeneity* applies to polynomial functions only. To enlarge the concept of the homogeneity so as to bring even other functions within its scope, we say that an expression in (x, y) is homogeneous of degree n, if it is expressible as

$$x^n f(y/x).$$

The polynomial function (i) which can be rewritten as

$$x^n \left[a_0 + a_1 \frac{y}{x} + a_2 \left(\frac{y}{x} \right)^2 + \ldots + a_n \left(\frac{y}{x} \right)^n \right]$$

is a homogeneous expression of order n according to the new definition also. The functions

$$f(x, y) = x^n \sin (y/x), f(x, y) = (\sqrt{y} + \sqrt{x})/(y + x)$$

are homogeneous according to the second definition only. The degree of the expression $x^n \sin (y/x)$ is n. Also

$$\frac{\sqrt{y} + \sqrt{x}}{y + x} = \frac{\sqrt{x} \left[1 + \dfrac{\sqrt{y}}{\sqrt{x}} \right]}{x \left[1 + \dfrac{y}{x} \right]} = x^{-1/2} \frac{1 + \sqrt{\dfrac{y}{x}}}{1 + \dfrac{y}{x}}$$

so that it is of degree $-\dfrac{1}{2}$.

11.8.1 Euler's theorem on Homogeneous Functions.

If $z = f(x, y)$ be a homogeneous function of x, y of degree n then

$$x \frac{\partial z}{\partial x} + y \frac{\partial z}{\partial y} = nz, \forall x, y \in \text{the domain of the function.}$$

(Kanpur, 2001, 2003; Kumaon, 1997; Vikram, 1999; Jiwaji, 1997, 1999, 2000; Ravishankar, 1996; Sagar, 1999; G.N.D.U. Amritsar, 2004; Indore, 2000; Avadh 2002; Gorakhpur 2003; MDU Rohtak 2000; Manipur 2002; Kuvempu 2005)

We have $\quad z = x^n f\left[\dfrac{y}{x}\right]$

$\Rightarrow \qquad \dfrac{\partial z}{\partial x} = nx^{n-1} f\left[\dfrac{y}{x}\right] + x^n\, f'\left[\dfrac{y}{x}\right] \times \dfrac{-y}{x^2}$

$\qquad\qquad\qquad = nx^{n-1} f\left[\dfrac{y}{x}\right] - yx^{n-2}\, f'\left[\dfrac{y}{x}\right]$

and $\qquad\qquad \dfrac{\partial z}{\partial y} = x^n f'\left[\dfrac{y}{x}\right]\cdot\dfrac{1}{x} = x^{n-1}\, f'\left[\dfrac{y}{x}\right]$

Thus, we have

$$x \dfrac{\partial z}{\partial x} + y \dfrac{\partial z}{\partial y} = nx^n f\left[\dfrac{y}{x}\right] = nz.$$

Hence the result.

Cor. *If $z = f(x, y)$ is a homogeneous function of x, y of degree n, then*

$$x^2 \dfrac{\partial^2 z}{\partial x^2} + 2xy \dfrac{\partial^2 z}{\partial x \partial y} + y^2 \dfrac{\partial^2 z}{\partial y^2} = n(n-1)\, z.$$

(Kumaon, 1999; Indore, 1995, 2000; Vikram, 2001; Jiwaji, 1995, 2001; MDU Rohtak 1998)

Differentiating

$$x \dfrac{\partial z}{\partial x} + y \dfrac{\partial z}{\partial y} = nz, \qquad\qquad\qquad\qquad\qquad ...(1)$$

partially with respect to x and y separately, we obtain

$$\dfrac{\partial z}{\partial x} + x \dfrac{\partial^2 z}{\partial x^2} + y \dfrac{\partial^2 z}{\partial x \partial y} = n \dfrac{\partial z}{\partial x} \qquad\qquad\qquad ...(2)$$

$$x \dfrac{\partial^2 z}{\partial y \partial x} + \dfrac{\partial z}{\partial y} + y \dfrac{\partial^2 z}{\partial y^2} = n \dfrac{\partial z}{\partial y}. \qquad\qquad\qquad ...(3)$$

Multiplying (2), (3) by x, y respectively and adding, we obtain

$$x^2 \dfrac{\partial^2 z}{\partial x^2} + 2xy \dfrac{\partial^2 z}{\partial x \partial y} + y^2 \dfrac{\partial^2 z}{\partial y^2} = n(n-1)\, z,$$

where we have assumed the equality of $\partial^2 z/\partial x \partial y$ and $\partial^2 z/\partial y \partial x$.

EXAMPLES

1. If $u = \cot^{-1} \dfrac{x+y}{\sqrt{x} + \sqrt{y}}$, *show that*

$$x \dfrac{\partial u}{\partial x} + y \dfrac{\partial u}{\partial y} + \dfrac{1}{4} \sin 2u = 0.$$

Let $z = \cot u = \dfrac{x+y}{\sqrt{x} + \sqrt{y}}$ ...(1)

z is a homogeneous function of x and y of degree 1/2. Therefore by Euler's Theorem,

$$x \dfrac{\partial z}{\partial x} + y \dfrac{\partial z}{\partial y} = \dfrac{1}{2}\, z \qquad\qquad\qquad\qquad\qquad ...(2)$$

From (1),

$$\frac{\partial z}{\partial x} = - \operatorname{cosec}^2 u \frac{\partial u}{\partial x}$$

and

$$\frac{\partial z}{\partial y} = - \operatorname{cosec}^2 u \frac{\partial u}{\partial y}$$

From (2)

$$- x \operatorname{cosec}^2 u \frac{\partial u}{\partial x} - y \operatorname{cosec}^2 u \frac{\partial u}{\partial y} = \frac{1}{2} \cot u$$

$$\Rightarrow \qquad x \frac{\partial u}{\partial x} + y \frac{\partial u}{\partial y} = \frac{-1}{2} \frac{\cot u}{\operatorname{cosec}^2 u}$$

$$\Rightarrow \qquad x \frac{\partial u}{\partial x} + y \frac{\partial u}{\partial y} + \frac{1}{4} \sin 2u = 0$$

2. *If* $u = \tan^{-1} \dfrac{x^3 + y^3}{x - y}$, $x \neq y$ *show that*

(i) $x \dfrac{\partial u}{\partial x} + y \dfrac{\partial u}{\partial y} = \sin 2u$

(ii) $x^2 \dfrac{\partial^2 u}{\partial x^2} + 2xy \dfrac{\partial^2 u}{\partial x \partial y} + y^2 \dfrac{\partial^2 u}{\partial y^2} = (1 - 4 \sin^2 u) \sin 2u.$

(Gorakhpur, 2002; Kumaon, 2003; Rohilkhand, 2003; Vikram, 2000; Ravishankar, 1997, 1998, 2000; Jabalpur, 1997; Rewa, 2001; Bhopal, 1997; Banglore 2004; 2005)

(i) Here u is not a homogeneous function. We, however, write

$$z = \tan u = \frac{x^3 + y^3}{x - y} = x^2 \frac{1 + (y/x)^3}{1 - (y/x)},$$

so that z is a homogeneous function of x, y of degree 2.

$$\therefore \quad x \frac{\partial z}{\partial x} + y \frac{\partial z}{\partial y} = 2z. \qquad \qquad \qquad \qquad \dots(1)$$

But

$$\frac{\partial z}{\partial x} = \sec^2 u \frac{\partial z}{\partial x}; \quad \frac{\partial z}{\partial y} = \sec^2 u \frac{\partial u}{\partial y}. \qquad \qquad \dots(2)$$

Substituting in (1), we obtain

$$\sec^2 u \left(x \frac{\partial u}{\partial x} + y \frac{\partial u}{\partial y} \right) = 2z = 2 \tan u$$

$$\Rightarrow x \frac{\partial u}{\partial x} + y \frac{\partial u}{\partial y} = \frac{2 \sin u}{\cos u} \cos^2 u = \sin 2u. \qquad \qquad \dots(3)$$

(ii) From (2)

$$\frac{\partial^2 z}{\partial x^2} = \sec^2 u \frac{\partial^2 u}{\partial x^2} + 2 \sec^2 u \tan u \left(\frac{\partial u}{\partial x} \right)^2,$$

$$\frac{\partial^2 z}{\partial y^2} = \sec^2 u \frac{\partial^2 u}{\partial y^2} + 2 \sec^2 u \tan u \left(\frac{\partial u}{\partial y} \right)^2$$

and $\qquad \dfrac{\partial^2 z}{\partial x \partial y} = \sec^2 u \dfrac{\partial^2 u}{\partial x \partial y} + 2 \sec^2 u \tan u \dfrac{\partial u}{\partial x} \dfrac{\partial u}{\partial y}$

Also by corollary of Euler's Theorem,

$$x^2 \dfrac{\partial^2 z}{\partial x^2} + 2xy \dfrac{\partial^2 z}{\partial x \partial y} + y^2 \dfrac{\partial^2 z}{\partial y^2} = 2 \, (2 - 1) \, z$$

$$\Rightarrow \quad \sec^2 u \left(x^2 \dfrac{\partial^2 u}{\partial x^2} + 2xy \dfrac{\partial^2 u}{\partial x \partial y} + y^2 \dfrac{\partial^2 u}{\partial y^2} \right)$$

$$+ 2 \sec^2 u \tan u \left(x^2 \left(\dfrac{\partial u}{\partial x} \right)^2 + 2xy \dfrac{\partial u}{\partial x} \dfrac{\partial u}{\partial y} + y^2 \left(\dfrac{\partial u}{\partial y} \right)^2 \right) = 2 \tan u$$

$$\Rightarrow \quad x^2 \dfrac{\partial^2 u}{\partial x^2} + 2xy \dfrac{\partial^2 u}{\partial x \partial y} + y^2 \dfrac{\partial^2 u}{\partial y^2} + 2 \tan u \left(x \dfrac{\partial u}{\partial x} + y \dfrac{\partial u}{\partial y} \right)^2 = 2 \sin u \cos u$$

$$\Rightarrow \quad x^2 \dfrac{\partial^2 u}{\partial x^2} + 2xy \dfrac{\partial^2 u}{\partial x \partial y} + y^2 \dfrac{\partial^2 u}{\partial y^2} = \sin 2u - 2 \tan u \sin^2 2u \qquad \text{(by 3)}$$

$$= (1 - 2 \tan u \sin 2u) \sin 2u$$

$$= (1 - 4 \sin^2 u) \sin 2u$$

3. *If* $z = x^m f(y/x) + x^n g(x/y)$, *prove that*

$$x^2 \dfrac{\partial^2 z}{\partial x^2} + 2xy \dfrac{\partial^2 z}{\partial x \partial y} + y^2 \dfrac{\partial^2 z}{\partial y^2} + mnz = (m + n - 1) \left(x \dfrac{\partial z}{\partial x} + y \dfrac{\partial z}{\partial y} \right).$$

Let $\qquad\qquad u = x^m f(y/x) \text{ and } v = x^n g(x/y).$

Then $\qquad\qquad z = u + v.$ $\qquad\qquad\qquad\qquad\qquad\qquad\qquad\qquad$ ...(i)

Now $u = x^m f(y/x)$ is a homogeneous function in x and y of degree m.

Therefore with the help of Euler's Theorem we can

Prove that $\qquad x^2 \dfrac{\partial^2 u}{\partial x^2} + 2xy \dfrac{\partial^2 u}{\partial x \, \partial y} + y^2 \dfrac{\partial^2 v}{\partial y^2} = m \, (m - 1) \, u$ $\qquad\qquad$ ..(ii)

[See cor §. 11.8.1]

Also $v = x^n g(x/y)$ is a homogeneous function in x and y of degree n, so we have

$$x^2 \dfrac{\partial^2 v}{\partial x^2} + 2xy \dfrac{\partial^2 v}{\partial x \, \partial y} + y^2 \dfrac{\partial^2 v}{\partial y^2} = n \, (n - 1) \, v. \qquad\qquad \text{...(iii)}$$

Adding (*ii*) and (*iii*), we have

$$x^2 \dfrac{\partial^2}{\partial x^2} (u + v) + 2xy \dfrac{\partial^2}{\partial x \, \partial y} (u + v) + y^2 \dfrac{\partial^2}{\partial y^2} (u + v) = m \, (m - 1) \, u + n \, (n - 1) \, v.$$

or $\quad x^2 \dfrac{\partial^2 z}{\partial x^2} + 2xy \dfrac{\partial^2 z}{\partial x \, \partial y} + y^2 \dfrac{\partial^2 z}{\partial y^2} = m \, (m - 1) \, u + n \, (n - 1) \, v.$ $\qquad$ ...(iv)

Again from Euler's Theorem we have for u and v, which are homogeneous functions in x and y of degree m and n respectively,

$$x \frac{\partial u}{\partial x} + y \frac{\partial u}{\partial y} = mu \text{ and } x \frac{\partial v}{\partial x} + y \frac{\partial v}{\partial y} = nv.$$

Adding these two with the help of (i), we get

$$x \frac{\partial z}{\partial x} + y \frac{\partial z}{\partial y} = mu + nv. \qquad \ldots(v)$$

Again
$$m(m-1)u + n(n-1)v$$
$$= (m^2 u + n^2 v) - (mu + nv)$$
$$= m(m+n)u + n(m+n)v - mn(u+v) - (mu + nv)$$
$$= (m+n)(mu + nv) - mn(u+v) - (mu + nv)$$
$$= (m+n-1)(mu + nv) - mn(z), \qquad \text{from } (i)$$

Substituting this value in (iv), we get

$$x^2 \frac{\partial^2 z}{\partial x^2} + \frac{\partial^2 z}{\partial x \partial y} + y^2 \frac{\partial^2 z}{\partial y^2} = (m+n-1)(mu+nv) - mnz$$

or $\quad x^2 \dfrac{\partial^2 z}{\partial x^2} + 2xy \dfrac{\partial^2 z}{\partial x \partial y} + y^2 \dfrac{\partial^2 z}{\partial y^2} + mnz = (m+n-1)(mu+nv)$

$$= (m+n-1)\left(x \frac{\partial z}{\partial x} + \frac{\partial z}{\partial y} \right) \quad \text{from } (v)$$

4. If $V = \log_e \sin \left\{ \dfrac{\pi (2x^2 + y^2 + xz)^{1/2}}{2(x^2 + xy + 2yz + z^2)^{1/3}} \right\}$, *find the value of*

$$x \frac{\partial V}{\partial x} + y \frac{\partial V}{\partial y} + z \frac{\partial V}{\partial z} \quad when\ x = 0, y = 1, z = 2.$$

Let $\qquad u = \dfrac{\pi (2x^2 + y^2 + xz)^{1/2}}{2(x^2 + xy + 2yz + z^2)^{1/3}}$,

then $\qquad V = \log_e \sin u$

$\therefore \qquad \dfrac{\partial V}{\partial x} = \cot u \dfrac{\partial u}{\partial x}, \dfrac{\partial V}{\partial y} = \cot u \dfrac{\partial u}{\partial y}$

and $\qquad \dfrac{\partial V}{\partial z} = \cot u \dfrac{\partial u}{\partial z}$

$\therefore \quad x \dfrac{\partial V}{\partial x} + y \dfrac{\partial V}{\partial y} + z \dfrac{\partial V}{\partial z} = \cot u \left(x \dfrac{\partial u}{\partial x} + y \dfrac{\partial u}{\partial y} + z \dfrac{\partial u}{\partial z} \right) \qquad \ldots(i)$

But u is a homogeneous function of degree $\dfrac{1}{3}$.

$\therefore \qquad x \dfrac{\partial u}{\partial x} + y \dfrac{\partial u}{\partial y} + z \dfrac{\partial u}{\partial z} = \dfrac{1}{3} u$

$\therefore$ From (i)

$$x \frac{\partial V}{\partial x} + y \frac{\partial V}{\partial y} + z \frac{\partial V}{\partial z} = \frac{1}{3} u \cot u .$$

When $x = 0, y = 1, z = 2$, we have $u = \dfrac{1}{4}\pi$, and then

$$x\frac{\partial V}{\partial x} + y\frac{\partial V}{\partial y} + z\frac{\partial V}{\partial z} = \frac{1}{3} \cdot \frac{\pi}{4} \cot \frac{\pi}{4} = \frac{\pi}{12}.$$

5. *If* $u = \log\left\{\dfrac{(x^4 + y^4)}{(x + y)}\right\}$, *show by Euler's Theorem that*

$$x\frac{\partial u}{\partial x} + y\frac{\partial u}{\partial y} = 3.$$

(*Jiwaji, 1997; Vikram, 1999; Rewa, 1999; Jabalpur, 1999, 2002*)

Here,

$$e^u = \frac{(x^4 + y^4)}{(x + y)} = z \qquad \text{(say)}$$

Now z is a homogeneous function of degree $(4 - 1) = 3$. Hence by Euler's Theorem

$$x\frac{\partial z}{\partial x} + y\frac{\partial z}{\partial y} = 3z$$

or $\qquad x\dfrac{\partial}{\partial x}(e^u) + y\dfrac{\partial}{\partial y}(e^u) = 3e^u$

or $\qquad e^u \cdot x\dfrac{\partial u}{\partial x} + e^u \cdot y\dfrac{\partial u}{\partial y} = 3e^u$

$\therefore \qquad x\dfrac{\partial u}{\partial x} + y\dfrac{\partial u}{\partial y} = 3.$

6. *If* $u = ze^{ax + by}$, *where z is a homogeneous function in x and y of degree n, prove that*

$$x\frac{\partial u}{\partial x} + y\frac{\partial u}{\partial y} = (ax + by + n)\,u.$$

If z is a homogeneous function of degree n in x and y, then from Euler's Theorem,

we get $\qquad x\dfrac{\partial z}{\partial x} + y\dfrac{\partial z}{\partial y} = nz.$ $\hfill ...(i)$

Now $\qquad x\dfrac{\partial u}{\partial x} + y\dfrac{\partial u}{\partial y} = x\dfrac{\partial}{\partial x}(ze^{ax + by}) + y\dfrac{\partial}{\partial y}(ze^{ax + by})$

$$= x\left[\frac{\partial z}{\partial x}e^{ax + by} + z\frac{\partial}{\partial x}e^{ax + by}\right] + y\left[\frac{\partial z}{\partial y}e^{ax + by} + z\frac{\partial}{\partial y}e^{ax + by}\right]$$

$$= \left(x\frac{\partial z}{\partial x} + y\frac{\partial z}{\partial y}\right)e^{ax + by} + z\left[x\frac{\partial}{\partial x}e^{ax + by} + y\frac{\partial}{\partial y}e^{ax + by}\right]$$

$$= nze^{ax + by} + z\,[xae^{ax + by} + bye^{ax + by}]$$

$$= (n + ax + by)\,ze^{ax + by} = (ax + by + n)\,u.$$

7. *If* $z = (x + y)\,\phi\,(y/x)$, *where ϕ is any arbitrary function prove that* $x\dfrac{\partial z}{\partial x} + y\dfrac{\partial z}{\partial y} = z.$

$$\frac{\partial z}{\partial x} = \phi\,(y/x) + (x + y)\,\phi'\,(y/x)\,(-y/x^2)$$

or $\qquad x\dfrac{\partial z}{\partial x} = x\phi\left(\dfrac{y}{x}\right) - \dfrac{y}{x}(x + y)\,\phi'\left(\dfrac{y}{x}\right)$ $\hfill ...(i)$

Also $\qquad \dfrac{\partial z}{\partial y} = \phi\left(\dfrac{y}{x}\right) + (x + y)\; \phi'\left(\dfrac{y}{x}\right)\left(\dfrac{1}{x}\right)$

or $\qquad y\dfrac{\partial z}{\partial y} = y\phi\left(\dfrac{y}{x}\right) + \dfrac{y}{x}\,(x + y)\; \phi'\left(\dfrac{y}{x}\right).$ ...(ii)

Adding (i) and (ii), we get

$$x\dfrac{\partial z}{\partial y} + y\dfrac{\partial z}{\partial y} = (x + y)\; \phi\left(\dfrac{y}{x}\right) = z.$$

Example 10. *If* $\sin v = (x + 2y + 3z)/\sqrt{(x^8 + y^8 + z^8)},$ *show that*

$$x\dfrac{\partial v}{\partial x} + y\dfrac{\partial v}{\partial y} + z\dfrac{\partial v}{\partial z} + 3\tan v = 0.$$

We have

$$\sin v = \dfrac{x + 2y + 3z}{\sqrt{x^8 + y^8 + z^8}} = f \text{ (say)}$$

Then f is a homogeneous function in x and of degree $1 - 4 = 3$.

∴ By Euler's Theorem

$$x\dfrac{\partial f}{\partial x} + y\dfrac{\partial f}{\partial y} + z\dfrac{\partial f}{\partial z} = -3f$$

$\Rightarrow \quad x\cos v\, \dfrac{\partial v}{\partial x} + y\cos v\, \dfrac{\partial v}{\partial y} + z\cos v\, \dfrac{\partial v}{\partial z} = -3\sin v$

$\Rightarrow \quad x\dfrac{\partial v}{\partial x} + y\dfrac{\partial v}{\partial y} + z\dfrac{\partial v}{\partial z} = -3\tan v.$

EXERCISES

1. Verify Euler's theorem for

   (i) $z = ax^2 + 2hxy + by^2.$ **(Kuvempu 2005)**   (ii) $z = (x^2 + xy + y^2)^{-1}.$

   (iii) $z = \sin^{-1}\dfrac{x}{y} + \tan^{-1}\dfrac{y}{x}.$ (iv) $z = x^n \log\dfrac{y}{x}.$

   (v) $z = \dfrac{x^{1/4} + y^{1/4}}{x^{1/5} + y^{1/5}}$ **(Ravishankar 1997; G.N.D.U. Amritsar, 2004)**

2. If $u = \log\dfrac{x^2 + y^2}{x + y}$, prove that $x\dfrac{\partial u}{\partial x} + y\dfrac{\partial u}{\partial y} = 1.$ **(Kanpur 2006)**

3. (a) If $u = \sin^{-1}\dfrac{x^2 + y^2}{x + y}$, show that $x\dfrac{\partial u}{\partial x} + y\dfrac{\partial u}{\partial y} = \tan u.$

   (Garhwal, 2000; Bhopal, 1996, 1999, 2000; Bilaspur, 1998; Vikram, 2001; Jiwaji, 1997; Jabalpur, 1996, 1998; Rewa, 2000; Gorakhpur, 2003; Patna, 2003; Indore, 1998, 2001)

   (b) If $u = \sin^{-1}\dfrac{\sqrt{x} - \sqrt{y}}{\sqrt{x} + \sqrt{y}}$, show that $x\dfrac{\partial u}{\partial x} + y\dfrac{\partial u}{\partial y} = 0.$ **(Garhwal, 2002)**

4. If $u = \cos^{-1}\dfrac{x + y}{\sqrt{x} + \sqrt{y}}$, show that $x\dfrac{\partial u}{\partial x} + y\dfrac{\partial u}{\partial y} + \dfrac{1}{2}\cot u = 0.$

5. If $z = \sec^{-1} \dfrac{x^3 + y^3}{x + y}$, show that $x \dfrac{\partial z}{\partial x} + y \dfrac{\partial z}{\partial y} = 2 \cot z$.   (**Banglore, 2004; 2005**)

6. If $u = \tan^{-1} \left(\dfrac{x + y}{\sqrt{x} + \sqrt{y}} \right)$, show that $x \dfrac{\partial u}{\partial x} + y \dfrac{\partial u}{\partial y} = \dfrac{1}{4} \sin 2u$.

7. If $V = r^m$ where $r^2 = x^2 + y^2 + z^2$, show that

$$\frac{\partial^2 V}{\partial x^2} + \frac{\partial^2 V}{\partial y^2} + \frac{\partial^2 V}{\partial z^2} = m (m + 1) r^{m - 2}.$$

8. If $u = \sin^{-1} \left\{ \dfrac{x^{1/3} + y^{1/3}}{x^{1/2} + y^{1/2}} \right\}^{1/2}$,

show that

$$x^2 \frac{\partial^2 u}{\partial x^2} + 2xy \frac{\partial^2 u}{\partial x \partial y} + y^2 \frac{\partial^2 u}{\partial y^2} = \frac{\tan u}{144} (13 + \tan^2 u).$$ (**Jiwaji, 1998**)

9. If $z = xyf(y/x)$, show that $x \dfrac{\partial z}{\partial x} + y \dfrac{\partial z}{\partial y} = 2z$,

and if z is a constant, then $\dfrac{f'(y/x)}{f(y/x)} = \dfrac{x \left(y + x \dfrac{dy}{dx} \right)}{y \left(y - x \dfrac{dy}{dx} \right)}$.

10. If $u = \dfrac{x^2 y^2}{x^2 + y^2}$, show that

(a) $x \dfrac{\partial^2 u}{\partial x^2} + y \dfrac{\partial^2 u}{\partial y \partial x} = \dfrac{\partial u}{\partial x}$

(b) $x \dfrac{\partial^2 u}{\partial x \partial y} + y \dfrac{\partial^2 u}{\partial y^2} = \dfrac{\partial u}{\partial y}$

(c) $x^2 \dfrac{\partial^2 u}{\partial x^2} + 2xy \dfrac{\partial^2 u}{\partial x \partial y} + y^2 \dfrac{\partial^2 u}{\partial y^2} = 2u$.   (**Utkal 2003**)

11.9. For the following developments, it will be assumed that f possesses continuous partial derivative $w. r.$ to x and y for every point of domain of the function.

11.9.1. Theorem on Total Differentials. We consider a function

$$z = f(x, y) \qquad \qquad \qquad \ldots(i)$$

Let (x, y), $(x + \Delta x, y + \Delta y)$ be any two points so that Δx, Δy are the changes in the independent variables x and y. Let Δz be the consequent change in z.

We have

$$z + \Delta z = f(x + \Delta x, y + \Delta y) \qquad \qquad \ldots(ii)$$

From (i) and (ii), we get

$$\Delta z = f(x + \Delta x, y + \Delta y) - f(x, y)$$
$$= [f(x + \Delta x, y + \Delta y) - f(x + \Delta x, y)] + [f(x + \Delta x, y) - f(x, y)] \qquad \ldots(iii)$$

so that we have subtracted and added $f(x + \Delta x, y)$.

Here the change Δz has been expressed as the sum of two differences, to each of which we shall apply Lagrange's mean value theorem.

We consider the function $f(y) = f(x + \Delta x, y)$; where $x + \Delta x$ is constant, so that by the mean value theorem,

$$f(x + \Delta x, y + \Delta y) - f(x + \Delta x, y) = \Delta y f_y(x + \Delta x, y + \theta_1 \Delta y)$$

We write

$$f_y(x + \Delta x, y + \theta_1 \Delta y) - f_y(x, y) = \varepsilon_2 \qquad \qquad ...(iv)$$

so that ε_2 depends on Δx, Δy and because of the assumed continuity of f_y tends to zero as Δx, Δy both tend to 0.

Again by considering $f(x) = f(x, y)$, keeping y constant, we have by the mean value theorem,

$$f(x + \Delta x, y) - f(x, y) = \Delta x f_x(x + \theta_2 \Delta x, y),$$

We write

$$f_x(x + \theta_2 \Delta x, y) - f_x(x, y) = \varepsilon_1 \qquad \qquad ...(v)$$

so that ε_1 depends upon Δx and because of the assumed continuity of f_x tends to 0 as Δx tends to 0.

From (iii), (iv) and (v), we get

$$\Delta z = \Delta x f_x(x, y) + \Delta y f_y(x, y) + \varepsilon_1 \Delta x + \varepsilon_2 \Delta x$$

$$= \underbrace{\frac{\partial z}{\partial x} \Delta x + \frac{\partial z}{\partial y} \Delta y} + \underbrace{\varepsilon_1 \Delta x + \varepsilon_2 \Delta y}.$$

Thus the change Δz in z consists of two parts as marked. Of these the first is called the **differential** of z and is denoted by dz.

Thus

$$dz = \frac{\partial z}{\partial x} \Delta x + \frac{\partial z}{\partial y} \Delta y \qquad \qquad ...(vi)$$

Now

$$z = x \Rightarrow dx = dz = 1 . \Delta x = \Delta x.$$

Similarly by taking $z = y$ we have $\Delta y = dy$.

Thus (vi) takes the form

$$\boxed{dz = \frac{\partial z}{\partial x} dx + \frac{\partial z}{\partial y} dy.}$$

It should be carefully noted that the *differentials dx and dy of the independent variables x and y are the* **actual** *changes Δx and Δy, but the differential dz of the dependent variable z is not the same as the change Δz; it being the* **principal part** *of the increment Δz.*

Cor. Approximate Calculations. From above, we *see that the approximate change dz in z corresponding to the small changes Δx and Δy in x, y is*

$$\frac{\partial z}{\partial x} dx + \frac{\partial z}{\partial y} dy$$

which has been denoted by dz.

11.9.2 Composite functions

Let

$$z = f(x, y) \qquad \qquad ...(i)$$

and let

$$x = \varphi(t) \qquad \qquad ...(ii)$$

$$y = \psi(t) \qquad \qquad ...(iii)$$

so that x, y are themselves functions of a third variable t.

The equations (i), (ii), (iii) are said to define z as a *composite function* of t.

Again, let $x = \phi (u, v)$ ...(iv)

$y = \psi (u, v)$ ...(v)

so that x, y are functions of the variables u, v. Here the functional equations (i), (iv), (v) define z as a function of u, v, which is called *a composite function* of u and v.

11.9.3 Differentiation of composite functions.

Let $z = f(x, y)$

possess continuous partial derivatives and let

$x = \varphi (t)$

$y = \psi (t)$

possess continuous derivatives.

Then $\dfrac{dz}{dt} = \dfrac{\partial z}{\partial x} \cdot \dfrac{dx}{dt} + \dfrac{\partial z}{\partial y} \dfrac{dy}{dt}.$

Let $t, t + \Delta t$ be any two values. Let $\Delta x, \Delta y, \Delta z$ be the changes in x, y, z consequent to the change Δt in t. We have

$x + \Delta x = \varphi (t + \Delta t), y + \Delta y = \psi (t + \Delta t)$

$z + \Delta z = f(x + \Delta x, y + \Delta y)$

$\Rightarrow$ $\Delta z = f(x + \Delta x, y + \Delta y) - f(x, y)$

$= [f(x + \Delta x, y + \Delta y) - f(x, y + \Delta y)] + [f(x, y + \Delta y) - f(x, y)].$

As in § 7.2, we apply Lagrange's mean value theorem to the two differences on the right and obtain

$\Delta z = \Delta x f_x (x + \theta_1 \Delta x, y + \Delta y) + \Delta y f_y (x, y + \theta_2 \Delta y)$ $(0 < \theta_1, \theta_2 < 1).$

$\Rightarrow$ $\dfrac{\Delta z}{\Delta t} = \dfrac{\Delta x}{\Delta t} f_x (x + \theta_1 \Delta x, y + \Delta y) + \dfrac{\Delta y}{\Delta t} f_y (x, y + \theta_2 \Delta y)$ ...(i)

Let $\Delta t \to 0$ so that Δx and $\Delta y \to 0$.

Because of the continuity of partial derivatives, we have

$\lim\limits_{(\Delta x, \Delta y) \to (0, 0)} f_x (x + \theta_1 \Delta x, y + \Delta y) = f_x (x, y) = \dfrac{\partial z}{\partial x},$

$\lim\limits_{\Delta y \to 0} f_y (x, y + \theta_2 \Delta y) = f_y (x, y) = \dfrac{\partial z}{\partial y}.$

Hence in the limit, (i) becomes

$\dfrac{dz}{dt} = \dfrac{\partial z}{\partial x} \cdot \dfrac{dx}{dt} + \dfrac{\partial z}{\partial y} \dfrac{dy}{dt}.$ ...(ii) (*Nagpur 2005*)

Cor 1. Let $z = f(x, y),$

possess continuous first order partial derivatives w. r. to x, y.

Let $x = \varphi (u, v)$

$y = \psi (u, v)$

possess continuous first order partial derivatives.

To obtain $\partial z/\partial u$, we regard v as a constant so that x and y may be supposed to be functions of u only. Then, by the above theorem, we have

$$\frac{\partial z}{\partial u} = \frac{\partial z}{\partial x} \cdot \frac{\partial x}{\partial u} + \frac{\partial z}{\partial y} \cdot \frac{\partial y}{\partial u}.$$

It may similarly be shown that

$$\frac{\partial z}{\partial v} = \frac{\partial z}{\partial x} \cdot \frac{\partial x}{\partial v} + \frac{\partial z}{\partial y} \cdot \frac{\partial y}{\partial v}.$$

EXAMPLES

1. *Find dz/dt when*
$$z = xy^2 + x^2 y, x = at^2, y = 2at.$$
Verify by direct substitution.

Now $\quad \dfrac{\partial z}{\partial x} = y^2 + 2xy, \dfrac{\partial z}{\partial y} = 2xy + x^2,$

$$\frac{dx}{dt} = 2at, \frac{dy}{dt} = 2a.$$

Substituting these values in (*ii*), § 11.9.3, we get

$$\frac{dz}{dt} = (y^2 + 2xy)\, 2at + (2xy + x^2)\, 2a$$

$$= (4a^2\, t^2 + 4a^2\, t^3)\, 2at + (4a^2\, t^3 + a^2\, t^4)\, 2a$$

$$= a^3\, (16t^3 + 10t^4)$$

Again $\quad z = x^2\, y + xy^2 = 2a^3\, t^5 + 4a^3\, t^4.$

$\Rightarrow \qquad \dfrac{dz}{dt} = 10a^3\, t^4 + 16a^3\, t^3 = a^3\, (16t^3 + 10t^4).$

Hence the verification.

2. *z is a function of x and y. Prove that if*
$$x = e^u + e^{-v}, y = e^{-u} - e^v,$$

then $\quad \dfrac{\partial z}{\partial u} - \dfrac{\partial z}{\partial v} = x\, \dfrac{\partial z}{\partial x} - y\, \dfrac{\partial z}{\partial y}.$

We look upon z as a composite function of u, v. We have

$$\frac{\partial z}{\partial u} = \frac{\partial z}{\partial x} \cdot \frac{\partial x}{\partial u} + \frac{\partial z}{\partial y}\frac{\partial y}{\partial u}$$

$$= \frac{\partial z}{\partial x} \cdot e^u - \frac{\partial z}{\partial y} \cdot e^{-u};$$

$$\frac{\partial z}{\partial v} = \frac{\partial z}{\partial x}\frac{\partial x}{\partial v} + \frac{\partial z}{\partial y} \cdot \frac{\partial y}{\partial v}$$

$$= -\frac{\partial z}{\partial x} \cdot e^{-v} - \frac{\partial z}{\partial y} \cdot e^v.$$

Subtracting, we get

$$\frac{\partial z}{\partial u} - \frac{\partial z}{\partial v} = \frac{\partial z}{\partial x}(e^u + e^{-v}) - \frac{\partial z}{\partial y}(e^{-u} - e^v)$$

$$= x\frac{\partial z}{\partial x} - y\frac{\partial z}{\partial y}.$$

3. *If* $H = f(y - z, z - x, x - y)$, *prove that*

$$\frac{\partial H}{\partial x} + \frac{\partial H}{\partial y} + \frac{\partial H}{\partial z} = 0.$$

(Kumaon, 1998, 2000; Garhwal, 2001;

MDU Rohtak 1999; Manipur 2000)

Let $\qquad u = y - z, v = z - x, w = x - y,$

$\Rightarrow \qquad H = f(u, v, w).$

We have expressed H as a composite function of x, y, z. We have

$$\frac{\partial H}{\partial x} = \frac{\partial H}{\partial u}\cdot\frac{\partial u}{\partial x} + \frac{\partial H}{\partial v}\cdot\frac{\partial v}{\partial x} + \frac{\partial H}{\partial w}\cdot\frac{\partial w}{\partial x}$$

$$= \frac{\partial H}{\partial u}\cdot 0 + \frac{\partial H}{\partial v}(-1) + \frac{\partial H}{\partial w}\cdot 1$$

$$= -\frac{\partial H}{\partial v} + \frac{\partial H}{\partial w}$$

Similarly

$$\frac{\partial H}{\partial y} = -\frac{\partial H}{\partial w} + \frac{\partial H}{\partial u}$$

$$\frac{\partial H}{\partial z} = -\frac{\partial H}{\partial u} + \frac{\partial H}{\partial v}.$$

Adding, we get the result.

4. *H is a homogeneous function of* x, y, z *of degree n, prove that*

$$x\frac{\partial H}{\partial x} + y\frac{\partial H}{\partial y} + z\frac{\partial H}{\partial z} = nz.$$

[Thus is Euler's theorem for a homogeneous function of three independent variables.] We have

$$H = x^n f\left[\frac{y}{x}, \frac{z}{x}\right] = x^n f(u, v) \text{ where } y/x = u, z/x = v.$$

$\therefore \qquad \dfrac{\partial H}{\partial x} = nx^{n-1} f(u, v) + x^n\left[\dfrac{\partial f}{\partial u}\cdot\dfrac{\partial u}{\partial x} + \dfrac{\partial f}{\partial v}\cdot\dfrac{\partial v}{\partial x}\right]$

But $\qquad \dfrac{\partial u}{\partial x} = -\dfrac{y}{x}, \dfrac{\partial v}{\partial x} = -\dfrac{z}{x^2}.$

Hence $\qquad \dfrac{\partial H}{\partial x} = nx^{n-1} f(u, v) - x^{n-2}\left[y\dfrac{\partial f}{\partial u} + z\dfrac{\partial f}{\partial v}\right].$

Again $\qquad \dfrac{\partial H}{\partial y} = x^n\left[\dfrac{\partial f}{\partial u}\cdot\dfrac{\partial u}{\partial y} + \dfrac{\partial f}{\partial v}\cdot\dfrac{\partial v}{\partial y}\right]$

$$= x^{n-1}\frac{\partial f}{\partial u}, \text{ for } \frac{\partial u}{\partial y} = \frac{1}{x}, \frac{\partial v}{\partial y} = 0.$$

Similarly $\qquad \dfrac{\partial H}{\partial z} = x^{n-1} \dfrac{\partial f}{\partial v}$

$\Rightarrow \qquad x \dfrac{\partial H}{\partial x} + y \dfrac{\partial H}{\partial y} + z \dfrac{\partial H}{\partial z} = nx^n f(u, v) = nz.$

5. *In a triangle ABC, the angles and sides a and b are made to vary in such a way that the area remains constant and the side c also remains constant. Show that a and b vary by small amounts δa, δb respectively, then $\cos A\ \delta a + \cos B\ \delta b = 0$.*

Let Δ denote the area of the triangle ABC. Then

$$\Delta = \dfrac{1}{2} ac \sin B \qquad\qquad\qquad\qquad\text{...(i)}$$

where Δ is constant,

Also, we know that

$$b^2 = c^2 + a^2 - 2ac \cos B \qquad\qquad\qquad\text{...(ii)}$$

Differentiating (i), we get

$$0 = \dfrac{1}{2} c\ [a \cos B\ \delta B + \sin B\ \delta a]$$

$\Rightarrow \qquad a \cos B\ \delta B = -\sin B\ \delta a \qquad\qquad\qquad\text{...(iii)}$

Differentiating (ii), we get

$$2b\ \delta b = 2a\ \delta a - 2c\ \{a\ (-\sin B)\ \delta B + \cos B\ \delta a\}$$

$\Rightarrow \qquad b\ \delta b = (a - a \cos B)\ \delta a + ac \sin B\ \delta B$

$$= (a - a \cos B)\ \delta a + ac \sin B\ \left[\dfrac{-\sin B\ \delta a}{a \cos B}\right]$$

$$= \left[\dfrac{(a - c \cos B) \cos B - c \sin^2 B}{\cos B}\right] \delta a$$

$$= \left(\dfrac{a \cos B - c}{\cos B}\right) \delta a = \left\{\dfrac{a \cos B - (a \cos B + b \cos A)}{\cos B}\right\} \delta a$$

$$= -\left(\dfrac{b \cos A}{\cos B}\right) \delta a$$

$\Rightarrow \quad \cos A\ \delta a + \cos B\ \delta b = 0.$

6. *If the sides and the angles of a plane triangle ABC vary in such a way that its circum radius remains constant, prove that $(\delta a/\cos A) + (\delta b/\cos B) + (\delta c/\cos C) = 0$, where δa, δb and δc denote small increments in the sides a, b and c respectively.*

Let R be the circum radius of the $\triangle ABC$, then

$$R = \dfrac{a}{2 \sin A} = \dfrac{b}{2 \sin B} = \dfrac{c}{2 \sin C}$$

$\Rightarrow \qquad a = 2R \sin A, b = 2R \sin B, c = 2R \sin C$

Differentiating, we get

$$\delta a = 2R \cos A\ \delta A, \delta b = 2R \cos B\ \delta B$$

and $\quad \delta c = 2R \cos C\ \delta C$

$\Rightarrow \qquad \dfrac{\delta a}{\cos A} + \dfrac{\delta b}{\cos B} + \dfrac{\delta c}{\cos C} = 2R\ (\delta A + \delta B + \delta C)$

$$= 2R\ \delta\ (A + B + C) = 2R\ \delta\ (\pi)$$

$$= 0.$$

11.9.4 Implicit functions. Let f be any function of two variables. Ordinarily, we say that, since on solving the equation

$$f(x, y) = 0 \qquad \qquad ...(i)$$

we can obtain y as a function of x, the equation (i) defines y as an *implicit* function of x.

There arises a theoretical difficulty here. Without investigation we cannot say that corresponding to each value of x, the equation (i) must determine one and only one value of y so that the equation (i) always determines, y, as a function of x and in fact, this is not the case, in general. The investigation of the conditions under which the equation (i) does define y as a function of x is not, however, within the scope of this book.

Assuming that the conditions under which the equation (i) defines y as a derivable function of x are satisfied, we shall *now obtain the values of dy/dx and d^2y/dx^2 in terms of the partial derivatives* $\partial f/\partial x$, $\partial f/\partial y$, $\partial^2 f/\partial x^2$, $\partial^2 f/\partial x \partial y$, $\partial^2 f/\partial y^2$ of 'f' w.r. to x and y.

Now, f is a function of two variables x, y and y is again a function of x so that we may regard f as a composite function of x. Its derivative with respect to x is

$$\frac{\partial f}{\partial x} \cdot \frac{dx}{dx} + \frac{\partial f}{\partial y} \cdot \frac{dy}{dx} = \frac{\partial f}{\partial x} + \frac{\partial f}{\partial y} \cdot \frac{dy}{dx}.$$

As f considered as a function of x alone, is identically equal to 0. As such we have

$$\frac{\partial f}{\partial x} + \frac{\partial f}{\partial y} \cdot \frac{dy}{dx} = 0$$

$$\frac{dy}{dx} = -\frac{\partial f/\partial x}{\partial f/\partial y} = -\frac{f_x}{f_y} \text{ if } f_y \neq 0.$$

Differentiating again w.r. to x, regarding $\partial f/\partial x$ and $\partial f/\partial y$ as composite functions of x, we get

$$\frac{d^2y}{dx^2} = -\frac{\left[\dfrac{\partial^2 f}{\partial x^2} + \dfrac{\partial^2 f}{\partial y \partial x} \dfrac{dy}{dx}\right]\dfrac{\partial f}{\partial y} - \dfrac{\partial f}{\partial x}\left[\dfrac{\partial^2 f}{\partial x \partial y} + \dfrac{\partial^2 f}{\partial y^2}\dfrac{dy}{dx}\right]}{\left[\dfrac{\partial f}{\partial y}\right]^2}$$

$$= -\frac{\dfrac{\partial^2 f}{\partial x^2}\left[\dfrac{\partial f}{\partial y}\right]^2 - 2\dfrac{\partial^2 f}{\partial y \partial x}\dfrac{\partial f}{\partial x}\dfrac{\partial f}{\partial y} + \dfrac{\partial^2 f}{\partial y^2}\left[\dfrac{\partial f}{\partial x}\right]^2}{\left[\dfrac{\partial f}{\partial y}\right]^3}$$

Hence
$$\frac{dy}{dx} = -\frac{f_x}{f_y}, \qquad \qquad ...(ii)$$

and
$$\frac{d^2y}{dx^2} = -\frac{f_x^{\,2}\,(f_y)^2 - 2f_{yx}\,f_x\,f_y + f_{y^2}\,(f_x)^2}{(f_y)^3} \qquad ...(iii) \;\; \textit{(MDU Rohtak 1999)}$$

Without making use of § 11.9.3, we may obtain dy/dx also as follows :

Now $f(x, y) = 0.$

Let Δx be the increment in x and Δy the consequent increment in y so that

$$f(x + \Delta x, y + \Delta y) = 0$$

$\Rightarrow \quad f(x + \Delta x, y + \Delta y) - f(x, y) = 0$

$\Rightarrow \quad f(x + \Delta x, y + \Delta y) - f(x, y + \Delta y) + f(x, y + \Delta y) - f(x, y) = 0$

$\Rightarrow \quad \Delta x f_x (x + \theta_1 \Delta x, y + \Delta y) + \Delta y f_y (x, y + \theta_2 \Delta y) = 0$

$\Rightarrow \quad \dfrac{\Delta y}{\Delta x} = - \dfrac{f_x (x + \theta_1 \Delta x, y + \Delta y)}{f_y (x, y + \theta_2 \Delta y)}.$

Let $\Delta x \to 0$. We obtain

$$\frac{dy}{dx} = - \frac{f_x}{f_y} \text{ if } f_y \neq 0.$$

EXAMPLES

1: *Prove that if* $y^3 - 3ax^2 + x^3 = 0$, *then*

$$\frac{d^2 y}{dx^2} + \frac{2a^2 x^2}{y^5} = 0.$$

Let $\qquad f(x, y) = y^3 - 3ax^2 + x^3 = 0$

We have $\qquad f_x (x, y) = -6ax + 3x^2, f_y (x, y) = 3y^2$;

$\qquad\qquad f_x^2 (x, y) = -6a + 6x, f_{xy} (x, y) = 0, f_y^2 (x, y) = 6y$

Substituting these values in (*iii*), on p. 468, we get

$$\frac{d^2 y}{dx^2} = - \frac{6(x - a) 9y^4 + (3x^2 - 6ax)^2 6y}{27 y^6}$$

$$= -2 \cdot \frac{(x - a)(3ax^2 - x^3) + (x^2 - 2ax)^2}{y^5}$$

$$= -2 \frac{a^2 x^2}{y^5}.$$

Thus $\qquad \dfrac{d^2 y}{dx^2} + \dfrac{2a^2 x^2}{y^5} = 0.$

Or, directly, differentiating the given relation *w.r.* to *x*,

$$3y^2 \frac{dy}{dx} = 6ax - 3x^2$$

$\Rightarrow \qquad\qquad \dfrac{dy}{dx} = \dfrac{2ax - x^2}{y^2}$

$\Rightarrow \qquad\qquad \dfrac{d^2 y}{dx^2} = \dfrac{(2a - 2x) y^2 - 2y (2ax - x^2) \, dy/dx}{y^4}$

$$= \frac{(2a - 2x) y^2 - 2y (2ax - x^2)(2ax - x^2)/y^2}{y^4}$$

$$= \frac{2(a - x) y^3 - 2(2ax - x^2)^2}{y^5} = - \frac{2a^2 x^2}{y^5}, \text{ as before.}$$

2. *If* $ax^2 + 2hxy + by^2 + 2gx + 2fy + c = 0$, *prove that*

$$\frac{d^2 y}{dx^2} = \frac{abc + 2fgh - af^2 - bg^2 - ch^2}{(hx + by + f)^3}.$$

$$f(x, y) = ax^2 + 2hxy + by^2 + 2gx + 2fy + c$$

$$f_x(x, y) = 2(ax + hy + g)$$

$$f_y(x, y) = 2(hx + by + f)$$

$$f_{xy} = 2h = f_{yx}$$

$$f_{x^2} = 2a$$

$$f_{y^2} = 2b$$

$$\therefore \quad \frac{d^2y}{dx^2} = \frac{\{8a(hx + by + f)^2 - 16h(ax + hy + g)}{8(hx + by + f)^3}$$

$$\frac{(hx + by + f) + 8b(ax + hy + g)^2\}}{8(hx + by + f)^3}$$

$$= \frac{-ab(ax^2 + 2hxy + by^2 + 2gx + 2fy) + 2fgh - af^2 - bg^2}{(hx + by + f)^3}$$

$$= \frac{+ h^2(ax^2 + 2hxy + by^2 + 2gx + 2fy)}{(hx + by + f)^3}$$

$$= \frac{abc + 2fgh - af^2 - bg^2 - ch^2}{(hx + by + f)^3}$$

3. *If A, B, C are the angles of a triangle such that*

$$\sin^2 A + \sin^2 B + \sin^2 C = constant$$

prove that $\quad \dfrac{dA}{dB} = \dfrac{\tan B - \tan C}{\tan C - \tan A}.$

We have $\quad \sin C = \sin(\pi - (A + B))$

$$= \sin(A + B)$$

Let $\quad f(A, B) \equiv \sin^2 A + \sin^2 B + \sin^2(A + B) - constant$

$$f_A(A, B) = 2\sin A \cos A + 2\sin(A + B)\cos(A + B)$$

$$f_B(A, B) = 2\sin B \cos B + 2\sin(A + B)\cos(A + B)$$

$$\therefore \quad \frac{dA}{dB} = -\frac{f_B}{f_A}$$

$$= -\frac{2\sin B \cos B + 2\sin(A + B)\cos(A + B)}{2\sin A \cos A + 2\sin(A + B)\cos(A + B)}$$

$$= -\frac{\sin 2B - \sin 2C}{\sin 2A - \sin 2C}$$

$$= \frac{-\cos A \{\sin C \cos B - \cos C \sin B\}}{-\cos B \{\sin A \cos C - \cos A \sin C\}}$$

$$= \frac{\cos A \sin B \cos C - \cos A \cos B \sin C}{\cos A \cos B \sin C - \sin A \cos B \cos C}$$

$$= \frac{\tan B - \tan C}{\tan C - \tan A}$$

EXERCISES

1. If $u = x^2 - y^2$, $x = 2r - 3s + 4$, $y = -r + 8s - 5$, find $\partial u/\partial r$.

2. If $z = (\cos y)/x$ and $x = u^2 - v$, $y = e^v$, find $\partial z/\partial v$.

3. If $z = \dfrac{\sin u}{\cos v}$, $u = \dfrac{\cos y}{\sin x}$, $v = \dfrac{\cos x}{\sin y}$, find $\partial z / \partial x$.

4. If $u = (x + y)/(1 - xy)$; $x = \tan(2r - s^2)$, $y = \cot(r^2 s)$, find $\partial u/\partial s$.

5. Find dy/dx in the following cases:—

    (i) $x \sin(x - y) - (x + y) = 0$.   (ii) $y^{x^y} = \sin x$.

    (iii) $(\cos x)^y - (\sin y)^x = 0$.   (iv) $x^y = y^x$.

    (v) $(\tan x)^y + y^{\cot x} = a$.

6. If $F(x, y, z) = 0$, find $\partial z/\partial x$, $\partial z/\partial y$.

7. If $z = xyf(y/x)$ and z is a constant, show that

$$\frac{f'(y/x)}{f(y/x)} = \frac{x[y + x(dy/dx)]}{y[y - x(dy/dx)]}.$$

8. If $f(x, y) = 0$, $\varphi(y, z) = 0$, show that

$$\frac{\partial f}{\partial y} \cdot \frac{\partial \varphi}{\partial z} \cdot \frac{dz}{dx} = \frac{\partial f}{\partial x} \cdot \frac{\partial \varphi}{\partial y}.$$

                                                           (Kumaon, 1999; Rohilkhand, 2001)

9. If $x\sqrt{(1 - y^2)} + y\sqrt{(1 - x^2)} = a$, show that

$$\frac{d^2 y}{dx^2} = \frac{a}{(1 - x^2)^{3/2}}.$$

10. If u and v are functions of x and y defined by

$$x = u + e^{-v}\sin u,\ y = v + e^{-v}\cos u,$$

prove that $\dfrac{\partial u}{\partial y} = \dfrac{\partial v}{\partial x}.$

11. Find $d^2 y/dx^2$ in the following cases :—

    (i) $x^3 + y^3 = 3axy$.   (ii) $x^4 + y^4 = 4a^2 xy$.

    (iii) $x^5 + y^5 = 5a^3 xy$.   (iv) $x^5 + y^5 = 5a^3 x^2$.

12. If the curves $f(x, y) = 0$ and $\phi(x, y) = 0$ touch each other, show that at the point of contact

$$\frac{\partial f}{\partial x} \cdot \frac{\partial \phi}{\partial y} - \frac{\partial f}{\partial y} \cdot \frac{\partial \phi}{\partial x} = 0.$$

ANSWERS

1. $4x + 2y$.

2. $(\cos y - xy \sin y)/x^2$.

3. $-(u \cot x \cos u \sin y + z \sin v \sin x)/(\cos v \sin y)$.

4. $-(1 + x^2)(1 + y^2)(2s + r^2)/(1 - xy)^2$.

5. (i) $[y + x^2 \cos(x - y)]/[x + x^2 \cos(x - y)]$.

    (ii) $(\cot x - yx^{y-1} \log y) y/x^y (y \log x \log y + 1)$.

    (iii) $\dfrac{\sin y (y \sin x + \cos x \log \sin y)}{\cos x (\sin y \log \cos x - x \cos y)}.$

    (iv) $y(y - x \log y)/x(x - y \log x)$.

(v) $-\dfrac{y\,(\tan x)^{y-1}\,\sec^2 x - \operatorname{cosec}^2 x \, \log y \, (y)^{\cot x}}{\log \tan x \, (\tan x)^y + \cot x \, (y)^{\cot x - 1}}.$

6. $\dfrac{\partial z}{\partial x} = -\dfrac{\partial F / \partial x}{\partial F / \partial z};\ \dfrac{\partial z}{\partial y} = -\dfrac{\partial F / \partial y}{\partial F / \partial z}.$

11. (i) $\dfrac{2a^3\,xy}{(ax - y^2)^3},$ (ii) $\dfrac{2a^2\,xy\,(3a^4 + x^2 y^2)}{(a^2 x - y^3)^3},$

(iii) $\dfrac{6a^3\,xy\,(2a^6 + x^3 y^3)}{(a^3 x - y^4)^3},$ (iv) $\dfrac{6a^3 x^2\,(a^3 + x^3)}{y^9}.$

11.10. Equality of f_{xy} (a, b) and f_{yx} (a, b). If has been seen that the two repeated second order partial derivatives are generally equal. They are not, however, *always* equal as is shown below by considering two examples. It is easy to see a *priori* also why $f_{yx}\,(a, b)$ may be different from $f_{xy}\,(a, b)$.

We have $f_{xy}\,(a, b) = \lim\limits_{h\to 0} \dfrac{f_y\,(a + h, b) - f_y\,(a, b)}{h}$

Also $f_y\,(a + h, b) = \lim\limits_{k\to 0} \dfrac{f\,(a + h, b + k) - f\,(a + h, b)}{k},$

and $f_y\,(a, b) = \lim\limits_{k\to 0} \dfrac{f\,(a, b + k) - f\,(a, b)}{k}.$

$\therefore$ $f_{xy}\,(a, b) = \lim\limits_{h\to 0} \lim\limits_{k\to 0} \dfrac{f\,(a+h, b+k) - f\,(a+h, b) - f\,(a, b+k) + f\,(a, b)}{hk}.$

$= \lim\limits_{h\to 0} \lim\limits_{k\to 0} \dfrac{\phi\,(h, k)}{hk},$ say.

It may similarly be shown that

$f_{yx}\,(a, b) = \lim\limits_{k\to 0} \lim\limits_{h\to 0} \dfrac{\phi\,(h, k)}{hk}.$

Thus we see that $f_{xy}\,(a, b)$ and $f_{yx}\,(a, b)$ are *repeated* limits of the same expression taken in *different orders*. Also the two repeated limits may not be equal, as, for example

$$\lim\limits_{h\to 0} \lim\limits_{k\to 0} \dfrac{h - k}{h + k} = \lim\limits_{h\to 0} \dfrac{h}{h} = 1,$$

and $\lim\limits_{k\to 0} \lim\limits_{h\to 0} \dfrac{h - k}{h + k} = \lim\limits_{k\to 0} \dfrac{-k}{k} = -1.$

EXAMPLES

1. *Prove that $f_{xy}\,(0, 0) \neq f_{yx}\,(0, 0)$ for the function f given by*

$$f(x, y) = \dfrac{xy\,(x^2 - y^2)}{x^2 + y^2};\ (x, y) \neq (0, 0)$$

$f(0, 0) = 0.$

We have $f_{xy}\,(0, 0) = \lim\limits_{h\to 0} \dfrac{f_y\,(0 + h, 0) - f_y\,(0, 0)}{h}$ ...(1)

Also $f_y\,(0, 0) = \lim\limits_{k\to 0} \dfrac{f\,(0, 0 + k) - f\,(0, 0)}{k}$ ...(2)

$$= \lim_{k \to 0} \frac{f(0, k) - f(0, 0)}{k} = \lim_{k \to 0} \frac{0}{k} = 0$$

and $\quad f_y(h, 0) = \lim_{k \to 0} \dfrac{f(h, 0 + k) - f(h, 0)}{k}$

$$= \lim_{k \to 0} \frac{hk(h^2 - k^2)}{k(h^2 + k^2)} = h. \qquad \qquad ...(3)$$

Thus from (1), (2) and (3)

$$f_{xy}(0, 0) = \lim_{h \to 0} \frac{h - 0}{h} = 1.$$

Again $\quad f_{yx}(0, 0) = \lim_{k \to 0} \dfrac{f_x(0, 0 + k) - f_x(0, 0)}{k} \qquad \qquad ...(4)$

Also $\quad f_x(0, 0) = \lim_{h \to 0} \dfrac{f(0 + h, 0) - f(0, 0)}{h} = \lim_{h \to 0} \dfrac{0}{h} = 0 \qquad ...(5)$

and $\quad f_x(0, k) = \lim_{h \to 0} \dfrac{f(0 + h, k) - f(0, k)}{h}$

$$= \lim_{h \to 0} \frac{hk(h^2 - k^2)}{h(h^2 + k^2)} = -k. \qquad \qquad ...(6)$$

From (4), (5) and (6),

$$f_{yx}(0, 0) = \lim_{k \to 0} \frac{-k - 0}{k} = -1.$$

Thus $\quad f_{xy}(0, 0) = 1 \ne -1 = f_{yx}(0, 0)$.

2. *Show that* $f_{xy}(0, 0) \ne f_{yx}(0, 0)$

where $\qquad f(x, y) = 0$ *if* $xy = 0$,

$$f(x, y) = x^2 \tan^{-1} \frac{y}{x} - y^2 \tan^{-1} \frac{x}{y}, \text{ if } xy \ne 0.$$

We have $f_{xy}(0, 0) = \lim_{h \to 0} \dfrac{f_y(h, 0) - f_y(0, 0)}{h}$

$$f_y(h, 0) = \lim_{k \to 0} \frac{f(h, k) - f(h, 0)}{k}$$

$$= \lim_{k \to 0} \frac{1}{k} \left(h^2 \tan^{-1} \frac{k}{h} - k^2 \tan^{-1} \frac{h}{k} \right)$$

$$= \lim_{k \to 0} \left[h \left(\frac{\tan^{-1} k/h}{k/h} \right) - k \tan^{-1} \frac{h}{k} \right]$$

$$= h \cdot 1 - 0 = h, \text{ for } [\tan^{-1} t/t] \to 1 \text{ as } t \to 0.$$

Also $\qquad \lim_{k \to (0 + 0)} \tan^{-1} \frac{h}{k} = \frac{\pi}{2}, \lim_{k \to (0 - 0)} \tan^{-1} \frac{h}{k} = -\frac{\pi}{2}$

$$f_y(0, 0) = \lim_{k \to 0} \frac{f(0, k) - f(0, 0)}{k} = 0.$$

$\therefore \qquad f_{xy}(0, 0) = \lim_{h \to 0} \dfrac{h - 0}{h} = 1.$

We may similarly show that

$$f_{yx}(0, 0) = -1.$$

Hence the result.

11.10.1. Equality of f_{xy} and f_{yx}. Theorem. *If $f(x, y)$ possesses continuous second order partial derivatives f_{xy} and f_{yx}, then*

$$f_{xy} = f_{yx}.$$

Consider the expression

$$\varphi(h, k) = f(x + h, y + k) - f(x + h, y) - f(x, y + k) + f(x, y).$$

For the sake of brevity, we write

$$\psi(x) = f(x, y + k) - f(x, y)$$

$$\Rightarrow \quad \varphi(h, k) = \psi(x + h) - \psi(x).$$

By Lagrange's mean value theorem,

$$\psi(x + h) - \psi(x) = h\,\psi'(x + \theta_1 h), \quad 0 < \theta_1 < 1.$$

Also

$$\psi'(x) = f_x(x, y + k) - f_x(x, y)$$

$$\Rightarrow \quad \varphi(h, k) = h[f_x(x + \theta_1 h, y + k) - f_x(x + \theta_1 h, y)]. \quad \text{...(1)}$$

Again applying the mean value theorem to the right side of (1), we obtain

$$\varphi(h, k) = hk\,f_{yx}(x + \theta_1 h, y + \theta_2 k) \quad 0 < \theta_2 < 1.$$

Thus

$$\frac{\varphi(h, k)}{hk} = f_{yx}(x + \theta_1 h, y + \theta_2 k).$$

Again considering $F(y) = f(x + h, y) - f(x, y)$ in place of $\psi(x)$ and proceeding as before, we may prove that

$$\frac{\varphi(h, k)}{hk} = f_{xy}(x + \theta_3 h, y + \theta_4 k).$$

$$\Rightarrow \quad f_{xy}(x + \theta_1 h, y + \theta_2 k) = f_{xy}(x + \theta_3 h, y + \theta_4 k).$$

Let $(h, k) \to (0, 0)$. Then, because of the assumed continuity of the partial derivatives, we obtain

$$f_{yx}(x, y) = f_{xy}(x, y).$$

$$\Rightarrow \quad f_{yx} = f_{xy}.$$

11.11. Taylor's theorem for a function of two variables.

As in the case of function of a single variable, we can prove Taylor's theorem with remainder after n terms. We shall however consider the case of Taylor's theorem with remainder after 3 terms only.

11.11.1. Theorem. *If f possesses continuous partial derivatives of the third order in a neighbourhood of a point (a, b) and if $(a + h, b + k)$ be a point of this neighbourhood, then there exists a positive number θ which is less than 1, such that*

$$f(a + h, b + k) = f(a, b) + [hf_x(a, b) + kf_y(a, b)]$$

$$+ \frac{1}{2!}[h^2 f_{x^2}(a, b) + 2hk f_{xy}(a, b) + k^2 f_{y^2}(a, b)] + \frac{1}{3!}[h^2 f_{x^3}(u, v)$$

$$+ 3h^2 k f_{y^2 x}(u, v) + 3hk^2 f_{y^2 x}(u, v) + k^3 f_{y^2 x}(u, v)]$$

where $\qquad u = a + \theta h, v = b + \theta k.$ $\qquad\qquad$ **(Utkal 2003)**

We write $\qquad z = f(x, y)$ and $x = a + ht, y = b + kt$

so that z is a function of t which are denote by $g(t)$.

We have $\qquad g'(t) = f_x(x, y) \dfrac{dx}{dt} + f_y(x, y) \dfrac{dy}{dt}$

$$= hf_x(x, y) + kf_y(x, y)$$

$$g''(t) = h\left[f_{x^2}(x, y)\frac{dx}{dt} + f_{yx}\frac{dy}{dt}\right] + k\left[f_{xy}\frac{dx}{dt} + f_{y^2}\frac{dy}{dt}\right]$$

$$= h^2 f_{x^2}(x, y) + 2hkf_{xy}(x, y) + k^2 f_{y^2}(x, y)]$$

$$g'''(t) = h^2\left[f_{x^3}(x, y)\frac{dx}{dt} + f_{yx^2}\frac{dy}{dt}\right] + 2hk\left[f_{x^2 y}(x, y)\frac{dy}{dx} + f_{y^2 x}(x, y)\frac{dy}{dx}\right]$$

$$+ k^2\left[f_{xy^2}(x, y)\frac{dx}{dt} + f_{y^3}(x, y)\frac{dy}{dt}\right]$$

$$= h^3 f_{x^3}(x, y) + 3h^2 kf_{x^2 y}(x, y) + 3hk^2 f_{y^2 x}(x, y) + k^3 f_{y^3}(x, y)$$

Now we have $\quad g(t) = g(0) + tg'(0) + \dfrac{t^2}{2!}g''(0) + \dfrac{t^3}{3!}g'''(\theta t), \qquad 0 < \theta < 1$

$\Rightarrow \qquad\qquad g(1) = g(0) + g'(0) + \dfrac{1}{2!}g''(0) + \dfrac{1}{3!}g'''(\theta).$

Thus, we have

$$f(a + h, b + k) = f(a, b) + [hf_x(a, b) + kf_y(a, b)] + \frac{1}{2!}[h^2 f_{x^2}(a, b) + 2hkf_{xy}(a, b)$$

$$+ k^2 f_{y^2}(a, b)] + \frac{1}{3!}[h^3 f_{x^3}(u, v) + 3h^2 kf_{x^2 y}(u, v)$$

$$+ 3hk^2 f_{y^2 x}(u, v) + k^3 f_{y^3}(u, v)]$$

where $\qquad u = a + \theta h, v = b + \theta k, 0 < \theta < 1.$

Other Form.

The theorem can be stated in still another form.

$$f(x, y) = f(a, b) + \left[(x - a)\frac{\partial}{\partial x} + (y - b)\frac{\partial}{\partial y}\right]f(a, b)$$

$$+ \frac{1}{2!}\left[(x - a)\frac{\partial}{\partial x} + (y - b)\frac{\partial}{\partial y}\right]^2 f(a, b)$$

$$+ \dots + \frac{1}{(n - 1)!}\left[(x - a)\frac{\partial}{\partial x} + (y - b)\frac{\partial}{\partial y}\right]^{n-1} f(a, b) + R_n,$$

where $\qquad R_n = \dfrac{1}{n!}\left[(x - a)\dfrac{\partial}{\partial x} + (y - b)\dfrac{\partial}{\partial y}\right]^n$

$$f(a + (x - a)\theta, b + (y - b)\theta),$$

$0 < \theta < 1$, called the Taylor's expansion of $f(x, y)$ about the point (a, b) in powers of $x - a$ and $y - b$.

EXAMPLES

1. *Expand the function* $f(x, y) = x^2 + xy - y^2$ *by Taylor's theorem in powers of* $(x - 1)$ *and* $(y + 2)$.   (*Indore, 1995, 1999; Bhopal, 1998, 2002; Vikram, 1999*)

We have $f(x, y) = x^2 + xy - y^2$

$\Rightarrow \qquad f(1, -2) = -5,$

$$\frac{\partial f}{\partial x} = 2x + y, \ \frac{\partial f}{\partial y} = x - 2y; \ \frac{\partial^2 f}{\partial x^2} = 2,$$

$$\frac{\partial^2 f}{\partial y^2} = -2, \ \frac{\partial^2 f}{\partial x \partial y} = 1,$$

and the values of third and higher order partial derivatives of $f(x, y)$ are zero.

$\therefore$ At $(1, -2)$

$$\frac{\partial f}{\partial x} = 0, \ \frac{\partial f}{\partial y} = 5, \ \frac{\partial^2 f}{\partial x^2} = 2, \ \frac{\partial^2 f}{\partial x \partial y} = 1,$$

$$\frac{\partial^2 f}{\partial y^2} = -2, \ \frac{\partial^3 f}{\partial x^3} = 0, \text{ etc.}$$

Hence the required Taylor's expansion of the given function about the point $(1, -2)$ is :

$$f(x, y) = f(1, -2) + \left[(x - 1)\frac{\partial}{\partial x} + (y + 2)\frac{\partial}{\partial y} \right] f(1, -2)$$

$$+ \frac{1}{2!} \left[(x-1)^2 \frac{\partial^2}{\partial x^2} + 2(x-1)(y+2)\frac{\partial^2}{\partial x \partial y} + (y+2)^2 \frac{\partial^2}{\partial x^2} \right] f(1, -2)$$

all other terms are zero.

Hence,

$$x^2 + xy - y^2 = -5 + \{(x - 1)\,0 + (y + 2).\,5\}$$

$$+ \frac{1}{2!} \ \{(x - 1)^2.\,2 + 2(x - 1)(y + 2).\,1 + (y + 2)^2(-2)\}$$

$$= -5 + (y + 2) + (x - 1)^2 + (x - 1)(y + 2) - (y + 2)^2.$$

2. *Obtain Taylor's formula for* $f(x, y) = \cos(x + y)$; $n = 3$ *at* $(0, 0)$.

                                                              (*Jabalpur, 1998, 1999; Sagar 2003*)

For $\qquad n = 3$; the Taylor's theorem is

$$f(x, y) = f(0, 0) + \left(x \frac{\partial}{\partial x} + y \frac{\partial}{\partial y} \right) f(0, 0)$$

$$+ \frac{1}{2!} \left(x^2 \frac{\partial^2}{\partial x^2} + 2xy \frac{\partial^2}{\partial x \partial y} + y^2 \frac{\partial^2}{\partial y^2} \right) f(0, 0) + R_3$$

where $\qquad R_3 = \frac{1}{3!}\left(x^3 \frac{\partial^3}{\partial x^3} + 3x^2 y \frac{\partial^3}{\partial x^2 \partial y} + 3xy^2 \frac{\partial^3}{\partial x \partial y^2} + y^3 \frac{\partial^3}{\partial y^3} \right) f(\theta x, \theta y), \ 0 < \theta < 1.$

Hence $\qquad f(x, y) = \cos(x + y)$

$\therefore \qquad\quad f(0, 0) = 1.$

$$\frac{\partial f}{\partial x} = \frac{\partial f}{\partial y} = -\sin(x + y)$$

$$\frac{\partial^2 f}{\partial x^2} = \frac{\partial^2 f}{\partial y^2} = \frac{\partial^2 f}{\partial x \, \partial y} = -\cos(x + y),$$

$$\frac{\partial^3 f}{\partial x^3} = \frac{\partial^3 f}{\partial x^2 \, \partial y} = \frac{\partial^3 f}{\partial x \, \partial y^2} = \frac{\partial^3 f}{\partial y^3} = \sin(x + y)$$

∴ At $(0, 0)$

$$\frac{\partial f}{\partial x} = \frac{\partial f}{\partial y} = 0, \frac{\partial^2 f}{\partial x^2} = \frac{\partial^2 f}{\partial y^2} = \frac{\partial^2 f}{\partial x \, \partial y} = -1;$$

and on substituting θx for x and θy for y,

$$\frac{\partial^3 f}{\partial x^3} = \frac{\partial^3 f}{\partial x^2 \, \partial y} = \frac{\partial^3 f}{\partial x \, \partial y^2} = \frac{\partial^3 f}{\partial x^3} = \sin\theta \,(x + y)$$

∴ $\cos(x + y) = 1 + 0 + \dfrac{1}{2!}(x^2 + 2xy + y^2)(-1) + \dfrac{1}{3!}(x^3 + 3x^2y + 3xy^2 + y^3)\sin\theta\,(x + y)$

or $\cos(x + y) = 1 - \dfrac{(x + y)^2}{2!} + \dfrac{(x + y)^3}{3!}\sin\theta\,(x + y).$

3. *Expand $\sin xy$ in powers of $(x - 1)$ and $\left(y - \dfrac{\pi}{2}\right)$ upto second degree terms.*

(Ravishankar, 1999; Sagar, 1998)

Solution. Here $f(x, y) = \sin xy$;

∴ $f\left(1, \dfrac{\pi}{2}\right) = 1.$ Also

$$\frac{\partial f}{\partial x} = y \cos xy; \frac{\partial f}{\partial y} = x \cos xy;$$

$$\frac{\partial^2 f}{\partial x^2} = -y^2 \sin xy;$$

$$\frac{\partial f}{\partial x \, \partial y} = \cos xy - xy \sin xy;$$

$$\frac{\partial^2 f}{\partial y^2} = -x^2 \sin xy, \text{ etc.}$$

∴ At the point $\left(1, \dfrac{\pi}{2}\right)$,

$$\frac{\partial f}{\partial x} = 0; \frac{\partial f}{\partial y} = 0$$

$$\frac{\partial^2 f}{\partial x^2} = -\frac{\pi^2}{4}; \frac{\partial^2 f}{\partial x \, \partial y} = \frac{\pi}{2}; \frac{\partial^2 f}{\partial y^2} = -1$$

Hence by Taylor's theorem,

$$f(x, y) = f\left(1, \frac{\pi}{2}\right) + \left[(x - 1)\frac{\partial}{\partial x} + \left(y - \frac{\pi}{2}\right)\frac{\partial}{\partial y}\right]f\left(1, \frac{\pi}{2}\right)$$

$$+ \frac{1}{2!}\left\{(x - 1)^2\frac{\partial^2}{\partial x^2} + 2(x - 1)\left(y - \frac{\pi}{2}\right)\frac{\partial^2}{\partial x \, \partial y} + \left(y - \frac{\pi}{2}\right)^2\frac{\partial^2}{\partial y^2}\right\}f\left(1, \frac{\pi}{2}\right)$$

upto second degree term.

$$\therefore \quad \sin x = 1\left\{(x-1)\cdot 0 + \left(y-\frac{\pi}{2}\right)\cdot 0\right\} + \frac{1}{2}\left\{(x-1)^2\cdot\left(\frac{\pi^2}{4}\right)\right.$$

$$\left. + 2(x-1)\left(y-\frac{\pi}{2}\right)\cdot\frac{\pi}{2} + \left(y-\frac{\pi}{2}\right)^2\cdot(-1)\right\}$$

$$= 1 - \frac{\pi^2}{8}(x-1)^2 - \frac{\pi}{2}(x-1)\left(y-\frac{\pi}{2}\right) - \frac{1}{2}\left(y-\frac{\pi}{2}\right)^2.$$

EXERCISES

1. Show that

    (i) $\sin x \sin y = xy - \dfrac{1}{6}\;[(x^3+3xy^2)\cos\theta x\cos\theta y + (y^3+3x^2y)\sin\theta x\cos\theta y]\;;\;0<\theta<1.$

    (ii) $e^{ax}\sin by = by + abxy + \dfrac{1}{6}\;[e^u(a^3x^3-3ab^2xy^2)\sin v + (3a^2bx^2y-b^3y^3)\cos v]$

    where　　　　　$u = a\,\theta\,x,\, v = b\,\theta\,y.$

2. Expand $x^2y+3y-2$ in powers of $x-1$ and $y+2$.

    (Rewa, 1999; Ravishankar, 2001; Bhopal, 2001; Bilaspur, 1998; Sagar, 2001; Jiwaji, 1999, 2002; Jabalpur, 2000)

3. Expand $f(x,y) = \log(x+e^y)$ by Taylor's series in powers of $(x-1)$ and y such that it includes all terms up to second degree.

4. Expand $f(x,y) = x^2 + xy + y^2$ in power of $(x-2)$ and $(y-3)$.

    (Vikram, 1996, 2001; Bhopal, 1997, 1999; Ravishankar, 1997, 2002; Jiwaji, 1998, 2003)

5. Obtain Taylor's formula for the function e^{x+y} at $(0,0)$ for $n=3$.　　　　　**(Bhopal, 2000)**

6. Expand $f(x,y) = e^{-2y^2+2xy}$ by Taylor's series in powers of x and y such that it includes all terms up to degree 4.

7. Expand $(x^2y+\sin y + e^x)$ in powers of $(x-1)$ and $(y-\pi)$ through quadratic terms and write the remainder, without computing θ.

8. Expand $f(x,y) = e^x\cos y$ by Taylor's series in powers of x and y such that it includes all terms up to third degree.　　　　　**(Bilaspur, 2002)**

ANSWERS

2. $x^2y+3y-2 = -10 - 4(x-1) + 4(y+2) - 2(x-1)^2 + 2(x-1)(y+2) + (x-1)^2(y+2)$

3. $\log(x+e^x) = \log 2 + \dfrac{1}{8}\,[(x-1)+y]\; - \dfrac{1}{8}\,[(x-2)^2 + 2(x-1)y - y^2] + ...$

4. $f(x,y) = 12 + [7(x-2)+8(y-3)] + \dfrac{1}{2}\,[2(x-2)^2 + 2(x-2)(y-3) + 2(y-3)^2]$

5. $e^{x+y} = 1 + (x+y) + \dfrac{1}{2!}(x+y)^2 + \dfrac{1}{3!}(x+y)^3\,e^{\theta(x+y)}$

6. $e^{-y^2+2xy} = 1 + (2xy-y^2)(2x^2y^2 - 2xy^3 + \dfrac{1}{2}y^4) + ...$

7. $x^2 y + \sin y + e^x = (\pi + e) + (2\pi + e)(x - 1) + \left(\pi + \dfrac{1}{2} e\right)(x-1)^2 + 2(x-1)(y-\pi)$

$$+ \dfrac{1}{6}(x-1)^3 e^{\theta x - \theta + 1} + (x-1)^2 (y-\pi) + \dfrac{1}{6}(y-\pi)^3 \cos(\theta y - \theta \pi).$$

8. $e^x \cos y = 1 + x + \dfrac{1}{2!}(x^2 - y^2) + \dfrac{1}{3!}(x^3 - 3xy^2) + \cdots$

OBJECTIVE QUESTIONS

For each of the following questions, four alternatives are given for the answer. Only one of them is correct. Choose the correct alternative.

1. $\lim\limits_{h \to 0} \dfrac{f(x+h,\, y) - f(x,\, y)}{h}$, if exists, is called the partial derivative of f with respect to :

   (a) x at (a, b) (b) x at (x, y)

   (c) y at (a, b) (d) y at (x, y)

2. If $u = \sin^{-1}\left(\dfrac{\sqrt{x} - \sqrt{y}}{\sqrt{x} + \sqrt{y}}\right)$, then :

   (a) $\dfrac{\partial u}{\partial x} = -\dfrac{x}{y}\dfrac{\partial u}{\partial y}$ (b) $\dfrac{\partial u}{\partial y} = -\dfrac{x}{y}\dfrac{\partial u}{\partial x}$

   (c) $\dfrac{\partial u}{\partial x} = \dfrac{x}{y}\dfrac{\partial u}{\partial y}$ (d) $\dfrac{\partial u}{\partial y} = \dfrac{x}{y}\dfrac{\partial u}{\partial x}$

3. If $f(x, y) = c$, then $\dfrac{\partial y}{\partial x}$ is :

   (a) $\dfrac{\partial f}{\partial x}$ (b) $\dfrac{\partial f}{\partial y}$ (c) $-\dfrac{\partial f / dx}{\partial f / \partial y}$ (d) $\dfrac{\partial f / dx}{\partial f / \partial y}$

                                                                      (Garhwal 2001)

4. If $z = \sin(x^2 y^2)$, then $\left(\dfrac{\partial z}{\partial x}\right)^2 + \left(\dfrac{\partial z}{\partial y}\right)^2 =$

   (a) $4x^2 y^2 \cos^2(x^2 y^2)$

   (c) $4x^2 y^2 (x^2 + y^2) \cos^2(x^2 y^2)$ (b) $4x^2 y^2 (x^2 + y^2)$

                                                            (d) None of these

5. If $z = (x+y)/(x-y);\; x \neq y$, then $\dfrac{\partial z}{\partial x} + \dfrac{\partial z}{\partial y} =$

   (a) $2(x+y)/(x-y)^2$ (b) $2/(x-y)$

   (c) $-2/(x-y)$ (d) None of these

6. If $z = x^2 \tan^{-1}(y/x) - y^2 \tan^{-1}(x/y)$, then $\partial^2 z / \partial x\, \partial y =$

   (a) $(y^2 - x^2)/(x^2 + y^2)$ (b) $(x^2 - y^2)/(x^2 + y^2)^2$

   (c) $(x - y^2)/(x^2 + y^2)$ (d) $(x^2 - y^2)/(x^2 + y^2)^2$

7. If $z = \cos(xy^3)$, then $\partial^2 z / \partial x\, \partial y =$

   (a) $-6xy \sin(xy^3) + 9x^2 y^4 \cos(xy^3)$

   (b) $6xy \sin(xy^3) - 9x^2 y^4 \cos(xy^3)$

   (c) $-6xy \sin(xy^3) - 9x^2 y^4 \cos(xy^3)$

   (d) $6xy \sin(xy^3) + 9x^2 y^4 \cos(xy^3)$

8. If $f(x, y) = x^3 + y^3 - 2x^2y^2$, then $(f_{xx})_{x=y=1} =$
   (a) 1   (b) -1 (c) 0   (d) None of these

9. If $f(x, y) = \log(x \tan^{-1} y)$, then $f_{xy} =$
   (a) $-1/x^2$ (b) 0   (c) $1/x^2$ (d) None of these

10. If $f(x, y) = \sin(e^{ax} + e^{by})$, then $f_{xy} =$
    (a) $-ab\, e^{(ax+by)} \sin(e^{ax} + e^{by})$
    (b) $ab\,(e^{ax} + e^{by}) \sin(e^{ax} + e^{by})$
    (c) $a^2 e^{ax} [\cos(e^{ax} + e^{by}) - e^{ax} \sin(e^{ax} + e^{by})]$
    (d) None of these

11. If $f(x, y, z) = (x^2 + y^2 + z^2)^{-1/2}$, then $f_{xx} + f_{yy} + f_{zz} =$
    (a) 0   (b) 1   (c) -1 (d) None of these

12. If $z = \tan^{-1}(y/x)$, then $dz =$

(a) $\dfrac{x\, dy + y\, dx}{x^2 + y^2}$ (b) $\dfrac{x\, dx - y\, dy}{x^2 + y^2}$ (c) $\dfrac{x\, dx + y\, dy}{x^2 + y^2}$ (d) $\dfrac{x\, dy - y\, dx}{x^2 + y^2}$

13. If $u = f(y - z, z - x, x - y)$, then value of $\dfrac{\partial u}{\partial x} + \dfrac{\partial u}{\partial y} + \dfrac{\partial u}{\partial z}$ is

(a) 3   (b) 0   (c) $\dfrac{\partial f}{\partial x} + \dfrac{\partial f}{\partial y} + \dfrac{\partial f}{\partial z}$ (d) None of these

14. If $z = e^{ax+by} f(ax - by)$, then $\dfrac{1}{z}\left[b \dfrac{\partial z}{\partial x} + a \dfrac{\partial z}{\partial y} \right] =$
    (a) ab (b) $2ab$ (c) $ab/2$ (d) None of these

15. If u be a homogeneous function of degree n, then

$x \dfrac{\partial^2 u}{\partial x^2} + y \dfrac{\partial^2 u}{\partial x \partial y} =$ *(Garhwal, 2003)*

(a) $n \dfrac{\partial u}{\partial x}$ (b) $(n-1) \dfrac{\partial u}{\partial x}$ (c) $(n+1) \dfrac{\partial u}{\partial x}$ (d) None of these

16. If $u = \sin^{-1}\left(\dfrac{x^2 + y^2}{x + y} \right)$, then $x \dfrac{\partial u}{\partial x} + y \dfrac{\partial u}{\partial y} =$

(a) u (b) $\sin u$ (c) $\tan u$ (d) None of these

17. If $z = xy\, f(x/y)$, then $x \dfrac{\partial z}{\partial x} + y \dfrac{\partial z}{\partial y} =$ *(Avadh 2002)*

(a) z (b) 0   (c) $1/z$ (d) $2z$

18. If $z = \tan^{-1}\left(\dfrac{\sqrt{x} - \sqrt{y}}{\sqrt{x} + \sqrt{y}} \right)$, then $x \dfrac{\partial z}{\partial x} + y \dfrac{\partial z}{\partial y} =$
    (a) $\sin z \cos z$ (b) 0   (c) $\tan z$ (d) None of these

19. If $\sin^{-1}\left(\dfrac{x + y}{\sqrt{x} + \sqrt{y}} \right) = z$, then $x \cdot \dfrac{\partial z}{\partial x} + y \dfrac{\partial z}{\partial y} =$

(a) $\tan z$ (b) $\dfrac{1}{2} \sin z$ (c) $2 \tan z$ (d) $\dfrac{1}{2} \tan z$

20. If $u = (x^{1/4} + y^{1/4}) / (x^{1/6} + y^{1/6})$ and $x \dfrac{\partial u}{\partial x} + y \dfrac{\partial u}{\partial y} = ku$, then $k =$

(a) $1/4$ (b) $1/12$ (c) $1/24$ (d) $1/6$

21. If $u = \log(x^3 + y^3 + z^3 - 3xyz)$, then $(\partial/\partial x + \partial/\partial y + \partial/\partial z)^2 \, u = \dfrac{k}{(x+y+z)^2}$, then $k =$

(a) 9   (b) -9 (c) 3   (d) -3

22. If $V = (x^2 + y^2 + z^2)^{-1/2}$, then $x \dfrac{\partial V}{\partial x} + y \dfrac{\partial V}{\partial y} + z \dfrac{\partial V}{\partial z} =$

(a) V (b) $-V$ (c) $2V$ (d) None of these

23. If $z = f(y/x^2)$, then $x \dfrac{\partial z}{\partial x} + 2y \dfrac{\partial z}{\partial y} =$

(a) $\dfrac{4y}{x^2} f'(y/x^2)$ (b) $\dfrac{4y}{x^3} f'(y/x^2)$ (c) 0   (d) None of these

ANSWERS							
1. (b)	2. (b)	3. (c)	4. (c)	5. (b)	6. (d)	7. (c)	
8. (d)	9. (b)	10. (a)	11. (a)	12. (d)	13. (b)	14. (b)	
15. (b)	16. (c)	17. (d)	18. (b)	19. (d)	20. (b)	21. (b)	
22. (b)	23. (c).						

CHAPTER
12
JACOBIANS

12.1. Definition

If $u_1, u_2, u_3, \dots, u_n$ be the functions of n variables $x_1, x_2, x_3, \dots, x_n$ then the determinant

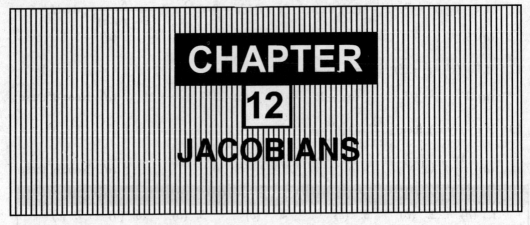

$$\begin{vmatrix} \dfrac{\partial u_1}{\partial x_1} & \dfrac{\partial u_1}{\partial x_2} & \dfrac{\partial u_1}{\partial x_3} & \cdots & \dfrac{\partial u_1}{\partial x_n} \\[2mm] \dfrac{\partial u_2}{\partial x_1} & \dfrac{\partial u_2}{\partial x_2} & \dfrac{\partial u_2}{\partial x_3} & \cdots & \dfrac{\partial u_2}{\partial x_n} \\[2mm] \dfrac{\partial u_3}{\partial x_1} & \dfrac{\partial u_3}{\partial x_2} & \dfrac{\partial u_3}{\partial x_3} & \cdots & \dfrac{\partial u_3}{\partial x_n} \\[2mm] \vdots & \vdots & \vdots & \vdots & \vdots \\[2mm] \dfrac{\partial u_n}{\partial x_1} & \dfrac{\partial u_n}{\partial x_2} & \dfrac{\partial u_n}{\partial x_3} & \cdots & \dfrac{\partial u_n}{\partial x_n} \end{vmatrix}$$

is called the *Jacobian* of $u_1, u_2, \dots, u_n$ with respect to $x_1, x_2, x_3, \dots, x_n$. We shall generally denote the Jacobian of $u_1, u_2, u_3, \dots, u_n$ by

$$\frac{\partial (u_1, u_2, u_3, \dots, u_n)}{\partial (x_1, x_2, x_3, \dots, x_n)} \text{ or } J (u_1, u_2, u_3, \dots, u_n).$$

12.1.1 Particular case.

If functions $u_1, u_2, \dots, u_n$ of $x_1, x_2, \dots, x_n$ are of the form

$$u_1 = f_1(x_1) \,;\, u_2 = f_2(x_1, x_2), \dots, u_n = f_n(x_1, x_2, \dots x_n),$$

then it is clearly seen that

$$\frac{\partial u_1}{\partial x_2} = 0 = \frac{\partial u_1}{\partial x_3} = \dots = \frac{\partial u_1}{\partial x_n}$$

and

$$\frac{\partial u_2}{\partial x_3} = 0 = \frac{\partial u_2}{\partial x_4} = \dots = \frac{\partial u_2}{\partial x_n} \,;\, \text{etc.}$$

and therefore

$$\frac{\partial (u_1, u_2, u_3, \dots, u_n)}{\partial (x_1, x_2, x_3, \dots, x_n)}$$

$$
= \begin{vmatrix} \partial u_1 / \partial x_1 & 0 & 0 & \ldots & 0 \\ \partial u_2 / \partial x_1 & \partial u_2 / \partial x_2 & 0 & \ldots & 0 \\ \ldots & \ldots & \ldots & \ldots & \ldots \\ \ldots & \ldots & \ldots & \ldots & \ldots \\ \partial u_n / \partial x_1 & \partial u_n / \partial x_2 & \ldots & \ldots & \partial u_n / \partial x_n \end{vmatrix}
$$

$$
= \frac{\partial u_1}{\partial x_1} \cdot \frac{\partial u_2}{\partial x_2} \cdot \frac{\partial u_3}{\partial x_3} \ldots \frac{\partial u_n}{\partial x_n}
$$

i.e., the Jacobian reduces to its leading terms.

EXAMPLES

1. If $x = r \cos \theta$, $y = r \sin \theta$, *then find* $\dfrac{\partial (x, y)}{\partial (r, \theta)}$ *and* $\dfrac{\partial (r, \theta)}{\partial (x, y)}$.

(*Garhwal 96, 2002; Indore 98; Sagar 98; Rewa 2002; Nagpur 2005*)

We have $\qquad \dfrac{\partial (x, y)}{\partial (r, \theta)} = \begin{vmatrix} \dfrac{\partial x}{\partial r} & \dfrac{\partial x}{\partial \theta} \\ \dfrac{\partial y}{\partial r} & \dfrac{\partial y}{\partial \theta} \end{vmatrix}$

$$
= \begin{vmatrix} \cos \theta & -r \sin \theta \\ \sin \theta & r \cos \theta \end{vmatrix}
$$

$$
= r (\cos^2 \theta + \sin^2 \theta) = r
$$

Again we have

$$
r = \sqrt{x^2 + y^2} \text{ and } \theta = \tan^{-1} \frac{y}{x}
$$

$$
\therefore \qquad \frac{\partial r}{\partial x} = \frac{1}{2} (x^2 + y^2)^{-1/2} \cdot 2x = \cos \theta
$$

$$
\frac{\partial r}{\partial y} = \frac{1}{2} (x^2 + y^2)^{-1/2} \cdot 2y = \sin \theta
$$

$$
\frac{\partial \theta}{\partial x} = \frac{1}{1 + \dfrac{y^2}{x^2}} \cdot \left(-\frac{y}{x^2} \right) = -\frac{\sin \theta}{r}
$$

and $\qquad \dfrac{\partial \theta}{\partial y} = \dfrac{\cos \theta}{r}$

$$
\therefore \qquad \frac{\partial (r, \theta)}{\partial (x, y)} = \begin{vmatrix} \dfrac{\partial r}{\partial x} & \dfrac{\partial r}{\partial y} \\ \dfrac{\partial \theta}{\partial x} & \dfrac{\partial \theta}{\partial y} \end{vmatrix}
$$

$$
= \begin{vmatrix} \cos \theta & \sin \theta \\ -\dfrac{\sin \theta}{r} & \dfrac{\cos \theta}{r} \end{vmatrix}
$$

$$
= 1/r (\cos^2 \theta + \sin^2 \theta) = 1/r
$$

2. *If* $x = r \cos \theta . \cos \phi$, $y = r \sin \theta \sqrt{(1 - m^2 \sin^2 \phi)}$,

$z = r \sin \phi (1 - n^2 \sin^2 \theta)$, *where* $m^2 + n^2 = 1$, *prove that*

$$\frac{\partial (x, y, z)}{\partial (r, \theta, \phi)} = \frac{r^2 (m^2 \cos^2 \phi + n^2 \cos^2 \theta)}{\left[(1 - m^2 \sin^2 \phi) (1 - n^2 \sin^2 \theta)\right]^{1/2}}.$$

From the given relations we have on squaring and adding,

$$x^2 + y^2 + z^2 = r^2 \cos^2 \theta \cos^2 \phi + r^2 . \sin^2 \theta - m^2 r^2 \sin^2 \theta \sin^2 \phi + r^2 \sin^2 \phi$$
$$- n^2 r^2 \sin^2 \phi \sin^2 \theta$$
$$= r^2 [\cos^2 \theta \cos^2 \phi + \sin^2 \theta + \sin^2 \phi - \sin^2 \phi \cos^2 \theta] \qquad \because m^2 + n^2 = 1$$
$$= r^2 [\cos^2 \theta \cos^2 \phi + \sin^2 \theta + \sin^2 \phi \cos^2 \theta]$$
$$= r^2 [\cos^2 \theta (\cos^2 \phi + \sin^2 \phi) + \sin^2 \theta]$$
$$= r^2$$

Differentiating above partially with respect to r, θ and ϕ we get

$$x \frac{\partial x}{\partial r} + y \frac{\partial y}{\partial r} + z \frac{\partial z}{\partial r} = r \qquad \qquad \dots(i)$$

$$x \frac{\partial x}{\partial \theta} + y \frac{\partial y}{\partial \theta} + z \frac{\partial z}{\partial \theta} = 0 \qquad \qquad \dots(ii)$$

$$x \frac{\partial x}{\partial \phi} + y \frac{\partial y}{\partial \phi} + z \frac{\partial z}{\partial \phi} = 0 \qquad \qquad \dots(iii)$$

Now,

$$J = \begin{vmatrix} \dfrac{\partial x}{\partial r} & \dfrac{\partial x}{\partial \theta} & \dfrac{\partial x}{\partial \phi} \\[2mm] \dfrac{\partial y}{\partial r} & \dfrac{\partial y}{\partial \theta} & \dfrac{\partial y}{\partial \phi} \\[2mm] \dfrac{\partial z}{\partial r} & \dfrac{\partial z}{\partial \theta} & \dfrac{\partial z}{\partial \phi} \end{vmatrix}$$

$$= \frac{1}{x} \begin{vmatrix} x \dfrac{\partial x}{\partial r} & x \dfrac{\partial x}{\partial \theta} & x \dfrac{\partial x}{\partial \phi} \\[2mm] \dfrac{\partial y}{\partial r} & \dfrac{\partial y}{\partial \theta} & \dfrac{\partial y}{\partial \phi} \\[2mm] \dfrac{\partial z}{\partial r} & \dfrac{\partial z}{\partial \theta} & \dfrac{\partial z}{\partial \phi} \end{vmatrix}$$

Applying $R_1 + y R_2 + z R_3$, we get

$$= \frac{1}{x} \begin{vmatrix} x \dfrac{\partial x}{\partial r} + y \dfrac{\partial y}{\partial r} + z \dfrac{\partial z}{\partial r} & x \dfrac{\partial x}{\partial \theta} + y \dfrac{\partial y}{\partial \theta} + z \dfrac{\partial z}{\partial \theta} & x \dfrac{\partial x}{\partial \phi} + y \dfrac{\partial y}{\partial \phi} + z \dfrac{\partial z}{\partial \phi} \\[2mm] \dfrac{\partial y}{\partial r} & \dfrac{\partial y}{\partial \theta} & \dfrac{\partial y}{\partial \phi} \\[2mm] \dfrac{\partial z}{\partial r} & \dfrac{\partial z}{\partial \theta} & \dfrac{\partial z}{\partial \phi} \end{vmatrix}$$

$$= \frac{1}{x} \begin{vmatrix} r & 0 & 0 \\ \dfrac{\partial y}{\partial r} & \dfrac{\partial y}{\partial \theta} & \dfrac{\partial y}{\partial \phi} \\ \dfrac{\partial z}{\partial r} & \dfrac{\partial z}{\partial \theta} & \dfrac{\partial z}{\partial \phi} \end{vmatrix}$$

from (i), (ii) and (iii)

$$= \frac{r}{x} \left(\frac{\partial y}{\partial \theta} \cdot \frac{\partial x}{\partial \phi} - \frac{\partial y}{\partial \phi} \cdot \frac{\partial z}{\partial \theta} \right)$$

$$= \frac{r}{r \cos \theta \cos \phi} \left[r \cos \theta \sqrt{(1 - m^2 \sin^2 \phi)} \cdot r \cos \phi \sqrt{(1 - n^2 \sin^2 \theta)} \right.$$

$$\left. - \frac{r \sin \theta}{2 \sqrt{(1 - m^2 \sin^2 \phi)}} (-2 m^2 \sin \phi \cos \phi) \frac{r \sin \phi}{2 \sqrt{(1 - n^2 \sin^2 \theta)}} (-2 n^2 \sin \theta \cos \theta) \right]$$

$$[\cos \theta \cos \phi (1 - m^2 \sin^2 \phi - n^2 \sin^2 \theta + m^2 n^2 \sin^2 \theta \sin^2 \phi)]$$

$$= \frac{r^3}{r \cos \theta \cos \phi} \frac{- m^2 n^2 \cdot \sin^2 \theta \cos \theta \sin^2 \phi \cos \phi]}{\sqrt{(1 - m^2 \sin^2 \phi) (1 - n^2 \sin^2 \theta)}}$$

$$= r^2 \left[\frac{(m^2 + n^2) - m^2 \sin^2 \phi - n^2 \sin^2 \theta}{[(1 - m^2 \sin^2 \phi) (1 - n^2 \sin^2 \theta)]^{1/2}} \right]$$

$$= r^2 \frac{[m^2 \cos^2 \phi + n^2 \cos^2 \theta]}{[(1 - m^2 \sin^2 \phi) (1 - n^2 \sin^2 \theta)]^{1/2}} \qquad \because \; m^2 + n^2 = 1$$

EXERCISES

1. If $u_1 = \dfrac{x_2 \, x_3}{x_1}$, $u_2 = \dfrac{x_3 \, x_1}{x_2}$ and $u_3 = \dfrac{x_1 \, x_2}{x_3}$, then prove that
$$J(u_1, u_2, u_3) = 4.$$
   (*Bilaspur 2000; Ravishankar 2002; Jabalpur 96, 2000; Kumaon 2001; Vikram 2000; Rohilkhand 2000; Kanpur 2001; Jiwaji 2000, 2001; Indore 2001; Avadh 2003*)

2. If $x = c \cos h \, \xi \cos \eta$, $y = c \sin h \, \xi \sin \eta$, show that
$$\frac{\partial (x, y)}{\partial (\xi, \eta)} = \frac{1}{2} c^2 (\cos h \, 2\xi - \cos 2\eta)$$
   (*Rohilkhand 2001; 2004; Ravishankar 2000, 01, 02; Vikram 2002*)

3. If $x = r \sin \theta \cos \phi$, $y = r \sin \theta \sin \phi$, $z = r \cos \theta$, show that
$$\frac{\partial (x, y, z)}{\partial (r, \theta, \phi)} = r^2 \sin \theta.$$
   (*Kanpur 2002; Rohilkhand 99; Garhwal 98; Kumaon 2004; Indore 99; Ravishankar 99; Rewa 98; Jabalpur 2000; Jiwaji 99*)

4. If $y_1 = (1 - x_1)$, $y_2 = x_1 (1 - x_2)$, $y_3 = x_1 x_2 (1 - x_3)$, $\dots$, $y_n = x_1 x_2 \dots x_{n-1} (1 - x_n)$, then prove that
$$\frac{\partial (y_1, y_2, \dots, y_n)}{\partial (x_1, x_2, \dots, x_n)} = (-1)^n \, x_1^{n-1} \, x_2^{n-2} \dots x_{n-1}. \qquad (\textit{Garhwal 97; Reva 99})$$

5. If $x = \sin \theta \sqrt{(1 - c^2 \sin^2 \phi)}$, $y = \cos \theta \cos \phi$, then show that
$$\frac{\partial (x, y)}{\partial (\theta, \phi)} = -\sin \phi \, \frac{(1 - c^2) \cos^2 \theta + c^2 \cos^2 \phi}{\sqrt{(1 - c^2 \sin^2 \phi)}}.$$

6. If $y_1 = r \sin \theta_1 \sin \theta_2$; $y_2 = r \sin \theta_1 \cos \theta_2$, $y_3 = r \cos \theta_1 \sin \theta_3$, $y_4 = r \cos \theta_1 \cos \theta_3$, then show that

$$J(y_1, y_2, y_3, y_4) = r^3 \sin \theta_1 \cos \theta_1.$$

12.2. Jacobian of Function of Function.

Theorem. *If* $u_1, u_2, u_3, \ldots u_n$ *are functions of* $y_1, y_2, y_3, \ldots , y_n$ *and* $y_1, y_2, y_3, \ldots , y_n$ *are functions of* $x_1, x_2, x_3, \ldots , x_n$ *then*

$$\frac{\partial (u_1, u_2, \ldots , u_n)}{\partial (x_1, x_2, \ldots , x_n)} = \frac{\partial (u_1, u_2, \ldots , u_n)}{\partial (y_1, y_2, \ldots , y_n)} \cdot \frac{\partial (y_1, y_2, \ldots , y_n)}{\partial (x_1, x_2, \ldots , x_n)}$$

Proof. Since it is given that $u_1, u_2, \ldots , u_n$ are functions of $y_1, y_2, y_3, \ldots , y_n$ which are themselves functions of $x_1, x_2, \ldots, x_n$, thus

$$\frac{\partial u_1}{\partial x_1} = \frac{\partial u_1}{\partial y_1} \cdot \frac{\partial y_1}{\partial x_1} + \frac{\partial u_1}{\partial y_2} \cdot \frac{\partial y_2}{\partial x_1} + \ldots + \frac{\partial u_1}{\partial y_n} \cdot \frac{\partial y_n}{\partial x_1}$$

$$= \sum \frac{\partial u_1}{\partial y_r} \cdot \frac{\partial y_r}{\partial x_1}$$

$$\frac{\partial u_1}{\partial x_2} = \frac{\partial u_1}{\partial y_1} \cdot \frac{\partial y_1}{\partial x_2} + \frac{\partial u_1}{\partial y_2} \cdot \frac{\partial y_2}{\partial x_2} + \ldots + \frac{\partial u_1}{\partial y_n} \cdot \frac{\partial y_n}{\partial x_2}$$

$$= \sum \frac{\partial u_1}{\partial y_r} \cdot \frac{\partial y_r}{\partial x_2}$$

$$\ldots\ldots\ldots\ldots\ldots\ldots\ldots\ldots\ldots\ldots$$
$$\ldots\ldots\ldots\ldots\ldots\ldots\ldots\ldots\ldots\ldots \text{ etc.}$$

Therefore

$$\frac{\partial (u_1, u_2, \ldots , u_n)}{\partial (y_1, y_2, \ldots , y_n)} \times \frac{\partial (y_1, y_2, \ldots , y_n)}{\partial (x_1, x_2, \ldots , x_n)}$$

$$= \begin{vmatrix} \dfrac{\partial u_1}{\partial y_1} & \dfrac{\partial u_1}{\partial y_2} & \cdots & \dfrac{\partial u_1}{\partial y_n} \\[2mm] \dfrac{\partial u_2}{\partial y_1} & \dfrac{\partial u_2}{\partial y_2} & \cdots & \dfrac{\partial u_2}{\partial y_n} \\[2mm] \cdots & \cdots & \cdots & \cdots \\[1mm] \cdots & \cdots & \cdots & \cdots \\[1mm] \dfrac{\partial u_n}{\partial y_1} & \dfrac{\partial u_n}{\partial y_2} & \cdots & \dfrac{\partial u_n}{\partial y_n} \end{vmatrix} \times \begin{vmatrix} \dfrac{\partial y_1}{\partial x_1} & \dfrac{\partial y_1}{\partial x_2} & \cdots & \dfrac{\partial y_1}{\partial x_n} \\[2mm] \dfrac{\partial y_2}{\partial x_1} & \dfrac{\partial y_2}{\partial x_2} & \cdots & \dfrac{\partial y_2}{\partial x_n} \\[2mm] \cdots & \cdots & \cdots & \cdots \\[1mm] \cdots & \cdots & \cdots & \cdots \\[1mm] \dfrac{\partial y_n}{\partial x_1} & \dfrac{\partial y_n}{\partial x_2} & \cdots & \dfrac{\partial y_n}{\partial x_n} \end{vmatrix}$$

$$= \begin{vmatrix} \sum \dfrac{\partial u_1}{\partial y_r} \cdot \dfrac{\partial y_r}{\partial x_1} & \sum \dfrac{\partial u_1}{\partial y_r} \cdot \dfrac{\partial y_r}{\partial x_2} & \cdots & \sum \dfrac{\partial u_1}{\partial y_r} \cdot \dfrac{\partial y_r}{\partial x_n} \\[3mm] \sum \dfrac{\partial u_2}{\partial y_r} \cdot \dfrac{\partial y_r}{\partial x_1} & \sum \dfrac{\partial u_2}{\partial y_r} \cdot \dfrac{\partial y_r}{\partial x_2} & \cdots & \sum \dfrac{\partial u_2}{\partial y_r} \cdot \dfrac{\partial y_r}{\partial x_n} \\[3mm] \cdots\cdots & \cdots\cdots & \cdots & \cdots\cdots \\[1mm] \cdots\cdots & \cdots\cdots & \cdots & \cdots\cdots \\[1mm] \sum \dfrac{\partial u_n}{\partial y_r} \cdot \dfrac{\partial y_r}{\partial x_1} & \sum \dfrac{\partial u_n}{\partial y_r} \cdot \dfrac{\partial y_r}{\partial x_2} & \cdots & \sum \dfrac{\partial u_n}{\partial y_r} \cdot \dfrac{\partial y_r}{\partial x_n} \end{vmatrix}$$

$$= \begin{vmatrix} \dfrac{\partial u_1}{\partial x_1} & \dfrac{\partial u_1}{\partial x_2} & \cdots & \dfrac{\partial u_1}{\partial x_n} \\[2mm] \dfrac{\partial u_2}{\partial x_1} & \dfrac{\partial u_2}{\partial x_2} & \cdots & \dfrac{\partial u_2}{\partial x_n} \\[2mm] \cdots & \cdots & \cdots & \cdots \\ \cdots & \cdots & \cdots & \cdots \\ \dfrac{\partial u_n}{\partial x_1} & \dfrac{\partial u_n}{\partial x_2} & \cdots & \dfrac{\partial u_n}{\partial x_n} \end{vmatrix}$$

$$= \frac{\partial (u_1, u_2, \dots, u_n)}{\partial (x_1, x_2, \dots, x_n)}.$$

Note. The above theorem is generalised result of the elementary theorem

$$\frac{d u}{d x} = \frac{d u}{d y} \cdot \frac{d y}{d x}$$

Cor. If we take $u_1 = x_1, u_2 = x_2, \dots, u_n = x_n$ then the above result becomes

$$\frac{\partial (x_1, x_2, \dots, x_n)}{\partial (y_1, y_2, \dots, y_n)} \times \frac{\partial (y_1, y_2, \dots, y_n)}{\partial (x_1, x_2, \dots, x_n)} = 1.$$

12.3. Jacobian of Implicit Functions.

If $u_1, u_2, u_3, \dots, u_n$ and $x_1, x_2, \dots, x_n$ are implicitly connected by n equations as :

$$f_1 (u_1, u_2, \dots, u_n, x_1, x_2, \dots, x_n) = 0$$
$$f_2 (u_1, u_2, \dots, u_n, x_1, x_2, \dots, x_n) = 0$$
$$\dots \dots \dots \dots \dots \dots \dots \dots \dots \dots \dots \dots$$
$$\dots \dots \dots \dots \dots \dots \dots \dots \dots \dots \dots \dots$$
$$f_n (u_1, u_2, \dots, u_n, x_1, x_2, \dots, x_n) = 0$$

then $\dfrac{\partial (f_1, f_2, \dots, f_n)}{\partial (u_1, u_2, \dots, u_n)} \cdot \dfrac{\partial (u_1, u_2, \dots, u_n)}{\partial (x_1, x_2, \dots, x_n)} = (-1)^n \dfrac{\partial (f_1, f_2, \dots, f_n)}{\partial (x_1, x_2, \dots, x_n)}.$

Proof. Differentiating the above given relations with respect to $x_1, x_2, \dots, x_n$ we get

$$\frac{\partial f_1}{\partial x_1} + \frac{\partial f_1}{\partial u_1} \cdot \frac{\partial u_1}{\partial x_1} + \frac{\partial f_1}{\partial u_2} \cdot \frac{\partial u_2}{\partial x_1} + \dots + \frac{\partial f_1}{\partial u_n} \cdot \frac{\partial u_n}{\partial x_1} = 0$$

or $\quad \Sigma \dfrac{\partial f_1}{\partial u_r} \cdot \dfrac{\partial u_r}{\partial x_1} = - \dfrac{\partial f_1}{\partial x_1}$

and $\quad \Sigma \dfrac{\partial f_1}{\partial u_r} \cdot \dfrac{\partial u_r}{\partial x_2} = - \dfrac{\partial f_1}{\partial x_2}$

$$\dots \dots \dots \dots \dots \dots \dots \dots \dots \dots$$
$$\dots \dots \dots \dots \dots \dots \dots \dots \dots \dots$$

and $\quad \Sigma \dfrac{\partial f_1}{\partial u_2} \cdot \dfrac{\partial u_r}{\partial x_n} = - \dfrac{\partial f_1}{\partial x_n}$

.... etc.

Now, $\quad \dfrac{\partial (f_1, f_2, \dots, f_n)}{\partial (u_1, u_2, \dots, u_n)} \times \dfrac{\partial (u_1, u_2, \dots, u_n)}{\partial (x_1, x_2, \dots, x_n)}$

$$
= \begin{vmatrix}
\dfrac{\partial f_1}{\partial u_1} & \dfrac{\partial f_1}{\partial u_2} & \cdots & \dfrac{\partial f_1}{\partial u_n} \\[2mm]
\dfrac{\partial f_2}{\partial u_1} & \dfrac{\partial f_2}{\partial u_2} & \cdots & \dfrac{\partial f_2}{\partial u_n} \\[2mm]
\cdots & \cdots & \cdots & \cdots \\
\cdots & \cdots & \cdots & \cdots \\
\dfrac{\partial f_n}{\partial u_1} & \dfrac{\partial f_n}{\partial u_2} & \cdots & \dfrac{\partial f_n}{\partial u_n}
\end{vmatrix}
\times
\begin{vmatrix}
\dfrac{\partial u_1}{\partial x_1} & \dfrac{\partial u_1}{\partial x_2} & \cdots & \dfrac{\partial u_1}{\partial x_n} \\[2mm]
\dfrac{\partial u_2}{\partial x_1} & \dfrac{\partial u_2}{\partial x_2} & \cdots & \dfrac{\partial u_2}{\partial x_n} \\[2mm]
\cdots & \cdots & \cdots & \cdots \\
\cdots & \cdots & \cdots & \cdots \\
\dfrac{\partial u_n}{\partial x_1} & \dfrac{\partial u_n}{\partial x_2} & \cdots & \dfrac{\partial u_n}{\partial x_n}
\end{vmatrix}
$$

$$
= \begin{vmatrix}
\Sigma \dfrac{\partial f_1}{\partial u_r} \cdot \dfrac{\partial u_r}{\partial x_1} & \Sigma \dfrac{\partial f_1}{\partial u_r} \cdot \dfrac{\partial u_r}{\partial x_2} & \cdots & \Sigma \dfrac{\partial f_1}{\partial u_r} \cdot \dfrac{\partial u_r}{\partial x_n} \\[3mm]
\Sigma \dfrac{\partial f_2}{\partial u_r} \cdot \dfrac{\partial u_r}{\partial x_1} & \Sigma \dfrac{\partial f_2}{\partial u_r} \cdot \dfrac{\partial u_r}{\partial x_2} & \cdots & \Sigma \dfrac{\partial f_2}{\partial u_r} \cdot \dfrac{\partial u_r}{\partial x_n} \\[3mm]
\cdots \cdots & \cdots \cdots & \cdots & \cdots \cdots \\
\cdots \cdots & \cdots \cdots & \cdots & \cdots \cdots \\
\Sigma \dfrac{\partial f_n}{\partial u_r} \cdot \dfrac{\partial u_r}{\partial x_1} & \Sigma \dfrac{\partial f_n}{\partial u_r} \cdot \dfrac{\partial u_r}{\partial x_2} & \cdots & \Sigma \dfrac{\partial f_n}{\partial u_r} \cdot \dfrac{\partial u_r}{\partial x_n}
\end{vmatrix}
$$

$$
= \begin{vmatrix}
- \dfrac{\partial f_1}{\partial x_1} & - \dfrac{\partial f_1}{\partial x_2} & \cdots & - \dfrac{\partial f_1}{\partial x_n} \\[2mm]
- \dfrac{\partial f_2}{\partial x_1} & - \dfrac{\partial f_2}{\partial x_2} & \cdots & - \dfrac{\partial f_2}{\partial x_n} \\[2mm]
\cdots & \cdots & \cdots & \cdots \\
\cdots & \cdots & \cdots & \cdots \\
- \dfrac{\partial f_n}{\partial x_1} & - \dfrac{\partial f_n}{\partial x_2} & \cdots & - \dfrac{\partial f_n}{\partial x_n}
\end{vmatrix}
$$

$$
= (-1)^n \dfrac{\partial (f_1, f_2, \dots, f_n)}{\partial (x_1, x_2, \dots, x_n)}
$$

The above result can be written as

$$
\dfrac{\partial (u_1, u_2, \dots, u_n)}{\partial (x_1, x_2, \dots, x_n)} = (-1)^n \dfrac{\dfrac{\partial (f_1, f_2, \dots, f_n)}{\partial (x_1, x_2, \dots, x_n)}}{\dfrac{\partial (f_1, f_2, \dots, f_n)}{\partial (u_1, u_2, \dots, u_n)}}
$$

Note. The above result is a generalised result of the elementary theorem

$$\frac{dy}{dx} = -\frac{\partial f/\partial x}{\partial f/\partial y}.$$

12.3.1 Particular case.   *(Garhwal 96)*

If the implicit relations in § 12.3 are given as

$$f_1(x_1, x_2, \ldots, x_n, u_1) = 0$$
$$f_2(x_2, x_3, \ldots, x_n, u_1, u_2) = 0$$
$$f_3(x_3, x_4, \ldots, x_n, u_1, u_2, u_3) = 0$$
$$\ldots \ldots \ldots \ldots \ldots \ldots \ldots \ldots$$
$$f_n(x_n, u_1, u_2, \ldots, u_n) = 0$$

then it is easily seen that

$$\frac{\partial(f_1, f_2, \ldots, f_n)}{\partial(u_1, u_2, \ldots, u_n)} = \frac{\partial f_1}{\partial u_1} \cdot \frac{\partial f_2}{\partial u_2} \cdots \frac{\partial f_n}{\partial u_n}$$

and

$$\frac{\partial(f_1, f_2, \ldots, f_n)}{\partial(x_1, x_2, \ldots, x_n)} = \frac{\partial f_1}{\partial x_1} \cdot \frac{\partial f_2}{\partial x_2} \cdots \frac{\partial f_n}{\partial x_n}$$

$$\therefore \quad \frac{\partial(u_1, u_2, \ldots, u_n)}{\partial(x_1, x_2, \ldots, x_n)} = (-1)^n \frac{\dfrac{\partial f_1}{\partial x_1} \cdot \dfrac{\partial f_2}{\partial x_2} \cdots \dfrac{\partial f_n}{\partial x_n}}{\dfrac{\partial f_1}{\partial u_1} \cdot \dfrac{\partial f_2}{\partial u_2} \cdots \dfrac{\partial f_n}{\partial u_n}}$$

EXAMPLES

1. *Prove that* $\dfrac{\partial(y_1, y_2, y_3)}{\partial(x_1, x_2, x_3)} \cdot \dfrac{\partial(x_1, x_2, x_3)}{\partial(y_1, y_2, y_3)} = 1.$ *(Sagar 2003; Bhopal 2003)*

Let $\quad y_1 = f_1(x_1, x_2, x_3), \qquad y_2 = f_2(x_1, x_2, x_3)$

$$y_3 = f_3(x_1, x_2, x_3) \qquad \ldots(i)$$

The above relations can be solved to give us the values of x_1, x_2, x_3 as

$$x_1 = F_1(y_1, y_2, y_3), \qquad x_2 = F_2(y_1, y_2, y_3),$$
$$x_3 = F_3(y_1, y_2, y_3) \qquad \ldots(ii)$$

$$\therefore \quad \frac{\partial(y_1, y_2, y_3)}{\partial(x_1, x_2, x_3)} \cdot \frac{\partial(x_1, x_2, x_3)}{\partial(y_1, y_2, y_3)}$$

$$= \begin{vmatrix} \dfrac{\partial y_1}{\partial x_1} & \dfrac{\partial y_1}{\partial x_2} & \dfrac{\partial y_1}{\partial x_3} \\[2mm] \dfrac{\partial y_2}{\partial x_1} & \dfrac{\partial y_2}{\partial x_2} & \dfrac{\partial y_2}{\partial x_3} \\[2mm] \dfrac{\partial y_3}{\partial x_1} & \dfrac{\partial y_3}{\partial x_2} & \dfrac{\partial y_3}{\partial x_3} \end{vmatrix} \begin{vmatrix} \dfrac{\partial x_1}{\partial y_1} & \dfrac{\partial x_1}{\partial y_2} & \dfrac{\partial x_1}{\partial y_3} \\[2mm] \dfrac{\partial x_2}{\partial y_1} & \dfrac{\partial x_2}{\partial y_2} & \dfrac{\partial x_2}{\partial y_3} \\[2mm] \dfrac{\partial x_3}{\partial y_1} & \dfrac{\partial x_3}{\partial y_2} & \dfrac{\partial x_3}{\partial y_3} \end{vmatrix}$$

Change the rows into columns and columns into rows in the second determinant, we have

$$\text{L.H.S.} = \begin{vmatrix} \dfrac{\partial y_1}{\partial x_1} & \dfrac{\partial y_1}{\partial x_2} & \dfrac{\partial y_1}{\partial x_3} \\[2mm] \dfrac{\partial y_2}{\partial x_1} & \dfrac{\partial y_2}{\partial x_2} & \dfrac{\partial y_2}{\partial x_3} \\[2mm] \dfrac{\partial y_3}{\partial x_1} & \dfrac{\partial y_3}{\partial x_2} & \dfrac{\partial y_3}{\partial x_3} \end{vmatrix} \begin{vmatrix} \dfrac{\partial x_1}{\partial y_1} & \dfrac{\partial x_2}{\partial y_1} & \dfrac{\partial x_3}{\partial y_1} \\[2mm] \dfrac{\partial x_1}{\partial y_2} & \dfrac{\partial x_2}{\partial y_2} & \dfrac{\partial x_3}{\partial y_2} \\[2mm] \dfrac{\partial x_1}{\partial y_3} & \dfrac{\partial x_2}{\partial y_3} & \dfrac{\partial x_3}{\partial y_3} \end{vmatrix}$$

$$
= \begin{vmatrix}
\Sigma \dfrac{\partial y_1}{\partial x_r} \cdot \dfrac{\partial x_r}{\partial y_1} & \Sigma \dfrac{\partial y_1}{\partial x_r} \cdot \dfrac{\partial x_r}{\partial y_2} & \Sigma \dfrac{\partial y_1}{\partial x_r} \cdot \dfrac{\partial x_r}{\partial y_2} \\[2mm]
\Sigma \dfrac{\partial y_2}{\partial x_r} \cdot \dfrac{\partial x_r}{\partial y_1} & \Sigma \dfrac{\partial y_2}{\partial x_r} \cdot \dfrac{\partial x_r}{\partial y_2} & \Sigma \dfrac{\partial y_2}{\partial x_r} \cdot \dfrac{\partial x_r}{\partial y_3} \\[2mm]
\Sigma \dfrac{\partial y_3}{\partial x_r} \cdot \dfrac{\partial x_r}{\partial y_1} & \Sigma \dfrac{\partial y_3}{\partial x_r} \cdot \dfrac{\partial x_r}{\partial y_2} & \Sigma \dfrac{\partial y_3}{\partial x_r} \cdot \dfrac{\partial x_r}{\partial y_3}
\end{vmatrix}
\qquad \ldots(iii)
$$

Differentiating every relation in (i) partially $w.r.t. y_1, y_2, y_3$ we get by the help of relation (ii)

$$
1 = \frac{\partial y_1}{\partial x_1} \cdot \frac{\partial x_1}{\partial y_1} + \frac{\partial y_1}{\partial x_2} \cdot \frac{\partial x_2}{\partial y_1} + \frac{\partial y_1}{\partial x_3} \cdot \frac{\partial x_3}{\partial y_1} = \Sigma \frac{\partial y_1}{\partial x_r} \cdot \frac{\partial x_r}{\partial y_1}
$$

$$
0 = \frac{\partial y_1}{\partial x_1} \cdot \frac{\partial x_1}{\partial y_2} + \frac{\partial y_1}{\partial x_2} \cdot \frac{\partial x_2}{\partial y_2} + \frac{\partial y_1}{\partial x_3} \cdot \frac{\partial x_3}{\partial y_2} = \Sigma \frac{\partial y_1}{\partial x_r} \cdot \frac{\partial x_r}{\partial y_2}
$$

$$
0 = \frac{\partial y_1}{\partial x_1} \cdot \frac{\partial x_1}{\partial y_3} + \frac{\partial y_1}{\partial x_2} \cdot \frac{\partial x_2}{\partial y_3} + \frac{\partial y_1}{\partial x_3} \cdot \frac{\partial x_3}{\partial y_3} = \Sigma \frac{\partial y_1}{\partial x_r} \cdot \frac{\partial x_r}{\partial y_3}
$$

Similarly, we shall have $\qquad \Sigma \dfrac{\partial y_2}{\partial x_r} \cdot \dfrac{\partial x_r}{\partial y_2} = 1, \ \Sigma \dfrac{\partial y_3}{\partial x_r} \cdot \dfrac{\partial x_r}{\partial y_3} = 1$

and rest of all sigmas in the above relation will be zero. Putting the values of these sigmas in (iii), we get

$$
\text{L.H.S.} = \begin{vmatrix} 1 & 0 & 0 \\ 0 & 1 & 0 \\ 0 & 0 & 1 \end{vmatrix} = 1.
$$

2. *Given the roots of the equation in λ,*
   $(\lambda - x)^3 + (\lambda - y)^3 + (\lambda - z)^3 = 0$ are u, v, w, then prove that

   $$
   \frac{\partial (u, v, w)}{\partial (x, y, z)} = -2 \, \frac{(y - z)(z - x)(x - y)}{(v - w)(w - u)(u - v)}.
   $$

   (Garhwal 97; Kanpur 2000)

   Equation in λ can be written as

   $$
   3\lambda^3 - 3\lambda^2 (x + y + z) + 3\lambda (x^2 + y^2 + z^2) - (x^3 + y^3 + z^3) = 0
   $$

   If the roots of the above equation be u, v, w then

   $$
   u + v + w = x + y + z,
   $$
   $$
   uv + vw + wu = x^2 + y^2 + z^2
   $$

   and $\qquad uvw = \dfrac{1}{3}(x^3 + y^3 + z^3)$

   These relations can now be written as

   $$
   f_1 \equiv u + v + w - x - y - z = 0
   $$
   $$
   f_2 \equiv uv + vw + wu - x^2 - y^2 - z^2 = 0
   $$
   $$
   f_3 \equiv uvw - \frac{1}{3}(x^3 + y^3 + z^3) = 0
   $$

   $\therefore \qquad \dfrac{\partial (u, v, w)}{\partial (x, y, z)} = (-1)^3 \, \dfrac{\partial (f_1, f_2, f_3)}{\partial (x, y, z)} \div \dfrac{\partial (f_1, f_2, f_3)}{\partial (u, v, w)}$

$$= - \begin{vmatrix} \dfrac{\partial f_1}{\partial x} & \dfrac{\partial f_1}{\partial y} & \dfrac{\partial f_1}{\partial z} \\[2mm] \dfrac{\partial f_2}{\partial x} & \dfrac{\partial f_2}{\partial y} & \dfrac{\partial f_2}{\partial z} \\[2mm] \dfrac{\partial f_3}{\partial x} & \dfrac{\partial f_3}{\partial y} & \dfrac{\partial f_3}{\partial z} \end{vmatrix} \div \begin{vmatrix} \dfrac{\partial f_1}{\partial u} & \dfrac{\partial f_1}{\partial v} & \dfrac{\partial f_1}{\partial w} \\[2mm] \dfrac{\partial f_2}{\partial u} & \dfrac{\partial f_2}{\partial v} & \dfrac{\partial f_2}{\partial w} \\[2mm] \dfrac{\partial f_3}{\partial u} & \dfrac{\partial f_3}{\partial v} & \dfrac{\partial f_3}{\partial w} \end{vmatrix}$$

$$= - \begin{vmatrix} -1 & -1 & -1 \\ -2x & -2y & -2z \\ -x^2 & -y^2 & -z^2 \end{vmatrix} \div \begin{vmatrix} 1 & 1 & 1 \\ v+w & w+u & u+v \\ vw & uw & uv \end{vmatrix}.$$

$$= - \begin{vmatrix} -1 & 0 & 0 \\ -2x & 2(x-y) & 2(x-z) \\ x^2 & x^2-y^2 & x^2-z^2 \end{vmatrix} \div \begin{vmatrix} 1 & 0 & 0 \\ v+w & u-v & u-w \\ vw & w(u-v) & v(u-w) \end{vmatrix}$$

by $C_2 - C_1$ in both determinants

$$= - 2 \frac{(y-z)(z-x)(x-y)}{(v-w)(w-u)(u-v)}$$

3. *If* $u = x(1-r^2)^{-1/2}$, $v = y(1-r^2)^{-1/2}$, $w = z(1-r^2)^{-1/2}$
   where $r^2 = x^2 + y^2 + z^2$,
   show that $J(u, v, w) = (1-r^2)^{-5/2}$.   (*Garhwal 95*)

Given $u = x(1-r^2)^{-1/2}$

or $x^2 = u^2(1-x^2-y^2-z^2)$ etc.

∴ Let $f_1 \equiv x^2 - u^2(1-x^2-y^2-z^2) = 0,$
              $f_2 \equiv y^2 - v^2(1-x^2-y^2-z^2) = 0$

and $f_3 \equiv z^2 - w^2(1-x^2-y^2-z^2) = 0$

∴ $\dfrac{\partial(u, v, w)}{\partial(x, y, z)} = (-1)^3 \cdot \dfrac{\partial(f_1, f_2, f_3)}{\partial(x, y, z)} \div \dfrac{\partial(f_1, f_2, f_3)}{\partial(u, v, w)}$

Now, $\dfrac{\partial(f_1, f_2, f_3)}{\partial(x, y, z)} = \begin{vmatrix} 2x(1+u^2) & 2yu^2 & 2zu^2 \\ 2xv^2 & 2y(1+v^2) & 2zv^2 \\ 2xw^2 & 2yw^2 & 2z(1+w^2) \end{vmatrix}$

$$= 8xyz \begin{vmatrix} 1+u^2 & u^2 & u^2 \\ v^2 & 1+v^2 & v^2 \\ w^2 & w^2 & 1+w^2 \end{vmatrix}$$

$$= 8xyz(1+u^2+v^2+w^2)$$

Also,

$$\dfrac{\partial(f_1, f_2, f_3)}{\partial(u, v, w)} = \begin{vmatrix} -2u(1-r^2) & 0 & 0 \\ 0 & -2v(1-r^2) & 0 \\ 0 & 0 & -2w(1-r^2) \end{vmatrix}$$

$$= -8 u v w (1 - r^2)^3$$

$$\therefore \quad \frac{\partial (u, v, w)}{\partial (x, y, z)} = \frac{8 x y z (1 + u^2 + v^2 + w^2)}{-8 u v w (1 - r^2)^3}$$

(Garhwal 96)

$$= (1 - r^2)^{-5/2}$$

as
$$1 + u^2 + v^2 + w^2 = 1 + \frac{r^2}{1 - r^2} = \frac{1}{1 - r^2}$$

and
$$u v w = x y z (1 - r^2)^{-3/2}$$

4. *If* $u^3 + v^3 + w^3 = x + y + z$, $u^2 + v^2 + w^2 = x^3 + y^3 + z^3$, $u + v + w = x^2 + y^2 + z^2$, *then show that*

$$\frac{\partial (u, v, w)}{\partial (x, y, z)} = \frac{(y - z) (y - x) (x - y)}{(u - v) (v - w) (w - u)}.$$

(Garhwal 97)

Let
$$f_1 \equiv u^3 + v^3 + w^3 - x - y - z = 0,$$
$$f_2 \equiv u^2 + v^2 + w^2 - x^3 - y^3 - z^3 = 0$$

and
$$f_3 \equiv u + v + w - x^2 - y^2 - z^2 = 0$$

We have

$$\frac{\partial (u, v, w)}{\partial (x, y, z)} = (-1)^3 \cdot \frac{\partial (f_1, f_2, f_3)}{\partial (x, y, z)} \div \frac{\partial (f_1, f_2, f_3)}{\partial (u, v, w)}$$

$$= - \begin{vmatrix} \dfrac{\partial f_1}{\partial x} & \dfrac{\partial f_1}{\partial y} & \dfrac{\partial f_1}{\partial z} \\[2mm] \dfrac{\partial f_2}{\partial x} & \dfrac{\partial f_2}{\partial y} & \dfrac{\partial f_2}{\partial z} \\[2mm] \dfrac{\partial f_3}{\partial x} & \dfrac{\partial f_3}{\partial y} & \dfrac{\partial f_3}{\partial z} \end{vmatrix} \div \begin{vmatrix} \dfrac{\partial f_1}{\partial u} & \dfrac{\partial f_1}{\partial v} & \dfrac{\partial f_1}{\partial w} \\[2mm] \dfrac{\partial f_2}{\partial u} & \dfrac{\partial f_2}{\partial v} & \dfrac{\partial f_2}{\partial w} \\[2mm] \dfrac{\partial f_3}{\partial u} & \dfrac{\partial f_3}{\partial v} & \dfrac{\partial f_3}{\partial w} \end{vmatrix}$$

$$= - \begin{vmatrix} -1 & -1 & -1 \\ -3 x^2 & -3 y^2 & -3 z^2 \\ -2 x & -2 y & -2 z \end{vmatrix} \div \begin{vmatrix} 3 u^2 & 3 v^2 & 3 w^2 \\ 2 u & 2 v & 2 w \\ 1 & 1 & 1 \end{vmatrix}$$

$$= 6 \begin{vmatrix} 1 & 1 & 1 \\ x^2 & y^2 & z^2 \\ x & y & z \end{vmatrix} \div 6 \begin{vmatrix} u^2 & v^2 & w^2 \\ u & v & w \\ 1 & 1 & 1 \end{vmatrix}$$

$$= \frac{(y - z) (z - x) (x - y)}{(u - v) (v - w) (w - u)}$$

5. *Given* $x = f(u, v, w)$; $y = \phi (u, v, w)$, $z = \psi (u, v, w)$;

and
$$J = \frac{\partial (x, y, z)}{\partial (u, v, w)} \neq 0.$$

If, when x, y, z *are taken as the coordinates of a point referred to rectangular axes, the three surfaces* $u = const.$, $v = const.$, $w = const.$, *intersect orthogonally, prove that*

$$J^2 = P_1^2 \, P_2^2 \, P_3^2,$$

where $P_1^2 = f_u^2 + \phi_u^2 + \psi_u^2$; $P_2^2 = f_v^2 + \phi_v^2 + \psi_v^2$;
$$P_3^2 = f_w^2 + \phi_w^2 + \psi_w^2.$$

The result is evident if $f_u, \phi_u, \psi_u; f_v, \phi_v, \psi_v$ and f_w, ϕ_w, ψ_w can be proved to be proportional to the direction cosines of three mutually perpendicular lines.

Let $l_1, m_1, n_1 \, ; \, l_2, m_2, n_2 \, ; \, l_3, m_3, n_3$ be the direction cosines of the normals to the surfaces $u =$ const., $v =$ const., $w =$ const., passing through x, y, z. We have then

$$l_1 = \frac{u_x}{\lambda}, \; m_1 = \frac{u_y}{\lambda}, \; n_1 = \frac{u_z}{\lambda}, \text{where } \lambda = \Sigma \, u_x^2$$

$$l_2 = \frac{v_x}{\mu}, \; m_2 = \frac{v_y}{\mu}, \; n_2 = \frac{v_z}{\mu}, \text{where } \mu = \Sigma \, v_x^2$$

and
$$l_3 = \frac{w_x}{\nu}, \; m_3 = \frac{w_y}{\nu}, \; n_3 = \frac{w_z}{\nu}, \text{where } \nu = \Sigma \, w_x^2$$

Now, $l_1 \, dx + m_1 \, dy + n_1 \, dz = \dfrac{1}{\lambda} (u_x \, dx + u_y \, dy + u_z \, dz) = \dfrac{1}{\lambda} \, du$ ...(1)

Similarly, $l_2 \, dx + m_2 \, dy + n_2 \, dz = \dfrac{1}{\mu} \, dv$ ...(2)

$$l_3 \, dx + m_3 \, dy + n_3 \, dz = \frac{1}{\nu} \, dw \qquad\qquad ...(3)$$

Multiplying (1), (2), (3) by l_1, l_2, l_3 respectively and adding

$$dx = \frac{l_1}{\lambda} \, du + \frac{l_2}{\mu} \, dv + \frac{l_3}{\nu} \, dw \qquad\qquad ...(4)$$

Similarly, $dy = \dfrac{m_1}{\lambda} \, du + \dfrac{m_2}{\mu} \, dv + \dfrac{m_3}{\nu} \, dw$ ...(5)

and $dz = \dfrac{n_1}{\lambda} \, du + \dfrac{n_2}{\mu} \, dv + \dfrac{n_3}{\nu} \, dw$ ...(6)

Comparing (4), (5) and (6) with

$$dx = x_u \, du + x_v \, dv + x_w \, dw \text{ etc.}$$

We get $\dfrac{l_1}{\lambda} = x_u = f_u, \dfrac{l_2}{\mu} = x_v = f_v, \dfrac{l_3}{\nu} = x_w = f_w, \text{etc.}$

We then have

$$f_u \, f_v + \phi_u \, \phi_v + \psi_u \, \psi_v = \frac{1}{\lambda \mu} (l_1 \, l_2 + m_1 \, m_2 + n_1 \, n_2) = 0 \text{ etc.} \qquad ...(7)$$

Now, $J^2 = \begin{vmatrix} f_u & f_v & f_w \\ \phi_u & \phi_v & \phi_w \\ \psi_u & \psi_v & \psi_w \end{vmatrix} \begin{vmatrix} f_u & \phi_u & \psi_u \\ f_v & \phi_v & \psi_v \\ f_w & \phi_w & \psi_w \end{vmatrix}$

$$= \begin{vmatrix} \Sigma f_u^2 & \Sigma f_u f_v & \Sigma f_w f_u \\ \Sigma f_v f_u & \Sigma f_v^2 & \Sigma f_v f_w \\ \Sigma f_w f_u & \Sigma f_v f_w & \Sigma f_w^2 \end{vmatrix}$$

$$= \begin{vmatrix} \Sigma f_u^2 & 0 & 0 \\ 0 & \Sigma f_v^2 & 0 \\ 0 & 0 & \Sigma f_w^2 \end{vmatrix}, \text{ using (7)}$$

$$= P_1^2 \, P_2^2 \, P_3^2 \qquad \text{where} \qquad P_1^2 = \Sigma f_u^2 \text{ etc.}$$

EXERCISES

1. If J be the Jacobian of the system u, v with regard to x, y and J' the Jacobian of x, y with regard to u, v then prove that $J J' = 1$.

2. If $u^3 + v^3 = x + y$, $u^2 + v^2 = x^3 + y^3$, show that

$$\frac{\partial (u, v)}{\partial (x, y)} = \frac{1}{2} \frac{y^2 - x^2}{uv (u - v)} \cdot \text{(Garhwal 99, 2003; Kumaon 96, 2002; Sagar 2003)}$$

3. If $x + y + z = u$, $y + z = uv$, $z = u v w$, find the value of the Jacobian of x, y, z with respect to u, v, w. **(Kumaon 2005)**

4. If $u^3 + v + w = x + y^2 + z^2$,

   $u + v^3 + w = x^2 + y + z^2$,

   $u + v + w^3 = x^2 + y^2 + z$,

   then prove that $\dfrac{\partial (u, v, w)}{\partial (x, y, z)} = \dfrac{1 - 4 (xy + yz + zx) + 16 x y z}{2 - 3 (u^2 + v^2 + w^2) + 27 u^2 v^2 w^2}$. **(Kumaon 97)**

5. If $\quad u_1 = x_1 + x_2, x_3 + x_4,$

   $u_1 u_2 = x_2 + x_3 + x_4,$

   $u_1 u_2 u_3 = x_3 + x_4,$

   $u_1 u_2 u_3 u_4 = x_4,$

   show that $\dfrac{\partial (x_1, x_2, x_3, x_4)}{\partial (u_1, u_2, u_3, u_4)} = u_1^3 u_2^2 u_3$. **(Kumaon 2000)**

6. If $u^3 = x y z$, $\dfrac{1}{v} = \dfrac{1}{x} + \dfrac{1}{y} + \dfrac{1}{z}$, $w^2 = x^2 + y^2 + z^2$, then prove that

   $$\frac{\partial (u, v, w)}{\partial (x, y z)} = \frac{v (y - z) (z - x) (x - y) (x + y + z)}{3 u^2 w(yz + zx + xy)}.$$

7. If λ, μ, v are the roots of the equation in R,

   $$\frac{x}{a + R} + \frac{y}{b + R} + \frac{z}{c + R} = 1, \text{ prove that}$$

   $$\frac{\partial (x, y, z)}{\partial (\lambda, \mu, v)} = - \frac{(\mu - v) (v - \lambda) (\lambda - \mu)}{(b - c) (c - a) (a - b)}.$$

8. If $\quad y_1 = r \cos \theta_1$

   $y_2 = r \sin \theta_1 \cos \theta_2$

   $y_3 = r \sin \theta_1 \sin \theta_2 \cos \theta_3$

   $\dots \dots \dots \dots \dots \dots \dots \dots \dots$

   and $\quad y_n = r \sin \theta_1 \sin \theta_2 \dots \sin \theta_{n-1} \cos \theta_n$, prove that

   $$\frac{\partial (y_1, y_2, \dots, y_n)}{\partial (r, \theta_1, \dots, \theta_{n-1})} = r^{n-1} \sin^{n-2} \theta_1 \cdot \sin^{n-3} \theta_2 \dots, \sin \theta.$$

3. $u^2 v$.

12.4. Theorem.

Let $u_1, u_2, ..., u_n$ be functions of n independent variables $x_1, x_2, ..., x_n$. The necessary and sufficient condition that the functions be connected by a relation $f(u_1, u_2, ..., u_n) = 0$ is that the Jacobian $\dfrac{\partial(u_1, u_2, ..., u_n)}{\partial(x_1, x_2, ..., x_n)}$ vanishes identically.

The condition is necessary.

We have $f(u_1, u_2, u_3, ..., u_n) = 0$.

Differentiating this with respect to $x_1, x_2, ..., x_n$, we get

$$\frac{\partial f}{\partial u_1} \cdot \frac{\partial u_1}{\partial x_1} + \frac{\partial f}{\partial u_2} \cdot \frac{\partial u_2}{\partial x_1} + ... + \frac{\partial f}{\partial u_n} \cdot \frac{\partial u_n}{\partial x_1} = 0$$

$$\frac{\partial f}{\partial u_1} \cdot \frac{\partial u_1}{\partial x_2} + \frac{\partial f}{\partial u_2} \cdot \frac{\partial u_2}{\partial x_2} + ... + \frac{\partial f}{\partial u_n} \cdot \frac{\partial u_n}{\partial x_2} = 0$$

....

....

$$\frac{\partial f}{\partial u_1} \cdot \frac{\partial u_1}{\partial x_n} + \frac{\partial f}{\partial u_2} \cdot \frac{\partial u_2}{\partial x_n} + ... + \frac{\partial f}{\partial u_n} \cdot \frac{\partial u_n}{\partial x_n} = 0$$

Eliminating $\dfrac{\partial f}{\partial u_1}, \dfrac{\partial f}{\partial u_2}, ..., \dfrac{\partial f}{\partial u_n}$ we obtain

$$\begin{vmatrix} \dfrac{\partial u_1}{\partial x_1} & \dfrac{\partial u_2}{\partial x_1} & \cdots & \dfrac{\partial u_n}{\partial x_1} \\ \dfrac{\partial u_1}{\partial x_2} & \dfrac{\partial u_2}{\partial x_2} & \cdots & \dfrac{\partial u_n}{\partial x_2} \\ \cdots & \cdots & \cdots & \cdots \\ \cdots & \cdots & \cdots & \cdots \\ \dfrac{\partial u_1}{\partial x_n} & \dfrac{\partial u_2}{\partial x_n} & \cdots & \dfrac{\partial u_n}{\partial x_n} \end{vmatrix} = 0$$

i.e.,

$$\frac{\partial(u_1, u_2, ..., u_n)}{\partial(x_1, x_2, ..., x_n)} = 0$$

The condition is sufficient.

If $J(u_1, u_2, ..., u_n) = 0$ then we will prove that there exists a relation between the functions.

The equation connecting $u_1, u_2, ..., u_n$ and $x_1, x_2, ..., x_n$ are always capable, by elimination, of being transformed into the following form :

$$F_1(x_1, x_2, ..., x_n, u_1) = 0$$

$$F_2(x_2, x_3, ..., x_n, u_1, u_2) = 0$$

....

....

$$F_n(x_n, u_1, u_2, ... u_n) = 0$$

Now we know that

$$J(u_1, u_2, \dots, u_n) = (-1)^n \frac{\begin{vmatrix} \dfrac{\partial F_1}{\partial x_1} \cdot \dfrac{\partial F_2}{\partial x_2} \dots \dfrac{\partial F_n}{\partial x_n} \end{vmatrix}}{\begin{vmatrix} \dfrac{\partial F_1}{\partial u_1} \cdot \dfrac{\partial F_2}{\partial u_2} \dots \dfrac{\partial F_n}{\partial u_n} \end{vmatrix}}$$

But it is given that $J(u_1, u_2, \dots, u_n) = 0$

$$\therefore \qquad \frac{\partial F_1}{\partial x_1} \cdot \frac{\partial F_2}{\partial x_2} \dots \frac{\partial F_n}{\partial x_n} = 0$$

i.e., we have $\dfrac{\partial F_r}{\partial x_r} = 0$, for same value of r between 1 and n.

Hence for that particular value of r the function F_r must not contain x_r. Therefore F_r becomes

$$F_r(x_{r+1}, x_{r+2}, \dots, x_n, u_1, u_2, \dots, u_r) = 0$$

Thus, between this and the remaining equation

$$F_{r+1} = 0, F_{r+2} = 0, \dots, F_n = 0,$$

The variables $x_{r+1}, x_{r+2}, \dots, x_n$ can be eliminated and consequently we get a final relation between $u_1, u_2, \dots, u_n$ alone.

EXAMPLES

1. If $u = \dfrac{x+y}{z}$, $v = \dfrac{y+z}{x}$, $w = \dfrac{y(x+y+z)}{xz}$, *show that u, v, w are not independent and find the relation between them.*

We have

$$J = \begin{vmatrix} \dfrac{1}{z} & \dfrac{1}{z} & -\dfrac{(x+y)}{z^2} \\[2ex] -\dfrac{(y+z)}{x^2} & \dfrac{1}{x} & \dfrac{1}{x} \\[2ex] -\dfrac{y^2-yz}{x^2 z} & \dfrac{x+2y+z}{xz} & \dfrac{-xy-y^2}{xz^2} \end{vmatrix}$$

Taking $\dfrac{1}{z^2}, \dfrac{1}{x^2}$ and $\dfrac{1}{x^2 z^2}$ common from R_1, R_2 and R_3 respectively, we get

$$J = \frac{1}{x^4 z^4} \begin{vmatrix} z & z & -(x+y) \\[1ex] -(y+z) & x & x \\[1ex] -(y^2 z + y z^2) & x^2 z + 2xyz + xz^2 & -(x^2 y + x y^2) \end{vmatrix}$$

Applying $C_2 - C_1$ and $C_3 - C_1$,

$$= \frac{1}{x^4 z^4} \begin{vmatrix} z & 0 & -(x+y+z) \\[1ex] -(y+z) & (x+y+z) & (x+y+z) \\[1ex] -yz(y+z) & z(x^2+y^2+2xy)+z^2(x+y) & -xy(x+y)+yz(y+z) \end{vmatrix}$$

$$= \frac{1}{x^4 \, z^4} \begin{vmatrix} z & 0 & -(x+y+z) \\ -(y+z) & (x+y+z) & (x+y+z) \\ -yz\,(y+z) & z\,(x+y)\,(z+x+z) & (x+y+z)\,(yz-xy) \end{vmatrix}$$

$$= \frac{(x+y+z)^2}{x^4 \, z^4} \begin{vmatrix} z & 0 & -1 \\ -(y+z) & 1 & 1 \\ -yz\,(y+z) & z\,(x+y) & (yz-xy) \end{vmatrix}$$

Applying $C_1 + z\,C_3$, we get

$$= \frac{(x+y+z)^2}{x^4 \, z^4} \begin{vmatrix} 0 & 0 & -1 \\ -y & 1 & 1 \\ -y^2 z - yz^2 + yz^2 - xyz & z\,(x+y) & (yz-xy) \end{vmatrix}$$

$$= \frac{(x+y+z)^2}{x^4 \, z^4}(-1) \begin{vmatrix} -y & 1 \\ -yz\,(y+x) & z\,(x+y) \end{vmatrix}$$

$$= \frac{(x+y+z)^2}{x^4 \, z^4}[-yz\,(x+y) + yz\,(x+y)] = 0$$

Since $J = 0$, hence the given functions are not independent.

Again $uv = \dfrac{xy + y^2 + yz + zx}{zx} = \dfrac{y\,(x+y+z)}{xz} + 1 = w + 1$

∴ $uv = w + 1$ is the required relation between them.

2. If $u = x^3 + x^2 y + x^2 z - z^2 (x + y + z)$, $v = x + z$, $w = x^2 - z^2 + xy - zy$, prove that u, v and w are connected by a functional relation.

We have

$$\frac{\partial(u,\, v,\, w)}{\partial(x,\, y,\, z)} = \begin{vmatrix} 3x^2 + 2xy + 2xz - z^2 & x^2 - z^2 & x^2 - 2zx - 2zy - 3z^2 \\ 1 & 0 & 1 \\ 2x + y & x - z & -2z - y \end{vmatrix}$$

Apply $C_3 - C_1$

$$J = \begin{vmatrix} 3x^2 + 2xy + 2xz - z^2 & x^2 - z^2 & \begin{array}{c} -2x^2 - 2xy \\ -4zx - 2yz - 2z^2 \end{array} \\ 1 & 0 & 0 \\ 2x + y & x - z & -2\,(x + y + z) \end{vmatrix}$$

$$= -1 \begin{vmatrix} (x-z)(x+z) & (-2x - 2z)(x+y+z) \\ (x-z) & -2\,(x+y+z) \end{vmatrix}$$

$$= 2\,(x-z)(x+y+z) \begin{vmatrix} x+z & x+z \\ 1 & 1 \end{vmatrix} = 0$$

Since $J = 0$; therefore the functions are not independent.

Clearly, $u = x^2 (x+y+z) - z^2 (x+y+z)$

$\qquad = (x+y+z)(x+z)(x-z)$

$w = (x-z)(x+z) + y\,(x-z)$

$\qquad = (x-z)(x+y+z)$

and $v = x + z$

∴ $u = w \cdot v$ is the required relation.

3. If $u = y \sqrt{(1 - x^2)} + x \sqrt{(1 - y^2)}$, $v = \sqrt{(1 - x^2)(1 - y^2)} - xy$ prove that u and v are not independent, and find the relation between them.

$$J = \frac{\partial(u, v)}{\partial(x, y)}$$

$$= \begin{vmatrix} \dfrac{xy}{\sqrt{(1 - x^2)}} + \sqrt{(1 - y^2)} & \sqrt{(1 - x^2)} - \dfrac{xy}{\sqrt{(1 - y^2)}} \\[3mm] -\dfrac{x\sqrt{(1 - y^2)}}{\sqrt{(1 - x^2)}} - y & -\dfrac{y\sqrt{(1 - x^2)}}{\sqrt{(1 - y^2)}} - x \end{vmatrix}$$

$$= \frac{1}{\sqrt{(1 - x^2)(1 - y^2)}} \begin{vmatrix} \sqrt{(1 - y^2)(1 - y^2)} - xy & \sqrt{(1 - x^2)(1 - y^2)} - xy \\[2mm] -x\sqrt{(1 - y^2)} - y\sqrt{(1 - x^2)} & -x\sqrt{(1 - y^2)} - y\sqrt{(1 - x^2)} \end{vmatrix}$$

The above determinant is zero as column No. 1 and 2 are identical. Hence the functions are not independent.

Again

$$u^2 + v^2 = y^2(1 - x^2) + x^2(1 - y^2) + 2xy\sqrt{(1 - x^2)(1 - y^2)} + (1 - x^2)(1 - y^2) + x^2y^2$$

$$- 2xy\sqrt{\{(1 - x^2)(1 - y^2)\}}$$

$$= y^2(1 - x^2 + x^2) + (1 - y^2)(x^2 + 1 - x^2)$$

$$= 1$$

Hence the required relation is $u^2 + v^2 = 1$.

EXERCISES

1. If $u = x + 2y + z$, $v = x - 2y + 3z$; $w = 2xy - xz + 4yz - 2z^2$, show that they are not independent, and find the relation between them.

2. Prove that the functions

   $u = x(y + z)$, $v = y(x + z)$, $w = z(x - y)$ are not independent, and find the relation between them.

3. Prove that $ax^2 + 2hxy + by^2$ and $Ax^2 + 2Hxy + By^2$ are independent unless

   $$\frac{a}{A} = \frac{h}{H} = \frac{b}{B}.$$

ANSWERS

1. $u^2 - v^2 = 4w$;   2. $u - v = w$.

OBJECTIVE QUESTIONS

For each of the following questions, four alternatives are given for the answer. Only one of them is correct. Choose the correct alternative.

1. If each of u, v, w is a function of the variables x, y, z then the Jacobian $\dfrac{\partial(u, v, w)}{\partial(x, y, z)}$ is determinant of order :

   (a) 9   (b) 3   (c) 1   (d) n

2. If $x = r \cos \theta, y = r \sin \theta$, then $\dfrac{\partial (x, y)}{\partial (r, \theta)}$ is equal to :

   (a) x (b) y (c) r (d) θ

                        (Rohilkhand 2002, 2003; Garhwal 2004; Avadh 2002; Kanpur 2001)

3. If $x = r \cos \theta, y = r \sin \theta$, then $\dfrac{\partial (r, \theta)}{\partial (x, y)} \times \dfrac{\partial (x, y)}{\partial (r, \theta)}$ is equal to :

   (a) 0 (b) 1 (c) 2 (d) ∞

4. If $u = f_1(x_1), u_2 = f_2(x_1, x_2), \ldots, u_n = f_n(x_1, x_2, \ldots, x_n)$, then $\dfrac{\partial (u_1, u_2, \ldots, u_n)}{\partial (x_1, x_2, \ldots, x_n)}$ is equal to :

   (a) $\dfrac{\partial u_1}{\partial x_1}$ (b) $\dfrac{\partial u_1}{\partial x_1} \cdot \dfrac{\partial u_n}{\partial x_n}$ (c) $\dfrac{\partial u_1}{\partial x_1} \dfrac{\partial u_2}{\partial x_2} \cdots \dfrac{\partial u_n}{\partial x_n}$ (d) None of these

5. If $x = u(1 + u), y = v(1 + u)$, then the Jacobian of x, y with respect to u, v is :

   (a) $u + v$ (b) $1 + u + v$ (c) $1 - u + v$ (d) $1 - u - v$

6. $\dfrac{\partial (u, v)}{\partial (x, y)} \times \dfrac{\partial (x, y)}{\partial (u, v)} =$

   (a) 1 (b) -1 (c) 0 (d) None of these

ANSWERS

1. (b)	2. (c)	3. (b)	4. (c)	5. (b)	6. (a)

CHAPTER
13
CONCAVITY AND POINTS OF INFLEXION

13.1. Introduction

In this chapter, we shall consider special types of points on a curve *viz.* those where the curve crosses the tangent. For example, the x-axis is the tangent to the curve $y = x^3$ at $(0, 0)$ and the curve which lies above x-axis on the right of the origin and below x-axis on the left, crosses the tangent at $(0, 0)$. A point where a curve crosses the tangent is known as a point of **Inflexion** on the curve.

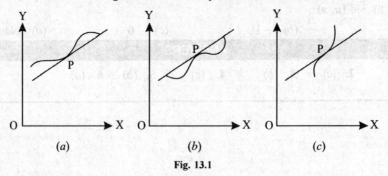

(a)   *(b)*   *(c)*

Fig. 13.1

In case a curve does not cross the tangent at a point, we have the following two possibilities :

(*i*) There is a portion of the curve on both sides of P, however small it may be, which lies *above* the tangent at P (*i.e.*, towards the positive direction of y-axis). In this case, we say that the curve is *concave upwards or convex downwards at P* [Fig. 13.1 (*a*)].

(*ii*) There is a portion of the curve on both sides of P, however small it may be, which lies *below* the tangent at P (*i.e.*, towards the negative direction of y-axis). In this case, we say that the curve is *concave downwards or convex upwards at P* [Fig. 13.1 (*b*)].

Thus the curve is concave upwards at P in Fig. 13.1 (*a*) and is concave downwards at P in Fig. 13.1 (*b*). Also the curve in Fig. 13.1 (*c*) has a point of inflexion at P.

A portion of a curve is said to be concave upwards (downwards) if it is concave upwards (downwards) at every point thereof.

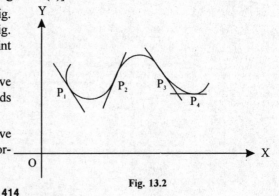

Thus the portions $P_1 P_2$ and $P_3 P_4$ of the curve in the Fig. 13.2 are concave upwards and the portion $P_2 P_3$ is concave downwards.

Also P_2 and P_3 are points of inflexion.

Fig. 13.2

414

At a point of inflexion, the curve changes from concavity upwards to concavity downwards or *vice-versa*.

Ex. Estimate through looking at the graphs the

    (*i*)   points of inflexion,

    (*ii*)   intervals of concavity upwards,

    (*iii*)   intervals of concavity downwards for the curves

$$y = \sin x, y = \tan x, y = \sec x, y = \cos^{-1} x, y = \cot^{-1} x, y = \operatorname{cosec}^{-1} x.$$

13.2 Criteria for (*i*) Concavity Upwards, (*ii*) Concavity Downwards, (*iii*) Inflexion at a given point.

Consider a curve $y = f(x)$ and a point $P\,[c, f(c)]$ thereof. We suppose that the tangent at P is *not* parallel to y-axis.

We take a point $Q\,[x, f(x)]$ on the curve near the point $P[c, f(c)]$. The point Q will lie to the right or the left of the point P according as $x > c$ or $x < c$.

Draw $QM \perp x$-axis meeting the tangent at P in R. The equation of the tangent at P being

$$y - f(c) = f'(c)\,(x - c),$$

the ordinate MR of the tangent corresponding to the abscissa x is

$$f(c) + f'(c)\,(x - c).$$

Also, the ordinate MQ of the curve for the abscissa x is $f(x)$.

$\therefore \qquad\qquad RQ = MQ - MR = f(x) - f(c) - f'(c)\,(x - c).$

(*i*) For concavity upwards at P, [Fig. 13.3 (*a*)], RQ is positive when Q lies on *either side* of P.

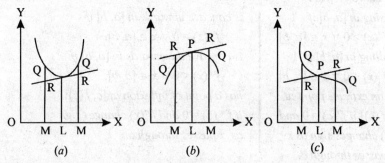

    (*a*)   (*b*)   (*c*)

Fig. 13.3

(*ii*) For concavity downwards at P, [Fig. 13.3 (*b*)], RQ is negative when Q lies on *either side* of P.

(*iii*) For inflexion at P, [Fig. 13.3 (*c*)], RQ is positive when Q lies on one side of P and negative when Q lies on the other side of P.

Thus, we have to examine the behaviour of the sign of RQ for values of x such that $|x - c|$ is sufficiently small. We write

$$RQ = \phi(x) = f(x) - f(c) - f'(c)\,(x - c)$$

so that $\qquad\qquad \phi'(x) = f'(x) - f'(c), \qquad\qquad \phi''(x) = f''(x).$

Case I. *Let $f''(c) > 0$.*

As $\phi''(c) = f''(c)$ is positive, there exists an interval $[c - \delta, c + \delta]$ around c for every point x of which $\phi''(x)$ is positive. Thus, $\phi'(x)$ is strictly increasing in $[c - \delta, c + \delta]$.

Also as $\phi'(c) = 0$, we see that

$$\phi'(x) < 0 \ \forall \ x \in [c - \delta, c[, \text{ and } \phi'(x) > 0 \ \forall \ x \in \,]x, c + \delta].$$

Again it follows that ϕ is strictly decreasing in $[c-\delta, c]$ and strictly increasing in $[c, c+\delta]$. Also $\phi(c) = 0$. Finally, it follows that

$$RQ = \phi(x) > 0 \ \forall \, x \in [c-\delta, c \ [\cup] \ c, c+\delta].$$

Thus, *the curve* $y = f(x)$ *is concave upwards at* $[c, f(c)]$ *if* $f''(c) > 0$.

Case II. *Let* $f''(c) < 0$.

In this case, we may show that there exists an interval $[c-\delta, c+\delta]$ around c such that

$$RQ = \phi(x) < 0 \ \forall \, x \in [c-\delta, c \ [\cup] \ c, c+\delta].$$

Thus, *the curve* $y = f(x)$ *is concave downwards at* $[c, f(c)]$ *if* $f''(c) < 0$.

Case III. *Let* $f''(c) = 0$.

We know that a curve has inflexion at P if it changes from concavity upwards to concavity downwards, or *vice-versa*, as a point moving along the curve passes through P, *i.e.*, if there is a complete neighbourhood of P such that at every point on one side of P lying in this neighbourhood, the curve is concave upwards and at every point on the other side of P the curve is concave downwards.

Thus, the curve has inflexion at $[c, f(c)]$, *if* $f''(x)$ *changes sign as* x *passes through* c.

Since f'' is assumed to be continuous, it follows that at a point of inflexion $f''(c) = 0$.

Note. To determine the criteria for an extreme value for $x = c$, we have to examine the behaviour of the sign of

$$f(x) - f(c)$$

and for determining the criteria for concavity upwards (downwards), point of inflexion for $x = c$, we examine the behaviour of the sign of

$$f(x) - f(c) - f'(c)(x - c).$$

Thus the curve $y = f(x)$

(a) is rising in $[a, b]$ if	is concave upwards in $[a, b]$ if
$f'(x) > 0 \ \forall \, x \in [a, b]$	$f''(x) > 0 \ \forall \, x \in [a, b]$.
(b) is falling in $[a, b]$ if	is concave downwards in $[a, b]$ if
$f'(x) < 0 \ \forall \, x \in [a, b]$	$f''(x) < 0 \ \forall \, x \in [a, b]$.
(c) has an extreme point at	has a point of inflexion at $[c, f(c)]$
$[c, f(c)]$ if $f'(c) = 0$ and	if $f''(c) = 0$ and $f''(x)$ changes sign
$f'(x)$ changes sign as x	as x passes through c.
passes through c.	

EXAMPLES

1. *Find the ranges of values of* x *for which the curve*

$$y = x^4 - 6x^3 + 12x^2 + 5x + 7$$

is concave upwards or downwards. Also determine the points of inflexion.

We have $\dfrac{dy}{dx} = 4x^3 - 18x^2 + 24x + 5,$

$\dfrac{d^2y}{dx^2} = 12x^2 - 36x + 24 = 12(x-1)(x-2)$

$\dfrac{d^2y}{dx^2} > 0 \ \forall \, x \in]-\infty, 1[, \dfrac{d^2y}{dx^2} < 0 \ \forall \, x \in]1, 2[$

$$\frac{d^2y}{dx^2} > 0 \; \forall \; x \in] \, 2, \, + \infty [.$$

Thus, the curve is concave upwards in the intervals $[-\infty, 1]$ and $[2, +\infty]$ and concave downwards in the interval $[1, 2]$. Also the curve has inflexions for $x = 1$ and $x = 2$ in that d^2y/dx^2 changes sign as x passes through 1 and 2. It follows that $(1, 19)$ and $(2, 33)$ are the two points of inflexion on the curve.

2. *Show that the curve $(a^2 + x^2)\, y = a^2 x$ has three points of inflexion.*

We have
$$\frac{dy}{dx} = \frac{a^2 (a^2 + x^2) - 2a^2 x^2}{(a^2 + x^2)^2} = \frac{a^2 (a^2 - x^2)}{(a^2 + x^2)^2}$$

$$\frac{d^2y}{dx^2} = a^2 \cdot \frac{-2x(a^2 + x^2)^2 - 4x(a^2 + x^2)(a^2 - x^2)}{(a^2 + x^2)^4}$$

$$= a^2 \cdot \frac{-2x(3a^2 - x^2)}{(a^2 + x^2)^3} = 2a^2 \frac{(x - \sqrt{3}a)x(x + \sqrt{3}\,a)}{(a^2 + x^2)^3}$$

$$\therefore \qquad \frac{d^2y}{dx^2} = 0 \text{ for } x = \sqrt{3}a, \; 0, \; -\sqrt{3}\,a.$$

It may be seen that d^2y/dx^2 changes sign as, x passes through each of these three values of x. Hence, the curve has inflexions at the corresponding points. Thus,

$$(\sqrt{3}\,a, \; \sqrt{3}\,a/4), \; (0,0) \; (-\sqrt{3}\,a, \; -\sqrt{3}\,a/4)$$

are the three points of inflexion on the curve.

3. *Find the points of inflexion on the curve $y = (\log x)^3$.*

We have
$$\frac{dy}{dx} = 3(\log x)^2 \cdot \frac{1}{x}$$

$$\frac{d^2y}{dx^2} = 6 \log x \cdot \frac{1}{x^2} - \frac{3}{x^2} (\log x)^2$$

$$= \frac{3 \log x}{x^2} (2 - \log x) = \frac{3 \log x}{x^2} \log \frac{e^2}{x}$$

Thus, $\dfrac{d^2y}{dx^2} = 0$ if $\log x = 0$ or $\log x = 2$.

Now, $\log x = 0 \Rightarrow x = e^0 = 1$ and $\log x = 2 \Rightarrow x = e^2$.

Thus, we expect points of inflexion on the curve for $x = 1$ and $x = e^2$.

Now, d^2y/dx^2 changes sign from negative to positive as x passes through 1 and changes sign from positive to negative as x passes through e^2.

Thus, $(1, 0)$ and $(e^2, 8)$ are the two points of inflexion of given curve.

EXERCISES

1. Show that $y = e^x$ is everywhere concave upwards and the curve $y = \log x$ is everywhere concave downwards.

2. Examine the curve $y = \sin x$ for concavity upwards, concavity downwards and for points of inflexion in the interval $[-2\pi, 2\pi]$. Also indicate the positions of the points of inflexion on the curve.

3. Find the ranges of values of x in which the curves

    (i) $y = 3x^5 - 40x^3 + 3x - 20$ **(Banglore 2006)**   (ii) $y = (x^2 + 4x + 5) e^{-x}$

    are concave upwards or downwards. Also find their points of inflexion.

4. Find the intervals in which the curve $y = (\cos x + \sin x) e^x$ is concave upwards or downwards; x varying in the interval $]\,0, 2\pi\,[$.

5. Find the points of inflexion on the curves.

    (i) $y = ax^3 + bx^2 + cx + d$ (ii) $y = (x^3 - x)/(3x^2 + 1)$

    (iii) $x = 3y^4 - 4y^3 + 5$ (iv) $x = (y - 1)(y - 2)(y - 3)$

    (v) $y = \dfrac{a^2(a - x)}{a^2 + x^2}$ (vi) $y = \dfrac{x^3}{a^2 + x^2}$

    (vii) $y = \dfrac{x}{x^2 + 2x + 2}$ (viii) $y = be^{-(x/a)^2}$

    (ix) $xy = a^2 \log(y/a)$ (x) $y = x^2 \log(x^2 / e^3)$

    (xi) $y^2 = x(x + 1)^2$ (xii) $a^2y^2 = x^2(a^2 - x^2)$

    (xiii) $54y = (x + 5)^2(x^3 - 10)$

    Also, obtain the equations of the inflexional tangents to the curves (ii), (iii) and (xi).

6. Show that the curve $ay^2 = x(x - a)(x - b)$ has two and, only two points of inflexion.

7. Show that the points of inflexion of the curve $y^2 = (x - a)^2(x - b)$ lie on the line $3x + a = 4b$.

    (MDU Rohtak 1998)

8. Find the points of inflexion on the curves :

    (i) $x = a(2\theta - \sin\theta), y = a(2 - \cos\theta)$ (ii) $x = a \tan t, y = a \sin t \cos t$

    (iii) $x = 2a \cot\theta, y = 2a \sin^2\theta$ (iv) $y = xe^{-x^2}$

    (v) $y = x^2 e^{-x}$ (vi) $y = e^{x/2} - 2e^{-x/2}$

    (vii) $y = x(x - 3) e^{2x}$ (viii) $y = e^{-2x} \cos x$

    (ix) $y = e^{(x - 1)/x^2}$ (x) $y = x + e^x$

    (xi) $y = 2e^x + e^{-x}$ (xii) $y = x^2 - 6 + e^x$

9. Find the points of inflexion on the curves :

    (i) $r = a\theta^2 / (\theta^2 - 1)$ (ii) $r^2 \theta = a^2$

    (iii) $r = ae^\theta / (1 + \theta)$

    (Transform to the cartesian parametric forms of equations).

10. Show that $r = a\theta^n$ has points of inflexion if and only if n lies between 0 and -1 and these are given by $\theta = \pm\sqrt{[-n(n + 1)]}$.

11. Show that the curve $r e^\theta = a(1 + \theta)$ has no point of inflexion.

ANSWERS

2. Concave downwards in $[-2\pi, -\pi]$ and $[0, \pi]$; upwards in $[-\pi, 0]$ and $[\pi, 2\pi]$.

3. (i) Concave upwards in $[-2, 0]$ and $[2, \infty\,[$; concave downwards in $]-\infty, -2]$ and $[0, 2]$. Inflexions at $(-2, 198)$, $(0, -20)$, $(2, -238)$.

    (ii) Concave upwards in $]-\infty, -1]$ and $[1, \infty[$; concave downwards in $[-1, 1]$. Inflexions at $(-1, 2e)$ and $(1, 10/e)$.

4. Concave upwards in $[0, \pi/4]$ and $[5\pi/4, 2\pi]$ and concave downwards in $[\pi/4, 5\pi/4]$.

5. (i) For $x = -b/3a$.

   (ii) $(0, 0), (1, 0), (-1, 0)$.

   (iii) $(5, 0), \left(\dfrac{119}{27}, \dfrac{2}{3}\right)$.

   (iv) $(0, 2)$.

   (v) For $x = -a$ and $a(2 \pm \sqrt{3})$

   (vi) For $x = 0$ and $\pm a\sqrt{3}$.

   (vii) For $x = -2$ and $1 \pm \sqrt{3}$.

   (viii) For $x = \pm a/\sqrt{2}$.

   (ix) $\left(\dfrac{3}{2} ae^{-3/2}, ae^{3/2}\right)$.

   (x) $(1, 3)$.

   (xi) $\left(\dfrac{1}{3}, \pm \dfrac{4}{3\sqrt{3}}\right)$.

   (xii) $(0, 0)$.

   (xiii) For $x = -2, -4 \pm \sqrt{18}$.

   Inflexional tangents to (ii) are $x + y = 0, x - 2y \pm 1 = 0$.

   Inflexional tangents to (iii) are $x = 5$ and $3(9x + 6y) = 151$.

   Inflexional tangents to (xi) are $9x \pm 3\sqrt{3}\sqrt{3y + 1} = 0$.

8. (i) $\left[a\left(4\pi n \pm \dfrac{2\pi}{3} \pm \dfrac{\sqrt{3}}{2}\right), \dfrac{3a}{2}\right]$

   (ii) $\left(\pm \sqrt{3a}, \pm \dfrac{\sqrt{3a}}{4}\right), (0, 0)$.

   (iii) $\left(\pm \dfrac{2a}{\sqrt{3}}, \dfrac{3a}{2}\right)$

   (iv) For $x = 0, \pm \dfrac{\sqrt{3}}{2}$.

   (v) For $x = 2 \pm \sqrt{2}$.

   (vi) For $x = \log 2$.

   (vii) For $x = \dfrac{1}{2} \pm \dfrac{1}{2}\sqrt{11}$.

   (viii) For $x : \tan x = -\dfrac{3}{4}$.

   (ix) For values of x which are the roots of the cubic $2x^3 - 5x^2 - 4x + 4 = 0$.

   (x) No inflexion.

   (xi) No inflexion.

   (xii) No inflexion.

9. (i) $(3a/2, \pm\sqrt{3})$.

   (ii) $\left(\pm \sqrt{2a}, \dfrac{1}{2}\right)$.

   (iii) $(a, 0)$.

OBJECTIVE QUESTIONS

For each of the following questions, four alternatives are given for the answer. Only one of them is correct. Choose the correct alternative.

1. A curve is concave upwards if

   (a) $\dfrac{d^2y}{dx^2} > 0$

   (b) $\dfrac{d^2y}{dx^2} < 0$

   (c) $\dfrac{d^2y}{dx^2} = 0$ and $\dfrac{d^3y}{dx^3} \neq 0$

   (d) $\dfrac{dy}{dx}$ becomes infinite

2. A curve is convex upwards if

   (a) $\dfrac{d^2y}{dx^2} > 0$

   (b) $\dfrac{d^2y}{dx^2} < 0$

   (c) $\dfrac{d^2y}{dx^2} = 0$ and $\dfrac{d^3y}{dx^3} \neq 0$

   (d) $\dfrac{dy}{dx}$ becomes infinite

3. At the point of inflexion :

(a) $\dfrac{d^2y}{dx^2} > 0$ (b) $\dfrac{d^2y}{dx^2} < 0$

(c) $\dfrac{d^2y}{dx^2} = 0, \dfrac{d^3y}{dx^3} \neq 0$ (d) $\dfrac{d^3y}{dx^3} = 0$.

4. The curve $y = x^3 - 3x^2 - 9x + 9$ has a point of inflexion at

(a) $x = -1$ (b) $x = 1$

(c) $x = -3$ (d) $x = 3$.

5. In the curve $a^{m-1}y = x^m$, origin is a point of inflexion if

(a) m is zero   (b) m is even and greater than 2

(c) m is odd and greater than 2   (d) Origin is not a point of inflexion.

ANSWERS

1. (a)   2. (b)   3. (c)   4. (b)   5. (c)

CHAPTER
14
CURVATURE AND EVOLUTES

14.1. Introduction. Definition of Curvature

In our every day language, we make statements which involve the comparison of bending or curvature of a road at two of its points. For instance, at times, we say, "The bend of the road is, sharper at this place than at that."

Here we depend upon intuitive means of comparing the curvature at the two points and the same may be reliable only when the difference is fairly marked. But *we are far from assigning* on the basis of intuition only any definite numerical measure to the curvature at any given point of a curve. In order to make 'Curvature' a subject of mathematical investigation, we have to assign, by means of some suitable definition, a numerical measure to it in a manner which may be in agreement with out intuitive notion of curvature. Thus we shall now proceed to do.

Consider a curve and a point P thereon.

We take a point Q on the curve lying near P.

Let A be a fixed point on the curve.

Let

arc AP = s, arc $AQ = s + \Delta s$

so that arc $PQ = \Delta s$.

Let ψ, $\psi + \Delta \psi$ be the angles which the tangents at P and Q make with some fixed line.

The symbol $\Delta \psi$ denotes the angle through which the tangent turns as a point moves along the curve from P to Q through a distance Δs. According to our intuitive feeling, $\Delta \psi$ will be large or small, as compared with Δs, depending on the degree of sharpness of the bend.

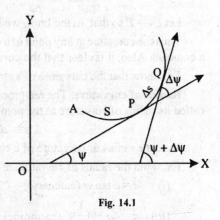

Fig. 14.1

This suggests the following definitions :–

(*i*) the **total bending or total curvature** or the arc PQ is defined to be the angle $\Delta \psi$.

(*ii*) the **average curvature** of the arc PQ is defined to be the ratio $\Delta \psi / \Delta s$,

and (*iii*) the **curvature** of the curve at P is defined to be

$$\lim_{Q \to P} \frac{\Delta \psi}{\Delta s} = \frac{d\psi}{ds}.$$

Thus, by def., **d** ψ**/ds** is the curvature of the curve at the point P. **(Banglore 2005)**

421

Curvature of a circle. *To prove that the curvature of a circle is constant.*

Intuitively, we feel that the curvature of a circle is the same at every point and that the larger the radius of the circle, the smaller will be its curvature. It will now be shown that these intuitive conclusions are consequences of our formal definition of curvature.

Consider a circle with radius, r and centre O. Let P, Q be two points on the circle and let arc $PQ = \Delta s$.

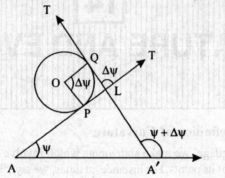

Fig. 14.2

Also let L be the point where the tangents PT, QT at P and Q meet.

We have $\angle POQ = \angle TLT' = \Delta \psi$.

From Trigonometry

$$\frac{arc\ PQ}{OP} = \angle POQ,$$

$$\Rightarrow \qquad \frac{\Delta s}{r} = \Delta \psi \Rightarrow \frac{\Delta \psi}{\Delta s} = \frac{1}{r}.$$

Let $Q \to P$ so that in the limit, we have $d\psi/ds = 1/r$.

Thus the curvature at any point of a circle is the reciprocal of the radius of the circle and, is, thus, a constant. Also, it is clear that the curvature, $1/r$, decreases as the radius increases.

Ex. Show that the curvature of a straight line at every point thereof is zero.

Radius of curvature. The reciprocal of the curvature of a curve at any point, in case it is $\neq 0$, is called its radius of curvature at the point and is generally denoted by, ρ, so that we have

$$\rho = ds/d\psi.$$

Thus the radius of curvature of a circle at any point, is equal to the radius of the circle.

Ex. Find the radius of curvature at any point of the following :–

(*i*) $s = c \tan \psi$ (catenary),   (*ii*) $s = 4a \sin \psi$ (cycloid),

(*iii*) $s = 4a \sin \dfrac{1}{3} \psi$ (cardioide),   (*iv*) $s = c \log \sec \psi$ (tractrix),

(*v*) $s = a \log (\tan \psi + \sec \psi) + a \tan \psi \sec \psi$ (Parabola).

ANSWERS		
(*i*) $c \sec^2 \psi$	(*ii*) $4\,a \cos \psi$	(*iii*) $\dfrac{4}{3} a \cos \dfrac{1}{3} \psi$
(*iv*) $c \tan \psi$	(*v*) $2a \sec^3 \psi.$	

The expression $ds/d\psi$ for the radius of curvature is suitable only for those curves whose equations are given by means of a relation between s and ψ. We must, therefore, transform the expression so that it may be applicable to the curves whose equations are given in the Cartesian or Polar form.

For this transformation we require the expression for ds/dx in terms of dy/dx which we now proceed to obtain.

14.2. Length of arc as a function. Derivative of arc.

Let $y = f(x)$ be the equation of a given curve on which we take a fixed point A. To any given value of x corresponds a value of y viz., $f(x)$ and to this pair of numbers x and $f(x)$ corresponds a point $P(x, y)$ on the curve, and this point P has some arcual length 's' from A. Thus, we have a function s of x for the curve $y = f(x)$. We shall now prove that,

$$\frac{ds}{dx} = \sqrt{\left[1 + \left(\frac{dy}{dx} \right)^2 \right]}$$

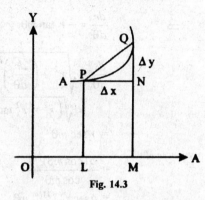

Fig. 14.3

We take a point $Q(x + \Delta x, y + \Delta y)$ on the curve near P.

Let arc $AQ = s + \Delta s$ so that arc

$$PQ = \Delta s.$$

From the rt. angled $\triangle PQN$, we have

$$PQ^2 = PN^2 = NQ^2,$$
$$= (\Delta x)^2 + (\Delta y)^2$$

$$\Rightarrow \quad \left(\frac{PQ}{\Delta x} \right)^2 = 1 + \left(\frac{\Delta y}{\Delta x} \right)^2$$

$$\Rightarrow \quad \left[\frac{\text{chord } PQ}{\text{arc } PQ} \right]^2 \left(\frac{\Delta s}{\Delta x} \right)^2 = 1 + \left(\frac{\Delta y}{\Delta x} \right)^2.$$

Assuming, on an intuitive basis that

$$\lim_{Q \to P} \frac{\text{chord } PQ}{\text{arc } PQ} = 1,$$

we obtain in the limit

$$\left(\frac{ds}{dx} \right)^2 = 1 + \left(\frac{dy}{dx} \right)^2$$

We make a convention that for the curve $y = f(x)$, 's' is measured positively in the direction of x increasing, so that, s, increase with x.

Hence ds/dx is positive.

Thus we have

$$\frac{ds}{dx} = \sqrt{\left[1 + \left(\frac{dy}{dx} \right)^2 \right]}$$

taking positive sign before the radical.

Cor. 1. $dx/ds = \cos \psi$, $dy/ds = \sin \psi$.

Cor. 2. For parametric cartesian equations with parameter t,

$$\left(\frac{ds}{dt} \right)^2 = \left(\frac{dx}{dt} \right)^2 + \left(\frac{dy}{dt} \right)^2.$$

EXAMPLES

1. *In the curve* $r^m = a^m \cos m\theta$, *prove that*

$$\frac{ds}{d\theta} = a \sec^{\frac{m-1}{m}} m\theta \text{ and } a^{2m} \frac{d^2 r}{ds^2} + m\, r^{2m-1} = 0.$$

We have $\quad r^m = a^m \cos m\theta$

$\Rightarrow \quad m \log r = m \log a + m \log \cos m\theta$

$\Rightarrow \qquad\qquad \dfrac{dr}{d\theta} = -r \tan m\theta$

$\therefore \qquad \dfrac{ds}{d\theta} = \sqrt{r^2 + \left(\dfrac{dr}{d\theta}\right)^2}$

$\qquad\qquad\quad = \sqrt{\left(r^2 + r^2 \tan^2 m\theta\right)}$

$\qquad\qquad\quad = r \sec m\theta$

$\qquad\qquad\quad = \dfrac{a(\cos m\theta)^{1/m}}{\cos m\theta}$

$\qquad\qquad\quad = a \sec^{(m-1)/m} m\theta.$

Again,

$$\frac{ds}{d\theta} = r \sec m\theta = \frac{r}{\cos m\theta} = \frac{r\, a^m}{r^m}$$

$$= a^m\, r^{1-m}$$

and $\qquad \dfrac{dr}{d\theta} = -r \tan m\theta$

$\therefore \qquad \dfrac{dr}{ds} = \dfrac{dr}{d\theta} \cdot \dfrac{d\theta}{ds} = -\dfrac{r \tan m\theta}{a^m\, r^{1-m}}$

$\qquad\qquad = -a^{-m}\, r^m \tan m\theta$

$\dfrac{d^2 r}{ds^2} = \dfrac{d}{ds}\left(\dfrac{dr}{ds}\right) = \dfrac{d}{dr}\left(\dfrac{dr}{ds}\right)\dfrac{dr}{ds}$

$\qquad = -a^{-m}\, r^m \tan m\theta \cdot \dfrac{d}{dr}\left\{-a^{-m}\, r^m \tan m\theta\right\}$

$\qquad = -a^{-2m}\, r^m \tan m\theta \left[m r^{m-1} \tan m\theta + m r^m \sec^2 m\theta \dfrac{d\theta}{dr}\right]$

$\qquad = m a^{-2m}\, r^m \tan m\theta \left[r^{m-1} \tan m\theta - r^m \sec^2 m\theta \cdot \dfrac{1}{r} \cot m\theta\right]$

$\qquad = m a^{-2m}\, r^{2m-1} \left[\tan^2 m\theta - \tan m\theta \sec^2 m\theta \cot m\theta\right]$

$\qquad = m a^{-2m}\, r^{2m-1} \left[\tan^2 m\theta - \sec^2 m\theta\right]$

$\qquad = m a^{-2m}\, r^{2m-1} (-1)$

$\therefore \quad \dfrac{d^2 r}{ds^2} + m a^{-2m}\, r^{2m-1} = 0$

or $\qquad a^{2m}\dfrac{d^2r}{ds^2} + m\, r^{2m-1} = 0.$

2. *Prove that for any curve*

$$\sin^2\phi\,\dfrac{d\phi}{d\theta} + r\,\dfrac{d^2r}{ds^2} = 0.$$

We know that $\cos\phi = \dfrac{dr}{ds}$

then $-\sin\phi \cdot \dfrac{d\phi}{ds} = \dfrac{d^2r}{ds^2}$, differentiating w.r.t.s,

or $\qquad -\sin\phi \cdot \dfrac{d\phi}{d\theta} \cdot \dfrac{d\theta}{ds} = \dfrac{d^2r}{ds^2}$

$\Rightarrow \qquad -\sin\phi\,\dfrac{d\phi}{d\theta} \cdot \dfrac{\sin\phi}{r} = \dfrac{d^2r}{ds^2}$ $\qquad\qquad \left[\because r\,\dfrac{d\theta}{ds} = \sin\phi \right]$

$\Rightarrow \qquad r\,\dfrac{d^2r}{ds^2} + \sin^2\phi\,\dfrac{d\phi}{d\theta} = 0.$

EXERCISES

1. Find ds/dx for the curves:

   (i) $y = \cosh x/c,$ (ii) $a\log[\,a^2/(a^2-x^2)]$

2. Show that for the parametric equation $x = f(t), y = F(t)$

$$\dfrac{ds}{dt} = \sqrt{\left[\left(\dfrac{dx}{dt}\right)^2 + \left(\dfrac{dy}{dt}\right)^2 \right]}$$

and use this result to find ds/dt for

   (i) $x = a\cos^3 t, y = b\sin^3 t.$ (ii) $x = a(t-\sin t), y = a(1-\cos t).$

   (iii) $x = a\,e^t\sin t, y = ae^t\cos t.$ **(Banglore 2005)**

   (iv) $x = a(\cos t + t\sin t), y = a(\sin t - t\cos t).$

3. Show that for polar equation $r = f(\theta),$

$$\dfrac{ds}{d\theta} = \sqrt{\left[r^2 + \left(\dfrac{dr}{d\theta}\right)^2 \right]}$$

and use this result to find $ds/d\theta$ for

   (i) $r = a(1+\cos\theta).$ (ii) $r^2 = a^2\cos 2\theta.$

ANSWERS

1. (i) $\cosh\dfrac{x}{c}.$ (ii) $\dfrac{a^2+x^2}{a^2-x^2}$

2. (i) $3\sin t\cos t\sqrt{(a^2\cos^2 t + b^2\sin^2 t)}.$

   (ii) $2a\sin\dfrac{1}{2}t.$ (iii) $\sqrt{2}ae^t.$ (iv) $at.$

3. (i) $2a\cos\theta/2$ (ii) $a\sqrt{(\sec 2\theta)}.$

14.3. Radius of Curvature -Cartesian Equations

(*Kanpur 2002; Kuvempu 2005*)

Consider the curve $y = f(x)$.

Differentiating

$$\tan \psi = \frac{dy}{dx}$$

w.r. to s, we obtain

$$\sec^2 \psi \, \frac{d\psi}{ds} = \frac{d^2 y}{dx^2} \cdot \frac{dx}{ds} \Rightarrow \frac{ds}{d\psi} = \frac{(1 + \tan^2 \psi) \dfrac{ds}{ds}}{\dfrac{d^2 y}{dx^2}}.$$

Also, we have

$$\frac{ds}{dx} = \sqrt{\left[1 + \left(\frac{dy}{dx} \right)^2 \right]},$$

where the positive sign is to be taken before the radical.

Hence $\qquad \rho = \dfrac{ds}{d\psi} = \dfrac{[1 + (dy/dx)^2]^{3/2}}{d^2 y / dx^2} = \dfrac{(1 + y_1^2)^{3/2}}{y_2}$ ...(A)

Cor. The radius of curvature, ρ, is positive or negative according as d^2y/dx^2 is positive, or negative, *i.e.* according as the curve is concave upwards or downwards. Also, (A) shows that the curvature is zero at a point of inflexion.

Note. Since the value of ρ is independent of the choice of x-axis and y-axis, interchanging x and y, we see that, ρ, is also given by

$$\left[1 + \left(\frac{dx}{dy} \right)^2 \right]^{3/2} \bigg/ \frac{d^2 x}{dy^2}.$$

This formula is a specially useful when the tangent is perpendicular to x-axis in which case $\dfrac{dx}{dy} = 0$.

Radius of Curvature—Parametric Equations:

$$x = f(t), y = F(t).$$

At point where $f'(t) \neq 0$, we have

$$\frac{dy}{dx} = \frac{F'(t)}{f'(t)}, \quad \frac{d^2 y}{dx^2} = \frac{f'(t) F''(t) - F'(t) f''(t)}{[f'(t)]^2} \cdot \frac{1}{f'(t)}.$$

Substituting these values in the formula (A) above, we get

$$\rho = \pm \frac{[f'^2(t) + F'^2(t)]^{3/2}}{f'(t) F''(t) - F'(t) f''(t)}$$ ...(B)

where the sign is $+$ or $-$ according as $f'(t)$ is positive or negative

Note. The formula (B) shows that the expression for ρ, will retain the same form for the point where $f'(t) = 0$ but $F'(t) \neq 0$.

Radius of Curvature—Polar Equations :

(*Bhopal 96; Banglore 2005*)

Let $r = f(\theta)$ be the given curve in polar co-ordinates. Now

$$x = r \cos \theta, \ y = r \sin \theta \text{ where } r = f(\theta),$$

may be thought of as the parametric cartesian equations of the curve, θ being the parameter. We have

$$\frac{dx}{d\theta} = r_1 \cos\theta - r\sin\theta, \quad \frac{dy}{d\theta} = r_1 \sin\theta + r\cos\theta$$

where $\quad r = f(\theta), r_1 = f'(\theta).$

Thus, $\quad y_1 = \dfrac{dy}{dx} = \dfrac{r_1 \sin\theta + r\cos\theta}{r_1 \cos\theta - r\sin\theta}.$

Again, $\quad y_2 = \dfrac{d^2 y}{dx^2} = \dfrac{d}{d\theta}\left(\dfrac{r_1 \sin\theta + r\cos\theta}{r_1 \cos\theta - r\sin\theta}\right) \cdot \dfrac{d\theta}{dx}$

$$= \frac{r^2 + 2r_1^2 - r r_2}{(r_1 \cos\theta - r\sin\theta)^3}.$$

Thus, $\quad \rho = \dfrac{(1 + y_1^2)^{3/2}}{y_2} = \dfrac{(r^2 + r_1^2)^{3/2}}{r^2 + 2r_1^2 - r r_2}$

where $\quad r_1 = f'(\theta), r_2 = f''(\theta).$

Radius of Curvature (Pedal Equations)

(Rohilkhand 1999; Gorakhpur 2003; Utkal 2003)

From the figure,

$$\psi = \theta + \phi$$

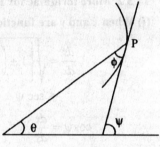

Fig. 14.4

Differentiating both sides with respect to s, we obtain

$$\frac{d\psi}{ds} = \frac{d\theta}{ds} + \frac{d\phi}{ds}$$

$$= \frac{d\theta}{ds} + \frac{d\phi}{dr} \cdot \frac{dr}{ds}$$

$\therefore \quad \dfrac{1}{\rho} = \dfrac{1}{r}\sin\phi + \cos\phi \cdot \dfrac{d\phi}{dr} \qquad [\because \dfrac{d\theta}{ds} = \dfrac{1}{r}\sin\phi \text{ and } \dfrac{dr}{ds} = \cos\phi]$

$\therefore \quad \dfrac{1}{\rho} = \dfrac{1}{r}\left(\sin\phi + r\cos\phi \cdot \dfrac{d\phi}{dr}\right)$

$$= \frac{1}{r}\frac{d}{dr}(r\sin\phi)$$

$$= \frac{1}{r}\frac{dp}{dr}$$

$\therefore \quad \rho = r\dfrac{dr}{dp}$

14.3.1 Radius of curvature when the equation of the curve is given in p and ψ or to prove that

$$\rho = p + \frac{d^2 p}{d\psi^2}$$

We have $\quad \dfrac{dp}{d\psi} = \dfrac{dp}{dr} \cdot \dfrac{dr}{ds} \cdot \dfrac{ds}{d\psi}$

$$= \dfrac{dp}{dr} \cdot \cos\phi \cdot \rho \qquad\qquad \left(\because \dfrac{dr}{ds} = \cos\phi \right)$$

$$= \dfrac{dp}{dr} \cos\phi \cdot r \dfrac{dr}{dp} \qquad\qquad \left(\because \rho = r \dfrac{dr}{dp} \right)$$

$$= r \cos\phi$$

Now again $p^2 + \left(\dfrac{dp}{d\psi} \right)^2 = r^2 \sin^2\phi + r^2 \cos^2\phi = r^2.$

Differentiating with respect to p, we obtain

$$2p + 2 \cdot \dfrac{dp}{d\psi} \cdot \dfrac{d^2 p}{d\psi^2} \cdot \dfrac{d\psi}{dp} = 2r \dfrac{dr}{dp}$$

or $\qquad p + \dfrac{d^2 p}{d\psi^2} = \rho.$

This is known as tangential polar formula.

14.3.2 More formulae for Radius of Curvature

(i) When x and y are functions of s.

$$\dfrac{ds}{dx} = \sqrt{\left\{ 1 + \left(\dfrac{dy}{dx} \right)^2 \right\}} = \sqrt{(1 + \tan^2\psi)}$$

$$= \sec\psi$$

$\therefore \qquad \cos\psi = \dfrac{dx}{ds}.$

Differentiating with respect to s, we obtain

$$-\sin\psi \cdot \dfrac{d\psi}{ds} = \dfrac{d^2 x}{ds^2}$$

or $\qquad -\sin\psi \cdot \dfrac{1}{\rho} = \dfrac{d^2 x}{ds^2}$

$\therefore \qquad \rho = -\dfrac{\sin\psi}{d^2 x / ds^2}$

while $\sin\psi = dy/ds$

Hence $\qquad \rho = -\dfrac{dy/ds}{\dfrac{d^2 x}{ds^2}}$

(ii) Similarly, we obtain

$$\rho = +\dfrac{dx/ds}{\dfrac{d^2 y}{ds^2}}$$

by taking $\quad \sin\psi = \dfrac{dy}{ds}.$

(iii) Now again, we know

$$\rho = -\frac{dy/ds}{d^2x/ds^2} = -\frac{\sin\psi}{d^2x/ds^2}$$

$$\therefore \qquad \frac{\sin\psi}{\rho} = \frac{d^2x}{ds^2} \qquad\qquad\qquad\qquad \text{...}(i)$$

and also $\qquad \dfrac{\cos\psi}{\rho} = \dfrac{d^2y}{ds^2} \qquad\qquad\qquad\qquad \text{...}(ii)$

Squaring and adding (i) and (ii)

we obtain $\qquad \dfrac{1}{\rho^2} = \left(\dfrac{d^2x}{ds^2}\right)^2 + \left(\dfrac{d^2y}{ds^2}\right)^2.$

EXAMPLES

1. *For the cycloid* $x = a(t + \sin t),\ y = a(1 - \cos t),$ *prove that*

$$\rho = 4a\cos\left(\frac{1}{2}t\right).$$

(Kumaon 2002; Utkal 2003; Vikram 97; Bhopal 2000;
Indore 99, 2000; Jabalpur 97; Sagar 98; Bilaspur 2000, 2001;
Devi Ahilya 2001; Banglore 2004)

We have $\qquad \dfrac{dx}{dt} = a(1+\cos t),\ \dfrac{dy}{dt} = a\sin t.$

$$\therefore \qquad \frac{dy}{dx} = \frac{a\sin t}{a(1+\cos t)} = \frac{2\sin t/2\,\cos t/2}{2\cos^2 t/2} = \tan t/2$$

$$\frac{d^2y}{dx^2} = \frac{1}{2}\sec^2\frac{t}{2}\cdot\frac{dt}{dx}$$

$$= \frac{1}{2}\sec^2\frac{t}{2}\cdot\frac{dt}{dx}\frac{1}{2a\cos^2\dfrac{t}{2}} = \frac{1}{4a}\cdot\frac{1}{\cos^4\dfrac{t}{2}}.$$

$$\therefore \qquad \left(\rho = \frac{[1+(dy/dx)^2]^{3/2}}{d^2y/dx^2} = \frac{\left(1+\tan^2\dfrac{1}{2}t\right)^{3/2}}{\dfrac{1}{4a}\cdot\dfrac{1}{\cos^4(t/2)}} = 4a\cos\frac{t}{2}.\right)$$

2. *For the curve* $r^m = a^m\cos m\theta,$ *prove that*

$$\rho = \frac{a^m}{(m+1)r^{m-1}}.$$

(Patna 2003)

We have $\qquad r^m = a^m\cos m\theta \Leftrightarrow m\log r = m\log a + \log\cos m\theta$

$$\Rightarrow \qquad \frac{m}{r}\cdot\frac{dr}{d\theta} = -m\frac{\sin m\theta}{\cos m\theta}$$

$$\Rightarrow \qquad r_1 = \frac{dr}{d\theta} = -r\tan m\theta$$

$$\Rightarrow \qquad r_2 = \frac{d^2 r}{d\theta^2} = -rm\sec^2 m\theta - \tan m\theta \frac{dr}{d\theta}$$

$$= -rm\sec^2 m\theta + r\tan^2 m\theta.$$

Hence $\qquad \rho = \dfrac{(r^2 + r^2 \tan^2 m\theta)^{3/2}}{r^2 + 2r^2 \tan^2 m\theta + r^2 m\sec^2 m\theta - r^2 \tan^2 m\theta}$

$$= \frac{r^3 \sec^3 m\theta}{r^2 \sec^2 m\theta + mr^2 \sec^2 m\theta}$$

$$= \frac{r}{(m+1)(\cos m\theta)} = \frac{1}{m+1} \cdot \frac{a^m}{r^{m-1}}.$$

3. *Show that the curvature of the point* $\left(\dfrac{3a}{2}, \dfrac{3a}{2}\right)$ *on the Folium* $x^3 + y^3 = 3\,axy$ *is* $-8\sqrt{2}/3a$.

(MDU Rohtak 2001)

Differentiating, we get

$$3x^2 + 3y^2 \frac{dy}{dx} = 3ay + 3ax \frac{dy}{dx} \qquad\qquad ...(i)$$

or $\qquad \dfrac{dy}{dx} = \dfrac{ay - x^2}{y^2 - ax}$

$\therefore \qquad \dfrac{dy}{dx}\bigg]_{\left(\frac{3a}{2}, \frac{3a}{2}\right)} = -1$

Again differentiating (i), we get

$$2x + 2y\left(\frac{dy}{dx}\right)^2 + y^2 \frac{d^2 y}{dx^2} = a\frac{dy}{dx} + a\frac{dy}{dx} + ax\frac{d^2 y}{dx^2}$$

Substituting $\dfrac{3a}{2}, \dfrac{3a}{2}, -1$ for x, y, $\dfrac{dy}{dx}$ respectively, we get

$$\frac{d^2 y}{dx^2}\bigg]_{\left(\frac{3a}{2}, \frac{3a}{2}\right)} = -\frac{32}{3a}$$

Hence the curvature at $\left(\dfrac{3a}{2}, \dfrac{3a}{2}\right) = \dfrac{d^2 y/dx^2}{\left[1 + \left(\dfrac{dy}{dx}\right)^2\right]^{3/2}} = \dfrac{-32/3a}{(2)^{3/2}} = \dfrac{-8\sqrt{2}}{3a}$

4. *Prove that the radius of curvature at any point of the catenary* $y = c\cosh\dfrac{x}{c}$ *varies as the square of the ordinate.*

(Kumaon 2001; Sagar 99, 2002; Jawaji 2001; Jabalpur 2001)

The given curve is

$$y = c\cosh\frac{x}{c}.$$

Differentiating it with respect to x, we have

$$\frac{dy}{dx} = c \cdot \sinh\frac{x}{c} \cdot \frac{1}{c} = \sinh\frac{x}{c}$$

and $\qquad \dfrac{d^2 y}{dx^2} = \cosh\dfrac{x}{c} \cdot \dfrac{1}{c}$

$$\therefore \qquad \rho = \frac{\left[1 + \left(\dfrac{dy}{dx}\right)^2\right]^{3/2}}{\dfrac{d^2 y}{dx^2}}$$

$$= \frac{c\left[1 + \sinh^2 \dfrac{x}{c}\right]^{3/2}}{\cosh \dfrac{x}{c}}$$

$$= c \cosh^2 x/c = c \cdot \frac{y^2}{c^2} = \frac{y^2}{c}$$

Hence $\rho \propto$ (ordinate)^2.

5. *If a curve is defined by the equation $x = f(t)$ and $y = \phi(t)$ prove that*

$$\rho = \frac{(x'^2 + y'^2)^{3/2}}{x' y'' - y' x''} , \qquad\qquad \textbf{\textit{(Bilaspur 2001; Garhwal 2001)}}$$

where dashes denote differentiations with respect to t.

We have $\qquad \dfrac{dx}{dt} = f'(t) = x'$ and $\dfrac{dy}{dt} = \phi'(t) = y'$

Now $\dfrac{dy}{dx} = \dfrac{dy/dt}{dx/dt} = \dfrac{y'}{x'}$ where y' and x' are functions of t.

$$\therefore \qquad \frac{d^2 y}{dx^2} = \frac{d}{dx}\left(\frac{y'}{x'}\right) = \frac{d}{dt}\left(\frac{y'}{x'}\right) \cdot \frac{dt}{dx}$$

$$= \frac{x' y'' - y' x''}{x'^2} \cdot \frac{1}{x'}$$

$$= \frac{x' y'' - y' x''}{x'^3}$$

$$\therefore \qquad \rho = \frac{\left[1 + \left(\dfrac{dy}{dx}\right)^2\right]^{3/2}}{d^2 y/dx^2} = \frac{\left(1 + \dfrac{y'^2}{x'^2}\right)^{3/2}}{\dfrac{x' y'' - y' x''}{x'^3}}$$

$$= \frac{(x'^2 - y'^2)^{3/2}}{x' y'' - y' x''}$$

6. *In the ellipse $\dfrac{x^2}{a^2} + \dfrac{y^2}{b^2} = 1$, show that the radius of curvature at an end of the major axis is equal to semi-latus rectum of the ellipse.*

$$\textbf{\textit{(Ravishankar 2000)}}$$

The given curve is

$$\frac{x^2}{a^2} + \frac{y^2}{b^2} = 1.$$

Differentiating it with respect to x, we get

$$\frac{2x}{a^2} + \frac{2y}{b^2} \cdot \frac{dy}{dx} = 0$$

or

$$\frac{dy}{dx} = -\frac{b^2 x}{a^2 y}$$

and

$$\frac{d^2 y}{dx^2} = -\frac{b^2}{a^2} \left[\frac{y - x\left(\dfrac{-b^2 x}{a^2 y}\right)}{y^2} \right]$$

$$= \frac{b^2}{a^4 y^3} \cdot (a^2 y^2 + b^2 x^2)$$

$$= -\frac{b^2}{a^4 y^3} \cdot a^2 b^2 \left(\frac{x^2}{a^2} + \frac{y^2}{b^2} \right)$$

$$= -\frac{b^4}{a^2 y^3}$$

Now,

$$\rho = \frac{\left[1 + \left(\dfrac{dy}{dx}\right)^2\right]^{3/2}}{d^2 y / dx^2}$$

$$= -\frac{\left(1 + \dfrac{b^4 x^2}{a^4 y^2}\right)^{3/2}}{b^4 / a^2 y^3} = -\frac{(a^4 y^2 + b^4 x^2)^{3/2}}{a^4 b^4}$$

Now, ρ at one end of major axis, $i.e.$, $(a, 0)$

$$= \frac{(a^4 \cdot 0 + b^4 \cdot a^2)^{3/2}}{a^4 b^4} = \frac{b^6 a^3}{a^4 b^4}$$

$$= \frac{b^2}{a} = \text{semi-latus rectum.}$$

7. *If CP and CD be a pair of conjugate semi-diameters of an ellipse, prove that the radius of curvature at P is* $\dfrac{CD^3}{ab}$ *, a and b being the lengths of the semi- axes of the ellipse.*

(*Ravishankar 1999*)

We know from coordinate geometry that if P and D are the ends of two conjugate diameters; then their coordinates shall be $(a \cos \phi, b \sin \phi)$ and $(-a \sin \phi, b \cos \phi)$ respectively where C is the centre of ellipse $\dfrac{x^2}{a^2} + \dfrac{y^2}{b^2} = 1$, whose coordinates are $(0, 0)$.

In example 6, we have just shown

$$\rho = (a^4 y^2 + b^4 x^2)^{3/2} / a^4 b^4$$

$$\therefore \qquad \rho \text{ at } P = (a^4 b^2 \sin^2 \phi + b^4 a^2 \cos^2 \phi)^{3/2} / a^4 b^4$$

$$= \frac{(a^2 \sin^2 \phi + b^2 \cos^2 \phi)^{3/2}}{ab}$$

But
$$CD = \sqrt{\left\{ (-a\sin\phi - 0)^2 + (b\cos\phi - 0)^2 \right\}}$$

$$= \sqrt{(a^2 \sin^2\phi + b^2 \cos^2\phi)}$$

Therefore ρ at $P = \dfrac{\left\{ (a^2 \sin^2\phi + b^2 \cos^2\phi)^{1/2} \right\}^3}{ab}$

$$= \frac{CD^3}{ab}.$$

8. *Find the value of ρ for $r = a\, e^{\theta\cot\alpha}$ and show that the radius of curvature subtends a right angle at the pole.* **(Sagar 98)**

We are given that

$$r = a\, e^{\theta\cot\alpha}$$

$$\frac{dr}{d\theta} = a\, e^{\theta\cot\alpha}.\cot\alpha = r\cot\alpha$$

and
$$\tan\phi = r\,/\,\frac{dr}{d\theta} = \frac{r}{r\cot\alpha} = \tan\alpha$$

$\therefore \qquad \phi = \alpha,$

also
$$p = r\sin\phi = r\sin\alpha.$$

Now differentiating with respect to r, we have

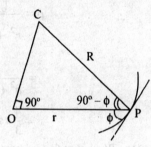

Fig. 14.5

$$\frac{dp}{dr} = \sin\alpha$$

But
$$\rho = \frac{r}{\dfrac{dp}{dr}} = \frac{r}{\sin\alpha} = r\operatorname{cosec}\alpha \qquad\qquad ...(i)$$

Now from $\triangle COP$

$$\frac{CP}{\sin COP} = \frac{r}{\sin PCO},$$

$$\frac{\rho}{\sin 90°} = \frac{r}{\sin\left\{ 180° - 90° - (90° - \phi) \right\}}$$

$$\rho = \frac{r}{\sin\phi} = \frac{r}{\sin\alpha} \qquad\qquad ...(ii)$$

The results (i) and (ii) are the same. Hence CP subtends a right angle at pole.

9. *Show that for any curve*

$$\frac{r}{\rho} = \sin\phi\left(1 + \frac{d\phi}{d\theta} \right).$$

We have

R. H. S. $= \sin\phi\left(1 + \dfrac{d\phi}{d\theta} \right)$

$= r\,\dfrac{d\theta}{ds}\left(1 + \dfrac{d\phi}{d\theta} \right)$

$$= r \left\{ \frac{d\theta}{ds} + \frac{d\theta}{ds} \cdot \frac{d\phi}{d\theta} \right\}$$

$$= r \left\{ \frac{d\theta}{ds} + \frac{d\phi}{ds} \right\}$$

$$= r \frac{d}{ds} (\theta + \phi) = r \frac{d\psi}{ds} = \frac{r}{\rho}.$$

10. *Find the radius of curvature for the curve* $r = a \, (1 - \cos \theta)$.

(Garhwal 96, 2000; Gorakhpur 2003; Ravishanker 98; Indore 97; Vikram 98; Jiwaji 99; Bilaspur 98, 2000)

We have $\dfrac{dr}{d\theta} = a \sin \theta, \; \dfrac{d^2 r}{d\theta^2} = a \cos \theta$

$\therefore$

$$\rho = \frac{\left\{ r^2 + \left(\dfrac{dr}{d\theta} \right)^2 \right\}^{3/2}}{r^2 + 2 \left(\dfrac{dr}{d\theta} \right)^2 - r \dfrac{d^2 r}{d\theta^2}}$$

$$= \frac{\left\{ a^2 (1 - \cos \theta)^2 + a^2 \sin^2 \theta \right\}^{3/2}}{a^2 (1 - \cos \theta)^2 + 2 \, a^2 \sin^2 \theta - a^2 \cos \theta \, (1 - \cos \theta)}$$

$$= \frac{4}{3} a \sin \frac{\theta}{2} = \frac{2}{3} \sqrt{(2 a r)}$$

11. *Show that for any curve* $r = f(\theta)$ *the curvature is given by*

$$\left(u + \frac{d^2 u}{d\theta^2} \right) \sin^3 \phi, \; where \; u = 1/r.$$

We have $$\rho = \frac{\left[u^2 + \left(\dfrac{du}{d\theta} \right)^2 \right]^{3/2}}{u^3 \left[u + \dfrac{d^2 u}{d\theta^2} \right]}$$

$\therefore$

$$\frac{1}{\rho} = \frac{u^3 \left(u + \dfrac{d^2 u}{d\theta} \right)}{\left[u^2 + \left(\dfrac{du}{d\theta} \right)^2 \right]^{3/2}}$$

$$= \frac{u^3}{\left[u^2 + \left(\dfrac{du}{d\theta} \right)^2 \right]^{3/2}} \left[u + \dfrac{d^2 u}{d\theta^2} \right]$$

$$= \frac{1/r^3}{\left[\dfrac{1}{r^2} + \dfrac{1}{r^4} \left(\dfrac{dr}{d\theta} \right)^2 \right]^{3/2}} \left[u + \dfrac{d^2 u}{d\theta^2} \right]$$

$$= \frac{1/r^3}{1/p^3} \left[u + \frac{d^2 u}{d\theta^2} \right] = \frac{p^3}{r^3} \left[u + \frac{d^2 u}{d\theta^2} \right]$$

$$= \frac{r^3 \sin^3 \phi}{r^3} \left[u + \frac{d^2 u}{d\theta^2} \right]$$

$$= \left[u + \frac{d^2 u}{d\theta^2} \right] \sin^3 \phi .$$

12. *Prove that for any curve* $\dfrac{1}{\rho} = \dfrac{d}{dx} \left(\dfrac{dy}{ds} \right).$ **(Rohilkhand 2001, 2004)**

We are given that

$$\frac{d}{dx} \left(\frac{dy}{ds} \right) = \frac{d}{dx} (\sin \psi)$$

$$= \frac{d}{d\psi} (\sin \psi) . \frac{d\psi}{dx}$$

$$= \cos \psi . \frac{d\psi}{ds} . \frac{ds}{dx}$$

$$= \frac{d\psi}{ds} = \frac{1}{\rho} .$$

13. *Show that for the curve* $s^2 = 8\,a\,y,\ \rho = 4a \sqrt{\left(1 - \dfrac{y}{2a} \right)}.$

The given curve is

$$s^2 = 8\,a\,\text{y}.$$

Differentiating with respect to s, we get

$$2\,s = 8\,a . \frac{dy}{ds}$$

But $\dfrac{dy}{ds} = \sin \psi$

$\therefore$ $2\,s = 8\,a \sin \psi$

or $s = 4\,a \sin \psi$

Now again differentiating with respect to ψ, we have

$$\frac{ds}{d\psi} = 4a \cos \psi$$

$\therefore$ $\rho = 4\,a \cos \psi$

$$= 4\,a \sqrt{(1 - \sin^2 \psi)}$$

$$= 4\,a \sqrt{\left(1 - \frac{s^2}{16a^2} \right)} = 4\,a \sqrt{\left(1 - \frac{8\,a\,y}{16\,a^2} \right)}$$

$$= 4\,a \sqrt{\left(1 - \frac{y}{2\,a} \right)}.$$

EXERCISES

1. Find the radius of curvature at any point on the curves:

   (i) $y = c \cosh x/c$ (catenary),   (*Kumaon 2001*)

   (i) $x = a(\cos t + t \sin t), y = a(\sin t - t \cos t)$,

   (iii) $x = (a \cos t)/t, \ y = (a \sin t)/t$.

2. Show that radius of curvature at any point of the astroid $x = a \cos^3\theta, y = a \sin^3\theta$ is equal to three times the length of the perpendicular from the origin to the tangent.

   (*MDU Rohtak 1999; Banglore 2006*)

3. Show that for the curve $x = a \cos \theta (1 + \sin \theta), y = a \sin \theta (1 + \cos \theta)$, the radius of curvature is a, at the point for which the value of the parameter θ is $-\pi/4$.

4. Show that the radius of curvature at any point of the curve

$$x = t - c \ \sinh \frac{t}{c} \ \cosh \frac{t}{c}, y = 2c \cosh \frac{t}{c}.$$

   is $-2c \ \cosh^2\left(\dfrac{t}{c}\right) \sinh\left(\dfrac{t}{c}\right)$,

   where t is the parameter.

5. Show that the radius of curvature at a point of the curve

$$x = ae^\theta (\sin\theta - \cos\theta), \ y = ae^\theta (\sin \theta + \cos \theta)$$

   is twice the distance of the tangent at the point from the origin.

6. Prove that the radius of curvature at the point $(-2a, 2a)$ on the curve $x^2 y = a(x^2 + y^2)$ is, $-2a$.

7. Find the radii of curvature of the curve $y^2(a - x) = x^2(a + x)$ at the origin.

8. Find the radius of curvature for $\sqrt{(x/a)} - \sqrt{(y/b)} = 1$ at the points where it touches the co-ordinate axes.

9. Show that the ratio of the radii of curvature of the curves $xy = a^2, x^3 = 3a^2y$, at points which have the same abscissae varies as the square root of the ratio of the ordinates.

10. Show that the radius of curvature at each point of the curve $x = a\left(\cos t + \log \tan\frac{1}{2}t\right)$,

    $y = a \sin t$, is inversely proportional to the length of the normal intercepted between the point on the curve and the x-axis.

11. Show that $3\sqrt{3}/2$ is the least value of $|\rho|$ for $y = \log x$.

12. Show that the radius of curvature of the curve given by

$$x^2 y = a(x^2 + a^2/\sqrt{5})$$

    is least for the point $x = a$ and its value there is $9a/10$.

13. Prove that for the ellipse

$$\frac{x^2}{a^2} + \frac{y^2}{b^2} = 1, \ \rho = \frac{a^2 b^2}{p^3};$$

    p, being the perpendicular from the centre on the tangent at the point (x, y).

   (*Reewa 96, 97, 98; Sagar 99; Kumaon 96, 2005; Garhwal 99, 2002; Indore 98; Jabalpur 99: Jiwaji 97, Avadh 2005*)

14. The tangents at two points P, Q on the cycloid $x = a(\theta - \sin\theta)$, $y = a(1 - \cos\theta)$, are at right angles; show if ρ_1, ρ_2 be the radii of curvature at these points, then $\rho_1^2 + \rho_2^2 = 16a^2$.

15. Find the points on the parabola $y^2 = 8x$ at which the radius of curvature is $7\dfrac{13}{16}$.

16. (a) Prove that, if ρ be the radius of curvature at any point P on the parabola $y^2 = 4ax$ and S be its focus, then, ρ^2 varies as $(SP)^3$.

(Kumaon 2003)

(b) ρ_1, ρ_2 are the radii of curvature at the extremities of a focal chord of a parabola whose semi-latus rectum is l; prove that

$$(\rho_1)^{-2/3} + (\rho_2)^{-2/3} = (l)^{-2/3}$$

17. Show that for the curve $x = \dfrac{3}{2}a(\sinh u \cosh u + u)$, $y = a\cosh^2 u$, if the normal at $P(x, y)$ meets the axis of x in G, the radius of curvature at P is equal to $3\,PG$.

18. Find the radius of curvature at the point of the curve $r = a\cos n\theta$. Also show that at the point where $r = a$, its value is $a/(1 + n^2)$.

19. Find the radius of curvature at the point (r, θ) on each of the following curves:

   (i) $r = a$,

   (ii) $r\theta = a$.

   (iii) $r^n = a^n \sin n\theta$,

   (iv) $\theta = \dfrac{\sqrt{(r^2 - a^2)}}{a} - \cos^{-1}\dfrac{a}{r}$,

   (v) $r(1 + \cos\theta) = a$.

(MDU Rohtak 2000)

20. Find the radius of curvature of the curve $r = a\sin n\theta$ at the origin.

21. Prove that for the cardioide $r = a(1 + \cos\theta)$, ρ^2/r is constant.

(Kumaon 98; Garhwal 98, 2004; Rohilkhand 2000; Banglore 2004; Bilaspur 96; Avadh 2003)

22. Show that at the points in which the Archimedean spiral $r = a\theta$ intersects the hyperbolical spiral $r\theta = a$, their curvatures are in the ratio $3 : 1$. *(Ravishankar 2000, Garhwal 2005)*

23. If ρ_1, ρ_2 be the radii of curvature at the extremities of any chord of the cardioide $r = a(1 + \cos\theta)$, which passes through the pole, then,

$$\rho_1^2 + \rho_2^2 = 16a^2/9.$$ *(Rewa 97; Vikram 2002)*

24. Find the radius of curvature of the curve $r = a(1 + \cos\theta)$ at the point where the tangent is parallel to the initial line.

25. A line is drawn through the origin meeting the cardioide $r = a(1 - \cos\theta)$ in the points P, Q and the normals at P, Q meet in C; show that the radii of curvatures at the points P and Q are proportional to PC and QC.

26. Prove that for the curve $r^2 = a^2 \sin 2\theta$, the tangent turns three times as fast as the radius vector and that the curvature varies as the radius vector.

27. Show for the curve $s = f(x)$,

$$\rho = \dfrac{\left[\left(\dfrac{ds}{dx}\right)^2 \sqrt{\left\{\left(\dfrac{ds}{dx}\right)^2 - 1\right\}}\right]}{\dfrac{d^2x}{ds^2}}.$$

28. For any curve, prove that

$$\frac{d^2r}{ds^2} = \frac{\sin^2\theta}{r^2} - \frac{\sin\phi}{\rho}.$$

29. Find the radius of curvature of $\sqrt{x} + \sqrt{y} = \sqrt{a}$, at the point where the line $y = x$ cuts it.

(Garhwal 97)

30. If ρ_1, ρ_2 be the radii of curvature at the extremities of two conjugate diameters on an ellipse, prove that

$$(\rho_1^{2/3} + \rho_2^{2/3})\, a^{2/3}\, b^{2/3} = a^2 + b^2.$$

(Rohilkhand 1999; Gorakhpur 2000; Bilaspur 99)

31. Show that for the curve $y = \dfrac{ax}{a+x}$

$$\left(\frac{2\rho}{a}\right)^{2/3} = \left(\frac{y}{x}\right)^2 + \left(\frac{x}{y}\right)^2.$$

(Kumaon 2004)

32. The coordinates of a point on a curve are given by

$$x = a\sin t - b\sin\frac{at}{b},\ y = a\cos t - b\cos\frac{at}{b}.$$

Show that the equation of the tangent at the point whose parameter is t, is

$$x\cos\frac{a+b}{2b}t - y\sin\frac{a+b}{2b}t + (a+b)\sin\frac{a-b}{2b}t = 0$$

and the radius of curvature at this point is

$$\frac{4ab}{a+b}\sin\frac{a-b}{2b}t.$$

33. Prove that for the curve $s = m(\sec^3 x - 1)$, $\rho = 3m\tan\psi\sec^3\psi$, and hence show that $3m\,.$
$$\frac{dy}{dx}\cdot\frac{d^2y}{dx^2} = 1\,.$$

34. Show that for the curve in which $s = ce^{x/e}$, $c\rho = s(s^2 - c^2)^{1/2}$. **(Kumaon 97; Bhopal 97)**

ANSWERS

1. (i) y^2/c. (ii) at. (iii) $-a(t^2+1)^{3/2}/t^4$.

7. $4\sqrt{2a}$. 8. $2a^2/b,\ 2b^2/a$.

15. $\left(\dfrac{9}{8}, \pm 3\right)$. 18. $\dfrac{(r^2 + a^2 n^2 - n^2 r^2)^{3/2}}{r^2 - r^2 n^2 + 2a^2 n^2}$.

19. (i) $\dfrac{a(1+\theta^2)^{3/2}}{(2+\theta^2)}$. (ii) $\dfrac{a(1+\theta^2)^{3/2}}{\theta^4}$. (iii) $a^n/(n+1)r^{n-1}$.

    (iv) $\sqrt{r^2 - a^2}$. (v) $\sqrt{(8r^3/a)}$.

20. $an/2$. 24. $\dfrac{2\sqrt{3}}{3}a$. 29. $a/\sqrt{2}$.

14.4. Newtonian Method. *If a curve passes through the origin and the axis of x is tangent at the origin, then*

$$\lim \frac{x^2}{2y}, \text{ as } x \to 0$$

gives the radius of curvature at the origin.

Here we obtain the values of y_1 and y_2 at the origin.

$$y_1(0) = \left(\frac{dy}{dx}\right)_{(0/0)} = 0.$$

Now, $x^2/2y$, assumes the indeterminate form $0/0$ as $x \to 0$.

$$\lim_{x \to 0} \frac{x^2}{2y} = \lim_{x \to 0} \frac{2x}{2y_1} \qquad \left(\frac{0}{0}\right)$$

$$= \lim_{x \to 0} \frac{1}{y_2} = \frac{1}{y_2(0)}.$$

Also, from the formula $(,A)$ of § 14.3, we have at the origin

$$\rho = \frac{(1 + 0)^{3/2}}{y_2(0)} = \frac{1}{y_2(0)}.$$

Thus at the origin where x- axis is a tangent,

$$\rho = \lim_{x \to 0} \left(\frac{x^2}{2y}\right).$$

It can similarly be shown *that at the origin where y – axis is a tangent*

$$\rho = \lim_{x \to 0} \left(\frac{y^2}{2x}\right).$$

These two formulae are due to Newton.

14. 4.1 Generalised Newtonian Formula. If a curve passes through the origin and x-axis is the tangent at the origin, we have

$$\lim \frac{x^2 + y^2}{2y} = \lim \left(\frac{x^2}{2y} + \frac{y}{2}\right)$$

$$= \lim \frac{x^2}{2y}$$

$$= \rho, \text{ at the origin.}$$

Here, $x^2 + y^2 = OP^2$, is the square of the distance of any point $P(x, y)$ on the curve form the origin O and, y, is the distance of the point P from the tangent x - axis at O.

Interpreted in general terms this conclusion can be stated as follows: (*see* Fig. **14.7**)

If OT be the tangent at any given point O of a curve, and PM, the length of the perpendicular drawn from any point P to the tangent at O, then the *radius of curvature at O*

$$= \lim \frac{OP^2}{2PM}$$

when the point P tends to O as its limit.

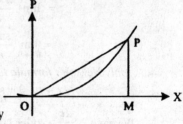

Fig. 14.6

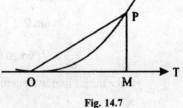

Fig. 14.7

EXAMPLES

1. *Find the radius of curvature at the origin of the curve:*
$$x^3 - 2x^2y + 3xy^2 - 4y^3 + 5x^2 - 6xy + 7y^2 - 8y = 0.$$
It is easy to see that x- axis is the tangent at the origin (Refer Chap. 16) Dividing by y, we get
$$x \cdot \frac{x^2}{y} - 2x^2 + 3xy - 4y^2 + 5\frac{x^2}{y} - 6x + 7y - 8 = 0$$

Let $x \to 0$ so that $\lim\limits_{x \to 0} \left(\dfrac{x^2}{y} \right) = 2\rho$

$\therefore \qquad 0.2\,\rho + 5.2\,\rho - 8 = 0$

$\Rightarrow \qquad \rho = 4/5$

2. *Find the radius of curvature at the origin for the curve*
$$x^3 + y^3 - 2x^2 + 6y = 0.$$
The curve passes through $(0, 0)$ and x - axis is tangent to the curve at $(0, 0)$.

Hence $\qquad \lim\limits_{\substack{x \to 0 \\ y \to 0}} \dfrac{x^2}{2y} = 0 \qquad$ or $\qquad \lim\limits_{\substack{x \to 0 \\ y \to 0}} \left(\dfrac{x^2}{y} \right) = 2\rho.$

Now dividing the curve by y, we obtain
$$x \cdot \frac{x^2}{y} + y^2 - \frac{2x^2}{y} + 6 = 0$$
Taking limit, we get
$$0.2\,\rho + 0 - 4\,\rho + 6 = 0$$

$\therefore \qquad\qquad\qquad \rho = \dfrac{3}{2}.$

3. *Find ρ at the pole for the curve $r = a \sin n\theta$.* **(Guwahati 2005)**
The curve is $r = a \sin n\theta$. Here r and θ are 0. Hence initial line is tangent to the curve at origin.

We know, $\qquad \rho = \lim\limits_{\theta \to 0} \dfrac{r}{2\theta}$

$$= \lim\limits_{\theta \to 0} \frac{a \sin n\theta}{2\,\theta}$$

$$= \lim\limits_{\theta \to 0} \frac{\sin n\theta}{n\,\theta} \cdot \frac{na}{2} = \frac{na}{2}.$$

4. *Apply Newton's formula to find the radius of curvature at the origin for the cycloid*
$$x = a(\theta + \sin\theta), \ y = a(1 - \cos\theta).$$

We have $\qquad \dfrac{dx}{d\theta} = a(1 + \cos\theta), \ \dfrac{dy}{d\theta} = a\sin\theta$

$\therefore \qquad \dfrac{dy}{dx} = \dfrac{dy/d\theta}{dx/d\theta} = \dfrac{a\sin\theta}{a(1 + \cos\theta)}$

$$= \tan\theta/2$$

$\therefore \qquad \dfrac{dy}{dx} = 0$, when $\theta = 0$

Hence initial line is tangent to origin.

Then $\qquad \rho = \lim\limits_{\theta \to 0} \dfrac{x^2}{2y} = \lim\limits_{\theta \to 0} \dfrac{a^2(\theta + \sin\theta)^2}{2a(1 - \cos\theta)}$

$$= \lim\limits_{\theta \to 0} \dfrac{a}{2}\left[\dfrac{2(\theta + \sin\theta)(1 + \cos\theta)}{\sin\theta}\right]$$

$$= \lim\limits_{\theta \to 0} \dfrac{a}{2}\left[\dfrac{2\cdot(1 + \cos\theta)^2 - \sin\theta\,(\theta + \sin\theta)}{\cos\theta}\right]$$

$$= 4a. \qquad\qquad\qquad\qquad\qquad\qquad\qquad \textbf{[By Hospital's rule]}$$

EXERCISES

Find ρ at origin of the curves:

1. $2x^4 + 3y^4 + 4x^2y + xy - y^2 + 2x = 0$

2. $y = x^4 - 4x^3 - 18x^2$

3. $3x^3 + y^3 + 5y^2 + 3yx^2 + 2x = 0$

4. $x^4 - y^4 + x^3 - y^3 + x^2 - y^2 + y = 0.$

5. Find the radius of curvature of the cardioid $r = a(1 - \cos\theta)$ at the pole (origin).

ANSWERS

1. 1;   **2.** $\dfrac{1}{36}$;   **3.** $\dfrac{1}{5}$;   **4.** $\dfrac{1}{2}$;   **5.** 0.

14.5. Centre of Curvature

The centre of curvature at any point P of a curve is the point which lies on the positive direction of the normal at P and is at a distance, ρ, from it.

The distance ρ, must be taken with a proper sign.

The positive direction of the normal is obtained by rotating the positive direction of the tangent (the positive direction of the tangent is the one in which x increases) through $\pi/2$ in the anti-clockwise direction.

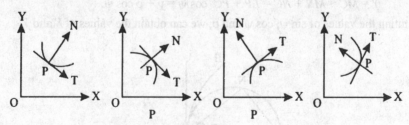

Fig. 14.8

From an examination of the above figures we see that the *centre of curvature at any point of a curve lies on the side towards which the curve is concave.*

To find the co-ordinates of the centre of curvature for any point $P(x, y)$ of the curve $y = f(x)$.

Let the positive direction of the tangent make angle, ψ, with x-axis so that the positive direction of the normal makes angle $\psi + \pi/2$ with x-axis.

The equation of the normal at (x, y) is

$$\dfrac{X - x}{\cos(\psi + \pi/2)} = \dfrac{Y - y}{\sin(\psi + \pi/2)} = r,$$

$$\Rightarrow \qquad \frac{X - x}{-\sin \psi} = \frac{Y - y}{\cos \psi} = r,$$

where $X,\ Y$ are the co-ordinates of any point on the normal and, r the variable distance of the variable point $(X,\ Y)$ from $(x,\ y)$.

Thus, the co-ordinates $(X,\ Y)$ of the point on the normal at a distance, r from $(x,\ y)$ are $(x - r \sin \psi, y + r \cos \psi)$.

For the centre of curvature, we have $r = \rho$. Hence, if $(x,\ y)$ be the centre of curvature, we have
$$X = x - \rho \sin \psi, Y = y + \rho \cos \psi.$$

But we know that

$$\sin \psi = \frac{y_1}{\sqrt{[1 + y_1^2]}}, \quad \cos \psi = \frac{1}{\sqrt{[1 + y_1^2]}}, \quad \rho = \frac{(1 + y_1^2)^{3/2}}{y_2}.$$

$$\therefore \qquad X = x - \frac{y_1(1 + y_1^2)}{y_2}; \ Y = y + \frac{1 + y_1^2}{y_2}.$$

Another method. If C be the centre of curvature for P, we have $PC = \rho$.

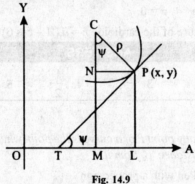

Fig. 14.9

Also $\angle NCP = \angle XTP = \psi.$

$$\therefore \qquad X = OM = OL - ML = OL - NP = x - \rho \sin \psi.$$
$$Y = MC = MN + NC = LP + PC \cos \psi = y + \rho \cos \psi.$$

Substituting the values of $\sin \psi$, $\cos \psi$ and ρ, we can obtain the values of X and $Y.$

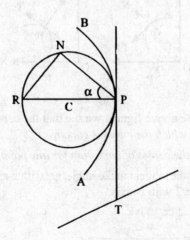

Fig. 14.10

14.6. Chord of curvature

If there is any point P at the given curve and a circle having the radius of curvature ρ is drawn passing through P, then any chord PN is called the **chord of curvature**. If C be the centre of the circle then

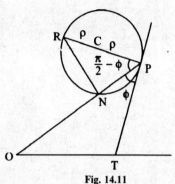

$$PN = R\,P \cos\,R\,PN = 2\,\rho \cos RPN$$
$$= 2\rho \cos \alpha.$$

Different cases of chord of curvature

Case I. Chord of curvature through pole (origin).

In the figure PN is the required chord of curvature

$$PN = RP \, \cos \, RPN$$

$$= 2\,\rho \cos\left(\frac{\pi}{2} - \phi\right)$$

Hence *chord of curvature through pole* $= 2\,\rho \sin \phi$.

Case II. *Chord of curvature through the pole for the curve* $p = f(r)$.

Fig. 14.11

We know $p = r \sin \phi$; $\quad \therefore \quad \sin \phi = \dfrac{p}{r}$

Hence chord of curvature $= 2\,\rho \cdot \dfrac{p}{r}$

$$= 2 \cdot r\,\frac{dr}{dp} \cdot \frac{p}{r} = 2p\,\frac{dr}{dp}$$

$$= 2f(r)\,\frac{dr}{dp}, \qquad i.e., \qquad \frac{2f(r)}{f'(r)} \qquad\qquad \textbf{\textit{(Jabalpur 96, 98)}}$$

Case III. Chord of curvature perpendicular to radius vector.

Here $\qquad PN$ = chord of curvature perpendicular to radius vector

$$= RP \, \cos \, RPN$$

$$= 2\,\rho \cos \phi$$

Case IV. Chord of curvature parallel to x - axis *and* y-axis.

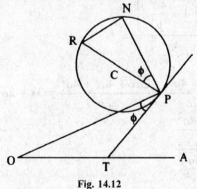

Fig. 14.12

Chord of curvature parallel to x-axis

$$= PN = P\,R \cos RPN$$

$$= 2\rho \, \cos\left(\frac{\pi}{2} - \psi\right) = 2\rho \sin \psi$$

and chord of curvature parallel to y-axis

$$= PN' = RP \cos RPN' = 2\,\rho \cos \psi$$

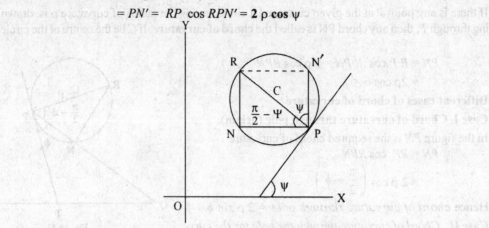

Fig 14.13

Circle of Curvature.

The circle with radius equal to radius of curvature ρ and its centre the centre of curvature (X, Y) is called the *circle of curvature.*

Then the equation of the circle of curvature is

$$(x - X)^2 + (y - Y)^2 = \rho^2.$$

EXAMPLES

1. *Find the co-ordinates of the centre of curvature at a point (x, y) of the parabola $y^2 = 4\,ax$.*

Differentiating, we get

$$2y\frac{dy}{dx} = 4a$$

$$\Rightarrow \qquad \frac{dy}{dx} = \frac{2a}{y} \Rightarrow \frac{d^2y}{dx^2} = -\frac{2a}{y^2} \cdot \frac{dy}{dx} = \frac{-4a^2}{y^3}.$$

If (X, Y) be the centre of curvature, we have

$$X = x - \frac{\dfrac{2a}{y}\left(1 + \dfrac{4a^2}{y^2}\right)}{-\dfrac{4a^2}{y^3}}$$

$$= x + \frac{y^2 + 4a^2}{2a} = \frac{2ax + 4ax + 4a^2}{2a} = 3x + 2a \qquad\qquad ...(i)$$

$$Y = y + \frac{1 + \dfrac{4a^2}{y^2}}{-\dfrac{4a^2}{y^3}}$$

$$= y - \frac{y(y^2 + 4a^2)}{4a^2} = \frac{-y^3}{4a^2} = \mp\frac{(4ax)^{3/2}}{4a^2} = \mp\frac{2x^{3/2}}{a^{1/2}} \qquad\qquad ...(ii)$$

2. *Find the centre of curvature of the four cusped hypocycloid*

$$x = a\cos^3\theta,\ \ y = a\sin^3\theta,\ i.e.,\ x^{2/3} + y^{2/3} = a^{2/3}.$$

(Guwahati 2005)

We have $x = a \cos^3\theta$, then $\dfrac{dx}{d\theta} = -3a \cos^2\theta \sin\theta$

$y = a \sin^3\theta$, then $\dfrac{dy}{d\theta} = -3a \sin^2\theta \cos\theta$

Then $y_1 = \dfrac{dy}{dx} = \dfrac{dy/d\theta}{dx/d\theta} = -\tan\theta$

and $y_2 = \dfrac{d^2y}{dx^2} = -\sec^2\theta \dfrac{d\theta}{dx}$

$= \dfrac{1}{3a} \sec^4\theta \,\operatorname{cosec}\theta$.

Now, if (X, Y) be the coordinates of the centre of curvature, then

$$X = x - \frac{y_1(1 + y_1^2)}{y_2} = a \cos^3\theta + \frac{\tan\theta(1 + \tan^2\theta)}{\dfrac{1}{3a}\sec^4\theta.\operatorname{cosec}\theta}$$

$$= a \cos^3\theta + 3a \sin^2\theta \cos\theta$$

and $$Y = y + \frac{1 + y_1^2}{y_2} = a \sin^3\theta + \frac{(1 + \tan^2\theta)3a}{\sec^4\theta . \operatorname{cosec}\theta}$$

$$= a \sin^3\theta + 3a \cos^2\theta \sin\theta.$$

3.   *Prove that the coordiantes of the centre of curvature at any point (x, y) can be expressed in the*

form $x - \dfrac{dy}{d\psi}$ *and* $y + \dfrac{dx}{d\psi}$.

We know if (X, Y) be the centre of curvature of any point (x, y) of a curve then

$$X = x - \frac{\left[1 + \left(\dfrac{dy}{dx}\right)^2\right] \cdot \dfrac{dy}{dx}}{\dfrac{d^2y}{dx^2}}$$

$$Y = \frac{y + \left\{1 + \left(\dfrac{dy}{dx}\right)^2\right\}}{\dfrac{d^2y}{dx^2}}$$

Also we know $\dfrac{dy}{dx} = \tan\psi$ and

$$\rho = \frac{\left[1 + \left(\dfrac{dy}{dx}\right)^2\right]^{3/2}}{\dfrac{d^2y}{dx^2}} = \frac{\sec^3\psi}{d^2y/dx^2}$$

or $\dfrac{d^2y}{dx^2} = \dfrac{1}{\rho}\sec^3\psi$

∴ $$X = x - \frac{(1 + \tan^2\psi)\tan\psi}{\left(\dfrac{1}{\rho}\right)\sec^3\psi}$$

$$= x - \rho \sin \psi = x - \frac{ds}{d\psi} \cdot \frac{dy}{ds} \qquad\qquad \left[\because \rho = \frac{ds}{d\psi} \text{ and } \sin\psi = \frac{dy}{ds} \right]$$

$$= x - \frac{dy}{d\psi}$$

Similarly,

$$Y = y + \frac{1 + \tan^2 \psi}{\left(\dfrac{1}{\rho}\right) \sec^3 \psi} = y + \rho \cos \psi$$

$$= y + \frac{ds}{d\psi} \cdot \frac{dx}{ds} = y + \frac{dx}{d\psi}$$

4. *Prove that the points on the curve $r = f(\theta)$, the circle of curvature at which passes through the origin are given by the equation*

$$f(\theta) + f''(\theta) = 0.$$

Let $P(r, \theta)$ be a point on the given curve, the circle of curvature, with centre C, at which passes through the origin O.

Let C be the centre of curvature. Join OP and let PC produced meet the circle of curvature at P in D. Let PT be the tangent to the curve at P, then

Fig. 14.14

$$\angle OPT = \phi = \angle PDO$$

Now, $OP =$ radius vector r and $PD = 2\rho$.

Hence in $\triangle POD$, we get $PO = PD \sin\phi$

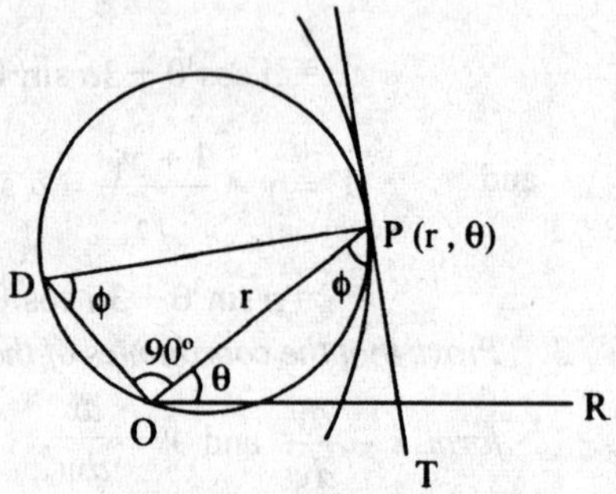

or $\qquad\qquad r = 2\rho \sin\phi \qquad\qquad\qquad ...(i)$

Also $\qquad \tan\phi = r \dfrac{d\theta}{dr} = \dfrac{r}{r_1}, \qquad$ where $r_1 = \dfrac{dr}{d\theta}$

$\therefore \qquad\qquad \sin\phi = \dfrac{r}{(r^2 + r_1^2)^{1/2}} \qquad\qquad\qquad ...(ii)$

Also $\qquad\qquad \rho = \dfrac{(r^2 + r_1^2)^{3/2}}{r^2 + 2r_1^2 - r r_2} \qquad\qquad\qquad ...(iii)$

Hence from (i), (ii) and (iii), we get

$$r = \frac{2(r^2 + r_1^2)^{3/2}}{r^2 + 2r_1^2 - r r_2} \cdot \frac{r}{\sqrt{(r^2 + r_1^2)}}$$

$$= \frac{2r(r^2 + r_1^2)}{r^2 + 2r_1^2 - r r_2}$$

or $\qquad r^2 + 2r_1^2 - r r_2 = + 2r^2 + 2r_1^2$

or $\qquad r^2 + r r_2 = 0$

or $\qquad r + r_2 = 0, \qquad$ where $r = f(\theta)$

or $\qquad f(\theta) + f''(\theta) = 0$

5. *Show that the chord of curvature through the pole of the equiangular spiral* $r = a\, e^{\theta \cot \alpha}$ *is* $2r$.

We have

$$r = a\, e^{\theta \cot \alpha}$$

$$\therefore \frac{dr}{d\theta} = a \cot \alpha \,.\, e^{\theta \cot \alpha}$$

$$= r \cot \alpha$$

Now, $\tan \phi = r \dfrac{d\theta}{dr} = \dfrac{r}{r \cot \alpha} = \tan \alpha$

$\therefore \qquad \phi = \alpha,$

Hence $p = r \sin \phi = r \sin \alpha.$

or $\dfrac{dp}{dr} = \sin \alpha$

$\therefore \qquad \rho = r \dfrac{dr}{dp} = \dfrac{r}{\sin \alpha}$

$\therefore$ But the chord of curvature through the pole

$$= 2\rho \sin \phi = \frac{2r}{\sin \alpha} \,.\, \sin \alpha = 2r \,.$$

6. *Show that the chord of curvature parallel to the axis of y for the curve* $y = c \cosh \dfrac{x}{c}$ *is double the ordinate whereas the chord parallel to x-axis is of length* $c \sinh \dfrac{2x}{c}$.

(Gorakhpur 2000; Bilaspur 97)

We have

$$y = c \cosh \frac{x}{c} \,; \qquad \therefore \frac{dy}{dx} = \sinh \frac{x}{c}$$

and $\dfrac{d^2 y}{dx^2} = \dfrac{1}{c} \cosh \dfrac{x}{c}$

$$\therefore \quad \rho = \frac{\left[1 + \left(\dfrac{dy}{dx} \right)^2 \right]^{3/2}}{d^2 y / dx^2} = \frac{\left(1 + \sinh^2 \dfrac{x}{c} \right)^{3/2}}{\dfrac{1}{c} \cosh \dfrac{x}{c}}$$

$$= c \cosh^2 \frac{x}{c}$$

Now, $\tan \psi = \sinh \dfrac{x}{c}$

then $\cos \psi = \dfrac{1}{\sqrt{1 + \tan^2 \psi}} = \dfrac{1}{\sqrt{\left(1 + \sinh^2 \dfrac{x}{c} \right)}} = \operatorname{sech} \dfrac{x}{c}$

and $\sin \psi = \sqrt{1 - \cos^2 \psi} = \sqrt{1 - \operatorname{sech}^2 \dfrac{x}{c}}$

$$= \tanh \frac{x}{c} \,.$$

Hence the chord of curvature parallel to y-axis

$$= 2\rho \cos \psi = 2\,c \cosh^2 \frac{x}{c} \cdot \operatorname{sech} \frac{x}{c}$$

$$= 2\,c \cosh \frac{x}{c} = 2y,$$

and the chord of curvature parallel to x-axis is

$$= 2\rho \sin \psi$$

$$= 2c \cosh^2 \frac{x}{c} \cdot \tanh \frac{x}{c} = 2c \cosh \frac{x}{c} \sinh \frac{x}{c}$$

$$= c \sinh \frac{2x}{c}.$$

14.7. Evolutes and Involutes

If the centre of curvature for each point on a curve be taken, we get a new curve called the *evolute* of the original one.

Also the original curve, when considered with respect to its evolute, is called an *involute*.

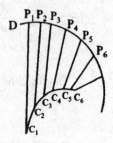

The connection between the curve is represented in the adjoining figure. Let $P_1, P_2, P_3 \ldots$ etc. represent a series of infinitely near points on a curve; $C_1, C_2, C_3 \ldots$ etc. the corresponding centres of curvature; then lines $P_1 C_1, P_2 C_2 , P_3, C_3$ etc. are normals to the curve and the lines $C_1 C_2, C_2 C_3, C_3 C_4$ etc, may be regarded in the limit and consecutive elements of the evolute ; also each of the normals $P_1 C_1, P_2 C_2, P_3 C_3$ etc. passes through two consecutive points on the evolute, they are tangents to that curve on the limit. Again if $P_1, P_2, P_3 P_4$ etc. denote the length of the radii of curvature at the points P_1, P_2, P_3 , etc.; we have

Fig. 14.15

$$\rho_1 = P_1 C_1, \rho_2 = P_2 C_2, \ \rho_3 = P_3 C_3 , \ \ldots\ldots$$

Therefore $\rho_1 - \rho_2 = P_1 C_1 - P_2 C_2$

$$= P_2 C_1 - P_2 C_2 = C_1 C_2 \qquad\qquad \text{(since the points } P_1, P_2 \text{ are nearer.)}$$

Also $\quad \rho_2 - \rho_3 = C_2 C_3, \rho_3 - \rho_4 = C_3 C_4 \ldots$

$$\rho_{n-1} - \rho_n = C_{n-1} C_n$$

Hence by addition, we have

$$\rho_1 - \rho_n = C_1 C_2 + C_2 C_3 + C_3 C_4 + \ldots\ldots + C_{n-1} C_n$$

This result still holds when the number n is increased indefinitely and we infer the length of any arc of the evolute is equal, in general to the difference between the radii of curvature at its extremities. In other words, *the evolute is the locus of the centre of curvature.*

14.7.1 To determine the equation of an evolute

We know that the coordinates of the centre of curvature are given by (X, Y)

whereas

$$X = x - \frac{y_1 (1 + y_1^2)}{y_2}$$

and

$$Y = y + \frac{(1 + y_1^2)}{y_2}.$$

Now we eliminate x and y (parameters in this case), and then a relation between X and Y is the required equation of the evolute.

1. *Obtain the evolute of the parabola $y^2 = 4ax$.*

   (Rohilkhand 2002; Kanpur 2000, 2006; Banglore 2004; Garhwal 2004; Kuvempu 2005)

   Here the point (x, y) may be taken as $x = at^2$, $y = 2at$, which satisfy the equation of the parabola. Hence

   $$\frac{dx}{dt} = 2at \text{ and } \frac{dy}{dt} = 2a$$

   $$\therefore \quad \frac{dy}{dx} = \frac{1}{t} \text{ and } \frac{d^2y}{dx^2} = -\frac{1}{t^2}\frac{dt}{dx}$$

   $$= -\frac{1}{2at^3}$$

   Then if (X, Y) be the coordinates of the centre of curvature, we have

   $$X = x - \frac{y_1(1 + y_1^2)}{y_2} = at^2 - \frac{\dfrac{1}{t}\left(1 + \dfrac{1}{t^2}\right)}{-\dfrac{1}{2at^3}}$$

   $$= at^2 + 2at^2\left(1 - \frac{1}{t^2}\right) = 3at^2 + 2a$$

   and
   $$Y = y + \frac{(1 + y_1^2)}{y_2} = 2at + \frac{\left(1 + \dfrac{1}{t^2}\right)}{-\dfrac{1}{2at^3}}$$

   $$= 2at + (t^2 + 1)(-2at)$$
   $$= -2at^3$$

   Now
   $$X = 3at^2 + 2a,$$
   $$Y = -2at^3$$

   Eliminating t between these relations, we obtain

   $$\left(\frac{X - 2a}{3a}\right)^{1/2} = \left(-\frac{Y}{2a}\right)^{1/3}$$

   or $\quad 4(X - 2a)^3 = 27 aY^2$

   Hence the locus is
   $$27 ay^2 = 4(x - 2a)^3.$$

2. *Find the evolute of the astroid $x = a\cos^3\theta$, $y = a\sin^3\theta$.*

   We have $\quad \dfrac{dy}{dx} = -\tan\theta; \quad \dfrac{d^2y}{dx^2} = \dfrac{1}{3a}\sec^4\theta \operatorname{cosec}\theta$

   $$X = a\cos^3\theta + \frac{\tan\theta(1 + \tan^2\theta)}{\sec^4\theta \operatorname{cosec}\theta}\cdot 3a$$

   $$= a\cos^3\theta + 3a\sin^2\theta\cos\theta \qquad \qquad ...(i)$$

   $$Y = a\sin^3\theta + \frac{1 + \tan^2\theta}{\sec^4\theta \operatorname{cosec}\theta}\cdot 3a$$

   $$= a\sin^3\theta + 3a\cos^2\theta\sin\theta \qquad \qquad ...(ii)$$

To eliminate θ, we separately add and subtract (i) , (ii), so that we have $(X + Y) = a$
$(\cos \theta + \sin \theta)^3 \Leftrightarrow (X + Y)^{1/3} = a^{1/3} (\cos \theta + \sin \theta)$

$$(X - Y) = a (\cos \theta - \sin \theta)^3 \Leftrightarrow (X - Y)^{1/3} = a^{1/3} (\cos \theta - \sin \theta)$$

On squaring and adding, we obtain

$$(X + Y)^{2/3} + (X - Y)^{2/3} = 2a^{2/3},$$

as the required evolute.

EXERCISES

1. Show that the evolute of the ellipse $x = a \cos\theta, y = b \sin\theta$ is $(ax)^{2/3} + (by)^{2/3} = (a^2 - b^2)^{2/3}$.

2. (a) Show that for the hyperbola $x^2/a^2 - y^2/b^2 = 1$, $a^4 X = (a^2 + b^2) x^3$, $b^4 Y = -(a^2 + b^2) y^3$ and
    the equation of the evolute is
    $$(ax)^{2/3} - (by)^{2/3} = (a^2 + b^2)^{2/3}.$$
    (b) Prove that the evolute of the hyperbola $2xy = a^2$.
    is $\qquad\qquad (x + y)^{2/3} - (x - y)^{2/3} = 2a^{2/3}$.

3. Show that the evolute of the tractrix
    $$x = a \left(\cos t + \log \tan \frac{1}{2} t \right), y = a \sin t$$
    is the catenary $y = a \cosh (x/a)$.

4. Show that the parabolas $y = -x^2 + x + 1, x = -y^2 + y + 1$ have the same circle of curvature
    at the point (1, 1).

5. Show that $\left(x - \frac{3}{4} a \right)^2 + \left(y - \frac{3}{4} a \right)^2 = \frac{1}{2} a^2$, is the circle of curvature of the curve
    $\sqrt{x} + \sqrt{y} = \sqrt{a}$, at the point ($a/4, a/4$).

6. Show that the circle of curvature at the parabola
    $$y = mx + x^2 \text{ is } x^2 + y^2 = (1 + m^2)(y - mx).$$

7. (a) Show that the circle of curvature, at the point (am^2, $2am$) of the parabola $y^2 = 4ax$, has for
    its equation
    $$x^2 + y^2 - 6am^2 x - 4ax + 4am^3 y = 3a^2 m^4.$$
    (b) Find the equation of the circle of curvature at the point (0, b) of the ellipse $x^2/a^2 + y^2/b^2 = 1$.

8. Show that the radii of curvatures of the curve $x = ae^\theta (\sin\theta - \cos\theta)$ $y = ae^\theta$,
    ($\sin \theta + \cos \theta$) and its evolute at corresponding points are equal.

9. Find the radius of curvature at any point P of the catenary $y = c \cosh (x/c)$ and show that $PC = PG$
    where C is the centre of curvature at P and G the point of intersection of the normal at P with x-
    axis.

10. For the lemniscate $r^2 = a^2 \cos 2\theta$, show that the length of the tangent from the origin to the
    circle of curvature at any point is $r\sqrt{3}/3$.

11. If P is a point on the curve $r^2 = a^2 \cos 2\theta$ and Q is the intersection of the normal at P with the
    line through the pole O perpendicular to OP, prove that the centre of curvature at P is a point
    of trisection of PQ.

12. If P is a point on the curve $r = a (1 + \cos \theta)$ and Q is the intersection of the normal at P with
    the line through the pole O perpendicular to OP, prove that the centre of curvature at P is a
    point of trisection of PQ remote from P.

13. The circle of curvature at any point P of the lemniscate $r^2 = a^2 \cos 2\theta$ meets the radius vector
    OP at A, show that $OP : AP = 1 : 2$; O being the pole.

14. ρ_1, ρ_2 are the radii of curvature at the corresponding points of a cycloid and its evolute; prove that $\rho_1{}^2 + \rho_2{}^2$ is a constant.

15. In the curve $y = c \cosh\left(\dfrac{x}{c}\right)$ show that coordinates of the centre of curvature are

$$X = x - y\left(\frac{y^2}{c^2} - 1\right)^{1/2}, \quad Y = 2y.$$

16. If (X, Y) be the coordinates of the centre of curvature of the parabola $\sqrt{x} + \sqrt{y} = \sqrt{a}$ at (x, y), then prove that
$$X + Y = 3(x + y).$$

17. Find the equation of circle of curvature at the point $(1, -1)$ of the curve $y = x^3 - 6x^2 + 3x + 1$.

18. Find the chord of curvature through the pole for the cardioid $r = a(1 + \cos\theta)$.

19. In the curve $y = a \log \sec\left(\dfrac{x}{a}\right)$ prove that the chord of curvature parallel to the axis of y is of constant length. **(Bhopal 1998; Jiwaji 96; Vikram 2000)**

20. Show that the chord of curvature through the pole of the curve $r^n = a^n \cos n\theta$ is
$$\frac{2r}{n+1}. \qquad\qquad\text{(Gorakhpur 2001; Avadh 2002)}$$

21. Find the chord of curvature through the pole of the curve $r^2 \cos 2\theta = a^2$.

22. If in the cardioid $r = a(1 + \cos\theta)$, C_r and C_θ be the chords of curvature respectively along and perpendicular to the radius vector, show that
$$C_r + C_\theta = \frac{8}{3}a\, C_r.$$

23. If C_x and C_y be the chords of curvature parallel to the axes at any point of the curve $y = ae^{x/a}$, prove that
$$\frac{1}{C_x^2} + \frac{1}{C_y^2} = \frac{1}{2aC_x}.$$

24. Show that the chord of curvature through the focus of a parabola is four times the focal distance of the point, and the chord of curvature parallel to the axis has the same length.

25. Prove that distance between the pole and the centre of curvature corresponding to any point on the curve $r^n = a^n \cos n\theta$ is
$$\frac{[a^{2n} + (n^2 - 1)r^{2n}]^{1/2}}{(n+1)r^{n-1}}.$$

ANSWERS

7. (b) $x^2 + y^2 - \dfrac{2(b^2 - a^2)}{b}y + (b^2 - 2a^2) = 0$ \qquad 9. y^2/c ;

17. $(x + 36)^2 + \left(y + \dfrac{43}{6}\right)^2 = \dfrac{37^3}{36}$ \qquad 18. $\dfrac{4}{3}r$; \qquad 21. $2r$.

14.8. Properties of the evolute.

If $P'(X, Y)$ be the centre of curvature for any point $P(x, y)$ on the given curve, we have

$$X = x - \rho \sin \psi, \quad Y = y + \rho \cos \psi.$$

Differentiating *w. r. to x*, we obtain

$$\frac{dX}{dx} = 1 - \rho \cos \psi \frac{d\psi}{dx} - \sin \psi \frac{d\rho}{dx}$$

$$= 1 - \frac{ds}{d\psi} \frac{dx}{ds} \frac{d\psi}{dx} - \sin \psi \frac{d\rho}{dx} = -\sin \psi \frac{d\rho}{dx} \qquad \ldots(i)$$

$$\left(\frac{dY}{dx} = \frac{dy}{dx} - \rho \sin \psi \frac{d\psi}{dx} + \cos \psi \frac{d\rho}{dx} \right)$$

$$= \frac{dy}{dx} - \frac{dx}{d\psi} \frac{dy}{ds} \frac{d\psi}{dx} + \cos \psi \frac{d\rho}{dx} = \cos \psi \frac{d\rho}{dx} \qquad \ldots(ii)$$

From (i) and (ii), we obtain

$$\frac{dY}{dX} = -\cot \psi \qquad \ldots(iii)$$

Now dY/dX is the slope of the tangent to the evolute at P' and $-\cot \psi$ is the slope of the normal PP' to the original curve at P.

By (iii) the slopes of two lines, which have a point P' in common, are equal, and as such they coincide.

Thus the normals to a curve are the tangents to its evolute.

Again, we square (i) and (ii) and add to obtain

$$\left(\frac{dX}{dx} \right)^2 + \left(\frac{dY}{dy} \right)^2 = \left(\frac{d\rho}{dx} \right)^2 \qquad \ldots(iv)$$

Let S be the length of the arc of the evolute measured from a fixed point on it upto (X, Y) so that

$$\left(\frac{dS}{dx} \right)^2 = \left(\frac{dX}{dx} \right)^2 + \left(\frac{dY}{dx} \right)^2 \qquad \ldots(v)$$

Here, x is a parameter for the evolute.

From (iv) and (v)

$$\left(\frac{dS}{dx} \right)^2 = \left(\frac{d\rho}{dx} \right)^2 \Rightarrow \frac{dS}{dx} = \frac{d\rho}{dx} \Rightarrow S = \rho + c,$$

where c is a constant.

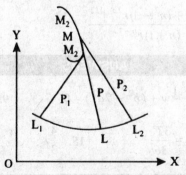

Fig. 14.16

Let M_1, M_2 be the two points on the evolute corresponding to the points L_1, L_2 on the original curve. Let ρ_1, ρ_2 be the values of, ρ, for L_1, L_2 and S_1, S_2 be the values of S for M_1, M_2. We have

$$S_1 = \rho_1 + c, \; S_2 = \rho_2 + c$$
$$\Rightarrow \qquad\qquad S_2 - S_1 = \rho_2 - \rho_1,$$

i.e., arc $M_1 M_2$ = difference between the radii of curvatures at L_1, L_2.

Thus, we have shown that *the difference between the radii of curvatures at two points of a curve is equal to the length of the arc of the evolute between the two corresponding points.*

Note. We suppose that S is measured positively in the direction of x increasing, so that dS/dx is positive or negative according as, ρ, increases or decreases as x increases. Thus

$$\frac{dS}{ds} = \frac{d\rho}{dx} \text{ or } -\frac{d\rho}{dx}$$

according as, ρ, increases or decreases for the values of x, under consideration.

It is easy to see that the conclusion arrived at in this section remains the same if we consider $dS/dx = -d\rho/dx$ instead of $dS/dx = d\rho/dx$. It should, however, be noted that conclusion holds good only for that case of a curve for which ρ, constantly increases or decreases so that $d\rho/dx$ keeps the same sign.

EXAMPLE

Find the length of the arc of the evolute of the parabola $y^2 = 4ax$ which is intercepted between the parabola.

The evolute is $27\,ay^2 = 4\,(x - 2a)^3$.

Let L, M be the points of intersection of the evolute LAM and the parabola. To find the co-ordinates of the points of intersection L, M, we solve the two equations simultaneously, so that

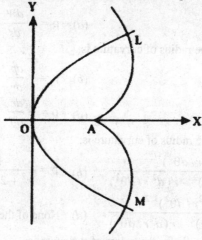

Fig. 14.17

$$27a \cdot 4ax = 4(x - 2a)^3,$$

$$\Rightarrow \qquad x^3 - 6ax^2 - 15\,a^2x - 8a^3 = 0.$$

Now, $8a$, $-a$, $-a$ are the roots of this cubic equation of which $x = 8a$ is the only admissible value; $-a$ being negative.

$$\therefore \quad (8a, \, 4\sqrt{2a}), \, (8a, \, -4\sqrt{2a}) \text{ are the co-ordinates of } L, M.$$

If (X, Y) be the centre of curvature for any point (x, y) on the parabola, we have

$$X = 3x + 2a, \quad Y = -y^3/4a^2.$$

Thus $A\,(2a, 0)$ is the centre of curvature for $O\,(0,0)$ and $L\,(8a, \, 4\sqrt{2a}\,)$ is the centre of curvature for $P\,(2a, -2\sqrt{2a}\,)$.

The radius of curvature for $Q = OA = 2a$.

The radius of curvature at $P = PL = 6\sqrt{3a}$.

$\therefore$ arc $AL = PL - OA = 2a\left(3\sqrt{3} - 1\right)$

Hence the required length $MAL = 4a\left(3\sqrt{3} - 1\right)$.

Ex. Show that the whole length of the evolute of the

(i)   ellipse $x^2/a^2 + y^2/b^2 = 1$ is $4\left(a^2/b - b^2/a\right)$,

(ii) astroid $x = a\cos^3\theta, y = x\sin^3\theta$ is $12a$.

OBJECTIVE QUESTIONS

For each of the following questions, four alternatives are given for the answer. Only one of them is correct. Choose the correct alternative.

1. The radius of curvature of any point (x, y) on the curve $y = f(x)$ is given by

(a) $\dfrac{\{1 + (dy/dx)^2\}^{2/3}}{d^2y/dx^2}$ (b) $\dfrac{\{1 + (dx/dy)^2\}^{2/3}}{d^2y/dx^2}$

(c) $\dfrac{\{1 + (dy/dx)^2\}^{3/2}}{d^2y/dx^2}$ (d) $\dfrac{\{1 + (dy/dx)^2\}^3}{d^2y/dx^2}$

2. The intrinsic formula for the radius of curvature is:

(a) $\rho = \dfrac{dy}{dx}$ (b) $\rho = \dfrac{ds}{d\Psi}$

(c) $\rho = \dfrac{1}{s}\dfrac{ds}{d\Psi}$ (d) $\rho = s\dfrac{d\Psi}{ds}$ **(Kumaon 2005)**

3. The pedal formula for the radius of curvature is:

(a) $\rho = r + \dfrac{dr}{dp}$ (b) $\rho = r\dfrac{dp}{dr}$

(c) $\rho = r\dfrac{dr}{dp}$ (d) $\rho = \dfrac{1}{r}\dfrac{dr}{dp}$

4. The polar formula for the radius of curvature is:

(a) $\rho = \dfrac{(1 + d^2r/d\theta^2)^{3/2}}{r^2 + 2(dr/d\theta)^2 - r(d^2r/d\theta^2)}$ (b) $\rho = \dfrac{\{r^2 + (dr/d\theta)^2\}^{3/2}}{r^2 + 2(dr/d\theta)^2 - r(d^2r/d\theta^2)}$

(c) $\rho = \dfrac{(r^2 + d^2r/d\theta^2)^{3/2}}{r^2 + 2(dr/d\theta)^2 - r(d^2r/d\theta^2)}$ (d) None of these

5. The polar tangential formula for the radius of curvature is:

(a) $\rho = p + \left(\dfrac{dp}{d\Psi}\right)^2$ (b) $\rho = p^2 + \dfrac{d^2p}{d\Psi^2}$

(c) $\rho = p + \dfrac{d^2p}{d\Psi^2}$ (d) $\rho = p + \dfrac{dp}{d\Psi}$

6. The radius of curvature of the origin, if y-axis is the tangent at the origin, is given by:

(a) $\lim\limits_{x\to 0}\dfrac{x^2}{2y}$ (b) $\lim\limits_{x\to 0}\dfrac{x^2}{y}$ (c) $\lim\limits_{x\to 0}\dfrac{y^2}{x}$ (d) $\lim\limits_{x\to 0}\dfrac{y^2}{2x}$

7. The radius of curvature at the point (s, Ψ) on the curve $s = c\log\sec\Psi$ is :

(a) $c\sin\Psi$ (b) $c\cos\Psi$ (c) $c\tan\Psi$ (d) $c\cot\Psi$

8. The radius of curvature at any point on the hyperbola $pr = a^2$ is :

(a) r^3/a^2 (b) r^2/a^2 (c) r/a^2 (d) r

9. Radius of curvature of any point (s, Ψ) on the curve $s = 4a \sin \Psi$ is:

   (a) $4a \cos \Psi$ (b) $4a \sin \Psi$ (c) $4a \sin 2\Psi$ (d) $4a \cos 2\Psi$

10. The radius of curvature at (x, y) of the curve $y = c \cos h (x/c)$ is:

    (a) y/c (b) y^2/c (c) y^3/c (d) cy

11. Radius of curvature of the curve $p^2 = ar$ is:

    (a) p^2/a^2 (b) $2p/a^2$ (c) $2p^3/a^2$ (d) $p^3/2a^2$

12. For the curve $s = ae^{x/a}$:

    (a) $\rho = s/\sqrt{(s^2 - a^2)}$ (b) $a\rho = s(s^2 - a^2)^{1/2}$

    (c) $\rho = (s^2 - a^2)^{1/2}$ (d) $\rho = s(s^2 - a^2)^{1/2}$

13. The radius of curvature for $y = c \log \sec \left(\dfrac{x}{c} \right)$ is

    (a) $c \sec \dfrac{x}{c}$ (b) $c \tan \dfrac{x}{c}$ (c) x/c (d) $\log \sec \left(\dfrac{x}{c} \right)$

    (Rohilkhand 2002, 2003)

14. The radius of curvature at the point (s, φ) on the curve $s = c \tan \varphi$ is :

    (a) $c \tan^2 \varphi$ (b) $c \sec^2 \varphi$ (c) c (d) None of these

    (Avadh 2002, 2005)

15. Radius of curvature of the curve $y = e^x$ at the point $(0,1)$ is :

    (a) $2\sqrt{2}$ (b) $3\sqrt{2}$ (c) 0 (d) None of these

    (Avadh 2003)

16. For any curve $\dfrac{d}{dx}\left(\dfrac{dx}{ds} \right)$:

    (a) $1/\rho$ (b) $-\dfrac{1}{\rho}$ (c) ρ (d) ρ^2

    (Kanpur 2001)

17. Locus of centre of curvature is known as:

    (a) circle of curvature   (b) chord of curvature

    (c) Evolute   (d) Envelope

    (Rohilkhand 2002, 2004)

18. Normals of a curve are

    (a) Normals to its evolute   (b) Tangents to its evolute

    (c) Chords to its evolute   (d) None of these

ANSWERS					
1. (c)	2. (b)	3. (c)	4. (b)	5. (c)	6. (d)
7. (c)	8. (a)	9. (a)	10. (a)	11. (c)	12. (b)
13. (a)	14. (b)	15. (b)	16. (a)	17. (c)	18. (b)

CHAPTER

15

ASYMPTOTES

15.1 Definition

*A straight line is said to be an **Asymptote** of an infinite branch of a curve, if, as the point P recedes to infinity along the branch, the perpendicular distance of P from the straight line tends to 0.*

(Kumaon 2004)

Illustration. The line $x = a$ is an asymptote of the *Cissoid*

$$y^2 (a - x) = x^3$$

It is easy to see that as $P(x, y)$ moves to infinity, its distance from the line $x = a$ tends to zero.

Ex. What are the asymptotes of the curves.

$$y = \tan x; \ y = \cot x; \ y = \sec x \text{ and } y = \operatorname{cosec} x.$$

15.2 Determination of Asymptotes.

We know that the equation of a line which is not *parallel to X-axis* is of the form

$$y = mx + c \qquad \qquad ...(i)$$

The abscissa, x, must tend to infinity as the point $P(x, y)$ recedes to infinity along this line.

We shall now determine, m, and c, so that the line (i) may be an asymptote of the given curve.

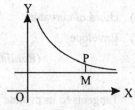

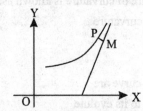

Fig. 15.1   Fig. 15.2

If $p = MP$ be the perpendicular distance of any point $P(x, y)$ on the infinite branch of a given curve from the line (i), we have

$$p = \frac{y - mx - c}{\sqrt{(1 + m^2)}}$$

Now $p \to 0$ as $x \to \infty$

$\therefore \qquad \lim (y - mx - c) = 0,$

which means that, when $x \to \infty$

$$\lim (y - mx) = c$$

Also,

456

$$y/x - m = (y - mx) . (1/x).$$

$\therefore$ $\qquad$ $\lim (y/x - m) = \lim (y - mx). \lim 1/x = c. 0 = 0$

or $\qquad$ $\lim (y/x) = m,$

when $x \to \infty$

Hence, $\qquad$ $m = \lim_{x \to \infty} (y/x), \quad c = \lim_{x \to \infty} (y - mx).$

We have thus the following method to determine asymptotes which are not parallel to y-axis:

(i) Find $\lim_{x \to \infty} (y/x)$; let $\lim_{x \to \infty} (y/x) = m.$

(ii) Find $\lim_{x \to \infty} (y - mx)$; let $\lim_{x \to \infty} (y - mx) = c$

Then $y = mx + c$ is an asymptote.

The values of y will be different according to the different branches along which P recedes to infinity, and so we expect several values of $\lim (y/x)$ corresponding to the several values of y and also several corresponding values of $\lim (y - mx)$. Thus a curve may have more than one asymptote.

This method will determine all the asymptotes except those which are parallel to Y-axis. To determine such asymptotes, we start with the equation $x = my + d$ which can represent every straight line not parallel to X-axis and show, that when $y \to \infty$

$$m = \lim (x/y) \text{ and } d = \lim (x - my).$$

The asymptotes not parallel to any axis can be obtained either way.

Ex. 1. *Examine the Folium*

$$x^3 + y^3 - 3\,axy = 0, \qquad\qquad\qquad ...(i)$$

for asymptotes.

The given equation is of the third degree.

To find $\lim (y/x)$, divide the equation (i) by x^3, so that

$$1 + \left(\frac{y}{x}\right)^3 - 3a\frac{y}{x} . \frac{1}{x} = 0.$$

Let $x \to \infty$. We then get

$$1 + m^3 = 0 \text{ or } (m + 1)(m^2 - m + 1) = 0.$$

$\therefore$ $\qquad\qquad\qquad$ $m = -1.$

The roots of $m^2 - m + 1 = 0$ are not real.

To find $\lim (y - mx)$ when $m = -1$, *i.e.*, to find $\lim (y + x)$, we put $y + x = p$ so that, p is a variable which $\to c$ when $x \to \infty$.

Putting $p - x$ for y in the equation (i), we get

$$x^3 + (p - x)^3 - 3ax (p - x) = 0,$$

or $\quad 3(p + a)x^2 - 3(p^2 + ap)x + p^3 = 0,$

which is of the *second* degree in x.

Dividing by x^2, we get

$$3(p + a) - 3(p^2 + ap). \frac{1}{x} + p^3 . \frac{1}{x^2} = 0.$$

Let $x \to \infty$. We then have

$$3(c + a) = 0 \text{ or } c = -a.$$

Hence, $\qquad$ $y = -x - a$ or $x + y + a = 0,$

is the only asymptote of the given curve.

If we start with $x = my + d$, we get no new asymptotes. Thus
$$x + y + a = 0$$
is the only asymptote of the given curve.

Ex. 2. *Find the asymptotes of the following curves:*

(i) $x^2 (x - y) + ay^2 = 0$.   (ii) $x^3 + y^3 = 3ax^2$.

(iii) $y^3 = x^3 + ax^2$.   *(Garhwal 2003)*

ANSWERS

(i) $x - y + a = 0$ (ii) $x + y = a$ (iii) $y = x + a/3$

15.3. Working rules for determining asymptotes. Shorter Methods. In practice the rules obtained below for determining asymptotes are found more convenient than the method which involves direct determination of
$$\lim (y/x) \text{ and } \lim (y - mx).$$

Firstly we shall consider the case of asymptotes parallel to the co-ordinate axes and then that of oblique asymptotes.

15.3.1 Determination of the aysmptotes parallel to the co-ordinate axes.

Asymptotes parallel to Y-axis.

Let
$$x = k \qquad \qquad ...(i)$$
be an asymptote of the curve, so that we have to determine k.

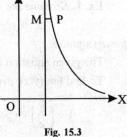

Here, y alone tends to infinity as a point $P(x, y)$ recedes to infinity along the curve.

The distance PM of any point $P(x, y)$ on the curve from the line (i) is equal to $x - k$.
$$\lim (x - k) = 0 \text{ when } y \to \infty,$$
or $\lim x = k$ when $y \to \infty$.

Fig. 15.3

which gives k.

Thus to find the asymptotes paralel to Y-axis, *we find the definite value or values* k_1, k_2, *etc., to which* x *tends as* y *tends to* ∞. *Then* $x = k_1$, $x = k_2$, *etc. are the required asymptotes.*

We will now obtain a simple rule to obtain the asymptotes of a rational *Algebraic* curve which are parallel to Y-axis.

We arrange the equation of the curve in descending powers of y, so that it takes the form
$$y^m \phi (x) + y^{m-1} \phi_1 (x) + y^{m-2} \phi_2 (x) + ... = 0; \qquad \qquad ...(ii)$$
where $\phi (x), \phi_1 (x), \phi_2 (x), \ etc.,$

are polynomials in x.

Dividing the equation (ii) by y^m, we get
$$\phi (x) + (1/y) \cdot \phi_1 (x) + (1/y^2) \cdot \phi_2 (x) + ... = 0 \qquad \qquad ...(iii)$$

Let $y \to \infty$. We write
$$\lim x = k$$

The equation (iii) gives
$$\phi (k) = 0,$$
so that, k is a root of the equation $\phi (x) = 0$.

Let k_1, k_2 etc., be the roots of $\phi(x) = 0$. Then the asymptotes parallel to Y-axis are

$$x = k_1, x = k_2, \text{ etc.}$$

From algebra, we know that $(x - x_1)$, $(x - k_2)$, etc., are the factors of $\phi(x)$ which is the co-efficient of the highest power y^m of y in the given equation.

Hence we have the rule : *The asymptotes parallel to Y-axis are obtained by equating to zero the real linear factors in the co-efficient of the highest power of y, in the equation of the curve.*

The curve will have no asymptote parallel to Y-axis, if the co-efficient of the highest power of y, is a constant or if its linear factors are all imaginary.

Asymptotes parallel to X-axis. As above it can be shown that *the asymptotes, which are parallel to X-axis, are obtained by equating to zero the real linear factors in the co-efficient of the highest power of x, in the equation of the curve.*

15. 3. 2 *To determine the asymptotes of the* **general rational algebraic equation**

$$U_n + U_{n-1} + U_{n-2} + \dots + U_2 + U_1 + U_0 = 0, \qquad \dots(i)$$

where, U_r, is a homogeneous expression of degree r in x, y.

We write U_r in the form

$$U_r \equiv x^r \, \phi_r(y/x)$$

where $\phi_r(y/x)$ is a polynomial in y/x of degree, r, at most.

So we write (i) in the form

$$x^n \phi_n\left(\frac{y}{x}\right) + x^{n-1} \phi_{n-1}\left(\frac{y}{x}\right) + x^{n-2} \phi_{n-2}\left(\frac{y}{x}\right) + \dots + x\phi_1\left(\frac{y}{x}\right) + \phi_0\left(\frac{y}{x}\right) = 0.$$

$$\dots(ii)$$

Dividing by x^n, we get

$$\phi_n\left(\frac{y}{x}\right) + \frac{1}{x}\phi_{n-1}\left(\frac{y}{x}\right) + \frac{1}{x^2}\phi_{n-2}\left(\frac{y}{x}\right) + \dots + \frac{1}{x^{n-1}}\phi_1\left(\frac{y}{x}\right) + \frac{1}{x^n}\phi_0\left(\frac{y}{x}\right) = 0.$$

$$\dots(iii)$$

On taking limits, as $x \to \infty$, we obtain the equation

$$\phi_n(m) = 0 \qquad \dots(iv)$$

which determines the slopes of the asymptote.

Let m_1 be one of the roots of this equation so that $\phi_n(m_1) = 0$.

We write

$$y - m_1 x = p_1, \text{ i.e., } \frac{y}{x} = m_1 + \frac{p_1}{x}.$$

Substituting this value of y/x in (ii), we get

$$x^n \phi_n\left(m_1 + \frac{p_1}{x}\right) + x^{n-1} \phi_{n-1}\left(m_1 + \frac{p_1}{x}\right) + x^{n-2} \phi_{n-2}\left(m_1 + \frac{p_1}{x}\right) +$$

$$\dots + x\phi_1\left(m_1 + \frac{p_1}{x}\right) + \phi_0\left(m_1 + \frac{p_1}{x}\right) = 0$$

Expanding each term by Taylor's theorem, we get

$$x^n \left[\phi_n(m_1) + \frac{p_1}{x}\phi_n'(m_1) + \frac{p_1^2}{2x^2}\phi_n''(m_1) + \dots\right]$$

$$+ x^{n-1}\left[\phi_{n-1}(m_1) + \frac{p_1}{x}\phi_{n-1}'(m_1) + \dots\right]$$

$$+ x^{n-2}\left[\phi_{n-2}(m_1) + \frac{p_1}{x}\phi_{n-2}'(m_1) + \dots\right] + \dots = 0$$

Arranging terms according to descending powers of x, we get

$$x^n \phi_n(m_1) + x^{n-1}[p_1 \phi_n'(m_1) + \phi_{n-1}(m_1)]$$

$$+ x^{n-2}\left[\frac{p_1^2}{2}\phi_n''(m_1) + p_1 \phi_{n-1}'(m_1) + \phi_{n-2}(m_1)\right] + \ldots = 0$$

Putting $\phi_n(m_1) = 0$ and then dividing by x^{n-1}, we get

$$[p_1 \phi_n'(m_1) + \phi_{n-1}(m_1)] + \left[\frac{p_1^2}{2}\phi_n''(m_1) + p_1 \phi_{n-1}'(m_1) + \phi_{n-2}(m_1)\right]\frac{1}{x} + \ldots = 0 \qquad \ldots(v)$$

Let $x \to \infty$. We write $\lim p_1 = c_1$. Therefore,

$$c_1 \phi_n'(m_1) + \phi_{n-1}(m_1) = 0, \qquad \ldots(vi)$$

or $\qquad c_1 = -\dfrac{\phi_{n-1}(m_1)}{\phi_n'(m_1)}$, if $\phi_n'(m_1) \neq 0$.

Therefore $\qquad y = m_1 x - \dfrac{\phi_{n-1}(m_1)}{\phi_n'(m_1)}$

is the asymptote corresponding to the slope m_1, if $\phi_n'(m_1) \neq 0$.

Similarly, $\qquad y = m_2 x - \dfrac{\phi_{n-1}(m_2)}{\phi_n'(m_2)}$; $y = m_3 x - \dfrac{\phi_{n-1}(m_3)}{\phi_n'(m_3)}$, etc.,

are the asymptotes of the curve corresponding to the slopes m_2, m_3 etc., which are the roots of $\phi_n(m)$ = 0, if $\phi_n'(m_2), \phi_n'(m_3)$, etc., are not 0.

Exceptional case. *Let* $\phi_n'(m_1) = 0$.

If $\phi_n'(m_1) = 0$ but $\phi_{n-1}(m_1) \neq 0$, then the equation (vi) does not determine any value of c_1 and, therefore, there is no asymptote corresponding to the slope m_1.

Now suppose

$$\phi_n'(m_1) = 0 = \phi_{n-1}(m_1).$$

In this case, (vi) becomes an identity and we have to reexamine the equation (v) which now becomes

$$\left[\frac{p_1^2}{2}\phi_n''(m_1) + p_1 \phi_{n-1}'(m_1) + \phi_{n-2}(m_1)\right] + [\ldots]\frac{1}{x} + \ldots = 0$$

On taking limits, as $x \to \infty$, we see that c_1 is a root of the equation

$$(c_1^2 / 2)\phi_n''(m_1) + c_1 \phi_{n-1}'(m_1) + \phi_{n-2}(m_1) = 0,$$

which determines two values of c_1, say c_1', c_1'', provided that

$$\phi_n''(m_1) \neq 0.$$

Thus

$$y = m_1 x + c_1', y = m_1 x + c_1'',$$

are *the two* asymptotes corresponding to the slope m_1, These are clearly parallel.

This is known as the *case of parallel asymptotes.*

Important Note. *The polynomial* $\phi_n(m)$ *is obtained by putting* $x = 1$ *and* $y = m$ *in the highest degree terms* $x^n\phi_n(y/x)$, *and* $\phi_{n-1}(m), \phi_{n-2}(m)$, *etc., are obtained from* $x^{n-1}\phi_{n-1}(y/x), x^{n-2}\phi_{n-2}(y/x)$, *etc., in a similar way.*

EXAMPLES

1. *Find the asymptotes parallel to co-ordinate axes, of the curves:*

   (*i*) $(x^2 + y^2)\, x - ay^2 = 0$ (*ii*) $x^2 y^2 - a^2 (x^2 + y^2) = 0.$

                                                              (*Banglore 2005, 2006*)

   (*i*) The co-efficient of the highest power y^2 of y is $x - a$. Hence, the asymptote parallel to y-axis is $x - a = 0$.

   The co-efficient of the highest power x^3 of x is 1 which is a constant. Hence there is no asymptote parallel to x-axis.

   (*ii*) The co-efficient of the highest power y^2 of y is $x^2 - a^2$.

   Also, $x^2 - a^2 = (x - a)(x + a)$

   Hence $x - a = 0,\ x + a = 0$

   are the two asymptotes parallel to y-axis.

   It may similarly be shown that $y - a = 0, y + a = 0$ are the two asymptotes parallel to x-axis.

2. *Find the asymptotes of the cubic curve*

   $$2x^3 - x^2 y + 2xy^2 + y^3 - 4x^2 + 8xy - 4x + 1 = 0$$

                                                              (*G.N.D.U. Amritsar 2004*)

   Putting $x = 1, y = m$ in the third degree and second degree terms separately, we get

   $$\phi_3(m) = 2 - m - 2m^2 + m^3,\ \phi_2(m) = -4 + 8m$$

   The slopes of the asymptotes are given by

   $$\phi_3(m) = m^3 - 2m^2 - m + 2 = 0$$

   or $(m + 1)(m - 1)(m - 2) = 0$

   ∴ $m = -1,\ 1,\ 2$

   Again, c is given by

   $$c\,\phi_3'(m) + \phi_2(m) = 0$$

   i.e., $c(-1 - 4m + 3m^2) + (-4 + 8m) = 0$

   Putting $m = -1, 1, 2$, we get $c = 2, 2, -4$ respectively.

   Therefore the asymptotes are

   $$y = -x + 2, y = x + 2, y = 2x - 4.$$

3. *Find the asymptotes of*

   $$x^3 - x^2 y - xy^2 + y^3 + 2x^2 - 4y^2 + 2xy + x + y + 1 = 0$$

                                                              (*Rohilkhand 2001*)

   Here $\phi_3(m) = 1 - m - m^2 + m^3 = (1 - m) - m^2(1 - m)$

                        $= (1 - m^2)(1 - m) = (1 - m)^2 (1 + m).$

             $\phi_2(m) = 2 + 2m - 4m^2.$

   The slopes of the asymptotes, given by $\phi_3(m) = 0$, are $1, 1, -1$.

   To determine c, we have

   $c\,\phi_3'(m) + \phi_2(m) = 0;$

   i.e., $c(-1 - 2m + 3m^2) + (2 + 2m - 4m^2) = 0.$ ...(*i*)

   For $m = -1$, this gives c $= 1$ and therefore, $y = -x + 1$ is the corresponding asymptote.

   For $m = 1$, the equation (*i*) becomes $0.\,c + 0 = 0$ which is identically true. In this exceptional case, c is determined from the equation

   $$(c^2 / 2).\ \phi_3''(m) + c\,\phi_2'(m) + \phi_1(m) = 0.$$

i.e.,

$$(c^2/2)(-2+6m) + c(2-8m) + 1(1+m) = 0.$$

Fore $m = 1$, this becomes

$$2c^2 - 6c + 2 = 0, \quad i.e., \quad c = (3 \pm \sqrt{5})/2.$$

Hence, $y = x + (3 \pm \sqrt{5})/2$ are the two parallel asymptotes corresponding to the slope 1. We have thus obtained all the asymptotes of the curve.

4. *Find the asymptotes of* $x^3 + 2x^2 y - xy^2 - 2y^3 + xy - y^2 - 1 = 0$

(Sagar 99; Kumaon 2004; Gorakhpur 2002)

The equation of the curve is

$$x^3 + 2x^2 y - xy^2 - 2y^3 + xy - y^2 - 1 = 0. \qquad \ldots(i)$$

Substituting $(mx + c)$ for y in (i), we get

$$x^3 - 2x^2(mx+c) - x(mx+c)^2 - 2(mx+c)^3 + x(mx+c) - (mx+c)^2 - 1 = 0$$

or $\qquad x^3(1+2m-m-2m^3) + x^2(2c-2mc-6m^2c+m-m^2) + \ldots = 0.$

∴ $\qquad$ m and c are given by $1 + 2m - m^2 - 2m^3 = 0$ $\qquad \ldots(ii)$

and $\qquad 2c - 2mc - 6m^2 c + m - m^2 = 0$ $\qquad \ldots(iii)$

From (ii), $(1 - m^2) + 2m(1 - m^2) = 0$ $\quad$ or $\quad$ $(1 - m^2)(1 + 2m) = 0$

or $\qquad\qquad m = 1, -1, -1/2.$

From (iii). $\qquad 2c(1 - m - 3m^2) = m^2 - m$ or $c = \dfrac{m^2 - m}{2(1 - m - 3m^3)}$

∴ when $\qquad m = 1, c = \dfrac{1-1}{2(1-1-3)} = 0,$

when $\qquad m = -1, c = \dfrac{1+1}{2(1+1-3)} = -1$ and

when $\qquad m = -1/2, c = \dfrac{(1/4)+(1/2)}{2\left(1+\frac{1}{2}-\frac{3}{4}\right)} = \dfrac{(3/4)}{3/2} = \dfrac{1}{2}$

Substituting these pairs of values of m and c in $y = mx + c$ one by one we have required asymptotes as

$$y = x, \quad y = -x - 1 \quad \text{and} \quad y = -\tfrac{1}{2}x + \tfrac{1}{2} \text{ or } 2y + x = 1.$$

5. *Find all the asymptotes of the curve* $y^3 - 6xy^2 + 11x^2 y - 6x^3 + x + y = 0.$

(Gorakhpur 2000)

Let $y = mx + c$ be an asymptote, then substituting $mx + c$ for y in the equation of the curve, we get

$$(mx + c)^3 - 6x(mx+c)^2 + 11x^2(mx+c) - 6x^3 + x + (mx+c) = 0.$$

or $\qquad x^3(m^3 - 6m^2 + 11m - 6) + x^2(3m^2 c - 12mc + 11c)$

$$+ x(3mc^2 - 6c^2 + 1 + m) + (c^3 + c) = 0$$

Equating the coefficients of x^3 and x^2 to zero we get

$$m^3 - 6m^2 + 11m - 6 = 0 \qquad \ldots(i)$$

and $\qquad 3m^2 c - 12mc + 11c = 0$ $\qquad \ldots(ii)$

From (i) we get

$$(m-1)(m-2)(m-3) = 0 \text{ or } m = 1, 2, 3$$

Also from (ii) we get $\qquad (3m^2 - 12m + 11)c = 0$

or $\qquad c = 0$, for all the three values of m viz. 1, 2, 3.

Hence for these values of m and c we get the required asymptotes as

$$y = x + 0, y = 2x + 0, y = 3x + 0$$

i.e., $\qquad y = x, y = 2x \text{ and } y = 3x.$

6. *Find the asymptotes of the cubic curve*

$$y^3 - 5xy^2 + 8x^2y - 4x^3 - 3y^2 + 9xy - 6x^2 + 2y - 2x = 1.$$

(*Kumaon 96; Bhopal 97; Avadh 2000*)

Putting $mx + c$ for y and rearranging;

$$(m^3 - 5m^2 + 8m - 4) x^3 + (3m^2 c - 13 mc + 8c - 3m^2 + 9m - 6) x^2$$

Equating to zero the coefficients of x^3 and x^2,

$$m^3 - 5m^2 + 8m - 4 = 0$$

and $\qquad 3m^2 c - 10 mc + 8c - 3m^2 + 9m - 6 = 0$

From the first equation, we get

$$(m - 1)(m - 2)^2 = 0$$

Hence $\qquad m = 1, 2, 2.$

If $m = 1$, the second equation gives $c = 0$ and the corresponding asymptote is $y = x$.

If $m = 2$, the left hand side of the second equation vanishes identically for all finite values of c. In such cases, to find c, equate the next coefficient to zero. hence

$$3 mc^2 - 5c^2 - 6mc + 9c + 2m - 2 = 0$$

Putting $m = 2$, we have

$$c^2 - 3c + 2 = 0, i.e., c = 1 \text{ or } 2.$$

Hence there are two parallel asymptotes.

$$y = 2x + 1 \text{ and } y = 2x + 2.$$

Hence the required asymptotes are

$$y = x, y = 2x + 1, y = 2x + 2$$

7. *Find the asymptotes of the curve* $x^3 - x^2y - xy^2 + y^3 + 2x^2 - 4y^2 + 2xy + x + y + 1 = 0.$

Substituting $mx + c$ for y in the given equation, we get

$$x^3 - x^2 (mx + c) - x (mx + c)^2 + (mx + c)^3 + 2x^2 - 4 (mx + c)^2$$

$$+ 2x (mx + c) + x + (mx + c) + 1 = 0$$

or $\quad x^3 (1 - m - m^2 + m^3) + x^2 (-c - 2mc + 3m^2c + 2 - 4m^2 + 2m)$

$$+ x (-c^2 + 3mc^2 - 8mc + 2c + 1 + m) + ... = 0. \qquad ...(i)$$

Equating the coefficient of highest power of x in (i) to zero, we have

$$1 - m - m^2 + m^3 = 0 \text{ or } (1 - m) - m^2 (1 - m) = 0$$

or $\quad (1 - m)(1 - m^2) = 0 \text{ or } (1 - m)^2 (1 + m) = 0 \text{ or } m = 1, 1, -1$

Equating the coefficient of x^3 in (i) to zero, we get

$$c (1 + 2m - 3m^2) = 2 - 4m^2 + 2m. \qquad ...(ii)$$

When , $m = -1$, from (*ii*) we have $c = 1$ and hence the corresponging asymptote is $y = -x + 1$ or $x + y = 1.$

When $m = 1$ from (*ii*) we have $0.c = 0$, where from c cannot be determined.

$\therefore \qquad$ Equating to zero the coefficient of x in (*i*), we get

$-c^2 + 3mc^2 - 8mc + 2c + 1 + m = 0$, which reduces to

$$2c^2 - 6c + 2 = 0 \text{ or } c^2 - 3c + 1 = 0, \text{ when } m = 1$$

or $\qquad c = \frac{1}{2} [3 \pm \sqrt{(9 - 4)}] = \frac{1}{2} (3 \pm \sqrt{5})$

∴ Corresponding asymptotes are $y = x + \frac{1}{2} (3 \pm \sqrt{5})$

or $\qquad 2y - 2x = 3 \pm \sqrt{5}$.

∴ The required asymptotes are $x + y = 1,\ 2x - 2y + 3 \pm \sqrt{5} = 0$.

EXERCISES

Find the asymptotes parallel to co-ordinate axes of the following curves:

1. $y^2x - a^2 (x - a) = 0$ 2. $x^2y - 3x^2 - 5xy + 6y + 2 = 0$

3. $y = x/(x^2 - 1)$ 4. $a^2/x^2 + b^2/y^2 = 1$ *(Banglore 2004)*

Find the asymptotes of the following curves:

5. $x (y - x)^2 = x (y - x) + 2$

6. $x^2 (x - y)^2 + a^2 (x^2 - y^2) = a^2xy$ *(Gorakhpur 2000)*

7. $(x - y)^2 (x^2 + y^2) - 10 (x - y) x^2 + 12y^2 + 2x + y = 0$

8. $x^2y + xy^2 + xy + y^2 + 3x = 0$ *(Banglore 2006)*

9. $(x - y + 1) (x - y - 2) (x + y) = 8x - 1$

10. $y^3 - x^2y + 2xy^2 - y + 1 = 0$ *(Jabalpur 1998, 2000; Ravishankar 1998; Bhopal 1999)*

11. $y (x - y)^2 = x + y$

12. $x^2y^2 (x^2 - y^2)^2 = (x^2 + y^2)^2$

13. $y (y - 1)^2 - x^2 = 1$

14. $xy^2 = (x + y)^2$

15. $(x + y) (x - y) (2x - y) - 4x (x - 2y) + 4x = 0$

16. $xy^2 - x^2y - 3x^2 - 2xy + y^2 + x - 2y + 1 = 0$

17. $2x (y - 3)^2 = 3y (x - 1)^2$

18. $(y - a)^2 (x^2 - a^2) = x^4 + a^4$

19. $(x + y)^2 (x^2 + xy + y^2) = a^2 x^2 + a^2 (y - x)$

20. $y^2 + 3y^2x - x^2y - 3x^2 + y^2 - 2xy + 3x^2 + 4y + 5 = 0$

21. $(x - y)^2 (x - 2y) (x - 3y) - 2a (x^3 - y^3) - 2a^2 (x - 2y) (x + y) = 0$

22. $y^2 = x^4 /(a^2 - x^2)$

Show that the following curves have no asymptotes :–

23. $x^4 + y^4 = a^2 (x^2 - y^2)$ 24. $y^2 = x (x + 1)^2$

25. $a^4y^2 = x^5 (2a - x)$ 26. $x^2 (y^2 + x^2) = a^2 (x^2 - y^2)$

27. Find the equation of the tangent to the curve $x^3 + y^3 = 3ax^2$ which is parallel to its asymptote.

28. An asymptote is sometimes defined as a straight line which cuts the curve in two points at infinity. Criticise this definition and replace it by a correct definition.

ANSWERS

1. $x = 0, y = \pm a$.   2. $x = 2, x = 3, y = 3$.

3. $x = \pm 1, y = 0$ 4. $x = \pm a, y = \pm b$.

5. $x = 0, y = x, y = x + 1$.   6. $y = x \pm a, x = \pm a$.

7. $y = x - 2, y = x - 3$.   8. $x + 1 = 0, y = 0, x + y = 0$

9. $y = x - 3, \ y = x + 2, x + y = 0.$ 10. $y = 0, y + x = \pm 1.$

11. $y = 0, y = x \pm \sqrt{2}$

12. $x + y = \pm \sqrt{2}, \ x = \pm 1, \ y = \pm 1, \ y - x = \pm \sqrt{2}$

13. $y = 0, y \pm x = 1.$ 14. $x = 1.$

15. $x + y = 2, y = x + 2, y = 2x - 4.$

16. $y + 3 = 0, x + 1 = 0, y = x + 4.$

17. $x = 0, y = 0, 2y - 3x = 6.$ 18. $x = \pm a, y \pm x = a.$

19. $x + y \pm a = 0.$

20. $4(y - x) + 1 = 0, 2(y + x) = 3, 4(y + 3x) + 9 = 0.$

21. $x = y + a, x = y + 2a, 2y = x + 14a, 3y = x - 13a.$

22. $x = \pm a.$ 27. $x + y = 4a.$

15.3.3 Some deduction from § 15.3.2

(*i*) *The number of asymptotes of an algebraic curve of the nth degree cannot exceed n.*

The slopes of the asymptotes which are not parallel to *y*-axis are given as the roots of the equation $\phi_n(m) = 0$ which is of degree *n* at the most.

In case the curve possesses one or more asymptotes parallel to *y*-axis, then it is easy to see that the degree of $\phi_n(m) = 0$ will be smaller than *n* by at least the same number.

Hence the result.

(*ii*) *The asymptotes of an algebraic curve are parallel to the lines obtained by equating to zero the factors of the highest degree terms in its equation.*

Let, m_1 be a root of the equation $\phi_n(m) = 0$, so that the line $y - m_1 x = 0$ is parallel to an asymptote.

By elementary algebra, $(y/x - m_1)$ is a factor of $\phi_n(y/x)$ and hence, $y - m_1 x$ is a factor of $x^n \phi_n(y/x)$, *i.e.*, U_n. Also conversely, we see that if, $y - m_1 x$ is a factor of U_n then m_1 is a root of $\phi_n(m) = 0$.

In case the highest degree terms contain, *x*, as a factor, then a little consideration will show that the curve will possess asymptotes parallel to $x = 0$, *i.e.*, to *y*-axis.

(*iii*) **Case of parallel asymptotes.**

In this case, m_1 satisfies the three equations

$$\phi_n(m) = 0, \ \phi'_n(m) = 0, \ \phi_{n-1}(m) = 0.$$

Since $\phi_n(m)$ and its derivative $\phi'_n(m)$ vanish for $m = m_1$, therefore, by elementary algebra, m_1 is a double root of $\phi_n(m) = 0$ and therefore, $(y - m_1 x)^2$ is a factor of the highest degree terms U_n.

Also, since m_1 is a root of $\phi_{n-1}(m) = 0$, $y - m_1 x$ is a factor of the $(n-1)$th degree terms U_{n-1}.

Thus, we see that in the exceptional case a twice repeated linear factor of U_n is also a non-repeated factor of U_{n-1}.

There will be *no* asymptote with slope m_1, if m_1 is a root of $\phi_n(m) = 0$, $\phi'_n(m) = 0$ but *not* of $\phi_{n-1}(m) = 0$, *i.e.*, if $(y - m_1 x)^2$ is a factor of U_n and $y - m_1 x$ is not a factor of U_{n-1}.

Note. The results obtained in the paragraphs (*ii*) and (*iii*) above enable us to shorten the process of determining the asymptotes as shown in the following examples.

The first step will always consist in factorising the expression formed of the highest degree in the given equation.

EXAMPLES

1. *Find the asymptotes of the Folium*
$$x^3 + y^3 - 3axy = 0. \qquad \textbf{\textit{(Bhopal 2000; Sagar 1998; Vikram 1998; Gorakhpur 2001)}}$$
The curve has no asymptotes parallel to co-ordinate axes.

Factorizing the highest degree terms, we get
$$(x + y)(x^2 - xy + y^2) - 3axy = 0.$$
so that, $y + x$, is the only real linear factor of the highest degree terms.

Hence, the curve has only one real asymptote which is parallel to the line $y + x = 0$ whose slope is -1.

We have, now to find, $\lim (y + x)$, when $x \to \infty$ and $y/x \to -1$.

We have
$$y + x = \frac{3axy}{x^2 - xy + y^2} = \frac{3a \cdot (y/x)}{1 - (y/x) + (y/x)^2}.$$

In the limit, we have

$$\lim (y + x) = \frac{-3a}{1 + 1 + 1} = -a \qquad \qquad ...(i)$$

$$\therefore \qquad y = -x - a, \textit{ i.e., } y + x + a = 0,$$

is the only asymptote of the folium.

It is easy to see that we could have eliminated the step (*i*), and simplified the process by saying that the required asymptote is

$$(y + x) = \lim \frac{3a (y/x)}{1 - (y/x) + (y/x)^2},$$

when $x \to \infty$ and $y/x \to -1$.

2. *Find the asymptotes of*
$$x^3 + 4x^2y + 4xy^2 + 5x^2 + 15xy + 10y^2 - 2y + 1 = 0.$$

Equating to zero the co-efficient of the highest power y^2 of y, we see that
$$4x + 10 = 0, \textit{ i.e., } 2x + 5 = 0,$$
is one asymptote.

Factorising the highest degree terms, we get
$$x (2y + x)^2 + 5x^2 + 15xy + 10y^2 - 2y + 1 = 0.$$

Here $2y + x$ is a repeated linear factor of highest degree terms, *i.e.*, 3rd degree. There will, therefore, be no asymptote parallel to $2y + x = 0$ if $(2y + x)$ is not a factor of the 2nd degree terms also. But this is not the case. In fact, the equation is
$$x (2y + x)^2 + 5 (x + y) (x + 2y) - 2y + 1 = 0.$$

Therefore, the curve has two asymptotes parallel to
$$2y + x = 0.$$

We have now to find him $\left(y + \dfrac{1}{2} x \right)$ when $x \to \infty$ and $y/x \to -\dfrac{1}{2}$. Let $\lim \left(y + \dfrac{1}{2} x \right) = c$ so that $\lim (2y + x) = 2c$.

Dividing by x, the equation becomes
$$(2y + x)^2 + 5 (2y + x) (1 + y/x) - 2y/x + 1/x = 0.$$

In the limit,

$$4c^2 + 5.2c \left(1 - \frac{1}{2} \right) - 2 \left(-\frac{1}{2} \right) + 0 = 0, \qquad \qquad ...(i)$$

or $\qquad 4c^2 + 5c + 1 = 0$

$\therefore \qquad\qquad c = -\dfrac{1}{4}, -1.$

Hence $\qquad\qquad y = -\dfrac{1}{2}x - \dfrac{1}{4}$ and $y = -\dfrac{1}{2}x - 1,$

i.e., $\qquad 4y + 2x + 1 = 0$ and $2y + x + 2 = 0,$

are the two more asymptotes.

It is easy to see that we could have eliminated the step (i) and simplified the process by saying that the asymptotes are

$$(2y + x)^2 + 5(2y + x) . \lim(1 + y/x) + \lim(-2y/x + 1/x) = 0,$$

i.e., $(2y + x)^2 + 5(2y + x) . \dfrac{1}{2} + 1 = 0$ or $2(2y + x)^2 + 5(2y + x) + 2 = 0,$

which gives

$$2y + x + 2 = 0 \text{ and } 4y + 2x + 1 = 0.$$

3. *Find the asymptotes of*

$$(x - y)^2(x^2 + y^2) - 10(x - y)x^2 + 12y^2 + 2x + y = 0.$$

The asymptotes parallel to the two imaginary lines $x^2 + y^2 = 0$ are imaginary. To obtain the two asymptotes parallel to the lines $x - y = 0$, we re-write the equation, on dividing it by $(x^2 + y^2)$, as

$$(x - y)^2 - 10(x - y)\dfrac{1}{1 + (y/x)^2} + \dfrac{12(y/x)^2 + 2/x + y/x . 1/x}{1 + (y/x)^2} = 0.$$

We take the limits when $x \to \infty$ and $y/x \to 1$. Therefore the asymptotes are

$$(x - y)^2 - 10(x - y)\lim\dfrac{1}{1 + (y/x)^2} + \lim\dfrac{12(y/x)^2 + 2/x + y/x . 1/x}{1 + (y/x)^2} = 0.$$

i.e., $\qquad (x - y)^2 - 5(x - y) + 6 = 0.$

i.e., $\qquad x - y - 2 = 0, x - y - 3 = 0.$

4. *Find the asymptotes of*

$$(x - y + 2)(2x - 3y + 4)(4x - 5y + 6) + 5x - 6y + 7 = 0.$$

The asymptote parallel to $x - y + 2 = 0$ is

$$x - y + 2 + \lim\dfrac{5x - 6y + 7}{(2x - 3y + 4)(4x - 5y + 6)} = 0,$$

when $x \to \infty$ and $y/x \to 1$,

i.e., $x - y + 2 + \lim\left[\dfrac{5 - 6y/x + 7/x}{(2 - 3y/x + 4/x)(4 - 5y/x + 6/x)} . \dfrac{1}{x}\right] = 0.$

or $\qquad\qquad x - y + 2 = 0,$

as the limit is zero because of the factor $1/x$.

Similarly, we can show that

$$2x - 3y + 4 = 0, 4x - 5y + 6 = 0$$

are also the asymptotes of the curve.

15.4. Asymptotes by Inspection. *If the equation of a curve of the nth degree can be put in the form*

$$F_n + F_{n-2} = 0,$$

where F_{n-2} is of degree $(n-2)$ at the most, then every linear, factor of F_n, when equated to zero will give an asymptote, provided that no straight line obtained by equating to zero any other linear factor of F_n is parallel to it or coincident with it.

Let $ax + by + c = 0$ be a non-repeated factor of F_n. We write
$$F_n = (ax + by + c) F_{n-1}$$
where F_{n-1} is of degree $(n-1)$. The asymptote parallel to $ax + by + c = 0$ is
$$ax + by + c + \lim \frac{F_{n-2}}{F_{n-1}} = 0,$$
when $x \to \infty$ and $y/x \to -a/b$.

To determine the limit (F_{n-2}/F_{n-1}), we divide the numerator as well as the denominator by x^{n-1} and see that $1/x$ appears as a factor so that $F_{n-2}/F_{n-1} \to 0$ as $x \to \infty$.

Thus $ax + by + c = 0$

is an asymptote.

EXERCISES

Find the asymptotes of the following curves :—

1. $xy (x + y) = a (x^2 - a^2)$
2. $(x - 1) (x - 2) (x + y) + x^2 + x + 1 = 0$
3. $y^3 - x^3 + y^2 + x^2 + y - x + 1 = 0$
4. $x (y^2 - 3by + 2b^2) = y^3 - 3bx^2 + b^3$
5. $x^3 + 6x^2 y + 11xy^2 + 6y^3 + 3x^2 + 12xy + 11y^2 + 2x + 3y + 5 = 0$
6. $x^2 (3y + x)^2 + (3y + x) (x^2 + y^2) + 9y^2 + 6xy + 9y - 6x + 9 = 0$
7. $(y^2 + xy - 2x^2)^2 + (y^2 + xy - 2x^2) (2y + x) - 7y^2 - 19xy - 28x^2 + x + 2y + 3 = 0$
8. $x (y - 3)^3 = 4y (x - 1)^3$
9. $(a + x)^2 (b^2 + x^2) = x^2 y^2$
10. $y^3 - 5xy^2 + 8x^2 y - 4x^3 - 3y^2 + 9xy - 6x^2 + 2y - 2x + 1 = 0$

ANSWERS

1. $y = a, x = 0, x + y + a = 0.$ 2. $x = 1, x = 2, x + y + 1 = 0.$
3. $3 (y - x) + 2 = 0.$ 4. $y - x = 0.$
5. $x + y + 1 = 0, 3y + x - 1 = 0, 2y + x + 1 = 0.$
6. $3y + x = (-5 \pm \sqrt{106})/9 .$
7. $y + 2x + 2 = 0, y + 2x - 1 = 0; y - x + 3 = 0, y - x - 2 = 0.$
8. $x = 0, y = 0, 2y = 4x + 3, 4x + 2y = 15.$
9. $x \pm y + a = 0, x = 0.$
10. $y = x, y = 2x + 1, y = 2x + 2.$

15.5. Intersection of a curve and its asymptotes.

Any asymptote of a curve of the nth degree cuts the curve in $(n-2)$ points.

Let $y = mx + c$ be an asymptote of the curve
$$x^n \phi_n (y/x) + x^{n-1} \phi_{n-1} (y/x) + x^{n-2} \phi_{n-2} (y/x) + ... = 0.$$

To find the points of intersection, we have to solve the two equations simultaneously.

The abscissae of the points of intersection are the roots of the equation
$$x^n \phi_n (m + c/x) + x^{n-1} \phi_{n-1} (m + c/x) + x^{n-2} \phi_{n-2} (m + c/x) + ... = 0. \qquad ...(i)$$

Expanding each term by Taylor's theorem and arranging according to descending powers of x, we get

$$x^n \phi_n (m) + [c \phi'_n (m) + \phi_{n-1} (m)] x^{n-1} +$$

$$\left[\frac{1}{2} c^2 \phi''_n (m) + c \phi'_{n-1} (m) + \phi_{n-2} (m)\right] x^{n-2} + ... = 0. ...(ii)$$

As $y = mx + c$ is an asymptote, the co-efficients of x^n and x^{n-1} are both zero.

Thus the equation (ii) reduces to that of $(n-2)$th degree and, therefore, determines $(n-2)$ values of x. Hence the result.

Cor. 1. *The, n, asymptotes of a curve of the nth degree cut it in n $(n-2)$ points.*

Cor. 2. *If the equation of a curve of the nth degree can be put in the form $F_n + F_{n-2} = 0$ where F_{n-2} is of degree $(n-2)$ at the most and F_n consists of, n, non-repeated linear factors, then the n (n − 2) points of intersection of the curve and its asymptotes lies on the curve*

$$F_{n-2} = 0.$$

The result follows at once from the fact that $F_n = 0$ is the joint equation of the, n, asymptotes. At the points of intersection of the curve and its asymptotes, the two equations $F_n = 0$ and $F_n + F_{n-2} = 0$ hold simultaneously and therefore at such points we have $F_{n-2} = 0$.

Particular cases

(i) *For a cubic*, $n = 3$, and therefore the asymptotes cut the curve in $3 (3 - 2) = 3$ points which lie on a curve of degree $3 - 2 = 1$, *i.e.*, on a straight line.

(ii) *For a quartic*, $n = 4$, and, therefore the asymptotes cut the curve in $4 (4 - 2) = 8$ points which lie on a curve of degree $4 - 2 = 2$ *i.e.*, on a conic.

EXAMPLES

1. *Find the asymptotes of the curve*

$$x^2y - xy^2 + xy + y^2 + x - y = 0$$

and show that they cut the curve again in three points which lie on the line

$$x + y = 0.$$ *(Devi Ahilya 2001)*

The asymptotes of the given curve, as may be easily shown are

$$y = 0, x = 1, x - y + 2 = 0.$$

The joint equation of the asymptotes is

$$y (x - 1) (x - y + 2) = 0,$$

i.e., $$x^2y - xy^2 + xy + y^2 - 2y = 0.$$

The equation of the curve can be written as

$$x^2y - xy^2 + xy + y^2 - 2y + (x + y) = 0.$$

Here $$F_3 = x^2y - xy^2 + xy + y^2 - 2y, F_1 = x + y.$$

Hence, the points of intersection lie on the line

$$F_1 \equiv x + y = 0.$$

2. *Show that the asymptotes of the quartic*

$$(x^2 - 4y^2) (x^2 - 9y^2) + 5x^2y - 5xy^2 - 30y^3 + xy + 7y^2 - 1 = 0,$$

cut the curve in the eight points which lie on a circle.

The asymptotes of the curve are

$$x + 2y = 0, x - 2y + 1 = 0, x - 3y = 0, x + 3y - 1 = 0,$$

so that their joint equation is

$$(x + 2y) (x - 2y + 1) (x - 3y) (x + 3y - 1) = 0,$$

i.e., $$(x^2 - 4y^2) (x^2 - 9y^2) + 5x^2y - 5xy^2 - 30y^3 - x^2 + xy + 6y^2 = 0.$$

The equation of the curve can be written as

$$(x^2 - 4y^2)(x^2 - 9y^2) + 5x^2y - 5xy^2 - 30y^3 - x^2 + xy + 6y^2 + (x^2 + y^2 - 1) = 0.$$

Hence, the points of intersection lie on the circle

$$x^2 + y^2 - 1 = 0,$$

3. *Find the equation of the cubic which has the same asymptotes as the curve*

$$x^3 - 6x^2y + 11xy^2 - 6y^3 + x + y + 1 = 0.$$ **(Guwahati 2005)**

and which passes through the points $(0, 0), (1, 0)$ *and* $(0, 1)$.

We write $F_3 \equiv x^3 - 6x^2y + 11xy^2 - 6y^3$

$$= (x - y)(x - 2y)(x - 3y).$$

$$F_1 \equiv x + y + 1.$$

The equation of the curve can be written in the form $F_3 + F_1 = 0$ where F_3 has non-repeated linear factors. Thus $F_3 = 0$ is the joint equation of the asymptotes of the cubic.

The general equation of the cubic is of the form

$$F_3 + ax + by + c = 0,$$

or $x^3 - 6x^2y + 11xy^2 - 6y^3 + ax + by + c = 0,$

where $ax + by + c$ is the general linear expression.

In order that it may pass through the points $(0, 0), (1, 0)$ and $(0, 1)$, we must have

$$c = 0$$

$$1 + a = 0 \text{ or } a = -1,$$

$$-6 + b = 0 \text{ or } b = 6.$$

Thus the required cubic is

$$x^3 - 6x^2y + 11xy^2 - 6y^3 - x + 6y = 0.$$

4. *Show that asymptotes of the cubic*

$$x^3 - 2y^3 + xy(2x - y) + y(x - y) + 1 = 0$$

(Indore 1998; Jiwaji 1998; Vikram 1999; MDU Rohtak 1998)

cut the curve again in three points which lie on the straight line $x - y + 1 = 0$.

The equation of the curve is

$$(x^3 - 2y^3 + 2x^2y - xy^2) + (xy - y^2) + 1 = 0 \qquad \qquad \dots(i)$$

$$f_0(m) = 1 - 2m^3 + 2m - m^2 = 0$$

i.e., $m = 1, -1, -\dfrac{1}{2}$

and $f_0'(m) = -6m^2 + 2 - 2m$

and $f_1(m) = m - m^2.$

Now $c = -\dfrac{f_1(m)}{f_0'(m)} = -\dfrac{m - m^2}{2 - 2m - 6m^2}$

For $m = 1, c = 0; \ m = -1, c = -1; \ m = -\dfrac{1}{2}, c = \dfrac{1}{2}$

Hence the asymptotes are

$$(y - x)(y + x + 1)(x + 2y - 1) = 0$$

i.e., $x^3 - 2y^3 + 2x^2y - xy^2 + xy - y^2 - x + y = 0$ $\dots(ii)$

The asymptotes cut the curve in

$n(n - 2),$ *i.e.,* $3(3 - 2) = 3$ points.

Subtracting (*ii*) from (*i*), we observe that points of intersection of the curve and asymptotes lie on

$$x - y + 1 = 0$$

5. *Find the equation of the straight line on which lie the three points of intersection of the curve* $(x + a) y^2 = (y + b) x^2$ *and its asymptotes.*

The asymptotes parallel to *x*-axis and *y*-axis are $y + b = 0; x + a = 0$.

Now putting $y - x = k$ or $x = -k + y$, the curve is

$$(-k + y) y^2 + a y^2 = (y + b) (-k + y)^2$$

or $\qquad y^2 (-k + a - b + 2k) + ... = 0$

If $y - x = k$ is asymptote, the coefficient of y^2 is zero,

i.e., $\qquad\qquad k = -a + b$.

Hence third asymptote is $y - x = b - a$,

The combined equation of asymptotes is

$$(x + a) (y + b) (-y + x + b - a) = 0$$

or $\quad x y (y - x) + a y^2 - b x^2 + a^2 y - b^2 x + a b (a - b) = 0$.

Subtracting the combined equation to the asymptotes from the curves, we obtain

$$a^2 y - b^2 x + a b (a - b) = 0 \qquad\qquad ...(i)$$

Hence the points of intersection of the curve and asymptotes are $3 (3 - 2) = 3$ and they lie on (*i*).

6. *Find the equation of the conic on which lie the eight points of intersection of the quartic curve*

$$x y (x^2 - y^2) + a^2 y^2 + b^2 x^2 - a^2 b^2 = 0,$$

with its asymptote.

The given equation of curve is

$$x^3 y - x y^3 + a^2 y^2 + b^2 x^2 - a^2 b^2 = 0$$

By putting $x = 1$ and $y = m$ in the highest degree term, we get

$$f_0 (m) = m (1 - m^2) \text{ and } f_0' (m) = 1 - 3 m^2$$

Also, $f_1 (m) = 0$ as there is no third degree term in the equation of curve.

By putting $f_0 (m) = 0$, we get $m = 0, 1, -1$.

Also, $\qquad c = - \dfrac{f_1 (m)}{f_0' (m)} = 0$

∴ The asymptotes are $y = 0, y = x$ and $y = -x$.

Also equating the coefficient of highest power of *y* to zero, we have $x = 0$ as the asymptote.

Thus we have the four asymptotes of curve as $x = 0, y = 0, x + y = 0$ and $x - y = 0$.

Their combined equation is

$$x y (x + y) (x - y) = 0 \qquad \text{or} \qquad x y (x^2 - y^2) = 0.$$

Subtracting this equation from the gives equation we find that the point of intersection of the curve and asymptotes lie on

$$a^2 y^2 + b^2 x^2 - a^2 b^2 = 0 \qquad \text{or} \qquad \dfrac{x^2}{a^2} + \dfrac{y^2}{b^2} = 1$$

Which is an ellipse.

Also the asymptotes cut the curve in $n (n - 2)$, i.e., $4 (4 - 2)$ or 8 points.

7. *Find the equation of the cubic curve whose asymptotes are* $x + a = 0, y - a = 0$ *and* $x + y + a = 0$ *and which touches the axis of x at the origin and passes through the point* $(-2a, -2a)$.

The combined equation of the asymptotes is

$$F_3 \equiv (x + a)(y - a)(x + y + a) = 0$$

∴ The equation of the curve having $F_3 = 0$ as its asymptotes can be written as $F_3 + Q_1 = 0$, where Q_1 is an expression of degree $n - 2$, i.e., $3 - 2$ or 1.

Let $Q_1 \equiv bx + cy + d = 0$, then the equation of the curve is

$$(x + a)(y - a)(x + y + a) + (bx + cy + d) = 0. \qquad \qquad ...(i)$$

∵ This passes through the origin, so we have $d = a^3$, substituting $x = 0$ and $y = 0$ in (i).

So the equation of the curve reduces to

$$(x - a)(y - a)(x - y + a) + bx + cy + a^3 = 0.$$

or $\quad xy(x + y) + a(y^2 - x^2 + xy) - 2a^2x + bx + cy = 0$

and the tangent at the origin is $(-2a^2 + b)x + cy = 0$. obtained by equating the lowest degree terms to zero. But we are given that x-axis, i.e., $y = 0$ is the tangent at the origin. Hence $-2a^2 + b = 0$ and the equation of the curve reduces is

$$xy(x + y) + a(y^2 - x^2 + xy) + cy = 0. \qquad \qquad ...(ii)$$

Also this passes through $(-2a, -2a)$, hence

$$4a^2(-4a) + a(4a^2 - 4a^2 + 4a^2) + c(-2a) = 0 \qquad \text{or} \quad c = -6a^2.$$

Hence from (ii), the required equation is

$$xy(x + y) + a(y^2 - x^2 + xy) - 6a^2y = 0$$

8. *Find the equation of the quartic curve which has $x = 0, y = 0, y = x$ and $y = -x$ for asymptotes and which passes through (a, b) and which cuts its asymptotes again in eight points that lie on a circle whose centre is origin and radius a.*

The combined equation of the asymptotes is

$$xy(x - y)(y + x) = 0 \text{ or } xy(y^2 - x^2) = 0. \qquad \qquad ...(i)$$

Also the equation of the circle with origin as centre and radius a is

$$x^2 + y^2 = a^2. \qquad \qquad ...(ii)$$

Let the equation of the curve whose asymptotes are given by (i) be

$$xy(y^2 - x^2) + \lambda(x^2 + y^2 - a^2) = 0. \qquad \qquad ...(iii)$$

If it passes through (a, b), then

$$ab(b^2 - a^2) + \lambda(a^2 + b^2 - a^2) = 0, \text{ or } \lambda = a(a^2 - b^2)/b.$$

∴ The required equation of the curve is

$bxy(y^2 - x^2) + a(a^2 - b^2)(x^2 + y^2 - a^2) = 0$, putting the value of λ in (iii).

EXERCISES

1. Show that the asymptotes of the cubic

$$x^3 - xy^2 - 2xy + 2x - y = 0$$

cut the curve again in points which lie on the line

$$3x - y = 0.$$

2. If a right line is drawn through the point $(a, 0)$ parallel to the asymptote of the cubic $(x - a)^3 - x^2 y = 0$, prove that the portion of the line intercepted by the axes is bisected by the curve.

3. Through any point P on the hyperbola $x^2 - y^2 = 2ax$, a straight line is drawn parallel to the only asymptote of the curve $x^3 + y^3 = 3ax^2$ meeting the curve in A and B; show that P is the mid-point of AB.

4. Show that $y = x + a$ is the only asymptote of the curve
$$x^2(x - y) + ay^2 = 0.$$
A straight line parallel to the asymptote meets the curve in P, Q; show that the mid-point of PQ lies on the hyperbola
$$x(x - y) + ay = 0.$$

5. Find the equation of the straight line on which lie three points of intersection of the cubic
$$x^3 + 2x^2y - xy^2 - 2y^3 + 4y^2 + 2xy + y - 1 = 0$$
and its asymptotes.

6. Find the asymptotes of the curve
$$4x^4 - 13x^2y^2 + 9y^4 + 32x^2y - 42y^3 - 20x^2 + 74y^2 - 56y + 4x + 16 = 0,$$
and show that they pass through the intersection of the curve with $y^2 + 4x = 0$.

7. Find all the asymptotes of the curve
$$3x^3 + 2x^2y - 7xy^2 + 2y^3 - 14xy + 7y^2 + 4x + 5 = 0.$$
Show that the asymptotes meet the curve again in three points which lie on a straight line, and find the equation of this line.

8. Find the equation of the cubic which has the same asymptotes as the curve
$$x^3 - 6x^2y + 11xy^2 - 6y^3 + x + y + 1 = 0,$$
and which touches the axis of y at the origin and passes through the point $(3, 2)$.

9. Find the asymtotes of the curve
$$4(x^4 + y^4) - 17x^2y^2 - 4x(4y^2 - x^2) + 2(x^2 - 2) = 0.$$
and show that they pass through the points of intersection of the curve with the ellipse $x^2 + 4y^2 = 4$.

10. Find the asymptotes of the curve
$$(2x - 3y + 1)^2(x + y) - 8x + 2y - 9 = 0$$
and show that they intersect the curve again in three points, which lie on a straight line. Obtain the equation of the line.

ANSWERS

5. $3y + x = 1.$ 6. $y = \pm x + 1,\ y = \pm\dfrac{2}{3}x + \dfrac{4}{2}.$

7. $6(y - x) + 7 = 0,\ 2(y - 3x) + 3 = 0.$
$3(2y + x) + 5 = 0,\qquad 106y - 381x + 105 = 0.$

8. $x^3 - 6x^2y + 11xy^3 - 6y^2 - x = 0.$

9. $y = 2x^3 + 1,\ y = -2x - 1,\ x = 2y,\ x + 2y = 0.$

10. $x + y = 0,\ 2x - 3y - 1 = 0,\ 2x - 3y + 3 = 0,\ 4x - 6y + 9 = 0.$

15.6. Asymptotes by expansion.

To show that
$$y = mx + c$$
is an asymptote of the curve
$$y = mx + c + A/x + B/x^2 + C/x^3 + \ldots \qquad \ldots(1)$$
Dividing by x, we have
$$y/x = m + c/x + A/x^2 + B/x^3 + C/x^4 + \ldots$$
so that when $x \to \infty$.
$$\lim(y/x) = m. \qquad \ldots(2)$$
Also from (1) we have

$$(y - mx) = c + A/x + B/x^2 + C/x^3 + ...$$

so that when $x \to \infty$.

$$\lim (y - mx) = c. \qquad\qquad ...(3)$$

From (2) and (3), we deduce that

$$y = mx + c$$

is an asymptote of the curve (1).

15.7. Position of a curve with respect to an asymptote.

To find the position of the curve

$$y = mx + c + A/x + B/x^2 + C/x^3 + ...$$

with respect to its asymptote

$$y = mx + c.$$

Let $A \neq 0$. Let y_1 and y_2 denote the ordinates of the curve and the asymptote corresponding to the same abscissa x. We have

$$y_1 - y_2 = A/x + B/x^2 + C/x^3 + ...$$
$$= (1/x)(A + B/x + C/x^2 + ...) \qquad\qquad ...(1)$$

By taking x sufficiently large, we can make

$$B/x + C/x^2 + ...$$

as small as we like. We suppose that x is so large numerically that this expression is numerically less than A. Thus, for sufficiently large values of x, the expression

$$A + B/x + C/x^2 + D/x^3 + ... \qquad\qquad ...(2)$$

has the sign of A.

Thus if A be positive, the expression (1) is positive for sufficiently large values of x so that, from (1) we deduce that when x is *positively* sufficiently large, then $y_1 - y_2$ is positive, *i.e.*, the curve lies above the asymptote and when x is *negative* but numerically large, $y_1 - y_2$ is negative, *i.e.*, the curve lies below the asymptote.

Similarly, we may deduce that if A be negative, then the curve lies below the asymptote when, x, is positive but sufficiently large and lies above the asymptote when, x, is negative but sufficiently large numerically.

Let $\qquad\qquad A = 0, B \neq 0$; we have

$$y_1 - y_2 = (1/x^2)(B + C/x + D/x^2 + ...)$$

As above we can show that for numerically sufficiently large values of x, the expression

$$B + C/x + D/x^2 + ...$$

has the sign of B.

In this case, the curve lies on the same side of the asymptote both for positive and negative values of x; it will be above or below the asymptote according as B is positive or negative.

It $B = 0$ and $C \neq 0$, we will have a situation similar to that of case (1).

Ex. 1. *Find the asymptotes of the curve*

$$y^2 = x(x - a)(x - 2a)/(x + 3a)$$

and determine on which side of the asymptotes the curve lies.

We have $\qquad y = \pm \sqrt{\left[\dfrac{x(x - a)(x - 2a)}{x + 3a} \right]}$

$$= \pm x \left(1 - \frac{a}{x} \right)^{1/2} \left(1 - \frac{2a}{x} \right)^{1/2} \left(1 - \frac{3a}{x} \right)^{-1/2}$$

$$= \pm x \left(1 - \frac{a}{2x} - \frac{a^2}{8x^2} \ldots\right)\left(1 - \frac{a}{x} - \frac{a^2}{2x^2} \ldots\right)\left(1 - \frac{3a}{2x} + \frac{27a^2}{8x^2} \ldots\right)$$

$$= \pm x \left(1 - \frac{3a}{x} + \frac{11a^2}{2x^2} \ldots\right)$$

Thus we have two values of y, viz.,

$$y = x - 3a + \frac{11}{2} a^2 x \ldots$$

$$y = -x + 3a - \frac{11}{2} a^2 x \ldots$$

Therefore $y = x - 3a,\ y = -x + 3a$

are the two asymptotes.

The difference between the ordinate of the curve and that of the asymptote $y = x - 3a$ being

$$\frac{11}{2} a^2 x - \ldots,$$

we see that the curve lies above the asymptote when x is positive and below it when x is negative.

It may similarly be seen that the curve lies below the second asymptote when x is positive and above it when x is negative.

It is easy to see that

$$x = -3a$$

is also an asymptote of the curve. To find the position of the curve relative to this asymptote, we suppose that

$$x = -3a + A/y + B/y^2 + C/y^3 + \ldots$$

Substituting this value of x in the equation of the curve, we have

$$y^2 \left[\frac{A}{y} + \frac{B}{y^2} + \frac{C}{y^3} + \ldots\right] = \left[-3a + \frac{A}{y} + \ldots\right]\left[-4a + \frac{A}{y} + \ldots\right]\left[-5a + \frac{A}{y} + \ldots\right]$$

Equating the co-efficients of like powers of y, we have

$$A = 0.$$
$$B = -60a^3,\ \text{etc.}$$

$$\therefore \qquad x = -3a - \frac{60a^3}{y^2} + \ldots$$

The difference between the abscissae of the curve and the asymptote $x = -3a$, for the same value of y, being

$$-60a^3/y^2 + \ldots,$$

which is negative whether y be positive or negative, we see that the curve lies towards the negative side of x-axis.

Ex. 2. *Find the asymptotes and their position with regard to the following curves* :

   (i) $x^3 + y^3 = 3ax^2$.   (ii) $x^3 + y^3 = 3axy$.   (iii) $x^2(x - y) + y^2 = 0$.

ANSWERS

  (i) $x + y = a$. The curve lies above or below the asymptote according as x is positive or negative.

 (ii) $x + y + a = 0$. The curve lies above the asymptote both for positive as well as negative values of x.

(iii) $y = x + 1$. The curve lies above or below the asymptote according as x is positive or negative.

15.8. Asymptotes in polar co-ordinates.

Lemma. The Polar Equation of a Line. *The polar equation of any line is*

$$p = r \cos(\theta - \alpha).$$

where, p, is the length of the perpendicular from the pole to the line and α, is the angle which this perpendicular makes with the initial line.

Let OY be the perpendicular on the given line; Y being its foot.

We are given that

$$OY = p; \angle XOY = \alpha.$$

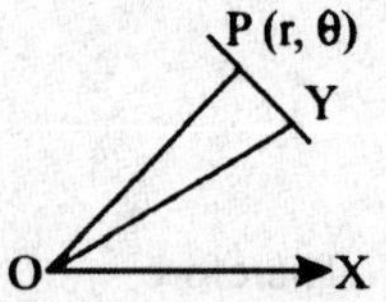

If $P(r, \theta)$ be any point on the line, we have

$$\angle YOP = \theta - \alpha.$$

Now, $$\frac{OY}{OP} = \cos \angle YOP.$$

Fig. 15.4

$\therefore$ $p/r = \cos(\theta - \alpha)$, *i.e.*, $p = r \cos(\theta - \alpha)$,

which is the required equation of the line.

To determine the asymptotes of the curve

$$r = f(\theta), \qquad \qquad \qquad ...(i)$$

we have to obtain the constants, p and, α, so that any line

$$p = r \cos(\theta - \alpha), \qquad \qquad ...(ii)$$

is the asymptote of the given curve.

Let $P(r, \theta)$ be any point on the curve (*i*).

Draw $OY \perp$ the line (*ii*).

Draw $PL \perp OY$ and $PM \perp$ the line (*ii*).

Now,

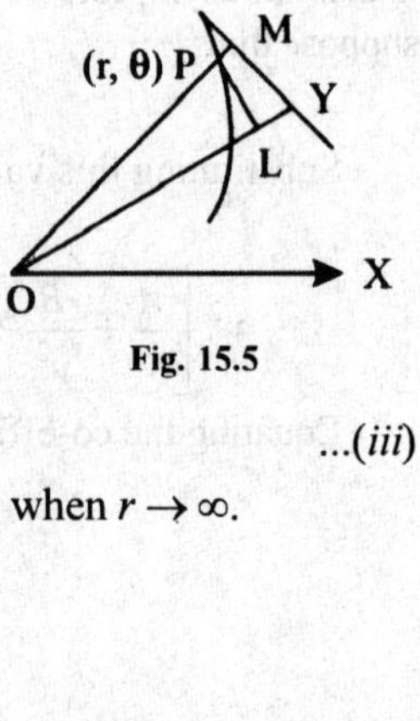

Fig. 15.5

$$PM = LY$$
$$= OY - OL$$
$$= p - OP \cos(\theta - \alpha)$$
$$= p - r \cos(\theta - \alpha). \qquad \qquad ...(iii)$$

Now $r \to \infty$ as the point recedes to infinity along the curve. Let $\theta \to \theta_1$ when $r \to \infty$.

We have $$\frac{PM}{r} = \frac{p}{r} - \cos(\theta - \alpha).$$

Now when $r \to \infty$, $PM \to 0$ so that

$$\frac{PM}{r} = PM \cdot \frac{1}{r} \to 0 \text{ and } \frac{p}{r} \to 0.$$

$\therefore$ $$\lim \cos(\theta - \alpha) = 0$$

or $$\lim (\theta - \alpha) = \pi/2,$$

or $\theta_1 - \alpha = \pi/2$, *i.e.*, $\alpha = \theta_1 - \pi/2$.

This gives α.

Again, $p = OY$ *is the polar sub-tangent of the point at which the asymptote touches the curve,* *i.e., the point at infinity on the curve.* This may be seen as follows :—

Join the pole O to the point at infinity on the curve *i.e.*, draw through O a line parallel to the asymptote. This line is the radius vector of the point at ∞.

Draw through O a line perpendicular to the asymptote meeting it at Y. Then, by def. OY is the polar sub-tangent of the point at infinity on the curve.

Thus $\qquad p = -\left[\dfrac{d\theta}{du}\right]_{\theta=\theta_1}$, where $u = \dfrac{1}{r}$.

Note. Without employing the notion of the polar sub-tangent and the point at infinity, the value of p, may also be obtained as follows :—

From (iii) we have, when $r \to \infty$,

$$p = \lim\, [r \cos(\theta - \alpha)]$$
$$= \lim\, [r \cos(\theta - \theta_1 + \pi/2)]$$
$$= \lim\, [r \cos(\theta_1 - \theta)] = \lim_{\theta \to \theta_1} \frac{\sin(\theta_1 - \theta)}{1/r}$$

which is of the form (0/0).

$$p = \lim_{\theta \to \theta_1} \frac{\cos(\theta_1 - \theta)}{\dfrac{1}{r^2}\dfrac{dr}{d\theta}}$$

$$= \lim_{\theta \to \theta_1} \left[r^2 \frac{d\theta}{dr}\right] = \lim \left[-\frac{d\theta}{du}\right] \text{ where } u = \frac{1}{r}$$

Hence, the asymptote is

$$\lim \left(-\frac{d\theta}{du}\right) = r \cos(\theta - \alpha)$$

$$= r \cos\left(\theta - \theta_1 + \frac{\pi}{2}\right)$$

$$= r \sin(\theta_1 - \theta).$$

where θ_1 is the limit of θ as $r \to \infty$ i.e., as $u \to 0$.

Working rule for obtaining asymptotes to polar curves.

Change r to 1/u in the given equation and find out the limit of θ as $u \to 0$.

Let θ_1, be any one of the several possible limits of θ.

Determine $(-d\theta/du)$ and its limit as $u \to 0$ and $\theta \to \theta_1$.

Let this limit be p.

Then $\qquad p = r \sin(\theta_1 - \theta).$

is the corresponding asymptote.

To draw the asymptote.

Through the pole O draw a line making angle $\left(\theta_1 - \dfrac{1}{2}\pi\right)$ with the initial line; on this line take a point Y such that

$$OY = \lim\,(-d\theta/du).$$

The line drawn through Y perpendicular to OY is the required asymptote.

EXAMPLES

1. *Find the asymptote of the hyperbolic spiral $r\theta = a$.*

Here $\qquad \theta = a/r = au$ so that $\theta \to 0$ as $u \to 0$.

Here $\qquad \theta_1 = \lim \theta = 0$.

Since $\qquad u = \theta/a,$

we have $\qquad du/d\theta = 1/a$ or $d\theta/du = a$.

Therefore $\qquad -a = r \sin(0 - \theta) = -r \sin \theta$.

i.e., $\qquad r \sin \theta = a,$

is the asymptote.

2. *Find the asymptote of the curve*

$$r = \frac{a}{\frac{1}{2} - \cos \theta}.$$

Here

$$u = \frac{1}{r} = \frac{1}{a}\left(\frac{1}{2} - \cos \theta\right).$$

When $u \to 0$, $\left(\frac{1}{2} - \cos \theta\right) \to 0$ so that $\cos \theta \to \frac{1}{2}$.

$\therefore \qquad \theta_1 = \pm \pi/3.$

Now, $\qquad \dfrac{du}{d\theta} = \dfrac{1}{a} \sin \theta$ or $- \dfrac{d\theta}{du} = - \dfrac{a}{\sin \theta}.$

$\therefore \qquad - \dfrac{d\theta}{du} \to - \dfrac{2a}{\sqrt{3}}$ as $\theta \to \dfrac{\pi}{3},$

and $\qquad \dfrac{-d\theta}{du} \to \dfrac{2a}{\sqrt{3}}$ as $\theta \to - \dfrac{\pi}{3}.$

$\therefore \qquad - \dfrac{2a}{\sqrt{3}} = r \sin\left(\dfrac{\pi}{3} - \theta\right)$, i.e., $4a = r\,(\sqrt{3} \sin \theta - 3 \cos \theta),$

and $\qquad \dfrac{2a}{\sqrt{3}} = r \sin\left(-\dfrac{\pi}{3} - \theta\right)$, i.e., $-4a = r\,(\sqrt{3} \sin \theta + 3 \cos \theta),$

are the two asymptotes.

EXERCISES

Find the asymptotes of the following curves :

1. $r = \dfrac{a\,\theta}{\theta - 1}.$

2. $r = \dfrac{3a \sin \theta \cos \theta}{\sin^3 \theta + \cos^3 \theta}.$

3. $r = a \sec \theta + b \tan \theta.$

4. $r^2 = a^2\,(\sec^2 \theta + \operatorname{cosec}^2 \theta).$

5. $r \sin 2\theta = a \cos 3\theta.$

6. $2r^2 = \tan 2\theta.$

7. $r\theta \cos \theta = a \cos 2\theta.$

8. $r \sin n\theta = a.$

9. $r^n \sin n\theta = a^n.$

10. $r = a \tan \theta.$

11. $r = a \log \theta.$

12. $r \log \theta = a.$

13. $r\,(1 - e^\theta) = a.$

14. $r\,(\theta^2 - n^2) = 2a\theta.$

15. $r \sin \theta = ae^\theta$

16. $r\,(\pi + \theta) = ae^\theta.$

17. Find the equation of the asymptotes of the curve given by the equation
$$r^n f_n(\theta) + r^{n-1} f_{n-1}(\theta) + \dots + f_0(\theta) = 0$$

18. Show that all the asymptotes of the curve $r \tan n\theta = a,$
    touch the circle $r = a/n.$

19. Find the asymptotes of the curve $r \cos 2\theta = a \sin 3\theta$ *(Kumaon 2003)*

ANSWERS

1. $-a = r \sin(1 - \theta).$

2. $a + \sqrt{2r}\,\sin\left(\theta + \dfrac{1}{4}\pi\right) = 0.$

3. $r \cos \theta \pm b = a.$

4. $r \cos \theta \pm a = 0,\ r \sin \theta \pm a = 0.$

5. $2r \sin \theta = a,\ 2\theta = \pi.$

6. $\theta = \pm \pi/4.$

7. $r \sin \theta = a$, $\pi (r \cos \theta) + 2a = 0$. 8. $r \sin \left(\theta - \dfrac{m\pi}{n} \right) = \dfrac{a}{n \cos m\pi}$ where m is any integer.

9. $n\theta = m\pi$ where m is any integer. 10. $r \cos \theta = \pm a$.

11. $\theta = 0$. 12. $a = r \sin (\theta - 1)$. 13. $y + a = 0$.

14. $y + a = 0$.

15. System of parallel lines $y = ae^{n\pi}$ where n is any integer or zero.

16. $ye^{\pi} + a = 0$.

17. $\dfrac{f_{n-1}(\theta)}{f_n(\theta_1)} = r \sin (\theta_1 - \theta)$, where θ_1 is any root or the equation $f_n(\theta) = 0$.

19. $a = 2r (\cos \theta - \sin \theta)$, $a + 2r (\cos \theta + \sin \theta) = 0$.

OBJECTIVE QUESTIONS

For each of the following questions, four alternatives are given for the answer, only one of them is correct. Choose the correct alternative.

1. The number of asymptotes of a curve of nth degree is :
   (a) at least one (b) at least n
   (c) at most n (d) at most 1 (**Avadh, 2002, 2005**)

2. The parabola $y^2 = 4ax$ possesses
   (a) one real asymptote (b) two real asymptotes
   (c) many real asymptotes (d) no real asymptote (**Garhwal 2004**)

3. A closed curve has
   (a) no asymptote (b) one asymptote
   (c) infinitely many asymptotes (d) n asymptotes (**Kumaon 2005**)

4. The asymptotes of the hyperbola $xy - 2y = 0$ are given by
   (a) $x - 2 = 0, y + 3 = 0$ (b) $x - 3 = 0, y + 2 = 0$
   (c) $x - 2 = 0, y - 3 = 0$ (d) $x - 3 = 0, y - 2 = 0$

5. Let $f(x, y) = 0$ be an algebraic equation of degree 8. The asymptotes of this curve will intersect in 18 points only if the number of asymptotes is :
   (a) 3 (b) 8 (c) 18 (d) 2

6. Which of the following is true ? An algebraic curve of an
   (a) even degree has at least one real asymptotes
   (b) even degree may\or may not have real asymptote
   (c) odd degree has at least two imaginary asymptotes
   (d) odd degree has at least two real asymptotes

7. The asymptotes of the curve $x^2y^2 - a^2 (x^2 + y^2) = 0$ form a :
   (a) circle (b) triangle (c) square (d) pentagon

ANSWERS

| 1. (c) | 2. (d) | 3. (a) | 4. (c) | 5. (a) | 6. (b) | 7. (c). |

CHAPTER

16

SINGULAR POINTS

MULTIPLE POINTS, DOUBLE POINTS

16.1 Introduction. Cusps, Nodes and Conjugate points. The cases of curves considered in § 17.4 p. 614 show that curves with implicit equations of the form $f(x, y) = 0$ exhibit some peculiarities which are not possessed by the curves with explicit equations of the form $y = F(x)$. These peculiarities arise from the fact that the equations $f(x, y) = 0$ may not define, y, as a single valued function of x. In fact, to each value of x correspond as many values of y as is the degree of the equation in y, and these different values of y give rise to different branches of the curve.

We recall to ourselves the following three curves considered in § 13.4.

Fig. 16.1   Fig. 16.2

(*i*) Origin is a point common to the two branches of the Cissoid
$$y^2 (a - x) = x^3,$$
and the two branches have a *common* tangent there.

Such a point on a curve is called a **cusp.**

(*ii*) Origin is a point common to the two branches of the Strophoid (Fig. 16.2)
$$(x^2 + y^2) x - a (x^2 - y^2) = 0,$$ (§ 17.4)
and the two branches have *different* tangents there.

Such a point on a curve is called a **node.**

(*iii*) $(- a, 0)$ is a point common to the two branches of the curve
$$ay^2 - x (x + a)^2 = 0$$ (§ 17.5)
and the two branches have *imaginary* tangents there. There is no point in the immediate neighbourhood of the point $(- a, 0)$ which lies on the curve. Here, a, is positive.

Such a point on a curve is a called an **isolated** or **conjugate** point.

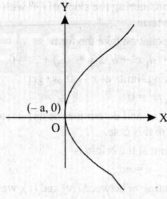

(-a, 0)

Fig. 16.3

16.2. Definitions.

Double points. Cusp. Node. Conjugate point. *A point through which there pass* **two** *branches of a curve is called a* **double point.**

A curve has two tangents at a double point, one for each branch.

The double point will be a **node,** *a* **cusp** *or an* **isolated** *point according as the two tangents are different and real, coincident or imaginary.*

Multiple point. *A point through which there pass, r, branches of a curve is called a* **multiple point of the rth order** *so that a curve has, r, tangents at a multiple point of the rth order.*

(Banglore 2005)

Thus a double point is a multiple point of the second order. A multiple point of the third order is also called a *triple* point. A multiple point is also, sometimes, called a **singular point.**

(Banglore 2005)

16.3. A simple rule for writing down the *tangent or tangents at the origin* to rational algebraic curves is obtained in the following article.

16.3.1 Tangents at the origin. The general equation of *rational algebraic* curve of the nth degree which passes through the origin O, when arranged according to ascending powers of x and y, is of the form

$$(b_1 x + b_2 y) + (c_1 x^2 + c_2 xy + c_3 y^2) + (d_1 x^3 + d_2 x^2 y + ...) + ... = 0 \qquad ...(i)$$

where the constant term is absent.

Let $P(x, y)$ be any point on the curve. The slope of the chord OP is y/x. Limiting position of the chord OP, when $P \to O$, is the tangent at O so that when $x \to 0$ and $y \to 0$.

$$\lim (y/x) = m,$$

the slope of this tangent.

From (i), we have, after dividing by x,

$$\left(b_1 + b_2 \frac{y}{x}\right) + \left(c_1 x + c_2 y + c_3 y \cdot \frac{y}{x}\right) + (d_1 x^2 + d_2 xy + ...) + ... = 0$$

On taking limits, when $x \to 0$, we get

$$b_1 + b_2 m = 0 \text{ so that } m = -b_1/b_2, \text{ if } b_2 \neq 0.$$

Hence $\qquad\qquad y/x = -b_1/b_2,$

i.e., $\qquad\qquad b_1 x + b_2 y = 0 \qquad\qquad\qquad ...(ii)$

is the tangent at the origin. This may be written down by equating to zero the lowest degree (first degree) terms in the equation (i).

If $b_2 = 0$ but $b_1 \neq 0$, then, considering the slope of OP with reference to Y-axis, it can be shown that the tangent retains the same form.

Let $b_1 = b_2 = 0$ so that the equation take the form

$$(c_1 x^2 + c_2 xy + c_3 y^2) + (d_1 x^3 + d_2 x^2 y + ...) + ... = 0. \qquad ...(iii)$$

Dividing by x^2 and then taking limits as $x \to 0$, we get

$$c_1 + c_2 m + c_3 m^2 = 0, \qquad ...(iv)$$

which is a quadratic equation in, m, and determines as its *two* roots the slopes of the *two tangents* so that the origin is a double point in this case.

The equation of either tangent at the origin is

$$y = mx, \qquad ...(v)$$

when m is a root of (iv). Eliminating m between (iv) and (v), we obtain

$$c_1 x^2 + c_2 xy + c_3 y^2 = 0, \qquad ...(vi)$$

as the joint equation of the two tangents at the origin. This can be written down by equating to zero the lowest degree terms in (iii).

The equation (vi) becomes an identity if $c_1 = c_2 = c_3 = 0$. In this case the second degree terms, also, do not appear in the equation of the curve. It can now be similarly shown that the equation of the tangents can still be written down by equating to zero the terms of the lowest degree which is third in this case.

In general, we see *that the equation of the tangent or tangents at the origin is obtained by equating to zero the terms of the lowest degree in the equation of the curve.*

The origin will be a multiple point on a curve whose equation does not, at least, contain the constant and the first degree terms.

Illustrations.

(i) The origin is a node on the curve

$$x^3 + y^3 - 3axy = 0,$$

and $x = 0, y = 0$ are the two tangents thereat.

(ii) The origin is a cusp on the curve

$$(x^2 + y^2) x - 2ay^2 = 0,$$

and $y = 0$ is the cuspidal tangent.

(iii) The origin is an isolated point on the curve

$$a^2 x^2 + b^2 y^2 = (x^2 + y^2)^2,$$

and $ax \pm iby = 0$ are the two imaginary tangents thereat.

(iv) The origin is a triple point on the curve

$$2y^5 + 5x^5 - 3x (x^2 - y^2) = 0,$$

and $x = 0, x = y, x = -y$ are the three tangents thereat.

EXAMPLE

Find the tangent at the origin of the curve $x^2 (x^2 + y^2) = a (x - y)$.

The given curve is

$$x^2 (x^2 + y^2) = a (x - y)$$

Equating to zero the lowest degree terms we obtain the equation of the tangents at origin as $x - y = 0$.

EXERCISES

Find the tangents at the origin to the following curves:–

1. $(x^2 + y^2)^2 = 4a^2xy$.

2. $y^2(a^2 - x^2) = x^2(b - x)^2$.

3. $(x^2 + y^2)(2a - x) = b^2x$.

4. $a^2(x^2 - y^2) = x^2y^2$.

5. $(x^2 + y^2)^3 = a^2(x^2 - y^2)^2$.

ANSWERS

1. $x = 0, y = 0$.

2. $bx = \pm ay$.

3. $x = 0$.

4. $y = \pm x$.

5. $y = \pm x$.

EXAMPLE

Find the equation of the tangent at $(-1, -2)$ *to the curve*

$$x^3 + 2x^2 + 2xy - y^2 + 5x - 2y = 0,$$

and show that this point is a cusp.

We will shift the origin to the point $(-1, -2)$. To do so we have to write

$$x = X - 1, y = Y - 2,$$

where X, Y are the current co-ordinates of a point on the curve with reference to the new axes. The transformed equation is

$$(X - 1)^3 + 2(X - 1)^2 + 2(X - 1)(Y - 2) - (Y - 2)^2 + 5(X - 1) - 2(Y - 2) = 0,$$

or

$$X^3 - X^2 + 2XY - Y^2 = 0.$$

Equating to zero the lowest degree terms, we get

$$-X^2 + 2XY - Y^2 = 0, i.e., (Y - X)^2 = 0,$$

which are two coincident lines, and, therefore, the point is a cusp and the cuspidal tangent, *i.e.*, the tangent at the cusp with reference to the new axes is

$$Y - X = 0.$$

To find the equation of the cuspidal tangent with reference to the given system of axes, we write

$$X = x + 1, Y = y + 2.$$

Hence the tangent at $(-1, -2)$ is

$$(y + 2) - (x + 1) = 0, i.e., y = x - 1.$$

EXERCISES

Find the equations of the tangents to the following curves :

1. $y^2(a^2 + x^2) = x^2(a^2 - x^2)$ at $(\pm a, 0)$.

2. $(x - 2)^2 = y(y - 1)^2$ at $(2, 1)$.

3. $x^4 - 4ax^3 - 2ay^3 + 4a^2x^2 + 3a^2y^2 - a^4 = 0$ at $(a, 0)$ and $(2a, a)$.

4. Show that the origin is a node; a cusp or a conjugate point on the curve

$$y^2 = ax^2 + ax^3,$$

according as, a is positive, zero or negative.

ANSWERS

1. $x = \pm a$.

2. $(y - 1) = \pm(x - 2)$.

3. $\pm\sqrt{3}y = \sqrt{2(x - a)}; 2(x - 2a) = \pm\sqrt{3(y - a)}$.

16.4. Conditions for any point (x, y) to be a multiple point of the curve
$$f(x, y) = 0.$$

In § 11.9.4, we have seen that at a point (x, y) of the curve

$$f(x, y) = 0,$$

the slope of the tangent, dy/dx is given by the equation

$$f_x + f_y \frac{dy}{dx} = 0. \qquad \qquad ...(i)$$

At a multiple point of a curve, the curve has at least two tangents and accordingly dy/dx must have at least two values at multiple point.

The equation (i), being of the first degree in dy/dx, can be satisfied by more than one value of dy/dx, if and only if,

$$f_x = 0, f_y = 0.$$

Thus, we see that *the necessary and sufficient conditions for any point (x, y) on $f(x, y) = 0$ to be a multiple point are that*

$$f_x(x, y) = 0, f_y(x, y) = 0.$$

To find multiple points (x, y), we have therefore to find the values of (x, y) which simultaneously satisfy the three equations

$$f_x(x, y) = 0, f_y(x, y) = 0, f(x, y) = 0.$$

16.4.1 To find the slopes of the tangents at a double point.

Differentiating (i), *w.r.* to x, we have

$$f_x^2 + f_{xy} \frac{dy}{dx} + \left(f_{yx} + f_{y^2} \frac{dy}{dx} \right) \frac{dy}{dx} + f_y \frac{d^2 y}{dx^2} = 0,$$

so that at the multiple point, where $f_y = 0, f_x = 0$, the values of dy/dx are the roots of the quadratic equation

$$f_{y^2} \left(\frac{dy}{dx} \right)^2 + 2 f_{xy} \frac{dy}{dx} + f_{x^2} = 0. \qquad \qquad ...(ii)$$

In case f_{x^2}, f_{xy}, f_{y^2}, are not all zero and $f_x = 0 = f_y$, the point (x, y) will be a double point and will be a node, cusp or conjugate according as the values of dy/dx are real and distinct, equal or imaginary i.e., according as

$$(f_{xy})^2 - f_{x^2} f_{y^2} > 0, = 0, < 0.$$

If $f_{x^2} = f_{xy} = f_{y^2} = 0$; the point (x, y) will be multiple point of order higher than the second.

EXAMPLES

1. *Find the multiple points on the curve*

$$x^4 - 2ay^3 - 3a^2 y^2 - 2a^2 x^2 + a^4 = 0.$$

Also, find the tangents at the multiple point.   *(Gorakhpur 2000, 2001)*

Let $f(x, y) = x^4 - 2ay^3 - 3a^2 y^2 - 2a^2 x^2 + a^4$

∴ $f_x(x, y) = 4x^3 - 4a^2 x.$

              $f_y(x, y) = -6ay^2 - 6a^2 y.$

              $f_x(x, y) = 0$ gives $x = 0, a, -a.$

              $f_y(x, y) = 0$ gives $y = 0, -a.$

Hence, the two partial derivatives vanish for the points

   $(0, 0), (0, -a), (a, 0), (a, -a), (-a, 0), (-a, -a)$

Of these the only points on the curve are
$$(a, 0), (-a, 0), (0, -a).$$
Hence, these are the only three multiple points, on the curve.

To find the tangents at the multiple points, we proceed as follows :–

We have
$$f_{x^2} = 12x^2 - 4a^2, f_{xy} = 0, f_{y^2} = -12ay - 6a^2.$$
Since at $(a, 0)$,
$$f_x^2 = 8a^2, f_{xy} = 0, f_y^2 = -6a^2.$$
Therefore, by the equation (ii), the values of dy/dx at $(a, 0)$ are given by
$$- 6a^2 (dy/dx)^2 + 8a^2 = 0.$$

i.e., $\qquad dy/dx = \pm 2/\sqrt{3}.$

The two values being both real, the point $(a, 0)$ is a node. The tangents at $(a, 0)$ are
$$y = \pm (2/\sqrt{3}) (x - a).$$
It may similarly be shown that the tangents at $(-a, 0)$ and $(0, -a)$ are
$$y = \pm \sqrt{\frac{4}{3}} (x + a), \qquad y + a = \pm \sqrt{\frac{2}{3}} x.$$

Second method. Differentiating the given equation *w. r.* to x, we get,
$$4x^3 - 6ay^2y_1 - 6a^2yy_1 - 4a^2x = 0,$$
which identically vanishes for the multiple points.

Differentiating again, we get
$$12x^2 - 12ayy_1{}^2 - 6ay^2y_2 - 6a^2y_1{}^2 - 6a^2yy_2 - 4a^2 = 0.$$
From this we see that

(i) for $(a, 0)$, $y_1{}^2 = \dfrac{3}{4}$, *i.e.*, $y_1 = \pm \sqrt{\dfrac{3}{4}}$.

(ii) for $(-a, 0)$, $y_1{}^2 = \dfrac{3}{4}$, *i.e.*, $y_1 = \pm \sqrt{\dfrac{3}{4}}$.

(iii) for $(0, -a)$, $y_1{}^2 = \dfrac{2}{3}$, *i.e.*, $y_1 = \pm \sqrt{\dfrac{2}{3}}$.

Knowing the slopes of the tangents, we can now put down their equations.

Third method. To find the tangents at $(a, 0)$, we shift the origin to this point. The transformed equation is
$$(x = X + a, y = Y + 0)$$
$$(X + a)^4 - 2aY^3 - 3a^2Y^2 - 2a^2 (X + a)^2 + a^4 = 0,$$
or $X^4 + 4X^3a - 2aY^3 + 4a^2X^2 - 3a^2Y^2 = 0.$

The tangents at the new origin are
$$4a^2X^2 - 3a^2Y^2 = 0,$$
$$Y = \pm \sqrt{(4/3)}\ X.$$

The tangents at the multiple point $(a, 0)$, therefore, are

$$Y = \pm \sqrt{(4/3)\,(x - a)}.$$

It may similarly be shown that

$$Y = \pm \sqrt{(4/3)\,(x + a)} \text{ and } y + a = \pm \sqrt{(2/3)\,x},$$

are the tangents at the multiple points $(-a, 0)$ and $(0, -a)$ respectively.

The three multiple points on the curve are nodes.

2. *Show that the origin of the curve* $y^2 = b\,x \sin \dfrac{x}{a}$, *there is a node, or a conjugate point according as a and b have like or unlike signs.*

The given equation of the curve is

$$y^2 = b\,x \sin \frac{x}{a}$$

or

$$y^2 = b\,x \left(\frac{x}{a} - \frac{x^3}{a^3 . 3!} + ... \right)$$

Then the approximate equation of the curve will ι

$$y^2 = \frac{b}{a}\,x^2 - \frac{b}{6\,a^3}\,x^4.$$

Now the tangents at the origin are given by

$$y^2 = \frac{b}{a}\,x^2.$$

If a and b have like signs, then two tangents at the origin are real and non-concident; so there is a node which can be confirmed by giving small values of positive or negative.

If a and b have unlike signs the two tangents at the origin will be imaginary and so we will have a conjugate point there.

Hence, origin will be a node or a conjugate point according as a and b have like or unlike signs.

3. *Locate the double points of the curve* $y\,(y - 6) = x^2\,(x - 2)^3 - 9$, *and ascertain their nature.*

(Avadh, 2000; Gorakhpur, 2003)

Here
$$f(x, y) = x^2\,(x - 2)^3 - y\,(y - 6) - 9 = 0 \qquad \qquad ...(i)$$

$$\frac{\partial f}{\partial x} = 2\,x\,(x - 2)^3 + x^3 . 3\,(x - 2)^2$$

$$= x\,(x - 2)^2\,(5\,x - 4) \qquad \qquad ...(ii)$$

$$\frac{\partial f}{\partial y} = 6 - 2\,y \qquad \qquad ...(iii)$$

$$\frac{\partial^2 f}{\partial x^2} = (x - 2)^2\,(5\,x - 4) + 2\,x\,(x - 2)\,(5\,x - 4) + 5\,x\,(x - 2)^2 \qquad \qquad ...(iv)$$

$$\frac{\partial^2 f}{\partial y^2} = -2 \qquad \qquad ...(v)$$

and
$$\frac{\partial^2 f}{\partial x \partial y} = 0 \qquad \qquad ...(vi)$$

Now
$$\frac{\partial f}{\partial x} = 0 \text{ gives } \quad x = 0,\ 2,\ \frac{4}{5}$$

and $\qquad \dfrac{\partial f}{\partial y} = 0$ gives $\qquad y = 3.$

Hence possible double points are $(0, 3)$, $(2, 3)$ and $\left(\dfrac{4}{5}, 3\right)$.

The point $\left(\dfrac{4}{5}, 3\right)$ does not satisfy $f(x, y) = 0$, and therefore there are only two double points.

Now at $(0, 3)$, $\dfrac{\partial^2 f}{\partial x^2} = -16$, $\dfrac{\partial^2 f}{\partial y^2} = -2$, $\dfrac{\partial^2 f}{\partial x \partial y} = 0.$

$\therefore \qquad \left(\dfrac{\partial^2 f}{\partial x \partial y}\right)^2 < \dfrac{\partial^2 f}{\partial x^2} \cdot \dfrac{\partial^2 f}{\partial y^2}.$

Hence $(0, 3)$, is a conjugate point.

At $(2, 3)$, $\qquad \dfrac{\partial^2 f}{\partial x^2} = 0$, $\dfrac{\partial^2 f}{\partial y^2} = -2$ and $\dfrac{\partial^2 f}{\partial x \partial y} = 0$

$\therefore \qquad \left(\dfrac{\partial^2 f}{\partial x \partial y}\right)^2 = \dfrac{\partial^2 f}{\partial x^2} \cdot \dfrac{\partial^2 f}{\partial y^2}$

Therefore the point $(2, 3)$ is a cusp.

Further to decide the nature of the cusp, transfer the origin to the point $(2, 3)$. The equation then becomes

$$(y + 3)(y - 3) = x^3 (x + 2)^2 - 9$$

i.e., $\qquad\qquad y^2 = x^3 (x + 2)^2.$

The tangents at the new origin are $y^2 = 0$, i.e., x–axis is the common tangent.

Solving for y, we get

$$y = \pm (x + 2) \sqrt{x^3},$$

which shows that when x is negative, y is imaginary and when x is positive, y has two values one positive other negative. Thus near the new origin the curve lies on both sides of x–axis (tangent) and only on one side of y–axis (normal).

Hence the new origin $(2, 3)$ is a single cusp of first species.

4. *Find the position and nature of the double points on the curve* $x^2 y^2 = (a + y)^2 (b^2 - y^2)$ *distinguishing between the cases* $a > b$ *or* $< b$.

Here $f(x, y) = x^2 y^2 - (a + y)^2 (b^2 - y^2) = 0$

$\therefore \qquad \dfrac{\partial f}{\partial x} = 2xy^2$, $\dfrac{\partial f}{\partial y} = 2x^2 y - 2(a + y)(b^2 - y^2) + 2y(a + y)^2$

Also at a double point $(\partial f / \partial x) = 0$; $(\partial f / \partial y) = 0$ and $f = 0.$

$\qquad (\partial f / \partial x) = 0 \Rightarrow x = 0$ or $\qquad y = 0$

and $\qquad (\partial f / \partial y) = 0 \Rightarrow x^2 y - (a + y)(b^2 - y^2) + y(a + y)^2 = 0 \qquad\qquad ...(i)$

Putting $x = 0$ in this result we have

$$(a + y)\{y(a + y) - (b^2 - y^2)\} = 0 \text{ or } (a + y)\{2y^2 + ay - b^2\} = 0$$

or $\qquad\qquad y = -a$ or $\dfrac{1}{4}[-a \pm \sqrt{(a^2 + 8b^2)}].$

Putting $y = 0$ in (i) we do not have any significant result.

Hence we get three set of values $x = 0, y = -a, \dfrac{1}{4}[-a \pm \sqrt{(a^2 + 8b^2)}]$ of which only one set viz. $x = 0, y = -a$ satisfies the curve.

∴ $(0, -a)$ is a double point.

Shifting the origin to $(0, -a)$ we get equation of the curve as

$$x^2 (y - a)^2 = y^2 \{b^2 - (y - a)^2\}$$

or $y^4 + x^2y^2 - 2ay^2 - 2ax^2y + (a^2 - b^2) y^2 + a^2x^2 = 0$ ...(ii)

Tangents at the new origin are given by

$(a^2 - b^2) y^2 + a^2x^2 = 0$, equating the lowest degree terms to zero.

If $a < b$, (iii) gives two imaginary tangents. Hence the new origin is a conjugate point, i.e., $(0, -a)$ is a conjugate point.

If $a > b$, (iii) gives two real and distinct tangents

$$\sqrt{(b^2 - a^2)} \; y = \pm\, ax.$$

∴   The new origin may be a node or a conjugate point. Neglecting y^4 and y^3 from (ii), we get

$(y^2 - 2ay + a^2) x^2 = (b^2 - a^2) y^2$ or $(y - a)^2 x^2 = (b^2 - a^2) y^2$

or $x = \pm \sqrt{(b^2 - a^2)} \; [y/(y - a)].$

This shows that the value of x near the origin are real when $b > a$ or $a < b$.

Hence the point $(0, -a)$ is a node.

[If $a = b$, (iii) gives two real and coincident tangents $x^2 = 0$.

∴   The new origin may be a cusp or a conjugage point.

Neglecting y^4 from (ii), we get $(y^2 - 2ay + a^2) x^2 = 2ay^3$

or $x = \pm [\sqrt{(2ay^3)}] (y - a) = \pm\, y \{\sqrt{(2ay)}\}/(y - a).$

This shows that for small positive values of y, the values of x are real. Hence the curve has real branches through the new origin. Hence $(0, -a)$ is a cusp.]

5. *Find the double points of $x^3 + y^3 = 3axy$.*   **(Garhwal, 1998, Kuvempu 2005)**

Let $f(x, y) \equiv x^3 + y^3 - 3axy = 0.$ ...(i)

∴ $(\partial f / \partial x) = 3x^2 - 3ay; (\partial f / \partial y) = 3y^2 - 3ax$

At a double point we know $(\partial f / \partial x) = 0, (\partial f / \partial y) = 0$, and $f = 0$

∴ $\dfrac{\partial f}{\partial x} = 0 \Rightarrow x^2 = ay$ and $\dfrac{\partial f}{\partial y} = 0 \Rightarrow y^2 = ax.$

Solving $x^2 = ay$ and $y^2 = ax$ we get $x^4 = a^2y^2 = a^3x$

or $x (x^3 - a^3) = 0$ or $x = 0, a$ (only real values being considered)

∴ $y = 0, \pm a$

∴   We have the following sets of values of x and y :

$x = 0, y = 0; x = 0, y = a; x = 0, y = -a; x = a, y = 0; x = a, y = a$ and $x = a, y = -a.$

Out of these only $x = 0, y = 0$ satisfy (i). Hence $(0, 0)$ is a double point.

Equating the lowest degree terms in (i) to zero we get $xy = 0$ or $x = 0, y = 0$ as the tangents to the given curve at $(0, 0)$. These tangents being real and distinct we conclude that $(0, 0)$ is a node.

6. *Show that the curve $(a^2 + x^2) y = a^2x$ has three points of inflexion. Find them.*

The curve is $y = a^2 x / (a^2 + x^2)$ ...(i)

$$\therefore \quad \frac{dy}{dx} = \frac{(a^2 + x^2)\, a^2 - a^2 x\,(2x)}{(a^2 + x^2)^2} = \frac{a^2\,(a^2 - x^2)}{(a^2 + x^2)^2}$$

$$\frac{d^2 y}{dx^2} = a^2 \left[\frac{(a^2 + x^2)^2\,(-2x) - (a^2 - x^2)\cdot 2\,(a^2 + x^2)\,2x}{(a^2 + x^2)^4} \right]$$

$$= a^2 \left[\frac{-2x\,\{(a^2 + x^2) + 2\,(a^2 - x^2)\}}{(a^2 + x^2)^3} \right] = \frac{2a^2 x\,(x^2 - 3a^2)}{(a^2 + x^2)^3}$$

Similarly $\quad \dfrac{d^3 y}{dx^3} = 6a^2 \left[\dfrac{(a^2 + x^2)\,(x^2 - a^2) - 2x\,(x^2 + 3a^2)}{(a^2 + x^2)^4} \right]$

For the point of inflexion, $d^2 y / dx^2 = 0$

i.e., $\quad x\,(x^2 - 3a^2) = 0$ or $x = 0, \pm a\sqrt{3}$.

For these values of x we find $d^3 y / dx^3 \neq 0$.

Also from (i) when $x = 0, y = 0$

and when $\quad x = \pm a\sqrt{3},\ y = \dfrac{\pm a^3 \sqrt{3}}{a^2 + 3a^2} = \pm \dfrac{\sqrt{3}}{4}\, a.$

$\therefore$ Points of inflexion are $\quad (0, 0),\ \left[a\sqrt{(3)},\ \dfrac{1}{4}\sqrt{(3)}\, a \right]$

and $\qquad\qquad\qquad \left[-a\sqrt{(3)},\ -\dfrac{1}{4}\sqrt{(3)}\, a \right].$

7. *Find the points of inflexion on the curve* $x = a\,(2\theta - \sin\theta), y = a\,(2 - \cos\theta).$

The curve is

$$x = a\,(2\theta - \sin\theta),\ y = a\,(2 - \cos\theta) \qquad\qquad ...(i)$$

$$\therefore \quad \frac{dx}{d\theta} = a\,(2 - \cos\theta) \Rightarrow \frac{d\theta}{dx} = \frac{1}{a\,(2 - \cos\theta)} \qquad\qquad ...(ii)$$

And $\quad \dfrac{dy}{d\theta} = a\,(\sin\theta)$

$$\therefore \quad \frac{dy}{dx} = \frac{dy/d\theta}{dx/d\theta} = \frac{a\sin\theta}{a\,(2 - \cos\theta)} = \frac{\sin\theta}{2 - \cos\theta}$$

$$\frac{d^2 y}{dx^2} = \frac{d}{dx}\left(\frac{dy}{dx} \right) = \frac{d}{d\theta}\left(\frac{dy}{dx} \right)\frac{d\theta}{dx}$$

$$= \left[\frac{d}{d\theta}\left(\frac{\sin\theta}{2 - \cos\theta} \right) \right]\cdot \frac{1}{a\,(2 - \cos\theta)},\ \text{from (ii)}$$

$$= \frac{(2 - \cos\theta)\cos\theta - \sin\theta\,(\sin\theta)}{(2 - \cos\theta)^2}\cdot \frac{1}{a\,(2 - \cos\theta)}$$

$$\Rightarrow \quad \frac{d^2 y}{dx^2} = \frac{2\cos\theta - 1}{a\,(2 - \cos\theta)^2}$$

Again differentiating, *w.r. to* x, we get

$$\frac{d^3 y}{dx^3} = \frac{d}{d\theta}\left(\frac{d^2 y}{dx^2} \right)\cdot \frac{d\theta}{dx} = \frac{d}{d\theta}\left[\frac{2\cos\theta - 1}{a\,(2 - \cos\theta)^2} \right]\cdot \frac{d\theta}{dx}$$

$$= \frac{(2 - \cos\theta)^2 \, (-\, 2\sin\theta) - (2\cos\theta - 1) \cdot 3 \, (2 - \cos\theta)^2 \, \sin\theta}{a \, (2 - \cos\theta)^6 \; a \, (2 - \cos\theta)} , \text{from } (ii)$$

$$= -\, [(1 + 4\cos\theta) \sin\theta] \, / \, \{a^2 \, (2 - \cos\theta)^5\}. \qquad\qquad ...(iii)$$

Now equating $\dfrac{d^2 y}{dx^2}$ to zero, we get $2\cos\theta - 1 = 0$

$\Rightarrow \qquad \cos\theta = 1/2 \Rightarrow \theta = \pi/3,\ 5\pi/6 \text{ in } 0 \le \theta \le 2\pi$

For these values of θ we find from (iii) that $\dfrac{d^3 y}{dx^3} \ne 0$.

Now obtain the coordinates of the points of inflexion from (i), putting $\theta = \pi/3$ and $5\pi/6$.

EXERCISES

Find the position and nature of the multiple points on the following curves :–

1. $x^2 \, (x - y) + y^2 = 0$

2. $y^3 = x^3 + ax^2$

3. $x^4 + y^3 - 2x^3 + 3y^2 = 0.$

4. $xy^2 - ax^2 + 2a^2 x - a^2 = 0.$

5. $y^2 = (x - 1) \, (x - 2)^2.$

6. $ay^2 = (x - a)^2 \, (x - b)^2.$

7. $x^4 - 4ax^3 + 2ay^3 + 4a^2 x^2 - 3a^2 y^2 - a^4 = 0.$

8. $x^3 + y^3 - 12x - 27y + 70 = 0.$

9. $x^4 + 4ax^3 + 4a^2 x^2 - b^2 y^2 - 2b^3 y - a^4 - b^4 = 0.$

10. $x^4 + y \, (y + 4a)^3 + 2x^2 \, (y - 5a)^2 = 5a^2 x^2.$

11. $x^3 + 2x^2 + 2xy - y^2 + 5x - 2y = 0.$

(Banglore 2006)

12. $(2y + x + 1)^2 - 4 \, (1 - x)^5 = 0.$

13. $(x + y)^3 - \sqrt{2} \, (y - x + 2)^2 = 0.$

(Garhwal, 1996)

14. $(y^2 - a^2)^3 + x^4 \, (2x + 3a)^2 = 0.$

Find the equations of the tangents at the multiple points of the following curves :–

15. $x^4 - 4ax^3 - 2ay^3 + 4a^2 x^2 + 3a^2 y^2 - a^4 = 0.$

16. $x^4 - 8x^3 + 12x^2 y + 16x^2 + 48xy + 4y^2 - 64y = 0.$

17. $(y - 2)^2 = x \, (x - 1)^2.$

18. Show that each of the curves

$(x \cos\alpha - y \sin\alpha - b)^3 = c \, (x \sin\alpha + y \cos\alpha)^2,$

for all different values of α, has a cusp ; show also that all the cusps lie on a circle.

ANSWERS

1. Cusp at $(0, 0)$.

2. Cusp at $(0, 0)$.

3. Node at $(0, 0)$.

4. Node at $(a, 0)$.

5. Node at $(2, 0)$.

6. $(a, 0)$ is a node, cusp or an isolated point according as b/a is less than, equal to or greater than 1.

7. Conjugate point at $(a, 0)$.

8. Conjugate point at $(2, 3)$.

9. Conjugate point at $(-a, -b)$.

10. Cusp at $(0, -4a)$.

11. Cusp at $(-1, -2)$.

12. Cusp at $(1, -1)$.

13. No multiple point.

14. $(0, \pm a)$ are triple point ; $\left(-\dfrac{3}{2}\, a, \pm a\right)$ are cusps ; node at $(-a, 0)$.

15. $y - a = \pm (2/\sqrt{3}) \, x ; y = \pm \sqrt{(2/3)} \, (x - a) ; \quad y - a = \pm (2/3) \, (x - 2a)$.

16. $y - 2 = \pm (2/\sqrt{2}) \, (x - 2)$.   17. $x - y + 1 = 0, x + y = 3$.

16.5. Types of cusps.

We know that two branches of a curve have a common tangent at a cusp. There are *five* different ways in which the two branches stand in relation to the common tangent and the common normal as illustrated by the following figures :–

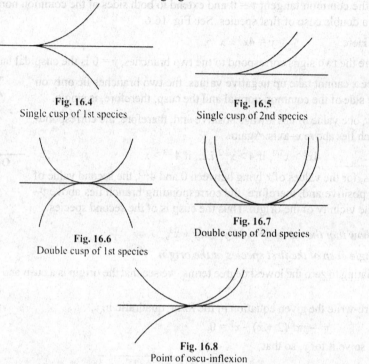

Fig. 16.4
Single cusp of 1st species

Fig. 16.5
Single cusp of 2nd species

Fig. 16.6
Double cusp of 1st species

Fig. 16.7
Double cusp of 2nd species

Fig. 16.8
Point of oscu-inflexion

In Fig. 16.4, the two branches lie on *the same side* of the common normal and on the *different sides* of the tangent.

In Fig. 16.5, the two branches lie on *the same side* of the normal and on *the same side* of the tangent.

In Fig. 16.6, the two branches lie on *the different sides* of the normal and on the *different sides* of the tangent.

In Fig. 16.7, the two branches lie on the *different sides* of the normal and on the *same side* of the tangent.

In Fig. 16.8, the two branches lie on the *different sides* of the normal but on one side they lie on the same, and on the other on opposite sides of the common tangent. One branch has inflexion at the point.

It will thus be seen that *the cusp is single or double according as the two branches lie on the same or different sides of the common normal. Also it is of the first or second species according as the branches lie on the different or the same side of the common tangent.*

EXAMPLES

1. *Find the nature of the cusps on the following curves :–*

    (i) $y^2 = x^3$.   (ii) $y^2 - x^4 = 0$.   (iii) $(y - 4x^2)^2 = x^7$.

(*i*) $y = 0$ is the cuspidal tangent. Since x cannot be negative, the two branches lie only on the same side of the common normal so that the cusp is single. See Fig. 16.4.

Again, $y = \pm x^{3/2}$ so that to each positive value of x correspond two values of y which are of opposite signs and hence the two branches lie on different sides of the common tangent and the cusp is of first species.

(*ii*) Two branches of $y^2 - x^4 = 0$, are the two parabolas $y - x^2 = 0$ and $y + x^2 = 0$ which lie on different sides of the common tangent $y = 0$ and extend to both sides of the common normal $x = 0$. Thus the origin is a double cusp of first species. See Fig. 16.6.

(*iii*) Here $y = 4x^2 \pm x^{7/2},$

where the two signs correspond to the two branches; $y = 0$ is the cuspidal tangent.

Since x cannot take up negative values, the two branches lie only on the same side of the common normal and the cusp, therefore, is single.

Now, one value of y is always positive and, therefore, the corresponding branch lies above x–axis. Again

$$4x^2 > x^{7/2}, \text{ if } 4 > x^{3/2} \text{ i.e., if } 4^{2/3} > x.$$

Thus, for the values of x lying between 0 and $4^{2/3}$, the second value of y is also positive and, therefore, the corresponding branch lies above y–axis in the vicinity of the origin. Thus the cusp is of the second species.

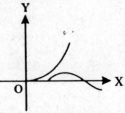

Fig. 16.9

2.   *Show that the curve* $y^2 = 2x^2 y + x^3 y + x^3,$

has a single cusp of the first species at the origin.

Equating to zero the lowest degree terms, we see that the origin is a cusp and $y = 0$ is the cuspidal tangent.

We re-write the given equation in the form, quadratic in y,

$$y^2 - yx^2 (2 + x) - x^3 = 0,$$

and solve it for y, so that

$$y = \frac{x^2 (2 + x) \pm \sqrt{[x^4 (2 + x)^2 + 4x^3]}}{2}$$

For positive values of x, we have
$$x^4 (2 + x)^2 + 4x^3 > x^4 (2 + x)^2,$$

or $$\sqrt{[x^4 (2 + x)^2 + 4x^3]} > x^2 (2 + x),$$

so that two positive values of x correspond two values of y with opposite signs.

Thus the two branches lie on opposite sides of x–axis when x is positive.

Again, we have
$$x^4 (2 + x)^2 + 4x^3 = x^3 (4 + 4x + 4x^2 + x^3).$$

For values of x which are sufficiently small in numerical value,
$$4 + 4x + 4x^2 + x^3$$
is positive, for the same $\to 4$ when $x \to 0$.

Thus, for negative values of x which are sufficiently small in numerical value,
$$x^3 (4 + 4x + 4x^2 + x^3),$$

is negative so that the values of y are imaginary. Thus x cannot take up negative values.

Hence, the curve has a single cusp of first species at the origin.

EXERCISES

Find the nature of the cusps on the following curves :–

1. $x^2 (x - y) + y^2 = 0$.
2. $x^2 (x + y) - y^2 = 0$.
3. $x^3 + y^3 - 2ay^2 = 0$.
4. $a^4 y^2 = x^5 (2a - x)$.
5. $(y - x)^2 + x^6 = 0$.
6. $x^6 - ayx^4 - a^3 x^2 y + a^4 y^2 = 0$.
7. $x^5 - ax^3 y - a^2 x^2 y + a^2 y^2 = 0$.

8. Examine the curve

$$x^5 + 16x^2 y - 64y^2 = 0 \quad \text{for singularities.}$$

9. Prove that the curve $x^3 + y^3 = ax^2$ has a cusp of the first species at the origin and a point of inflexion where $x = a$.

10. Show that the curve

$$y^3 = (x - a)^2 (2x - a)$$

has a single cusp at $(a, 0)$.

ANSWERS

1. Single cusp of first species.
2. Single cusp of first species.
3. Single cusp of first species.
4. Single cusp of first species.
5. Isolated point.
6. Double cusp of second species.
7. Oscu-inflexion.
8. Oscu-inflexion.

16.6 Radii of curvature at multiple points. The formula for the radius of curvature at any point (x, y) on the curve $f(x, y) = 0$, as obtained in § 14.4, becomes meaningless at a multiple point where $f_x = f_y = 0$. At a multiple point we expect as many values of, ρ, as its order. Of course, these values of ρ may not be all distinct.

The following examples will illustrate the method of determining the values of ρ at such points.

EXAMPLES

1. *Find the radii of curvature at the origin of the branches of the curve*

$$y^4 + 2axy^2 = ax^3 + x^4.$$

Here, $2xy^2 = x^3$, *i.e.*, $x = 0, y = \pm (1/\sqrt{2})x$ are the three tangents at the origin so that it is a *triple* point.

To find, ρ, for the branch which touches $x = 0$, we find lim $(y^2/2x)$. To do this, we write

$$y^2/2x = \rho_1, \ i.e., \ x = y^2/2 \rho_1,$$

and substitute this value of x in the given equation. Lim $\rho_1 = \rho$ is the radius of curvature of the corresponding branch at the origin. We get

$$y^4 + \frac{2ay^4}{2 \rho_1} = a \frac{y^6}{8 \rho_1{}^3} + \frac{y^8}{16 \rho_1{}^4}$$

or

$$1 + \frac{a}{\rho_1} = a \frac{y^2}{8 \rho_1{}^3} + \frac{y^4}{16 \rho_1{}^4}.$$

Let $y \to 0$ so that we have $1 + a/\rho = 0$.

Thus, ρ, for this branch $= -a$.

To find, ρ, for the other branches we proceed as follows.

Suppose that the equation of either branch is given by

$$y = f(0) + xf'(0) + \frac{x^2}{2!} f''(0) + ...$$

We have, there

$$f(0) = 0.$$

Also we write $f'(0) = p, f''(0) = q$. Thus we have

$$y = px + \frac{1}{2} qx^2 + ...$$

Making substitution in the given equation, we get

$$\left(px + \frac{1}{2} qx^2 + ... \right)^4 + 2ax \left(px + \frac{1}{2} qx^2 + ... \right)^2 = ax^3 + x^4$$

Equating co-efficients of x^3 and x^4, we get

$$2ap^2 = a, p^4 + 2apq = 1.$$

These give $p = \dfrac{1}{\sqrt{2}}, q = \dfrac{3\sqrt{2}}{8a};$

$$p = -\frac{1}{\sqrt{2}}, q = -\frac{3\sqrt{2}}{8a}.$$

∴ $\rho = \dfrac{(1 + p^2)^{3/2}}{q} = \pm 2\sqrt{3}\,a,$

for the two branches.

2.   *Show that the pole is a triple point on the curve*

$$r = a(2\cos\theta + \cos 3\theta),$$

and that the radii of curvature of the three branches are

$$\sqrt{3}\,a/2, a/2, \sqrt{3}\,a/2.$$

The radius vector, r, vanishes for the values of, θ, given by

$$2\cos\theta + \cos 3\theta = 0,$$

i.e., $2\cos\theta + 4\cos^2\theta - 3\cos\theta = 0,$

or $\cos\theta (4\cos^2\theta - 1) = 0,$

or $\cos\theta = 0, \cos\theta = \dfrac{1}{2}, \cos\theta = -\dfrac{1}{2}.$

Thus, $r = 0$ when $\theta = \pi/3, \pi/2, 2\pi/3$ so that the pole is a triple point.
We now proceed to find ρ. We have

$$r_1 = a(-2\sin\theta - 3\sin 3\theta), r_2 = a(-2\cos\theta - 9\cos 3\theta).$$

For $\theta = \pi/3,$

$$r_1 = -\sqrt{3}\,a, \qquad r_2 = 8a;$$

For $\theta = \pi/2,$

$$r_1 = a \qquad\qquad r_2 = 0;$$

For $\theta = 2\pi/3,$

$$r_1 = -\sqrt{3}\,a, \qquad r_2 = -8a.$$

Also, $r = 0$ for each of these branches.

Putting these values in the formula

$$\rho = \frac{(r^2 + r_1^2)^{3/2}}{r^2 + 2r_1^2 - rr_2},$$

we get the required result.

EXERCISES

1. Show that the radius of curvature at the origin for both the branches of the curve

$$y^2(a-x) = x^2(a+x) \qquad \text{is } \sqrt{2}\,a.$$

2. Find the radius of curvature at the point $(1, 2)$ for the curve

$$(y-2)^2 = x(x-1)^2.$$

3. Find the radii of curvature at the origin of the two branches of the curve given by the equations

$$y = t - t^2, x = 1 - t^2.$$

   (For the origin $t = \pm 1$, so that the two branches correspond to these two values of t).

   Find the radii of curvature at the origin of the following curves :

4. $x^3 + y^3 = 3axy.$ (*Folium*).

5. $x^2 - 3xy - 4y^2 + y^3 + y^4x + x^5 = 0.$

6. $x^5 + ax^2y^2 - ax^3y - 2a^2xy^2 + a^2y^2 = 0.$

7. Show that $(a, 0)$, in polar co-ordinates, is a triple point on the curve

$$r = a\left(1 + 2\sin\frac{1}{2}\theta\right),$$

   and find the radii of curvature at the point.

ANSWERS

2. $2\sqrt{2}.$ 3. $2\sqrt{2}$ for each.   4. $3a/2$ for each.

5. $\dfrac{85}{2}\sqrt{17}, 5\sqrt{2}.$ 6. $a, -a/2, -5^{3/2}\,a.$

7. $\dfrac{2}{3}\sqrt{2}a, \dfrac{2}{3}\sqrt{2}a, \dfrac{2}{3}a.$

OBJECTIVE QUESTIONS

For each of the following questions, four alternatives are given for the answer. Only one of them is correct. Choose the correct alternative.

1. If two branches of a curve pass through a point P, and the two tangents at P are real and distinct, then P is a :

   (*a*) node   (*b*)   cusp

   (*c*) conjugate point   (*d*)   none

2. An isolated point on the curve is one at which :

   (*a*) two branches are real

   (*b*) two branches are real and distinct

   (*c*) two branches are real and coincident

   (*d*) no branch of the curve lies in the neighbourhood of the point.

3. For the curve $y^2(1+x) = x^2(1-x)$, the origin is a :
    (a) node   (b) cusp
    (c) point of inflexion   (d) none of these

4. The curve $y = x^3 - 3x^2 - 9x + 9$ has a point of inflexion at :
    (a) $x = -1$ (b) $x = 1$
    (c) $x = -3$ (d) $x = 3$

5. A double point on the curve is a cusp if tangents are :
    (a) real and distinct   (b) imaginary and distinct
    (c) imaginary and coincident   (d) real and coincident.

                                               (*Garhwal, 2001; Avadh 2002; 2003*)

6. For the curve $x^4 - ax^2y + axy^2 + a^2y^2 = 0$, the origin is :
    (a) node   (b) cusp
    (c) conjugate point   (d) point of inflexion

7. Nature of thed origin for the curve $y^2(a^2 + x^2) = x^2(a^2 - x^2)$ is :
    (a) Node   (b) Cusp
    (c) Conjugate point   (d) Point of inflexion

8. Curve $x^3 + x^2y = ay^2$ has at origin a :
    (a) Node   (b) Cusp
    (c) Conjugate Point   (d) Point of inflexion

9. Nature of origin on the curve $\dfrac{a^2}{x^2} - \dfrac{b^2}{y^2} = 1$ is :

    (a) Node   (b) Cusp
    (c) Conjugate point   (d) None of these.

ANSWERS

1. (a)   2. (d)   3. (a)   4. (b)   5. (d)   6. (c)   7. (a)
8. (b)   9. (a).

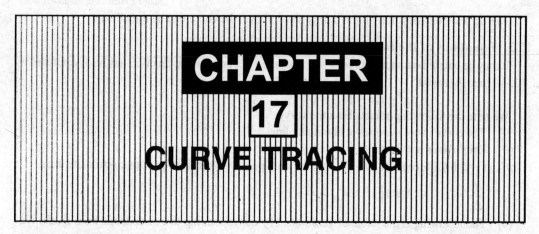

CHAPTER 17
CURVE TRACING

17.1. Introduction

The general problem of curve tracing in its elementary aspects, will be taken up in this chapter.

It will be seen that the equations of curves which we shall trace are generally solvable for y, x and r. Some equations which are not solvable for y or x may be rendered solvable for r, on transformation from Cartesian to Polar system.

17.2. Procedure for tracing Cartesian Equations

I. *Find out if the curve is symmetrical about any line.* In this connection, the following rules, whose truth is evident are helpful:-

(i) A curve is symmetrical about x-axis if the powers of y which occur in its equation are all even;

(ii) A curve is symmetrical about y-axis if the powers of x which occur in its equation are all even;

(iii) A curve is symmetrical about the line $y = x$ if, on inter-changing x and y its equation does not change.

(iv) A curve is symmetrical about the line $y = -x$ if the equation of the curve remains unchanged when x and y are replaced by $-y$ and $-x$ respectively.

(v) The curve is symmetrical in opposite quadrants if the equation of the curve remains unchanged when x and y are replaced by $-x$ and $-y$ respectively.

II. *Find out if the origin lies on the curve.* If it does, write down the tangent or tangents thereat. In case the origin is a multiple point, find out its nature.

III. *Find out the points common to the curve and the co-ordinate axes if there be any.* Also obtain the tangents at such points.

IV. *Find out the asymyptotes* and the points in which each asymptote meets the curve.

V. *Find out if there is any region of the plane such that no part of the curve lies in it.*

Such a region is generally obtained on solving the equation for one variable in terms of the other, and find out the set of values of one variable which make the other imaginary.

VI. Find out $\dfrac{dy}{dx}$ and the points where the tangent is parallel to the co-ordinate axes.

VII. Finally find out the intervals in which y is increasing and decreasing and in particular, the values of x for which y has maximum or minimum values.

This investigation depends upon the determination of $\dfrac{dy}{dx}$ and of the intervals in which it is $+ve$ or $-ve$. Also find the intervals in which the curve is concave upwards or downwards and in particular the values for which the curve has points of inflexion. This investigation depends upon $\dfrac{d^2y}{dx^2}$.

Taking into consideration all the points given above, the curve can be traced out.

17.3. Equations of the form y = f (x).

It may be pointed out that a curve $y = f(x)$ where f is a polynomial function has no asymptotes.

1. *Trace the curve*
$$y = x^3 - 12x - 16.$$

We note the following points about the curve.

(*i*)   The curve is not symmetric about any line.

(*ii*)   Origin does not lie on it.

(*iii*) $(-2, 0), (4, 0)$ and $(0, -16)$ are the only points of intersection with co-ordinate axes.

(*iv*) $\dfrac{dy}{dx} = 3x^2 - 12 = 3(x^2 - 4), \dfrac{d^2y}{dx^2} = 6x$.

(*v*)   The following is the table of variations of y corresponding to x as deduced from dy/dx. We also note that

$$\lim_{x \to -\infty} y = -\infty \text{ and } \lim_{x \to +\infty} y = +\infty.$$

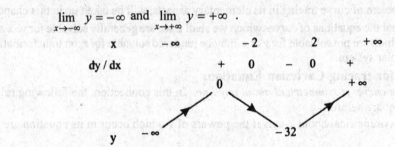

Fig 17.1

Thus, y is strictly increasing in the intervals $]-\infty, -2]$ and $[2, \infty[$. Also y is strictly decreasing in the interval $[-2, 2]$.

Finally, y is maximum for $x = -2$ and minimum for $x = 2$.

(*vi*) $d^2y/dx^2 = 0$ for $x = 0$.

Also, $d^2y/dx^2 < 0 \ \forall \ x \in]-\infty, 0[$ and $> 0 \ \forall \ x \in]0, \infty[$.

Thus, the curve is concave downwards in $]-\infty, 0]$ and concave upwards in $[0, \infty[$. Since d^2y/dx^2 change sign as x passes through 0, the curve has a point of inflexion for $x = 0$, *i.e.*, at the point $(0, -16)$.

We have the curve as given in the following figure 17.2

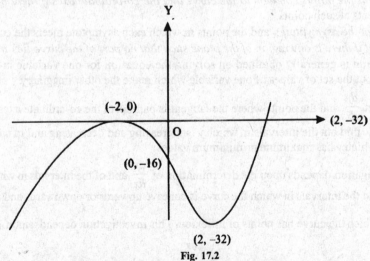

(2, −32)

Fig. 17.2

2. *Trace the curve*

$$y = -\frac{3}{2}x^4 + 4x^3 + 3x^2 - 12x.$$

We note the following points about the curve.

(*i*) The curve is not symmetrical about any line.

(*ii*) Origin lies on it and it meets co-ordinate axes in origin only.

(*iii*) $\dfrac{dy}{dx} = -6x^3 + 12x^2 + 6x - 12 = -6\,(x-2)\,(x-1)\,(x+1),$

$\dfrac{d^2y}{dx^2} = -18x^2 + 24x + 6 = 6(-3x^2 + 4x + 1)$

The following is the table of variations:

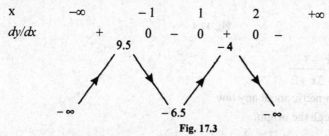

Fig. 17.3

Thus, y is strictly increasing in $]-\infty, -1\,]$ and $[\,1, 2\,]$ and decreasing in $[-1, 1\,]$ and $[\,2, \infty\,[$. Also, y is maximum for $x = -1$ and 2 and minimum for $x = 1$.

(*iv*) $\quad d^2y/dx^2 = 0$ if and only if $x = \dfrac{1}{3}\,(2 \pm \sqrt{7})$.

Since $\sqrt{7} = 2.6.....$, we may see that

$$\frac{1}{3}\,(2 - \sqrt{7}) = -.2..... \text{ and } \frac{1}{3}\,(2 + \sqrt{7}) = 1.5\,...$$

Also, d^2y/dx^2 changes sign as x passes through

$$\frac{1}{3}(2 - \sqrt{7}) \text{ and } \frac{1}{3}(2 + \sqrt{7}).$$

We have $\quad d^2y/dx^2 < 0 \ \forall \ x \in \ ]-\infty, \dfrac{1}{3}\,(2 - \sqrt{7})\,[$

and $\quad d^2y/dx^2 < 0 \ \forall \ x \in [\,\dfrac{1}{3}\,(2 + \sqrt{7}), \infty\,[$

$\quad d^2y/dx^2 > 0 \ \forall \ x \in \ ]\,\dfrac{1}{3}(2 - \sqrt{7}), \dfrac{1}{3}\,(2 + \sqrt{7})\,[.$

Thus, the curve is concave downwards in $]-\infty, \dfrac{1}{3}\,(2 - \sqrt{7})\,]$, upwards in $\left[\dfrac{1}{3}(2 - \sqrt{7}), \dfrac{1}{3}(2 + \sqrt{7})\right]$ and downwards in $[\dfrac{1}{3}\,(2 + \sqrt{7}), +\infty\,[$.

Also the curve has two points of inflexion corresponding to

$$x = \frac{1}{3}\,(2 - \sqrt{7}) \text{ and } x = \frac{1}{3}\,(2 + \sqrt{7}).$$

We thus have the curve as drawn in Fig. 17.4.

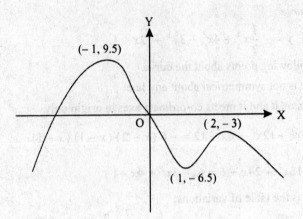

(− 1, 9.5)

O

(2, − 3)

(1, − 6.5)

Fig. 17.4

3. *Trace the curve*
$$y = \frac{2x - 3}{x^2 - 3x + 2}$$

(*i*)   The curve is not symmetric about any line.

(*ii*)   It does not pass through the origin.

(*iii*)   It meets co-ordinate axes in $\left(\dfrac{3}{2}, 0\right)$ and $\left(0, -\dfrac{3}{2}\right)$ only.

(*iv*) y is defined for $x \in R \sim \{1, 2\} = \,]-\infty \, 1 \, [\, \cup \,] \, 1, 2 \, [\, \cup \,] \, 2 + \infty \, [$

(*v*) $\dfrac{dy}{dx} = \dfrac{-2x^2 + 6x - 5}{(x^2 - 3x + 2)^2} = \dfrac{-2\left[(x - 3/2)^2 + 1/4\right]}{(x^2 - 3x + 2)^2}$

The following is the table of variations of y:

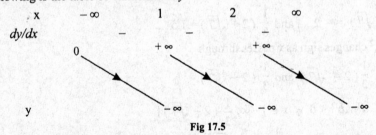

x	− ∞		1		2		∞
dy/dx		−		−		−	
	0		+ ∞		+ ∞		
y			− ∞		− ∞		− ∞

Fig 17.5

Thus, y is strictly decreasing in each of the intervals $]-\infty, 1\,[, \,] \, 1, 2 \, [$ and $] \, 2, +\infty \, [$.

Also y has no maximum or minimum

(*vi*) $x = 1$, $x = 2$ and $y = 0$ are the asymptotes to the curve

(*vii*) $\dfrac{d^2 y}{dx^2} = \dfrac{2(2x - 3)(x^2 - 3x + 3)}{(x^2 - 3x + 2)^3}$ so that

$\dfrac{d^2 y}{dx^2} = 0$ when $x = \dfrac{3}{2}$ and changes sign as x passes through $\dfrac{3}{2}$. Thus $\left(\dfrac{3}{2}, 0\right)$ is a point of inflexion and the slope of the tangent at this point is − 8.

We have the curve as drawn in Fig. 17.6.

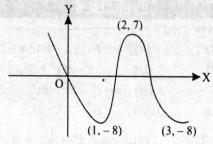

Fig. 17.6.

4. *Trace the curve* $y = \dfrac{8a^3}{x^2 + 4a^2}$

(*i*)   The curve is symmetrical about *y*-axis.

(*ii*)   It does not pass through the origin.

(*iii*)   It meets *y-axis* in $(0, 2a)$ and $y = 2a$ is the tangent thereat.

(*iv*) $y = 0$ is the only asymptote to the curve

(*v*) y is positive for all values of x. Therefore the curve lies in 1st and 2nd quadrant only.

(*vi*) $\dfrac{dy}{dx} = \dfrac{-16\,x\,a^3}{(x^2 + 4a^2)^2}.$

The following is the table of variations of y

x	$-\infty$	$-2a$	$-a$	0	a	$2a$	∞
dy/dx		$+$	$+$	0	$-$	$-$	

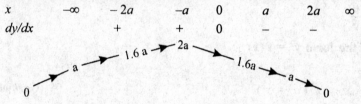

Fig 17.7

$\therefore$ $\dfrac{dy}{dx} < 0 \ \forall\, x > 0$

and $\dfrac{dy}{dx} > 0 \ \forall\, x < 0$

Thus, y is strictly increasing in $]-\infty, 0\,]$ and strictly decreasing in $[\,0, \infty\,[$. Also y is maximum for $x = 0$ and its value there is $2a$.

Thus, we have the curve as given in the figure Figure 17.8

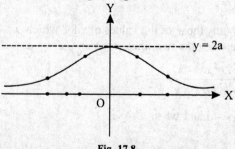

Fig. 17.8

EXERCISES

1. Trace the following curves:

   (i) $y = 3x^4 - 16x^3 + 24x^2$,   (ii) $y = (x^2 - x - 6)(x - 7)$,

   (iii) $y = (x + 1)^2 (x - 3)$,   (iv) $y = x^4 - 24x^2 + 80$,

   (v) $y = x^4 - 2x + 10$,   (vi) $y = x^4 - 5x^3 + 6x^2$,

   (vii) $y = 4 - 8x + 5x^2 - x^3$,   (viii) $y = 2x^5 - 3x^4 + 112$,

   (ix) $y = \dfrac{1}{4} x^4 - 2x^2 + 2$,   (x) $y = \dfrac{3}{2} x^4 + \dfrac{1}{2} x^2 - 4$,

   (xi) $y = \dfrac{1}{5} x^5 - \dfrac{1}{3} x^3 + 2$,   (xii) $y = \dfrac{1}{6} x^6 - x^5 + x^4$.

2. Trace the following curves:

   (i) $y = \dfrac{x^2 - 3x}{x - 1}$ (ii) $y = \dfrac{2x^2 - 7x + 5}{x^2 - 5x + 7}$

   (iii) $y = \dfrac{1 + x^2}{2 - 3x + x^2}$ (iv) $y = \dfrac{2x - 3}{3x^2 - 4x}$

   (v) $y = \dfrac{x^2 - 12x + 27}{x^2 - 4x + 5}$ (vi) $y = \dfrac{x^2 - 1}{(x - 2)^2}$

   (vii) $y = \dfrac{x - 1}{x^2}$ (viii) $y = \dfrac{x^2 + 4x - 5}{(x + 2)^2}$

   (ix) $y = \dfrac{3x + 6}{(x - 1)^2}$.

17.4. Equations of the form $y^2 = f(x)$

1. *Trace the curve*

$$y^2 (a^2 + x^2) = x^2 (a^2 - x^2)$$ *(Garhwal 1999)*

We note the following particulars about this curve:

(i) It is symmetrical about both the axes.

(ii) It passes through the origin and $y = \pm x$ are the two tangents thereat. Thus the origin is a node.

(iii) It meets x - axis at $(a, 0)$, $(0, 0)$ and $(-a, 0)$ and meets y-axis at $(0, 0)$ only. The tangents at $(a, 0)$ and $(-a, 0)$ are $x = a$ and $x = -a$ respectively.

(iv) The curve has no asymptotes

(v) $y = \pm x \sqrt{\dfrac{a^2 - x^2}{a^2 + x^2}}$

y is defined for only those of the values of x for which $a^2 - x^2 \geq 0 \Leftrightarrow -a \leq x \leq a$ so that the curve lies between the two lines $x = a$ and $x = -a$.

(vi) $\dfrac{dy}{dx} = \dfrac{a^4 - 2a^2 x^2 - x^4}{(a^2 + x^2)^{3/2} (a^2 - x^2)^{1/2}}$

$\dfrac{dy}{dx} \to \infty$ as $x \to -a$ and when $x \to a$

Also $\dfrac{dy}{dx} = 0$ when $a^4 - 2a^2 x^2 - x^4 = 0$

Again $a^4 - 2a^2 x^2 - x^4 = -[x^4 + 2a^2 x^2 - a^4]$

$$=-[x^2-(-1+\sqrt{2})a^2][x^2-(-1-\sqrt{2})a^2]$$

$$=-[x-\sqrt{(-1+\sqrt{2})}a][x+\sqrt{(-1+\sqrt{2})}a][x^2+(1+\sqrt{2})a^2]$$

Thus $\dfrac{dy}{dx}=0$ for $x=\sqrt{(-1+\sqrt{2})}a$ and $x=-\sqrt{(-1+\sqrt{2})}a$

$\dfrac{dy}{dx}<0$ for $x\in\,]-a,-\sqrt{(-1+\sqrt{2})}a\,[\cup]\,\sqrt{(-1+\sqrt{2})}a,a\,[$

and $\dfrac{dy}{dx}>0$ when $x\in\,]-\sqrt{(-1+\sqrt{2})}a,\,[$

Thus y is strictly decreasing in $[-a,-\sqrt{(-1+\sqrt{2})}a]$ and $[(-1+\sqrt{2})a,a]$ and strictly increasing in $[-\sqrt{(-1+\sqrt{2})}a,\sqrt{(-1+\sqrt{2})}a]$

We may note that $\sqrt{(-1+\sqrt{2})}=.6$

Thus the curve

$$y^2=\frac{x^2(a^2-x^2)}{a+x^2}$$

is as given in the Fig. 17.9

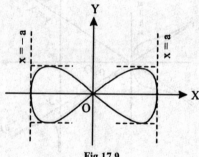

Fig 17.9

2. *Trace the curve*

$$y^2(x-a)=x^2(x+a)$$

(*Garhwal 2001*)

(i) The curve is symmetrical about x-axis only.

(ii) It passes through the origin and $y^2+x^2=0$ *i.e.* $y\pm ix=0$ are the two imaginary tangents thereat. Thus the origin is an isolated point.

(iii) It meets x-axis $(-a,0)$ and $(0,0)$ and y-axis at the origin only; $x=-a$ is the tangent at $(-a,0)$.

(iv) $y=\pm(x+a)$ and $x=a$ are its three asymptotes.

(v) $y^2=\dfrac{x^2(x+a)}{x-a}$.

$\Rightarrow\qquad y=\pm x\sqrt{\dfrac{x+a}{x-a}}$.

y is defined only for those of the values of x for which $x^2-a^2\geq0$ and $x-a\neq0$ or equivalently when $x\leq-a$ and $x>a$.

Thus, no part of the curve lies between the lines $x=-a$ and $x=a$ except the point $(0,0)$.

(vi) $\dfrac{dy}{dx}=\pm\dfrac{x^2-ax-a^2}{(x-a)^{3/2}(x+a)^{1/2}}$

$$\frac{dy}{dx} = 0 \Leftrightarrow x^2 - ax - a^2 = 0 \Leftrightarrow x = \frac{1}{2}(1 \pm \sqrt{5})a,$$

For $x = (1 - \sqrt{5})\frac{a}{2}$, y is not real.

Now $\frac{dy}{dx} > 0 \ \forall \ x \in \ ]-\infty, -a]$ and $\forall \ x \in \left[\frac{1}{2}(1+\sqrt{5})a, \infty\right[$ so that y is strictly increasing

in $]-\infty, -a]$ and $\left[\frac{1}{2}(1+\sqrt{5})a, \infty\right[$

Again $\frac{dy}{dx} < 0 \ \forall \ x \in \ \left]a, \frac{1}{2}(1+\sqrt{5})a\right]$ so that y is strictly decreasing in

$\left]a, \frac{1}{2}(1+\sqrt{5})a\right[$.

Taking the above points into consideration, we can trace the curve as given in the figure.

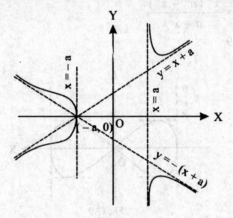

Fig 17.10

3. *Trace the curve*

$$y^2 x^2 = x^2 - a^2$$

(*i*)   The curve is symmetrical about both the axes.

(*ii*) It does not pass through the origin.

(*iii*) It meets x- axis at $(-a, 0)$ but it does not meet y-axis at any point; $x = a$ and $x = -a$ are the tangents at $(a, 0)$ and $(-a, 0)$ respectively.

(*iv*) $y = \pm 1$ are the two asymptotes of the curve.

(*v*) $y^2 = \frac{x^2 - a^2}{x^2} \Rightarrow y = \pm\frac{\sqrt{x^2 - a^2}}{x}$

y is defined for only those values of x for which $x \neq 0$ and $x^2 - a^2 \geq 0 \Leftrightarrow x \leq -a$ or $x \geq a$ so that no part of the curve lies between the lines $x = a$ and $x = -a$.

(*vi*) $\frac{dy}{dx} = \pm\frac{a^2}{x^2\sqrt{x^2 - a^2}}$

which is never zero.

$$\frac{dy}{dx} > 0 \ \forall \ x \in \ ]-\infty, -a[\text{ and } [a, \infty[.$$

so that y is strictly increasing in $]-\infty, -a]$ and $[a, \infty[$.

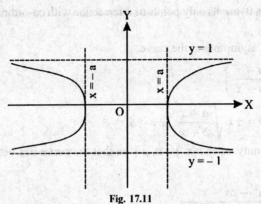

Fig. 17.11

Combining the above facts, we see that the curve $y^2 = \dfrac{x^2 - a^2}{x^2}$ is as given in the figure.

4. *Trace the curve*

$$(x^2 + y^2)x - ay^2 = 0 \ (a > 0)$$

(*i*)   The curve is symmetrical about x-axis.

(*ii*) Origin lies on it and x-axis is the tangent thereat.

(*iii*) Origin is the only point of intersection with axes.

(*iv*) $x = a$ is the only asymptote to the curve.

(*v*) $y^2 = \dfrac{x^3}{a - x}, \ x \neq a$

$\Rightarrow \quad y = \pm x \sqrt{\dfrac{x}{a - x}}$

y is defined only for those of the values of x for which $\dfrac{x}{a - x}$ is non-negative, *i.e.*, when

$$0 \leq x < a \Leftrightarrow \forall x \in [0, a[$$

Thus, the curve is situated between the lines $x = 0$ and $x = a$

(*vi*) $\dfrac{dy}{dx} = \dfrac{\left(\dfrac{3a}{2} - x\right)\sqrt{x}}{(a - x)\sqrt{a - x}}$

$\dfrac{dy}{dx} = 0$ for $x = 0$ where y is 0 and $0 < x < a \Rightarrow \dfrac{dy}{dx} > 0$

so that y is strictly increasing in $[0, a[$.

The curve is as given in Fig 17.11.

Note: This curve is known as the cissoid of Diocles.

5. *Trace the curve*

$$(x^2 + y^2)x - a(x^2 - y^2) = 0, \ (a > 0)$$

(*i*)   The curve is symmetrical about x-axis.

(*ii*) Origin lies on it. $y = \pm\, x$ are the two tangents thereat so that origin is a node.

(*iii*) $(0, 0)$ and $(a, 0)$ are its only points of intersection with co-ordinate axes. $x = a$ is the tangent at $(a, 0)$.

(*iv*) $x = -a$ is an asymptote of the curve.

(*v*) $y^2 = x^2 \left(\dfrac{a - x}{a + x} \right)$

$$\Rightarrow \qquad y = \pm x \sqrt{\dfrac{a - x}{a + x}}, \; x \neq -a$$

y is defined only when $x \in\;]-a, a\,]$ so that the curve lies between the lines $x = -a$ and $x = a$.

(*vi*) $\dfrac{dy}{dx} = \pm\, \dfrac{a^2 - ax - x^2}{(a + x)\sqrt{a^2 - x^2}}$

$\dfrac{dy}{dx} = 0 \Leftrightarrow a^2 - ax - x^2 = 0$

$$\Rightarrow \qquad x = \dfrac{-1 \pm \sqrt{5}}{2}\, a$$

Now $\dfrac{-1 - \sqrt{5}}{2}\, a$ is not a member of the interval $]-a, a\,]$.

Also $\dfrac{dy}{dx} > 0 \;\; \forall\, x \in\;]-a, \dfrac{-1 + \sqrt{5}}{2}\, a\,[$ and $< 0 \;\; \forall\, x \in\; \left] \dfrac{-1 + \sqrt{5}}{2}\, a, a\, \right[$

Thus y is strictly increasing in $\left] -a, \dfrac{-1 + \sqrt{5}}{2}\, a \right]$ and strictly decreasing in $\left[\dfrac{-1 + \sqrt{5}}{2}\, a, a \right]$

The point corresponding to $x = \dfrac{-1 + \sqrt{5}}{2}\, a$ gives a maximum value of y.

Combining the above facts, the curve is as given in the Fig. 17.12

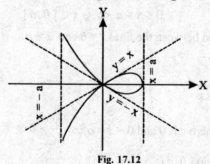

Fig. 17.12

Note: The curve referred to above is known as strophoid.

6. *Trace the curve*

$$y^2 = (x - 1)(x - 2)(x - 3)$$

(*i*) The curve is symmetrical about x-axis.

(*ii*) It does not pass through the origin.

(iii) It meets x- axis at (1, 0), (2, 0) and (3, 0) but it does not meet y-axis; $x = 1, x = 2, x = 3$ are the tangents at (1, 0), (2, 0), (3, 0) respectively.

(iv) It has no asymptotes.

(v) Now $(x - 1)(x - 2)(x - 3) < 0$ when $x < 1$ and $2 < x < 3$ and $(x - 1)(x - 2)(x - 3) > 0$ when $1 < x < 2$ and $x > 3$. Thus no point of the curve lies to the left of the line $x = 1$ and between the lines $x = 2, x = 3$.

(vi) $y = \pm \sqrt{(x - 1)(x - 2)(x - 3)}$

$$\frac{dy}{dx} = \pm \frac{3x^2 - 12x + 11}{2\sqrt{(x - 1)(x - 2)(x - 3)}}$$

$$= \pm \frac{3(x - \alpha)(x - \beta)}{2\sqrt{(x - 1)(x - 2)(x - 3)}}$$

Thus $\dfrac{dy}{dx} = 0$ when

$x = \alpha = \dfrac{6 - \sqrt{3}}{3} = 1.42$ approx.

and $x = \beta = \dfrac{6 + \sqrt{3}}{3} = 2.5$ approx. $(\alpha < \beta)$.

Also β does not belong to the domain of definition of y.

Now $\dfrac{dy}{dx} > 0 \ \forall \ x \in [1, \alpha], \dfrac{dy}{dx} < 0 \ \forall \ x \in [\alpha, 2]$

and $\dfrac{dy}{dx} > 0 \ \forall \ x > 3$

Thus, y is strictly increasing in [1, α] and [3, ∞ [and strictly decreasing in [α , 2]

Also y is maximum for $x = \alpha$.

We also notice that $\dfrac{dy}{dx} \to \infty$ as $x \to \infty$ so that the curve tends to be $\parallel$ to y-axis as $x \to \infty$. Since the curve, while departing from (3, 0) where the tangent is parallel to y-axis, must again tend to become parallel to y-axis, we see that it must change its direction of bending for some point whose abscissae is > 3 and where therefore we have a point of inflexion.

We thus have the curve as in Fig. 17.13.

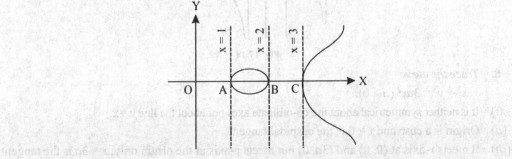

Fig. 17.13

7. *Trace the curve*

$$y^2(x^2 - 1) = 2x - 1$$

   (*i*)   The curve is symmetrical about x-axis.

   (*ii*) It does not pass through the origin.

   (*iii*) It meets x-axis in $\left(\dfrac{1}{2}, 0\right)$ and y-axis in $(0, 1)$ and $(0, -1)$ respectively.

   The line $x = \dfrac{1}{2}$ is the tangent to the curve at $\left(\dfrac{1}{2}, 0\right)$.

   (*iv*) $x = \pm 1$, and $y = 0$ are its only asymptotes.

   (*v*) $y^2 = \dfrac{2x - 1}{x^2 - 1}$

   For $x < -1$ and $\dfrac{1}{2} < x < 1$, y is imaginary, therefore the curve lies in the region $-1 < x < \dfrac{1}{2}$ and $x > 1$.

   (*vi*) $y = \pm \sqrt{\dfrac{2x - 1}{x^2 - 1}}$

   $$\dfrac{dy}{dx} = \pm \left(\dfrac{-x^2 + x + 1}{(2x - 1)^{1/2} (x^2 - 1)^{3/2}} \right)$$

   For $x > 1$, $\dfrac{dy}{dx} < 0$. Thus y is decreasing for $x > 1$.

   Again for $-1 < x < \dfrac{1}{2}$, $\dfrac{dy}{dx} < 0$, so that y is decreasing in the interval $\left[-1, \dfrac{1}{2}\right]$. Taking all these points into consideration, the curve is as shown in the figure.

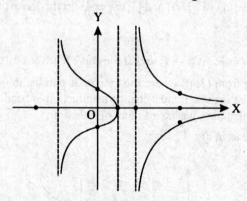

Fig. 17.14

8. *Trace the curve*

$$x^3 + y^3 = 3ax^2 \ (a > 0).$$

   (*i*)   It is neither symmetrical about the co-ordinate axes nor about the line $y = x$.

   (*ii*) Origin is a cusp and $x = 0$ is the cuspidal tangent.

   (*iii*) It meets x-axis at $(0, 0)$ and $(3a, 0)$ but meets y-axis at the origin only; $x = 3a$ is the tangent at $(3a, 0)$.

(*iv*) $y + x = a$ is its only asymptote and curve meets the asymptote at $\left(\dfrac{a}{3}, \dfrac{2a}{3}\right)$.

(*v*) x and y cannot both be negative. Therefore the curve does not lie in the 3rd quadrant.

(*vi*) $y^2 \dfrac{dy}{dx} = x(2a - x)$ and, therefore

$$\frac{dy}{dx} = 0 \ \text{ for } x = 2a$$

Solving for y, we obtain

$$y = [x^2(3a - x)]^{1/3}$$

If $x = 0$ then $y = 0$ and y-axis is the tangent there. If $0 < x < 3a$, then $y > 0$ and if x increases from

0, y also increases; y will go on increasing with x till $x = 2a$ where $\dfrac{dy}{dx} = 0$. When x increases

beyond $2a$, y will constantly be decreasing; $y = 0$ for $x = 3a$ and is negative for $x > 3a$.

We now consider the negative values of x. If x is negative, y is positive and constantly goes on increasing as x increases numerically, *i.e.* as x varies from 0 to $-\infty$.

Also $x + y = a$ is the only asymptote of the curve. Taking all the above facts into consideration, we see that the complete curve is as shown in Fig. 17.15

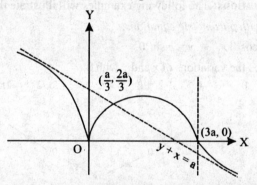

Fig. 17.15

EXERCISES

1. Trace the following curves :

(*i*) $3ay^2 = x^2(x - a)$

(*ii*) $y^2 = x(x + 1)^2$

(*iii*) $ay^2 = x(a^2 - x^2)$

(*iv*) $4a^4 y^2 = x^5(2a - x)$. (**Gorakhpur 2003**)

(*v*) $y^2 = x^2(4 - x^2)$

(*vi*) $x^2 y^2 = (a + y)^2(a^2 - y^2)$.

(*vii*) $a^2 y^2 = x^2(a^2 - x^2)$

(*viii*) $y^2(a^2 - x^2) = a^3 x$.

(*ix*) $y^2 x = a^2(a - x)$

(*x*) $ay^2 = x(a^2 + x^2)$

(*xi*) $y^2(a^2 + x^2) = a^2 x^2$

(*xii*) $y^2 x = a(x^2 + a^2)$.

(*xiii*) $y^2 x^2 = x^2 - 1$

(*xiv*) $y^2(a^2 - x^2) = x^4$

(*xv*) $y^2 x = a(x^2 - a^2)$

(*xvi*) $y^2 x^2 = x^2 + 1$

(*xvii*) $a^2 y^2 = x^2(2a - x)(x - a)$

(*xviii*) $y^2(x + 2) = x + 1$.

(*xix*) $y^2(2x - 1) = x(x - 1)$.

(*xx*) $y^2 = x^2(4 - x)$.

(*xxi*) $y^2 = x^2(x - 5)$.

(*xxii*) $y^2 = x(16 - x^2)$.

$(xxiii)$ $y^2 = (x-2)^2 (x-3)$. $(xxiv)$ $y^2 = x^3 - x^2$.

(xxv) $y^2 = (x-a)^2 (x-b)$ for different values of a and b.

$(xxvi)$ $y^2 (x-1) = x^2$. $(xxvii)$ $y^2 = x^2 (x^2 - 4)$.

$(xxviii)$ $y^2 = x^2 (25 - 4x^2)$ $(xxix)$ $y^2 = x^2 (x^2 - 4)$.

(xxx) $y^2 = x^2 (3-x)$. $(xxxi)$ $y^2 (a+x) = x^2 (a-x)$

(Avadh 96, 98; Garhwal 2001; Gorakhpur 99, 2002; Ravi Shankar 96; Jabalpur 98; Vikram 2000; Sagar 2000)

2. $y(a^2 + x^2) = a^2 x$ 3. $y^3 = a^2 x - x^3$

(Kanpur 2001, Jiwaji 96; Indore 96, 99; Bilaspur 96; Sagar 99)

4. $y(x^2 - 1) = x^2 + 1$ 5. $y(1 - x^2) = x^2$

6. $x = (y-1)(y-2)(y-3)$ **(Avadh 97; Ravi Shankar 2001; Bilaspur 2000)**

7. $y^2 (a+x) = x^2 (3a - x)$.   **(Garhwal 1996, 2000)**

8. $ay^2 = x^3$ 9. $ay^2 = (x-a)(x-5a)^2$.

10. $y^2 (a-x) = x^3, a > 0$.   **(Banglore 2006)**

17.5. Parametric Equations: The following examples will illustrate the process.

1. *Trace the curve with parametric equations.*

$$x = a \cos^3 \theta, \qquad y = b \sin^3 \theta.$$

The following table gives the variations of x and y with θ.

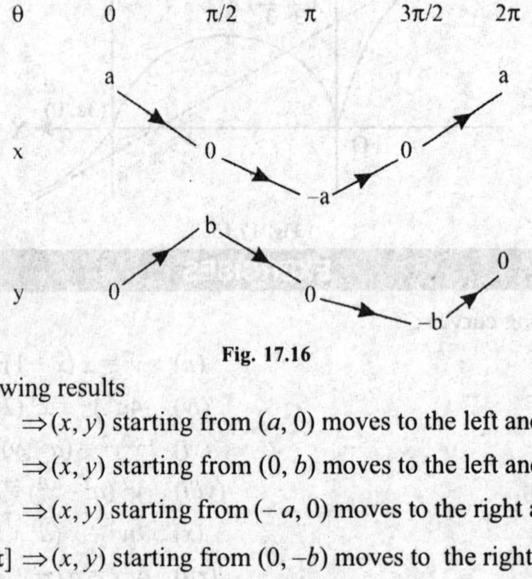

Fig. 17.16

so that we have the following results

θ varies in $[0, \pi/2]$ $\Rightarrow (x, y)$ starting from $(a, 0)$ moves to the left and upwards to $(0, b)$.

θ varies in $[\pi/2, \pi]$ $\Rightarrow (x, y)$ starting from $(0, b)$ moves to the left and downwards to $(-a, 0)$.

θ varies in $[\pi, 3\pi/2]$ $\Rightarrow (x, y)$ starting from $(-a, 0)$ moves to the right and downwards to $(0, -b)$.

θ varies in $[3\pi/2, 2\pi]$ $\Rightarrow (x, y)$ starting from $(0, -b)$ moves to the right and upwards to $(a, 0)$.

Again, we have

$$\frac{dx}{d\theta} = -3a \cos^2 \theta \sin \theta, \quad \frac{dy}{d\theta} = 3b \sin^2 \theta \cos \theta.$$

Now $dx/d\theta = 0$ if $\theta \in \{0, \pi/2, \pi, 3\pi/2, 2\pi\}$

It follows that

$$\frac{dy}{dx} = \frac{dy}{d\theta} \bigg/ \frac{dx}{d\theta} = -\frac{b}{a} \cdot \frac{\sin^2 \theta \cos \theta}{\cos^2 \theta \sin \theta} = -\frac{b}{a} \tan \theta$$

for the values of θ for which $dx/d\theta \neq 0$. Also

    (i) $\theta \to 0 \Rightarrow dy/dx \to 0$,   (ii) $\theta \to \pi/2 \Rightarrow dy/dx \to \infty$

    (iii) $\theta \to \pi \Rightarrow dy/dx \to 0$,   (iv) $\theta \to 3\pi/2 \Rightarrow dy/dx \to \infty$

Thus the tangent at each of the points $(a, 0)$, $(-a, 0)$ is the x-axis and the tangent at each of the points $(0, b)$, $(0, -b)$ is the y-axis.

The curve is as given in the following diagram.

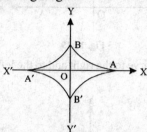

Fig. 17.17

The curve with parametric equations $x = a \cos^3\theta$, $y = b \sin^3\theta$ is known as **Astroid**.

2. *Trace the curve with parametric equations* θ

    $x = a\,(\theta + \sin\theta), \;\; y = a\,(1 + \cos\theta)$

The following table gives the variations of x and y with θ.

θ	$-\pi$	0	π
x	$-a\pi$	0	$a\pi$
y	0	$2a$	0

Fig. 17.18

so that we have the following results

    θ varies in $[-\pi, 0]$ $\Rightarrow$ (x, y) starting from $(-a\pi, 0)$ moves to the right and upwards to $(0, 2a)$.

    θ varies in $[0, \pi]$ $\Rightarrow$ the point (x, y) starting from $(0, 2a)$ moves to the right and downwards to $(a\pi, 0)$.

Again, we have

$$\frac{dx}{d\theta} = a(1 + \cos\theta), \;\; \frac{dy}{d\theta} = -a\sin\theta$$

Now $dx/d\theta = 0$ if $\theta = \pi$ or $-\pi$.

Thus $\dfrac{dy}{dx} = \dfrac{dy}{d\theta} \Big/ \dfrac{dx}{d\theta} = -\tan\dfrac{\theta}{2}$

except for the values $\pm \pi$ of θ for which $dx/d\theta = 0$.

Also it may be seen that at the points given by $\theta = \pi$ and $\theta = -\pi$, the tangents are parallel to y-axis.

Thus we have the curve as shown below

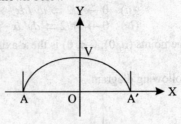

Fig. 17.19

We obtain identical portions of curve to the right and to the left if θ varies in $[\pi, 3\pi]$, $[3\pi, 5\pi]$ etc. and $[-3\pi, -\pi]$, $[-5\pi, -3\pi]$ etc.......

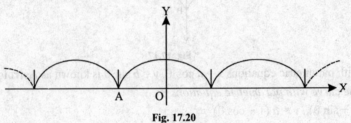

Fig. 17.20

3. *Trace the curve*

$$x = a \sin 2\theta (1 + \cos 2\theta), \; y = a \cos 2\theta (1 - \cos 2\theta).$$

Since, x, y are periodic functions of θ, with π as their period the values of x, y will repeat themselves as θ varies in the interval

$$]x, 2\pi [,] 2\pi, 3\pi [\text{ etc.}$$

We thus confine our attention to $\theta \in [0, \pi [$ only.

We have $\dfrac{dx}{d\theta} = 4a\cos 3\theta \cos \theta; \dfrac{dy}{d\theta} = 4a\cos 3\theta$

$\dfrac{dx}{d\theta} = 0$ for $\theta = \dfrac{\pi}{6}, \dfrac{\pi}{2}, \dfrac{5\pi}{6}.$

$\therefore$ $\dfrac{dy}{dx} = \tan \theta$ if $\theta \notin \left\{ \dfrac{\pi}{6}, \dfrac{\pi}{2}, \dfrac{5\pi}{6} \right\}$

Thus, the tangent at any point corresponding to the value θ of the parameter makes an angle θ with x-axis.

Table of Variations of x and y with θ.

θ	0	π/6	π/3	π/2	2π/3	5π/6	π
dx/dθ	+	–	–	–	–	+	
dy/dθ	+	–	–	+	+	–	

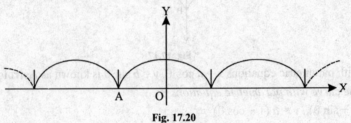

Fig. 17.21

The curve is as given in Fig. 17.22.

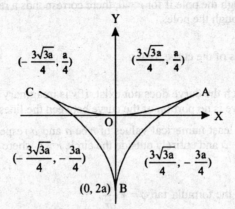

Fig. 17.22

EXERCISES

Trace the following curves:

(i) $x = a(\theta - \sin\theta), y = a(1 - \cos\theta)$.

(ii) $x = a(\theta + \sin\theta), y = a(1 - \cos\theta)$.

(iii) $x = a\cos^3\theta, y = a(\sin 3\theta + \sin\theta)$

(iv) $x = a(3\cos\theta - \cos^3\theta), y = a(3\sin\theta - \sin^3\theta)$

(v) $x = a(\cos\theta + \theta\sin\theta), y = a(\sin\theta - \theta\cos\theta)$

(vi) $x = a(\sin\theta + \dfrac{1}{3}\sin 3\theta), \quad y = a(\cos\theta - \dfrac{1}{3}\cos 3\theta)$

(vii) $x = \cos t, y = \cot t$.

(viii) $x = 1 - e^{-t}, y = t^2 + 1$,

(ix) $x = e^t + e^{-t}, y = e^t - e^{-t}$.

(x) $x = 1 + \sin t, y = 2\cos 2t$.

(xi) $x = \dfrac{1}{2\cos t}, y = 1 + \cot t$.

17.6. Tracing of Polar Curves

The following points may be taken into consideration while tracing polar curves of the type $r = f(\theta)$.

1. *Symmetry*

(i)   The curve $r = f(\theta)$ is symmetrical about the initial line $\theta = 0$ if the equation remains unchanged when θ is replaced by $-\theta$. The curve $r = a(1 - \cos\theta)$ is symmetrical about the line $\theta = 0$.

(ii)   The curve is symmetrical about the line $\theta = \pi/2$ (*i.e.* the line through the pole perpendicular to the initial line) if the equation remains unchanged when θ is replaced by $\pi - \theta$. The curve $r = a(1 - \sin\theta)$ is symmetrical about the line $\theta = \pi/2$.

(iii)   The curve is symmetrical about the pole

2. *Pole*

The curve passes through the pole if for $r = 0$, there corresponds a real value of θ. For example $r = a(1 + \cos \theta)$ passes through the pole.

3. *Asymptotes*

Find out the asymptotes of the curve, if any.

4. *Region*

Find the region in which the curve does not exist. If r is imaginary for some values of θ lying between θ_1 and θ_2 then there is no portion of the curve between the lines $\theta = \theta_1$ and $\theta = \theta_2$.

If the greatest and the least numerical values of r be a and b respectively then the curve lies entirely within the circle $r = a$ and entirely outside the circle $r = b$ where $a > 0$ and $b > 0$.

5. *Values of* φ

Find φ with the help of the formula $\tan \varphi = r \dfrac{d\theta}{dr}$.

If $\theta = \theta_1$ when $\varphi = 0$ then the line $\theta = \theta_1$ will be a tangent to the curve at the point $\theta = \theta_1$ and if $\theta = \theta_2$ when $\varphi = \pi/2$ then $\theta = \theta_2$, the tangent will $\perp$ to the radius vector $\theta = \theta_2$.

6. *Special points*

Trace the variations of r as θ varies. Also if $\dfrac{dr}{d\theta} > 0$ then r increases as θ increases and if $\dfrac{dr}{d\theta} < 0$, then r decreases as θ increases.

Taking the above points into consideration we can trace the requisite curve.

EXAMPLES

1. *Trace the curve*

$r = a(1 + \cos \theta)$ *(Rohilkhand 1999, 2000, 2002, Bhopal 98 Ravishankar 2000; Rewa 98; Banglore 2005)*

The curve is symmetrical about the initial line.

$r = 0 \Rightarrow \theta = \pi$. Therefore the curve passes through the pole, the tangent at the pole being $\theta = \pi$. There are no asymptotes to the curve.

The maximum value of r is $2a$. Therefore the curve lies within a circle whose centre is at the pole nd whose radius is $2a$.

$$\frac{dr}{d\theta} = -a \sin \theta$$

$$\therefore \qquad \tan \phi = r \frac{d\theta}{dr} = -\frac{(1 + \cos \theta)}{\sin \theta} = \cot \theta/2 = \tan(\pi/2 + \theta/2).$$

$$\Rightarrow \qquad \phi = \pi/2 + \theta/2.$$

Therefore the tangent to the curve at the point $(2a, 0)$ is perpendicular to the initial line.

As θ increases from 0 to $\pi/2$, r decreases from $2a$ to a. As θ increases from $\pi/2$ to π, r decreases from a to 0. The table of variations of r as θ varies from 0 to π is as follows:

θ	:	0	$\pi/3$	$\pi/2$	$2\pi/3$	π
r	:	$2a$	$3a/2$	a	$a/2$	0

Since the curve is symmetrical about the initial line, the complete graph is as shown in the figure:

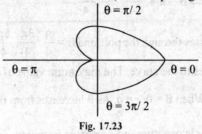

Fig. 17.23

2. *Trace the curve*:

$$r = a \sin 3\theta.$$
(Kumaon 2003; Gorakhpur 2000)

The curve is symmetrical about the line through the pole perpendicular to the initial line.

$$r = 0 \Rightarrow \theta = 0, \frac{\pi}{3}, \frac{2\pi}{3}, \pi, \frac{4\pi}{3}, \frac{5\pi}{3}.$$

Therefore the curve passes through the pole and $\theta = 0, \pi/3, \frac{2\pi}{3}, \pi, \frac{4\pi}{3}, \frac{5\pi}{3}$ are the tangents at the pole.

There are no asymptotes to the curve. The maximum value of r is a so that the curve lies wholly within a circle of radius a.

As θ increases from 0 to $\frac{\pi}{6}$, r increases from 0 to a. As θ increases from $\frac{\pi}{6}$ to $\frac{\pi}{3}$, r decreases form a to 0. As θ increases from $\frac{\pi}{3}$ to $\frac{\pi}{2}$, r is negative and numerically increases from 0 to a. This portion of the curve lies in the 3rd quadrant. The curve is symmetrical about the line $\theta = \pi/2$. Therefore we take the table of variations of r as θ increases from $-\pi/2$ to $\pi/2$

θ :	$-\dfrac{\pi}{2}$	$-\dfrac{\pi}{3}$	$-\dfrac{\pi}{6}$	0	$\dfrac{\pi}{6}$	$\dfrac{\pi}{3}$	$\dfrac{\pi}{2}$
r :	a	0	$-a$	0	a	0	$-a$

The complete figure is as follows:

$\theta = 2\pi/3$ $\theta = \pi/3$

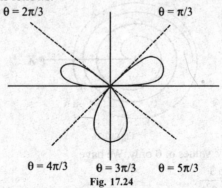

$\theta = 4\pi/3$ $\theta = 3\pi/3$ $\theta = 5\pi/3$

Fig. 17.24

3. *Trace the curve*:

$$r = a \cos 2\theta, \, a > 0$$

The curve is symmetrical about the initial line as well as about the line through the pole perpendicular to the initial line.

$$r = 0 \Rightarrow \theta = \pi/4, \frac{3\pi}{4}, \frac{5\pi}{4}, \frac{7\pi}{4}.$$

Therefore the curve passes through the pole and $\theta = \frac{\pi}{4}, \frac{3\pi}{4}, \frac{5\pi}{4}, \frac{7\pi}{4}$ are the tangents at the pole.

There are no asymptotes to the curve. The maximum value of r is a so that the curve lies wholly within a circle of radius a. When $\theta = 0$, $r = a$. As θ increases from 0 to $\frac{\pi}{4}$, r decreases from a to 0. As θ increases from $\frac{\pi}{4}$ to $\frac{\pi}{2}$, r is negative and decreases from 0 to $-a$. This portion of the curve lies in the 3rd quadrant. As θ increases from $\frac{\pi}{2}$ to $\frac{3\pi}{4}$, r increases from $-a$ to 0 and as θ increases from $\frac{3\pi}{4}$ to π, r increases from 0 to a. Since the curve is symmetrical about the initial line the figure is as follows:

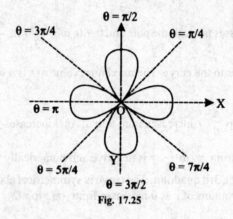

Fig. 17.25

4. *Trace the curve*:

$$r = \frac{a\theta^2}{1 + \theta^2}.$$

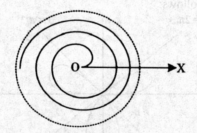

Fig. 17.26

We first consider *positive* values of θ only. We have

$$\frac{dr}{d\theta} = \frac{2a\theta}{(1 + \theta^2)},$$

which is always positive so that r, constantly increases as, θ , increases.

Also, $r = 0$ when $\theta = 0$.

Again we have

$$\frac{a\theta^2}{1 + \theta^2} = \frac{a}{\theta^{-2} + 1} \text{ which} \to a \text{ as } \theta \to \infty.$$

Thus r, starting from its initial value 0, constantly increases as θ increases and approaches, a, as $\theta \to \infty$ so that the point (r, θ) approaches nearer and nearer the circle whose centre is at the pole and radius equal to a.

The circle is shown dotted in the Fig. 17.26.

The figure shows the part of the curve corresponding to the positive values of θ only, and the part of the curve for negative values is its reflections in the initial line.

5. *Trace the curve*:

$$r = a (\sec \theta + \cos \theta) \hspace{4cm} \textbf{\textit{(Gorakhpur 2001)}}$$

Here $\qquad r = a\left(\dfrac{1}{\cos\theta} + \cos\theta\right) = a\dfrac{1 + \cos^2\theta}{\cos\theta}$

(*i*) The curve is symmetrical about the initial line.

(*ii*) $r \cos \theta = a$, *i.e.*, $x = a$ is its asymptote.

(*iii*) $\dfrac{dr}{d\theta} = a \dfrac{\sin^3\theta}{\cos^2\theta}$ so that $\dfrac{dr}{d\theta}$ is positive when θ lies between 0 and $\pi/2$.

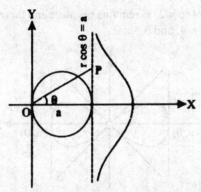

Fig. 17.27

When $\theta = 0$, $dr/d\theta = 0$ so that $\phi = \pi/2$ and, therefore, the tangent is perpendicular to the initial line at the point $(2a, 0)$.

Also, $r = 2a$, when $\theta = 0$ and $r \to \infty$ as $\theta \to \pi/2$.

Hence we see that when θ increases from 0 to $\dfrac{1}{2}\pi$, r monotonically increases from $2a$ to ∞ and the point $P(r, \theta)$ describes the part of the curve drawn in the first quadrant.

When θ increases from $\dfrac{1}{2}\pi$ to π, r, remains negative and decreases in numerical value from ∞ to $-2a$ and so the point $P(r, \theta)$ describes the part of curve as shown in the fourth quadrant.

As the curve is symmetrical about the initial line. no new point will be obtained when θ varies from π to 2π.

17.7. In the case of some curves, it is found convenient to make use of the polar as well as the Cartesian form of their equations. Some facts are obtained from the Cartesian and the others from the Polar form.

Curves whose cartesian equations are not solvable for x and y, but whose polar equations are solvable for r, are generally dealt with in this manner.

6. *Trace the curve*:
$$x^4 + y^4 = a^2(x^2 - y^2).$$

(*i*)   It is symmetrical about both the axes.

(*ii*)   Origin is a node on the curve and $y = \pm x$ are the nodal tangents.

(*iii*)   It meets x-axis at $(0, 0)$, $(a, 0)$ and $(-a, 0)$, but meets y-axis at $(0, 0)$ only; $x = a$ and $x = -a$ are the tangents at $(a, 0)$ and $(1, -a, 0)$.

(*iv*)   It has no asymptotes.

(*v*)   On changing to polar co-ordinates, the equation becomes
$$r^2 = \frac{a^2 \cos 2\theta}{\cos^4 \theta + \sin^4 \theta}.$$

We see that
$$r = \frac{dr}{d\theta} = -\frac{8a^2 \sin^3 \theta \cos^3 \theta}{\left(\cos^4 \theta + \sin^4 \theta\right)^2}$$

so that $dr/d\theta$ remains negative as θ varies from 0 to $\pi/4$ and therefore r decreases from a to 0 as θ increases from 0 to $\pi/4$.

(*vi*)   As θ changes from $\pi/4$ to $\pi/2$, r^2 remains negative and therefore no point on the curve lies between the lines $\theta = \pi/4$. and $\theta = \pi/2$.

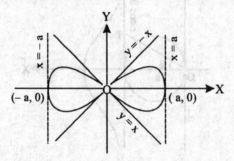

Fig. 17.28

As the curve is symmetrical about both the axes, we have its shape as shown. (Fig. 17.28).

7. *Trace the curve*:
$$y^4 - x^4 + xy = 0. \hspace{4cm} \textbf{(P.U.)}$$

(*i*)   It is neither symmetrical about the co-ordinate axes. nor about the line $y = x$.

(*ii*)   It passes through the origin : $x = 0, y = 0$ are the two tangents thereat so that the origin is a node.

(*iii*)   It cuts the co-ordinate axes at the origin only.

(*iv*) $y = x, y = -x$ are its asymptotes.

(*v*)   On transforming to polar co-ordinates, we get
$$r^2 = \frac{1}{2}\tan 2\theta.$$

When θ increases from 0 to $\pi/4$, 2θ increases from 0 to $\pi/2$, and, therefore, r^2 monotonically increases from 0 to ∞.

When θ increases from $\pi/4$ to $\pi/2$, tan 2θ and therefore also r^2 remains negative and, thus, there is no part of the curve lying between the lines $\theta = \pi/4$ and $\pi/2$.

When θ increases from π/2 to 3π/4, r^2 increases from 0 to ∞.

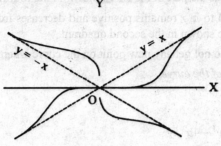

Fig. 17.29

When θ increases from 3π/4 to π, r^2 remains negative and so there is no part of the curve lying between the lines θ = 3π/4, and θ = π.

We can similarly consider the variations of r^2 as θ increases from π to 2π.

Hence we have the curve as drawn. (Fig. 17.29).

8. *Trace the Folium of Descartes*:
    $$x^3 + y^3 = 3axy. \qquad \textbf{(Gorakhpur 2000, Rohilkhand 2003; Poorvanchal 2004)}$$
    (*i*)   It is symmetrical about the line $y = x$ and meets it in the point $(3a/2, 3a/2)$.
    (*ii*) It passes through the origin and $x = 0, y = 0$ are the tangents there so that the origin is a node on the curve.
    (*iii*) It meets the co-ordinate axes at the origin only.
    (*iv*) $x + y + a = 0$ is its only asymptote.
    (*v*) x, y cannot both be negative so that no part of the curve lies in the third quadrant.
    On transforming to polar co-ordinates, we get
    $$r = \frac{3a\sin\theta\cos\theta}{\cos^3\theta + \sin^3\theta}$$
    Now, $r = 0$ for θ = 0 and θ = π/2.
    $$\frac{dr}{d\theta} = \frac{3a(\cos\theta - \sin\theta)(1 + \sin\theta\cos\theta + \sin^2\theta\cos^2\theta)}{(\cos^2\theta + \sin^2\theta)^2}$$
which vanishes only when
    $$\cos\theta - \sin\theta = 0, \; i.e., \; \tan\theta = 1. \; i.e., \; \theta = \pi/4 \text{ or } 5\pi/4.$$
For ——   θ = π/4, $r = 3a/\sqrt{2}$

Thus r monotonically increases from 0 to $3a/\sqrt{2}$, as θ increases from 0 to π/4, and monotonically decreases from $3a/\sqrt{2}$ to 0, as θ increases from π/4 to π/2.

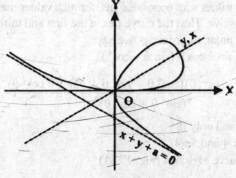

Fig. 17.30

Again as θ increases from $\pi/2$ to $3\pi/4$, r remains negative and numerically increases from 0 to ∞ so that the point (r, θ) describes the part of the curve shown in the fourth quadrant. (Fig. 17.30).

As θ increases from $3\pi/4$ to π, r remains positive and decreases from ∞ to 0 so that the point describes the part of the curve shown in the second quadrant.

It is easy to see that we do not get any new point on the curve when, θ, increases from π to 2π.

9. *Find the double point of the curve*:
$$x^4 + y^4 = 4a^2xy.$$
and trace it.

Let $f(x, y) = x^4 + y^4 - 4a^2\, xy.$

$\therefore$ $f_x(x, y) = 4\,(x^3 - a^2y),$

                $f_y(x, y) = 4\,(y^3 - a^2x).$

Putting $f_x = 0$ and $f_y = 0$, we have

$x^3 - a^2y = 0,\ y^3 - a^2x = 0.$

$\therefore$ $y^3 - a^3y = 0,$

or $y = 0, \pm a.$

Thus we see that f_x and f_y vanish at the points $(0, 0)\,(a, a)\,(-a, -a)$. Of these only $(0, 0)$ is a point of the curve so that the origin is the only double point of the curve.

To trace the curve we note the following particulars about it :—

(*i*)   It is symmetrical about the line $y = x$ and meets it at
$$(\pm\sqrt{2}a, \pm\sqrt{2}a).$$

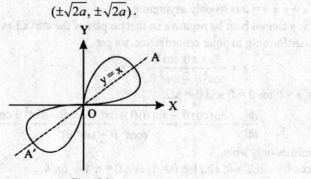

Fig. 17.31

(*ii*)   It passes through the origin and $x = 0, y = 0$, are the two tangents there.

(*iii*)   It meeets the co-ordinate axes at the origin only.

(*iv*)   It has no asymptotes.

(*v*) x and y cannot have values with opposite signs, for such values make $4a^2xy$ negative whereas $x^4 + y^4$ is always positive. Thus the curve lies in the first and third quadrants only.

(*vi*)   On transforming to polar co-ordinates, we have
$$r^2 = 4a^2 \sin\theta \cos\theta/(\sin^4\theta + \cos^4\theta).$$

$\therefore$ $r\dfrac{dr}{d\theta} = 2a^2 \dfrac{(\cos^2\theta - \sin^2\theta)(\cos^4\theta + \sin^4\theta + 4\sin^2\theta \cos^2\theta)}{(\sin^4\theta + \cos^4\theta)^2}$

so that $dr/d\theta = 0$ if and only if
tan $\theta = 1$, *i.e.*, $\theta = \pi/4$ and $5\pi/4$..

Thus we have the curve as drawn (Fig. 17.31)

10. *Trace the curve*:
$$x^5 + y^5 = 5a^2 xy^2$$

(i) The curve is symmetrical in opposite quadrants.

(ii) The curve passes through the origin and $x = 0, y = 0$ are the tangents there so that $(0, 0)$ is a node.

(iii) It meets the co-ordinate axes at the origin only.

(iv) $x + y = 0$ is an asymptote to the curve. There are no other asymptotes.

(v) On transforming to polar co-ordinates we get

$$r^2 = \frac{5a^2 \cos\theta \sin^2\theta}{\cos^5\theta + \sin^5\theta}$$

When $\theta = 0, r = 0$. Also when $\theta = \pi/2, r = 0$.

As θ increases from $\pi/2$ to $3\pi/4$, r^2 is negative and hence r is imaginary. Therefore no portiion of the curve lies in this region.

At $\theta = \dfrac{3\pi}{4}$, $r = \infty$, As θ increases from $\dfrac{3\pi}{4}$ to π, r decreases from ∞ to 0.

Since the curve is symmetric in opposite quadrants, the curve is as traced in the figure 17.32.

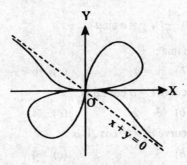

Fig. 17.32

EXERCISES

Trace the following curves:

1. $r = a + b \cos\theta$

2. $r \cos^3\theta = a \cos 2\theta$

3. $r \cos\theta = a \sin 3\theta$

4. $r^2 \cos\theta = a^2 \sin 3\theta$

5. $r = a \log\theta$

6. $r \log\theta = a$

7. $r = ae^\theta \sin\theta$

8. $r = a(\theta - \sin\theta)$

9. $r = 2(1 - 2\sin\theta)$

10. $r^2 \cos\theta = a^2 \sin^3\theta$

11. $r = a(1 + \sin\theta), a > 0$

12. $r = a\theta \sin\theta, a > 0$

13. $r = ae^{m\theta} (a > 0, m > 0)$

14. $r^2 = a^2 \cos 2\theta$

(*Kumaon 1997*; *Gorakhpur 99*; *Bilaspur 98*)

15. $r = a(1 - \cos\theta)$

16. $r \cos\theta = a \cos 2\theta$

17. $x^4 + y^4 = 4axy^2$

18. $x^5 + y^5 = 5ax^2y^2$

19. $x^5 + y^5 = 5a^2x^2y$

20. $x^6 + y^6 = 6a^2x^2y^2$

OBJECTIVE QUESTIONS

For each of the following questions, four alternatives are given for the answer. Only one of them is correct. Choose the correct alternative.

1. For the curve $y = \tan x$, which of the following is false?

    (a) The curve is symmetrical about the x-axis

    (b) The origin is a cusp

    (c) $x = 2a$ is an asymptote

    (d) curve exists for all $x \geq 0$

2. The number of loops in the curve $r = a \sin 5\theta$ is:

    (a) 2   (b) 5   (c) 10   (d) 1

    (Avadh 2005)

3. The folium of Descartes is given by the equation:

    (a) $x^2 + y^2 = 3ax^2y^2$ (b) $x^3 + y^3 = 3axy$ (c) $x^3 + y^3 = 3\,ax^2y^2$ (d) $x^4 + y^4 = 3axy$

4. The graph of the curve

    $$x = a\left(\cos t + \frac{1}{2}\log \tan^2 \frac{1}{2}t\right),\ y = a \sin t$$

    is symmetrical about the line :

    (a) $x = 0$ (b) $y = 0$ (c) $y = x$ (d) none of these

5. Number of loops in the curve $r = a \cos 2\theta$ is:

    (a) 2   (b) 3   (c) 4   (d) 6

6. Number of loops in the curve $r^2 = a^2 \cos 2\theta$ is :

    (a) 2   (b) 3   (c) 4   (d) 6

7. The curve $r = a + b \cos \theta$ is symmetrical about :

    (a) initial line   (b) y-axis   (c) line $\theta = \pi/2$ (d) line $\theta = \pi/4$

8. For the curve

    $$x = a(t + \sin t), y = a(1 - \cos t)$$

    (a) axis of y is tangent at origin   (b) curve is symmetrical about x-axis

    (c) axis of x is tangent at origin   (d) Curve is symmetrical about y-axis

ANSWERS

1. (a)   2. (b)   3. (b)   4. (b)   5. (c)   6. (a)   7. (a)   8. (c)

CHAPTER 18
ENVELOPES

18.1 One Parameter family of curves :

If $(x, y, \alpha) \to f(x, y, \alpha)$ be a function of three variables, then the equation

$$f(x, y, \alpha) = 0$$

determines a curve corresponding to each particular value of α.

The *totality* of these curves, obtained by assigning different values to α, is said to be a *one parameter family of curves*.

The variable α, which is different for different curves is said to be the *parameter* for the family.

Illustration :

(*i*) The equation

$$x^2 + y^2 - 2\,ax = 0,$$

determines a family of circles with their centres on X-axis and which pass through the origin. Here, a is the parameter.

(*ii*) The equation

$$y = mx - 2\,am - am^3,$$

determines a family of straight lines which are normal to the parabola

$$y^2 = 4ax.$$

Here the parameter is m.

We now introduce the concept of the *envelope of a one - parameter family of curves* by means of an example considered in the next article.

18.2 Consider the family of straight lines

$$y = mx + a / m, \qquad \qquad ...(i)$$

where, m is the *parameter* and, a is some given constant.

The two members of this family corresponding to the values m_1 and $m_1 + \Delta\,m$ of the parameter, m are

$$y = m_1 x + \frac{a}{m_1} \qquad \qquad ...(ii)$$

$$y = (m_1 + \Delta m)\,x + \frac{a}{m_1 + \Delta m} \qquad \qquad ...(iii)$$

We shall keep, m_1, fixed and regard Δm as a variable which tends towards 0 so that the line (*iii*) tends to coincide with the line (*ii*). The two lines (*ii*), (*iii*) intersect at the point x, y where

$$x = \frac{a}{m_1 (m_1 + \Delta m)}, \quad y = \frac{a(2m_1 + \Delta m)}{m_1 (m_1 + \Delta m)}.$$

As $\Delta m \to 0$, this point of intersection goes on changing its position on the line (*ii*) and, in the limit, tends to the point

$$\left(\frac{a}{m_1^2}, \frac{2a}{m_1} \right)$$

which lies on (*ii*).

This point is the limiting position of the point of intersection of the line (*ii*) with another line of the family when the latter tends to coincide with the former.

There will be a point similarly obtained, on every line of the family. The *locus of such points is called the envelope of the given family of lines.*

To find this locus for the family of lines (*ii*), we notice that the co-ordinates (*x, y*) of such a point lying on the line '*m*' are given by

$$x = \frac{a}{m^2}, \quad y = \frac{2a}{m}.$$

Eliminating *m*, we obtain

$$y^2 = 4ax,$$

as the envelope of the given family of lines.

18.3 Definition.

*The **envelope** of a one parameter of family of curves is the locus of the limiting positions of the points of intersection of any two curves of the family when one of them tends to coincide with the other which is kept fixed.*

18.4 Determination of Envelope

Let $f(x, y, \alpha) = 0$,   ...(*i*)

be any given family of curves.

Consider the two curves

$$f(x, y, \alpha) = 0 \text{ and } (x, y, \alpha + \Delta \alpha) = 0 \qquad \qquad ...(ii)$$

corresponding to the values α and $\alpha + \Delta \alpha$ of the parameter. The points common to the two curves satisfy the equation

$$f(x, y, \alpha + \Delta \alpha) - f(x, y, \alpha) = 0$$

$$\Leftrightarrow \qquad \frac{f(x, y, \alpha + \Delta \alpha) - f(x, y, \alpha)}{\Delta \alpha} = 0 \qquad \qquad ...(iii)$$

Let $\Delta \alpha \to 0$. Therefore the limiting positions of the points of intersection of the curves (*i*) satisfy the equation which is the limit of (*iii*) viz.,

$$f_\alpha (x, y, \alpha) = 0, \qquad \qquad ...(iv)$$

Thus the co-ordinates of the points on the envelope satisfy the equations

$$f(x, y, \alpha) = 0 \text{ and } f_\alpha (x, y, \alpha) = 0.$$

Let the elimination of α between (*i*) and (*iv*) lead to an equation

$$\phi (x, y) = 0.$$

This is, then, the required envelope.

Rule. *To obtain the envelope of the family of curves*

$$f(x, y, \alpha) = 0,$$

eliminate, α, *between*

$$f(x, y, \alpha) = 0, \text{ and } f_\alpha(x, y, \alpha) = 0,$$

where $f_\alpha(x, y, \alpha)$ *is the partial derivative of* $f(x, y, \alpha)$ *w.r. to* α.

The equations

$$x = \phi(\alpha), y = \psi(\alpha)$$

obtained on solving $f(x, y, \alpha) = 0, f_\alpha(x, y, \alpha) = 0$ are the parametric equations of the envelope; α being the parameter.

Illustration. To find the envelope of the family of lines

$$y - \alpha x - a / \alpha = 0, \qquad \qquad ...(i)$$

we eliminate, α, between (*i*) and

$$-x + a/\alpha^2 = 0,$$

which is obtained by differentiating (*i*) w.r. to α.

The eliminant is

$$y^2 = 4 \, ax,$$

which is the envelope of the given family of lines.

This conclusion agrees with the one already arrived at '*abinitio*' in § 18.2.

18.5 Theorem.

The evolute of a curve is the envelope of its normals.

PQ, QR. are the normals and PT, QT, the tangents at two point P, Q of a curve. L is the point of intersection of the tangents.

$$\angle PRQ = \angle TLT' = \Delta \psi$$

$$\text{arc } PQ = \Delta s$$

Applying the sine formula to the $\Delta \, PRQ$, we get

$$\frac{PR}{PQ} = \frac{\sin \angle RQP}{\sin \angle PRQ}$$

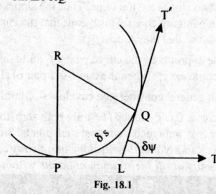

Fig. 18.1

$$\Rightarrow \qquad PR = \sin \angle RQP \, \frac{PQ}{\sin \Delta \, \psi}$$

$$= \sin \angle RQP \cdot \frac{\text{chord } PQ}{\text{arc } PQ} \cdot \frac{\Delta s}{\Delta \psi} \cdot \frac{\Delta \psi}{\sin \Delta \psi}$$

Let $Q \to P$ so that $\angle RQP \to \angle RPT = \pi/2$.

$$\therefore \qquad \lim_{Q \to P} PR = \sin\frac{\pi}{2} \cdot 1 \cdot \frac{ds}{d\psi} \cdot 1 = \rho.$$

Thus the limiting position of R which is the intersection of the normals at P and Q is the centre of curvature at P.

Hence the theorem.

18.6 *To prove that, in general, the envelope of a family of curves touches each member of the family.*

Let $\qquad f(x, y, \alpha) = 0$ $\qquad\qquad\qquad\qquad\qquad\qquad$ ...(i)

be a given family of curves.

Its envelope is obtained by eliminating α between (i) and

$$f_\alpha(x, y, \alpha) = 0. \qquad\qquad\qquad\qquad\qquad ...(ii)$$

Let $\qquad x = \phi(\alpha), y = \psi(\alpha)$ $\qquad\qquad\qquad\qquad\qquad\qquad$ ...(iii)

be the parametric equations of the envelope obtained by solving (i) and (ii) for x and y in terms of α.

The equation (iii) satisfy the equation (i) identically, *i.e.*, for every value of α.

We differentiate (i) *w. r. t* to α, regarding x, y as functions of x so that we obtain

$$\frac{\partial f}{\partial x} \cdot \frac{dx}{d\alpha} + \frac{\partial f}{\partial y} \cdot \frac{dy}{d\alpha} + \frac{\partial f}{\partial \alpha} = 0$$

which, with the help of (ii), becomes

$$\frac{\partial f}{\partial x}\phi'(\alpha) + \frac{\partial f}{\partial y}\psi'(\alpha) = 0 \qquad\qquad\qquad\qquad ...(iv)$$

We consider points (x, y) such that

$$f_x(x, y) \text{ and } f_y(x, y)$$

are not both zero.

At such a point the slope of the tangent at the curve 'α' of the family is

$$-\frac{\partial f/\partial x}{\partial f/\partial y}.$$

Also the slope of the tangent at the same point to the envelope (iii) is $\psi'(\alpha)/\phi'(\alpha)$.

We see from (iv) that these two slopes are the same. Thus the slopes of the tangents to the curve and the envelope at the common point are equal. This means that the curve at the envelope have the same tangent at the common point so that they touch.

This situation may not arise at points of the curve where f_x and f_y are both zero. This situation does not occur for a line and a conic except when the conic is a pair of straight lines.

In the following we shall, in general consider the envelopes of families of lines or conics only.

Note. If for any point on a curve, $\partial f/\partial x$ and $\partial f/\partial y$ are both zero, then the above argument will break down, so that the envelope may not touch a curve at such points. Since at no point on a line and a conic, we have such points, we can say that *the envelope of a family of straight lines or of conics touches each member of the family at all their common points without exception.*

EXAMPLES

1. *Find the envelope of the straight lines*

$$x \cos\alpha + y \sin\alpha = l \sin\alpha \cos\alpha,$$

α *being the parameter.* $\qquad\qquad\qquad\qquad$ **(*Kumaon* 2000; *Avadh* 99; *Rohilkhand* 99)**

The given equation of the family of straight lines can be written as

$$\frac{x}{\sin\alpha} + \frac{y}{\cos\alpha} = l. \qquad \qquad \dots(1)$$

Differentiating eq. (1) partially w.r.t. α, we have

$$-\frac{x}{\sin^2\alpha} \cdot \cos\alpha + \frac{y}{\cos^2\alpha} \cdot \sin\alpha = 0$$

i.e., $\qquad \tan^3\alpha = \frac{x}{y}$ or $\tan\alpha = \frac{x^{1/3}}{y^{1/3}}$

so that $\qquad \sin\alpha = \frac{x^{1/3}}{\sqrt{(x^{2/3} + y^{2/3})}}$ and $\cos\alpha = \frac{y^{1/3}}{\sqrt{(x^{2/3} + y^{2/3})}}.$

Substituting the values of $\sin\alpha$ and $\cos\alpha$ in (1), we get the required envelope as,

$$\frac{x\sqrt{(x^{2/3} + y^{2/3})}}{x^{1/3}} + \frac{y\sqrt{(x^{2/3} + y^{2/3})}}{y^{1/3}} = l$$

or $\qquad (x^{2/3} + y^{2/3})^{3/2} = l \qquad$ i.e., $\qquad x^{2/3} + y^{2/3} = l^{2/3}$

2. *Find the envelope of the parabolas*

$$y^2 = m^2 (x - m),$$

m being the parameter.

We have, $\qquad y^2 = m^2 (x - m).$ $\qquad\qquad \dots(1)$

Differentiating eq. (1) partially w.r.t. m, we have

$$0 = 2mx - 3m^2, \ i.e., \ m = 0 \ \text{or} \ 2x/3.$$

If we substitute $m = 0$ in eq. (1), we obtain $y = 0$, which is not a part of the envelope since it touches none of the given parabolas.

Substituting $m = 2x/3$ in eq. (1), we obtain $27y^2 = 4x^3$, which is the required envelope of the given parabolas.

3. *Find the envelope of $x^2 \sin\alpha + y^2 \cos\alpha = a^2$, α is a parameter.*

The given curve is

$$x^2 \sin\alpha + y^2 \cos\alpha = a^2 \qquad\qquad \dots(i)$$

Differentiating partially with respect to α, we have

$$x^2 \cos\alpha - y^2 \sin\alpha = 0,$$

$\therefore \qquad\qquad \tan\alpha = \frac{x^2}{y^2},$

or $\qquad\qquad \frac{x^2}{\sin\alpha} = \frac{y^2}{\cos\alpha} = \frac{\sqrt{x^4 + y^4}}{1}$

Putting the values of $\sin\alpha$ and $\cos\alpha$ in (i) we get

$$\frac{x^4}{(x^4 + y^4)^{1/2}} + \frac{y^4}{(x^4 + y^4)^{1/2}} = a^2$$

$\therefore \qquad\qquad x^4 + y^4 = a^4,$

is the required envelope.

4. *Find the envelope of the family of circles*

$$x^2 + y^2 - 2ax \cos\alpha - 2ay \sin\alpha = c^2 \qquad\qquad \textbf{\textit{(Kumaon 2004)}}$$

where α is the parameter, and interpret the result. $\qquad\qquad$ **_(Gorakhpur 1999; Agra 2001)_**

We have $x^2 + y^2 - 2ax \cos \alpha - 2ay \sin \alpha - c^2 = 0$

Differentiating partially with respect to α, we get

$$2ax \sin \alpha - 2ay \cos \alpha = 0 \qquad \qquad \qquad ...(i)$$

The curve is $2ax \cos \alpha + 2ay \sin \alpha = x^2 + y^2 - c^2$ ...(ii)

Squaring and adding, we have

$$4a^2 (x^2 + y^2) = (x^2 + y^2 - c^2)^2 \qquad \qquad ..(iii)$$

This is the required equation of the envelope. (iii) can also be written as

$$(x^2 + y^2)^2 - 2(2a^2 + c^2)(x^2 + y^2) + c^4 = 0$$

or $(x^2 + y^2) = [(a^2 + c^2) + a^2] \pm 2a\sqrt{(a^2 + c^2)}$

$$= \left[\sqrt{(a^2 + c^2)} \pm a \right]^2$$

represents circle with origin as centre.

5. *Find the envelope of the straight line*

$$x \cos t + y \sin t = a + a \cos t \log \tan \frac{t}{2},$$

where t is parameter.

We are given that

$$x \cos t + y \sin t = a + a \cos t \log \tan \frac{t}{2}$$

or $x + y \tan t = a \sec t + a \log \tan \dfrac{t}{2}$ ...(i)

Differentiating partially with respect to t, we obtain

$$y \sec^2 t = a \sec t \tan t + a \cdot \frac{1}{\tan t/2} \sec^2 t/2 \cdot \frac{1}{2}$$

$$= a \sec t \tan t + \frac{a}{\sin t}$$

$$= a \left(\frac{\sin t}{\cos^2 t} + \frac{1}{\sin t} \right)$$

or $y = \dfrac{a}{\sin t}$

Putting this value of y in (i), we have

$$x + a \sec t = a \sec t + a \log \tan \frac{t}{2}$$

or $\dfrac{x}{a} = \log \tan \dfrac{t}{2}$

or $e^{x/a} = \tan t/2.$

$$2 \cosh \frac{x}{a} = e^{x/a} + e^{-x/a} = \tan \frac{t}{2} + \cot \frac{t}{2}$$

or $\cosh \dfrac{x}{a} = \dfrac{1}{\sin t} = \dfrac{y}{a}$

Hence $y = a \cosh \dfrac{x}{a}$ is the equation of the envelope.

Find the envelope of the following curves, where α is the parameter:

1. $y = mx + a/m$, m being the parameter.

2. $y = mx + am^3$

3. $y = mx + am^p$

4. $a x \sec \alpha - b y \csc \alpha = a^2 - b^2$ (*Kumaon* 99)

5. $\dfrac{c^2}{x} \cos \alpha - \dfrac{b^2}{y} \sin \alpha = k$

6. $x \cos^n \alpha + y \sin^n \alpha = c$.

7. $x \cos^3 \alpha + y \sin^3 \alpha = c$.

8. $tx^3 + t^2 y = a$, parameter being t.   (*Kanpur* 2003)

9. $y = t^2 (x - t)$.

10. Show that the envelope of the straight line

$$y \cos \theta - x \sin \theta = a - a \sin \theta \log \tan \left(\frac{\theta}{2} + \frac{\pi}{4} \right),$$

where θ is parameter, is the curve $y = a \cosh a/c$.   (*Gorakhpur* 2001)

1. $y^2 = 4 a x$;   2. $4 x^3 + 27 a y^2 = 0$;

3. $(p - 1)^{p-1} x^p = -p^p a \cdot y^{p-1}$;   4. $a^{2/3} x^{2/3} + b^{2/3} y^{2/3} = (a^2 - b^2)^{2/3}$;

5. $\dfrac{a^4}{x^2} + \dfrac{b^4}{y^2} = k^2$;   6. $x^{2/(n-2)} + y^{2/(n-2)} = c^{2/(n-2)}$;

7. $\dfrac{1}{x^2} + \dfrac{1}{y^2} = \dfrac{1}{c^2}$;   8. $x^6 + 4 ay = 0$;

9. $4 x^3 = 27 y$

18.7 If A, B, C are functions of x and y and m is a parameter then the envelope of $Am^2 + Bm + C = 0$ is $B^2 = 4 AC$.   (*Rohilkhand* 98)

We have

$$A m^2 + B m + C = 0 \qquad \qquad \text{...(i)}$$

Differentiating with respect to m partially, we get

$$2 m A + B = 0$$

or $\qquad m = -\dfrac{B}{2 A}$

By putting the value of m in (i), we get

$$A \cdot \frac{B^2}{4 A^2} - \frac{B^2}{2 A} + C = 0$$

or $\qquad B^2 = 4 A C$.

1. *Find the envelope of the family of straight lines*

$$y = mx + \frac{a}{m}.$$

The given equation is

$$y = mx + \frac{a}{m}$$

which can be written as quadratic equation in m as

$$m^2 x - m y + a = 0$$

Thus $A = x, B = -y, C = a$

The required envelope is given by

$$B^2 = 4 A C$$

or $\qquad\qquad y^2 = 4 a x.$

2. *Find the envelope of the family of curves given by*

$$\frac{x^2}{\alpha^2} + \frac{y^2}{k^2 - \alpha^2} = 1, \text{ where } \alpha \text{ is a parameter.} \qquad \textbf{(\textit{Garhwal 97; Gorakhpur 2000})}$$

The given curve is

$$\frac{x^2}{\alpha^2} + \frac{y^2}{(k^2 - \alpha^2)} = 1$$

or $\qquad x^2 (k^2 - \alpha^2) + y^2 \alpha^2 = \alpha^2 (k^2 - \alpha^2)$

or $\qquad \alpha^4 - (x^2 - y^2 + k^2) \alpha^2 + k^2 x^2 = 0$

This is a quadratic equation in α^2,

$$A = 1, B = -(x^2 - y^2 + k^2), C = k^2 x^2$$

The required envelope is $B^2 = 4 A C$

or $\qquad (x^2 - y^2 + k^2)^2 = 4 k^2 x^2.$

3. *Find the envelope of the family of trajectories*

$$y = x \tan \alpha - \frac{g x^2}{2 u^2 \cos^2 \alpha}, \text{ where } \alpha \text{ is the parameter.} \qquad \textbf{(\textit{Avadh 98, 2000})}$$

The given curve is

$$y = x \tan \alpha - \frac{g x^2}{2 u^2} \sec^2 \alpha$$

or $\qquad\qquad y = x \tan \alpha - \frac{g x^2}{2 u^2} (1 + \tan^2 \alpha)$

or $\qquad \tan^2 \alpha \cdot \frac{g x^2}{2 u^2} - x \tan \alpha + \frac{g x^2}{2 u^2} + y = 0$

which is quadratic in $\tan \alpha$.

$$A = \frac{g x^2}{2 a^2}, B = -x, C = \frac{g x^2}{2 u^2} + y$$

Hence the required envelope is

$$B^2 = 4 A C$$

or $\qquad\qquad x^2 = \frac{4 g x^2}{2 u^2} \left(\frac{g x^2}{2 u^2} + y \right).$

On simplification, we obtain

$$\frac{u^2}{2 g} = \frac{g x^2}{2 u^2} + y$$

EXERCISES

1. Show that the envelope of the lines $\dfrac{x}{\alpha} = \dfrac{y}{a-\alpha} = 1$ is the parabola $\sqrt{x} + \sqrt{y} = \sqrt{a}$.

2. Find the envelope of the family of $y = mx + \sqrt{(a^2 m^2 + b^2)}$.

   (Kanpur 2002; Rohilkhand 2002, 01, 04)

3. Find the envelope of the family of $y = mx + a\sqrt{(1 + m^2)}$ where m is the parameter.

4. Find the envelope of the curve $(x - \alpha)^2 + y^2 = 4\,\alpha$.   *(Banglore 2006)*

5. Find the envelope of the family of straight lines

   $$y = mx + 1/m.$$ *(Avadh 97; Gorakhpur 2003)*

6. Find the envelope of family of straight lines

   $$y = m^2 x + 1/m^2.$$

7. Find the envelope of $(x - \alpha)^2 + (y - \alpha)^2 = 2\,\alpha$, α being the parameter.

8. Find the envelope of $\dfrac{x}{m} + \dfrac{my}{c^2} = 1$, c is a constant.

9. Find the envelope of the ellipses

   $$x = a \sin(\theta - \alpha),\ y = b \cos \theta$$

   where α is the parameter.   *(Rohilkhand 97; Agra 99, 2000)*

ANSWERS

2. $\dfrac{x^2}{a^2} + \dfrac{y^2}{b^2} = 1$;   3. $x^2 + y^2 = a^2$;   4. $y^2 - 4x - 4 = 0$;

5. $y^2 = 4ax$;   6. $y^2 = 4x$;   7. $(x + y - 1)^2 = 8(x^2 + y^2)$;

8. $xy = c^2/4$;   9. $x = \pm a$.

18.8 Two parameters connected by a relation

When two parameters are involved in the equation of the curve and one parameter is related to the second one, *i.e.*,

$$\phi(x, y, c_1, c_2) = 0 \qquad\qquad\qquad ...(i)$$

and $\qquad\qquad f(c_1, c_2) = 0 \qquad\qquad\qquad ...(ii)$

where c_1 is independent parameter. Then

$$\frac{\partial \phi}{\partial c_1} + \frac{\partial \phi}{\partial c_2} \cdot \frac{\partial c_2}{\partial c_1} = 0 \qquad\qquad\qquad ...(iii)$$

and $\qquad \dfrac{\partial f}{\partial c_1} + \dfrac{\partial f}{\partial c_2} \cdot \dfrac{\partial f}{\partial c_1} = 0 \qquad\qquad\qquad ...(iv)$

Eliminating c_1, c_2 from (*i*) and (*ii*) with the help of (*iii*) and (*iv*), we get the required equation of the envelope.

1. *Find the envelope of the curve*

$$\left(\frac{x}{a}\right)^m + \left(\frac{y}{b}\right)^m = 1,$$

when $a^n + b^n = c^n$.

The given curve is

$$\frac{x^m}{a^m} + \frac{y^m}{b^m} = 1.$$

Differentiating both the sides with respect to t when a and b are taken as functions of some arbitrary parameter t, we have

$$-\frac{x^m}{a^{m+1}} \cdot \frac{da}{dt} - \frac{y^m}{b^{m+1}} \cdot \frac{db}{dt} = 0 \qquad \qquad \dots(i)$$

Now, $a^n + b^n = c^n$,

Differentiating with respect to t, we have

$$n a^{n-1} \frac{da}{dt} + n b^{n-1} \frac{db}{dt} = 0 \qquad \qquad \dots(ii)$$

Comparing (i) and (ii), we have

$$\frac{\dfrac{x^m}{a^{m+1}}}{a^{n-1}} = \frac{\dfrac{y^m}{b^{m+1}}}{b^{n-1}}$$

$i..e,$

$$\frac{\dfrac{x^m}{a^m}}{a^n} = \frac{\dfrac{y^m}{a^m}}{b^n} = \frac{\left(\dfrac{x}{a}\right)^m + \left(\dfrac{y}{b}\right)^m}{a^n + b^n} = \frac{1}{c^n}$$

or

$$\frac{x^m}{a^{m+n}} = \frac{1}{c^n} = \frac{y^m}{b^{m+n}}$$

Therefore $a^{m+n} = c^n x^m$ and $b^{m+n} = c^n y^n$

$i.e.,$ $a = c^{n/(m+n)} x^{m/(m+n)}$

and $b = c^{n/(m+n)} y^{m/(m+n)}$

Putting these values in

        $a^n + b^n = c^n$,

we get $(c^{n/m+n})^n [x^{mn/(m+n)} + y^{mn/(m+n)}] = c^n$

or $x^{mn/(m+n)} + y^{mn/(m+n)} = c^{mn/(m+n)}$

2. *Find the envelope of the straight line* $\dfrac{x}{a} + \dfrac{y}{b} = 1$, *when* $ab = c^2$, *where c is constant.*

(Kumaon 2001)

Given that

$$\frac{x}{a} + \frac{y}{b} = 1,$$

Regarding a, b as functions of t and differentiating with respect to t, we have

$$-\frac{x}{a^2} \cdot \frac{da}{dt} - \frac{y}{b^2} \cdot \frac{db}{dt} = 0 \qquad \ldots(i)$$

Also $\qquad ab = c^2$

Differentiating $\qquad b\dfrac{da}{dt} + a\dfrac{db}{dt} = 0 \qquad \ldots(ii)$

Comparing (i) and (ii), we have

$$\frac{x/a^2}{b} = \frac{y/b^2}{a}$$

or $\qquad \dfrac{x/a}{ab} = \dfrac{y/b}{ab} = \dfrac{x/a + y/b}{2ab} = \dfrac{1}{2c^2}$

Therefore $\qquad \dfrac{x}{a} = \dfrac{y}{b} = \dfrac{1}{2}$

or $\qquad 2x = a$ and $2y = b$

Hence the required envelope is

or $\qquad 2x \cdot 2y = c^2$

or $\qquad xy = c^2/4.$

EXERCISES

1. Find the envelope of the lines $\dfrac{x}{a} + \dfrac{y}{b} = 1$, when the parameters a and b are connected by the relation $a + b = c$. **(Garhwal 2001, 2003; Avadh 2002)**

2. Find the envelope of the straight lines $\dfrac{x}{a} + \dfrac{y}{b} = 1$, where the parameters a and b are connected by the relation $a^n + b^n = c^n$. **(Rohilkhand 96; Garhwal 98; Gorakhpur 2002)**

3. Find the envelope of the family of ellipses $\dfrac{x^2}{a^2} + \dfrac{y^2}{b^2} = 1$, where the two parameters a and b are connected by the relation $a + b = c$, c being constant. **(Garhwal 2002; Banglore 2004)**

4. Find the envelope of the ellipses having the axes of coordinates as principal axes and sum of their semi-axes constant.

5. Find the envelope of the family of parabolas

$$\left(\frac{x}{a}\right)^{1/2} + \left(\frac{y}{b}\right)^{1/2} = 1 \qquad \text{where } a^n + b^n = c^n.$$

6. Find the envelope of the family of the parabolas

$$\left(\frac{x}{a}\right)^{1/2} + \left(\frac{y}{b}\right)^{1/2} = 1,$$

when $ab = c^2$. **(Garhwal 96, 2000, 2004; Kumaon 2002)**

7. Find the envelope of the straight line $\dfrac{x}{a} + \dfrac{y}{b} = 1$, when $a^m b^n = c^{m+n}$ where a and b are parameters and c is constant. **(Avadh 96; Gorakhpur 2003)**

8. Find the envelope of a system of concentric and coaxial ellipses of constant area.

9. Find the envelope of the family of curves

$$\sqrt{\frac{x}{a}} + \sqrt{\frac{y}{b}} = 1,$$

when $a + b = c$.

10. Find the envelope of the straight lines

$$\frac{x}{a} + \frac{y}{b} = 1, \text{ if } a^2 + b^2 = c.$$

11. Find the envelope of the circle whose diameter is a line of constant length which slides between two fixed straight lines at right angles.

12. Find the envelope of a straight line of given length which slides with its extremities on two fixed straight lines at right angles.

13. Find the envelope of the ellipse

$$\frac{x^2}{a^2} + \frac{y^2}{b^2} = 1$$

(i)   When $a^2 + b^2 = c^2$;   (*Avadh 2003*)

(ii) When $a^n + b^n = c^n$;

(iii) When $a^m b^n = c^n$;

where a and b are the parameters and c is a constant.

ANSWERS

1. $\sqrt{x} + \sqrt{y} = \sqrt{c}$;

2. $x^{n/(n+1)} + y^{n/(n+1)} = c^{n/(n+1)}$;

3. $x^{2/3} + y^{2/3} = c^{2/3}$;

4. $x^{2/3} + y^{2/3} = c^{2/3}$;

5. $x^{n/(2n+1)} + y^{n/(2n+1)} = c^{n/(2n+1)}$;

6. $16\,xy = c^2$;

7. $(m+n)^{m+n}\, x^m\, y^n / m^m\, n^n = c^{m+n}$;

8. $2\,xy = c$;

9. $x^{1/3} + y^{1/3} = c^{1/3}$;

10. $x^{2/3} + y^{2/3} = c^{2/3}$;

11. $x^2 + y^2 = c^2$;

12. $x^{2/3} + y^{2/3} = c^{2/3}$;

13. (i) $x \pm y = \pm c$;

(ii) $x^{2n/(n+2)} + y^{2n/(n+2)} = c^{2n/(n+2)}$;

(iii) $\dfrac{(m+n)^{m+n}}{m^m\, n^n}\, x^{2m}\, y^{2n} = c^{2m+2n}$.

18·9 *When the equation to a family of curves is not given, but the law is given in accordance with which any member of the family can be determined.*

EXAMPLES

1. *Find the envelope of the circles described on the radii vectors of the parabola $y^2 = 4\,a\,x$ as diameter.*   (*Avadh 96*)

The parametric coordinates of a point on parabola $y^2 = 4\,a\,x$ are $(at^2, 2\,a\,t)$.

Equation of the circle joining $(0, 0)$ and $(at^2, 2\,a\,t)$ as diameter is

$$x\,(x - a\,t^2) + y \cdot (y - 2\,a\,t) = 0$$

or $a\,x\,t^2 + 2\,a\,y\,t - (x^2 + y^2) = 0,$

which is quadratic equation in t.

Hence $\qquad A = ax, B = 2\,ay, C = -(x^2 + y^2).$

The envelope is

$$B^2 = 4\,A\,C.$$

Hence the required envelope is

$$4\,a^2 y^2 = -4\,ax\,(x^2 + y^2)$$

or $\quad x^2 + (a + x)\,y^2 = 0.$

2. *A series of circles have their centres on a given straight line and their radii are proportional to the distances of their corresponding centres from a given point in that line. Find the envelope.*

Let the x-axis be considered as straight line, on which the centres of the circles lie and $(\alpha, 0)$ be the centre of the circles where α is a parameter.

Then the radii $= k\,\alpha.$

Hence the equation of family of circles is

$$(x - \alpha)^2 + (y - 0)^2 = k^2\,\alpha^2$$

or $\qquad \alpha^2\,(1 - k^2) - 2\,x\,\alpha + (x^2 + y^2) = 0,$

which is a quadratic equation in α.

Hence $\qquad A = 1 - k^2, B = -2\,x, C = x^2 + y^2$

The envelope is $B^2 = 4\,A\,C.$

Therefore the required envelope is

$$4\,x^2 = 4\,(1 - k^2)\,(x^2 + y^2)$$

or $\qquad k^2\,x^2 = (1 - k^2)\,y^2,$

which represents a pair of lines if $k^2 < 1$.

3. *Show that the envelope of a circle whose centre lies on the parabola $y^2 = 4\,ax$ and which passes through its vertex is the cissoid*

$$y^2\,(2a + x) + x^3 = 0.$$

Now, $(at^2, 2at)$ is a point on the parabola other than the vertex $(0, 0)$ so that $t \neq 0$. Its distance from the vertex $(0, 0)$ is

$$\sqrt{(a^2\,t^4 + 4\,a^2\,t^2)}\,.$$

Thus, the equation of the given family of circles is

$$(x - at^2)^2 + (y - 2\,at)^2 = a^2\,t^4 + 4\,a^2\,t^2.$$

$\Leftrightarrow \qquad x^2 + y^2 - 2\,at^2 x - 4\,aty = 0.$ $\qquad\qquad$ ...(i)

Differentiating (i) w.r. to t, we get

$$-4\,atx - 4\,ay = 0 \text{ or } t = -y/x.$$

Substituting this value of t in (i), we get the required envelope.

EXERCISES

1. Find the envelope of the circles which pass through the origin and whose centres lie on

(a) $\dfrac{x^2}{a^2} + \dfrac{y^2}{b^2} = 1$ $\qquad\qquad\qquad\qquad\qquad$ (*Garhwal 2005*)

(b) $x^2 - y^2 = a^2.$

2. A circle moves with its centre on the parabola $y^2 = 4\,ax$ and always passes through the vertex of the parabola. Show that the envelope of the circle is the curve

$$x^3 + y^2\,(x + 2a) = 0. \qquad\qquad\qquad\qquad\qquad (\textit{Garhwal 95})$$

3. Show that the envelope of the family of circles whose diameters are double ordinates of the parabola $y^2 = 4ax$ is the parabola $y^2 = 4a(x + a)$.

4. From any point on the ellipse $\dfrac{x^2}{a^2} + \dfrac{y^2}{b^2} = 1$ perpendiculars are drawn to the axes and the feet of these perpendiculars are joined. Show that the straight lines thus formed always touch the curve $\left(\dfrac{x}{a}\right)^{2/3} + \left(\dfrac{y}{b}\right)^{2/3} = 1$.

5. Show that the envelope of the straight line joining the extremities of a pair of conjugate diameter of an ellipse is a similar ellipse.

6. Show that the envelope of the circles described on the central radii of the rectangular hyperbola $x^2 - y^2 = a^2$ is the lemniscate $r^2 = a^2 \cos 2\theta$.

7. Find the envelope of the circles drawn upon the central radii vectors of the ellipse

$$\frac{x^2}{a^2} + \frac{y^2}{b^2} = 1, \text{ as diameter,} \qquad\qquad (\textit{Nagpur 2005})$$

8. Show that the envelope of the polars of the point on the ellipse $\dfrac{x^2}{h^2} + \dfrac{y^2}{k^2} = 1$ with respect to the ellipse is

$$\frac{h^2 x^2}{a^4} + \frac{k^2 y^2}{b^4} = 1.$$

ANSWERS

1. (a) $(x^2 + y^2)^2 = 4(a^2 x^2 + b^2 y^2)$,

   (b) $(x^2 + y^2)^2 = 4a^2(x^2 - y^2)$.

7. $(x^2 + y^2)^2 = a^2 x^2 + b^2 y^2$.

18.10 Envelopes of Polar Curves

EXAMPLES

1. *Find the envelope of the straight lines at right angles to the radii vectors to the cardioid $r = a(1 + \cos\theta)$ drawn through their extremities.*

Let (r, θ) be any point on the cardioid. If (R, ϕ) be the polar co-ordinates of any point on the straight line through (r, θ) and at right angles to the radius vector, then the equation is

$$R \cos(\phi - \theta) = r,$$

i.e., $\qquad\qquad R \cos(\phi - \theta) = a(1 + \cos\theta) \qquad\qquad\qquad ...(i)$

Differentiating logarithmically with respect to the parameter θ, we get

$$\tan(\phi - \theta) = \frac{-\sin\theta}{1 + \cos\theta}$$

$$= -\tan\frac{\theta}{2}$$

$\therefore \qquad\qquad \phi - \theta = \pi - \dfrac{\theta}{2}$

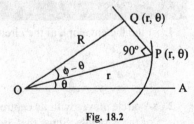

Fig. 18.2

or $\qquad\qquad \theta = 2\phi - 2\pi.$

Substituting this value of θ in (*i*),

we get $\qquad R \cos (\phi - 2\phi) = a (1 + \cos 2\phi)$

or $\qquad R \cos \phi = a (1 + \cos 2\phi)$

or $\qquad R \cos \phi = a (1 + \cos 2\phi) = 2 a \cos^2 \phi$

or $\qquad R = 2 a \cos \phi.$

Therefore in current co-ordinates (r, θ), the equation of the envelope is

$$r = 2 a \cos \theta.$$

2. *Find the envelope of the circles described on the radii vectors of the curve* $r^n = a^n \cos n\theta$ *as diameter.*

Let the polar co-ordinates of any point P on the curve $r^n = a^n \cos n\theta$ be (r, θ) and the point θ which lies on the circle described on OP as diameter (R, ϕ).

Now, $\qquad OQ = OP \cos (\phi - \theta),$

i.e. $\qquad R = r \cos (\phi - \theta)$

or $\qquad R^n = r^n \cos^n (\phi - \theta)$ $\qquad\qquad\qquad$ ...(*i*)

But $\qquad r^n = a^n \cos n\theta$; then (*i*) becomes

$\qquad R^n = a^n \cos n\theta \cos^n (\phi - \theta)$

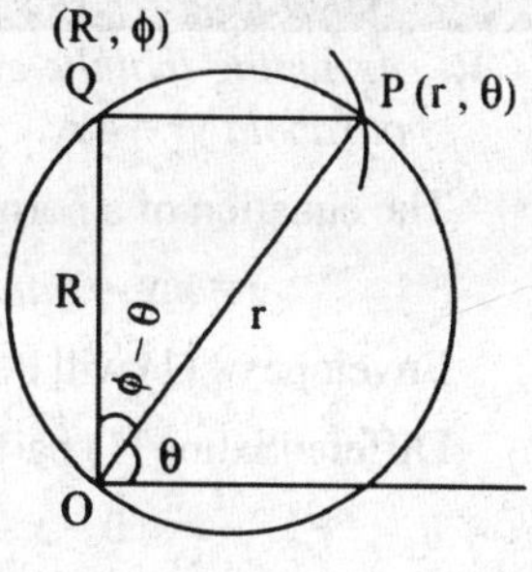

Taking log of both the sides, we obtain

$\qquad n \log R = n \log a + \log \cos n\theta + n \log \cos (\phi - \theta)$

Differentiating partially with respect to θ, we have

$\qquad 0 = - n \tan n\theta + n \tan (\phi - \theta)$

or $\qquad \tan (\phi - \theta) = \tan n\theta$

or $\qquad \phi - \theta = k\pi + n\theta$

or $\qquad \theta = \dfrac{(\phi - k\pi)}{(n + 1)}$

Then $\qquad R^n = a^n \cos n \left(\dfrac{\phi - k\pi}{n + 1} \right) \cdot \cos^n \left(\dfrac{k\pi - \phi}{1 + n} + \phi \right)$

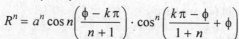

Therefore in current co-ordinates (r, θ), the equation of the envelope is

$$r^n a^n \cos n \left(\dfrac{\theta - k\pi}{n + 1} \right) \cos^n \left(\dfrac{k\pi - \theta}{1 + n} + \theta \right).$$

(R , φ)

Q $\qquad\qquad$ P (r , θ)

R $\qquad$ r

O $\qquad$ θ

Fig. 18.3

EXERCISES

1. Find the envelopes of circles described on the radii vectors of the following curves as diameters :

   (*a*) $\dfrac{l}{r} = 1 + e \cos \theta;$ $\qquad\qquad$ (*b*) $r^3 = a^3 \cos 3\theta;$ $\qquad\qquad$ (*c*) $r \cos^n (\theta/n) = a.$

2. Find the envelopes of the straight lines drawn through the extremities of and at right angles to the radii vectors of the following curves :

   (*a*) $r \cos (\theta + \alpha) = p;$ $\qquad\qquad$ (*b*) $r^2 \cos n\theta = a^n;$ $\qquad\qquad$ (*c*) $r^n = a^n \cos n\theta.$

ANSWERS

1. (a) $r^2(e^2 - 1) = 2\,e\,l\,r\cos\theta - l^2$; (b) $r^{3/4} = a^{3/4}\cos\left(\dfrac{3}{4}\theta\right)$;

 (c) $r\cos^{n-1}\dfrac{\theta}{n-1} = a$;

2. (a) $r\cos^2\dfrac{(\theta + \alpha)}{2} = p$; (b) $r^{n/(n+1)}\cos\left(\dfrac{n\theta}{n+1}\right) = a^{n/(n+1)}$;

 (c) $r^{n/(1-n)} = a^{n/(1-n)}\cos\left(\dfrac{n\theta}{n+1}\right)$.

18.11 Envelope of Normals (Evolutes)

The evolute of a curve has been defined as the locus of the centre of curvature, and it has been shown that the centre of curvature is the limiting position of the point of intersection of two consecutive normals. Hence, **the evolute of a curve is the envelope of the normals of that curve.**

Therefore, **the normals of a curve touch the evolute.**

EXAMPLES

1. *Assuming that the evolute of a curve is the envelope of its normals, find the evolute of the parabola $y^2 = 4ax$.* (*Kanpur* **2001**; *Rohilkhand* **98, 2002**; *Banglore* **2004**)

The equation of a normal to the parabola $y^2 = 4\,ax$ is given by

$$y = mx - 2\,am - am^3, \text{ where } m \text{ is the parameter} \qquad ...(1)$$

Envelope of (1) will be the evolute of the parabola $y^2 = 4\,ax$.

Differentiating (1) partially w.r.t m, we have

$$0 = x - 2a - 3am^2$$

∴ $$m = \left(\dfrac{x - 2a}{3a}\right)^{1/2}$$

Substituting the value of m in (1), we have

$$y = \left(\dfrac{x - 2a}{3a}\right)^{1/2}\left\{x - 2a - a\cdot\dfrac{(x - 2a)}{3a}\right\}$$

or $$27\,ay^2 = 4\,(x - 2a)^3,$$

which is the required equation of the evolute.

2. *Prove that the evolute of the tractrix*

$$x = a\left(\cos t + \frac{1}{2}\log\tan^2 t/2\right),\ y = a\sin t,$$

is the catenary $y = a\cosh(x/a)$.

We have $x = a(\cos t + \log\tan t/2),\ y = a\sin t$

Differentiating w.r.t. t, we have

$$\frac{dx}{dt} = a\left(-\sin t + \frac{\sec^2 t/2}{\tan t/2}\cdot\frac{1}{2}\right) = a\left(-\sin t + \frac{1}{\sin t}\right)$$

i.e., $$\frac{dx}{dt} = a\frac{\cos^2 t}{\sin t}$$

and $\qquad \dfrac{dy}{dt} = a\cos t$

$\therefore \qquad \dfrac{dy}{dx} = \dfrac{a\cos t \sin t}{a\cos^2 t} = \tan t.$

The equation of normal to the tractrix is

$\qquad \tan t\,(y - a\sin t) + x - a\cos t - a\log \tan t/2 = 0$

or $\qquad x + y\tan t - a\sec t - a\log \tan t/2 = 0,$ $\qquad\qquad$ ...(1)

where t is the parameter.

Differentiating (1) partially w.r.t. t, we have

$$y\sec^2 t - a\sec t \,\tan t - \dfrac{a\sec^2 t/2}{\tan t/2}\cdot\dfrac{1}{2} = 0$$

or $\qquad y\sec^2 t - \dfrac{a\sin t}{\cos^2 t} - \dfrac{a}{\sin t} = 0$

or $\qquad \dfrac{y}{\cos^2 t} - \dfrac{a}{\sin t \cos^2 t} = 0 \qquad$ or $\qquad y = \dfrac{a}{\sin t}$ $\qquad\qquad$ ...(2)

Substituting for y from (2) in (1), we have

$\qquad x + a\sec t - a\sec t - a\log \tan t/2 = 0$

$\qquad\qquad x/a = \log \tan t/2$

or $\qquad e^{x/a} = \tan t/2$ and $e^{-x/a} = \cot t/2$

$\therefore \qquad e^{x/a} + e^{-x/a} = \tan t/2 + \cot t/2 = \dfrac{\sin^2 t/2 + \cos^2 t/2}{\sin t/2 \cos t/2} = \dfrac{1}{\sin t}$

or $\qquad \cosh\left(\dfrac{x}{a}\right) = \dfrac{1}{\sin t}$ or $\cosh\dfrac{x}{a} = \dfrac{y}{a}$ $\qquad\qquad$ [from (2)]

or $\qquad y = a\cosh\left(\dfrac{x}{a}\right),$

which is the required evolute.

3. *Prove that the evolute of the cardioid* $r = a(1 + \cos\theta)$ *is cardioid* $r = \dfrac{1}{3}a(1 - \cos\theta),$ *the pole in the latter equation being at the point* $\left(\dfrac{2}{3}a, 0\right).$

Let $P(r, \theta)$ be any point on the given cardioid

$\qquad r = a(1 + \cos\theta)$

$\qquad\quad = 2a\cos^2 \theta/2$ $\qquad\qquad$ ...(1)

Let O' be the point $\left(\dfrac{2}{3}a, 0\right)$, C the centre of curvature at P, $O'C = R$, $\angle CO'X = \delta$.

Draw the perpendiculars CM and $O'M'$ upon OP. To find out the evolute of (1), we have to find the locus of C.

Now radius of curvature (ρ) at

$\qquad \rho = PC = \dfrac{4}{3}a\cos\dfrac{1}{2}\theta$

Also $\qquad \dfrac{dr}{d\theta} = -a\sin\theta$

$\therefore \qquad \tan\phi = r\,\dfrac{d\theta}{dr} = \dfrac{a(1 + \cos\theta)}{-a\sin\theta} = -\cot\dfrac{1}{2}\theta$

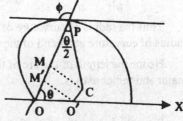

Fig. 18.4

or $\qquad \tan\phi = \tan\left(\dfrac{\pi}{2} + \dfrac{\theta}{2}\right)$

i.e., $\qquad \phi = \dfrac{\pi}{2} + \dfrac{\theta}{2}.$

$\therefore$ From the figure, $\angle MPC = \theta/2$

$\therefore \qquad CM = PC \sin\dfrac{1}{2}\theta = \dfrac{4}{3}a\cos\dfrac{1}{2}\theta\sin\dfrac{1}{2}\theta = \dfrac{2}{3}a\sin\theta$

and $\qquad O'M' = OO'\sin\theta = \dfrac{2}{3}a\sin\theta = CM.$

$\therefore$ $O'C$ is parallel to OP and consequently $\delta = \theta.$ $\qquad\qquad\qquad\qquad\qquad$ ...(2)

Now $\qquad\qquad R = O'C = M'M = OP - OM' - MP$

$\qquad\qquad\qquad\quad = r - \dfrac{2}{3}a\cos\theta - \dfrac{4}{3}a\cos^2\dfrac{1}{2}\theta$

or $\qquad\qquad R = a(1 + \cos\theta) - \dfrac{2}{3}a\cos\theta - \dfrac{2}{3}a(1 + \cos\theta)$

$\qquad\qquad\qquad\quad = \dfrac{1}{3}a(1 - \cos\theta)$ on simplification.

Hence the locus of C is $r = \dfrac{1}{3}a(1 - \cos\theta),$ which is the required evolute.

4. *Show that the whole length of the evolute of the ellipse*

$$\dfrac{x^2}{a^2} + \dfrac{y^2}{b^2} = 1 \text{ is } 4\left(\dfrac{a^2}{b} - \dfrac{b^2}{a}\right).$$

Transferring the origin to $(a, 0)$, the equation of the ellipse becomes

$$\dfrac{(x + a)^2}{a^2} + \dfrac{y^2}{b^2} = 1 \qquad i.e., \qquad \dfrac{x^2}{a^2} + \dfrac{y^2}{b^2} + \dfrac{2x}{a} = 0 \qquad\qquad ..(1)$$

Since $x = 0$ *i.e.*, y-axis is the tangent at origin.

$\therefore$ By Newton's method, the radius of curvature at the origin,

$$\rho = \lim_{x\to 0} \dfrac{y^2}{2x}$$

Eq. (1) can be written as

$$\dfrac{x}{2a} + \dfrac{1}{b^2}\cdot\dfrac{y^2}{2x} + \dfrac{1}{a} = 0$$

Taking limit when $x \to 0$, we have

$$\dfrac{1}{b^2}\rho + \dfrac{1}{a} = 0 \qquad i.e., \qquad \rho = \dfrac{b^2}{a} \qquad\qquad\qquad\qquad\qquad \text{(numerically)}$$

Thus the radius of curvature of the ellipse at an end of the major axis is (b^2 / a). Similarly, the radius of curvature at an end of the minor axis is (a^2/b).

Hence the length of the arc of the evolute between points corresponding to extremities of the major and minor axes is

$$a^2/b - b^2/a.$$

From the symmetry of the figure the whole length of the evolute is equal to $4(a^2/b - b^2/a)$.

1. Find the evolute of the ellipse

    $$\frac{x^2}{a^2} + \frac{y^2}{b^2} = 1.$$ *(Agra 2001; Rohilkhand 2004)*

2. Prove that the evolute of the hyperbola $2xy = a^2$ is

    $$(x+y)^{2/3} - (x-y)^{2/3} = 2a^{2/3}.$$ *(Rohilkhand 96; Agra 2000)*

3. Show that the evolute of an equiangular spiral is an equal equiangular spiral.

4. Prove that the ellipse $b^2x^2 + a^2y^2 = a^2b^2$ and $a^2x^2 \sec^4\alpha + b^2y^2 \csc^4 \alpha = (a^2 - b^2)^2$ are so related that the evolute of the first is the envelope of the second for all values of α.

5. Find the evolute of

    $$x^{2/3} + y^{2/3} = a^{2/3} \quad \text{or} \quad x = a\cos^3 t, \quad y = a\sin^3 t.$$

6. Find the evolute of $x^2 - y^2 = a^2$.

7. Show that in the parabola $y^2 = 4ax$, the length of the part of the evolute intercepted within the parabola is $4a(3\sqrt{3} - 1)$.

8. Prove that the whole length of the evolute of the astroid

    $$x = a\cos^3 \theta, y = a\sin^2 \theta \text{ is } 12\,a.$$

ANSWERS

1. $(ax)^{2/3} + (by)^{2/3} = (a^2 - b^2)^{2/3}$ 5. $(x+y)^{2/3} + (x-y)^{2/3} = 2a^{2/3}$
6. $(ax)^{2/3} - (ay)^{2/3} = (2a^2)^{2/3}$

OBJECTIVE QUESTIONS

For each of the following questions, four alternatives are given for the answer. Only one of them is correct. Choose the correct alternative.

1. The equation of the envelope of the family of curves $F(x, y, \alpha)$, where α is a parameter, is obtained by eliminating α between the equations $F(x, y, \alpha) = 0$ and: *(Kumaon 2005)*

    (a) $\dfrac{\partial}{\partial x} F(x, y, \alpha) = 0$ (b) $\dfrac{\partial}{\partial y} F(x, y, 0) = 0$

    (c) $\dfrac{\partial}{\partial \alpha} F(x, y, \alpha) = 0$ (d) None of these

2. The envelope of the family of curves $A\alpha^2 + B\alpha + C = 0$, where A, B and C are functions of x and y is given by:

    (a) $\dfrac{-B + \sqrt{(B^2 - 4AC)}}{2A} = 0$ (b) $\dfrac{-B - \sqrt{(B^2 - 4AC)}}{2A} = 0$

    (c) $B^2 + 4AC = 0$ (d) $B^2 - 4AC = 0$

                                                                    (Garhwal 2002; Avadh 2003)

3. Envelope of the family of straight lines $y = mx + a/m$ is :

    (a) $x^2 + y^2 = a^2$ (b) $xy = a^2$
    (c) $y^2 = 4ax$ (d) $x^2 = 4ay$

                                    (Kanpur 2002; Rohilkhand 2002, 2003; Avadh 2002)

4. Envelope of $x^2 \sin \alpha + y^2 \cos \alpha = a^2$ is:

    (a) $x^2 + y^2 = a^2$ (b) $x^4 + y^4 = a^4$

    (c) $x^3 + y^3 = a^3$ (d) $x^2 - y^2 = a^2$

5. Envelope of the system of circles $(x - \alpha)^2 + y^2 = 4\,\alpha$ is:

    (a) $y^2 - 4x - 4 = a$ (b) $y^2 + 4x - 4 = a$

    (c) $y^2 - 4x + 4 = a$ (d) $y^2 + 4x + 4 = a$

6. Envelope of the family of curve $tx^2 + t^2y = a$ is :

    (a) $x^2 = 4\,ay$ (b) $x^2 + 4\,ay = 0$

    (c) $x^2 + 4ay = 0$ (d) $x^2 - 4ay = 0$

7. Envelope of the family of curves $y^2 = t^2 (x - t)$ is :

    (a) $x^3 = y^2$ (b) $4x^3 + 27y^2 = 0$

    (c) $4x^3 = 27y^2$ (d) $x^3 = 4y^2$

ANSWERS

1. (c)	**2.** (c)	**3.** (d)	**4.** (c)	**5.** (a)	**6.** (c)	**7.** (c)

CHAPTER 19

CHANGE OF INDEPENDENT VARIABLES

19.1 To change the single independent variable into the dependent variable

In $y = f(x)$, let y be dependent variable where x is independent variable; now we want to change the independent variable, into dependent variable, *i.e.*,

$$\frac{dy}{dx} = \frac{1}{\dfrac{dx}{dy}} = \left(\frac{dx}{dy}\right)^{-1}$$

$$\frac{d^2y}{dx^2} = \frac{d}{dx}\left[\frac{dx}{dy}\right]^{-1} = -\left(\frac{dx}{dy}\right)^{-2} \cdot \frac{d^2x}{dy^2} \cdot \frac{dy}{dx} = -\frac{\dfrac{d^2x}{dy^2}}{\left(\dfrac{dx}{dy}\right)^3}$$

$$\frac{d^3y}{dx^3} = \frac{d}{dx}\left[\frac{\dfrac{d^2x}{dy^2}}{\left(\dfrac{dx}{dy}\right)^3}\right] = \frac{\dfrac{d^3x}{dy^3} \cdot \dfrac{dy}{dx} \cdot \left(\dfrac{dx}{dy}\right)^3 - 3\left(\dfrac{dx}{dy}\right)^2 \cdot \left(\dfrac{d^2x}{dy^2}\right)^2 \cdot \dfrac{dy}{dx}}{\left(\dfrac{dx}{dy}\right)^6}$$

$$= \frac{\dfrac{dy}{dx} \cdot \left(\dfrac{dx}{dy}\right)^2}{\dfrac{dx}{dy}} \cdot \left[\frac{\dfrac{d^3x}{dy^3} \cdot \dfrac{dx}{dy} - 3\left(\dfrac{d^2x}{dy^2}\right)^2}{\left(\dfrac{dx}{dy}\right)^5}\right]$$

$$= \frac{\dfrac{d^3x}{dy^3} \cdot \dfrac{dx}{dy} - 3\left(\dfrac{d^2x}{dy^2}\right)^2}{\left(\dfrac{dx}{dy}\right)^5}$$

and similarly we can change these independent variables to dependent variables.

19.2 To change the independent variable x into another variable t, where $x = f(t)$

We know, $\quad \dfrac{dy}{dx} = \dfrac{dy}{dt} \cdot \dfrac{dt}{dx} = \dfrac{\dfrac{dy}{dt}}{\dfrac{dx}{dt}}$

543

The operator $\dfrac{d}{dx}$ is therefore equivalent to the operator $\dfrac{1}{\dfrac{dx}{dt}} \cdot \dfrac{d}{dt}$.

Therefore $\dfrac{d^2 y}{dx^2} = \dfrac{1}{\dfrac{dx}{dt}} \cdot \dfrac{d}{dt}\left[\dfrac{\dfrac{dy}{dt}}{\dfrac{dx}{dt}}\right] = \dfrac{\dfrac{d^2 y}{dt^2} \cdot \dfrac{dx}{dt} - \dfrac{d^2 x}{dt^2} \cdot \dfrac{dy}{dt}}{\dfrac{dx}{dt} \cdot \left(\dfrac{dx}{dt}\right)^2}$

$$= \dfrac{\dfrac{d^2 y}{dt^2} \cdot \dfrac{dx}{dt} - \dfrac{d^2 x}{dt^2} \cdot \dfrac{dy}{dt}}{\left(\dfrac{dx}{dt}\right)^3}$$

and $\dfrac{d^3 y}{dx^3} = \dfrac{1}{\dfrac{dx}{dt}} \cdot \dfrac{d}{dt}\left[\dfrac{\dfrac{d^2 y}{dt^2} \cdot \dfrac{dx}{dt} - \dfrac{d^2 x}{dt^2} \cdot \dfrac{dy}{dt}}{\left(\dfrac{dx}{dt}\right)^3}\right]$

$$= \dfrac{1}{\dfrac{dx}{dt}}\left[\left\{\left(\dfrac{d^3 y}{dt^3} \cdot \dfrac{dx}{dt} + \dfrac{d^2 x}{dt^2} \cdot \dfrac{d^2 y}{dt^2}\right) - \left(\dfrac{d^3 x}{dt^3} \cdot \dfrac{dy}{dt} + \dfrac{d^2 x}{dt^2} \cdot \dfrac{d^2 y}{dt^2}\right)\right\}\left(\dfrac{dx}{dt}\right)^3\right.$$

$$\left. - 3\left(\dfrac{dx}{dt}\right)^2 \dfrac{d^2 x}{dt^2} \cdot \left\{\dfrac{d^2 y}{dt^2} \cdot \dfrac{dx}{dt} - \dfrac{d^2 x}{dt^2} \cdot \dfrac{dy}{dt}\right\}\left(\dfrac{dx}{dt}\right)^{-6}\right]$$

$$= \left(\dfrac{dx}{dt}\right)^{-7}\left[\left(\dfrac{dx}{dt}\right)^3\left\{\dfrac{d^3 y}{dt^3} \cdot \dfrac{dx}{dt} - \dfrac{d^3 x}{dt^3} \cdot \dfrac{dy}{dt}\right\}\right.$$

$$\left. - 3\left(\dfrac{dx}{dt}\right)^2 \cdot \dfrac{d^2 x}{dt^2}\left\{\dfrac{d^2 y}{dt^2} \cdot \dfrac{dx}{dt} - \dfrac{d^2 x}{dt^2} \cdot \dfrac{dy}{dt}\right\}\right]$$

$$= \left(\dfrac{dx}{dt}\right)^{-5}\left[\dfrac{dx}{dt}\left(\dfrac{d^3 y}{dt^3} \cdot \dfrac{dx}{dt} - \dfrac{d^3 x}{dt^3} \cdot \dfrac{dy}{dt}\right) - 3\dfrac{d^2 x}{dt^2}\left\{\dfrac{d^2 y}{dt^2} \cdot \dfrac{dx}{dt} - \dfrac{d^2 x}{dt^2} \cdot \dfrac{dy}{dt}\right\}\right]$$

$$= \dfrac{\dfrac{d^3 y}{dt^3}\left(\dfrac{dx}{dt}\right)^2 - 3\dfrac{d^2 x}{dt^2} \cdot \dfrac{d^2 y}{dt^2} \cdot \dfrac{dx}{dt} + 3\left(\dfrac{d^2 x}{dt^2}\right)^2 \cdot \dfrac{dy}{dt} - \dfrac{dx}{dt} \cdot \dfrac{dy}{dt} \cdot \dfrac{d^3 x}{dt^3}}{\left(\dfrac{dx}{dt}\right)^5}$$

and similar other higher order differential coefficients can be written.

Now x is a function of t; so the differential coefficients of 1st, 2nd and 3rd order are functions of t etc. and can be known.

19.3 Case of Single Independent Variable

Let it be required to change x to t, where

$$x = f(t) \tag{i}$$

We will express $\dfrac{dy}{dx}, \dfrac{d^2y}{dx^2}$ etc. in terms of $\dfrac{dy}{dt}, \dfrac{d^2y}{dt^2}$ etc.

We know that $\dfrac{dy}{dx} = \dfrac{\dfrac{dy}{dt}}{\dfrac{dx}{dt}}$ ...(ii)

Operator $\quad D \equiv \dfrac{d}{dx} \equiv \dfrac{1}{\dfrac{dx}{dt}} \cdot \dfrac{d}{dt}$ ...(iii)

Now, $\quad \dfrac{d^2y}{dx^2} = \dfrac{d}{dx} \cdot \left(\dfrac{dy}{dx} \right) = \dfrac{1}{\dfrac{dx}{dt}} \dfrac{d}{dt} \cdot \left[\dfrac{\dfrac{dy}{dt}}{\dfrac{dx}{dt}} \right]$

$= \dfrac{1}{\dfrac{dx}{dt}} \cdot \dfrac{\left\{ \dfrac{d^2y}{dt^2} \cdot \dfrac{dx}{dt} - \dfrac{d^2x}{dt^2} \cdot \dfrac{dy}{dt} \right\}}{\left(\dfrac{dx}{dt} \right)^2}$

$= \dfrac{\dfrac{d^2y}{dt^2} \cdot \dfrac{dx}{dt} - \dfrac{d^2x}{dt^2} \cdot \dfrac{dy}{dt}}{\left(\dfrac{dx}{dt} \right)^3}$ ...(iv)

Again, $\quad \dfrac{d^3y}{dx^3} = \dfrac{d}{dx} \left(\dfrac{d^2y}{dx^2} \right) = \dfrac{1}{\dfrac{dx}{dt}} \cdot \dfrac{d}{dt} \left[\dfrac{\dfrac{dx}{dt} \cdot \dfrac{d^2y}{dt^2} - \dfrac{dy}{dt} \cdot \dfrac{d^2x}{dt^2}}{\left(\dfrac{dx}{dt} \right)^3} \right]$

$= \dfrac{1}{\left(\dfrac{dx}{dt} \right)^7} \left[\left(\dfrac{dx}{dt} \right)^3 \left(\dfrac{dx}{dt} \cdot \dfrac{d^3y}{dt^3} - \dfrac{dy}{dt} \cdot \dfrac{d^3x}{dt^3} \right) \right.$

$\left. -3 \left(\dfrac{dx}{dt} \right)^2 \cdot \dfrac{d^2x}{dt^2} \left\{ \dfrac{dx}{dt} \cdot \dfrac{d^2y}{dt^2} - \dfrac{dy}{dt} \cdot \dfrac{d^2x}{dt^2} \right\} \right]$

Since $\dfrac{d^2x}{dt^2}, \dfrac{d^2y}{dt^2}$ cancel in the first bracket.

$= \dfrac{1}{\left(\dfrac{dx}{dt} \right)^5} \left[\left(\dfrac{dx}{dt} \right)^2 \cdot \dfrac{d^3y}{dt^3} - 3 \dfrac{dx}{dt} \cdot \dfrac{d^2x}{dt^2} \cdot \dfrac{d^2y}{dt^2} + \dfrac{dy}{dt} \left\{ 3 \left(\dfrac{d^2x}{dt^2} \right)^2 - \dfrac{dx}{dt} \cdot \dfrac{d^3x}{dt^3} \right\} \right]$ ...(v)

and so on.

In relations (ii), (iv), (v), we can substitute the value of $\dfrac{dx}{dt}, \dfrac{d^2x}{dt^2}, \dfrac{d^3x}{dt^3}$ from (i) as they are functions of t. Such transformations can be performed.

EXAMPLES

1. *Transform the equation* $\qquad x^2 \dfrac{d^2 y}{dx^2} + 2x \dfrac{dy}{dx} + \dfrac{a^2}{x^2} y = 0,$

into another when z is the independent variable, being given $x = \dfrac{1}{z}.$

We have $\qquad \dfrac{dx}{dz} = -\dfrac{1}{z^2}$

$\therefore \qquad \dfrac{dy}{dx} = \dfrac{1}{\dfrac{dx}{dz}} \cdot \dfrac{d}{dz}(y) = -z^2 \dfrac{dy}{dz}.$

or $\qquad \dfrac{1}{z} \dfrac{dy}{dx} = -z \dfrac{dy}{dx}$

or $\qquad x \dfrac{dy}{dx} = -z \dfrac{dy}{dz}$ $\qquad\qquad\qquad\qquad\qquad …(i)$

From (i), $\qquad x \dfrac{d}{dx}\left(x \dfrac{dy}{dx}\right) = -z \dfrac{d}{dz}\left(-z \dfrac{dy}{dz}\right)$

or $\qquad x^2 \dfrac{d^2 y}{dx^2} + x \dfrac{dy}{dx} = z^2 \dfrac{d^2 y}{dz^2} + z \dfrac{dy}{dz}$ $\qquad\qquad …(ii)$

Adding (i) and (ii), we get

$$x^2 \dfrac{d^2 y}{dx^2} + 2x \dfrac{dy}{dx} = z^2 \dfrac{d^2 y}{dz^2}$$

Hence the transformed equation becomes

$$z^2 \dfrac{d^2 y}{dz^2} + \dfrac{a^2}{x^2} y = 0$$

or $\qquad \dfrac{d^2 y}{dz^2} + a^2 y = 0$

Aliter. $\qquad \dfrac{dy}{dx} = \dfrac{dy}{dz} \cdot \dfrac{dz}{dx} = -z^2 \dfrac{dy}{dz}$

$$\dfrac{d^2 y}{dx^2} = \dfrac{d}{dx}\left(\dfrac{dy}{dx}\right) = \dfrac{1}{\dfrac{dx}{dz}} \cdot \dfrac{d}{dz}\left\{-z^2 \dfrac{dy}{dz}\right\}$$

$$= -z^2 \left[-2z \dfrac{dy}{dz} - z^2 \dfrac{d^2 y}{dz^2}\right]$$

$$= z^4 \dfrac{d^2 y}{dz^2} + 2z^3 \dfrac{dy}{dz}$$

Hence, the transformed equation,

$$x^2 \left[z^4 \dfrac{d^2 y}{dz^2} + 2z^3 \dfrac{dy}{dz}\right] - 2x \left[z^2 \dfrac{dy}{dz}\right] + a^2 z^2 y = 0$$

or $\qquad z^2 \dfrac{d^2 y}{dz^2} + 2z \dfrac{dy}{dz} - 2z \dfrac{dy}{dz} + a^2 z^2 y = 0$

or $\qquad\qquad\qquad\qquad\qquad\qquad\qquad \dfrac{d^2 y}{dz^2} + a^2 y = 0 .$

2. *Change the independent variable from x to θ in the equation*

$$(1 - x^2) \frac{d^2 y}{dx^2} - x \frac{dy}{dx} + y = 0.$$

Having given $x = \cos \theta$.

We have, $\qquad x = \cos \theta$

then $\qquad \dfrac{dx}{d\theta} = -\sin \theta, \ \dfrac{d^2 x}{d\theta^2} = -\cos \theta = -x.$

Now, $\qquad \dfrac{dy}{dx} = \dfrac{\dfrac{dy}{d\theta}}{\dfrac{dx}{d\theta}} = -\dfrac{1}{\sin \theta} \dfrac{dy}{d\theta},$

$$\frac{d^2 y}{dx^2} = \frac{d}{dx}\left(\frac{dy}{dx}\right) = \frac{1}{\dfrac{dx}{d\theta}} \cdot \frac{d}{d\theta}\left\{-\frac{1}{\sin \theta} \frac{dy}{d\theta}\right\}$$

$$= -\frac{1}{\sin \theta}\left\{\frac{-\sin\theta \dfrac{d^2 y}{d\theta^2} + \cos\theta \dfrac{dy}{d\theta}}{\sin^2 \theta}\right\}$$

$$= \frac{1}{\sin^3 \theta}\left\{\sin \theta \frac{d^2 y}{d\theta^2} - \cos \theta \frac{dy}{d\theta}\right\},$$

By putting these values the given equation becomes

$$(1 - \cos^2 \theta)\left\{\frac{\sin\theta \dfrac{d^2 y}{d\theta^2} - \cos\theta \dfrac{dy}{d\theta}}{\sin^3 \theta}\right\} + \frac{\cos\theta}{\sin\theta}\frac{dy}{d\theta} + y = 0$$

or $\qquad \dfrac{d^2 y}{d\theta^2} - \dfrac{\cos\theta}{\sin\theta}\dfrac{dy}{d\theta} + \dfrac{\cos\theta}{\sin\theta}\dfrac{dy}{d\theta} + y = 0$

or $\qquad \dfrac{d^2 y}{d\theta^2} + y = 0.$

3. *Show that the equation* $\dfrac{d^2 x}{ds^2} = a$ *may be written in the form*

$$\frac{d^2 s}{dx^2} + a\left(\frac{ds}{dx}\right)^3 = 0.$$

We have $\qquad \dfrac{dx}{ds} = \dfrac{1}{\dfrac{ds}{dx}}$

then, $\qquad \dfrac{d^2 x}{ds^2} = \dfrac{1}{\dfrac{ds}{dx}} \cdot \dfrac{d}{dx}\left\{\dfrac{1}{\dfrac{ds}{dx}}\right\}$

$$= \frac{-\dfrac{d^2 s}{dx^2}}{\left(\dfrac{ds}{dx}\right)^3}$$

Substituting this value in $\dfrac{d^2x}{ds^2} = a$,

we get
$$\dfrac{-\dfrac{d^2s}{dx^2}}{\left(\dfrac{ds}{dx}\right)^3} = a$$

or $\quad \dfrac{d^2s}{dx^2} + a\left(\dfrac{ds}{dx}\right)^3 = 0$

EXERCISES

Change the independent variable from x to t in the following :

1. $(1 - x^2)\dfrac{d^2y}{dx^2} = x\dfrac{dy}{dx} - x^2 y$, when $x = \sin t$.

2. $x^2\dfrac{d^2y}{dx^2} + x\dfrac{dy}{dx} + y = 0$, when $x = \dfrac{1}{t}$.

3. $\sin^2 2x\dfrac{d^2y}{dx^2} + \sin 4x\dfrac{dy}{dx} + 4y = 0$, when $\tan x = e^t$.

4. $(a + bx)^2\dfrac{d^2y}{dx^2} + A(a + bx)\dfrac{dy}{dx} + By = f(x)$, when $a + bx = e^t$.

5. $\dfrac{d}{dx}\left\{(1 - x^2)\dfrac{dP}{dx}\right\} + n(n + 1)P = 0$ when $x = \cot t$.

   or

   $$(1 - x^2)\dfrac{d^2P}{dx^2} - 2x\dfrac{dP}{dx} + n(n + 1)P = 0$$

   (i) when $x = \cos t$, $\qquad$ (ii) when $x = \dfrac{1}{2}\left(t + \dfrac{1}{t}\right)$.

6. $x^3\dfrac{d^3y}{dx^3} + 2x^2\dfrac{d^2y}{dx^2} + 3x\dfrac{dy}{dx} + 4y = 0$, when $x = e^t.0$

ANSWERS

1. $\dfrac{d^2y}{dt^2} + n^2 y = 0$; $\qquad\qquad$ 2. $t^2\dfrac{d^2y}{dt^2} + t\dfrac{dy}{dt} + y = 0$;

3. $\dfrac{d^2y}{dt^2} + y = 0$;

4. $b^2\dfrac{d^2y}{dt^2} + (Ab - b^2)\dfrac{dy}{dt} + By = f\left(\dfrac{e^t - a}{b}\right)$;

5. (i) $\dfrac{d^2P}{dt^2} + \cos t\dfrac{dP}{dt} + n(n + 1)P = 0$;

   (ii) $t^2(t^2 - 1)\dfrac{d^2P}{dt^2} + 2t^3\dfrac{dP}{dt} - n(n + 1)(t^2 - 1)P = 0$;

6. $\dfrac{d^3y}{dt^3} - \dfrac{d^2y}{dt^2} + 3\dfrac{dy}{dt} + 4y = 0$.

19.4 Transformation from Cartesian to Polars and vice-versa

It often happens that a result in cartesian is much simplified on reduction to polar or vice-versa.

Let
$$x = r \cos \theta$$
$$y = r \sin \theta.$$

Suppose θ to be independent variable, then

$$\frac{dy}{dx} = \frac{\dfrac{dy}{d\theta}}{\dfrac{dx}{d\theta}} = \frac{\dfrac{dr}{d\theta} \sin \theta + r \cos \theta}{\dfrac{dr}{d\theta} \cos \theta - r \sin \theta}$$

$$\frac{d^2 y}{dx^2} = \frac{1}{\dfrac{dx}{d\theta}} \cdot \frac{d}{d\theta} \left\{ \frac{\dfrac{dr}{d\theta} \sin \theta + r \cos \theta}{\dfrac{dr}{d\theta} \cos \theta - r \sin \theta} \right\}$$

$$= \left[\left(\frac{dr}{d\theta} \cos \theta + \frac{d^2 r}{d\theta^2} \sin \theta - r \sin \theta + \frac{dr}{d\theta} \cos \theta \right) \times \left(\frac{dr}{d\theta} \cos \theta - r \sin \theta \right) \right.$$

$$\left. - \left(\frac{dr}{d\theta} \sin \theta + r \cos \theta \right) \times \left(-\frac{dr}{d\theta} \sin \theta + \frac{d^2 r}{d\theta^2} \cos \theta - r \cos \theta - \frac{dr}{d\theta} \sin \theta \right) \right]$$

$$\div \left(\frac{dr}{d\theta} \cos \theta - r \sin \theta \right)^3$$

$$= \left[2 \left(\frac{dr}{d\theta} \right)^2 \cos^2 \theta - 3r \frac{dr}{d\theta} \sin \theta \cos \theta + \frac{d^2 r}{d\theta^2} \sin \theta \cos \theta \frac{dr}{d\theta} \right.$$

$$- r \frac{d^2 r}{d\theta^2} \sin^2 \theta + r^2 \sin^2 \theta + 2 \left(\frac{dr}{d\theta} \right)^2 \sin^2 0 + 3r \frac{dr}{d\theta} \sin \theta \cos \theta$$

$$\left. - \frac{d^2 r}{d\theta^2} \sin \theta \cos \theta \frac{dr}{d\theta} - r \frac{d^2 r}{d\theta^2} \cos^2 \theta + r^2 \cos^2 \theta \right] \div \left(\frac{dr}{d\theta} \cos \theta - r \sin \theta \right)^3$$

$$= \frac{2 \left(\dfrac{dr}{d\theta} \right)^2 + r^2 - r \dfrac{d^2 r}{d\theta^2}}{\left(\dfrac{dr}{d\theta} \cos \theta - r \sin \theta \right)^3}$$

$$= \frac{r^2 + 2 \left(\dfrac{dr}{d\theta} \right)^2 - r \dfrac{d^2 r}{d\theta^2}}{\left(\dfrac{dr}{d\theta} \cos \theta - r \sin \theta \right)^3}$$

19.5 x and y to be expressed in terms of some third variable

To show that

$$x \frac{dx}{dt} + y \frac{dy}{dt} = r \frac{dr}{dt} \qquad \qquad \text{...(1)}$$

$$x\frac{dy}{dt} - y\frac{dx}{dt} = r^2\frac{d\theta}{dt} \qquad \dots(2)$$

$$\left(\frac{dx}{dt}\right)^2 + \left(\frac{dy}{dt}\right)^2 = \left(\frac{dr}{dt}\right)^2 + r^2\left(\frac{d\theta}{dt}\right)^2 \qquad \dots(3)$$

We have $x = r \cos \theta$

$$y = r \sin \theta$$

$$x^2 + y^2 = r^2.$$

Differentitating, We get

$$2x\frac{dx}{dt} + 2y\frac{dy}{dt} = 2r\frac{dr}{dt}$$

or $x\dfrac{dx}{dt} + y\dfrac{dy}{dt} = r\dfrac{dr}{dt}.$

To prove result (3) we will find the value of $\left(\dfrac{ds}{dt}\right)^2$ in two forms.

$$\left(\frac{ds}{dt}\right)^2 = \left(\frac{ds}{dx}\right)^2 \cdot \left(\frac{dx}{dt}\right)^2$$

$$= \left\{1 + \left(\frac{dy}{dx}\right)^2\right\}\left(\frac{dx}{dt}\right)^2$$

$$= \left(\frac{dx}{dt}\right)^2 + \left(\frac{dy}{dt}\right)^2 \qquad \dots(i)$$

Again, $\left(\dfrac{ds}{dt}\right)^2 = \left(\dfrac{ds}{d\theta}\right)^2 \cdot \left(\dfrac{d\theta}{dt}\right)^2$

$$= \left\{r^2 + \left(\frac{dr}{d\theta}\right)^2\right\} \cdot \left(\frac{d\theta}{dt}\right)^2$$

$$= r^2\left(\frac{d\theta}{dt}\right)^2 + \left(\frac{dr}{dt}\right)^2 \qquad \dots(ii)$$

Equating the two values,

$$\left(\frac{dx}{dt}\right)^2 + \left(\frac{dy}{dt}\right)^2 = r^2\left(\frac{d\theta}{dt}\right)^2 + \left(\frac{dr}{dt}\right)^2$$

To prove 2nd relation, multiply (3) by r^2 and subtracting the square of relation (1) from it.

$$(x^2 + y^2)\left\{\left(\frac{dx}{dt}\right)^2 + \left(\frac{dy}{dt}\right)^2\right\} = r^2\left(\frac{dr}{dt}\right)^2 + r^4\left(\frac{d\theta}{dt}\right)^2 \qquad ..(iii)$$

Squarting 1st relation,

$$x^2\left(\frac{dx}{dt}\right)^2 + x^2\left(\frac{dy}{dt}\right)^2 + 2xy\frac{dx}{dt}\cdot\frac{dy}{dt} = r^2\left(\frac{d\theta}{dt}\right)^2 \qquad \dots(iv)$$

Subtracting (iv) from (iii) we have

$$y^2\left(\frac{dx}{dt}\right)^2 + x^2\left(\frac{dy}{dt}\right)^2 - 2xy\frac{dx}{dt}\cdot\frac{dy}{dt} = r^4\left(\frac{d\theta}{dt}\right)^2$$

or $$\left(y\frac{dx}{dt} - x\frac{dy}{dt}\right)^2 = \left(r^2\frac{d\theta}{dt}\right)^2$$

or $$y\frac{dx}{dt} - x\frac{dy}{dt} = r^2\frac{d\theta}{dt}.$$

EXAMPLES

1. *Transform* $p = \dfrac{x\dfrac{dy}{dx} - y}{\sqrt{\left\{1 + \left(\dfrac{dy}{dx}\right)^2\right\}}}$ *into polars.*

Multiplying numerator and denominator by $\dfrac{dx}{dt}$, we get

$$p = \frac{x\dfrac{dy}{dt} - y\dfrac{dx}{dt}}{\sqrt{\left\{\left(\dfrac{dx}{dt}\right)^2 + \left(\dfrac{dy}{dx}\right)^2\right\}}}$$

$$= \frac{r^2\dfrac{d\theta}{dt}}{\sqrt{\left\{\left(\dfrac{dr}{dt}\right)^2 + r^2\left(\dfrac{d\theta}{dt}\right)^2\right\}}}$$

or $$\frac{1}{p^2} = \frac{\left(\dfrac{dr}{dt}\right)^2 + r^2\left(\dfrac{d\theta}{dt}\right)^2}{r^4\left(\dfrac{d\theta}{dt}\right)^2}$$

$$= \frac{1}{r^2} + \frac{1}{r^4}\cdot\frac{\left(\dfrac{dr}{dt}\right)^2}{\left(\dfrac{d\theta}{dt}\right)^2}$$

$$= \frac{1}{r^2} + \frac{1}{r^4}\left(\frac{dr}{d\theta}\right)^2.$$

2. *Transform* $\tan\phi = r\dfrac{d\theta}{dr}$ *into cartesian form.*

We have $$\tan\phi = r\frac{d\theta}{dr} = \frac{r^2\dfrac{d\theta}{dt}}{r\dfrac{dr}{dt}}$$

$$= \frac{r\dfrac{d\theta}{dt}}{\dfrac{dr}{dt}} = \frac{x\dfrac{dy}{dt} - y\dfrac{dx}{dt}}{x\dfrac{dx}{dt} + y\dfrac{dy}{dt}}$$

$$= \frac{\left\{x\dfrac{dy}{dx} - y\right\}\dfrac{dx}{dt}}{\left\{x + y\dfrac{dy}{dx}\right\}\dfrac{dx}{dt}}$$

$$= \frac{x\dfrac{dy}{dx} - y}{x + y\dfrac{dy}{dx}}$$

3. *Transform* $\dfrac{d^2 y}{dx^2}$ *to the new variables u and v taking u as the independent variable, given* $x = v^{-1}, y = uv.$

From the given relations, we have

$$\frac{dx}{du} = -\frac{1}{v^2} \cdot \frac{dv}{du}; \text{ and } \frac{dy}{du} = v + u\frac{dv}{du}$$

Now,
$$\frac{dy}{dx} = \frac{\dfrac{dy}{du}}{\dfrac{dx}{du}} = \frac{v + u\dfrac{dv}{du}}{-\dfrac{1}{v^2}\dfrac{dv}{du}}$$

Therefore
$$\frac{d^2 y}{dx^2} = \frac{1}{\dfrac{dx}{du}} \cdot \frac{d}{du} \frac{v + u\dfrac{dv}{du}}{-\dfrac{1}{v^2}\dfrac{dv}{du}}$$

$$= \frac{v^2}{\dfrac{dv}{du}} \frac{d}{du}\left\{\frac{v^3 + uv^2\dfrac{dv}{du}}{\dfrac{dv}{du}}\right\}$$

$$= \frac{v^2}{\left(\dfrac{dv}{du}\right)^3}\left[\frac{dv}{du}\left\{3v^2 \cdot \frac{dv}{du} + v^2\frac{dv}{du} + 2uv\left(\frac{dv}{du}\right)^2 + uv^2\frac{d^2 v}{du^2}\right\} - \left(v^3 + uv^2\frac{dv}{du}\right)\frac{d^2 v}{du^2}\right]$$

$$= \frac{u^3}{\left(\dfrac{dv}{du}\right)^3}\left[4v\left(\frac{dv}{du}\right)^2 + 2u\left(\frac{dv}{du}\right)^3 - v^2\frac{d^2 v}{du^2}\right]$$

EXERCISES

1. Transform the formula $\rho = \dfrac{\left\{1 + \left(\dfrac{dy}{dx}\right)^2\right\}^{3/2}}{\dfrac{d^2 y}{dx^2}}$ into polar coordinates.

2. Given $x = $ Given $x = a\,(\theta + \sin \theta), y = a\,(1 - \cos \theta)$, prove that
$$\frac{d^2 y}{dx^2} = \frac{1}{4a} \sec^4 \frac{\theta}{2}.$$

3. Show that by putting $x^2 = s$ and $y^2 = t$, the equatio

$$A x y \left(\frac{dy}{dx}\right)^2 + (x^2 - A y^2 - B) \frac{dy}{dx} - xy = $$

is reduced to

$$t = s \frac{dt}{ds} - \frac{B \dfrac{dt}{ds}}{A \dfrac{dt}{ds} + 1}.$$

ANSWERS

1. $\rho = \dfrac{\left\{r^2 + \left(\dfrac{dr}{d\theta}\right)^2\right\}^{3/2}}{r^2 + 2\left(\dfrac{dr}{d\theta}\right)^2 - r \dfrac{d^2 r}{d\theta^2}}$

19.6 Two Independent Variables

We shall now consider the case in which there are two independent variables x and y.

Let $\qquad U = f(x, y)$ $\qquad\qquad\qquad$ ...(i)

$\qquad\qquad x = \phi_1(u, v)$ $\qquad\qquad\qquad$ ...(ii)

$\qquad\qquad y = \phi_2(u, v)$

be the proposed transformations ; then we have

$$\left. \begin{aligned} \frac{\partial U}{\partial u} &= \frac{\partial U}{\partial x} \cdot \frac{\partial x}{\partial u} + \frac{\partial U}{\partial y} \cdot \frac{\partial y}{\partial u} \\[2mm] \frac{\partial U}{\partial v} &= \frac{\partial U}{\partial x} \cdot \frac{\partial x}{\partial v} + \frac{\partial U}{\partial y} \cdot \frac{\partial y}{\partial v} \end{aligned} \right\} \qquad ...(iii)$$

These equations may be solved for $\dfrac{\partial U}{\partial x} \cdot \dfrac{\partial U}{\partial y}$.

$$\frac{\partial U}{\partial x} \cdot \frac{\partial x}{\partial y} + \frac{\partial U}{\partial y} \cdot \frac{\partial y}{\partial u} - \frac{\partial U}{\partial u} = 0$$

$$\frac{\partial U}{\partial x} \cdot \frac{\partial x}{\partial y} + \frac{\partial U}{\partial y} \cdot \frac{\partial y}{\partial v} - \frac{\partial U}{\partial v} = 0$$

Solving, we get

$$\frac{\dfrac{\partial U}{\partial x}}{\dfrac{\partial U}{\partial v} \cdot \dfrac{\partial y}{\partial u} + \dfrac{\partial U}{\partial u} \cdot \dfrac{\partial y}{\partial v}} = \frac{\dfrac{\partial U}{\partial y}}{-\dfrac{\partial U}{\partial u} \cdot \dfrac{\partial x}{\partial v} + \dfrac{\partial U}{\partial v} \cdot \dfrac{\partial x}{\partial u}}$$

$$= \frac{1}{\dfrac{\partial x}{\partial u} \cdot \dfrac{\partial y}{\partial v} - \dfrac{\partial y}{\partial u} \cdot \dfrac{\partial x}{\partial v}}$$

$$\frac{\partial U}{\partial x} = \frac{\dfrac{\partial U}{\partial u} \cdot \dfrac{\partial y}{\partial v} - \dfrac{\partial U}{\partial v} \cdot \dfrac{\partial y}{\partial u}}{\dfrac{\partial x}{\partial u} \cdot \dfrac{\partial y}{\partial v} - \dfrac{\partial y}{\partial u} \cdot \dfrac{\partial x}{\partial v}}$$

and
$$\frac{\partial U}{\partial y} = \frac{\dfrac{\partial U}{\partial v} \cdot \dfrac{\partial x}{\partial u} - \dfrac{\partial U}{\partial u} \cdot \dfrac{\partial x}{\partial v}}{\dfrac{\partial x}{\partial u} \cdot \dfrac{\partial y}{\partial v} - \dfrac{\partial y}{\partial u} \cdot \dfrac{\partial x}{\partial v}}$$

If, however, we could solve equations (*ii*) in *u*, *v*,

$$u = F_1(x, y)$$
$$v = F_2(x, y),$$

then substitute the values of $\dfrac{\partial u}{\partial x}, \dfrac{\partial u}{\partial y}, \dfrac{\partial v}{\partial x}, \dfrac{\partial v}{\partial y}$.

When the connecting relations are

$$V = \phi(u, v),\ u = F_1(x, y),\ v = F_2(x, y)$$

Then,
$$\frac{\partial V}{\partial x} = \frac{\partial V}{\partial u} \cdot \frac{\partial u}{\partial x} + \frac{\partial V}{\partial v} \cdot \frac{\partial v}{\partial x} \qquad \qquad \qquad ...(iii)$$

and
$$\frac{\partial V}{\partial y} = \frac{\partial V}{\partial u} \cdot \frac{\partial u}{\partial y} + \frac{\partial V}{\partial v} \cdot \frac{\partial v}{\partial y} \qquad \qquad \qquad ...(iv)$$

Differentiating (*iii*) with respect to *x*, we get

$$\frac{\partial^2 V}{\partial x^2} = \frac{\partial}{\partial x}\left[\frac{\partial V}{\partial u} \cdot \frac{\partial u}{\partial x} + \frac{\partial V}{\partial v} \cdot \frac{\partial v}{\partial x}\right]$$

$$= \frac{\partial v}{\partial u}\frac{\partial^2 u}{\partial x^2} + \frac{\partial u}{\partial x} \cdot \frac{\partial}{\partial x}\left(\frac{\partial V}{\partial u}\right) + \frac{\partial V}{\partial v} \cdot \frac{\partial^2 v}{\partial x^2} + \frac{\partial v}{\partial x} \cdot \frac{\partial}{\partial x}\left(\frac{\partial V}{\partial v}\right)$$

$$= \frac{\partial V}{\partial u} \cdot \frac{\partial^2 u}{\partial x^2} + \frac{\partial u}{\partial x}\left[\frac{\partial^2 V}{\partial v^2} \cdot \frac{\partial u}{\partial x} + \frac{\partial^2 V}{\partial u\,\partial v} \cdot \frac{\partial v}{\partial x}\right] + \frac{\partial V}{\partial v} \cdot \frac{\partial^2 v}{\partial x^2}$$

$$\qquad\qquad\qquad\qquad + \frac{\partial v}{\partial x}\left[\frac{\partial^2 V}{\partial u\,\partial v} \cdot \frac{\partial u}{\partial x} + \frac{\partial^2 V}{\partial v^2} \cdot \frac{\partial v}{\partial x}\right]$$

To get the bracketted expression, put $\dfrac{\partial V}{\partial u}$ and $\dfrac{\partial V}{\partial v}$ for *V* in (*iii*)

Similarly,

$$\frac{\partial^2 V}{\partial y^2} = \frac{\partial V}{\partial u} \cdot \frac{\partial^2 u}{\partial y^2} + \frac{\partial u}{\partial y}\left[\frac{\partial^2 V}{\partial u^2} \cdot \frac{\partial u}{\partial y} + \frac{\partial^2 V}{\partial u\,\partial v} \cdot \frac{\partial v}{\partial y}\right]$$

$$\qquad\qquad\qquad\qquad + \frac{\partial V}{\partial v} \cdot \frac{\partial^2 v}{\partial y^2} + \frac{\partial v}{\partial y}\left[\frac{\partial^2 V}{\partial u\,\partial v} \cdot \frac{\partial u}{\partial y} + \frac{\partial^2 V}{\partial v^2} \cdot \frac{\partial v}{\partial y}\right]$$

19.7 Transformation from cartesian to the polar

We have, $\qquad y = r\sin\theta, \qquad\qquad r = \sqrt{(x^2 + y^2)},$

$$x = r\cos\theta, \qquad\qquad \theta = \tan^{-1}\frac{y}{x},$$

$$\frac{\partial x}{\partial r} = \cos\theta, \quad \frac{\partial r}{\partial x} = \frac{1}{2} \cdot \frac{2x}{\sqrt{(x^2 + y^2)}} = \frac{x}{\sqrt{(x^2 + y^2)}}$$

$$= \frac{r \cos \theta}{r} = \cos \theta,$$

$$\frac{\partial y}{\partial r} = \sin \theta, \quad \frac{\partial r}{\partial y} = \frac{1}{2} \cdot \frac{2y}{\sqrt{(x^2 + y^2)}} = \frac{y}{\sqrt{(x^2 + y^2)}}$$

$$= \frac{r \sin \theta}{r} = \sin \theta$$

$$\frac{\partial x}{\partial \theta} = -r \sin \theta, \quad \frac{\partial \theta}{\partial x} = -\frac{1}{1 + \frac{y^2}{x^2}} \cdot \frac{y}{x^2} = -\frac{y}{x^2 + y^2}$$

$$= -\frac{r \sin \theta}{r^2} = -\frac{\sin \theta}{r},$$

$$\frac{\partial y}{\partial \theta} = r \cos \theta, \quad \frac{\partial \theta}{\partial y} = \frac{1}{1 + \frac{y^2}{x^2}} \cdot \frac{1}{x} = \frac{x}{x^2 + y^2} = \frac{r \cos \theta}{r^2} = \frac{\cos \theta}{r}.$$

Then

$$\frac{\partial V}{\partial x} = \frac{\partial V}{\partial r} \cdot \frac{\partial r}{\partial x} + \frac{\partial V}{\partial \theta} \cdot \frac{\partial \theta}{\partial x} = \cos \theta \frac{\partial V}{\partial r} - \frac{\sin \theta}{r} \frac{\partial V}{\partial \theta},$$

$$\frac{\partial V}{\partial y} = \frac{\partial V}{\partial r} \cdot \frac{\partial r}{\partial y} + \frac{\partial V}{\partial \theta} \cdot \frac{\partial \theta}{\partial y} = \sin \theta \frac{\partial V}{\partial r} - \frac{\cos \theta}{r} \frac{\partial V}{\partial \theta},$$

$$\frac{\partial V}{\partial r} = \frac{\partial V}{\partial x} \cdot \frac{\partial x}{\partial r} + \frac{\partial V}{\partial y} \cdot \frac{\partial y}{\partial r} = \cos \theta \frac{\partial V}{\partial x} + \sin \theta \frac{\partial V}{\partial y},$$

$$\frac{\partial V}{\partial \theta} = \frac{\partial V}{\partial x} \cdot \frac{\partial x}{\partial \theta} + \frac{\partial V}{\partial y} \cdot \frac{\partial y}{\partial \theta} = -r \sin \theta \frac{\partial V}{\partial x} + r \cos \theta \frac{\partial V}{\partial y},$$

$$\therefore \quad \frac{\partial}{\partial x} \equiv \cos \theta \frac{\partial}{\partial r} - \frac{\sin \theta}{r} \frac{\partial}{\partial \theta}.$$

$$\frac{\partial}{\partial y} \equiv \sin \theta \frac{\partial}{\partial r} - \frac{\cos \theta}{r} \frac{\partial}{\partial \theta}.$$

19.8 Transformation of $\dfrac{\partial^2 V}{\partial x^2}$ and $\dfrac{\partial^2 V}{\partial y^2}$ to polars

We have

$$\frac{\partial^2 V}{\partial x^2} = \left(\frac{\partial}{\partial x} \right)^2 V = \left(\cos \theta \frac{\partial}{\partial r} - \frac{\sin \theta}{r} \frac{\partial}{\partial \theta} \right) \left(\cos \theta \frac{\partial V}{\partial r} - \frac{\sin \theta}{r} \frac{\partial V}{\partial \theta} \right)$$

$$= \cos \theta \frac{\partial}{\partial r} \left(\cos \theta \frac{\partial V}{\partial r} - \frac{\sin \theta}{r} \frac{\partial V}{\partial \theta} \right) - \frac{\sin \theta}{r} \frac{\partial}{\partial \theta} \left(\cos \theta \frac{\partial V}{\partial r} - \frac{\sin \theta}{r} \frac{\partial V}{\partial \theta} \right)$$

$$= \cos^2 \theta \frac{\partial^2 V}{\partial r^2} - \frac{2 \cos \theta \sin \theta}{r} \frac{\partial^2 V}{\partial r \partial \theta} + \frac{\sin^2 \theta}{r^2} \cdot \frac{\partial^2 V}{\partial \theta^2}$$

$$+ \frac{\sin^2 \theta}{r} \frac{\partial V}{\partial r} + \frac{2 \cos \theta \sin \theta}{r^2} \frac{\partial V}{\partial \theta} \qquad \ldots (i)$$

Similarly,

$$\frac{\partial^2 V}{\partial y^2} = \left(\frac{\partial}{\partial y} \right)^2 V = \left(\sin \theta \frac{\partial}{\partial r} + \frac{\cos \theta}{r} \frac{\partial}{\partial \theta} \right) \left(\sin \theta \frac{\partial V}{\partial r} + \frac{\cos \theta}{r} \frac{\partial V}{\partial \theta} \right)$$

$$= \sin^2 \theta \, \frac{\partial^2 v}{\partial r^2} + \frac{2 \sin \theta \cos \theta}{r} \frac{\partial^2 V}{\partial r \, \partial \theta} - \frac{2 \sin \theta \cos 0}{r^2} \frac{\partial V}{\partial \theta}$$

$$+ \frac{\cos^2 \theta}{r} \frac{\partial V}{\partial r} + \frac{\cos^2 \theta}{r^2} \frac{\partial^2 V}{\partial \theta^2}, \qquad \ldots(ii)$$

Adding (*i*) and (*ii*), we get

$$\frac{\partial^2 V}{\partial x^2} + \frac{\partial^2 V}{\partial y^2} = \frac{\partial^2 V}{\partial r^2} (\cos^2 \theta + \sin^2 \theta) + \frac{\partial^2 V}{\partial r \, \partial \theta} \left(\frac{2 \sin \theta \cos \theta}{r} - \frac{2 \sin \theta \cos \theta}{r} \right)$$

$$+ \frac{\partial V}{\partial \theta} \left(\frac{2 \cos \theta \sin \theta}{r^2} - \frac{2 \sin \theta \cos \theta}{r^2} \right) + \frac{1}{r} \frac{\partial V}{\partial r} (\cos^2 \theta + \sin^2 \theta) + \frac{\partial^2 V}{\partial \theta^2} \cdot \frac{1}{r^2} (\cos^2 \theta + \sin^2 \theta)$$

or $\qquad \dfrac{\partial^2 V}{\partial x^2} + \dfrac{\partial^2 V}{\partial y^2} = \dfrac{\partial^2 V}{\partial r^2} + \dfrac{1}{r} \dfrac{\partial V}{\partial r} + \dfrac{1}{r^2} \dfrac{\partial^2 V}{\partial \theta^2}$

19.9 Transformation of $\nabla^2 V$

We know that

$$\frac{\partial^2 V}{\partial x^2} + \frac{\partial^2 V}{\partial y^2} + \frac{\partial^2 V}{\partial z^2} = \nabla^2 V \qquad \ldots(i)$$

If $x = r \cos \theta$, $y = r \sin \theta$, then we know

$$\frac{\partial^2 V}{\partial x^2} + \frac{\partial^2 V}{\partial y^2} = \frac{\partial^2 V}{\partial r^2} + \frac{1}{r} \frac{\partial V}{\partial r} + \frac{1}{r^2} \frac{\partial^2 V}{\partial \theta^2} \qquad \ldots(ii)$$

The operator ∇^2 stands for $\dfrac{\partial^2}{\partial x^2} + \dfrac{\partial^2}{\partial y^2} + \dfrac{\partial^2}{\partial z^2}$.

The transformation formulae are as follows :–

$x = r \sin \theta \cos \phi.$ If $r \sin \theta = u,$

$y = r \sin \theta \sin \phi,$ then $x = u \cos \phi,$

$z = r \cos \theta.$ $y = u \sin \phi.$

By (*ii*)

$$\frac{\partial^2 V}{\partial x^2} + \frac{\partial^2 V}{\partial y^2} = \frac{\partial^2 V}{\partial u^2} + \frac{1}{u} \frac{\partial V}{\partial u} + \frac{1}{u^2} \frac{\partial^2 V}{\partial \phi^2} \qquad \ldots(iii)$$

Where $z = r \cos \theta$, $u = r \sin \theta$,

$$\therefore \qquad \frac{\partial^2 V}{\partial u^2} + \frac{\partial^2 V}{\partial z^2} = \frac{\partial^2 V}{\partial r^2} + \frac{1}{r} \frac{\partial V}{\partial r} + \frac{1}{r^2} \frac{\partial^2 V}{\partial \theta^2} \qquad \ldots(iv)$$

Adding (*iii*) and (*iv*), we get

$$\frac{\partial^2 V}{\partial x^2} + \frac{\partial^2 V}{\partial y^2} + \frac{\partial^2 V}{\partial z^2} + \frac{\partial^2 V}{\partial u^2} = \frac{\partial^2 V}{\partial u^2} + \frac{1}{u} \frac{\partial V}{\partial u} + \frac{1}{u^2} \frac{\partial^2 V}{\partial \phi^2}$$

$$+ \frac{\partial^2 V}{\partial r^2} + \frac{1}{r} \frac{\partial V}{\partial r} + \frac{1}{r^2} \frac{\partial^2 V}{\partial \theta^2} \qquad \ldots(v)$$

We know that

$$\frac{\partial V}{\partial u} = \frac{\partial V}{\partial r} \cdot \frac{\partial r}{\partial u} + \frac{\partial V}{\partial \theta} \cdot \frac{\partial \theta}{\partial u}$$

$$= r \cos \theta, \ u = r \sin \theta$$

$$\therefore \qquad r = \sqrt{(z^2 + u^2)}, \qquad \frac{\partial r}{\partial u} = \frac{u}{\sqrt{(z^2 + u^2)}} = \sin \theta.$$

$$\theta = \tan^{-1} \frac{u}{z}, \qquad \therefore \frac{\partial \theta}{\partial u} = \frac{z}{u^2 + z^2} = \frac{\cos \theta}{r}$$

$$\therefore \qquad \frac{\partial V}{\partial u'} = \sin \theta \frac{\partial V}{\partial r} + \frac{\cos \theta}{r} \frac{\partial V}{\partial \theta}$$

i.e.,
$$\frac{1}{u} \left(\frac{\partial V}{\partial u} \right) = \frac{1}{r} \frac{\partial V}{\partial r} + \frac{1}{r^2} \cot \theta \frac{\partial V}{\partial \theta} .$$

Substituting this value in (*v*), we get

$$\frac{\partial^2 V}{\partial x^2} + \frac{\partial^2 V}{\partial y^2} + \frac{\partial^2 V}{\partial z^2} = \frac{2}{r} \frac{\partial V}{\partial r} + \frac{1}{r^2} \cot \theta \frac{\partial V}{\partial \theta} + \frac{1}{u^2} \frac{\partial^2 V}{\partial \phi^2} + \frac{\partial^2 V}{\partial r^2} + \frac{1}{r^2} \frac{\partial^2 V}{\partial \theta^2}$$

$$= \frac{\partial^2 V}{\partial r^2} + \frac{2}{r} \frac{\partial V}{\partial r} + \frac{1}{r^2} \frac{\partial^2 V}{\partial \theta^2} + \frac{1}{r^2} \cot \theta \frac{\partial V}{\partial \theta} + \frac{1}{r^2 \sin^2 \theta} \frac{\partial^2 V}{\partial \phi^2}$$

$$\therefore \frac{\partial^2 V}{\partial x^2} + \frac{\partial^2 V}{\partial y^2} + \frac{\partial^2 V}{\partial z^2} = \frac{\partial^2 V}{\partial r^2} + \frac{2}{r} \frac{\partial V}{\partial r} + \frac{1}{r^2} \frac{\partial^2 V}{\partial \theta^2} + \frac{1}{r^2} \cot \theta \frac{\partial V}{\partial \theta} + \frac{1}{r^2 \sin^2 \theta} \frac{\partial^2 V}{\partial \phi^2}$$

$$= \nabla^2 V.$$

EXAMPLES

1. *If V be a function of r alone, when $r^2 = x^2 + y^2$, show that*

$$\frac{\partial^2 V}{\partial x^2} + \frac{\partial^2 V}{\partial y^2} = \frac{\partial^2 V}{\partial r^2} + \frac{1}{r} \frac{\partial V}{\partial r}.$$

(Garhwal 2004)

We have $\qquad r^2 = x^2 + y^2.$

$$\therefore \qquad r \frac{dr}{dx} = 2x; \qquad \therefore \qquad \frac{dr}{dx} = \frac{x}{r}$$

$$\frac{\partial V}{\partial x} = \frac{\partial V}{\partial r} \cdot \frac{\partial r}{\partial x} = \frac{x}{r} \frac{\partial V}{\partial r}, \text{ so that}$$

$$\frac{\partial}{\partial x} \equiv \frac{x}{r} \frac{\partial}{\partial r}.$$

$$\therefore \qquad \frac{\partial^2 V}{\partial x^2} = \frac{\partial}{\partial x} \left(\frac{\partial V}{\partial x} \right) = \frac{x}{r} \frac{\partial}{\partial r} \left(\frac{x}{r} \frac{\partial V}{\partial r} \right)$$

$$= \frac{1}{r} \frac{\partial V}{\partial r} + \frac{x^2}{r} \frac{\partial}{\partial r} \left(\frac{1}{r} \frac{\partial V}{\partial r} \right)$$

$$= \frac{1}{r} \frac{\partial V}{\partial r} - \frac{x^2}{r^3} \frac{\partial V}{\partial r} + \frac{x^2}{r^2} \frac{\partial^2 V}{\partial r^2} \qquad \qquad ...(i)$$

Similarly,

$$\frac{\partial^2 V}{\partial y^2} = \frac{1}{r} \frac{\partial V}{\partial r} - \frac{y^2}{r^3} \frac{\partial V}{\partial r} + \frac{y^2}{r^2} \frac{\partial^2 V}{\partial r^2} \qquad \qquad ...(ii)$$

Adding (*i*) and (*ii*), we get

$$\frac{\partial^2 V}{\partial x^2} + \frac{\partial^2 V}{\partial y^2} = \frac{2}{r} \frac{\partial V}{\partial r} - \frac{1}{r^3} \frac{\partial V}{\partial r} (x^2 + y^2) + \frac{1}{r^2} \frac{\partial^2 V}{\partial r^2} (x^2 + y^2)$$

$$= \frac{2}{r} \frac{\partial V}{\partial r} - \frac{1}{r} \frac{\partial V}{\partial r} + \frac{\partial^2 V}{\partial r^2}$$

$$= \frac{\partial^2 V}{\partial r^2} + \frac{1}{r} \frac{\partial V}{\partial r}$$

2. *If* $u = e^x \sec y$, $v = e^x \tan y$ *and V is a function of u, v, show that*

$$\cos y \left(\frac{\partial^2 V}{\partial x \partial y} - \frac{\partial V}{\partial y} \right) = uv \left(\frac{\partial^2 V}{\partial u^2} + \frac{\partial^2 V}{\partial v^2} \right) + (u^2 + v^2) \frac{\partial^2 V}{\partial u \partial v}.$$

We are given that

$$u = e^x \sec y, \quad v = e^x \tan y,$$

$$\therefore \qquad u^2 = e^{2x} \sec^2 y = e^{2x} (1 + \tan^2 y)$$

$$v^2 = e^{2x} \tan^2 y$$

$$\therefore \qquad u^2 - v^2 = e^{2x}.$$

Now, $\quad \dfrac{\partial u}{\partial x} = \dfrac{\partial V}{\partial u} \cdot \dfrac{\partial u}{\partial x} + \dfrac{\partial V}{\partial v} \cdot \dfrac{\partial v}{\partial x}$

$$= e^x \sec y \frac{\partial V}{\partial u} + e^x \tan y \frac{\partial V}{\partial x}.$$

or $\qquad \dfrac{\partial V}{\partial x} = u \dfrac{\partial V}{\partial u} + v \dfrac{\partial V}{\partial v}$ ...(i)

Again, $\quad \dfrac{\partial V}{\partial y} = \dfrac{\partial V}{\partial u} \cdot \dfrac{\partial u}{\partial y} + \dfrac{\partial V}{\partial v} \cdot \dfrac{\partial v}{\partial y}$

We have $\quad \dfrac{\partial v}{\partial y} = e^x \sec^2 y, \quad \dfrac{\partial u}{\partial y} = e^x \sec y \tan y$

$$\therefore \qquad \frac{\partial V}{\partial y} = e^x \sec y \tan y \frac{\partial V}{\partial u} + e^x \sec^2 y \frac{\partial V}{\partial v}$$

$$= \sec y \left\{ v \frac{\partial V}{\partial u} + u \frac{\partial V}{\partial v} \right\}$$

or $\qquad \cos y \dfrac{\partial V}{\partial y} = v \dfrac{\partial V}{\partial u} + u \dfrac{\partial V}{\partial y}$ ...(ii)

Now, $\cos y \cdot \dfrac{\partial}{\partial y} \left(\dfrac{\partial V}{\partial x} \right) = \left(v \dfrac{\partial}{\partial u} + u \dfrac{\partial}{\partial v} \right) \left(u \dfrac{\partial V}{\partial u} + v \dfrac{\partial V}{\partial v} \right)$

$$= uv \frac{\partial^2 V}{\partial u^2} + v^2 \frac{\partial^2 V}{\partial u \partial v} + v \frac{\partial V}{\partial u} + u^2 \frac{\partial^2 V}{\partial u \partial v} + uv \frac{\partial^2 V}{\partial v^2} + u \frac{\partial V}{\partial v}$$

$$= uv \left\{ \frac{\partial^2 V}{\partial u^2} + \frac{\partial^2 V}{\partial v^2} \right\} + (u^2 + v^2) \frac{\partial^2 V}{\partial u \partial v} + u \frac{\partial V}{\partial v} + v \frac{\partial V}{\partial u} \qquad \text{from } (ii)$$

$$\therefore \quad \cos y \left\{ \frac{\partial^2 V}{\partial y \partial x} - \frac{\partial V}{\partial y} \right\} = uv \left\{ \frac{\partial^2 V}{\partial u^2} + \frac{\partial^2 V}{\partial v^2} \right\} + (u^2 + v^2) \frac{\partial^2 V}{\partial u \partial v}.$$

3. *If* $x = r \cos \theta$, $y = r \sin \theta$, *prove that*

(i) $\quad \dfrac{\partial^2 \theta}{\partial x \partial y} = \dfrac{\cos 2\theta}{r^2}$

(ii) $(x^2 - y^2)\left(\dfrac{\partial^2 u}{\partial x^2} - \dfrac{\partial^2 u}{\partial y^2}\right) + 4xy \dfrac{\partial^2 u}{\partial x \partial y} = r^2 \dfrac{\partial^2 u}{\partial^2 r} - r \dfrac{\partial u}{\partial \theta} - \dfrac{\partial^2 u}{\partial \theta^2}.$

(i) We have $\theta = \tan^{-1}(y/x)$

so that $\dfrac{\partial \theta}{\partial y} = \dfrac{1}{1 + y^2/x^2} \cdot \dfrac{1}{x} = \dfrac{x}{x^2 + y^2}$

and $\dfrac{\partial^2 \theta}{\partial x \partial y} = \dfrac{\partial}{\partial x}\left(\dfrac{x}{x^2 + y^2}\right) = \dfrac{x^2 + y^2 - 2x^2}{(x + y^2)^2} = \dfrac{y^2 - x^2}{(x^2 + y^2)^2}$

$= \dfrac{r^2(\sin^2 \theta - \cos^2 \theta)}{r^4} = -\dfrac{\cos 2\theta}{r^2}.$

(ii) We have, $r \dfrac{\partial u}{\partial r} = x \dfrac{\partial u}{\partial x} + y \dfrac{\partial u}{\partial y}$

$r^2 \dfrac{\partial^2 u}{\partial r^2} = x^2 \dfrac{\partial^2 u}{\partial x^2} + 2xy \dfrac{\partial^2 u}{\partial x \partial y} + y^2 \dfrac{\partial^2 u}{\partial y^2}$

and $\dfrac{\partial^2 u}{\partial^2 \theta} = y^2 \dfrac{\partial^2 u}{\partial x^2} - 2xy \dfrac{\partial^2 u}{\partial x \partial y} + x^2 \dfrac{\partial^2 u}{\partial y^2} - x \dfrac{\partial u}{\partial x} - y \dfrac{\partial u}{\partial y}$

$\therefore$ $r^2 \dfrac{\partial^2 u}{\partial r^2} - r \dfrac{\partial u}{\partial r} - \dfrac{\partial^2 u}{\partial^2 \theta} = (x^2 - y^2)\left(\dfrac{\partial^2 u}{\partial x^2} - \dfrac{\partial^2 u}{\partial y^2}\right) + 4xy \dfrac{\partial^2 u}{\partial x \partial y}.$

4. *If V be a function of r alone where $r^2 = x_1^2 + x_2^2 + \qquad x_n^2$, then show that*

$$\dfrac{\partial^2 V}{\partial x_1^2} + \dfrac{\partial^2 V}{\partial x_2^2} + \cdots + \dfrac{\partial^2 V}{\partial x_n^2} = \dfrac{d^2 V}{dr^2} + \dfrac{n-1}{r} \dfrac{dV}{dr}.$$

Since V is a function of r alone, differential coefficient of V w.r.t. r will be total.

Now $r^2 = x_1^2 + x_2^2 + \cdots + x_n^2$

$\therefore$ $2r \dfrac{\partial r}{\partial x_1} = 2x,$ *i.e.,* $\dfrac{\partial r}{\partial x_1} = \dfrac{x_1}{r}.$

Thus $\dfrac{\partial V}{\partial x_1} = \dfrac{dV}{dr} \cdot \dfrac{\partial r}{\partial x_1} = \dfrac{x_1}{r} \dfrac{dV}{dr}$

$\dfrac{\partial^2 V}{\partial x_1^2} = \dfrac{\partial}{\partial x_1}\left(\dfrac{\partial V}{\partial x_1}\right) = \dfrac{\partial}{\partial x_1}\left(\dfrac{x_1}{r} \cdot \dfrac{dV}{dr}\right)$

$= \dfrac{x_1}{r} \dfrac{d^2 V}{dr^2} \cdot \dfrac{\partial r}{\partial x_1} + \dfrac{1}{r} \dfrac{dV}{dr} - \dfrac{x_1}{r^2} \dfrac{\partial r}{\partial x_1} \cdot \dfrac{dV}{dr}$

$= \dfrac{x_1^2}{r^2} \dfrac{d^2 V}{dr^2} + \dfrac{1}{r} \dfrac{dV}{dr} - \dfrac{x_1^2}{r^3} \dfrac{dV}{dr},$ As $\dfrac{\partial r}{\partial x_1} = \dfrac{x_1}{r}.$

Similarly $\dfrac{\partial^2 V}{\partial x_2^2} = \dfrac{x_2^2}{r^2} \dfrac{d^2 V}{dr^2} + \dfrac{1}{r} \dfrac{dV}{dr} - \dfrac{x_2^2}{r^3} \dfrac{dV}{dr}$

....

and $\qquad \dfrac{\partial^2 V}{\partial x_n^2} = \dfrac{x_n^2}{r^2}\dfrac{d^2V}{dr^2} + \dfrac{1}{r}\dfrac{dV}{dr} - \dfrac{x_n^2}{r^3}\dfrac{dV}{dr}$

Adding these we get

$$\dfrac{\partial^2 V}{\partial x_1^2} + \dfrac{\partial^2 V}{\partial x_2^2} + \cdots + \dfrac{\partial^2 V}{\partial x_n^2} = \dfrac{x_1^2 + x_2^2 + \ldots + x_n^2}{dr^2} + \dfrac{n}{r}\dfrac{dV}{dr} - \dfrac{x_1^2 + x_2^2 + \ldots + x_n^2}{r^3}\dfrac{dV}{dr}$$

$$= \dfrac{d^2V}{dr^2} + \dfrac{n}{r}\dfrac{dV}{dr} - \dfrac{1}{r}\dfrac{dV}{dr} \qquad \text{as } r^2 = x_1^2 + x_2^2 + \ldots + x_n^2$$

$$= \dfrac{d^2V}{dr^2} + \dfrac{n-1}{r}\dfrac{dV}{dr}.$$

5. *If the equation* $\dfrac{\partial^2 \phi}{\partial x^2} + \dfrac{\partial^2 \phi}{\partial y^2} = 0$ *is satisfied when* ϕ *is a function of* x *and* y, *show that it will also be satisfied when* ϕ *is the same function of* u *and* v *where* $u = \dfrac{1}{2}\log(x^2 + y^2)$ *and* $v = \tan^{-1}(y/x)$.

We have $\quad u + iv = \dfrac{1}{2}\log(x^2 + y^2) + i\tan^{-1}(y/x)$

$$= \log(x + iy).$$

$\therefore \qquad\qquad x + iy = e^{u+iv} = e^u(\cos v + i\sin v).$

Hence $\qquad\quad x = e^u\cos v$ and $y = e^u\sin v$.

Now $\qquad \dfrac{\partial \phi}{\partial u} = \dfrac{\partial \phi}{\partial x}\cdot\dfrac{\partial x}{\partial u} + \dfrac{\partial \phi}{\partial y}\dfrac{\partial y}{\partial u} = x\dfrac{\partial \phi}{\partial x} + y\dfrac{\partial \phi}{\partial y}$ $\qquad\qquad\qquad$ ...(1)

and $\qquad \dfrac{\partial \phi}{\partial v} = \dfrac{\partial \phi}{\partial x}\dfrac{\partial x}{\partial v} + \dfrac{\partial \phi}{\partial y}\dfrac{\partial y}{\partial v} = -y\dfrac{\partial \phi}{\partial x} + x\dfrac{\partial \phi}{\partial y}$ $\qquad\qquad\qquad$ ...(2)

Now the equation will be satisfied when ϕ is the same function of u and v if $\dfrac{\partial^2 \phi}{\partial u^2} + \dfrac{\partial^2 \phi}{\partial v^2} = 0$.

But $\dfrac{\partial^2 \phi}{\partial u^2} + \dfrac{\partial^2 \phi}{\partial v^2} = \dfrac{\partial}{\partial u}\left(\dfrac{\partial \phi}{\partial u}\right) + \dfrac{\partial}{\partial v}\left(\dfrac{\partial \phi}{\partial v}\right)$

$$= x\left(\dfrac{\partial}{\partial x} + y\dfrac{\partial}{\partial y}\right)\left(x\dfrac{\partial \phi}{\partial x} + y\dfrac{\partial \phi}{\partial y}\right) + \left(-y\dfrac{\partial}{\partial x} + x\dfrac{\partial}{\partial y}\right)\left(-y\dfrac{\partial \phi}{\partial x} + x\dfrac{\partial \phi}{\partial y}\right)$$

$$= x^2\dfrac{\partial^2 \phi}{\partial x^2} + y^2\dfrac{\partial^2 \phi}{\partial y^2} + 2xy\dfrac{\partial^2 \phi}{\partial x\,\partial y} + x\dfrac{\partial \phi}{\partial x} + y\dfrac{\partial \phi}{\partial y} + y^2\dfrac{\partial^2 \phi}{\partial x^2} + x^2\dfrac{\partial^2 \phi}{\partial y^2}$$

$$- 2xy\dfrac{\partial^2 \phi}{\partial x\,\partial y} - x\dfrac{\partial \phi}{\partial y} + y\dfrac{\partial \phi}{\partial y}$$

$$= (x^2 + y^2)\left[\dfrac{\partial^2 \phi}{\partial x^2} + \dfrac{\partial^2 \phi}{\partial y^2}\right] = 0 \qquad \text{as } \dfrac{\partial^2 \phi}{\partial x^2} + \dfrac{\partial^2 \phi}{\partial y^2} = 0.$$

6. *If* $x = r\cos\theta$, $y = r\sin\theta$ *and* $r = e^z$, *prove that*

$$x^2\dfrac{\partial^2 u}{\partial x^2} + 2xy\dfrac{\partial^2 u}{\partial x\,\partial y} + y^2\dfrac{\partial^2 u}{\partial y^2} = r\dfrac{\partial}{\partial r}\left(r\dfrac{\partial}{\partial r} - 1\right)u = \dfrac{\partial}{\partial z}\left(\dfrac{\partial}{\partial z} - 1\right)u.$$

When $x = r \cos \theta$, $y = r \sin \theta$

$$\frac{\partial u}{\partial r} = \frac{\partial u}{\partial x} \frac{\partial x}{\partial r} + \frac{\partial u}{\partial y} \frac{\partial y}{\partial r} = \cos \theta \frac{\partial u}{\partial x} + \sin \theta \frac{\partial u}{\partial y}$$

so that $\quad r\dfrac{\partial u}{\partial r} = x\dfrac{\partial u}{\partial y} + y\dfrac{\partial u}{\partial y}.$ ...(1)

Thus $\quad r\dfrac{\partial r}{\partial r}\left(r\dfrac{\partial}{\partial r} - 1\right)u = \left(x\dfrac{\partial}{\partial x} + y\dfrac{\partial}{\partial y}\right)\left[x\dfrac{\partial}{\partial x} + y\dfrac{\partial}{\partial y} - 1\right]u$

$$= x^2 \frac{\partial^2 u}{\partial x^2} + y^2 \frac{\partial^2 u}{\partial y^2} + 2xy\frac{\partial^2 u}{\partial x\, \partial y} + x\frac{\partial u}{\partial x} + y\frac{\partial u}{\partial y} - x\frac{\partial u}{\partial x} - y\frac{\partial u}{\partial y}$$

$$= x^2 \frac{\partial^2 u}{\partial x^2} + 2xy\frac{\partial^2 u}{\partial x\, \partial y} + y^2 \frac{\partial^2 u}{\partial y^2}.$$

This proves one of the relations.

Second Part. From the relation $r = e^z$, $\dfrac{dr}{dz} = e^z = r,$

$\therefore \qquad \dfrac{\partial u}{\partial r} = \dfrac{\partial u}{\partial z} \cdot \dfrac{dz}{dr} = \dfrac{1}{r}\dfrac{\partial u}{\partial z}$

Thus $\quad r\dfrac{\partial u}{\partial r} = \dfrac{\partial u}{\partial z}$ or $r\dfrac{\partial}{\partial r} \equiv \dfrac{\partial}{\partial z}.$

There $\quad r\dfrac{\partial}{\partial r}\left(r\dfrac{\partial}{\partial r} - 1\right)u = \dfrac{\partial}{\partial z}\left(\dfrac{\partial}{\partial z} - 1\right)u.$

This proves the other part of the result.

7. *Show that*

$$x^2 \frac{\partial^3 u}{\partial x^3} + 3x^2 y\frac{\partial^3 u}{\partial x^2\, \partial y} + 3xy^2 \frac{\partial^2 u}{\partial x\, \partial y^2} + y^3 \frac{\partial^3 u}{\partial y^3} = \nabla(\nabla - 1)(\nabla - 2)u$$

where $\nabla = x\dfrac{\partial}{\partial x} + y\dfrac{\partial}{\partial y} = r\dfrac{\partial}{\partial r}$ *in polars.*

We have $\nabla . \nabla u = \left(x\dfrac{\partial}{\partial x} + y\dfrac{\partial}{\partial y}\right)\left(x\dfrac{\partial u}{\partial x} + y\dfrac{\partial u}{\partial y}\right)$

$$= x\left[x\frac{\partial^2 u}{\partial x^2} + \frac{\partial u}{\partial x} + y\frac{\partial^2 u}{\partial x\, \partial y}\right] + y\left[x\frac{\partial^2 u}{\partial x\, \partial y} + \frac{\partial u}{\partial y} + y\frac{\partial^2 u}{\partial y^2}\right]$$

$$= x^2 \frac{\partial^2 u}{\partial x^2} + 2xy\frac{\partial^2 u}{\partial x\, \partial y} + y^2 \frac{\partial^2 u}{\partial y^2} + x\frac{\partial u}{\partial x} + y\frac{\partial u}{\partial y}$$

$$= x^2 \frac{\partial^2 u}{\partial x^2} + 2xy\frac{\partial^2 u}{\partial x\, \partial y} + y^2 \frac{\partial^2 u}{\partial y^2} + \nabla u$$

i.e., $\nabla . \nabla u - \nabla u = x^2 \dfrac{\partial^2 u}{\partial x^2} + 2xy\dfrac{\partial^2 u}{\partial x\, \partial y} + y^2 \dfrac{\partial^2 u}{\partial y^2}$

i.e., $\nabla(\nabla - 1)u = x^2 \dfrac{\partial^2 u}{\partial x^2} + 2xy\dfrac{\partial^2 u}{\partial x\, \partial y} + y^2 \dfrac{\partial^2 u}{\partial y^2}$

Now operating on the both sides by operator ∇, we have

$$\nabla^2(\nabla - 1)u = \left(x\frac{\partial}{\partial x} + y\frac{\partial}{\partial y}\right)\left(x^2\frac{\partial^2 u}{\partial x^2} + 2xy\frac{\partial^2 u}{\partial x\,\partial y} + y^2\frac{\partial^2 u}{\partial y^2}\right)$$

$$= x\left[\frac{\partial^3 u}{\partial x^3} + 2x\frac{\partial^2 u}{\partial x^2} + 2xy\frac{\partial^3 y}{\partial x^2\,\partial y} + 2y\frac{\partial^2 u}{\partial x\,\partial y} + y^2\frac{\partial^3}{\partial y^2\,\partial x}\right]$$

$$+ y\left[x^2\frac{\partial^2 u}{\partial x^2\,\partial y} + 2xy\frac{\partial^3 u}{\partial x\,\partial y^2} + 2x\frac{\partial^2 u}{\partial x\,\partial y} + 2y\frac{\partial^2 u}{\partial y^2} + y^2\frac{\partial^3 u}{\partial x^3}\right]$$

$$= x^3\frac{\partial^3 u}{\partial x^3} + 3x^2 y\frac{\partial^3 u}{\partial x^2\,\partial y} + 3xy^2\frac{\partial^3 u}{\partial x\,\partial y^2} + y^3\frac{\partial^3 u}{\partial y^3}$$

$$+ 2\left(x^2\frac{\partial^3 u}{\partial x^2} + 2xy\frac{\partial^2 u}{\partial x\,\partial y} + y^2\frac{\partial^2 u}{\partial y^2}\right)$$

$$= x^3\frac{\partial^3 u}{\partial x^3} + 3x^2 y\frac{\partial^3 u}{\partial x^2\,\partial y} + 3xy^2\frac{\partial^3 u}{\partial x\,\partial y^2} + y^3\frac{\partial^3 u}{\partial y^3} + 2u\nabla(\nabla - 1)u$$

or $\quad x^3\dfrac{\partial^3 u}{\partial x^3} + 3x^2 y\dfrac{\partial^3 u}{\partial x^2\,\partial y} + 3xy^2\dfrac{\partial^3 u}{\partial x\,\partial y^2} + y^3\dfrac{\partial^3 u}{\partial y^3} + 2u\nabla(\nabla - 1)u = \nabla^2(\nabla - 1)u$

or $\quad x^3\dfrac{\partial^3 u}{\partial x^3} + 3x^2 y\dfrac{\partial^3 u}{\partial x^2\,\partial y} + 3xy^2\dfrac{\partial^3 u}{\partial x\,\partial y^2} + y^3\dfrac{\partial^3 u}{\partial y^3}$

$$= \nabla^2(\nabla - 1)u - 2\nabla(\nabla - 1)u$$

$$= \nabla(\nabla - 1)(\nabla - 2)u.$$

8. *Transform the equation*

$$\frac{\partial^2 z}{\partial x^2} + 2xy^2\frac{\partial z}{\partial x} + 2(x - y^3)\frac{\partial z}{\partial y} + x^2 y^2 z = 0$$

by the substitution $x = uv$, $y = \dfrac{1}{v}$ *and hence show that z is the same function of u and v as of x and y.*

We have from the given relations

$$u = xy, \quad v = \frac{1}{y}.$$

$$\therefore \quad \frac{\partial r}{\partial x} = \frac{\partial z}{\partial u}\frac{\partial u}{\partial x} + \frac{\partial z}{\partial v}\frac{\partial v}{\partial x} = y\frac{\partial z}{\partial u} + \frac{\partial z}{\partial v}\cdot 0 = \frac{1}{v}\frac{\partial z}{\partial u} \qquad \dots(1)$$

$$\therefore \quad \frac{\partial^2 z}{\partial x^2} = \frac{1}{v}\frac{\partial}{\partial u}\left(\frac{1}{v}\frac{\partial z}{\partial u}\right) = \frac{1}{v^2}\frac{\partial^2 z}{\partial u^2}$$

and $\quad \dfrac{\partial z}{\partial y} = \dfrac{\partial z}{\partial u}\dfrac{\partial u}{\partial y} + \dfrac{\partial z}{\partial v}\dfrac{\partial v}{\partial y} = x\dfrac{\partial z}{\partial u} - \dfrac{1}{y^2}\dfrac{\partial z}{\partial v}$

or $\quad y\dfrac{\partial z}{\partial y} = xy\dfrac{\partial z}{\partial u} - \dfrac{1}{y}\dfrac{\partial z}{\partial v} = u\dfrac{\partial z}{\partial u} - v\dfrac{\partial z}{\partial v}.$

Putting these values in the given equation, we get

$$\frac{1}{v^2}\frac{\partial^2 z}{\partial u^2} + 2u\cdot\frac{1}{v}\left(\frac{1}{v}\frac{\partial z}{\partial u}\right) + 2\left(1 - \frac{1}{v^2}\right)\left(u\frac{\partial z}{\partial u} - v\frac{\partial z}{\partial v}\right) + u^2 z = 0$$

or $\quad \dfrac{\partial^2 z}{\partial u^2} + \dfrac{\partial z}{\partial u}[2u + 2u(v^2 - 1)] - 2v(v^2 - 1)\dfrac{\partial z}{\partial v} + u^2 v^2 z = 0,$

or $\quad\quad \dfrac{\partial^2 z}{\partial u^2} = 2uv^2\dfrac{\partial z}{\partial u} + 2(u - v^2)\dfrac{\partial z}{\partial v} + u^2 v^2 z = 0,$

Since this equation is exactly similar to the given equation so z is the same function of u and v as $f x$ and y.

9. *If* $u = f(x, y)$, $x^2 = \xi\eta$ *and* $y^2 = \dfrac{\xi}{\eta}$, *change the independent variables to* ξ, η *in the equation*,

$$x^2\dfrac{\partial^2 u}{\partial x^2} - 2xy\dfrac{\partial^2 u}{\partial y^2} + y^2\dfrac{\partial^2 u}{\partial y^2} + 2y\dfrac{\partial u}{\partial y} = 0.$$

Solution. We have $x^2 = \xi\eta$, $y^2 = \dfrac{\xi}{\eta}$

so that $\quad\quad\quad\quad \xi = xy$ and $\eta = \dfrac{y}{x}.$

Now $\quad\quad \dfrac{\partial u}{\partial x} = \dfrac{\partial u}{\partial \xi}\cdot\dfrac{\partial \xi}{\partial x} + \dfrac{\partial u}{\partial \eta}\cdot\dfrac{\partial \eta}{\partial x} = y\dfrac{\partial u}{\partial \xi} - \dfrac{y}{x^2}\dfrac{\partial u}{\partial \eta}$

i.e., $\quad\quad x\dfrac{\partial u}{\partial x} = xy\dfrac{\partial u}{\partial \xi} - \dfrac{y}{x}\dfrac{\partial u}{\partial \eta} = \xi\dfrac{\partial u}{\partial \xi} - \eta\dfrac{\partial u}{\partial \eta}$ ...(1)

and $\quad\quad \dfrac{\partial u}{\partial y} = \dfrac{\partial u}{\partial \xi}\dfrac{\partial \xi}{\partial y} + \dfrac{\partial u}{\partial \eta}\cdot\dfrac{\partial \eta}{\partial y} = x\dfrac{\partial u}{\partial \xi} + \dfrac{1}{x}\dfrac{\partial u}{\partial \eta}$

i.e., $\quad\quad y\dfrac{\partial u}{\partial y} = xy\dfrac{\partial u}{\partial \xi} + \dfrac{y}{x}\dfrac{\partial u}{\partial \eta} = \xi\dfrac{\partial u}{\partial \xi} + \eta\dfrac{\partial u}{\partial \eta}.$ ...(2)

$\therefore \quad\quad x\dfrac{\partial}{\partial x} - y\dfrac{\partial}{\partial y} \equiv -2\eta\dfrac{\partial}{\partial \eta}$ ...(3)

Now $\left(x\dfrac{\partial}{\partial x} - y\dfrac{\partial}{\partial y}\right)\left(x\dfrac{\partial u}{\partial x} - x\dfrac{\partial u}{\partial y}\right)$

$= x^2\dfrac{\partial^2 u}{\partial x^2} - 2xy\dfrac{\partial^2 u}{\partial x\partial y} + y^2\dfrac{\partial^2 u}{\partial y^2} + x\dfrac{\partial u}{\partial x} + y\dfrac{\partial u}{\partial y}$

$= x^2\dfrac{\partial^2 u}{\partial x^2} - 2xy\dfrac{\partial^2 u}{\partial x\partial y} + y^2\dfrac{\partial^2 u}{\partial y^2} + 2y\dfrac{\partial u}{\partial y} + x\dfrac{\partial u}{\partial x} - y\dfrac{\partial u}{\partial y}$

adding and subtracting $y\dfrac{\partial u}{\partial y}$

so that $\quad\quad x^2\dfrac{\partial^2 u}{\partial x^2} - 2xy\dfrac{\partial^2 u}{\partial x\partial y} + y^2\dfrac{\partial^2 u}{\partial y^2} + 2y\dfrac{\partial u}{\partial y}$

$= \left(x\dfrac{\partial}{\partial x} - y\dfrac{\partial}{\partial y}\right)\left(x\dfrac{\partial u}{\partial x} - y\dfrac{\partial u}{\partial y}\right) - \left(x\dfrac{\partial u}{\partial x} - y\dfrac{\partial u}{\partial y}\right)$

$= \left(-2\eta\dfrac{\partial}{\partial \eta}\right)\left(-2\eta\dfrac{\partial u}{\partial \eta}\right) - \left(-2\eta\dfrac{\partial u}{\partial \eta}\right)$

$$= 4\eta^2 \frac{\partial^2 u}{\partial \eta^2} + 2\eta \frac{\partial u}{\partial \eta}.$$

Thus given equation becomes

$$4\eta^2 \frac{\partial^2 u}{\partial \eta^2} + 2\eta \frac{\partial u}{\partial \eta} = 0, \qquad i.e., \qquad 2\eta \frac{\partial^2 u}{\partial \eta^2} + \frac{\partial u}{\partial \eta} = 0.$$

10. *The position of a point in a plane is defined by the length r of the tangent from it to a fixed circle of radius a and the inclination θ of the tangent to a fixed line. Show that the continuity*

equation $\dfrac{\partial^2 \phi}{\partial x^2} + \dfrac{\partial^2 \phi}{\partial y^2} = 0$ *transforms into*

$$\frac{\partial^2 \phi}{\partial r^2} + \frac{1}{r}\frac{\partial \phi}{\partial r} + \frac{1}{r^2}\frac{\partial^2 \phi}{\partial \theta^2} + \frac{a^2}{r^2}\left(\frac{\partial^2 \phi}{\partial r^2} - \frac{1}{r}\frac{\partial \phi}{\partial r}\right) - \frac{a}{r^2}\left(2\frac{\partial^2 \phi}{\partial r \partial \theta} - \frac{1}{r}\frac{\partial \phi}{\partial r}\right) = 0.$$

Let (x, y) be the coordinates of a point P in the plane referred to Ox, Oy axes through the centre circle as axes of reference.

Draw PR tangent to the circle, then as given

$$r = PR$$

and θ is the angle RTO

$$x = ON + LP = a\sin\theta + r\cos\theta,$$

$$y = RN - RL$$

$$= a\cos\theta - r\sin\theta.$$

Now $\qquad \dfrac{\partial \phi}{\partial \theta} = \dfrac{\partial \phi}{\partial x}\dfrac{\partial x}{\partial \theta} + \dfrac{\partial \phi}{\partial y}\dfrac{\partial y}{\partial \theta}$

$$= (a\cos\theta - r\sin\theta)\frac{\partial \phi}{\partial x} + (-a\sin\theta - r\sin\theta)\frac{\partial \phi}{\partial x} \qquad \text{...(1)}$$

and $\qquad \dfrac{\partial \phi}{\partial r} = \dfrac{\partial \phi}{\partial x}\dfrac{\partial x}{\partial r} + \dfrac{\partial \phi}{\partial y}\dfrac{\partial y}{\partial r}$

$$= \cos\theta\,\frac{\partial \phi}{\partial x} - \sin\theta\,\frac{\partial \phi}{\partial y} \qquad \text{...(2)}$$

Solving these equations for $\dfrac{\partial \phi}{\partial x}$ and $\dfrac{\partial \phi}{\partial \theta}$ simultaneously, we get

$$\frac{\partial \phi}{\partial x} = \left(\frac{a\sin\theta}{r} + \cos\theta\right)\frac{\partial \phi}{\partial r} - \frac{\sin\theta}{r}\frac{\partial \phi}{\partial \theta}$$

and $\qquad \dfrac{\partial \phi}{\partial y} = \left(\dfrac{a\cos\theta}{r} - \sin\theta\right)\dfrac{\partial \phi}{\partial r} - \dfrac{\cos\theta}{r}\dfrac{\partial \phi}{\partial \theta}.$

$$\left[\frac{\partial \phi}{\partial y} \text{ may also be obtained by putting } \frac{1}{2}\pi = \theta \text{ for } \theta \text{ in } \frac{\partial \phi}{\partial x}\right]$$

Now

$$\frac{\partial^2 \phi}{\partial x^2} = \left[\left(\frac{a\sin\theta}{r} + \cos\theta\right)\frac{\partial}{\partial r} - \frac{\sin\theta}{\partial r} - \frac{\sin\theta}{r}\frac{\partial}{\partial\theta}\right]$$

$$\times \left[\left(\frac{a\sin\theta}{r} + \cos\theta\right)\frac{\partial\phi}{\partial r} - \frac{\sin\theta}{r}\frac{\partial\phi}{\partial\theta}\right]$$

$$= \left(\frac{a\sin\theta}{r} + \cos\theta\right)\left[\left(\frac{a\sin\theta}{r} + \cos\theta\right)\frac{\partial^2\phi}{\partial r^2} - \frac{a\sin\theta}{r^2}\frac{\partial\phi}{\partial r}\right.$$

$$\left. + \frac{\sin\theta}{r^2}\frac{\partial\phi}{\partial\theta} - \frac{\sin\theta}{r}\frac{\partial^2\phi}{\partial r\,\partial\theta}\right] - \frac{\sin\theta}{r}\left[\left(\frac{a\sin\theta}{r} + \cos\theta\right)\right.$$

$$\frac{\partial^2\phi}{\partial r\,\partial\theta}\left(\frac{a\cos\theta}{r} - \sin\theta\right)\frac{\partial\phi}{\partial r} - \frac{\sin\theta}{r}\frac{\partial^2\phi}{\partial\theta^2} - \frac{\cos\theta}{r}\frac{\partial\phi}{\partial\theta}\right]$$

Now putting $\frac{1}{2}\pi + \theta$ for θ, we get

$$\frac{\partial^2 \phi}{\partial y^2} = \left(\frac{a\cos\theta}{r} - \sin\theta\right)\left[\left(\frac{a\cos\theta}{r} - \sin\theta\right)\frac{\partial^2\phi}{\partial r^2} - \frac{a\cos\theta}{r^2}\frac{\partial\phi}{\partial r}\right.$$

$$\left. + \frac{\cos\theta}{r^2}\frac{\partial\phi}{\partial\theta} - \frac{\cos\theta}{r}\frac{\partial^2\phi}{\partial r\,\partial\theta}\right] - \frac{\cos\theta}{r}\left[\left(\frac{a\cos\theta}{r} - \sin\theta\right)\frac{\partial^2\phi}{\partial\theta\,\partial r}\right.$$

$$\left. + \left(-\frac{a\sin\theta}{r} - \cos\theta\right)\frac{\partial\phi}{\partial\theta} - \frac{\cos\theta}{r}\frac{\partial^2\phi}{\partial\theta^2} + \frac{\sin\theta}{r}\frac{\partial\phi}{\partial\theta}\right]$$

Adding these, term by term, we get

$$\frac{\partial^2\phi}{\partial x^2} + \frac{\partial^2\phi}{\partial y^2} = \left(\frac{a^2}{r^2} + 1\right)\frac{\partial^2\phi}{\partial r^2} - \frac{a^2}{r^3}\frac{\partial\phi}{\partial r} + \frac{a}{r^3}\frac{\partial\phi}{\partial\theta} - \frac{a}{r^2}\frac{\partial^2\phi}{\partial r\,\partial\theta} - \frac{a}{r^2}\frac{\partial^2\phi}{\partial r\,\partial\theta} + \frac{1}{r}\frac{\partial\phi}{\partial r} + \frac{1}{r^2}\frac{\partial^2\phi}{\partial\theta^2}.$$

$$= \frac{\partial^2\phi}{\partial r^2} + \frac{1}{r}\frac{\partial\phi}{\partial r} + \frac{1}{r^2}\frac{\partial^2\phi}{\partial\theta^2} + \frac{a^2}{r^2}\left(\frac{\partial^2\phi}{\partial r^2} - \frac{1}{r}\frac{\partial\phi}{\partial\theta}\right) - \frac{a}{r^2}\left(2\frac{\partial^2\phi}{\partial r\,\partial\theta} - \frac{1}{r}\frac{\partial\phi}{\partial\theta}\right),$$

As $\frac{\partial^2\phi}{\partial x^2} + \frac{\partial^2\phi}{\partial y^2} = 0$,

$$\therefore \frac{\partial^2\phi}{\partial r^2} + \frac{1}{r}\frac{\partial\phi}{\partial r} + \frac{1}{r^2}\frac{\partial^2\phi}{\partial\theta^2} + \frac{a^2}{r^2}\left(\frac{\partial^2\phi}{\partial r^2} - \frac{1}{r}\frac{\partial\phi}{\partial\theta}\right) - \frac{a}{r^2}\left(2\frac{\partial^2\phi}{\partial r\,\partial\theta} - \frac{1}{r}\frac{\partial\phi}{\partial\theta}\right) = 0$$

11. *Given that $f(x, y)$ has continuous partial derivatives of the first two orders and $x + y = (u + v)^3$, $x - y = (u - v)^3$, show that*

$$9(x^2 - y^2)\left(\frac{\partial^2 f}{\partial x^2} - \frac{\partial^2 f}{\partial y^2}\right) = (u^2 - v^2)\left(\frac{\partial^2 f}{\partial u^2} - \frac{\partial^2 f}{\partial v^2}\right).$$

We have $x + y = (u + v)^3 = u^3 + 3u^2v + 3uv^2 + v^3$

and $x - y = (u - v)^3 = u^3 - 3u^2v + 3uv^2 - v^3$.

So that on adding and subtracting, we get

$x = u^3 + 3uv^2$ and $y = 3u^2v + v^3$.

Now $\frac{\partial f}{\partial u} = \frac{\partial f}{\partial x} \cdot \frac{\partial x}{\partial u} + \frac{\partial f}{\partial y} \cdot \frac{\partial y}{\partial u}$

$$= 3(u^2 + v^2)\frac{\partial f}{\partial x} + 6uv\frac{\partial f}{\partial y}$$

and $\quad \dfrac{\partial f}{\partial v} = 6uv\dfrac{\partial f}{\partial x} + 3(u^2 + v^2)\dfrac{\partial f}{\partial y}.$

$\therefore \quad \dfrac{\partial}{\partial u} + \dfrac{\partial}{\partial v} \equiv 3(u^2 + v^2 + 2uv)\dfrac{\partial}{\partial x} + 3(u^2 + v^2 + 2uv)\dfrac{\partial}{\partial y}$

$$\equiv 3(u + v)^2\left(\frac{\partial}{\partial x} + \frac{\partial}{\partial y}\right)$$

and $\quad \dfrac{\partial}{\partial u} - \dfrac{\partial}{\partial v} \equiv 3(u - v)^2\left(\dfrac{\partial}{\partial x} - \dfrac{\partial}{\partial y}\right).$

So that $\quad (u + v)\left(\dfrac{\partial}{\partial x} + \dfrac{\partial}{\partial v}\right) = 3(u + v)^3\left(\dfrac{\partial}{\partial x} + \dfrac{\partial}{\partial y}\right)$

$$= 3(x + y)\left(\frac{\partial}{\partial x} + \frac{\partial}{\partial y}\right)$$

and $\quad (u - v)\left(\dfrac{\partial}{\partial u} - \dfrac{\partial}{\partial v}\right) = 3(u - v)^3\left(\dfrac{\partial}{\partial x} - \dfrac{\partial}{\partial y}\right)$

$$= 3(x - y)\left(\frac{\partial}{\partial x} - \frac{\partial}{\partial y}\right).$$

Multiplying the above two operators and then operating on f, we have

$$(u + v)\left(\frac{\partial}{\partial u} + \frac{\partial}{\partial v}\right)(u - v)\left(\frac{\partial}{\partial u} - \frac{\partial}{\partial u}\right)f$$

$$= 3\,(x + y)\left(\frac{\partial}{\partial x} + \frac{\partial}{\partial y}\right)3(x - y)\left(\frac{\partial}{\partial x} - \frac{\partial}{\partial y}\right)f$$

or $\quad (u^2 - v^2)\left(\dfrac{\partial^2 f}{\partial u^2} - \dfrac{\partial^2 f}{\partial v^2}\right) = 9(x^2 - y^2)\left(\dfrac{\partial^2 f}{\partial x^2} - \dfrac{\partial^2 f}{\partial y^2}\right).$

12. *In cartesian-polar transformation by the formulae*

$$x = r \sin\theta \cos\phi, \, y = r \sin\theta \sin\phi, \, z = r \cos\theta$$

prove that (i) $\dfrac{\partial x}{\partial r} = \dfrac{\partial r}{\partial x},$ (ii) $\dfrac{\partial x}{\partial \theta} = r^2\dfrac{\partial \theta}{\partial x},$ (iii) $\dfrac{\partial x}{\partial \phi} \cdot \dfrac{\partial \phi}{\partial x} + \dfrac{\partial y}{\partial \phi}\dfrac{\partial \phi}{\partial y} = 1.$

(*i*) We have from the relation $x = r \sin\theta \cos\phi$

$$\frac{\partial x}{\partial r} = \sin\theta \cos\phi. \qquad\qquad\qquad\qquad \dots(1)$$

Also from the relation $r^2 = x^2 + y^2 + z^2$

$$r\frac{\partial r}{\partial x} = x \quad \text{or} \quad \frac{\partial r}{\partial x} = \frac{x}{r} = \frac{r\sin\theta\cos\phi}{r} = \sin\theta\cos\phi. \qquad \dots(2)$$

Comparing (1) and (2), we get

$$\frac{\partial x}{\partial r} = \frac{\partial r}{\partial x}.$$

(*ii*) Again from $\quad x = r \sin\theta \cos\phi, \dfrac{\partial x}{\partial \theta} = r\cos\theta\cos\phi \qquad\qquad \dots(3)$

and from the relation $\quad \tan^2 \theta = \dfrac{\sin^2 \theta}{\cos^2 \theta} = \dfrac{x^2 + y^2}{z^2},$

we have $\quad 2 \tan \theta \sec^2 \theta \dfrac{\partial \theta}{\partial x} = \dfrac{2x}{z^2}$

or $\quad r^2 \dfrac{\partial \theta}{\partial x} = r^2 \cdot \dfrac{2r \sin \theta \cos \phi}{r^2 \cos^2 \theta \cdot 2 \tan \theta \sec^2 \theta}$

$$= r \cos \theta \cos \phi = \dfrac{\partial x}{\partial \theta} \text{ from (3)}.$$

(*iii*)   We have $x = r \sin \theta \cos \phi,\ \therefore\ \dfrac{\partial x}{\partial \phi} = -r \sin \theta \sin \phi$

and from the relation $y = r \sin \theta \sin \phi,\ \dfrac{\partial y}{\partial \phi} = r \sin \theta \cos \phi$.

Now from the relation $\tan \phi = \dfrac{y}{x}$ *i.e.,* $\phi = \tan^{-1} \left(\dfrac{y}{x} \right)$

$$\dfrac{\partial \phi}{\partial x} = \dfrac{1}{1 + y^2/x^2} \left(-\dfrac{y}{x^2} \right) = -\dfrac{y}{x^2 + y^2} = \dfrac{r \sin \theta \sin \phi}{r^2 \sin^2 \theta}$$

$$= -\dfrac{\sin \phi}{r \sin \theta}$$

and $\quad \dfrac{\partial \phi}{\partial y} = \dfrac{1}{1 + y^2/x^2} \cdot \left(\dfrac{1}{x} \right) = \dfrac{x}{x^2 + y^2} = \dfrac{r \sin \theta \cos \phi}{r^2 \sin^2 \theta} = \dfrac{\cos \theta}{r \sin \theta}.$

Substituting these values, we have $\dfrac{\partial x}{\partial \phi} \cdot \dfrac{\partial \phi}{\partial x} + \dfrac{\partial y}{\partial \phi} \cdot \dfrac{\partial \phi}{\partial y}$

$$= (-r \sin \theta \sin \phi) \left(-\dfrac{\sin \phi}{r \sin \theta} \right) + (r \sin \theta \cos \phi) \left(\dfrac{\cos \phi}{r \sin \theta} \right)$$

$$= \sin^2 \phi + \cos^2 \phi = 1.$$

13. *Show that* $x^n \dfrac{\partial^n u}{\partial x^n} + nx^{n-1} y \dfrac{\partial^n u}{\partial x^{n-1} \partial y} + ... + y^n \dfrac{\partial^n u}{\partial y^n} = \nabla (\nabla - 1)(\nabla - 2) ... (\nabla - n + 1)\, u$

where $\nabla \equiv x \dfrac{\partial}{\partial x} + y \dfrac{\partial}{\partial y} \equiv r \dfrac{\partial}{\partial r}$ *in polars.*

Let $\quad \phi_n = x^n \dfrac{\partial^n u}{\partial x^n} + nx^{-1} y \dfrac{\partial^n u}{\partial x^{n-1} \partial y} + ... + y^n \dfrac{\partial^n u}{\partial y^n}$

then $\quad \nabla \phi_n = \left(x \dfrac{\partial}{\partial x} + y \dfrac{\partial}{\partial y} \right) \left(x^n \dfrac{\partial^n u}{\partial x^n} + nx^{n-1} y^n \dfrac{\partial^n u}{\partial x^{n-1} \partial y} + ... + y^n \dfrac{\partial^n u}{\partial y^n} \right)$

$$= \phi_{n+1} + n\phi_n$$

i.e., $\quad \phi_{n+1} = (\nabla - n)\, \phi_n$

so that $\quad \phi_n = (\nabla - \overline{n-1})\phi_{n-1} = (\nabla - n + 1)\phi_{n-1}.$

Putting $n = 1, 2, 3, ..., n$, we have

$$\phi_1 = \nabla \phi_0 = \nabla u$$

$$\phi_2 = (\nabla - 1)\,\phi_1 = \nabla\,(\nabla - 1)\,u$$

$$\phi_3 = \nabla\,(\nabla - 1)\,(\nabla - 2)\,u$$

$$..................................etc.$$

and $$\phi_n = \nabla\,(\nabla - 1)\,(\nabla - 2)\,.\,(\nabla - n + 1)\,u.$$

14. *If* $\xi = y^2 + z^{-2}$, $\eta = z^2 + x^{-2}$, $\zeta = x^2 + y^{-2}$ *and* u *be a function of* x, y, z *show that*

$$x\frac{\partial u}{\partial x} + y\frac{\partial u}{\partial y} + z\frac{\partial u}{\partial z} = 2\left(\xi\frac{\partial u}{\partial \xi} + \eta\frac{\partial u}{\partial \eta} + \zeta\frac{\partial u}{\partial \zeta}\right)$$

$$= 4\left(y^2\frac{\partial u}{\partial \xi} + z^2\frac{\partial u}{\partial \eta} + x^2\frac{\partial u}{\partial \zeta}\right).$$

We have $$\frac{\partial u}{\partial x} = \frac{\partial u}{\partial \xi}\frac{\partial \xi}{\partial x} + \frac{\partial u}{\partial \eta}\frac{\partial \eta}{\partial x} + \frac{\partial u}{\partial \zeta}\frac{\partial \zeta}{\partial x}$$

$$= \frac{\partial u}{\partial \xi}0 + \frac{\partial u}{\partial \eta}(-2x^{-3}) + \frac{\partial u}{\partial \zeta}(2x).$$

Multiplying both the sides by x, we get

$$x\frac{\partial u}{\partial x} = -2x^{-2}\frac{\partial u}{\partial \eta} + 2x^2\frac{\partial u}{\partial \zeta} \qquad\qquad ...(1)$$

Similarly $$y\frac{\partial u}{\partial y} = -2y^{-2}\frac{\partial u}{\partial \xi} + 2y^2\frac{\partial u}{\partial \zeta} \qquad\qquad ...(2)$$

and $$z\frac{\partial u}{\partial z} = -2z^{-2}\frac{\partial u}{\partial \xi} + 2z^2\frac{\partial u}{\partial \eta}. \qquad\qquad ...(3)$$

Adding (1), (2) and (3), we get

$$x\frac{\partial u}{\partial x} + y\frac{\partial u}{\partial y} + z\frac{\partial u}{\partial z}$$

$$= 2(y^2 - z^{-2})\frac{\partial u}{\partial \xi} + 2(z^2 - x^{-2})\frac{\partial u}{\partial \eta} + 2(x^2 - y^{-2})\frac{\partial u}{\partial \zeta}$$

$$= 2(2y^2 - \xi)\frac{\partial u}{\partial \xi} + 2(2z^2 - \eta)\frac{\partial u}{\partial \eta} + 2(x^2 - \zeta)\frac{\partial u}{\zeta}$$

as $$\xi = y^2 + z^{-2}\ \text{etc.}$$

Suitable transposing, we get

$$x\frac{\partial u}{\partial x} + y\frac{\partial u}{\partial y} + z\frac{\partial u}{\partial z} = 2\left(\xi\frac{\partial u}{\partial \xi} + \eta\frac{\partial u}{\partial \eta} + \zeta\frac{\partial u}{\partial \zeta}\right)$$

$$= 4\left(y^2\frac{\partial u}{\partial \xi} + z^2\frac{\partial u}{\partial \eta} + x^2\frac{\partial u}{\partial \zeta}\right).$$

19.10 Orthogonal Transformation of $\nabla^2 V$.

To show that the expression $\dfrac{\partial^2 V}{\partial x^2} + \dfrac{\partial^2 V}{\partial y^2} + \dfrac{\partial^2 V}{\partial z^2}$ *transforms to* $\dfrac{\partial^2 V}{\partial \xi^2} + \dfrac{\partial^2 V}{\partial \eta^2} + \dfrac{\partial^2 V}{\partial \zeta^2}$ *by changing*

to any other set of axes $O\xi$, $O\eta$, $O\zeta$, *mutually at right angles, the origin being the same.*

Let (l_1, m_1, n_1), (l_2, m_2, n_2), (l_3, m_3, n_3) be the direction cosines of $O\xi$, $O\eta$, $O\zeta$, referred to Ox, Oy, Oz as axes. Then

$$l_1^2 + m_1^2 + n_1^2 = 1,\ l_2^2 + m_2^2 + n_2^2 = 1,\ l_3^2 + m_3^2 + n_3^2 = 1$$

$$l_1 l_2 + m_1 m_2 + n_1 n_2 = 0 \text{ etc.} \qquad ...(A)$$

Again $[l_1, l_2, l_3]$, $[m_1, m_2, m_3]$, $[n_1, n_2, n_3]$ are the direction cosines of Ox, Oy, Oz referred to $O\xi$, $O\eta$, $O\zeta$;

$$\therefore \quad l_1^2 + l_2^2 + l_3^2 = 1, \text{ etc. and } l_1 m_1 + l_2 m_2 + l_2 m_3 = 0, \text{ etc.} \qquad ...(B)$$

Now ξ = projection of line joining $(0, 0, 0)$ and (x, y, z) on $O\xi = l_1 x + m_1 y + n_1 z$

Similarly, $\qquad \eta = l_2 x + m_2 y + n_2 z, \; \zeta = l_3 x + m_3 y + n_3 z.$

Now $\qquad \dfrac{\partial V}{\partial x} = \dfrac{\partial V}{\partial \xi}\dfrac{\partial \xi}{\partial x} + \dfrac{\partial V}{\partial \eta}\dfrac{\partial \eta}{\partial x} + \dfrac{\partial V}{\partial \zeta}\dfrac{\partial \zeta}{\partial x}$

$$= l_1 \frac{\partial V}{\partial \xi} + l_2 \frac{\partial V}{\partial \eta} + l_3 \frac{\partial V}{\partial \zeta};$$

$$\therefore \quad \frac{\partial^2 V}{\partial x^2} = \left(l_1 \frac{\partial}{\partial \xi} + l_2 \frac{\partial}{\partial \eta} + l_3 \frac{\partial}{\partial \zeta}\right)\left(l_1 \frac{\partial V}{\partial \xi} + l_2 \frac{\partial V}{\partial \eta} + l_3 \frac{\partial V}{\partial \zeta}\right)$$

$$= l_1^2 \frac{\partial^2 V}{\partial \xi^2} + l_2^2 \frac{\partial^2 V}{\partial \eta^2} + l_3^2 \frac{\partial^2 V}{\partial \zeta^2} + 2l_2 l_3 \frac{\partial^2 V}{\partial \eta \partial \zeta} + 2l_1 l_3 \frac{\partial^3 V}{\partial \xi \partial \zeta} + 2l_1 l_2 \frac{\partial^2 V}{\partial \xi \partial \eta}$$

Similarly, $\qquad \dfrac{\partial^2 V}{\partial y^2} = m_1^2 \dfrac{\partial^2 V}{\partial \xi^2} + ... \text{ etc.}$

and $\qquad \dfrac{\partial^2 V}{\partial z^2} = n_1^2 \dfrac{\partial^2 V}{\partial \xi^2} + ... \text{ etc.}$

Adding these and using results (A) and (B) (namely

$$l_1^2 + m_1^2 + n_1^2 = 1 \text{ etc. and } l_2 l_3 + m_2 m_3 + n_2 n_3 = 0 \text{ etc.})$$

we get $\qquad \dfrac{\partial^2 V}{\partial x^2} + \dfrac{\partial^2 V}{\partial y^2} + \dfrac{\partial^2 V}{\partial z^2} = \dfrac{\partial^2 V}{\partial \xi^2} + \dfrac{\partial^2 V}{\partial \eta^2} + \dfrac{\partial^2 V}{\partial \zeta^2}.$

Example. *If the co-ordinates x, y be transformed orthogonally to ξ, η and V be any function of x, y then*

$$\frac{\partial^2 V}{\partial x^2}\frac{\partial^2 V}{\partial y^2} - \left(\frac{\partial^2 V}{\partial x\,\partial y}\right)^2 = \frac{\partial^2 V}{\partial \xi^2}\cdot\frac{\partial^2 V}{\partial \eta^2} - \left(\frac{\partial^2 V}{\partial \xi\,\partial \eta}\right)^2.$$

For orthogonal transformation in two dimensions we have

$$x = \xi \cos\theta - \eta \sin\theta, \; y = \xi \sin\theta + \eta \cos\theta.$$

$$\therefore \quad \frac{\partial V}{\partial \xi} = \frac{\partial V}{\partial x}\cdot\frac{\partial x}{\partial \xi} + \frac{\partial V}{\partial y}\frac{\partial y}{\partial \xi} = \cos\theta\frac{\partial V}{\partial x} + \sin\theta\frac{\partial V}{\partial y};$$

$$\therefore \quad \frac{\partial^2 V}{\partial \xi^2} = \left(\cos\theta\frac{\partial}{\partial x} + \sin\theta\frac{\partial}{\partial y}\right)\left(\cos\theta\frac{\partial V}{\partial x} + \sin\theta\frac{\partial V}{\partial y}\right)$$

$$= \cos^2\theta\frac{\partial^2 V}{\partial x^2} + 2\sin\theta\cos\theta\frac{\partial^2 V}{\partial x\,\partial y} + \sin^2\theta\frac{\partial^2 V}{\partial v^2}.$$

Similarly, $\qquad \dfrac{\partial V}{\partial \eta} = -\sin\theta\dfrac{\partial V}{\partial x} + \cos\theta\dfrac{\partial V}{\partial y} \qquad \qquad ...(1)$

and $\qquad \dfrac{\partial^2 V}{\partial \eta^2} = \sin^2\theta\dfrac{\partial^2 V}{\partial x^2} - 2\sin\theta\cos\theta\dfrac{\partial^2 V}{\partial x\,\partial y} + \cos^2\theta\dfrac{\partial^2 V}{\partial v^2} \qquad ...(2)$

Also, $\dfrac{\partial^2 V}{\partial\xi\,\partial\eta} = \left(\cos\theta\dfrac{\partial}{\partial x} + \sin\theta\dfrac{\partial}{\partial y}\right) + \left(-\sin\theta\dfrac{\partial V}{\partial x} + \cos\theta\dfrac{\partial V}{\partial y}\right)$

$$= \sin\theta\,\cos\theta\left(\dfrac{\partial^2 V}{\partial y^2} - \dfrac{\partial^2 V}{\partial x^2}\right) + (\cos^2\theta - \sin^2\theta)\dfrac{\partial^2 V}{\partial x\,\partial y} \qquad \text{...(3)}$$

Hence, $\dfrac{\partial^2 V}{\partial\xi^2}\cdot\dfrac{\partial^2 V}{\partial\xi^2} - \left(\dfrac{\partial^2 V}{\partial\xi\,\partial\eta}\right)^2$

$$= \left(\cos^2\theta\dfrac{\partial^2 V}{\partial x^2} + 2\sin\theta\,\cos\theta\dfrac{\partial^2 V}{\partial x\,\partial y} + \sin^2\theta\dfrac{\partial^2 V}{\partial y^2}\right)$$

$$\times\left(\sin^2\theta\dfrac{\partial^2 V}{\partial x^2} - 2\sin\theta\,\cos\theta\dfrac{\partial^2 V}{\partial x\,\partial y} + \cos^2\theta\dfrac{\partial^2 V}{\partial y^2}\right)$$

$$- \left[\sin\theta\,\cos\theta\left(\dfrac{\partial^2 V}{\partial y^2} - \dfrac{\partial^2 V}{\partial x^2}\right) + \cos 2\theta\dfrac{\partial^2 V}{\partial x\,\partial y}\right]^2$$

$$= \dfrac{\partial^2 V}{\partial\xi^2}\cdot\dfrac{\partial^2 V}{\partial\eta^2} - \left(\dfrac{\partial^2 V}{\partial\xi\,\partial\eta}\right)^2 \quad \text{after proper simplification.}$$

EXERCISES

1. If $u + v = 2\,e^x\cos y$, $u - v = 2\,i\,e^x\sin y$, prove that

   $\dfrac{\partial^2 V}{\partial x^2} + \dfrac{\partial^2 V}{\partial y^2} = 4uv\dfrac{\partial^2 V}{\partial u\,\partial v}$, where V is a function of u and v.

2. If $u + iv = f(x + iy)$, when x and y are independent and u, v, x, y are all real, prove that

   $\dfrac{\partial^2 V}{\partial x^2} + \dfrac{\partial^2 V}{\partial y^2} = \left(\dfrac{\partial^2 V}{\partial u^2} + \dfrac{\partial^2 V}{\partial v^2}\right)f'(x + iy)\,f'(x - iy).$

3. If V be a function of r alone when $r^2 = x^2 + y^2 + z^2$, show that

   $\dfrac{\partial^2 V}{\partial x^2} + \dfrac{\partial^2 V}{\partial y^2} + \dfrac{\partial^2 V}{\partial z^2} = \dfrac{\partial^2 V}{\partial r^2} + \dfrac{2}{r}\dfrac{\partial V}{\partial r}.$

4. In transforming any function u of x, y, z from cartesian to polar by the formulae $x = r\sin\theta\cos\phi$, $y = r\sin\theta\sin\phi$, $z = r\cos\theta$, prove that

   (a) $\dfrac{\partial x}{\partial r} = \dfrac{\partial r}{\partial x}$ 　　　(b) $\dfrac{\partial x}{\partial\theta} = r^2\dfrac{\partial\theta}{\partial x}$ 　　　(c) $\dfrac{\partial x}{\partial\phi}\cdot\dfrac{\partial\phi}{\partial x} + \dfrac{\partial y}{\partial\phi}\cdot\dfrac{\partial\phi}{\partial y} = 1$

   (d) $\left(\dfrac{\partial u}{\partial x}\right)^2 + \left(\dfrac{\partial u}{\partial y}\right)^2 + \left(\dfrac{\partial u}{\partial z}\right)^2 = \left(\dfrac{\partial u}{\partial r}\right)^2 + \dfrac{1}{r}\left(\dfrac{\partial u}{\partial\theta}\right)^2 + \left(\dfrac{1}{r\sin\theta}\dfrac{\partial u}{\partial\theta}\right)^2$

5. If $x = r\cos\theta$, $y = r\sin\theta$, show that $\dfrac{\partial^2\theta}{\partial x^2} + \dfrac{\partial^2\theta}{\partial y^2} = 0$.

6. If $x = r\cos\theta$, $y = r\sin\theta$, prove that the equation

   $xy\left(\dfrac{\partial^2 u}{\partial x^2} - \dfrac{\partial^2 y}{\partial y^2}\right) - (x^2 - y^2)\dfrac{\partial^2 u}{\partial x\,\partial y} = 0$

   becomes $\dfrac{\partial^2 u}{\partial r\,\partial\theta} - \dfrac{\partial u}{\partial\theta} = 0$.

7. If $x = r \cos \theta$, $y = r \sin \theta$ and $r = e^z$, prove that

$$x^2 \frac{\partial^2 y}{\partial y^2} - 2xy \frac{\partial^2 u}{\partial x \, \partial y} + y^2 \frac{\partial^2 u}{\partial x^2} = \frac{\partial^2 u}{\partial \theta^2} + r \frac{\partial u}{\partial r} = \frac{\partial^2 u}{\partial \theta^2} + \frac{\partial u}{\partial z}.$$

8. If $x = ae^\theta \cos \phi$, $y = ae^\theta \sin \phi$, prove that

$$y^2 \frac{\partial^2 u}{\partial x^2} - 2xy \frac{\partial^2 u}{\partial x \, \partial y} + x^2 \frac{\partial^2 u}{\partial y^2} = \frac{\partial^2 u}{\partial \phi^2} + \frac{\partial u}{\partial \theta}.$$

9. If $x = r \cos \theta$, $y = r \sin \theta$, pove that

$$r \frac{\partial^2 u}{\partial r \, \partial \theta} - \frac{\partial u}{\partial \theta} = xy \left(\frac{\partial^2 u}{\partial y^2} - \frac{\partial^2 u}{\partial x^2} \right) + (x^2 - y^2) \frac{\partial^2 u}{\partial x \, \partial y}$$

10. If $x + iy = f(\xi + i\eta)$, prove that the equation $\dfrac{\partial^2 V}{\partial r^2} + \dfrac{\partial^2 V}{\partial y^2} = 0$ becomes

$$\frac{\partial^2 V}{\partial \xi^2} + \frac{\partial^2 V}{\partial \eta^2} = 0.$$

11. If u is a function of x, y having continuous derivatives upto the second order, and variables x, y are changed to ξ, η by the transformation

$$x + y = (\xi + \eta)^n, \quad x - y = (\xi - \eta)^n,$$

prove that $(x^2 - y^2)\left(\dfrac{\partial^2 u}{\partial x^2} - \dfrac{\partial^2 u}{\partial y^2} \right) = \dfrac{1}{n^2}(\xi^2 - \eta^2)\left(\dfrac{\partial^2 u}{\partial \xi^2} - \dfrac{\partial^2 u}{\partial \eta^2} \right).$

12. By putting $G = x^n V$ and changing the independent variables x, y to u, v where $u = y/x$ and $v = xy$, transform the equations

$$x \frac{\partial G}{\partial x} + y \frac{\partial G}{\partial x} = xG, \quad x \frac{\partial V}{\partial x} - y \frac{\partial V}{\partial y} = 0.$$

Hence show that $G = x^n \phi(y/x)$ and $V = \phi(xy)$.

OBJECTIVE QUESTIONS

For each of the following questions,four alternatives are given for the answer. Only one of them is correct. Choose the correct alternative.

1. Equation $\dfrac{d^2 y}{dx^2} = a$ can be written as :

   (a) $\dfrac{d^2 x}{dy^2} + \dfrac{dx}{dy} = 0$

   (b) $\dfrac{d^2 x}{dy^2} + a\left(\dfrac{dx}{dy} \right)^3 = 0$

   (c) $\dfrac{d^2 x}{dy^2} = 0$

   (d) None of these

2. When $x = \dfrac{1}{z}$, then value of $\dfrac{dy}{dx}$ will be :

   (a) $z \dfrac{dy}{dz}$

   (b) $z^2 \dfrac{dy}{dz}$

   (c) $-z^2 \dfrac{dy}{dz}$

   (d) $\dfrac{1}{z} \dfrac{dy}{dz}$

3. When $x = \tan z$, then value of $\dfrac{dy}{dx}$ will be :

   (a) $\cos^2 z \dfrac{dy}{dz}$

   (b) $\sin^2 z \dfrac{dy}{dz}$

   (c) $\sec^2 z \dfrac{dy}{dz}$

   (d) $\cot z \dfrac{dy}{dz}$

4. When $x = \cos\theta$, $\dfrac{dy}{dx}$ will be:

(a) $\sin\theta\,\dfrac{dy}{d\theta}$

(b) $\text{cosec}\,\theta\,\dfrac{dy}{d\theta}$

(c) $-\sin\theta\,\dfrac{dy}{d\theta}$

(d) $-\text{cosec}\,\theta\,\dfrac{dy}{d\theta}$

5. In polar coordinates $x = r\cos\theta$, value of $\dfrac{dy}{dx}$ will be:

(a) $-\tan\theta$

(b) $\dfrac{r\cos\theta + \sin\theta\,\dfrac{dr}{d\theta}}{-r\sin\theta + \cos\theta\,\dfrac{dr}{d\theta}}$

(c) $\dfrac{r\cos\theta - \sin\theta\,\dfrac{dr}{d\theta}}{-r\sin\theta\,\cos\theta\,\dfrac{dr}{d\theta}}$

(d) $\dfrac{r\cos\theta + \sin\theta\,\dfrac{dr}{d\theta}}{r\sin\theta - \cos\theta\,\dfrac{dr}{d0}}$

6. Polar formula $\tan\phi = r\,\dfrac{d\theta}{dr}$ in cartesian system becomes:

(a) $\dfrac{x\dfrac{dy}{dx} - y}{x + y\dfrac{dy}{dx}}$

(b) $\dfrac{x\dfrac{dy}{dx} + y}{x - y\dfrac{dy}{dx}}$

(c) $\dfrac{x\dfrac{dy}{dx} - y}{x - y\dfrac{dy}{dx}}$

(d) $\dfrac{x\dfrac{dy}{dx} + y}{x + y\dfrac{dy}{dx}}$

7. If x and y be functions of t, then value of $x\dfrac{dx}{dt} + y\dfrac{dy}{dt}$ in polar coordinates will be:

(a) $r^2\dfrac{d\theta}{dt}$

(b) $r^2\left(\dfrac{d\theta}{dt}\right) + \left(\dfrac{dr}{dt}\right)$

(c) $r\dfrac{dr}{dt}$

(d) None of these

ANSWERS						
1. (b)	2. (c)	3. (a)	4. (d)	5. (b)	6. (a)	7. (c)